Radial Flow Turbocompressors

An introduction to the theory and engineering pra[illegible] ponent design and analysis of radial flow turbocompressors. Draw[illegible] tensive theoretical background and years of practical experience, the authors [illegible]vide the following:

- Descriptions of applications, concepts, component design, analysis tools, performance maps, flow stability and structural integrity, with numerous illustrative examples
- Wide coverage of all types of radial compressor over a range of applications unified by the consistent use of dimensional analysis
- The methods needed to analyse the performance, flow and mechanical integrity that underpin the design of efficient centrifugal compressors with good flow range and stability
- Explanation of the design of all radial compressor components, including inlet guide vanes, impellers, diffusers, volutes, return channels, deswirl vanes and side-streams

This volume is suitable as a reference for advanced students of turbomachinery, and a perfect tool for practising mechanical and aerospace engineers already within the field as well as those just entering it.

Michael Casey is a director of PCA Engineers Limited, and was previously a Professor of Thermal Turbomachinery at the University of Stuttgart.

Chris Robinson is managing director at PCA Engineers Limited and a specialist in the aeromechanical design of centrifugal compressors.

Radial Flow Turbocompressors

Design, Analysis, and Applications

MICHAEL CASEY
PCA Engineers Limited

CHRIS ROBINSON
PCA Engineers Limited

CAMBRIDGE
UNIVERSITY PRESS

University Printing House, Cambridge CB2 8BS, United Kingdom

One Liberty Plaza, 20th Floor, New York, NY 10006, USA

477 Williamstown Road, Port Melbourne, VIC 3207, Australia

314–321, 3rd Floor, Plot 3, Splendor Forum, Jasola District Centre, New Delhi – 110025, India

79 Anson Road, #06–04/06, Singapore 079906

Cambridge University Press is part of the University of Cambridge.

It furthers the University's mission by disseminating knowledge in the pursuit of education, learning, and research at the highest international levels of excellence.

www.cambridge.org
Information on this title: www.cambridge.org/9781108416672
DOI: 10.1017/9781108241663

First published 2021

A catalogue record for this publication is available from the British Library.

Library of Congress Cataloging-in-Publication Data
Names: Casey, Michael, 1948- author. | Robinson, Chris, 1959- author.
Title: Radial flow turbocompressors : design, analysis, and applications / Michael Casey, PCA Engineers Limited, Chris Robinson, PCA Engineers Limited.
Description: Cambridge, United Kingdom ; New York, NY, USA : Cambridge University Press, 2021. | Includes bibliographical references and index.
Identifiers: LCCN 2020046835 (print) | LCCN 2020046836 (ebook) | ISBN 9781108416672 (hardback) | ISBN 9781108241663 (epub)
Subjects: LCSH: Centrifugal compressors.
Classification: LCC TJ990 .C366 2021 (print) | LCC TJ990 (ebook) | DDC 621.5/1–dc23
LC record available at https://lccn.loc.gov/2020046835
LC ebook record available at https://lccn.loc.gov/2020046836

ISBN 978-1-108-41667-2 Hardback

The first author gratefully dedicates his part in this book to his parents, Jean and Eric Casey, who made it all possible by their struggles and sacrifices in the hard times of the 1950s and 1960s and for their endearing love and encouragement over the whole of their lives.

Contents

Credits

Unless otherwise stated, the simulations presented in the book have been obtained with ANSYS CFX and ANSYS mechanical software.

All photographs, simulation results, line diagrams and charts are supplied from PCA Engineers Limited with the exception of those where credit is given to another source. The copyright of the figures remains with the original sources.

Introduction

This book is about the engineering science that is most useful for the design and analysis of radial turbocompressors over a wide range of applications. The aim is to provide the most relevant information for designers, especially the pragmatic approaches that the authors find are quickest and the most effective. The book concentrates on one-dimensional (1D) and two-dimensional (2D) analyses of the fluid dynamic, thermodynamic and mechanical phenomena. These analyses provide rigorous, common-sense ways of looking at the problems and are usually the most insightful way to an engineering solution that is suitable for design purposes. These methods are also the most amenable and doable for engineers who study them. The 1D and 2D analyses often provide a good grounding for understanding the results arising from detailed analysis with three-dimensional (3D) simulations.

During their careers, the authors have been faced with many difficult questions with regard to radial compressors, and this book documents some of the approaches and shortcuts in the wide literature on this subject that they have found to be the most useful. The book provides a coherent description of the most useful theory and the necessary pragmatic experience associated with it in the design environment. It is hoped that the book provides a suitable introduction to engineers entering this field, enabling them to get up to speed when starting from scratch, and gives useful guidance for advanced students of turbomachinery. It should also act as a practical detailed reference for practitioners, specialists and academics working in this area, helping them to see the clusters in the forest rather than individual trees. For additional guidance, each chapter includes a brief list of the important learning objectives related to its content and gives extensive references for further study.

In Chapter 1, the basic definition of a turbomachine is considered together with a classification and history of different turbomachines. Different applications of radial flow turbocompressors are then outlined to show their wide range of aerodynamic duty. They are used in many different industries due to their inherent robustness, good efficiency and broad operating range. Meeting environmental goals with reduced carbon and other greenhouse gas emissions is likely to lead to more applications of radial compressors in the future. The focus of Chapters 2–9 is to take all readers to the same level of understanding in the essential principles and concepts of aerothermodynamics relevant to compressors before details of the design and performance of radial compressors are examined in the subsequent chapters. The background knowledge in these early chapters is also relevant for most other turbomachines, but is presented in a way that draws out specific information and useful insights relevant to radial compressors. Much of the content of these first chapters derives from lecture courses on fluid dynamics, gas dynamics and thermodynamics.

Chapter 2 considers energy transfer in radial compressors, both from the kinematic point of view through the velocity triangles and from the statements of the first and second laws of thermodynamics. Here some emphasis is placed on the definition and physics of the aerodynamic work, also known as the polytropic head, in isentropic, polytropic and isothermal compression processes. The importance of the centrifugal effect and the relevance of the degree of reaction are particularly emphasised. This is followed by Chapter 3 introducing different equations of state that can be used to model the gas behaviour for the many radial compressor applications where the gas deviates from ideal gas behaviour. Chapter 4 covers some thorny and oft-neglected concepts related to the definition of efficiency in compressors taking into account the dissipation losses, the aerodynamic work and the change in kinetic energy across a component. Chapters 3 and 4 might be omitted on a first reading of the book.

Chapters 5–9 provide fundamentals of the basic fluid dynamic theory needed to understand the complex flow in radial turbocompressors. Chapter 5 provides a physical description of the fluid dynamics of diffusing flow in curved blades and flow channels and how this is affected by the duty of the compressor. Chapter 6 provides an overview of gas dynamics for compressible flow including the important features of the shock system at the inlet to the blade rows in a transonic impeller. Chapter 7 highlights the important fluid dynamic principles of the flows around aerofoils, cascades of blades and the aerodynamic loading limits involved in the diffusing flow in radial compressor flow channels both in impellers and vaned diffusers. Chapter 8 introduces the nondimensional similarity parameters that are of most use in the analysis of turbocompressors and provides a description of their relevance and application. A consistent use of these dimensionless parameters is used throughout the book to categorise the different aspects and link the technology in different compressor applications. The effect of compressibility is considered from the concept of thermodynamic similarity leading to the definition of the tip-speed Mach number, based on the ratio of the inlet speed of sound and the mechanical blade speed at the impeller outlet. Chapter 9 discusses two other non-dimensional parameters known as the specific speed and specific diameter, together with their presentation in a Cordier diagram. The awkwardness of these dimensionless parameters is described in Chapter 9, where reasons for scepticism with regard to their use for radial compressors is explained. Despite its popularity in many radial compressor publications, the specific speed is otherwise hardly mentioned in the book.

These earlier chapters are a compendium of fundamental knowledge and concepts that are referred to in the subsequent chapters, which then examine details of the design, development, performance and testing of radial turbocompressors. Chapters 10–20 are derived from the lecture course on axial and radial compressors given by the first author at the Institute of Thermal Turbomachinery and Machinery Laboratory (ITSM) in Stuttgart University and from training material used by PCA Engineers Limited. Chapter 10 begins this process with a one-dimensional description of the losses and the prediction of the performance levels of radial compressors. The relevance of the duty required in terms of the important nondimensional parameters, the flow coefficient and the tip-speed Mach number in determining the level of the efficiency that can be expected is highlighted. Elementary considerations which determine the size and rotational speed of a compressor for a given duty are also described.

The background to the design of different components is covered in in the next three chapters. In Chapter 11, the design of the impeller inlet to achieve the most compact stage is emphasised. The crucial importance of the work coefficient in determining the steepness of the impeller characteristic is explained. The impeller and the diffuser are coupled together by the impeller outlet flow and its velocity triangle and a little-known diagram, first used by Mehldahl (1941), is presented in different forms to elucidate the matching of these components at different flow rates and for different styles of design. In Chapter 12, the pressure recovery of a vaneless diffuser and of various forms of vaned diffuser is described. The performance of the different zones of vaned diffuser is linked to the performance of planar diffusers. Chapter 13 considers other stator components, including the inlet nozzle and inlet guide vanes upstream of the impeller, and the different possible downstream components, including crossover channels, return channels, deswirl vanes, volutes, sidestream inlets and the rotor-stator cavities, and their influence on axial thrust.

The aerodynamic design tools that the authors have found to be most useful in practical applications are described in Chapter 14 (geometry definition), Chapter 15 (throughflow methods) and Chapter 16 (computational fluid dynamics). The first author was involved in the development of geometry definition methods and throughflow methods for radial machines in the early 1980s. The second author has spent most of his career using these methods. Both authors were separately involved in some of the first applications of computational fluid dynamics (CFD) methods in industry in the 1980s and early 1990s. Some of the teething difficulties of CFD technology during its development from infancy to maturity over the last 40 years left a lasting impression and has certainly coloured our reflection of this important technology.

The critical difference between compressors and turbines is that compressors are subject to rotating stall and to surge, which limit the safe operating range of the machines. The critical issue is the onset of instabilities in the flow and how to avoid them. Despite the advances in numerical methods, the stable operating range remains one of the least well-predicted aspects of performance. The current understanding of the unstable nature of the flow in compressors at low flow near the surge line is reviewed in Chapter 17 together with different control strategies to increase the operating range. Performance maps from surge to choke in single-stage and multistage compressors are described in Chapter 18, and their use in the matching of components in different compressor applications is described.

Chapter 19 offers an introduction to the key aspects of structural integrity and rotor dynamics. The objective here is to introduce the important mechanical concepts so that an aerodynamic designer is aware of the issues and can take advantage of the many specialist works in this field. Chapter 20 closes the book by providing some details of the typical design and development processes up to and including compressor testing.

An extensive list of references is provided to demonstrate that our views are well supported by the evidence in the technical literature. This is barely the tip of the iceberg of the relevant papers that are available on the subject, and these in turn give copious leads to other sources of useful information. The list includes most of our own technical publications, not that these have any special modicum of merit compared to other publications, but because these are the ones with which we are most familiar and are our personal anchor to the technology. Like all engineer scientists we are pleased with our contribution to design methods which were not there until we personally developed them.

Communication with the authors with regard to any aspects of the book can be made via the following website: www.pcaeng.co.uk.

Preface

This book is our homage to radial flow turbocompressors. These have not only presented us with many challenging engineering design problems but also provided us with a rewarding subject of academic study. They have also given us many friends, a network of great like-minded colleagues the world over and the major part of our income over many years.

The risk of writing a preface is that it can have the unwanted effect that some readers may read the preface and leave the rest of the book untouched. While working together as engineers on axial compressors, the authors experienced this effect with several doctoral and master theses. In many of these, the preface was invariably compelling reading, occasionally more immediately memorable than the theses themselves. So here we give our greetings to all those who, like us, prefer to start a book with the preface.

There is no single correct way to structure a book on radial turbocompressors, and our approach is outlined in the Introduction. The content is guided by our experience and knowledge of the subject. It gives a glimpse of the theoretical concepts in the order they are needed, much like an introductory lecture course, and provides references to more detailed coverage. As the story develops, important things get emphasised by appearing in many different chapters, often in different guises and usually increasing in complexity. The book is based on part of an introductory lecture course on axial and radial compressors given by the first author to master students in the University of Stuttgart. The content of the lecture course has been greatly expanded and tempered by practical experience both authors have gained in consultancy and design work for many compressor industries. It has been helped by notes made by the first author with regard to technical issues throughout his career, either jotted on paper or annotated on to copies of important technical papers. Without these, the book would have been a great deal poorer, as memories of the initial difficulty in understanding a new concept when it is still on the learning curve fade quickly when the concept has finally been grasped. The second author has been more than happy to let the first author lead the way in describing the theory, feeling himself to be more suited to describing the pragmatism in application.

The purpose of this book is to set out the basic principles and rules associated with the design of radial flow turbocompressors across a range of applications. Primarily the book is concerned with the aerodynamic design, but this cannot be divorced from the mechanical aspects, and so these are touched upon but are not gone into so deeply.

The book is written from the point of view of the designer and tries to cover most of the points that need to be considered in order to produce a successful radial flow turbocompressor. The emphasis is on understanding the theory behind the design process and on minimising the reliance on empirical rules. However, because of the complexity of the flow patterns, a lot of empiricism and some conjecture still remains.

This is a book the authors wished they had had when they first started work on centrifugals. We hope that it gives guidance to newcomers on their own road of discovery with this technology and conveys something of our own pleasure and excitement of working in this field. It is also intended to provide fresh insights for experienced engineers and specialists already in this field; we hope to incite reflection on the basic engineering science and the value and enduring contribution of 1D and 2D methods, often as a means of understanding modern 3D simulations.

Acknowledgements

Many individuals have helped in the writing of this book, and it is a pleasure to acknowledge this. Foremost among these are Hamid Hazby, Daniel Rusch, Paul Galpin and Maxine Backus. Dr Hazby provided the initial formulation of the content on transonic impeller design and was kind enough to review some of the early drafts of other chapters, making several useful contributions. Dr Rusch of ABB Turbocharging read several chapters and was always willing to help by explaining his own understanding of many issues. Paul Galpin provided a short description about modern CFD codes, which is the basis of some of Chapter 16 . Maxine Backus read and corrected the use of English in all the chapters in draft form and made the book better with her concise improvements to the style.

Closer to home within PCA Engineers Limited, we are grateful to all of our colleagues and partners who have supported our activity on this book. Specific thanks go to Peter Came, who provided the solid foundation for the evolution of compressor design at PCA. John Calvert kindly reviewed some of the content and made useful suggestions. Ian Woods, Mark Dempsey and Colin McFarlane helped with the technical content on mechanical integrity and rotor dynamics. Simon Welburn and Rob O'Donoghue did excellent work on many of the figures.

The first author would like to thank many graduate and postgraduate students in Stuttgart University who attended or helped with the lecture course on *Turboverdichter* (turbocompressors in German), which provided the initial structure for this book, Casey (2008).

For the stimulating discussions on many issues, feedback on different chapters, supply of offprints of technical papers, provision of initial versions of illustrations and technical data, provision of some excellent summaries of technical issues and help in various other ways, we are grateful to the following persons: Nicola Aldi of the University of Ferrara, Claudine Bargetzi of MAN Energy Solutions, Urs Baumann of MAN Energy Solutions, Yves Bidaut of MAN Energy Solutions, John Bolger of Rolls-Royce Ltd., Colin Casey, Professor Nick Cumpsty of Imperial College, Peter Dalbert, Fabian Dietmann, Ivor Day of Cambridge University, Professor Günther Dibelius of RWTH Aachen University (deceased), Markus Diehl, Professor John Denton formerly of Cambridge University, Professor Erik Dick of Ghent University, Fabian Dietmann of Celeroton AG, Dietrich Eckardt, Professor Phillip Epple of Friedrich Alexander University, John Fabian, Paul Galpin of ISIMQ, Daniel Grates of RWTH Aachen University, Professor Ed Greitzer of Massachusett Institute of

Technology (MIT) Gas Turbine Laboratory, Professor Georg Gyarmathy of the ETH Zürich (deceased), Tom Heuer of BorgWarner, Werner Jahnen of SWS Tech AG, Tony Jones of Pratt and Whitney, Professor Nicole Key and Fangyuan Lou of Purdue University, Franz Marty (deceased), Professor Rob Miller of Cambridge University, Professor Beat Ribi of FNHW, Colin Rodgers, Professor Isabelle Trébinjac of the École Centrale de Lyon, Professor Markus Schatz of Helmut Schmidt University, Georg Scheurerer of ISIMQ, Joachim Schmid of Delta JS AG, Christoph Schreiber, Om Sharma of United Technologies, Professor Zoltan Spakovsky of MIT Gas Turbine Laboratory, Professor Stephen Spence of Trinity College in Dublin, Bobby Sirakov of Garret Motion Inc., Professor Xinqian Zheng of Tsinghua University in Peking, Christof Zwyssig of Celeroton AG, and others who have helped but may have inadvertently been forgotten in this list.

For their patience and forbearance in guiding us through the production of the book, we would also like to thank all the team at Cambridge University Press.

We would also like to acknowledge the enormous debt we both owe to many engineers and teachers who, even when being unaware of exactly how they were helping us, encouraged us to find our own way and own solutions throughout our careers. The book is written partly in thanks to the many engineers, researchers and companies who have a similar interest in this technology and have shared their enthusiasm by publishing a wealth of technical literature describing the essential engineering science that explains the performance of radial flow turbocompressors. Hopefully the book does justice to their work.

For those parts of the book which are still unclear, and for any muddles in the text and blunders in the equations, the authors are directly responsible.

Conventions and Nomenclature

Conventions

1. Unless otherwise stated, all units in all equations are SI units.
2. It has not been possible to avoid the use of the same symbol for several different quantities. This is usually made clear by the context.
3. Parameter values and quantities used only very locally in the text are defined where they are used and are not all listed here.
4. Several slightly different conventions have been used for station locations. The normal notation uses locations with numerical values increasing from 1, 2 and 3 and so on through a compressor stage, as shown in Figure 10.6. In different situations, 1 and 2 can also indicate the inlet and exit of a stage or the inlet and outlet of a component. Subscript 2 is most often used to denote the impeller outlet. In some situations, i is used to denote the inlet and o is used to denote the outlet.
5. The subscripts h and c are used to denote the hub and the casing contour coordinates, and the subscript m is used for the arithmetic mean and rms for the root mean square.
6. All angles are given in degrees. The blade angles are defined relative to the meridional flow direction. A positive value indicates flow angles that are in the direction of rotation and negative values are against the direction of rotation, as shown in Figure 2.4. With this convention, the typical inlet and outlet angle of an impeller blade has a negative value. In such a situation, the statement that the backsweep or the inlet angle increases usually implies a more negative value.

Nomenclature

Letters

a	Speed of sound
$a_1, a_2, \ldots$	Polynomial coefficients in equation for specific heat
$a, b, c, d, \ldots$	Coefficients in equations
$A, B, C, D, \ldots$	Coefficients in equations
A	Area
b	Flow channel width
B	Bernstein polynomial

B	(1) Blockage
	(2) Greitzer B parameter
BL	Blade loading parameter
c	(1) Absolute velocity
	(2) Damping coefficient
c_u, c_r, c_z, c_m	Absolute velocity components
c_d	Dissipation coefficient
c_f	Skin friction coefficient
c_p	Specific heat at constant pressure
c_v	Specific heat at constant volume
C	Coefficient in Mallen and Saville method
C_D	Drag coefficient
C_L	Lift coefficient
C_M	Moment coefficient
C_p	Pressure coefficient
D	(1) Diameter
	(2) Drag
D_h	Hydraulic diameter
D_s	Specific diameter
DF	Diffusion factor
DH	The De Haller number
e	Eccentricity
E	(1) Energy
	(2) Young's modulus
f_D	Darcy friction factor
f_s	Schultz correction factor
F	Force component
g	(1) Staggered spacing
	(2) Acceleration due to gravity
h	(1) Specific enthalpy
	(2) Height
h_t	Specific total enthalpy
H	Boundary layer shape factor
i	Incidence angle
I	(1) Specific rothalpy
	(2) Moment of inertia
j_{12}	Specific dissipation losses
k, l, m, n	Coefficients in equations
k	(1) Blockage factor
	(2) Bending stiffness
k_c	The Casey coefficient for heat transfer to a compressor
K	Coefficient
KE	Kinetic energy

L	(1) Length
	(2) Lift
m	(1) Distance along the meridional direction
	(2) Mass
	(3) Exponent in gas dynamics equations
m'	Distance along the normalised meridional direction
$\dot{m}$	Mass flow rate
M	(1) Mach number
	(2) Torsional moment on a shaft
M_{u2}	Impeller tip-speed Mach number
M_w	Molecular weight
MR	Mach number ratio
n	(1) Rotational speed in rps
	(2) Polytropic exponent
	(3) Normal vector to a surface
	(4) Distance in the normal direction
	(5) Coefficient in Haaland equation
N	(1) Rotational speed in rpm
	(2) Amount of a constituent expressed in moles
$NPSH$	Net positive suction head
o	Throat width
p	(1) Pressure
	(2) s-shaped logistic function
p^*	Rotary stagnation pressure
P	(1) Power
	(2) Parameter value
PIF	Power input factor
P_h	Preheat factor
PE	Potential energy
q	Distance along a quasiorthogonal
q_{12}	Specific heat input
Q	Heat input
r	Radius
r, z, θ	Coordinates in cylindrical coordinate system
r_k	Kinematic degree of reaction
R	Specific gas constant
R_m	Universal gas constant
Ra	Centre-line average roughness
Re	Reynolds number
RF	Relaxation factor
s	(1) Specific entropy
	(2) Distance along a streamline or a blade
	(3) Spacing

S	(1) Entropy
	(2) Source terms in Chapter 16
St	Strouhal number
t	(1) Time
	(2) Blade thickness
T	Absolute temperature
$Trim$	Square of the impeller casing diameter ratio
u	(1) Blade speed
	(2) Specific internal energy
	(3) Fluid velocity in Chapter 16
U	Internal energy
v	Specific volume
V	(1) Volume
	(2) Velocity in Chapter 16
$\dot{V}$	Volume flow rate
w	Relative velocity
w_{s12}	Specific shaft work
w_{v12}	Specific displacement work
W	(1) Width of flow channel in diffusers
	(2) Work input
x	Mole fraction
x, y, z	Coordinates in Cartesian coordinate system
r, θ, z	Coordinates in cylindrical coordinate system
y_{12}	Specific aerodynamic work
y^+	Normalised distance from the wall in a boundary layer
Y	Stagnation pressure loss coefficient
Z	(1) Altitude above a datum level
	(2) Gas compressibility factor
	(3) Blade number

Symbols

α	Absolute flow angle
$\Delta\alpha$	Flow turning angle of a cascade
α'	Absolute blade angle
α, β	Reynolds-independent and Reynolds-dependent losses
β	(1) Relative flow angle
	(2) Reduced frequency
	(3) Core rotation factor in swirling flow
β'	Relative blade angle
γ	(1) Ratio of specific heats (and isentropic exponent of an ideal gas)
	(2) Blade lean angle and rake angle
δ	(1) Boundary layer thickness
	(2) Deviation angle

	(3) Shock wave deflection angle
	(4) Logarithmic decrement
δ^*	Boundary layer displacement thickness
δ, ε	Coefficients in real gas equations
Δ	Difference or change in a parameter, such as Δh_t
ε	(1) Tip clearance gap
	(2) Kinetic energy thickness of a boundary layer
	(3) Turbine expansion ratio
	(4) Shock wave angle
	(5) Inclination angle of meridional flow to axial direction
	(6) Diffusion coefficient
ζ	(1) Energy or enthalpy loss coefficient
	(2) Damping ratio
ζ_q	Heat transfer in a diabatic compression as a fraction of the aerodynamic work
η	Efficiency
θ	(1) Boundary layer momentum thickness
	(2) Circumferential coordinate
	(3) Planar diffuser divergence half-angle
	(4) Camber line slope angle
κ	Isentropic exponent
ω_s	Specific speed
Ω_s	Secondary circulation
λ	(1) Work coefficient
	(2) Fluid conductivity in Chapter 16
μ	Dynamic viscosity
ν	(1) Kinematic viscosity
	(2) Inlet dimeter to outlet diameter ratio
	(3) Poisson's ratio
ξ	(1) Kinetic energy loss coefficient
	(2) Ratio of local kinetic energy to the shaft work
π	Pressure ratio
ρ	Density
σ	(1) Solidity
	(2) Stress
	(3) Slip velocity ratio
τ	Shear stress
υ	Polytropic ratio
Φ	Mass-flow function
ϕ	(1) Flow coefficient
	(2) Phase angle
ϕ_{t1}	Global inlet flow coefficient
χ	Entropy loss coefficient

χ_d	Nondimensional slope of work characteristic at design point
ψ	(1) Pressure rise coefficient
	(2) Slope angle of a quasiorthogonal to the meridional direction
ω	(1) Rotational speed in radians per second
	(2) Pressure loss coefficient

Subscripts

0	At zero Mach number
1, 2, 3	(1) Station numbers
	(2) Component numbers
1	Inlet
2	Outlet
∞	Far upstream
a	Annular
act	Actual
b	Burst
$btob$	Blade-to-blade
c	(1) Compressor
	(2) Curvature
	(3) At critical conditions
	(4) Vortex centre
	(5) Relating to the camber line
	(6) Absolute velocity
ch	Choke
$corr$	Corrected
d	(1) Disturbance
	(2) Discharge
	(3) Design
dep	Departure function value
$diff$	Diffuser
ds	Deswirl
DF	Disc friction
e	(1) Value at the edge of a boundary layer
	(2) Engine
eff	Effective
eq	Equivalent
$exit$	Exit
$Euler$	From the Euler equation
f	(1) Friction
	(2) Fuel
$htoc$	Hub to casing
GT	Gas turbine
i, j	Station location

i	Isothermal
id	Ideal
igv	Inlet guide vanes
imp	Impeller
irr	Irreversible
j	Jet
l	Lower
le	Leading edge
lam	Laminar
leak	Leakage
m	(1) Meridional
	(2) Mechanical
max	Maximum value
min	Minimum value
n	(1) Normal to meridional direction
	(2) Natural
o	Outlet
opt	Optimum value
p	(1) Polytropic
	(2) at peak efficiency
ps	Pressure surface
q	Along the quasiorthogonal
r	(1) Rotor
	(2) Reduced
rc	Return channel
rec	Recirculation
ref	Reference
rev	Reversible
rp	At rated point
RS	Rotating stall
s	(1) Stator
	(2) Isentropic
	(3) Sand
	(4) Shroud
shutoff	Shutoff value at zero flow
ss	Suction surface
SW	Swept
t	Total conditions
te	Trailing edge
th	Throat
turb	(1) Turbulent
	(2) Turbine
TC	Turbocharger

u	(1) In the circumferential direction
	(2) Virtual station in Figure 2.18
	(3) Upper
	(3) Ultimate
v	Vapour pressure
vd	Vaned diffuser
vld	Vaneless diffuser
vol	Volumetric
w	(1) Wheel
	(2) Wake
	(3) relative velocity
θ	Circumferential direction

Superscripts

0	Inlet
*	Critical conditions at a throat
′	related to blade value
c	Compressor
$comp$	Compressible
d	Design
ss	Static-static
$stall$	At stall
t	turbine
ts	Total–static
tt	Total–total

1 Introduction to Radial Flow Turbocompressors

1.1 Overview

1.1.1 Introduction

In this chapter, the subject of the book is introduced to the reader. Classifications of different types of turbomachine and different types of compressor are reviewed, to define precisely what is meant by a radial flow turbocompressor. This first chapter also considers the basic operating principle of turbocompressors, known as the Euler turbine equation, which is valid for all turbomachines. The function of the components of a single-stage compressor are introduced, and the many different forms of single-stage and multistage compressors are described. An overview of different types of impeller is provided.

Centrifugal compressors have become ubiquitous over the twentieth century across a wide range of applications due to their inherent robustness, good efficiency and broad operating range. Numerous applications of radial flow turbocompressors are described in this chapter whereby the special aerodynamic features that are relevant to their design for each application are highlighted. The wide range of applications explains why the book is devoted to this relatively narrow part of the wide field of turbomachines. Even small gains of efficiency, improvements in operating range and an increase in swallowing capacity are welcomed across all industrial applications.

Some comments are provided in the chapter on the effect of renewable energy on the applications of radial compressors. The future of centrifugal compressors seems equally bright with the present focus on environmental goals, including the future decarbonisation of power generation and transportation, with a possible shift to a hydrogen economy. There is ever-increasing pressure to produce higher-efficiency gas turbines at increased cycle pressure ratios for aviation propulsion and ground-based power generation, which also favour the use of centrifugal or axial-centrifugal machines.

Aspects of the history of turbomachines are also given together with a short overview of some other useful books on this topic.

1.1.2 Learning Objectives

- Define what is meant by a turbomachine.
- How does a turbocompressor differ from other rotary compressors?

- Define the differences between turbines and compressors, axial and radial machines and thermal and hydraulic turbomachines.
- Know the basic operating principle of a turbomachine based on the Euler turbine equation.
- Be able to identify various applications of radial flow turbocompressors.
- Have a broad knowledge of the history of turbomachinery with emphasis on radial flow turbocompressors.
- Be aware of other useful books on the same subject matter.

1.2 Definition of Turbomachinery

Turbomachines are rotating machines used to change the state of a working fluid – liquids in pumps, gases in compressors and one or the other in turbines. Pumps, fans, ventilators and blowers are used to transport fluids; turbines and wind turbines extract energy from a fluid stream. Turbochargers, propellers and jet engines constitute part of propulsion or transportation devices. This wide remit means that turbomachines span almost all industrial sectors and play a vital role in many of them. They are often only a small part of a more complex system, which typically imposes the requirements for and constraints on their design. There is a huge range of types, sizes and speeds, with large economic significance in numerous applications. In addition, the technological and scientific interest of fluid dynamics, thermodynamics and mechanics makes turbomachinery a highly worthwhile subject of engineering study.

Turbomachinery is fundamentally linked to energy conversion in its many forms. Most electrical power is currently generated by steam turbines in nuclear and coal-fired plants, but even modern solar or biomass power plants use small steam or gas turbines for energy conversion. Natural gas is the fuel most commonly used for land-based gas turbine power plants, and nowadays these are often combined with a steam turbine which, in turn, obtains its heat source from the hot exhaust gases of the gas turbine. Water turbines in hydroelectric schemes and wind turbines use renewable energy sources for power generation.

Turbomachinery plays an equally important role in the transportation industry – the gas turbine jet engine is used for propulsion in nearly all aircraft; turbochargers are widely used in ground transportation and as an integral part of diesel propulsion in nearly all ships, lorries and diesel cars and, increasingly, in gasoline-fuelled vehicles. Pumps are the world's most ubiquitous turbomachines, found everywhere where liquid needs to be transported. Compressors and pumps are important to chemical processes, to the use of industrial gases and in the oil and gas industries. Compressors and ventilators are key components in air conditioning, cooling and refrigeration equipment. Turbomachinery can even be found in the medical industry, where pumps are deployed for blood circulation, ventilation-assist fans are used in the treatment of sleep apnoea and in other clinical applications. The drive powering a high-speed dental drill is, in fact, generally a tiny high-speed air-driven turbine.

The distinguishing feature of a turbomachine is that energy is transferred continuously to or from a fluid by the aerodynamic action of the flow around rotating blades in an open system. Different aspects of this are discussed later in this chapter.

1.2.1 Open System

The definition of an open system is discussed in Chapter 2 on the thermodynamics of energy transfer. To explain a turbomachine as part of an open system, it is useful to make a simple comparison between the work-producing processes in a gas turbine and those of an internal combustion engine. In the gas turbine, the energy transfer takes place through the continuous aerodynamic forces of the gas passing through the flow channels of a machine which is open at both ends. There are no valves or movable plates to force the flow to pass forwards through the machine or to stop the flow from reversing direction, as is the case in a reciprocating machine.

The different parts of the gas turbine process take place continuously in separate components which are open at both ends and are specifically designed to achieve high efficiency for each purpose: the intake, the compressor, the combustion chamber and the turbine and, in a jet engine, the outlet nozzle. Work is produced in the rotor blade rows of the turbine, and this is used as motive power to drive the compressor. The excess power not needed by the compressor is then available to drive a generator or to provide thrust for an aero engine.

In the internal combustion engine, similar processes take place but the energy is transferred by the intermittent operation of forces acting on the pistons in a closed cylinder, which are forced to change direction regularly. There are valves which open and close to trap gas in the cylinder and to allow it to enter and leave during the different expansion and compression processes. These variable processes take place in the same component but at different times, and the work transfer is intermittent. In a two-stroke engine, there is one power stroke every two strokes of the piston in a single revolution of a shaft; in a four-stroke engine, two shaft revolutions are needed for each power stroke.

In a gas turbine, there are no strokes, since each blade passage provides more or less continuous flow and ensures that the forces acting on the shaft, with the exception of the small variations due to the interaction of the moving and stationary blade rows, are steady. Because of this, a turbomachine basically has only one moving part. This rotates rather than oscillates, and has high reliability and a long lifetime. The continuous rotating motion applies relatively simple mechanical loads, and the rotors may achieve high rotational speeds, resulting in high power density – turbomachines that are generally light in weight for a given power.

1.2.2 Continuous Energy Transfer by Flow over Blades Rotating around an Axis

Turbomachines differ from positive displacement devices in that they effect a continuous energy transfer. In the case of positive displacement machines, the work is transferred discontinuously by changing the volume of a trapped mass of fluid, releasing it and then intermittently repeating the process. Most positive displacement machines are reciprocating – as in a piston engine – but some, such as screw compressors, liquid ring compressors and sliding vane blowers, may have rotary motion and be nearly continuous in operation.

The limitations of the intermittent power and mass flow capability associated with reciprocating machinery explains the fact that the largest diesel engine (for container ships) has a power output of about 80 MW, compared with typically 1600 MW for the largest single-shaft steam turbines. In contrast to discontinuous operation, a large energy flux can be obtained in turbomachines and hence high gas flow rates require only comparatively small components. In addition, rotary motion is a natural characteristic of turbomachinery and does not require complex oscillating connecting rods from the machine crankshaft to a rotating shaft.

A categorisation of different compressor types is given in Figure 1.1. The piston compressor comprises a reciprocating piston and associated valves and operates on the same principle as a bicycle pump. A diaphragm compressor comprises a flexible diaphragm which is moved by means of a hydraulic system. The movement of the diaphragm draws the process gas into a chamber enclosed by the diaphragm and then forces it under pressure through a valve. Another possibility, often used for vacuum applications, is to use a jet of gas at high pressure in an ejector to transfer its momentum to a stream of gas at low pressure.

Screw compressors use two helical screws, both known as rotors, which mesh to enclose a small volume of gas. Gas enters at the suction end of the casing and moves through the threads of the screws as they rotate and is then forced at pressure out of the exit of the compressor casing. The rotary lobe compressor also consists of two rotors that have two or three lobes. The rotors rotate such that air is sucked into the inlet and is forced out of the discharge by the lobes. The liquid-ring compressor has a vaned

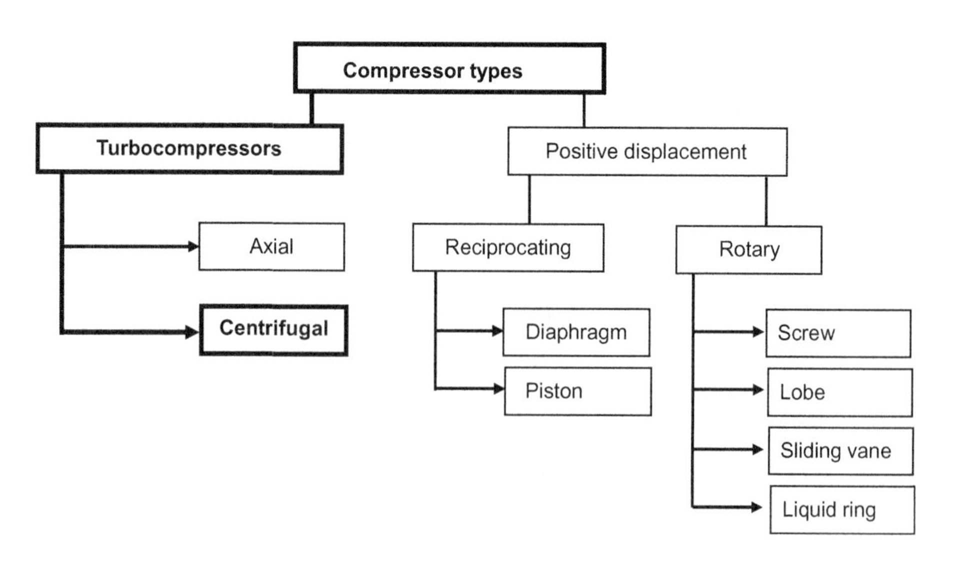

Figure 1.1 Categories of different compressor types.

impeller rotating eccentrically within a cylindrical casing. Liquid (usually water) within the casing is centrifuged outwards and provides a liquid ring that seals the space between the impeller blades and the casing. The eccentric impeller results in a periodic variation of the volume enclosed by the vanes and the liquid ring; this draws air into the casing and delivers it at a higher pressure. A sliding vane compressor also has an eccentric rotor, but the blades themselves are able to move within the rotor to seal the casing. The blades are held in close contact with the outer shell under the action of the centrifugal force, and the sealing between the vanes and the casing is improved by the injection of a lubricating oil over the entire length of the blade. The main drawback of positive displacement compressors is that the compression cycle is discrete: the gas comes in bursts rather than smoothly and continuously.

On a timescale of the order of seconds, the time-averaged properties within a turbomachine can be considered as a smooth and continuous steady flow process, at least at the usual stable operating points. However, the forces acting on a blade are not steady at very small timescales (milliseconds). The aerodynamic interaction of the motion of adjacent blade rows leads to small, unsteady variations in pressure at frequency with which the blades and vanes pass each other. In fact, these variations can force blade vibrations, and if they are not under control the mechanical damage can severely interfere with the life and aerodynamic function of the machine. In fact, a classical analysis by Dean (1959), clarified by Hodson et al. (2012), shows that the equations describing the change in enthalpy for a fluid particle in a turbomachine require a time variation in pressure. Conveniently, for most analyses the design engineer is concerned with the mean, time-averaged performance and may consider the smooth operation to be steady.

1.2.3 Aerodynamic

The truly key feature of a turbomachine is the aerodynamic forces acting on the blades. It is the motion of the fluid compared to the motion of the machine that is primary. The actual motion of the fluid in a positive displacement machine is, by contrast, generally less well ordered and is secondary to the motion of the components – such as pistons or valves. As the flow passes over the blade surface of a turbomachine, an aerodynamic lift force is generated, similar to the lift force on an aircraft wing. If the blades are moving in the direction of this force, then there is a transfer of work between the fluid and the shaft. This is the essence of turbomachinery, and a good understanding of fluid dynamics is therefore one of the key building blocks in the design of good turbomachines.

As an example of the importance of aerodynamic forces, it is useful to consider the work input into an Archimedean screw. This rotating machine acts as a pump to raise water to a higher level (or as a water turbine when the water is flowing downhill), but it is not a turbomachine because the work input is not related to an aerodynamic force. Instead, it is due to work done against gravitation to lift the weight of the water as it periodically fills and empties from a section of the screw.

Similarly, one can distinguish between undershot and overshot water wheels. In an overshot water wheel, the water enters at the top of the wheel and the weight of the water held in each bucket provides the necessary torque. This is not a turbomachine as long as the small impulse from the water entering and leaving the buckets can be neglected. An undershot wheel, by contrast, makes use of the kinetic motion of the fluid to provide a force on the paddle blades. The head of water must first be converted into kinetic energy or the wheel needs to be mounted in a fast-running stream. The undershot water wheel is considered to be a turbomachine as its operation is due to an aerodynamic force from the motion of the water acting on the blades. In this case, however, it is a drag force and not an aerodynamic lift force, which is more normal in turbomachines.

The range of types and sizes of turbomachines is vast. Common examples in rough order of increasing size are: high-speed dental drills, computer cooling fans, hair dryers, ventilation fans, turbochargers for cars, water pumps, blowers for air compression, compressors and turbines for helicopter engines, turbochargers for ships' diesel engines, turbines and compressors in aero engines and industrial gas turbines, water turbines, steam turbines and wind turbines in the power generation industry. If world records are considered, wind turbines are the largest (having a rotor tip diameter above 200 m), Francis turbines are the most powerful in terms of the energy from a single blade row (with a power of 800 MW from a single impeller with a diameter of between 5 and 10 m), and the steam turbine is the most powerful turbomachine on a single shaft.

Practical radial turbocompressors can be found in a range of applications with impeller diameters, D_2, of between 10 mm and 2000 mm, and an impeller tip-speed, u_2, of between 100 and 700 m/s. An overview is provided in Figure 1.2. Low-pressure compressors with low tip-speeds are known as fans and medium-pressure compressors as blowers. While

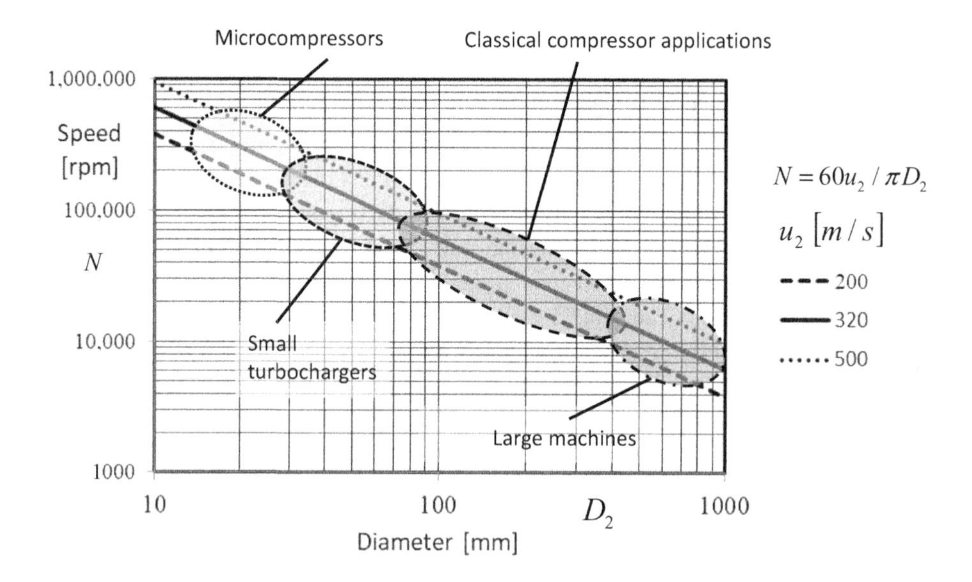

Figure 1.2 An overview of the range of size and rotational speed of radial flow turbocompressors.

most of the fundamental principles of design can be comfortably applied across the range of size and tip-speed, some aspects do not scale. For example, the tip clearance of open impellers is one of the key factors limiting the efficiency of small stages, and the blade thickness tends to become larger relative to the diameter at small size due to manufacturing constraints. In many applications at the larger sizes, it may be more efficient – and may reduce the outer dimensions – to use a smaller axial compressor or an axial-centrifugal compressor. Large centrifugal machines are often preferred, however, as they are more robust and have much wider operating characteristics.

1.2.4 Principle of Operation

An understanding of the fact that machines of such a broad range of power density and application operate on the same single physical principle is the key to effective turbomachinery design. The unifying theoretical system is the Euler turbine equation, first derived from Newton's second law by the polymath Leonhard Euler (1707–1783) in the eighteenth century.

Newton showed that a force is needed to generate a change in the speed of an object, and the change in speed is proportional to, and along the direction of, the applied force. Euler extended this and deduced that the torque acting on a turbomachinery shaft is related to the forces causing a change in the angular momentum of the gas by a change in its circumferential velocity between the inlet and outlet of a blade row. This is illustrated in Section 2.2 and discussed in detail in most other chapters. When the Euler turbine equation is applied to the gas passing from the inlet (station 1) to outlet (station 2) across a rotor blade row, it gives the specific shaft work (the mechanical work per unit mass flow on the shaft) in terms of the changes in the circumferential component of the gas velocity (c_u) and the blade speed (u) as

$$w_{s12} = u_2 c_{u2} - u_1 c_{u1} = \Delta(u c_u). \tag{1.1}$$

The individual blade rows guide the fluid as it passes through the flow channels and produce a change in direction of the swirling flow relative to the circumferential direction. It is this change in angular momentum across a rotor which creates work on the shaft: such that the principle of a turbomachine can be thought of as a machine in which the blades cause a change in the swirl of the flow. The addition of swirl in a rotor blade row of a compressor is usually followed by its removal in a downstream stator row, and so a turbomachine usually has at least two blade rows.

Not only is the Euler turbomachinery equation universally applicable, it also combines great simplicity and elegance. Furthermore, it is formidably potent: as it is based on a control volume approach, it determines the work of the machine simply from the changes between the mean conditions at inlet and outlet of a rotor with no knowledge of the inner workings of the blade row. It is immediately clear from the dimensions of this equation, for example, that the work done per unit mass on the fluid in a compressor is proportional to the product of two velocities and thus to the square of the blade speed. Therefore, for compressors with high pressure ratios, high blade tip-speeds are required.

In all turbomachines, at least one blade row rotates, which means that a fundamental aspect of the application of the Euler turbomachinery equation is the consideration of the flow in the blade rows, both in an inertial system relative to the moving blades and in the absolute frame of reference. The stator vanes guide the direction of the flow in the absolute frame, while the rotor blades guide the relative flow in the rotating coordinate system. An important aspect is then the relationship between flow in the absolute coordinate system, as seen by an observer outside the machine, and the flow in a relative coordinate system that rotates with the rotor. This approach was first formulated in the theory of steam turbines by Aurel Stodola (1859–1942) (Stodola, 1905). The transformation of velocity vectors from one system to the other requires vector addition of velocities. At each point in the flow through a turbomachine, the absolute velocity, c, and the relative velocity, w, can be identified and combined through the local circumferential blade speed, u, with a so-called Galilean transformation between the two inertial systems:

$$\vec{c} = \vec{w} + \vec{u}\,. \tag{1.2}$$

The Galilean transformation of velocities leading to velocity triangles is discussed in Section 2.2.5.

Another simple description of the principle of operation of a compressor is sketched in Figure 1.3, which depicts the change in the nondimensional absolute velocity and static pressure through a 2D compressor stage. The absolute gas velocity, c, increases through the impeller along together with the static pressure. At the impeller outlet, the absolute gas velocity almost reaches the tip-speed of the impeller. The deceleration of the gas in the downstream diffuser leads to a further rise in pressure. The proportion of the pressure rise in the impeller to that in the whole stage, in this case about 60%, is called the degree of reaction.

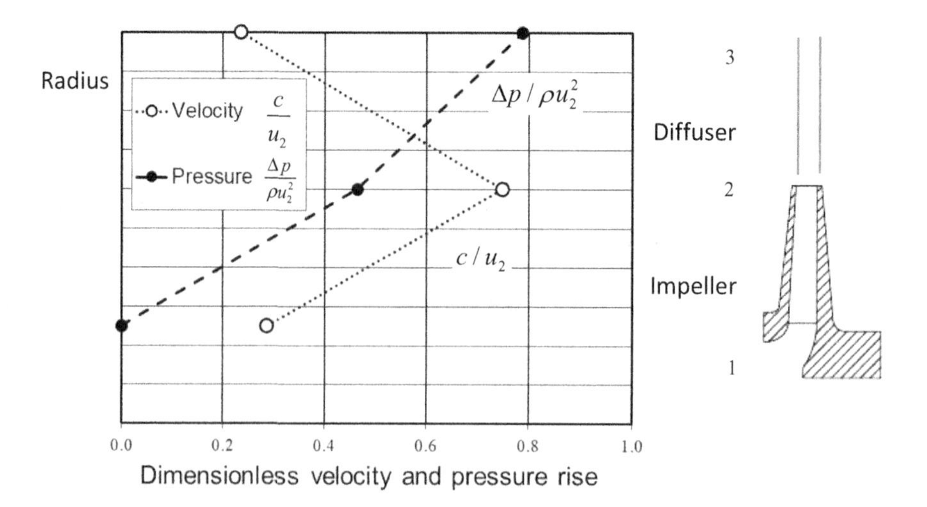

Figure 1.3 Dimensionless absolute gas velocity and pressure rise in a 2D compressor stage.

The aerodynamic forces and the stresses resulting from high-speed rotation must be safely carried by the blades and by the other mechanical elements. In addition, the blades need to stay attached to the rotor. Special attention must be given to the sealing between the rotating elements and the stationary elements. Some of the unsteady stresses in turbomachinery blade rows are associated with vibration phenomena from blade row interactions and flow instabilities, both of which may lead to material failure by fatigue. Stress analysis, vibration analysis, the selection of materials and the mechanical component design for safe operation are key aspects in the integration of turbomachinery components into actual products and are introduced in Chapter 19 of the book.

1.3 Classification of Turbomachines

1.3.1 Power Consuming and Power Producing Machines

There are many ways to classify turbomachines. This book covers only a fraction of the many categories of turbomachinery, as it deals with thermal and not hydraulic machines, radial and not axial machines, compressors and not turbines, and only includes machines enclosed within a casing, as explained in this section.

The most fundamental classification is into machines which add energy to a fluid and those which extract energy from a fluid: work-absorbing machines and work-producing machines. The addition or extraction of energy is usually achieved at the expense of the fluid pressure. Turbines have an output of shaft work from the machine which they obtain by converting the internal energy of the fluid into rotational energy of the shaft. In all turbines, except wind turbines, the pressure falls as rotational kinetic energy is generated in upstream components, and this energy is then extracted by the rotor blades. Compressors (and pumps) require work input to the shaft to generate a pressure (or head) rise. The work is first converted into static pressure or enthalpy rise and an increase in kinetic energy in a rotor blade row. The kinetic energy is then converted into a further pressure rise by a stator blade row. In some low-cost applications, such as ventilator fans, the downstream stator is not present and the exit kinetic energy is simply discarded.

1.3.2 Thermal and Hydraulic Machines

The distinguishing feature of a thermal turbomachine, as opposed to a hydraulic turbomachine, is that the working fluid is compressible. The gas undergoes a marked change in volume with the density and temperature changes that occur as it passes through the machine. In a hydraulic machine, the fluid (usually cold water) is effectively incompressible and the density remains constant: the fluid then has the same volume, and very nearly the same temperature, at the inlet and the outlet of the machine. Thermodynamics is not relevant to the performance analysis in this case.

A multistage pump has similar impellers in all of its stages as there is no difference in the volume flow at different positions in the machine. On the other hand, a multistage compressor requires that the flow channel of the stages is adapted to

Figure 1.4 The shaft of a three-stage air compressor with open impellers manufactured by Entenmach RPC LCC, Saint Petersburg, Russia. (image by courtesy of Entenmach)

account for the decrease in volume of the gas being compressed, which means that different impellers with narrower flow channels are needed on the passage through the machine. Figure 1.4 from Neverov and Liubimov (2018) shows this for a multistage machine with three different open impellers of the same diameter on the same shaft. The amount of volume compression for a given pressure rise is also a function of the gas properties of the fluid being compressed so that different impellers are also needed for different gases. The manufacturer of multistage pumps needs to develop a single stage that can be used at each position in a multistage machine as the volume flow does not change through the machine, whereas the compressor manufacturer needs to develop or adapt separate stages for each location.

A limiting case in respect of thermal turbomachinery is a low-speed ventilator where the density change is almost negligible and so these machines may be regarded as incompressible. The large changes in temperature and density in high-speed turbomachines can only be properly accounted for by the theory of compressible gas dynamics. This is another key building block to understanding the design and operation of turbocompressors and is covered in Chapter 6 of the book.

1.3.3 Acceleration and Deceleration of the Flow

The addition or extraction of energy is usually made at the expense of the fluid pressure; compressors generate a pressure rise, whereas turbines cause a pressure drop. Furthermore, whether the static pressure rises or falls through the machine is an important difference. In compressors and ventilators, the static pressure usually rises in the direction of flow, whereas it usually drops in the direction of flow in turbines. It is fundamentally difficult to persuade a fluid to move 'uphill' against rising pressure. This means that the basic aerodynamic design of compressors is generally more problematic and more affected by aerodynamic limits than that of turbines.

Compressor blade rows generally experience decelerating flow as the flow gives up its kinetic energy to produce a static pressure rise. This tends to cause the slow-moving fluid in the thin fluid boundary layers on the blades to become slower and, as a consequence, the boundary layers become thicker and are more liable to separate from the blade surface. The propensity for flow to separate means that the rate of reduction in velocity or the rise in pressure must be carefully managed, and this effectively results in a limit on the deceleration of the flow for compressor stages. Turbines, with accelerating flow in the direction of the pressure gradient, tend to have thin, stable boundary layers on the blades, and there is no limit to the amount of acceleration that the fluid encounters.

This fundamental difference between the two types of machine is evident from the cross section of any industrial single-shaft gas turbine – where there are usually three to four times as many compressor stages as turbine stages, although both have the same rotational speed and nearly the same pressure ratio.

Some turbomachines, such as a reversible pump-turbine for a pumped hydroelectric storage scheme, operate as both turbines and compressors. At times of low electrical demand, excess generation capacity is used to pump water into the higher reservoir. When there is higher demand, water is released into the lower reservoir through the same machine, which is now operating as a turbine and generating electricity. The design for pump operation takes priority; the machine will operate adequately as a turbine with accelerated flow when the water runs downhill, albeit nonoptimally.

1.3.4 Flow Direction

Another important criterion for the classification of turbomachinery is the flow direction relative to the axis of rotation. In purely axial rotors, the radius of the streamlines is approximately constant and the flow passes through the machine roughly parallel to the rotational axis. In purely radial machines, the radius changes significantly and the flow travels through the rotor perpendicular to the axis of rotation, as shown in Figure 1.5. In practice, axial machines tend to have some small radial velocity components as neither the inner nor the outer casing walls are perfectly cylindrical, and most radial machines have an axial component of velocity. Most radial flow compressors have axial flow at the inlet and are radial at the outlet (and vice versa for radial flow turbines), as shown in Figure 1.5.

In some situations, these may qualify as mixed-flow machines because of the axial inlet and radial outlet. In the present text, the adjectives *radial* and *centrifugal* are used for all machines with a significant change of radius across the impeller and with a radial flow at exit. The name *centrifugal impeller* implies that the effect of centrifugal force (radius change) in generating a pressure rise across the impeller is significant.

Defining the machine type based on the orientation of the leading and trailing edges can become confusing: for an axial inlet flow and radial outlet flow, the leading and trailing edges can also be designed with different sweeps with respect to the mean meridional flow. The term *diagonal* is sometimes used for rotors which have leading and trailing edges which are neither axial nor radial and are swept forwards or backwards in the flow channel. The term *mixed flow* can be linked to the direction

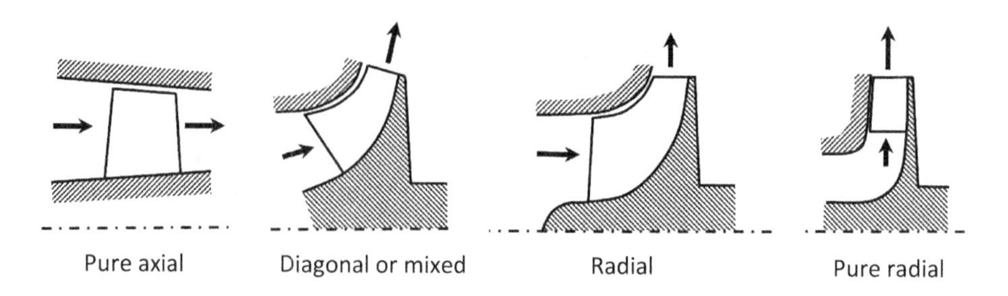

Figure 1.5 Sketches of axial, diagonal, mixed and radial flow compressors.

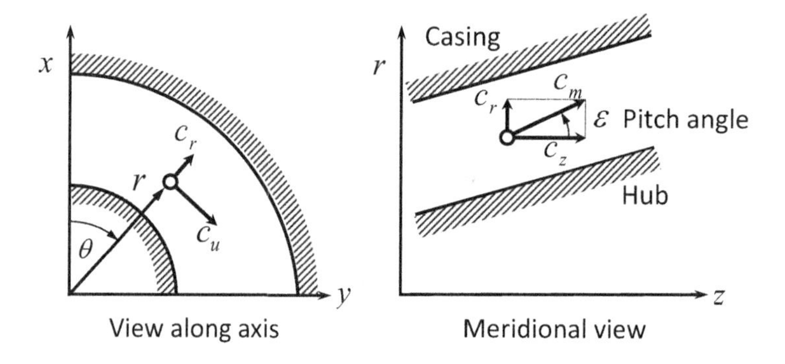

Figure 1.6 Axial, radial, circumferential and meridional velocity components.

of the mean flow at the impeller outlet, which then has a large axial component of velocity. This naming convention is mainly relevant to the impeller and not to the entire machine, as turning the meridional flow downstream of the rotor to the axial direction does not change the type of the machine.

In this book, the term *meridional direction* is often used. This refers to the mean direction of the flow in a meridional plane through the axis of the rotor and can be axial, radial or something in between. This formulation has the advantage that it allows the same equations to be used for the analysis of axial and radial machines; see the notation in Figure 1.6.

The key significance of the distinction between axial and radial flow arises owing to the centrifugal effect whereby an object travelling in a circle behaves as if it is experiencing an outward force. In an axial machine with swirling flow, this causes the static pressure to rise towards the casing to counteract the swirling motion of the fluid such that the pressure at the casing is higher than at the hub. This radial pressure gradient causes a natural pressure rise on a swirling fluid particle whenever its radius increases within a rotating blade row, and this pressure gradient accounts for 75% of the static pressure rise in centrifugal compressor impellers. This is sometimes considered to be a free pressure rise as it does not require flow deceleration and does not, in itself, result in any flow separations. The centrifugal effect makes centrifugal stages somewhat easier to design than axial stages for a given pressure rise. For this reason, the first effective gas turbine engines all made use of radial compressors; see Section 1.4.

As shown in Figure 1.3, compressor rotors increase the absolute velocity across the rotor blade rows by adding swirl to the circumferential flow, and they are equipped with a downstream diffuser or row of stator vanes to convert the rotor outlet kinetic energy into a useful pressure rise. Compressor rotors turn the flow from the meridional direction towards the circumferential direction in the direction of rotation. In contrast, turbines are equipped with an upstream stator vane or a casing to provide swirl in the direction of rotation, and this leads to high inlet kinetic energy, which is then extracted by the rotor. The turbine rotor turns the absolute flow velocities from the circumferential direction against the direction of rotation back towards the meridional direction.

1.3.5 Degree of Reaction

The degree of reaction for a turbomachinery stage describes the relative pressure rise (or fall) in the rotor compared to that of the stage. This is often discussed in the context of steam turbines where reaction turbines and impulse designs with no reaction can be found. In the history of steam turbines, both types were invented independently, the reaction turbine by Sir Charles Parsons (1854–1931) and the impulse turbine by Gustaf de Laval (1845–1913). The reaction turbine, as the name implies, is turned by reactive force rather than by a direct push or impulse. An impulse turbine has fixed nozzles (stator blades) that orient the fluid flow into high-speed jets. These jets contain significant kinetic energy, and the rotor blades, shaped symmetrically like buckets, convert this into shaft rotation as the jet changes direction. A pressure drop occurs in the nozzle, but the pressure is the same when the flow enters and leaves the rotor blade, and this gives rise to an impulse machine with zero degree of reaction. In a reaction turbine, the rotor makes use of the reaction force produced as the fluid accelerates through the rotor blades. The fluid is directed on to the rotor by the fixed vanes of the stator. It leaves the stator as a stream that fills the entire circumference of the rotor. In a reaction machine, the pressure change occurs across both the stator and the rotor.

Centrifugal compressors with no inlet swirl are reaction machines and generally have a degree of reaction close to 0.6 indicating that about 60% of the overall static pressure rise of the stage takes place in the rotor, Figure 1.3.

1.3.6 Boundary of the Flow Field

Turbomachines can also be classified with respect to configurations with and without a casing. In machines without a casing, such as windmills, wind turbines, propellers and some axial fans, where no walls exist to support a pressure difference against the ambient pressure, it is essentially the kinetic part of the change in the fluid energy which effects the energy transfer with no change in static pressure.

1.4 Short History of Thermal Turbomachines

1.4.1 The Prefix Turbo

The prefix *turbo* in the word *turbomachine* comes from the Latin source – *turbo, turbinis*, meaning whirlwind, whirlpool, vortex and, by extension, a child's spinning top. Given that whirlwinds can wreak havoc, a secondary meaning of turbo in Latin was confusion, or disorder, and this can still be identified in English words such as *disturb* and *turbulent*. The word *turbine* was first coined in French in 1828 by Claude Burdin (1790–1873) to describe his submission in a government-sponsored engineering competition for a new water power source. The French authorities of the time were keen on developing water power because they had few natural fossil fuel resources, whereas Britain had abundant energy in the form of coal, which could be used to power steam engines and had made more rapid progress in the early days of industrialisation.

In modern usage, *turbo* is a prefix for many types of turbomachine, such as turbopump, turbojet, turbofan, turboprop, turboshaft, turbocompressor and turbocharger. A turbopump often refers to a rocket fuel pump. Turbojet, turbofan, turboprop and turboshaft are different architectures of gas turbine aeroengines. Turbochargers are used in internal combustion engines to increase their power by forcing more air into the charge of a cylinder, on the principle that introducing more oxygen means that more fuel can be burned. The use of turbochargers in fast sports cars caused the word *turbo* to become synonymous with anything superlative and fast. Turbomachinery has a long history and can be considered to be a very mature technology. Despite this, new developments and applications continually take place by applying incremental pieces of knowledge discovered by the many researchers, engineers and companies in the field.

1.4.2 Historical Overview

Examples of turbines can be found in antique literature, such as Hero's turbine (or aeliopile) as designed by Hero of Alexandria (c. 10–70 AD). The centrifugal pump was invented by Denis Papin (1647–1710) in France at the end of the seventeenth century but was not brought to fruition (Harris, 1951). The decisive scientific steps to explain the operation of a pump or a turbine were taken by Leonhard Euler (1707–1783) with the derivation of the Euler turbomachinery equation; see Section 1.2.4. Developments in hydraulic turbomachinery throughout the nineteenth century involving James Francis (1815–1892), Lestor Pelton (1829–1908) and Osborne Reynolds (1842–1912) led to the pioneering work of Sir Charles Algernon Parsons (1854–1931) and to the development in 1884 of the first multistage reaction steam turbine for driving a generator as an electrical power source. Given that about 70% of the world's power at the end of the last century was still produced in large centralised steam turbine power stations, this is rightly considered to be one of the landmark technological achievements and a watershed leading to today's energy-intensive society.

John Barber (1734–1793) published the first patent for a gas turbine based on a reciprocating compressor which preceded Parsons' steam turbine by nearly 100 years, but the development of a practical gas turbine took rather longer. Even the turbocharger is older than the first practical gas turbine, with the first patent being awarded to the Swiss engineer Alfred Büchi (1879–1959) in 1905 for the effective demonstration of the use of radial turbocompressor technology for turbocharging a diesel engine. This became the standard for ship propulsion from around 20 years afterwards. The compressor technology used in his turbocharger was derived from the multistage inline radial compressors patented by Professor Auguste Rateau (1863–1930) in 1899 for providing high-pressure air for blast furnaces in steelworks and for mining ventilation.

The biggest obstacle to the practicable gas turbine at the beginning of its development was the low efficiency of the compressor. An analysis showing the large effect of the compressor efficiency on the efficiency of a simple gas turbine cycle is given in Section 18.9.1. All the first successful gas turbines made use of radial compressor technology: thanks to the centrifugal effect, this was able to produce a high-pressure rise more effectively than a multistage axial compressor at that time. The first functioning gas turbine was developed by Aegidius Elling (1861–1949). His first design from 1903 provided power in the form of compressed air from a radial compressor to drive other machines. The breakthrough in gas turbine technology leading to the modern jet engine for aircraft propulsion was made independently by Sir Frank Whittle (1907–1966) in England and Hans von Ohain (1911–1998) in Germany in the years just preceding the Second World War. Both of these inventors made use of radial turbocompressors in their first engines. The use of gas turbines for power generation also dates from this period, as the first marketable industrial gas turbine in the world was designed by Brown Boveri & Cie in 1939 for the Swiss city of Neuchatel, making use of axial compressors. The breakthrough for the gas turbine as an efficient power source for electricity generation only came during the last two decades of the twentieth century with the development of combined steam and gas turbine cycle power plant, operating on natural gas as a fuel. Such plants now achieve cycle efficiencies of over 60%.

More details with regard to the history of turbomachinery can be found in the books by Eckert and Schnell (1961), Jenny (1993), Wilson and Korakianitis (1998), St. Peter (1999), Lüedtke (2004) and Eckardt (2014) and historical overviews of the development of centrifugal compressors are provided by Engeda (1998) and Krain (2005).

1.5 Components of Radial Flow Turbocompressors

1.5.1 The Single-Stage Radial Flow Turbocompressor Stage

The compression process is carried out in a compressor stage which comprises four main elements: the inlet, the impeller, the diffuser and the outlet. These components are shown in Figure 1.7, which depicts a typical single-stage radial compressor,

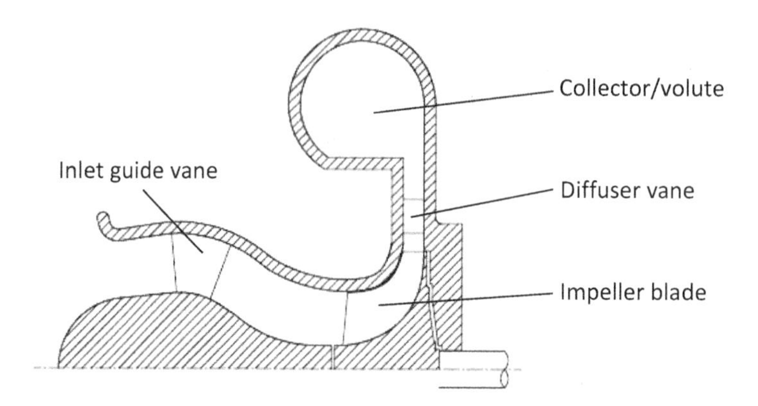

Figure 1.7 Sketch of a compressor stage components with a volute.

Figure 1.8 A single-stage compressor with a variable vaned diffuser and a volute with the inlet casing removed. (image courtesy of MAN Energy Solutions)

consisting of an intake with an inlet guide vane, an open (or unshrouded) impeller, a diffusing system (in this case a vaned diffuser) and a discharge volute. A single-stage machine with the front casing removed is shown in Figure 1.8.

The inlet casing, inlet nozzle or suction pipe, accelerates the flow from the low velocities at the compressor flange, or from the surroundings, to the impeller inlet. It is important that this is done with the minimum of distortion to the velocity profile. There is no energy transfer; the total enthalpy remains constant, but losses occur. The inlet duct may be axial or radial and it may be fitted with variable inlet guide vanes to change the swirl component of velocity at the impeller inlet in order to provide a means of controlling the performance; see Section 17.5.4.

The impeller transfers energy to the fluid, raising both the pressure of the gas and its absolute velocity. In impellers with an axial inlet, the axial section of the blading at the front of the impeller is often known as the inducer. This axial element is rather like the inlet of an axial compressor and in the past was often treated as such. Industrial compressors often have the leading edge of the impeller in the axial to radial bend and do not have an inducer. The outlet of the impeller generally has backswept blades, which means they are leaning backwards away from the direction of rotation. Impeller design is considered in Chapter 11 of the book.

The diffuser converts the high kinetic energy at impeller outlet into a rise in static pressure. There is no further energy transfer from the shaft, so the stagnation enthalpy remains constant but the total pressure falls due to the losses. In some situations, with heat transfer from the surroundings, the total enthalpy may rise slightly, for example in a turbocharger compressor in close proximity to a turbine driven by the hot engine exhaust gases. The diffuser can be vaned or vaneless. In a vaned diffuser there is a small vaneless region upstream of the diffuser blades. Chapter 12 considers the design of the diffuser.

The outlet casing takes the fluid either to the outlet flange, to the inlet of a subsequent stage or to a combustion chamber. The scroll, volute or spiral casing collects the flow leaving the diffuser and brings it to the outlet flange. In many designs, a further reduction of kinetic energy is carried out downstream of the scroll by means of a conical outlet diffuser, as in Figure 1.8. In multistage (inline) compressor applications, the scroll is replaced by a crossover bend and a vaned return channel to remove the swirl velocity and direct the flow inwards into the inlet of the next stage, as shown in Figure 1.9. In gas turbine applications, there are often axial exit guide vanes to condition the flow for the combustor, which requires low swirl velocity, typically with a swirl angle less than 20°, and low absolute Mach number, 0.25 or less,

Figure 1.9 A nine-stage compressor with shrouded impellers. (image courtesy of MAN Energy Solutions)

as shown in Figure 13.10. Chapter 13 considers the design of the volute, the return channel and other stationary components of the gas path.

1.5.2 Multistage Configurations

If a single-stage machine cannot reach a sufficiently high pressure, then multiple stages are used. Figure 1.4 shows the shaft of a three-stage compressor with open impellers. In this case, the lack of an impeller shroud cover requires that the mechanical design needs to include well-machined casing components to match with the impeller blade outer contour otherwise leakage flow over the tips of the blades lowers the efficiency. This leakage loss may be offset by the increased pressure ratio available due to the higher tip-speeds that are allowable when compared with shrouded impellers. Two-stage and three-stage compressors are possible with open impellers, but for more stages shrouded impellers are invariably used. The maximum number of impellers on a single shaft is approximately eight to ten. Figure 1.9 shows a process compressor with nine stages with a smaller outlet width of the impellers as the gas density increases.

1.5.3 Impeller Types

There are three broad categories that describe the different types of impeller. The first is whether the impeller is open or shrouded. The second is the value of the global flow coefficient, which is a nondimensional flow capacity defined in Chapter 8 and which determines the width of the flow channels needed. The third is whether the impeller has splitter blades between the main blades or not. Historically there is also a fourth distinction between impellers whose blades are radial at outlet and those with backswept blades. Nowadays nearly all impellers have some backsweep at the impeller outlet.

In single-stage applications, the impeller is usually unshrouded, or open, and has a high flow coefficient with an axial inlet, wide flow channels and three-dimensional blades with splitter blades, as shown in Figure 1.10, which shows an impeller with a low backsweep. There is usually no constraint on the axial length, so the impeller has an axial inlet with an inducer section. There is a small clearance gap between the tip of

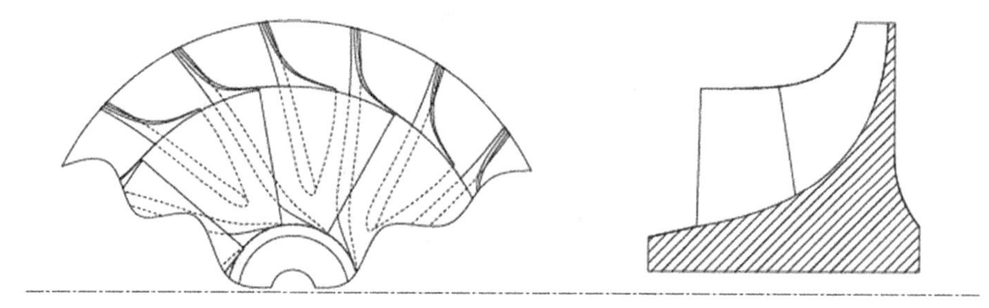

Figure 1.10 The open impeller of a high flow coefficient stage with splitter vanes.

the rotating blades and the casing. As in all compressor applications, a relatively large clearance between the ends of the rotor blades and the casing has a detrimental effect on efficiency, and may also affect flow stability, such that the gap needs to be as small as practicable. In impellers with a diameter greater than 250 mm, the practical operating gap is typically 1/1000th of the impeller diameter but with a minimum size of between 0.5 and 1.0 mm.

In multistage inline compressors, shrouded or closed impellers are most often used, as shown in Figure 1.11. This is because it is difficult to maintain such a small clearance between the impellers and the casing in all stages – the differential expansion between the rotor and the casing during startup causes an increase or decrease in the gaps and may lead to rubs between the impeller blades and the casing. In shrouded impellers axial labyrinth seals are used both in order to seal the flow over the shroud cover from the impeller outlet to inlet and also to resist leakage over the impeller backplate. In a multistage compressor, the volume flow decreases through the machine so that impellers of different flow coefficient are needed at different positions.

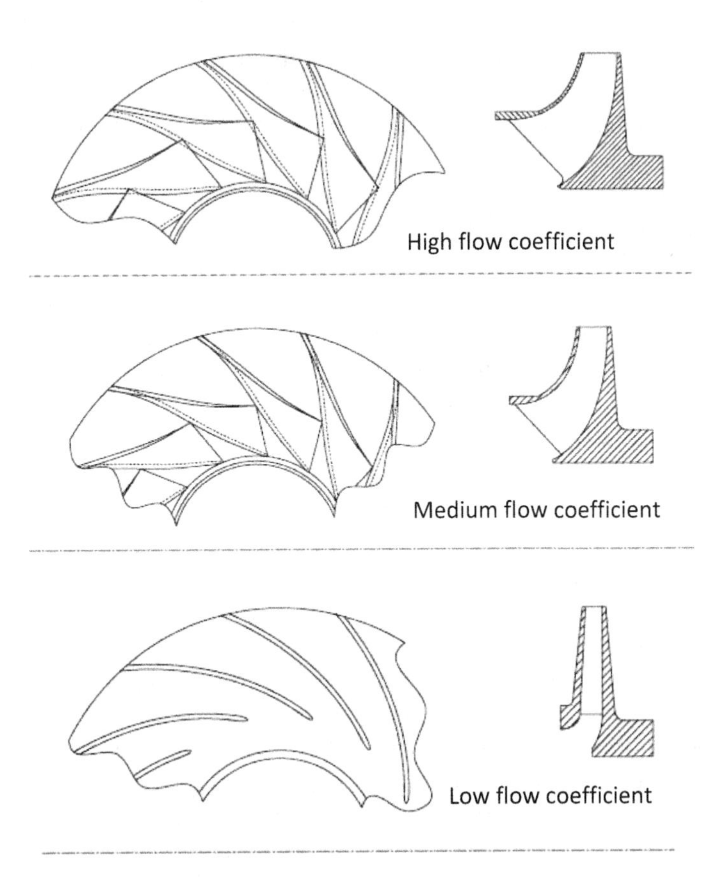

Figure 1.11 Shrouded impellers of high, medium and low flow coefficient process stages.

The leading edge of the impeller blades of multistage compressors is often inclined at an angle of about 30–45° to the radial direction in the axial-to-radial bend, in some cases even at 90°, that is, after the axial-to-radial bend. The reason for this inclination is purely mechanical. In multistage machines, thick shaft diameters are necessary for rotor stability. Since the root of the blade cannot be fixed to the shaft itself, the leading edge has to be inclined.

Shrouded impellers for high, medium and low flow coefficient process compressor stages are shown in Figure 1.11. High flow coefficient impellers are typically used in the inlet section of a multistage compressor and have three-dimensional blade shapes, an axial inlet and wide flow channels. Medium flow coefficient impellers have narrower flow channels, are shorter than those with high flow coefficients and are typically used downstream of the high flow impellers. They often have the leading edge in the bend to the radial direction and so do not have an axial inducer but still make use of three-dimensional blade shapes. Low flow coefficient impellers have narrow flow channels with the leading edge in a region of radial flow. They often have two-dimensional blades with a constant blade profile in the axial direction.

Shrouded impellers typically run at lower tip speeds than open wheels due to the higher mechanical stresses caused by the heavy shroud at a large radius, whereby the highest stress levels occur in the high flow coefficient impellers. However, shrouded impellers provide a more robust component that is less prone to blade vibration problems. As such, they are highly suitable for industrial and process applications.

In some industrial multistage compressors, the first stage may be positioned close to the axial thrust bearing, so that the impeller can be open but subsequent impellers are invariably shrouded. The absence of a cover means that open impellers can operate at higher tip speeds and thus produce greater pressure rise than shrouded stages. The disadvantage of open impellers forced to operate with a large clearance gap is their lower efficiency due to increased leakage flow between the blade tips and the casing. Furthermore, there is an increased risk of lower resonant blade natural frequencies resulting from the cantilevered attachment of the blades to the hub. Chapter 19 discusses mechanical design aspects (impeller stress, critical speeds and rotor stability) in more detail.

1.5.4 Features of Multistage Radial Turbocompressors

Multistage turbocompressors are mostly classified in relation to their mechanical design, method of construction and arrangement of the stages (Sandberg, 2016). The main distinctions are between single-stage and multistage machines, which may be arranged in series or in parallel to meet the flow and head requirements of a specific application, as shown schematically in Figure 1.12. Single-stage designs are invariably overhung from the bearing with a single impeller placed outside of the bearing span, and in some cases two-stage configurations are also overhung designs. In some older applications with a very high-volume flow for the technology of the time, a double-suction configuration may be used with two impellers operating in parallel

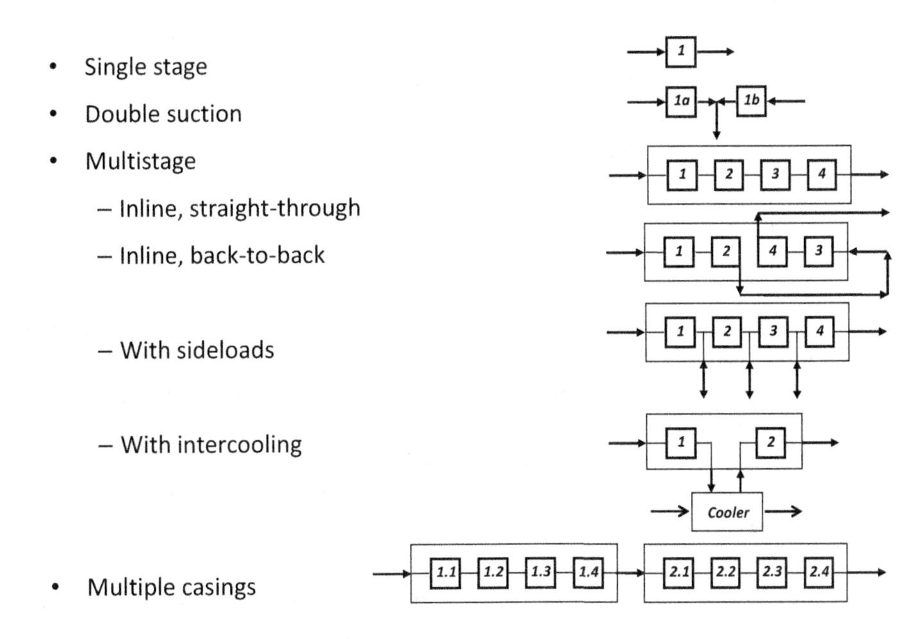

Figure 1.12 Schematic diagram of different compressor configurations.

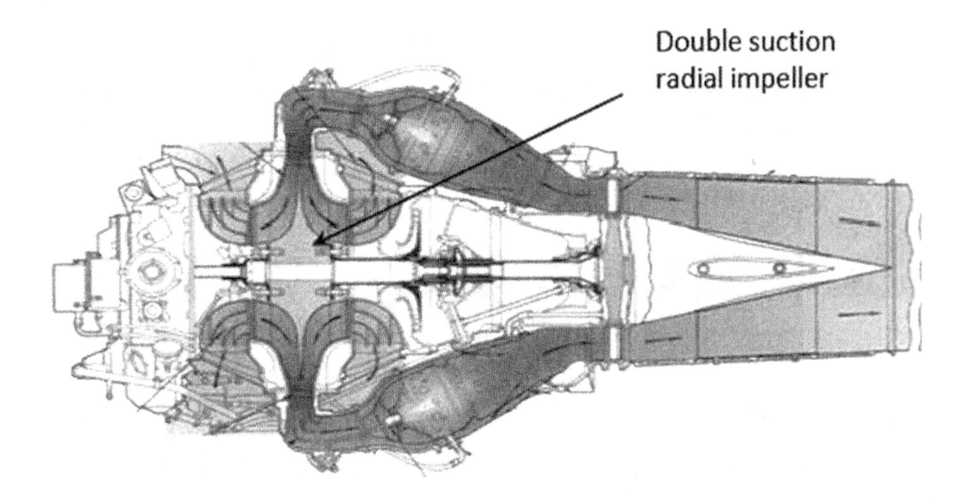

Figure 1.13 An early aero engine for propeller drive with a double-suction radial compressor, Rolls-Royce Nene developed in 1944. (image courtesy of Rolls-Royce plc., ©Rolls-Royce plc, all rights reserved)

back-to-back on a single shaft with two inlets but only one outlet. In this configuration, both impellers are of the same design and may make use of a single diffuser, as shown in Figure 1.13.

The basic multistage configuration comprises several impellers aligned on a single shaft between two radial bearings, with the flow in the same direction within a single casing and with single inlet and discharge flanges. The limitation on the number of impellers allowed on a single shaft is determined by the rotordynamics of the long shaft, which in turn is strongly influenced by the axial length of the impellers and by the temperature rise, due to mechanical or process considerations; see Section 19.7.

The pressure profile around each impeller results in an unbalanced axial thrust force from the back disc towards the eye of the impeller. The sum of these forces for each impeller is counteracted by the combination of a balance piston at one end of the rotor and a thrust bearing. The balance piston is provided with a cross-sectional area where discharge pressure is imposed on one end of the piston and suction pressure on the other, partially negating the unbalanced axial force due to the impellers.

As an alternative to the straight-through inline configuration a back-to-back inline configuration may be used. The flow in the stages of the first section is then towards the centre of the machine, where the flows leaves the casing. It reenters at the other end of the shaft and then also flows towards the centre of the machine so that the inducers of the impellers in each section are oriented towards the shaft ends at both ends of the casing. This configuration can provide for external intercooling and mass flow addition or extraction between the sections. One advantage of the back-to-back design is its inherent characteristic to reduce, and roughly balance, the axial thrust force generated in the stages of each section. Since the two sections are oriented in the opposite direction to each other, unbalanced axial thrust forces from the individual sections act in opposite directions.

Another multistage compressor configuration is known as an integral gear compressor, shown in Figure 1.14. In its simplest form, a single impeller is directly connected to the end of a pinion which is driven by a bull gear. These impellers are often of an open design. More complex versions of the integrally geared compressor exist in which impellers are attached to both ends of the pinion (usually one of these

Figure 1.14 An integral gear compressor. (image courtesy of MAN Energy Solutions)

impellers is open and one shrouded). Multiple pinions can be fixed to a single bull gear to develop a multistage machine. The rotational speed and size of each impeller can be optimised to enhance the overall efficiency of the compressor by using stages which are close to the optimum flow coefficient, as discussed in Chapter 10. In some process applications, one of the compressor stages on the pinion may be replaced by a turbine expander to recover process energy.

There are many applications where intercooling of the gas is required to maintain temperatures at acceptable levels or to provide the thermodynamic advantage of a compression process that is more nearly isothermal, as discussed in Section 4.7.4. Here it is possible to use multiple, single-section casings with intercooling provided between the casings. An alternative is to provide multiple sections within a single physical casing. In some applications, coolers are integrated into the casing of the compressor. These generally need to be specially adapted to particular applications, as shown in Figure 1.15.

Process compressors may require the addition or extraction of sidestream flows at some intermediate pressure of the overall application. The inlet and discharge flanges for the sideloads of each section must be attached to the casing. Adequate space is needed to accommodate these nozzles, which means that each section usually has at least two stages. For sidestream flows entering the compressor, special arrangements are needed in the casing to mix with the discharge flow of the preceding section. The sideload compressor is often used for refrigeration applications where the refrigerant gas can be obtained from economisers that operate at intermediate pressure levels in order to increase the overall efficiency of the refrigeration process.

Figure 1.15 An inline multistage compressor with two internal coolers. (image courtesy of MAN Energy Solutions)

Large pressure ratios cannot be handled by one casing alone as, for rotordynamic reasons, it is usually not possible to split the compression cycle into more than three sections within one casing. The major process compressor manufacturers therefore build compressor trains that may consist of up to four separate casings. The compressors are driven by a common driver, but gearing may be used between the stages to allow the sections to run at different speeds.

Additionally, there are two different modes of assembly for inline applications: horizontally and vertically split machines. In the horizontally split machines, which are used for low pressure service with up to about 75 bar, the casing is split horizontally along the centre line of the machine and the upper and lower halves are bolted together. Above this pressure, barrel compressors are used in which the horizontally split casing is inserted into a cylindrical casing which has a removable end cover bolted on the cylindrical casing. The increased strength available from the barrel casing allows these machines to be used up to 600–700 bar discharge pressure. Hanlon (2001), Brown (2005), Bloch (2006) and Brun and Kurz (2019) give many practical details. Some of the figures in the later parts of this chapter show some of the features discussed here.

1.6 Applications of Centrifugal Turbocompressors

1.6.1 Turbochargers and Superchargers

The largest application of radial compressors in terms of production numbers is in turbochargers, where around 25 million units are produced each year for automotive applications. Centrifugal compressors are used in turbochargers in conjunction with reciprocating internal combustion engines to provide an increased density of air for the engine. These are known as turbochargers if they are driven by the engine's exhaust gas through a turbine; as turbosuperchargers if the compressor is mechanically driven by the engine or by an electric motor; and as turbocompound turbochargers if the turbine of the turbocharger is also mechanically linked to the engine shaft to recover power.

Turbochargers have been in use in marine diesel engines for nearly a century. Exhaust-gas turbocharging was the greatest single technical contribution to diesel engine progress helping the diesel engine to replace the steam turbine in marine propulsion. Turbocharging increases the power for a given engine size simply by forcing more oxygen into the cylinder so that more fuel can be burned. In the early days of the great oceangoing liners of the Norddeutscher Lloyd, Cunard and White Star Line, steam power was the preferred motive power. *Kaiser Wilhelm der Grosse* (1897) had two screws driven by quadruple expansion steam engines, the *Lusitania* (1909) had twin screws driven by Parsons steam turbines and the *Titanic* (1912) and her sister ships had triple screws, the outer two being driven by driven by triple expansion steam engines from Harland and Wolf exhausting into a Parsons low-pressure steam turbine for the central screw. The first application for large diesel motors in ships were the *Preussen* (1923) and *Hansastadt Danzig* (1926), both equipped with turbochargers, following their patented invention by Alfred Büchi in

Switzerland in 1905. The high thermal efficiency of the turbocharged compression ignition engine gives high fuel economy, and the diesel engine became the primary power source for marine engines and for large engines for rail traction. Auguste Rateau extended the use of turbocharging to aircraft engines, allowing them to achieve higher power at altitude (Jenny, 1993).

Turbocharging of diesel engines is one of the most effective techniques for increasing their performance while reducing pollutants and fuel consumption (Galindo et al., 2007). Turbocharging is essential for modern diesel engines and is becoming more so for gasoline engines. Turbocharging allows acceptable air-to-fuel ratios to be maintained with high injected fuel levels. The increase in the air density inside the cylinders enhances combustion as it is an efficient technique for accelerating the atomisation and mixing of the injected fuel and decreasing the penetration of the fuel jets inside the cylinders. During the 1980s, turbochargers became commonplace in medium-size diesel engines for trucks and lorries, and in the 1990s they began to be used for small automobile diesel engines. Here they were first deployed to provide more torque and power from a given engine and then, more recently, to allow motor downsizing – providing a smaller engine with improved power density and reduced emissions (see Figures 1.16 and 18.19). Latterly, turbochargers have become more commonplace on automotive gasoline engines for the same reasons.

The key aerodynamic issues relating to the compressor in this application are the achievement of a wide operating range to match the variability of the engine flow at different speeds and power levels, and the maximisation of efficiency. Weak performance of a compressor in a turbocharged engine limits the behaviour of the drive train, particularly during transient conditions such as the drive-away process or acceleration

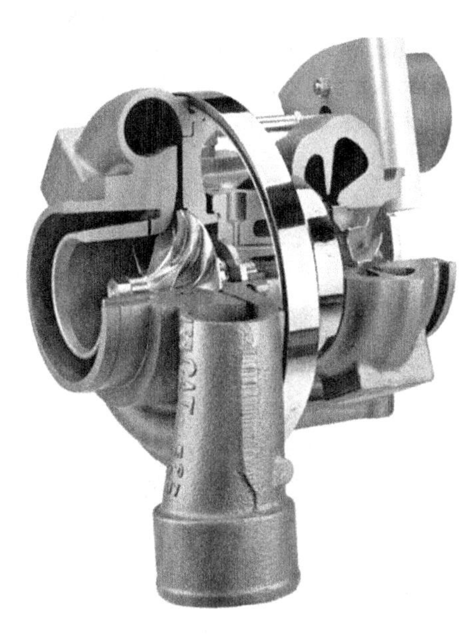

Figure 1.16 Radial compressor impeller in an automotive turbocharger driven by a radial turbine from the exhaust gas stream of the engine. (image courtesy of Garrett Motion Inc.)

from low engine speeds. These considerations come up against practical limits associated with compressor instability at low flow and the choking of transonic flows at high flow rates, both subjects covered in Section 18.10. In early automotive turbochargers, there was an issue of delayed transient response when power was needed which was related to the time taken for the turbocharger to accelerate to a suitable speed. This problem, known as turbo lag, has largely been solved using turbine bypass valves (wastegates), but higher efficiency is obtained with variable geometry turbines and, more recently, with multiple turbocharger arrangements with compressor stages in series or in parallel (Baines, 2005) and improvements derived from reduction in inertia of the turbine.

Specific issues associated with small turbochargers are the high rotational speeds of above 250,000 rpm in small passenger car applications, combined with variable geometry turbines for the control of the matching of the turbine with the engine, as discussed in Section 18.10.6. At low engine speeds the stator (nozzle) vanes of the turbine are closed to reduce the flow capacity of the stage, which provides a high pressure drop at low flow producing sufficient power for the compressor to supply low-speed boost. At high engine speeds, the vanes are opened as otherwise there would be too much power and excessive boost for the engine.

Turbochargers have a large range of sizes; a turbocharger for a marine diesel engine may have an impeller diameter close to 1000 mm, and a turbocharger for a small gasoline engine may have a diameter of less than 40 mm. The compressor used in turbochargers needs to achieve a pressure ratio of between two and four in automotive applications and above six for larger diesel engines (for ships and power generation). In both cases, this pressure ratio is usually achieved by a single radial compressor impeller made of aluminium or titanium, but two-stage boosting systems are also sometimes used. Standards set by the industry for turbochargers have been established by SAE (1995). Ideal gas properties often work well to support the design, test and analysis of turbocharger centrifugal compressor performance.

1.6.2 Compressed Air

Compressed air at a pressure greater than atmospheric is required for many domestic and industrial purposes. It is used for power tools, such as air hammers, drills, wrenches and others and is also used to atomise paint and to operate air cylinders for automation. A study by Radgen and Blaustein (2001) estimated that in Europe, 10% of all industrial electricity consumption is used in the production of compressed air. Typical system storage pressures are close to 1 bar (gauge), which means that a pressure ratio of two is required for the compressor, and a single-stage compressor as shown in Figure 1.8 is suitable.

1.6.3 Industrial Wastewater Treatment

Compressors are required in the treatment of sewage and wastewater. Air is needed to supply oxygen for the activation of bacteria to digest and break down waste material during the process. In addition, injecting air into a treatment tank provides a strong mixing and stirring movement of the fluid to keep solids and sludge in suspension in the liquid. A similar application is the provision of air for biochemical fermentation processes associated with yeast and pharmaceutical processes.

This is an interesting application as the air is forced into large digestion tanks with a fixed depth of water. The absolute pressure at the bottom of a 10 m tank is nearly 2 bar, and so a pressure ratio of two is needed to blow the ambient air into the water: a lower pressure rise is needed for shallower tanks. The same pressure rise is needed at different volume flow rates if the water depth remains the same. This is an interesting technical control problem for centrifugal compressors where the outlet pressure naturally increases at lower flows. The solution requires variable speed, variable guide vanes or variable diffuser vanes, or a combination of these features, as discussed in Section 17.5. Historically, machines used for this purpose were driven at constant speed through a gearbox by induction motors, but more recently advances in high-speed motor technology permit the use of direct drive with variable speed to assist in the achievement of broad range at constant pressure. Normally several compressors are available at any particular plant to give coarse increments in air supply. This drives the range requirement; each machine must be capable of operating at constant pressure to just below half of its maximum flow.

1.6.4 Air Conditioning, Air Extraction and Building Ventilation Applications

Ventilators are required to aerate and ventilate, to cool or heat processes and buildings and to dry products in many technical processes. The ventilators used range from large axial ventilators with a diameter of 20 m in forced convection cooling towers to small axial ventilators with a diameter of just a few centimetres as in a conventional laptop. Centrifugal machines are used where higher pressures are needed and, in these applications, they are sometimes called blowers.

There are several special forms of design in this market. The designs are generally at low absolute pressure levels and low pressure rise but with low blade speeds – mainly to avoid noise, which increases with the sixth power of the speed. This results in relatively high nondimensional aerodynamic loading (see Section 7.5), hence the attraction of a centrifugal stage. These machines superficially appear to be less technically advanced than other high-pressure radial compressor applications: as in many low-speed applications, there is no diffuser, but the impeller discharges directly into the volute casing. Nevertheless, the requirement of low speeds to avoid noise, and the general task of keeping costs as low as possible, leads to substantial design challenges.

1.6.5 Vacuum Compressors

In addition to providing compressed air, compressors can be used to evacuate air and so reduce the pressure in an enclosure. Examples can be found in machines developed for dewatering of pulp felt on paper, board and tissue machines. The

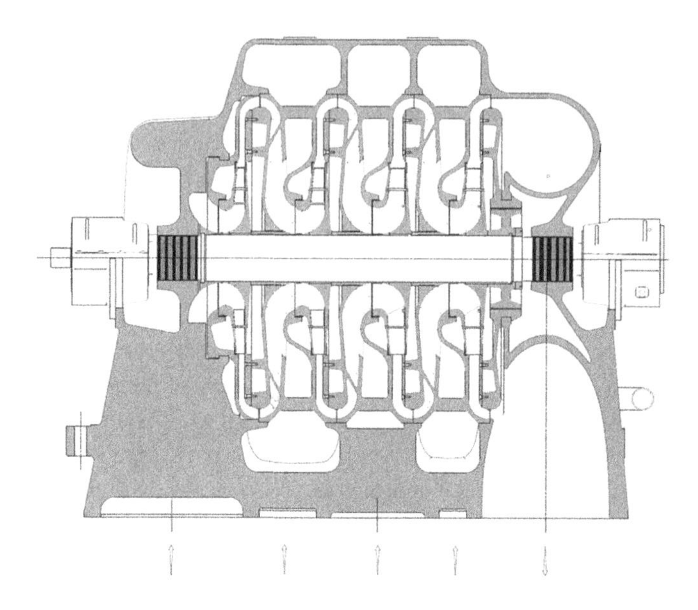

Figure 1.17 A vacuum blower with multiple sidestreams for dewatering paper pulp. (image courtesy of MAN Energy Solutions)

example shown in Figure 1.17 has several different inlet streams along. the axis to match the suction and vacuum requirements at different positions of the paper machine. Air at the highest vacuum level is compressed in the first stage of the blower and ducted via return channels to the second stage, where it is mixed with air from the medium vacuum level in the first sidestream. This process is repeated in the third and fourth compression stages. The design is such that four different vacuum levels can be generated in one unit. Thanks to the additional inlet flow upstream of each stage, the flow coefficients and the outlet width of successive stages do not reduce towards the end of the compressor but at first increase through the machine.

Similar applications with low inlet pressure requirements are found in vacuum cleaners, albeit on a much smaller scale using small single-stage compressors at high speed; see Section 1.6.19.

1.6.6 Vapour Compression Refrigeration and Heat Pumps

Refrigeration plants transfer heat to or from processes and buildings. The vapour cycle refrigeration process is the most common method of doing this. The vapour compression thermodynamic cycle is similar to a heat engine cycle but works in reverse. In the basic cycle, the liquid refrigerant at low pressure is evaporated to cool the evaporator at low temperature, then the evaporated gas is compressed to a higher pressure to the condenser, where heat is removed by cooling the gas to the liquid state. The liquid is then throttled back to the evaporator pressure through a valve. This is the standard process in a domestic refrigerator where the evaporator removes heat from within the enclosed space. Where the flow rate is small, reciprocating compressors are used. In larger heat pumps, it is the heat transferred to the condenser which is of interest for heating purposes. The evaporator can make use of different sources of heat for the evaporation (ambient air, rivers, lakes or the ground). The electric-driven heat pump is one of the most promising technologies in term of reducing greenhouse gas emissions in the future as it is an energy efficient way to provide space heating provided the power is from renewable sources (Schiffmann, 2008).

On a larger scale, a wide variety of vapour compression cycles are used to improve the coefficient of performance, involving multistage compressors with liquid suction heat exchangers, flash gas economisers, heat exchange economisers and cascade refrigeration systems with multiple refrigerants (van Gerner, 2014). The major focus today is the effective use of synthetic refrigerant gases that are not damaging to the environment. There is a wide range of new working gases (refrigerants) becoming available for application at different temperature levels. Carbon dioxide (CO_2) has been widely investigated for use as a working fluid in refrigeration cycles: it is inexpensive, nonexplosive and nonflammable and has both no ozone-depleting potential (ODP) and low global warming potential (GWP) compared to other refrigerants.

Centrifugal compressors are used in a wide range of sizes and configurations for larger heat pumps and refrigeration applications. The use of refrigerants means that real gas properties are needed to design, test and analyse the performance of these machines. The gases generally have a high molecular weight and consequently a low speed of sound, such that large pressure ratios and transonic flow occur even though the blade speeds remain moderate compared to compressors operating with air. Standards set by the industry for these compressors include those published by the American Society of Heating, Refrigerating and Air-Conditioning Engineers (ASHRAE), the American Society of Mechanical Engineers (ASME), the American Petroleum Institute (API) and the European standard EN-14511.

Air cycle refrigeration is a form of gas refrigeration in which air is used as the working fluid and does not change its state but remains as a gas throughout the process in a reversed Brayton or Joule cycle. If air is used as the working fluid, a closed system is not required and the cooled air can be used directly. This technology is used in most modern aircraft cooling applications for the supply of cabin air, in wing ice protection systems and in larger transport systems, such as train air conditioning and for cooling of electronics. An air cycle machine typically includes a centrifugal compressor and one or two radial expansion turbines. The air from the compressor is first cooled by a heat exchanger, then expanded through a turbine, thus producing cold air. In an aircraft cooling and ventilation system, there are often two radial turbines, a radial compressor and an axial fan all on the same shaft.

1.6.7 Gas Turbines for Power Generation

In power generation, the main application of gas turbines is in combined cycle gas and steam turbines, where the hot exhaust gases from the gas turbine (at around 600°C) are used in a heat recovery steam generator to produce steam for a so-called bottoming cycle steam power plant. Such high-power compressor applications need a large flow capacity and are invariably equipped with axial compressors. Centrifugal compressors, both in single and two-stage configurations, are suitable for application to relatively small simple-cycle gas turbines in the 2–5 MW power range at pressure ratios typically between seven and eleven.

The first gas turbine to provide a net power output was that of Aegidius Elling which made use of a six-stage centrifugal compressor with water injection as a means of intercooling (Bakken et al., 2004). The impeller was a radial bladed design and looks quite dated compared to modern designs, but the stage used vaned diffusers and the shape of these look quite modern in form, as can be seen by comparison of those in Chapter 12 (see Figure 1.18). Many other images of this machine are available on the website of the Norway Museum of Science and Technology.

In more recent years, small micro gas turbines (MGT) for distributed power have been developed, based largely on turbocharger technologies making use of natural gas to provide local sources of power and heat (Moore, 2002). These smaller machines make use of radial flow turbocompressors in a regenerative gas turbine cycle and are becoming widespread in distributed power and combined heat and power applications. They range from handheld units producing less than a kilowatt, to commercial-sized systems that produce tens or hundreds of kilowatts. Microturbine systems have many advantages over reciprocating engine generators, including a higher power-to-weight ratio, lower emissions and the fact that they have only one main moving part.

Normal gas turbine cycles are open and make use of air as the working fluid such that the combustion products pass directly through the turbine. An active area of

Figure 1.18 An i.mpeller and diffuser from a radial compressor stage of the 1904 gas turbine of Aegidius Elling. (image by Aegidius Elling courtesy of the Norway Museum of Science and Technology)

research is the development of closed cycle Brayton power units employing supercritical CO_2 with high density as the working fluid and radial compressors to pressurise the gas (Conboy et al., 2012). Supercritical CO_2 is a fluid state of carbon dioxide held above its critical point (i.e., critical pressure and temperature). The density is similar to that of a liquid and allows for the pumping power needed in a compressor for a given pressure ratio to be significantly reduced, as discussed in Section 2.8.2. Such cycles could provide significant efficiency gains in geothermal, coal, nuclear and solar thermal power production.

1.6.8 Automotive Gas Turbines

One almost obsolete application, included for completeness, is automotive gas turbines, which, because of their relatively small size and power requirement, made use of radial turbocompressors in a regenerative gas turbine cycle. Prototypes were developed by Allison, Chrysler, Daimler, Rover, Austin and others in the middle of the last century. The poor throttling response from the high rotational moment of inertia and the low thermal efficiency in such small machines given their low pressure ratio, plagued the further development for this application (Walsh and Fletcher, 2004).

The automotive relevance of radial compressors in small gas turbines has recently been revised: small micro gas turbines can be used to provide power to charge the batteries of hybrid or electric vehicles and so act as range extenders. As the gas turbine is not directly connected to the vehicle wheels, it can operate intermittently at its natural high rotational speed without throttling.

1.6.9 Jet Propulsion Gas Turbines

Over 25,000 gas turbines are in operation as aircraft jet engines, many of them in military applications. The key development in this area is towards lighter and quieter machines with lower specific fuel consumption and lower emissions. Normally a key criterion for engines for aviation propulsion is high flow per unit frontal area, and this tends to favour the axial compressor for most civil transport and modern military applications. In addition, at the physical size typically needed for aero engines in large passenger aeroplanes, the multistage axial compressor normally has an efficiency advantage over a centrifugal at a similar overall pressure ratio.

Historically, however, radial compressor technology was used in all early jet engines, for example in both of the first two turbojets, the Whittle engine of 1937 (Whittle, 1945) and the Ohain engine of a similar date (Von Ohain, 1989). This was partly due to the advantage given by the centrifugal effect producing compression via radius change. Furthermore, experience with suitable centrifugal stages had already been accrued at that time in the supercharging of piston engines for aircraft propulsion, such as the Rolls-Royce Merlin engine of the Spitfire aircraft. Figure 1.13 shows an early engine of this period. The advantage of the centrifugal compressor stage was the relative ease with which an adequate level of pressure ratio and efficiency could be reached. The requirement for high throughflow in the gas turbine was initially achieved by applying a double suction impeller, with two inlets and a single outlet. A double suction impeller, as shown in Figure 1.13, has no back-plate and so in this configuration there is the added advantage that there are no parasitic disc friction losses; see Section 10.6.2.

Major contributions to aerodynamic theory were made in the 1940s and 1950s, and these led to the development of satisfactory axial compressors with more power and higher efficiency, and axial compressors gradually became the standard designs for larger aeroengines (Nichelson, 1988). However, in the earliest versions of the first passenger jet aircraft, the de Havilland Comet, four Rolls-Royce Ghost engines with a single-stage centrifugal impeller were used. The Comet engines were later replaced by the Rolls-Royce Avon engine with a multistage axial compressor. Since that time, centrifugal compressor technology has only been retained for smaller aeroengines found, for example, in business jets, helicopters, drones and in auxiliary power units (APU). The specific applications with different engine architectures and power requirements necessitate different types of centrifugal compressor design (Rodgers, 1991; Lou and Key, 2019). Cross-sectional images of many different gas turbine configurations using centrifugal compressors are given by Grieb (2009).

One important application of centrifugal technology is in turboshaft engines for helicopters and propeller-driven aeroplanes. These may use a low-flow coefficient centrifugal stage for the final compression stage downstream of a multistage axial compressor on the same shaft, shown in Figure 1.19, known as an axicentrifugal configuration. The radial compressor in an axicentrifugal compressor has a very high hub radius compared with other radial machines designed for higher flow coefficients. At intermediate and lower thrust levels, a turboshaft engine may incorporate a single-stage centrifugal compressor, shown in Figure 1.19, or a two-stage centrifugal compressor with open impellers, as shown in Figure 1.20.

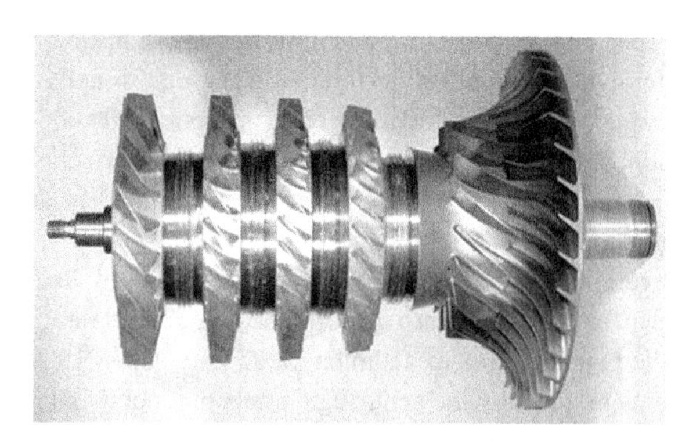

Figure 1.19 The rotor of an axicentrifugal compressor with a low-flow coefficient impeller and the rotor of a high-flow coefficient impeller for helicopter engines with a single radial compressor stage of a similar duty. (images courtesy of Rolls-Royce plc, ©Rolls-Royce plc, all rights reserved)

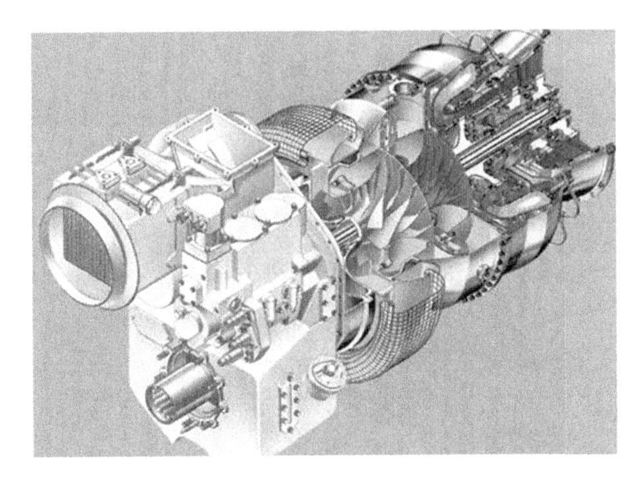

Figure 1.20 A cutaway view of the shaft of the MTR390 gas turbine engine with two centrifugal compressors. (image courtesy of MTR)

In the ongoing quest for reduced specific fuel consumption (SFC) and emissions, the thermodynamic cycles of gas turbines for aircraft application are pushed towards increased overall pressure ratio and a higher bypass ratio. The ramification for the core engine is a reduction in the volume flow at compressor delivery. For axial stages, this implies a high hub tip ratio, reduced blade heights and stronger effects of clearance which erode the efficiency benefit over centrifugal stages. Recent trends have seen axicentrifugal machines considered for engines for smaller regional jet transport aircraft, where at this smaller size a high efficiency centrifugal stage can replace four or five lower efficiency axial stages.

An example is given in Figure 1.21. This involves a high-flow coefficient centrifugal stage as the high-pressure compressor on the high-speed spool of a twin-spool jet engine, following the fan and a low-pressure axial compressor (also called a booster) on the low-speed spool. In this case, there is a swan-neck duct downstream of the booster to bring the flow down to the radius of the impeller eye so the impeller is not constrained to have a thick shaft. Larger engines remain all axial. The architecture for more adventurous cycles, which divert core flow through heat exchangers in the bypass duct to benefit from an intercooled compression process, as discussed in Section 4.7.4, may also suit a final centrifugal stage in the core engine.

1.6.10 The Auxiliary Power Unit

The APU is a small constant-speed gas turbine engine located in an aircraft's tail used to run an electric generator and provide pressurised air during operation on the ground. The design of such devices forces the choice of high-flow coefficient impellers to minimise the size. A cutaway view of the shaft of an APU of the type used in many Airbus aeroplanes is shown in Figure 1.22. The impeller on the left is part of the APU gas turbine, and the impeller on the right is known as the load compressor and provides compressed air output at a pressure ratio of four. The compressed air feeds

Figure 1.21 A cutaway view of the HF120 twin-spool aero engine for a business jet with a centrifugal compressor on the high-speed spool. (image courtesy of GE Honda Aero Engines)

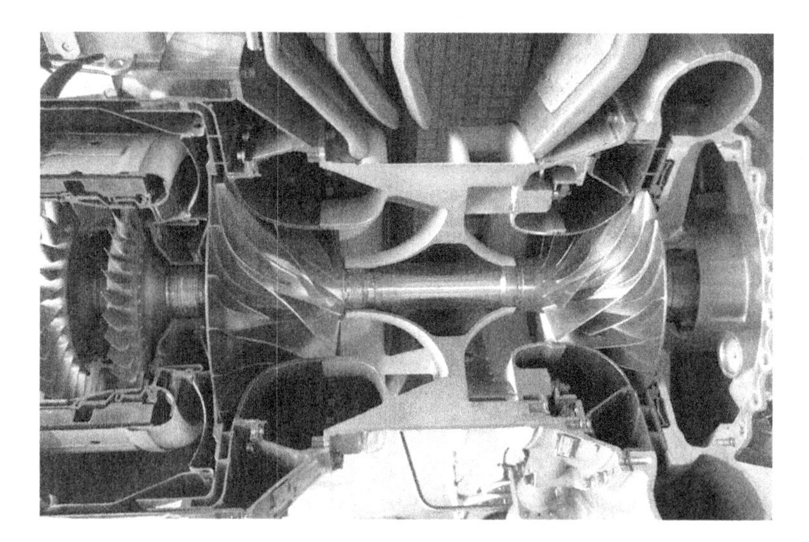

Figure 1.22 A cutaway view of the shaft of an auxiliary power unit with two centrifugal compressors driven by an axial turbine. (image courtesy of Pratt & Whitney)

the aircraft air conditioning system via pipes running through the fuselage and may also be employed in an air turbine starter to spin the main engines during starting.

1.6.11 Oil, Gas and Chemical Applications

There are innumerable applications of radial compressors in the oil, gas and chemical industries, and an excellent description is given by Brun and Kurz (2019) and a summary given here.

Pipeline compressors are installed along natural gas pipelines in compressor stations at intervals of roughly 100 km to pressurise the gas and so aid its transport along the pipeline. Typically, they make use of single-stage or multistage centrifugal compressors driven by a natural gas fired gas turbine, making use of the fuel in the pipeline. The special requirements in this application are high flow stages with high efficiency and a wide operating range.

Centrifugal compressors for natural gas processing, petrochemical and chemical plants are often single-shaft multistage machines driven by large steam or gas turbines. Their casings are often horizontally split at pressure levels below 100 bar, as shown in Figure 1.23, or are of the barrel type at high pressure levels, where a horizontally split unit is inserted into a vertically split casing acting as a pressure vessel, as shown in Figure 1.24.

Standards set by the industry (ANSI, API, ASME) result in large, thick casings to maximise safety. The impellers are mostly, if not always, of the covered style with a shroud attached to the blades. This type of compressor is also often called an API-style compressor, from the American Petroleum Institute standard API 617 (2002). Use of real gas properties is needed to properly design, test and analyse the performance of natural gas pipeline centrifugal compressors, and great care must be taken in the matching of the stages.

There is increasing industrial activity in the compression and liquefaction of natural gas (CNG and LNG) and ethylene. These more unusual working fluids have high molecular weights and low sonic velocities leading to high Mach numbers. This causes the performance maps to be very narrow with a limited range between surge and choke, exacerbating the difficulty in matching stages. Another design issue is that in many applications these compressors have sidestreams to accommodate flows

Figure 1.23 A horizontally split compressor with with two inline sections and a sidestream entering after two stages. (image courtesy of MAN Energy Solutions)

Figure 1.24 A barrel compressor showing the horizontally split internals extracted from the barrel casing. (image courtesy of MAN Energy Solutions)

entering or leaving the compressor at intermediate pressures. These pressures depend on the process for which they were intended or the requirements of intercooling. The standard approaches that have been used for machines working with air are no longer applicable to these compressors, and the importance of the Mach number on performance is given strong emphasis in the analysis presented by the current authors.

A range of different compressors are used in these processes as follows:

- The feed gas compressor compresses natural gas received at the plant to the required pressure for the condensate removal process.
- The flash gas compressor repressurises the vapour produced during the condensate removal process.
- The off-gas compressor compresses natural gas to the required pressure needed for acid gas removal, mercury removal and dehydration.
- The regeneration gas compressor feeds the main cryogenic exchanger with required refrigerant.
- The CO_2 injection compressor is used to inject carbon dioxide into a storage well.
- The lean gas compressor raises the lean natural gas pressure to a level suitable for the main cryogenic exchanger.
- The end flash gas compressor pressurises the vapour produced during the refrigeration process.
- The boil-off gas compressor repressurises boil-off gas for re-liquefaction or use in this gaseous state.
- The N_2 booster compressor adjusts the calorific value of the gas being sold.
- The fuel gas compressor or fuel-boost compressor elevates the pressure of the gas stream to the inlet pressure required by the turbine combustion chamber.

Gas reinjection compressors are used for the reinjection of natural gas into an underground reservoir to increase the pressure in the chamber, decrease the viscosity of the crude oil and increase the flow of oil to production, leading to enhanced oil recovery (EOR). Centrifugal compressors for such uses are often single-shaft, multistage machines driven by gas turbines. With discharge pressures approaching 700 bars, casings are of the barrel style, and the stages are special designs of low flow capacity with narrow flow channels.

In some applications, the gas recovered from the oil well contains significant amounts of hydrogen sulphide. The gas can be reinjected directly into the well without removing the hydrogen sulphide (so-called sour gas) and without removing the acid content such as carbon dioxide (acid gas). In oil refineries or natural gas processing plants, the removal of hydrogen sulphide and other compounds is referred to as 'sweetening' as it removes the sour smell of the gas. Similar applications involve the injection of CO_2 into oil wells. This is particularly used for compressors in subsea gas recovery applications, which are usually hermetically sealed and oil-free with magnetic bearings and direct-drive integrated motors (Fulton et al., 2001).

1.6.12 Fluid Catalytic Cracking Compressors

The growing demand for light hydrocarbons is driven by the worldwide need for kerosene, diesel oil and gasoline as combustion fuel in many transport industries and the need for benzine or propylene for plastic or synthetic fibres. The fluid catalytic cracking (FCC) process refines various heavy hydrocarbons to lighter, more valuable products via high-temperature catalytic cracking. In the first part of the process, high molecular weight hydrocarbons are cracked in the presence of a catalyser with air at high temperature and pressure, often supplied from a large axial-centrifugal compressor.

1.6.13 Air Separation Plants

Air separation plants separate air into its constituent components, such as nitrogen, oxygen, argon, helium and other inert gases. The process is known as cryogenic distillation, and the air must first be cooled and liquefied using the Joule–Thomson effect. Air is first compressed to typically between 5 and 10 bar in a multistage turbocompressor and cooled before starting the expansion process. The compression work is minimised by using intercooling to try to approximate an isothermal (constant temperature) process, as discussed in Section 2.7.4. The cooling can be internal to the inline compressor (see Figure 1.13), or the compressor may have external coolers and comprise multiple single stage compressors driven by a bull gear known as integral gear compressors (see Figure 1.12). The cooler can also be in the form of cooling jackets embracing the compressor flow channel. A brief history of intercooled compressors with internal cooling, going back to a machine with a water-cooled casing in 1906, is given by Strub (1974).

1.6.14 Synthetic Fuels, Coal Liquefaction and Gas to Liquid

The Fischer–Tropsch process describes a series of chemical reactions that convert a mixture of carbon monoxide and hydrogen into liquid hydrocarbons, which are then used as synthetic lubrication oils and synthetic fuels. These reactions occur in the presence of appropriate catalysts, typically at temperatures between 150 and 300°C and pressures of up to several tens of atmospheres, which means that suitable compressors are needed. The process was first developed in Germany (Stranges, 2000), and was used during the Second World War to provide liquid fuels from coal. It now serves as an important reaction in both coal liquefaction and gas-to-liquids technology as well as many other chemical processes aimed at producing compounds based on hydrocarbon chains from coal. In the decarbonised energy system of the future, synthetic fuels from renewable electricity are likely be an important supplement to the direct use of electricity from renewables.

1.6.15 Carbon Capture and Storage (CSS)

This gas reinjection technology may in the future be applied to reinjection compressors for subterranean sequestration of CO_2 in carbon capture and storage systems. The technology reduces greenhouse gas emissions from thermal power plants by injecting CO_2 into depleted oil and gas fields or in to other suitable geological formations. The pressure level required for reinjection is above 200 bar. Very large machines will be needed to deal with large volumes of CO_2.

1.6.16 Compressed Air Energy Storage

The exploitation of wind energy is an efficient way to reduce dependence on fossil fuels and reduce global warming, However, wind power is intermittent in nature. Compressed air energy storage (CAES) is a technology which uses available electrical power in times of surplus power to compress air and store it in a large void above or below ground. The stored air is then reheated and expanded through expanders which are attached to generators which convert the stored energy to electricity (Berman, 1978). In large systems, axial turbomachinery may be used, but in smaller systems radial turbomachinery technology may be appropriate (Zhang et al., 2017).

1.6.17 Centrifugal Steam Compressors for Mechanical Vapour Recompression

Thermal separation processes, such as evaporation and distillation, are energy intensive and often make use of mechanical vapour recompression systems. In a conventional evaporator, the vapour stream produced is condensed, meaning that most of its energy content is lost. A mechanical vapour recompression (MVR) system compresses the vapour to a higher pressure and therefore acts like a heat pump, adding energy to the stream which can be recovered at a higher temperature. The elevated temperature enables the use of the compressed vapour to heat the evaporator. In mechanical vapour recompression, a turbofan or blower (either single-stage or multistage) is used to compress the whole vapour flow to a pressure that is sufficient to heat the evaporator. In this way, the evaporator does not consume steam during operation.

The main fields of application are currently in the food and beverage industries (evaporation of milk, whey and sugar solutions), the chemical industry (evaporation of aqueous solutions), salt works and desalination (evaporation of saline solutions) and environmental technology (concentration of wastewater). Single-stage centrifugal compressors with impellers made of standard materials, for example steel, are capable of achieving a water vapour pressure increase of a factor of 1.8. If higher-quality materials such as titanium or multistage compressors are used, the vapour pressure increase may be up to 2.5. The technical challenges for this application are the high tip-speeds needed because of the high speed of sound in steam – which is around 475 m/s – and the erosion by water droplets, if the vapour falls below the saturation line.

1.6.18 Fuel Cells

Research and development of fuel cell systems for many applications has dramatically increased in the past few years. In particular, the use of fuel cells running on hydrogen in vehicles with an efficiency close to 50% could replace internal combustion engine running on gasoline or diesel fuels with an efficiency of 20–25%. This leads to a number of unique challenges, such as safety, physical packaging within the vehicle, durability and operation under extreme environmental conditions. In addition, there are demanding duty cycles that include high peak power requirements and a rapid response time (Zheng et al., 2008). Increasing the pressure of the fuel cell system can give higher efficiency, higher power density and better water balance characteristics for the fuel cell. The centrifugal compressor, driven by an auxiliary power source or from a turbine in the fuel cell exhaust, offers higher power with good system efficiency with the advantages of small volume, low cost and high reliability for this innovative application (Berenyi, 2006).

1.6.19 Microcompressors

New applications of small centrifugal compressors are associated with the development of high-speed motors allowing the rotor size for a similar mass flow and pressure ratio to be reduced. The application of ultra-high-speed miniature centrifugal compressors with an impeller diameter less than 30 mm and the use of ultra-high-speed electric motors to provide rotational speeds between 200,000 and 600,000 rpm are now possible. The enabling technology is the use of high-speed permanent-magnet synchronous motors up to 1,000,000 rpm, whereby the rotor is a diametrically magnetised samarium cobalt (Sm_2Co_{17}) permanent magnet which is encased in a retaining titanium sleeve.

Electromagnetic, thermal, rotordynamic and mechanical design are all a challenge for such machines. In a growing number of applications at low flow rates, such microcompressors can be successfully used to replace much larger positive displacement devices (such as liquid ring compressors) or to replace larger centrifugal

Figure 1.25 Ultra-high-speed radial microcompressor impellers of different sizes and flow coefficients. (image courtesy of Celeroton AG)

compressors operating at lower rotational speeds (Casey et al., 2013). A common application is in high-speed compact vacuum cleaners. A range of small impellers with different flow coefficients and diameters for such compressors is shown in Figure 1.25. There are also concepts for microgasturbines capable of achieving a high power density compared to batteries using compressor impellers with an impeller of 8 mm rotating at 1.2 million rpm at design (Epstein, 2004; Sirakov et al., 2004).

1.6.20 Medical Applications

At the lowest level of pressure rise and flow, centrifugal compressors and ventilators are becoming increasingly common in medical applications. They are used as the source of air supply for small continuously positive airway pressure (CPAP) devices for the treatment of sleep apnoea and in the provision of ventilation air for patients with critical lung infections. These particular devices must operate as quietly as possible. This means that although the power requirement for the flow and pressure rise necessary for even a large patient is modest, high efficiency and relatively high loading and low tip-speed to support low broadband noise are needed.

1.7 Some Other Publications

1.7.1 Books

There are many useful publications offering complementary cover of similar or related technical areas as this book. For completeness, a very short description of the most relevant books on radial compressors is given here.

Aungier, R. H. (2000) *Centrifugal Compressors – a Strategy for Aerodynamic Design And Analysis.* ASME Press

A useful guide to practical aspects and to design guidelines for radial flow industrial compressors but less depth on the theoretical aspects and high-speed applications.

Bloch, H. P. (2006) *A Practical Guide to Compressor Technology.* Wiley

Not theoretical, mainly intended for users and operators close to real applications.

Boyce, M. (2003) *Centrifugal Compressors: A Basic Guide.* PennWell

Not theoretical, mainly intended for users and operators close to real applications.

Brown, R. N. (2005) *Compressors: Selection and Sizing.* Elsevier

Not theoretical, mainly intended for compressor users. Includes all types of compressors.

Brun, K. and Kurz, R. (2019) *Compression Machinery for Oil and Gas.* Gulf Professional Publishing.

A useful and practical up-to-date review of centrifugal, reciprocating and screw compressors in oil and gas applications, with multiple contributors from industry and academic research.

Cumpsty, N. A. (2004) *Compressor Aerodynamics.* Krieger Publishing Company

A useful theoretical introduction to both axial and radial compressors but concentrating mainly on axial machines. Very strong on research issues, less pragmatic advice offered on design aspects.

Eckert, B. and Schnell, E. (1961), *Axial- und Radial-Kompressoren*, Springer, Berlin.

A good section on radial compressors but in German, out of date and out of print.

Ferguson, T. B. (1963) *The Centrifugal Compressor Stage*, Butterworth, London

A logical, well-written introductory book on basic principles, but out of date and out of print.

Gresh, T. M. (2018) *Compressor Performance.* Butterworth Heinemann, Elsevier.

A useful book explaining the practical aspects of aerodynamics of compressors for equipment users involved in selecting, monitoring and maintaining various types of compressors; uses American units.

Gülich, J. F. (2008) *Centrifugal Pumps.* Springer, Berlin.

A good and wide-ranging book on centrifugal pumps with good design guidelines, some of which are useful for radial compressors.

Hanlon, P. C. (2001) *Compressor Handbook.* McGraw-Hill.

Not at all theoretical, mainly practical and intended for compressor users. Includes all types of compressors and describes many applications from many different contributors.

Japikse, D. (1990) *Centrifugal Compressor Design and Performance.* Concepts/NREC.

A compendium of knowledge giving an introduction to centrifugal compressors and including useful information on the design of the various components.

Lüdtke, K. H. (2004) *Process centrifugal compressors.* Springer.

An excellent book on industrial multistage process compressors with useful design guidelines and practical information on these machines.

Traupel, W. (1962) *Die Theorie der Strömung durch Radialmaschinen*. G Braun.

An interesting description of the physics of radial compressors as understood in the early 1960s; in German and out of print.

Van den Braembussche, R. (2019) *Design and Analysis of Centrifugal Compressors*. ASME Press, Wiley.

A useful up-to-date book with some excellent complementary information, especially on secondary flows, vaneless diffuser instability, asymmetry from the volute flow and optimisation methods.

Watson, N. and Janota, M.S. (1982) *Turbocharging the Internal Combustion Engine*. Macmillan.

Despite being out of date, this includes an excellent introduction to radial flow compressors in the context of turbocharging; a reprint has recently been issued.

Whitfield, A. and Baines, N. C. (1990) *Design of Radial Turbomachines*. Longman, John Wiley and Sons.

A useful and pragmatic description of the technology and design methods for both centrifugal compressors and radial turbines working in compressible flow.

1.7.2 Technical Conferences and Journals

The major sources of up-to-date technical literature for radial compressors are the following technical conferences:

The ASME/IGTI Turbo Expo. The Turbomachinery Technical Conference and Exposition, organised annually by the ASME International Gas Turbine Institute (IGTI). The conference hosts an enormous number of turbomachinery sessions, including several technical sessions on radial turbomachinery aerodynamics, microturbines, turbochargers, and oil and gas applications. www.asme.org

The European Turbomachinery Conference. This is organised every two years by the European Turbomachinery Society, which represents many engineering associations from most countries in Europe and includes technical sessions on axial, radial and mixed-flow turbomachinery. www.euroturbo.eu

The International Conference on Turbochargers and Turbocharging. This is organised by the Institution of Mechanical Engineers, England, and addresses current and novel aspects of turbocharging system designs. www.events.imeche.org

The Turbomachinery and Pump Symposium. This is organised annually by the Texas A&M University, Houston, and is an industry event, offering a forum for the exchange of ideas between rotating equipment engineers and technicians worldwide.

https://tps.tamu.edu/event-info/

The American Institute of Aeronautics and Astronautics (AIAA) Science and Technology Forum and Exposition. The AIAA SciTech Forum is an important annual event for aerospace research, development and technology bringing together aerospace industry experts in a variety of technical disciplines

The major journals which include publications on centrifugal turbocompressors are the following:

***The Journal of Turbomachinery*, ASME.** *The Journal of Turbomachinery* publishes high-quality, peer-reviewed technical papers that advance the state of the art of turbomachinery technology, including gas-path technologies associated with axial compressors, centrifugal compressors and turbines.

https://turbomachinery.asmedigitalcollection.asme.org/journal.aspx

***The Journal of Engineering for Gas Turbines and Power*, ASME.** *The Journal of Engineering for Gas Turbines and Power* publishes high-quality papers in the broad technical areas of gas and steam turbines, internal combustion engines and power generation including centrifugal compressors.

https://gasturbinespower.asmedigitalcollection.asme.org/journal.aspx

***The Journal of Power and Energy*, Institution of Mechanical Engineers (IMechE).** *The Journal of Power and Energy, Part A of the Proceedings of the Institution of Mechanical Engineers*, is dedicated to publishing papers of high scientific quality on all aspects of the technology of energy conversion systems.

https://journals.sagepub.com/home/pia

***International Journal of Rotating Machinery*, Hindawi.** *The International Journal of Rotating Machinery* is a peer-reviewed, open access journal that publishes original research articles as well as review articles on all types of rotating machinery.

www.hindawi.com/journals/ijrm/

***The Journal of Propulsion and Power*, AIAA.** This journal provides readers with primary interests in propulsion and power access to papers spanning the range from research through development to applications in fluid dynamics and combustion, including many turbocompressor papers.

https://arc.aiaa.org/jpp/about

International Journal of Turbomachinery, Propulsion and Power(IJTPP). This is an international peer-reviewed open access journal focused on turbomachinery, propulsion and power. It is the official dissemination tool of *Euroturbo*, the European turbomachinery society, which runs the *European Turbomachinery Conferences*.

https://www.mdpi.com/journal/ijtpp

2 Energy Transfer

2.1 Overview

2.1.1 Introduction

Because of the fundamental importance of the Euler turbine equation, the chapter begins with a brief analysis of this equation. This provides some key insights into energy transfer in compressors, particularly with regard to the importance of the centrifugal effect on the pressure rise. To understand this, the basics of one-dimensional velocity triangles need to be introduced. The velocity triangle shows the blade tangential velocity, u, the gas absolute velocity, c, and the gas relative velocity, w, making up three sides of the triangle at a particular location in the machine. The velocity triangle provides an excellent method for representing the circumferential and meridional components of velocities of the working fluid at a particular location. In addition, the changes in the velocity triangle across a component are used to highlight the way in which the flow changes direction and the mean gas velocities change their magnitude along the flow path.

The role of thermodynamics in turbocompressor energy transfer is then considered. A brief introduction to thermodynamics leads to the description of the steady flow energy equation (SFEE), which is the first law of thermodynamics applied to a fixed region with steady flow. The SFEE is used to account for the changes in fluid properties along the flow path as a result of work and heat transfer. It shows that the bookkeeping of the energy transfer needs to be carried out using the total enthalpy or the rothalpy. The efficiency of processes concerned is a primary aspect of the study of compressors. The second law of thermodynamics provides a rigorous way to study this through the thermodynamic state variable known as entropy. In the context of energy transfer, the entropy production is equivalent to the lost work in the machine due to dissipation losses. The definition of the isentropic, isothermal and polytropic processes is introduced. The importance of the aerodynamic work and the value of a polytropic analysis are considered.

2.1.2 Learning Objectives

- Define the Euler turbine equation, its relationship to velocity triangles and the insights it provides
- What circumstances cause the classical Euler turbine equation to be slightly incorrect?

- Be able to draw the velocity triangles of a radial compressor stage and relate the velocity vectors to the geometry of the blades and the flow channel.
- Explain why centrifugal impellers really are mostly centrifugal!
- Know the circumstances in which the classical Euler turbine equation is slightly incorrect.
- Be familiar with the SFEE.
- Understand the differences between static and stagnation properties.
- Know the second law of thermodynamics in the Gibbs form with the so-called temperature entropy relationships.
- Be able to describe different types of work, such as flow work, displacement work, aerodynamic work and shaft work.
- Be able to explain the relevance of various enclosed areas on the *T-s* diagram.
- Define the different state changes known as polytropic, isentropic and isothermal.
- Understand the effect of the gas molecular weight on the aerodynamic work needed to achieve a certain pressure ratio.

2.2 The Euler Turbine Equation

2.2.1 Newton's Laws of Motion

Newton's first law of motion states that an object will remain at rest or in uniform motion in a straight line, unless acted upon by an external force. In a fluid, the application of a pressure force can change the velocity of the fluid, and the rate of change in velocity is proportional to, and along the direction of, the applied force. More precisely, according to Newton's second law of motion the sum of all forces acing on a body equals the change in linear momentum. For a steady process, the change in momentum is due only to the change in velocity and not to the change in mass, so that in the direction x of the applied force

$$\sum F_x = \frac{d}{dt}(mc_x) = \dot{m}(c_{x2} - c_{x1}). \tag{2.1}$$

The forces acting on the fluid are pressure forces, viscous forces and body forces. From Newton's third law of motion, these cause a resultant reaction force on the turbomachine. In any situation where the fluid accelerates, decelerates or does not continue in a straight path, we know that forces must be acting. In a turbocompressor, the strongest of these forces are the pressure forces.

Within a radial compressor impeller, for example, the meridional plane is a bend from the axial to the radial direction. This bend forces the fluid to change direction from axial to radial by generating radial pressure forces which act to turn the flow radially. The meridional streamlines are then curved, and the curvature causes higher pressure on the hub and lower pressure on the casing. This pressure difference provides the force to drive the fluid from the axial to the radial direction. A decrease in pressure in a flow with no losses involves an increase in velocity (see the Bernoulli equation in Section 5.3.1), so that in the inlet of a radial compressor impeller the meridional velocity on the casing is generally higher than that on the hub.

Other forces are, of course, also at play within the complex flow field of a radial compressor, and these are described in some detail in Section 15.4, which details the action of the different forces on the circumferentially averaged meridional flow.

2.2.2 The Euler Turbine Equation

Euler realised that in a turbomachine it is the moment of momentum that is of relevance. He derived that, for a steady process,

$$M = \sum rF_u = \frac{d}{dt}(mrc_u) = \dot{m}(r_2c_{u2} - r_1c_{u1}). \tag{2.2}$$

Only the force in the direction of rotation makes a contribution to the torsional moment, M. This force is denoted by suffix u to indicate that it is in the direction of the blade speed, u, but in some texts the suffix θ is used. The torque is the change in flux of angular momentum, usually known as the swirl, which is the product of radius and circumferential component of absolute velocity, rc_u. In general, for flow through a radial compressor rotor, the fluid enters at a radius of r_1 with tangential velocity c_{u1} and leaves at a higher radius of r_2 with tangential velocity c_{u2}; see Figure 2.1. When considering the whole rotor, it is the mass-averaged integral values of the velocities that are included in this equation. The rate of energy transfer, or power, in the rotor is then given by $P = \omega M$, and the specific shaft work done is given by

$$w_{s12} = P/\dot{m} = \omega M/\dot{m} = \omega(r_2c_{u2} - r_1c_{u1}) = u_2c_{u2} - u_1c_{u1}. \tag{2.3}$$

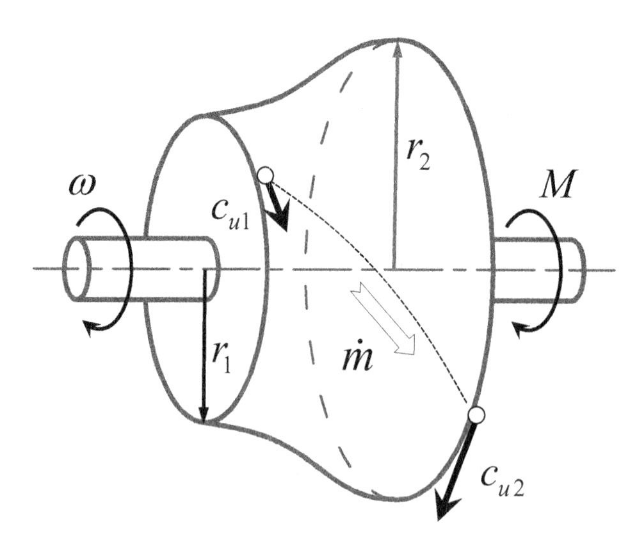

Figure 2.1 A stream tube passing through a turbocompressor with an increase in the radius and in the absolute tangential velocity of the gas.

The shaft work is transmitted across the boundary of the control volume to the rotor by a torque, or torsional moment, M, on the shaft. In some special circumstances, an additional term known as the viscous work term may be needed in this equation; this is discussed in Section 2.5.4.

2.2.3 First Insights from the Euler Turbine Equation

The derivation of the Euler turbine equation arises purely from mechanical arguments and imposes no conditions with respect to the type of fluid (compressible, incompressible, gas or liquid, perfect or real) and no assumptions are made about friction losses and heat transfer. The equation in this form is universal and applies without modification to all turbomachinery flows for all fluids, with or without frictional losses and heat transfer.

The mechanical work on the turbomachine blades takes place only in the rotor, where the applied circumferential force imposed on the blades by the shaft turns the rotor at a particular speed. The stator is stationary, with $\omega = 0$, and the change in swirl caused by the stator does no work on the fluid. In a compressor, work is input to the flow, $w_{s12} > 0$, so that the swirl increases across the rotor blade row, $r_2c_{u2} > r_1c_{u1}$. Most often, the impeller inlet flow enters in the axial direction and the circumferential component of velocity at inlet is zero, $c_{u1} = 0$. As a consequence, it is usually only the outlet of the impeller which determines the specific shaft work.

In a turbine, rotor work is extracted from the flow, $w_{s12} < 0$, so that the swirl decreases across the blade row, $r_2c_{u2} < r_1c_{u1}$. In a region of duct flow without blades, no work is done on the flow so that the swirl remains constant along a streamline. The Latin for swirl (or vortex) is 'turbo', and so turbomachines get their name from being swirl changing devices; see Section 1.4.1.

In a compressor, swirl is added by the rotor and is removed by the downstream stator component, so it is customary to think of the compressor stage as a rotor followed by a stator. In a turbine, an upstream stator generates the swirl, which the rotor removes to provide shaft work out of the control volume, and the turbine stage is usually denoted as a stator followed by a rotor. This leads to the generally applicable notation for the rotor and stator and the numbering of the locations between them as shown in Figure 10.6. The rotor inlet and outlet have the locations 1 and 2 respectively for both turbines and compressors in most systems of notation used in the literature. Other notations are also logical and possible.

The Euler turbine equation identifies that the energy transfer per unit mass is proportional to the change in the product of two velocity components, which has the units of the square of velocity. In a radial compressor, a fundamental dimensionless work coefficient can be defined by dividing the work input by the square of a reference speed, such as

$$\lambda = \frac{w_{s12}}{u_2^2} = \frac{c_{u2}}{u_2} - \frac{u_1}{u_2}\frac{c_{u1}}{u_2}, \tag{2.4}$$

where u_2 is the blade speed at the impeller outlet. This is invariably used as a reference speed for radial compressors and throughout this book. In other turbomachines, other reference blade speeds may be more suitable (at inlet, at the mean radius, at the hub radius etc.).

The change in swirl, which takes place between the blades in the impeller, arises because the blades are sufficiently closely spaced to guide the flow in the rotating coordinate system to follow the blade direction. The curvature of the blades applies a pressure force on the flow, as explained in Chapter 5. The flow leaves the rotor with a relative velocity, w, more or less in the direction of the impeller blade at the outlet, but with a small deviation from this. Nevertheless, the work is determined by the change in the circumferential component of the absolute velocity, c_u. An understanding of the flow field in terms of velocity magnitude and direction in both the rotating and the stationary frame of reference is axiomatic in the design of turbomachinery. This leads on to the issue of velocity triangles.

2.2.4 Velocity Triangles

To obtain a well-designed turbomachinery stage, it is essential to select a velocity field that acknowledges the principles of good aerodynamic design and to choose the shape of the blades in keeping with this. The essential part of the velocity field is best described in terms of the velocity vectors, or velocity triangles. The triangles represent a circumferential average perspective of the mean flow. These have a determining influence on the performance of a turbomachine, both at design and off-design, and so they are introduced here in considerable detail. Errors made in the selection of the velocity triangles used for a design are extremely difficult to correct by more sophisticated analysis in the later stages of the design process. Stodola (1905) was the first to use velocity triangles to explain the performance of turbomachinery.

First, consider the velocity triangle for a particular location in the flow, as shown in Figure 2.2. At any point in the flow the absolute velocity, c, the blade speed associated with the radial location of this point, $u = \omega r$, and the relative velocity, w, can be

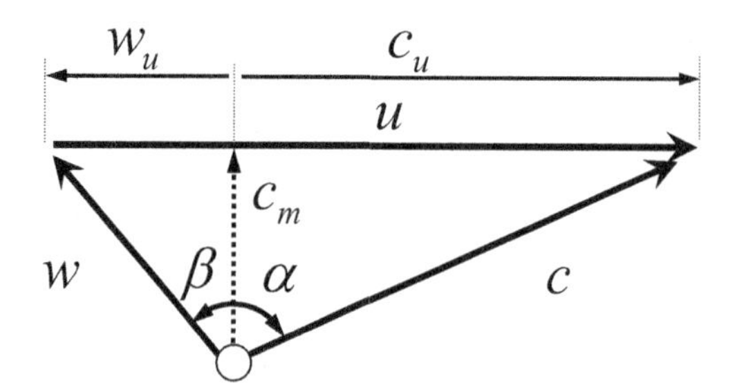

Figure 2.2 Velocity triangle showing flow angles and velocity components. In the notation of this book, the numerical values of w_u and β are negative in this velocity triangle as they are against the direction of rotation.

identified. Clearly, in a stator it is useful to look at the absolute velocities and, in a rotor, to look at the relative velocities, but in the gap between two blade rows, both views of the flow are of relevance as there is a need to switch attention from the rotor to the stator and vice versa. The velocity field is shown most simply and graphically by the algebraic expression for the vector addition of the relative and absolute velocities as

$$\vec{c} = \vec{w} + \vec{u}. \tag{2.5}$$

The velocity triangle can be considered for any single point in the flow. In the application of the Euler turbine equation on a one-dimensional circumferential-averaged basis across a stream-tube representing a complete flow channel of a compressor impeller it is sensible to consider the triangle at a representative point of the inlet and outlet of the blade row. The change between the velocity vectors between the inlet and outlet determine the terms in the Euler turbine equation.

The essence of this triangle is the addition of velocity vectors, and is the same as the relationship between the absolute velocity, apparent velocity and wind speed when, for example, riding on any vehicle, as shown in Figure 2.3. If there is no wind, the absolute velocity of the fluid is zero, $c = 0$, and the vehicle has a speed of u into still air. The driver will notice a headwind (a relative velocity of $w = -u$) relative to the motion. If there is a headwind of $-c$ and the vehicle rides into the wind at a speed u, then the driver feels a stronger apparent wind speed of $w = -(u + c)$. If there is a backwind, $c = u$, at the same speed as the vehicle is travelling, u, there is no relative velocity, $w = c - u = u - u = 0$. If there is a weaker backwind, or the vehicle is going at a higher velocity than the backwind, the driver still feels a headwind. If there is an even stronger backwind, the driver still feels a backwind. Figure 2.3 shows these cases and also the situation in which the wind speed includes a component of side wind. In these diagrams the component of the side wind normal to the vehicle motion is shown as constant in all of the triangles.

In a turbomachine, the blade speed, u, is always normal to the throughflow component of velocity, c_m. This leads to three different individual forms of velocity triangles that can occur in a turbomachine, as shown in Figure 2.4, depending on the magnitude and direction of the absolute flow vector, c, relative to the blade moving

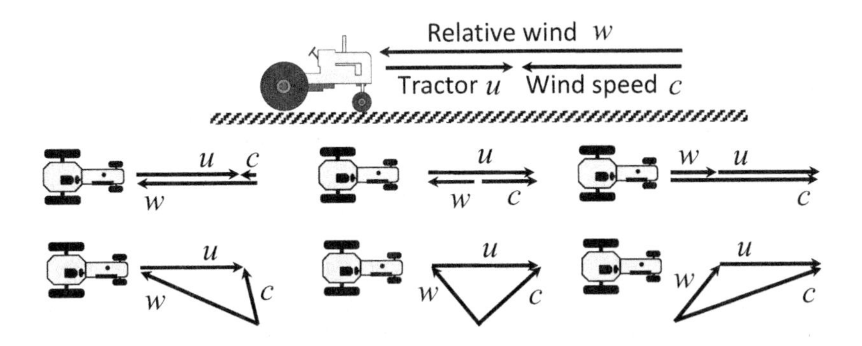

Figure 2.3 Velocity triangles for a vehicle in a headwind, backwind and a crosswind.

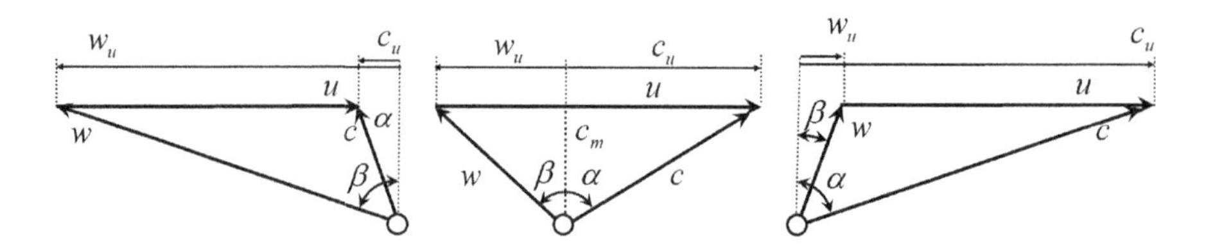

Figure 2.4 Three forms of velocity triangle.

with velocity $u = \omega\, r$. These forms, and only these forms, can occur in forward flow in a turbomachine. The middle triangle of Figure 2.4 is typical for a radial impeller outlet operating at its design point, and the diagram on the right is similar to the inlet of a radial compressor with swirl in the flow and on the left with counterswirl in the inlet flow.

In Figure 2.4, the horizontal base of the triangle represents the local blade speed and the blade is moving from the left to the right. The height of the velocity triangle (the length from the vertex along a normal to the blade speed vector at its base) is determined by the meridional component of velocity, whereby the meridional flow is up the page in the representation in this figure. The blade speed and the meridional component are the same in both the absolute and relative systems. Given that the base of the triangle represents the local blade speed, the height of the triangle immediately shows the value of the throughflow component of the flow relative to the blade speed. This can differ at different points in the flow and for different designs. An important nondimensional parameter to characterise the throughflow velocity is the ratio of the meridional velocity to the local blade speed, known as the local flow coefficient:

$$\phi = c_m/u. \tag{2.6}$$

From the velocity triangle, it is clear that a high value of the local flow coefficient implies low flow angles, and a low flow coefficient gives rise to large flow angles. Several useful algebraic relationships between the velocity components and the flow angles in a single velocity triangle which apply everywhere in the compressor are as follows:

$$\begin{aligned}
&\vec{c} = \vec{w} + \vec{u}, \quad && c_m = w_m, \quad && c_u = u + w_u, \quad && u = \omega r\\
&c = \sqrt{c_m^2 + c_u^2}, && c_m = c\cos\alpha, && c_u = c\sin\alpha, && \tan\alpha = c_u/c_m\\
&w = \sqrt{w_m^2 + w_u^2}, && w_m = w\cos\beta, && w_u = w\sin\beta, && \tan\beta = w_u/w_m\\
&c_m^2 = c^2 - c_u^2, \quad c_m^2 = w^2 - w_u^2 = w^2 - (u - c_u)^2\\
&c^2 - c_u^2 = w^2 - (u - c_u)^2 = w^2 - u^2 - c_u^2 + 2uc_u\\
&2uc_u = c^2 + u^2 - w^2.
\end{aligned} \tag{2.7}$$

The important step in constructing velocity triangles for a turbomachine is to consider the separate velocity triangles at the inlet and at the outlet of the rotor blade row. These show the circumferential component of the absolute velocity, c_u, as referred to in the Euler turbine equation (2.3). There are several conventions for combining the velocity triangles at different locations in axial machines, essentially differing in whether the tips of the vectors are united, or the base of the vectors (the blade speed) are united. For radial machines, the blade speed changes across the rotor such that the second of these options is most suitable and will be used here. The advantage of combining the inlet and outlet velocity triangles in this way is that it enables the work coefficient of the machine to be estimated from the velocity triangle on sight. The circumferential difference between the ends of the absolute velocity vectors before and after the rotor is the change in c_u and so is directly related to the work input or extraction of the machine, (2.4). Figure 2.5 and Figure 11.16 show visually the deceleration in the relative velocity across the impeller by comparing w_2 to w_1.

The key feature of the velocity triangles is that, at the design point, they also indicate the blade shapes associated with them. To avoid flow disturbances and high losses, the relative inlet flow on to a blade row needs to be close to zero incidence in order that the inlet angles of the blades are roughly aligned with the inlet velocity vectors. The blade row turns the flow in the circumferential direction, and the relative outlet flow direction is closely related to the direction of the blade profile at its outlet. For closely spaced blades, there is little deviation between the blade outlet angle and the flow direction. So, from a sketch of the velocity triangle, it is possible to sketch the blade forms that will be required. And, from inspection of the typical blade forms, it is possible to sketch the velocity triangles at the design point. This is shown in Figure 2.5 for a 3D radial impeller with an axial inlet and a radial outlet.

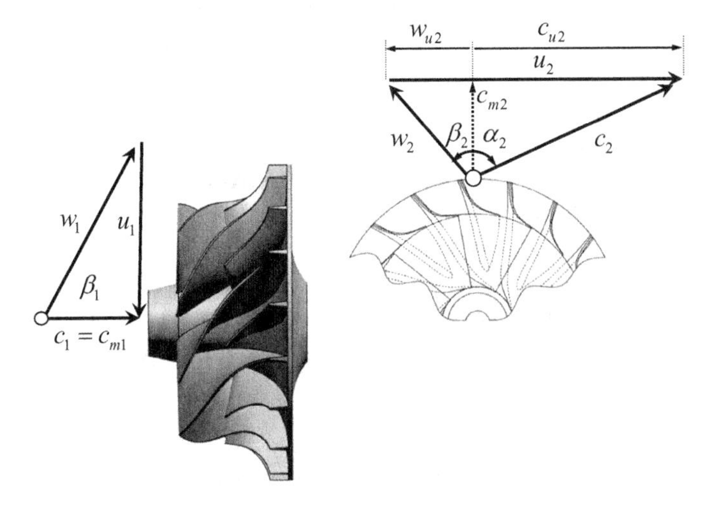

Figure 2.5 Velocity triangles and blade shape of a 3D radial compressor impeller.

The metal angle at the outlet of a blade row approximately determines the direction of the flow relative to the blade row, as no flow can pass through the blades. In a rotor, the metal angle at the outlet determines the value of the relative flow angle, β. In a stator, the metal outlet angle determines the value of the absolute flow angle, α. Thus, in both cases the flow direction at the outlet of a blade row is determined largely by the blade angle, although the flow deviates slightly from the blade metal angle. The flow angle at the inlet to a given blade row is not determined by that blade row but by the upstream blade row in the alternative coordinate system. This is explored in more detail later in this chapter.

At the rotor inlet, the upstream stator, if present, imposes the direction of the absolute flow. If no upstream blade row exists, as is the normal case in radial compressors, it is the design of the inlet duct which determines this absolute flow angle. This angle is normally zero. In combination with the meridional velocity (which is directly related to the volume flow and the area of the flow channel through the continuity equation, as described in Section 5.2.3) we can determine the relative flow angles at the inlet to the rotor as follows:

$$c_u = c_m \tan\alpha, \quad w_u = c_u - u, \quad \tan\beta = w_u/c_m$$
$$\tan\beta = \frac{c_u - u}{c_m} = \frac{c_u}{c_m} - \frac{u}{c_m} = \tan\alpha - \frac{u}{c_m} = \tan\alpha - \frac{1}{\phi}. \tag{2.8}$$

At a stator inlet, the upstream rotor imposes the direction of the relative flow angle, β, and we can determine the stator inlet flow angle as follows:

$$\tan\alpha = \frac{u}{c_m} + \tan\beta = \frac{1}{\phi} + \tan\beta. \tag{2.9}$$

An important aspect of these equations is that they show immediately that the inlet flow angle to a blade row is not constant but rather varies with the flow rate. The inlet angle can only be matched to the blade inlet angle at a specific flow coefficient, c_m/u, which is typically that at the design point. If the flow at the inlet to a blade row is not in the direction of the blade, then the inlet flow is not incidence free. Incidence often dominates the loss production in the blade row, as described in Section 10.4.2, and has an important effect on the efficiency and the pressure rise of a compressor stage, which are relatively intolerant to operation far off design. At high incidence, the flow can separate and break down to a reverse flow so that the compressor operates in an unstable manner. The velocity triangle also suggests a way of dealing with this problem through the use of variable geometry; see Section 17.5.

If the outlet flow angle in a radial compressor is known, together with the inlet flow angle of the downstream diffuser, the expected design flow coefficient at impeller outlet can be calculated from

$$\frac{1}{\phi} = \tan\alpha - \tan\beta. \tag{2.10}$$

This equation is used in Section 17.5 to consider the effect of inlet guide vane control and diffuser vane control on the flow capacity of a centrifugal stage.

2.2.5 Centrifugal Effect and Degree of Reaction

The key difference between a centrifugal rotor and an axial rotor can be identified when the Euler turbine equation is expressed in terms relative to the rotor (Lewis et al., 1972), as

$$\begin{aligned} w_{s12} &= u_2 c_{u2} - u_1 c_{u1} = u_2(u_2 + w_{u2}) - u_1(u_1 + w_{u1}) \\ &= (u_2 w_{u2} - u_1 w_{u1}) + (u_2^2 - u_1^2) \\ &= u_2\left(w_{u2} - \frac{r_1}{r_2} w_{u1}\right) + u_2^2\left(1 - \frac{r_1^2}{r_2^2}\right). \end{aligned} \tag{2.11}$$

The first term is the work done due to the deflection of the fluid by the blades, and the second term is the work done due to the centrifugal forces caused by the change of radius. In an axial machine, there is no radius change and the value of the second term is zero. The work is then entirely related to the change in the relative swirl velocity caused by the blade profile shape, which is the same as the change in the absolute swirl velocity. In a radial machine, the work done is strongly assisted by the centrifugal forces and the change in the swirl velocity is different in the relative and absolute systems.

A further insight from the trigonometry of velocity triangles in (2.7) can be obtained from a second form of the Euler turbine equation, with the help of the last line of (2.7), which identifies the work done with the kinematics of the velocities:

$$w_{s12} = u_2 c_{u2} - u_1 c_{u1} = ½(c_2^2 - c_1^2) + ½(w_1^2 - w_2^2) + ½(u_2^2 - u_1^2). \tag{2.12}$$

The three terms on the right-hand side are the increase in kinetic energy of the absolute flow through the rotor, the decrease in the kinetic energy of the relative flow through the rotor and the centrifugal effect which is the increase in blade speed with radius in the impeller. As there is positive work into the control volume in a compressor, all three terms should be positive to maximise the shaft work. This equation is used in Section 2.6.4 to define a characteristic conserved property of turbomachinery flow known as the rothalpy.

The absolute velocity of the flow through the impeller increases, which leads to a high kinetic energy at the rotor outlet. Downstream of the impeller, deceleration of this high-speed flow in the diffuser implies that the flow is moving into a region of higher pressure and that there is a risk of flow separation leading to high losses, as discussed in Section 10.3.3. In order to convert this outlet kinetic energy into a useful pressure rise, an effective diffusing system downstream of the impeller is needed.

To make a positive contribution to the work input, the relative velocity across the impeller must decrease through the blade passage. The hub streamline of a radial impeller has a larger change in blade speed than the casing streamline, with the result that the flow near the hub requires less deceleration than the flow in the casing streamline. In fact, the relative flow on the hub streamline often accelerates through the impeller. In order to decrease the relative velocity, the internal blade channels have an increase in area between the blades from inlet to outlet. In a turbine where the relative velocity has to increase across the impeller, the flow channels appear like nozzles with a decreasing area from the inlet to the outlet. There is a physical limit to the amount of deceleration in a turbomachinery blade row, named after de Haller (1953), who first suggested what is now known as the de Haller number, which sets a limit of $w_2/w_1 = 0.7$ for axial blade rows. A slightly lower number giving more deceleration is often used as the limit in radial compressors. The de Haller number is only a function of the velocity triangles, as no knowledge of the hardware is required, so it is particularly useful in the guidance of preliminary design; see Section 7.5.4.

In order to profit as much as possible from the increase of the blade speed, the impeller is centrifugal with a higher diameter at the outlet than at the inlet. This is the main reason why radial compressor impellers have a radial outlet flow. In a radial turbine impeller, the work is extracted from the rotor, and as a consequence the centrifugal term needs to be negative, leading to radial inflow turbines which are centripetal. In fact, approximately half of the pressure rise in a radial compressor arises from the centrifugal effect, which provides a high pressure increase without the need to decelerate the flow and has few direct losses associated with it. For this reason, the radial compressor can achieve relatively high efficiencies despite the high curvature of the meridional flow compared to an axial stage, which relies mostly on deflection of the circumferential flow to achieve pressure rise.

It is useful to consider the numerical value of the kinetic energy terms in (2.12). Considering state 2 to be the outlet of the impeller, a useful parameter to characterise the kinetic energy at this location is the so-called kinematic degree of reaction. This is usually defined as

$$r_k = \frac{\Delta h_r}{\Delta h_r + \Delta h_s} = \frac{h_2 - h_1}{h_2 - h_1 + h_3 - h_2} = \frac{h_2 - h_1}{h_3 - h_1} \approx \frac{\Delta p_r}{\Delta p_r + \Delta p_s}, \tag{2.13}$$

which is the ratio of the static enthalpy rise across the rotor, Δh_r, to the static enthalpy rise across the stage comprising the rotor and stator, $\Delta h_r + \Delta h_s$. Enthalpy is discussed in Section 2.4.2. The degree of reaction is sometimes also defined in terms of the changes in pressure across the components, (2.13), but in this case there is no direct link to the velocity triangles. In the present work, the degree of reaction is defined following Mehldahl (1941): the stage inlet and stage outlet conditions are taken to be the total conditions, and the impeller outlet conditions to be the static conditions, leading to

$$r_k = \frac{h_2 - h_{t1}}{h_{t3} - h_{t1}} = \frac{h_2 - h_{t1}}{h_{t2} - h_{t1}} = \frac{h_{t2} - \tfrac{1}{2}c_2^2 - h_{t1}}{h_{t2} - h_{t1}} = 1 - \frac{\tfrac{1}{2}c_2^2}{h_{t2} - h_{t1}}. \tag{2.14}$$

This form of the equation highlights the link between the kinetic energy at the impeller outlet and the degree of reaction. The numerical values from (2.13) and (2.14) are very similar, as the kinetic energy at the inlet and outlet flanges of the stage is usually small relative to that at the impeller outlet. The gas velocities at the flanges are about 10% of the impeller outlet velocity so that the kinetic energy in the flow at the flanges is only about 1% of that contained in the gas in the impeller outlet flow, $\tfrac{1}{2}c_2^2$. The degree of reaction lies between 0.6 and 0.65 for many designs of impeller, indicating that the kinetic energy at the impeller outlet is between 35 and 40% of the work input. Chapter 11 discusses this in more detail in connection with the design choices available for the selection of the impeller outlet velocity triangle.

A further rearrangement of (2.12) and (2.14) leads to

$$r_k = \frac{\tfrac{1}{2}\left(w_1^2 - w_2^2\right)}{h_{t2} - h_{t1}} + \frac{\tfrac{1}{2}\left(u_2^2 - u_1^2\right)}{h_{t2} - h_{t1}} = r_{\text{diffusion}} + r_{\text{centrifugal}}. \tag{2.15}$$

This shows that the degree of reaction is the sum of the fraction of the work produced by deceleration of the relative flow in the rotor, $r_{\text{diffusion}}$, and the fraction produced by the centrifugal effect, $r_{\text{centrifugal}}$. In a purely axial compressor, the mean radius of the rotor is nearly constant across the blade row, $u_2 = u_1$, and therefore $r_{\text{centrifugal}} = 0$ and $r_k = r_{\text{diffusion}}$. For an impeller with no inlet swirl, the centrifugal effect can be further expressed as

$$r_{\text{centrifugal}} = \frac{\tfrac{1}{2}\left(u_2^2 - u_1^2\right)}{h_{t2} - h_{t1}} = \frac{1 - u_1^2/u_2^2}{2c_{u2}/u_2} = \frac{1 - r_1^2/r_2^2}{2c_{u2}/u_2}. \tag{2.16}$$

This shows that for a typical radial compressor stage with $c_{u2}/u_2 = 0.65$ and a mean radius ratio of $r_1/r_2 = 0.5$, approximately 50% of the work input is produced by the centrifugal effect. This indicates that the breakdown of the enthalpy rise in (2.12) for a typical stage is approximately 35% due to the change in kinetic energy across the stator, 15% due to the decrease in the kinetic energy of the relative flow in the rotor and 50% due to the centrifugal effect. These proportions vary slightly for different impeller designs, but the centrifugal effect dominates the pressure rise in all radial compressor stages.

Considering the impeller alone, the proportion due to the centrifugal effect becomes even more dominant. For the static condition at the impeller outlet, the aforementioned values indicate that roughly 75% of the static enthalpy rise (equivalent to 75% of the static pressure rise) across the impeller is due to the centrifugal effect and only 25% to the deceleration in the relative impeller flow. In different designs of impeller with the same mean radius ratio, the work done by the centrifugal effect remains the same, with the result that through the choice of work coefficient the designer can typically influence only the 25% of the pressure rise due to diffusion. Equation (2.16) shows that reducing the work coefficient increases the proportion of the pressure rise due to the centrifugal effect. In impellers with a high radius ratio, such as impellers on a thick shaft in multistage process compressors, or impellers downstream of an axial compressor in an axial-centrifugal design, the centrifugal effect is reduced and this makes it more difficult to achieve a high work coefficient. Designs with a high work coefficient require a higher deceleration of the relative flow and so are more prone to separation in the impeller. This becomes important in the consideration of the jet-wake phenomenon in Section 5.8.

2.3 The First Law of Thermodynamics

2.3.1 Thermodynamic Concepts

A basic knowledge of thermodynamics is clearly needed to understand the processes in a radial turbocompressor, as the changes in state of the fluid are important in determining the actual local volume flow. The volume flow determines the local meridional flow velocity, and as a result the local velocity triangle and work input. The equations developed here are, of course, just as relevant to hydraulic machines, such as centrifugal pumps, but in hydraulic machines the density is constant and the flow can be understood without any reference to thermodynamics. Analysis of the different machine types requires the precise definition of the equations of state of the fluids in order to properly understand the nature of their differences. For example, thermodynamics explains why in a water turbine, which by definition has an incompressible flow, there is always a temperature rise, however small, as the water flows through the turbine. Or why in the expansion process of a compressible fluid in a gas or steam turbine the temperature of the gas always drops with the extraction of the work. This chapter describes the basics of the theory of thermodynamics for turbomachinery following closely the approach that can be found in many text books, including Traupel (2000), Moran and Shapiro (2007), Baehr and Kabelac (2012), Cengel and Boles (2015) and Geller (2015). It assumes that the equation of state of the fluid is that of an ideal gas. Chapter 3 considers the effect of real gases on the equation of state.

2.3.2 Definition of a System

In thermodynamics, the definition of the system of interest is important as it allows us to identify the boundaries with which it interacts with the surroundings. The system can be stationary or moving, and the intelligent choice of the appropriate system boundary simplifies the analysis of most problems. An isolated system has constant mass and constant energy with no heat, work or mass transfer across the boundary. An insulated system has no heat transfer across the boundary. A closed system has constant mass, but variable energy through work and heat transfers across the boundary.

Consider a closed system of a small element of fluid with fixed mass passing through a turbomachine: it would first enter the machine, pass through each blade row in turn and, at some point, leave the machine. Such an analysis is called the Lagrangian view after Joseph-Louis Lagrange (1736–1813). This analysis is, in fact, not without its difficulties as the fixed mass passes through all the components in an unsteady fashion and at any time the energy balance of the system of fixed mass differs from that of the internal space of the turbomachine through which it has already passed.

Clearly, a better method of analysis is needed, and the most useful is known as a Eulerian – or a control volume – analysis. This considers the turbomachine as an open system with a fixed region in space but with work, heat and mass transfers across the boundary. The control volume is a defined internal space within the machine, which allows mass flow across its boundaries to and from the surroundings together with heat and work transfer across the boundaries. If the turbomachine is operating steadily, the thermodynamic properties and the kinematic properties of the velocity triangles at the boundaries of the control volume are constant. This is why turbomachines are invariably analysed as open systems. The following discourse is limited to an analysis of a control volume with steady flow, where the mass flows into and out of the control volume are the same and the work input and any heat transfer are steady. Details of more complex situations are covered in the thermodynamics textbooks already mentioned in Section 2.3.1 and by Greitzer et al. (2004) for turbomachinery flows.

2.3.3 Thermodynamic Properties

The thermodynamic state of a system is defined by its properties, and different types of thermodynamic properties can be identified. Extensive properties are proportional to the system mass (such as volume, V; internal energy, E; enthalpy, H; and entropy, S). Intensive properties are independent of the system mass (such as pressure, p; ρ and temperature, T). Specific properties, which are used mostly in this text, are extensive properties per unit mass (specific volume, v; specific internal energy, u; specific enthalpy h; and specific entropy, s) and these are also intensive.

The thermodynamic state at a particular location within a control volume is defined by any two of its properties, and the equations relating these are defined as the equation of state of the fluid concerned. In this chapter, either a perfect gas with constant specific heat or an ideal incompressible liquid is considered. The extension to other equations of state is discussed in Chapter 3. The properties fall into one of two categories. The first category comprises those properties, such as pressure, temperature and volume, which are easily perceivable and relatively easily to measure directly with suitable measurement equipment. The second category covers thermodynamic properties, such as enthalpy, internal energy and entropy, which cannot be measured directly and are difficult to perceive directly and so need to be calculated from the first group. There are two types of equations of state: the thermal equation of state relates temperature, pressure and density to each other; and the calorific equation of state relates enthalpy, entropy and internal energy to the other properties.

2.3.4 Quasiequilibrium and Reversible Processes

An important aspect in turbomachinery is the change of the thermodynamic state of the gas as it passes through the machine. The fluid in a turbomachine is considered to be in a state of equilibrium at all locations. In addition, as the fluid passes through the machine it is assumed that it goes through a series of quasiequilibrium changes driven by a number of infinitesimal departures from equilibrium. This is a reasonable approximation for most engineering processes since the time to reach equilibrium is much less than that taken for any change of state. For compressors with steady operating conditions, this is a good approximation and is used throughout this book.

A reversible process is an idealised one in which the process can be reversed through the same series of equilibrium states so that the system and its surroundings return to the same conditions as at the outset.

2.3.5 Sign Convention

In the notation of this book, both the shaft work and the heat transfer acting on a system are considered to be positive when they are both inputs into the control volume. The advantage of this notation when considering compressors is that the work is a positive number as the shaft does work on the fluid. Some textbooks prefer to consider heat into the control volume as positive and the work out of this as positive. This has the advantage when considering heat engines with an output of shaft work that both heat and work have positive values. However, both the shaft work input and the heat transfer into the system lead to an increase in the internal energy and are here considered as positive.

2.3.6 The First Law Applied to a Fluid Element

The equations representing the first law of thermodynamics are of fundamental importance in turbomachinery flows. Because of this, it is worthwhile to derive them in the simplified form used most often in this book. This highlights the significance of the terms used in the equations, and of others yet to be defined.

As a first step, consider the first law applied to a fluid element, a closed system of fixed mass, as it passes through a thermodynamic device in a Lagrangian view. The device may include components with work addition and heat transfer, as shown in the schematic rendering of the compressor and combustion chamber of a gas turbine in Figure 2.6. In this case, shaft work is exchanged with the environment through the shaft via the blades of the compressor, and heat is added through combustion in the combustion chamber. An element of fluid mass, δm, exhibits variable energy as it undergoes the process from state 1 to state 2 through the machine due to the work and heat transfers. When applied to a fixed mass, the first law of thermodynamics determines that, if no work or heat is added, the total energy of the system remains constant throughout the process, that is, the sum of the internal energy, the kinetic energy and any other forms of energy of the fluid element does not change. When work or heat is added, the change in the energy of

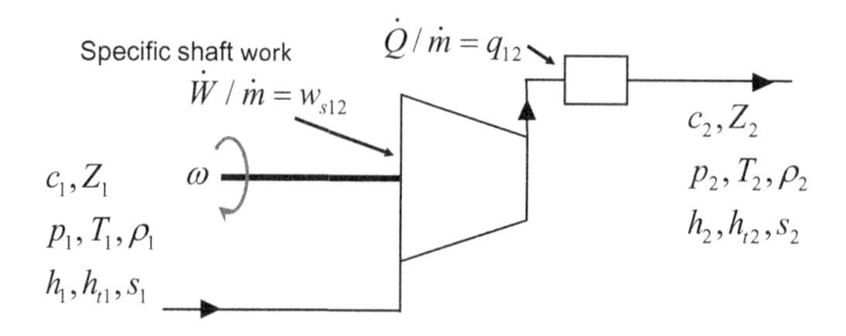

Figure 2.6 Schematic of a compressor and a combustion chamber of a gas turbine.

the fluid element is equal to the sum of the heat received and the work done on it as follows:

$$\delta E = \delta W + \delta Q, \quad \delta E = \delta U + \delta KE + \delta PE, \tag{2.17}$$

where W is the work input, Q is the heat input, E represents the different forms of energy, U is used to represent the internal energy, KE is the kinetic energy and PE is the potential energy of the control mass.

In this case, the work being added could be shaft work, electrical work, work done against gravity, work done against any other body forces or work done through a pressure force which moves the boundary of the control mass, as in a piston. The latter, if it takes place reversibly, is here called displacement work, but it could also be called boundary work, volume change work or *pdv* work, and this is calculated as

$$W_{v12,rev} = -\int_1^2 pdV = -m\int_1^2 pdv \quad or \quad w_{v12,rev} = -\int_1^2 pdv, \tag{2.18}$$

where the suffix, v, indicates volume change. This would be the work done by a piston on an enclosed mass as it decreases the volume of the fluid in a reversible compression process. A visualisation of the displacement work is given in Figure 2.7 as the area under the pressure-volume curve as the gas is compressed. Note that this equation has a negative sign in the notation used here because a reduction in volume (where dv is negative) occurs when work is done on the system.

2.3.7 First Law Applied to a Control Volume

A turbomachine has a continuous flow through it, so it is not a closed system of fixed mass but an open system with fluid mass passing through it. For a clearer perspective, it is necessary to derive an equation based on the Eulerian view that can be used for a fixed region in space. The control volume under consideration is shown in Figure 2.8 and includes the whole compressor and short parallel sections of the inlet and outlet

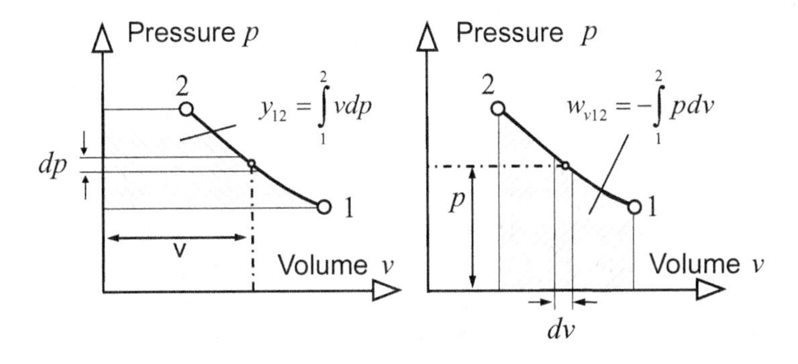

Figure 2.7 Aerodynamic work (left) and displacement work (right) in a *p-v* diagram.

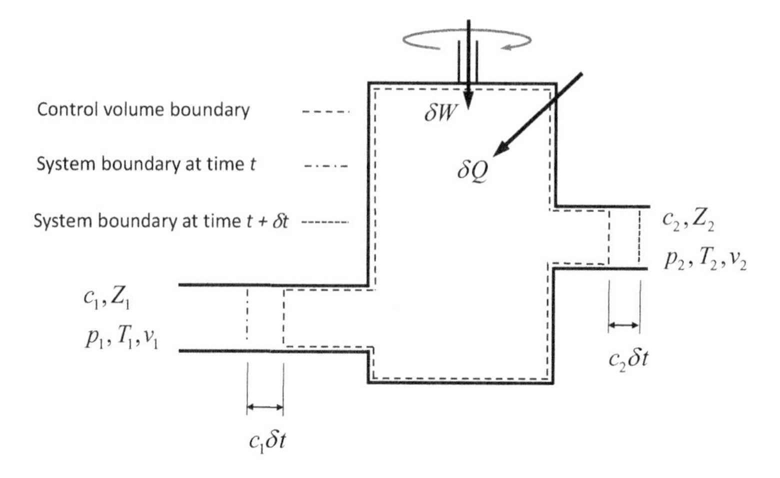

Figure 2.8 Control volume used for the derivation of the SFEE.

ducting. Shaft work is applied by the compressor blade rows to the control volume under consideration, and heat transfer occurs over the boundaries. One-dimensional mean values of the flow conditions are taken to be representative of the local state of the fluid at the inlet and outlet planes. In this case, we have one inlet and one outlet, but a more general derivation of the first law for a control volume can be made with multiple inlets and outlets. It is then the mass-averaged integral values of the properties over all the relevant inlets and outlets that are important (Greitzer et al., 2004). The flow properties within the control volume are considered to be steady. Even though unsteady processes occur, the time-average properties of the fluid at the boundaries of the control volume are assumed to remain constant.

Considering a fixed mass of fluid at time t and at time $t + \delta t$, it is clear that over this time step a small element of the fluid moves into the control volume at inlet and a small element leaves at the outlet boundary. Work is done by the external environment to bring this small element of mass at the inlet into the control volume, and at the same time the control volume does work on a small element of fluid to expel it from the system into the surroundings. At both locations, the force needed is the product of the pressure and the area, pA, and the distance moved is the velocity times the time taken, $c\,\delta t$. The work done on a small element of mass δm by the surroundings is then given by

$$p_1 A_1 c_1 \delta t - p_2 A_2 c_2 \delta t = p_1 v_1 \delta m - p_2 v_2 \delta m = (p_1 v_1 - p_2 v_2)\delta m, \tag{2.19}$$

where $Ac\,\delta t$ has been replaced by $v\,\delta m$, or $\delta m/\rho$, from the conservation of mass ($\delta m = \rho\, Ac\,\delta t$). The change in the product of the pressure and the specific volume which appears in this equation is called the *flow work*. The term arises in the Eulerian view out of the work needed at the boundaries to maintain a continuous flow through the control volume. In addition to this, work is needed to raise the altitude of a small element of mass from the inlet to the outlet height, $\delta m g(Z_1 - Z_2)$.

The total energy change of the element of mass during the time step δt is then given by

$$\delta W + \delta Q + p_1 v_1 \delta m - p_2 v_2 \delta m = \delta m\left[(u_2 - u_1) + ½\left(c_2^2 - c_1^2\right) + g(Z_2 - Z_1)\right], \tag{2.20}$$

where u is the specific internal energy. The rate of energy addition (rate of work done and rate of heat transfer) is given by

$$\dot{W} + \dot{Q} = \dot{m}\left[(u_2 - u_1) + (p_2 v_2 - p_1 v_1) + ½\left(c_2^2 - c_1^2\right) + g(Z_2 - Z_1)\right]. \tag{2.21}$$

On dividing by the mass flow rate, the SFEE is obtained as

$$w_{s12} + q_{12} = (u_2 - u_1) + ½\left(c_2^2 - c_1^2\right) + (p_2 v_2 - p_1 v_1) + g(Z_2 - Z_2), \tag{2.22}$$

where w_{s12} is the specific shaft work and q_{12} is the specific heat addition. At the outlet of the control volume, this equation includes the internal and kinetic energy that is transported out of the system by the flow, $u_2 + ½c_2^2$; the potential energy due to the altitude of the fluid, gZ_2; and the flow work done on the surroundings to expel the flow from the control volume, $p_2 v_2$. It also includes the equivalent terms at the inlet.

2.4 The Steady Flow Energy Equation

2.4.1 Application of the SFEE

Equation (2.22) represents one of the fundamental relationships used in thermal turbomachinery calculations. It is the first law of thermodynamics in the form of the SFEE for steady flow through a control volume including a turbomachine with inlet state 1 and outlet state 2. The specific shaft work, w_{s12}, done on the control volume together with the specific heat transferred to the control volume, q_{12}, produces a change of properties between the flow at the inlet and outlet of the control volume. The specific heat transfer and specific shaft work both have two suffices to indicate that they are not state variables and are only defined in terms of interactions with the system between the inlet and outlet. The use of u for internal energy is not to be confused with u denoting the blade speed in many equations. Fortunately, the u used for internal energy is combined into a property known as the enthalpy in Section 2.4.2, and so does not appear in equations involving the blade speed. In the notation used here, both work and heat added to the system are positive. These cause a change in the specific internal energy, (Δu), and in the kinetic energy, $\Delta(½c^2)$, of the fluid. Work is done by the surroundings to displace fluid from the inlet to the outlet in a region at a different pressure level (the flow work, $\Delta(pv)$), or against gravity by displacing the fluid to a different altitude (the potential energy, $g\Delta Z$).

Given that (2.22) is derived for a control volume as a fixed region of space, only knowledge of the values of properties explicitly on the inlet and outlet boundaries of the control volume is needed. No information is necessary with regard to the actual processes within the control volume. These can be irreversible with frictional dissipation and can even be unsteady, provided time-averaged values are used in the equation and remain constant. The equation can also be applied to individual stream tubes of the flow. It is, however, more generally used with the mean values of the properties at inlet and outlet, whereby the mass-flow average mean values are usually taken. In combination with the velocity triangles at the inlet and outlet and with the Euler turbine equation, the control volume analysis becomes a powerful tool for understanding compressors.

For the application of (2.22), it is useful to choose a control volume that is coincident with the bounding surfaces of the device under consideration, such as the hub and casing walls of the turbomachine. In the analysis of turbomachinery, the same equation can then be applied to different locations of the inlet and outlet boundary planes of the control volume. These planes can then be selected to examine individual stage components (inlet duct, rotor, stator, exit duct, diffuser, etc.), whole stages (inlet to outlet of a stage with a rotor and a stator), groups of stages or from flange to flange. It is useful to use a different number or letter to identify each location, and this is generally done when considering a radial compressor stage, using state 1 for the inlet, state 2 for the impeller exit and diffuser inlet and state 3 for the outlet. In some cases, it is useful to use the equation simply from states 1 to 2 representing the inlet and outlet of a generic component.

Most compressors may be also considered to be adiabatic with no heat transfer under steady flow conditions. There is typically so little heat transfer that this is negligible compared with the work transfer, and so q_{12} is generally taken as zero. One exception here is small turbochargers operating at low speed which pick up heat from the adjacent hot turbine where this heat is significant compared to the work input. A further exception is small gas turbines where the compressor, combustor and turbine are physically close together. In the stationary components, such as stator vanes, inlet and outlet casings, diffusers or heat exchangers and boilers, there is no work addition or extraction and so the shaft work, w_s, in these components is zero, although the aerodynamic work may differ from zero.

2.4.2 Definition of Static and Total Properties

In thermal turbomachinery, the potential energy term, gZ, can invariably be neglected and will only rarely be used here. For example, the specific kinetic energy, $\frac{1}{2}c^2$, associated with a gas velocity of only 50 m/s would be equivalent to an altitude change of 127 m in air, so that even in low-speed ventilators and fans, the potential term, gZ, is clearly negligible. The thermodynamic property which occurs naturally in a compressible flow process through an open system described by the SFEE is the enthalpy, which is the sum of the internal energy and the product of pressure and specific volume, $h = u + pv$. The analysis of all thermal turbomachines can then easily be unified by defining the local enthalpy as

$$h = u + pv = u + p/\rho \tag{2.23}$$

and the total (or stagnation) enthalpy as

$$h_t = u + pv + ½c^2 = h + ½c^2 \tag{2.24}$$

such that

$$h_{t2} - h_{t1} = (u_2 - u_1) + ½(c_2^2 - c_1^2) + (p_2v_2 - p_1v_1). \tag{2.25}$$

The total enthalpy accounts for the internal energy and the associated kinetic energy and flow work at the boundaries of an open system.

In the case of hydraulic turbomachines, it is not appropriate to neglect the potential energy term as the fluid density is much higher than in a gas, and the level of the inlet and outlet above a reference level, Z, needs to be taken into account. In flows of incompressible liquids, Traupel (1962) defines a generalised enthalpy, for which he uses the symbol $\boldsymbol{i}$, including the gZ term as

$$\boldsymbol{i} = u + pv + gz = h + gZ. \tag{2.26}$$

The use of this definition for the enthalpy allows the thermodynamic equations described in this book to be generalised for use with incompressible liquid flows so that many aspects of the book apply also to centrifugal pumps. The head of water in a pump multiplied by the gravitational constant can then be interpreted as an enthalpy.

With the simplification of using the total enthalpy to characterise the energy transfer, the first law for a steady flow through a control volume for all turbomachines then takes a very compact form,

$$w_{s12} + q_{12} = h_{t2} - h_{t1}, \tag{2.27}$$

and in an adiabatic machine with $q_{12} = 0$ even more so. Using the appropriate definition of total enthalpy, this equation is valid for all fluids. It provides a simple way to account for changes in work and heat input through bookkeeping of the total enthalpy and identifies the key role of total enthalpy as a measure of energy interactions in turbomachinery processes as open systems. In an adiabatic flow in a compressor, the total enthalpy changes across a rotor (with work transfer) and is conserved across a stator (with no work input or extraction).

2.4.3 Total Conditions

It is worthwhile examining what the concept of total enthalpy means physically. Considering a flow situation with no work or heat input and neglecting altitude changes, the total enthalpy remains constant between the inlet and outlet of the control volume. Applying this to a stream-tube in the flow with no work or heat transfer, (2.27) indicates that the quantity

$$h_t = h + ½c^2 \tag{2.28}$$

is conserved, so that the total enthalpy stays constant along the stream-tube. The total enthalpy can be imagined as a reference enthalpy for the flow whose numerical value is the enthalpy at a location where the velocity is reduced to zero (for example, at a stagnation point in the absolute flow with $c = 0$). In turbomachines, the inlet total conditions are often taken as the conditions in a large plenum chamber from which the fluid is drawn and where the velocity is zero. In compressor applications with atmospheric air, the inlet plenum is the ambient external environment. In Section 2.6.4, a new term will be introduced called *rothalpy*, which can be visualised as the total relative enthalpy of the flow when brought to rest on the axis of the rotating machine. Both the total and the static enthalpy can be shown in an *h-s* or Mollier diagram, shown in Figure 2.9, which is a convenient diagram to visualise the changes in the thermodynamic properties.

Associated with the total enthalpy, there is a total temperature, which using the calorific equation of state for an ideal gas, $dh = c_p dT$, can be written as

$$T_t = T + \frac{c^2}{2c_p} = const. \tag{2.29}$$

This equation can be applied in the situation where a body is moving into a flow with a certain velocity c into still air with an upstream static temperature of T. The increase in temperature between the upstream conditions and the stagnation point on the nose of the body can be calculated for different velocity levels. For example, 10 m/s (36 km/hour) leads to a temperature increase at the stagnation point of 0.05°C. This is hardly noticeable, but at 100 m/s the temperature rise is 5°C and in a high-speed radial air compressor with a gas velocity of 300 m/s at the impeller outlet, the temperature difference between the stagnation and the static states would be 45°C. Similarly, this equation shows that air drawn from the environment and accelerated into the inlet of an impeller will cool down, potentially leading to ice formation in inlet guide vanes.

Considering the equivalent in terms of pressure, and neglecting altitude changes, the total pressure is the pressure at a stagnation point in the flow where the velocity is reduced to zero. The deceleration is reversible and adiabatic, and therefore at constant entropy, so the total condition lies vertically above the static condition in an *h-s* diagram – the stagnation entropy is the same as the static entropy, and normally no distinction is made here. For a perfect gas with constant specific heats, the stagnation state can be calculated from the isentropic relations for an adiabatic reversible process, as

$$\frac{p_t}{p} = \left(\frac{T_t}{T}\right)^{\gamma/(\gamma-1)} = \left(1 + \frac{c^2}{2c_pT}\right)^{\gamma/(\gamma-1)}. \tag{2.30}$$

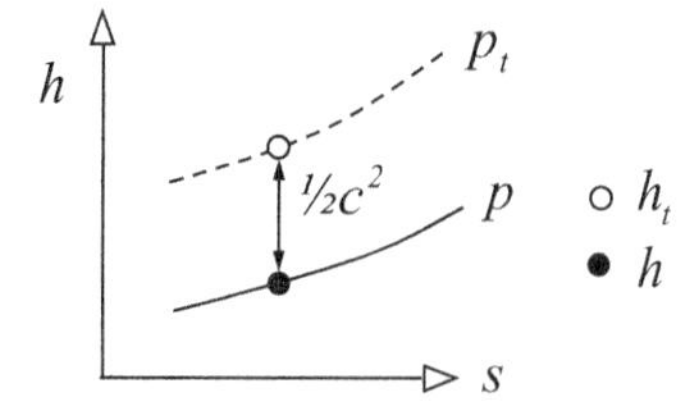

Figure 2.9 The total and static conditions represented in a Mollier *h-s* diagram.

This equation relates the total pressure and the static pressure for a compressible flow. Further elucidation of this equation and its relationship to the Mach number of the flow is given in Chapter 6 on gas dynamics.

2.4.4 Special Cases of the Steady Flow Energy Equation

The SFEE can be applied to a range of flow situations to describe the energy exchanges in the components considered. The simplest case is a flow process with no heat or work transfer, such that (2.28) applies. The total enthalpy remains constant but the static enthalpy depends on the velocity level. Two cases of this type are given in Figure 2.10 to represent the energy transfer in a simple pipe and a diffuser with decelerating flow. Note the coordinate x in these diagrams represents the distance through the particular component but, as is shown in the figure, it could be a representation of an increase in the entropy through the device. The special case of the pipe would also be valid for the flow through a restrictor or a valve. These special cases are also relevant to the flow in compressors; the lower part of Figure 2.10, for instance, represents the diffusing flow in diffuser vanes. The corresponding diagrams for a radial compressor impeller from state 1 to 2 at impeller outlet and a compressor stage to state 3 at the compressor outlet, with no heat addition but work added, are given in Figure 2.11.

The combination of work and heat transfer occurs in the analysis of complete power plant, but also in the special case of the radiation of heat from a turbomachine, or, as in a turbocharger compressor, with the transfer of heat by radiation and conduction from the turbine to the compressor. In many experimental tests on compressors, the inlet and outlet total temperature is measured and used to determine the inlet and outlet total enthalpy. Assuming that the machine is adiabatic, $q_{12} = 0$, then a measurement of the temperature change gives a direct link to the specific work done on the flow through (2.27).

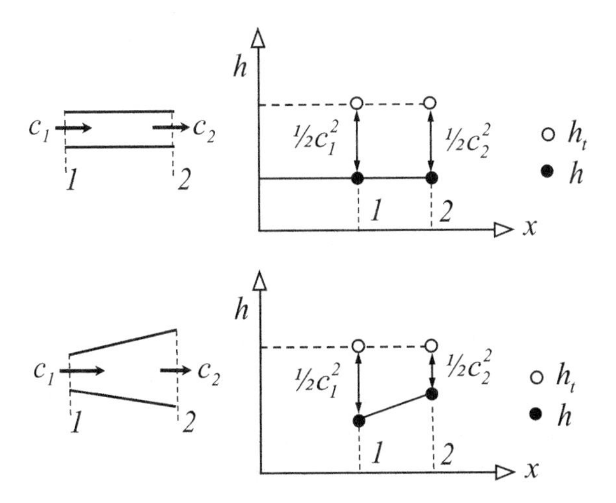

Figure 2.10 The flow process through a constant area duct and a duct of increasing area.

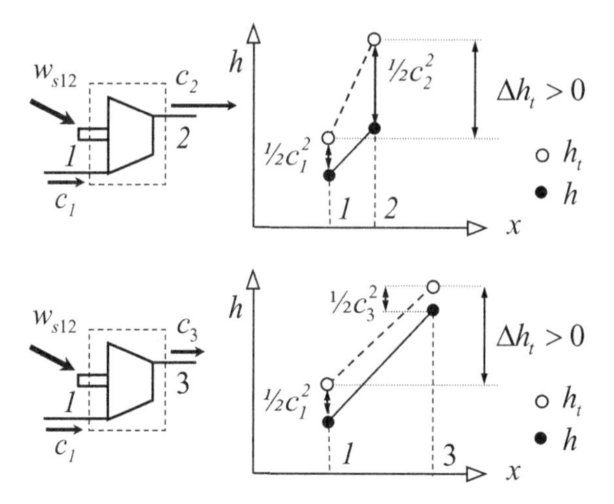

Figure 2.11 The flow process in an adiabatic impeller and a compressor stage.

Note, however, that if the compressor has just started operating and the metal components are cold, then heat may still be transferred from the fluid to the casing, and this means that a lower exit temperature is measured. An error of this type suggests a lower work input and a higher compression efficiency. In order to avoid errors due to this effect, it is crucial in a test campaign to make sure that the machine is in thermal equilibrium when data are taken and not still warming up. Conversely, a machine that has already warmed up at high speed would give up heat to the flow as the rotational speed is reduced. This increases the apparent temperature rise across the machine and may be misinterpreted as a higher work input and a lower efficiency. In the case of a turbocharger compressor, the heat transfer to the compressor from the turbine, which increases the total enthalpy at outlet, may be falsely interpreted as work input, and the efficiency of the compressor then appears to be lower than it really is (Sirakov and Casey, 2013).

2.5 The Second Law of Thermodynamics

2.5.1 Entropy

In addition to the bookkeeping of the energy transfer through the steady flow energy equation and the change in total enthalpy, it is necessary to track the frictional losses produced in the machine. The irreversibility is best monitored by using the entropy generated in the compression process. It is not possible to measure entropy directly, but it is a useful and rigorous description of the dissipative processes that occur. The full analytical development of the concept of entropy is derived from the second law of thermodynamics. This can be found in many thermodynamic textbooks and so is not covered in detail here. Its importance is that it defines entropy as a state variable and describes the direction in which any real adiabatic process takes place; in such a process, the entropy always increases. The theory considers the reversible heat transfer to a system and leads to a definition of the change in the specific entropy, s, the entropy per unit mass. For an adiabatic flow, an increase in the entropy is a measure of the work lost which is made unavailable due to the irreversibility (Greitzer, 2004). Thus, looking back at Figures 2.10 and 2.11, the parameter x on the axis of these sketches could be replaced by the entropy, s, which increases in the direction of flow, which enables the processes to be shown in an h-s diagram.

If heat is transferred reversibly across an infinitesimal temperature difference to a closed system of mass m at temperature T, then the change in the entropy per unit mass is defined by

$$ds = dq_{rev}/T, \tag{2.31}$$

where dq_{rev} is the heat per unit mass that has been added reversibly. Depending on whether the heat is added or removed, the entropy increases or decreases. In any real (irreversible) process, the change in entropy is always greater than this, such that the entropy change in a real process becomes

$$Tds = dq_{rev} + Tds_{irr}. \tag{2.32}$$

The product Tds_{irr} is known here as the specific dissipation and for any real process is always greater than zero. It is a measure of the increase in entropy due to dissipative processes and, as it occurs in many equations, following Traupel (2000) and Baehr and Kabelac (2012), it is given its own symbol, j:

$$dj = Tds_{irr} \geq 0. \tag{2.33}$$

For an adiabatic process, $dq = 0$, the change in entropy is always greater than zero. A common misconception of the second law is that it states that entropy always increases, but this is not the case for a system with heat extraction. The entropy of the flow falls across a turbine blade row with intensive cooling, but this reduction in entropy in the turbine flow is associated with an increase in entropy of the surroundings of the turbine blade. For the turbine and the surroundings taken together, the entropy always increases. When applied to an adiabatic control volume in a turbomachine, the equation states that the entropy between inlet and outlet increases and only for a fully reversible ideal adiabatic process does it remain constant. Another case where entropy would remain constant is a diabatic process with heat transfer in which the heat extraction exactly matches and cancels the effect of dissipation on the entropy change, and this becomes important when considering polytropic processes in Section 2.7.4.

The product Tds_{irr} is the sum of all the dissipation losses related to internal frictional dissipation in the flow and is called specific dissipation here. Alternative terms are the frictional work, the dissipation work, lost work or simply the dissipation. It represents the lost opportunity to obtain work during a frictional dissipative process. As it appears together in an equation with the term dq, it is also sometimes confusingly known as the frictional heat as it has the same effect on the entropy change as heat addition, as follows:

$$Tds = dq + dj. \tag{2.34}$$

In an adiabatic frictional flow, dj corresponds to the heat addition that would be needed in a frictionless flow to achieve the same change of state from inlet to outlet. For a process between an inlet at state 1 and an outlet at state 2 of a turbomachine, (2.34) can be integrated to give

$$\int_1^2 Tds = q_{12} + j_{12}. \tag{2.35}$$

Note that for the integration of the Tds term, the real process path from inlet to outlet must be known such that, just like the heat addition, the specific dissipation, j_{12}, is a process parameter depending on the path of the process and is not a state parameter.

2.5.2 The Gibbs Equation

The Gibbs equation is a combination of the two great principles of thermodynamics, the first and the second laws, and was derived by Josiah Gibbs (1839–1903). For a system of fixed mass undergoing an internally reversible process with internally reversible heat addition, $dq_{rev} = Tds$, and specific displacement work, on the boundary ($dw_{rev} = -pdv$) and with no change in kinetic or potential energy, the first law can be written as

$$du = dq_{rev} + dw_{rev} = Tds - pdv. \tag{2.36}$$

With the definition of entropy this relation combines the first and second law and has been derived for a reversible process. But it is a relationship which relates purely thermodynamic properties and so it must also be valid for any process, reversible or irreversible. From the definition of enthalpy, $h = u + pv$, so that, $dh = du + pdv + vdp$, (2.36) can be written in two forms as

$$\begin{aligned} Tds &= dh - vdp \\ Tds &= du + pdv. \end{aligned} \tag{2.37}$$

Equation (2.37) can also be applied to the gas state at the total conditions and hence

$$dh_t = T_t ds + v_t dp_t. \tag{2.38}$$

In turbomachinery applications, the second law of thermodynamics is mainly used in the first form defined here by (2.37), which includes enthalpy, which is directly related to the work and heat addition through the SFEE.

Applying (2.33) to a compression process between inlet and outlet states, 1 and 2, gives

$$h_2 - h_1 = \int_1^2 Tds + \int_1^2 vdp. \tag{2.39}$$

As explained previously, the entropy can change due to heat transfer or due to dissipation, (2.34), so this gives

$$h_2 - h_1 = q_{12} + j_{12} + y_{12}, \quad \text{where } y_{12} = \int_1^2 v dp. \tag{2.40}$$

A new term has been introduced, y_{12}, the integral of *vdp* from state 1 to 2, sometimes known as the pressure change work, or the *vdp* work. Baehr and Kabelac (2012) and Traupel (2000) call this the *Strömungsarbeit* (meaning flow work in German) but in English textbooks this is the name already given to the change in the *pv* products across the control volume, as in (2.22). In this book the term, y_{12}, is called the aerodynamic work. In practice a polytropic process may be used to integrate the static fluid conditions *vdp* along the compression path and then the term is generally called the polytropic head rise, polytropic head or simply head. Equation 2.40 states that for any steady process between the inlet and outlet of an open system defined as a control volume, the enthalpy change between inlet and outlet is the sum of the aerodynamic work, y_{12}, the heat transfer, q_{12}, and the dissipation, j_{12}. As with the heat transfer and the dissipation the aerodynamic work, y_{12}, is not a state property but is a process parameter which depends on the path of the compression process.

2.6 Energy Transfer in Radial Turbocompressors

2.6.1 Combination of the First and Second Laws

The steady flow energy equation, (2.23), can be combined with the Gibbs version of the second law in the form given by (2.40) across an impeller from state 1 to state 2, to obtain

$$w_{s12} = y_{12} + \tfrac{1}{2}\left(c_2^2 - c_1^2\right) + j_{12}. \tag{2.41}$$

This relationship shows that the shaft work in the impeller increases the kinetic energy of the flow, overcomes frictional dissipation and causes aerodynamic work to be done to increase the pressure of the fluid. Interestingly, the equation contains neither the change in static enthalpy nor the heat addition that takes place, although no assumption has been made about the adiabatic nature of the process. The heat addition does influence the equation, however, as the dissipation, j_{12}, and the aerodynamic work, y_{12}, are process variables and depend upon the integration of the state changes from state 1 to state 2, which themselves depend on the heat transfer.

Considering a stator blade row from its inlet to outlet, state 1 to state 2 with no work input, $w_{s12} = 0$, such as the diffuser downstream of an impeller, then (2.41) can be rearranged as

$$\tfrac{1}{2}\left(c_1^2 - c_2^2\right) = y_{12} + j_{12}. \tag{2.42}$$

This shows that in a diffuser it is the reduction in absolute kinetic energy that overcomes the dissipation and gives rise to aerodynamic work to increase the pressure.

2.6.2 The Reversible Steady Flow Shaft Work

Now considering a perfect reversible compression process with no dissipation, $j_{12} = 0$, from state 1 to state 2, then from (2.41)

$$w_{s12,rev} = y_{12} + \frac{1}{2}\left(c_2^2 - c_1^2\right). \tag{2.43}$$

This term is the reversible steady flow shaft work. It represents the best that can be done in a reversible process with no dissipation. In this case, the shaft work would either provide a contribution to the aerodynamic work, so increasing the pressure, or to the change in kinetic energy and nothing would be lost. This is the minimum work that would be needed without frictional dissipation and, as such, is very useful in defining the efficiency of the compression process. Note that the minimum work is the sum of the aerodynamic work and the change in kinetic energy so that both need to be properly accounted for in definitions of efficiency.

As noted previously, the perfect process is assumed to be reversible. However, no assumption has been made in this derivation as to the adiabatic nature of the process. This condition, therefore, also applies to (2.43). A reversible adiabatic process between the same inlet conditions and outlet pressure would not finish at state 2, but at state 2s (the state with the outlet pressure p_2 and the inlet entropy, s_1; see Figure 2.12). It is clear, then, that the only way a reversible process from state 1 to 2 can really be achieved is if it is diabatic. A reversible diabatic process can have the same end states as a real irreversible adiabatic process if the heat removed exactly matches the dissipation in the real process. This can be seen from (2.34) as, if $dq = -dj$, then $ds = 0$.

2.6.3 Use of the First and the Second Laws in Efficiency Definition

The important use of (2.43) is in the definition of efficiency. This will be discussed in detail in Chapter 4 but is introduced here. The efficiency is the key parameter used to denote the quality of a machine as it has a direct influence on the power consumption of a compressor and on the operating and environmental costs. It is defined in such a way that it would take the value of unity for a perfect machine and less than unity for an imperfect machine. The efficiency gives a guide to the magnitude of any thermodynamic imperfections. For this purpose, an expression is needed for the minimum possible effort that could have been expended in the

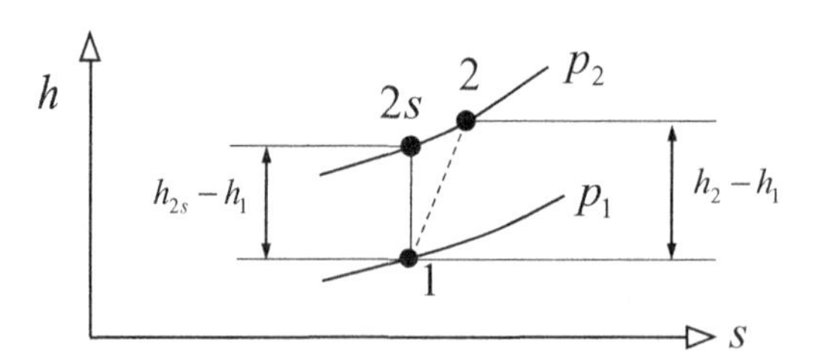

Figure 2.12 A comparison of the actual and an isentropic compression process.

compression process as a theoretical basis of comparison, and this is exactly what is given by (2.43).

For simplicity, consider an adiabatic machine with no change in kinetic energy between the inlet and outlet flange; then, from (2.41), in a process from state 1 to 2, the shaft work is the sum of the aerodynamic work and the dissipation, giving $w_{s12} = y_{12} + j_{12}$. Further, from (2.43), the minimum work to achieve this duty from a compressor with no dissipation in a reversible process is the aerodynamic work for the actual process, y_{12}. From this, a concise expression for the compressor efficiency can be derived from the first and second laws, from the inlet flange, state 1, to the outlet flange, state 2, as follows:

$$\eta_c = \frac{y_{12}}{w_{s,12}} = \frac{w_{s,12} - j_{12}}{w_{s,12}} = 1 - \frac{j_{12}}{w_{s,12}}. \tag{2.44}$$

An ideal machine with no dissipation (i.e., reversible with $j_{12} = 0$) would have an efficiency of unity. The common engineering use of the value 1-η as a convenient statement of the losses is then more or less self-evident, in that from (2.44) a compressor with an efficiency of 80% has 20% dissipation or lost work.

The concepts of dissipation and aerodynamic work are fundamentally related to the efficiency of a compressor. Chapter 4 considers many further issues related to different forms of efficiency definition, which take into account any changes in the kinetic energy and heat transfer. In this chapter, further discussion is provided of other thermodynamic concepts.

2.6.4 Rothalpy

Combining the steady flow energy in the form of (2.27) with the Euler turbine equation in the form of (2.3) gives

$$h_{t2} - h_{t1} = u_2 c_{u2} - u_1 c_{u1}. \tag{2.45}$$

This can be rearranged to identify that a particular function, known as the rothalpy, I, which was first used by Wu (1952), has a value that remains constant through a rotating blade row:

$$h_{t2} - u_2 c_{u2} = h_{t1} - u_1 c_{u1} = I \quad \text{or} \quad I_2 = I_1. \tag{2.46}$$

On this basis, rothalpy appears to be conserved through a rotor. In terms of a stator, this equation reverts to the conservation of total enthalpy.

The precise conditions under which this equation is valid are discussed in detail by Lyman (1993), who shows that the equation does not apply exactly in rotors where the viscous fluid on the walls of a stationary casing extract shear work from the impeller control volume. This additional shear work is not included in the derivation of the Euler turbine equation. The shaft has to supply this extra work, which causes a small increase in the rothalpy. In fact, taking into account the shear work term (2.45) should be modified to

$$h_{t2} - h_{t1} = u_2 c_{u2} - u_1 c_{u1} + w_v, \tag{2.47}$$

where w_v is the work out of the control volume to the stationary components due to the viscous work term. This term is zero for the case with a shroud attached to the impeller and moving with it, and rothalpy is only conserved through a rotor if it is shrouded. In the calculations of Moore et al. (1984) of an open impeller, this additional viscous work term amounts to an additional 1% of the shaft work; the rothalpy also increases by this amount.

A rearrangement of (2.46) using the expressions for velocities given in (2.12) shows that

$$I = h + \tfrac{1}{2}w^2 - \tfrac{1}{2}u^2 = h_{t,rel} - \tfrac{1}{2}u^2, \tag{2.48}$$

where $h_{t,rel}$ is the stagnation enthalpy in the relative frame of reference. Note that the temperature and a pressure associated with the rothalpy are known as rotary stagnation states.

Following Cumpsty (2004), the changes in enthalpy across the stage are shown schematically in Figure 2.13. The flow at inlet enters the compressor from the ambient environment with a total enthalpy of h_{t1}. The flow accelerates in the intake to the static conditions at the impeller inlet, h_1. Here the absolute velocity is low but the relative velocity is much higher, so the relative total enthalpy is high. Subtracting the kinetic energy associated with the blade speed from the relative total enthalpy results in the rothalpy at inlet, I_1. This remains constant across the rotor, so $I_2 = I_1$. Then adding the term representing the centrifugal effect at impeller outlet, which with a typical radius ratio between inlet and outlet of two is typically four times that at inlet, leads to the relative total enthalpy at impeller outlet. Finally subtracting the relative kinetic energy at outlet, which is less than half that at inlet, due to the typical deceleration of the relative velocity in the impeller $w_2/w_1 = 0.6$, the static conditions at impeller outlet are obtained. The absolute flow has accelerated across the impeller, with the result that

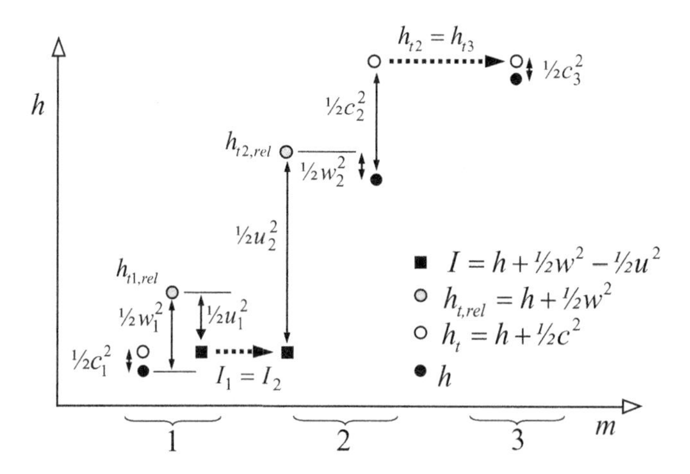

Figure 2.13 Variation of enthalpy, h, total enthalpy, h_t, relative total enthalpy, $h_{t,rel}$, and rothalpy, I, in a typical radial compressor stage.

adding the outlet kinetic energy results in the total conditions at the impeller outlet. These remain constant across the diffuser, and the diffuser decelerates the absolute flow to the static conditions at the stage outlet.

For incompressible flow, the analogous quantity to the rothalpy is the rotary stagnation pressure, defined as

$$p^* = p + \tfrac{1}{2}\rho w^2 - \tfrac{1}{2}\rho u^2. \tag{2.49}$$

This is conserved along streamlines in an inviscid flow. Section 5.6 describes the use of this parameter by Johnson and Moore (1983) to show the development of secondary flows in radial compressors.

The process between the impeller inlet and impeller discharge according to (2.12) implies a rise in static enthalpy given by

$$h_2 = h_1 + \tfrac{1}{2}\left(w_1^2 - w_2^2\right) + \tfrac{1}{2}\left(u_2^2 - u_1^2\right). \tag{2.50}$$

Figure 2.13 shows that most of the static enthalpy rise across an impeller results from the large radius change in the impeller, which in turn produces the centrifugal effect, described in Section 2.2.5. It is often stated that the centrifugal effect is loss free as it causes no direct increase in entropy and the centrifugal pressure rise does not increase the tendency for the boundary layers to separate. Nevertheless, it seems unreasonable to say that this is really loss free as the compression process producing the change in radius includes frictional losses, tip clearance losses and secondary flow. The deceleration in the diffuser and in the relative flow in the impeller are a smaller proportion of the enthalpy rise, but these may still have additional losses associated with them due to flow separation if recommended limits for the deceleration are exceeded; see Section 7.5.

Vavra (1970a) produced an interesting sketch of the property changes in a *T*-*s* diagram across a radial compressor stage with no backsweep, giving another useful insight into the centrifugal effect in the impeller. This is brought up to date in the form of an *h*-*s* diagram for a backswept impeller in Figure 2.14. In this diagram, the virtual

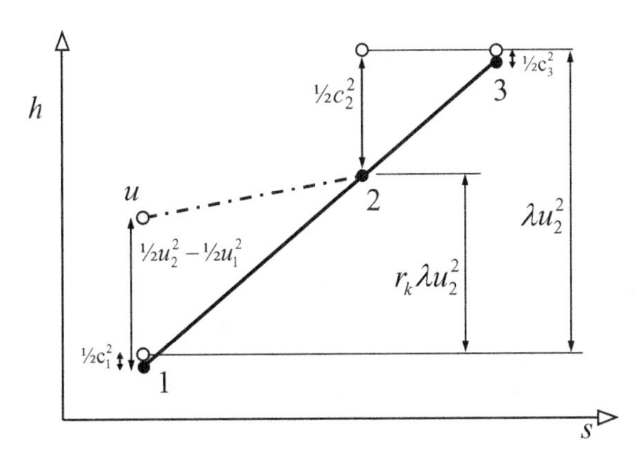

Figure 2.14 Variation of enthalpy and entropy in a compressor stage with virtual state u.

state u is introduced. This has the same entropy as at state 1, but has a static enthalpy given by

$$h_u = h_1 + ½(u_2^2 - u_1^2). \tag{2.51}$$

This is the state that would be achieved if the centrifugal effect were the only cause of the static enthalpy change and if this were considered to be free of losses so it has the same entropy as the inlet condition. The temperatures and pressures associated with this intermediate state are given by

$$T_u = T_1 + (h_u - h_1)/c_p = T_1 + ½(u_2^2 - u_1^2)/(2c_p), \quad p_u/p_1 = (T_u/T_1)^{\gamma/(\gamma-1)}. \tag{2.52}$$

This condition is a state that represents the enthalpy rise that is produced in the impeller purely by the change in blade speed between the impeller inlet and outlet. It is used to define the impeller wheel efficiency in Section 4.5. The remaining effects on the enthalpy rise in the impeller are due to the change in the relative kinetic energy in the impeller, as given by (2.50). As can be seen in Figure 2.14, the centrifugal effect accounts for about 75% of the static enthalpy rise in the impeller; see Section 2.2.5. The decreased slope of the line from state u to state 2, compared with that from state 1 to state 2, indicates that this process has a lower efficiency. The similar slopes of the lines from the impeller inlet to static conditions at impeller outlet and from there to the stage outlet indicate that these processes have similar efficiencies.

2.7 Different Ideal Compression Processes

2.7.1 The Isentropic Process

For an isentropic process in an ideal gas with constant specific heats, the Gibbs equations give

$$\begin{aligned} dh - vdp &= 0, \quad \Rightarrow c_p dT = vdp \\ du + pdv &= 0, \quad \Rightarrow c_v dT = -pdv \end{aligned} \tag{2.53}$$

and as a result, the isentropic process of an ideal gas can be defined as

$$\begin{aligned} &\gamma = c_p/c_v = -vdp/pdv, \quad \gamma pdv + vdp = 0, \quad p\gamma v^{\gamma-1}dv + v^{\gamma}dp = 0 \\ &pv^{\gamma} = constant. \end{aligned} \tag{2.54}$$

Therefore, for an ideal gas the exponent in an isentropic process is the same as γ, the ratio of the specific heats. Considering an ideal gas, with $pv = RT$, we obtain the relationship between the temperature and pressure in an isentropic process from state 1 to state 2 as

$$T_2/T_1 = (p_2/p_1)^{\frac{\gamma-1}{\gamma}}. \tag{2.55}$$

The isentropic process for a real gas is considered in Section 3.5.1, where the isentropic exponent is given the symbol, κ, to distinguish it from the ratio of the specific heats, γ, as these differ for a real gas.

2.7.2 Diagrams Describing Changes of State in Radial Compressors

In this section, several important features of the change of state in a turbocompressor are described by areas in an appropriate diagram. The two most common forms are the diagram showing the displacement work, which is the area under the pressure-volume curve in a p-v diagram, and the aerodynamic work, which can be identified as the area to the left of the pressure-volume curve in a p-v diagram, shown in Figure 2.7. The diagrams given in Figure 2.7 can be combined into a single diagram, Figure 2.15, to show the nature of the aerodynamic work, y_{12}, based on the following equation:

$$y_{12} = w_{v,in} + w_{v,12} + w_{v,out} = -p_1 v_1 - \int_1^2 p dv + p_2 v_2 = \int_1^2 v dp. \tag{2.56}$$

Other useful diagrams are the heat transfer in a reversible process, which is the area under the temperature-entropy curve in a T-s diagram. For heat addition, the area is positive, and for heat removal with entropy reduction the area is negative (as $ds < 0$), as shown in the top of Figure 2.16 This consideration of the area under the temperature-entropy curve in a T-s diagram can be extended to the case of an irreversible flow. In an adiabatic flow, the area under the temperature-entropy curve represents the dissipation work, and in a nonadiabatic flow it represents the sum of the heat addition and the frictional losses.

Interestingly, other process transitions can be represented as areas under the T-s curve in a T-s diagram. The Gibbs equation, as given in (2.37), suggests that for an isobar ($dp = 0$), the area under a T-s diagram joining two points at a constant pressure represents the enthalpy change related to the process, as shown in the middle of

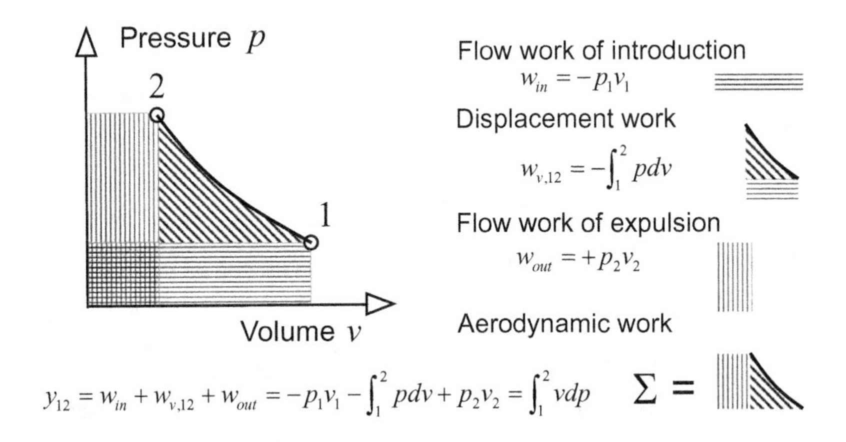

Figure 2.15 Flow work, displacement work and aerodynamic work.

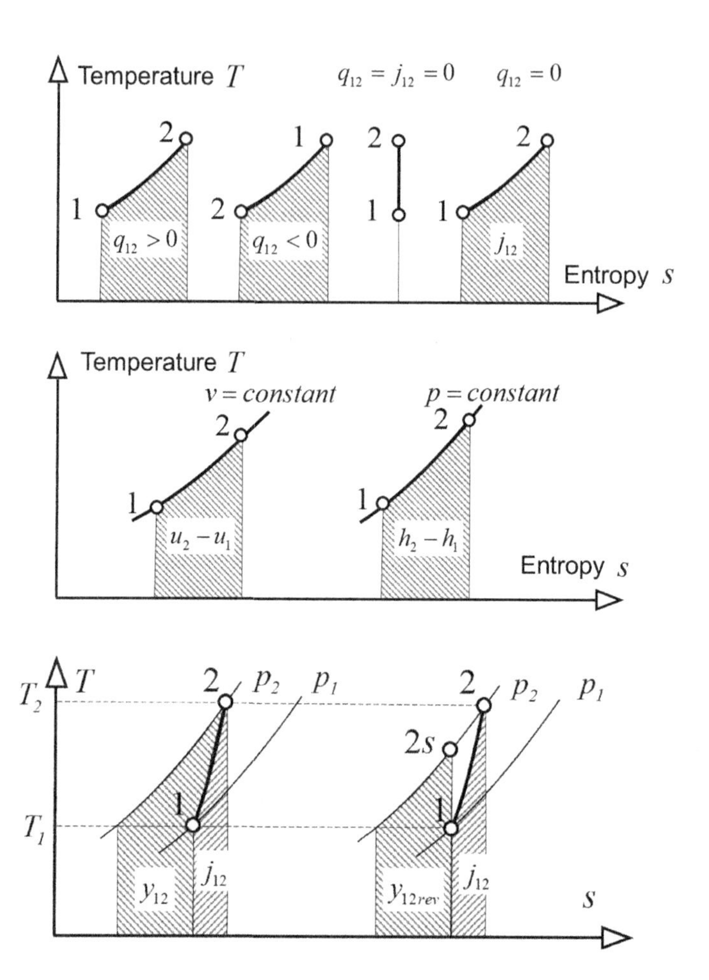

Figure 2.16 Diabatic and adiabatic processes shown as areas in a temperature entropy diagram.

Figure 2.16. Similarly, the other form of the Gibbs equation identifies that the internal energy along a constant volume process from state 1 to 2 is also an area under a constant volume line.

Considering real compression processes, the integrated form of the Gibbs equation, (2.40), for an adiabatic process identifies that

$$h_2 - h_1 = j_{12} + y_{12}. \tag{2.57}$$

As explained previously, the area under the constant pressure line joining two points represents the actual enthalpy change of the process. The dissipation in an adiabatic process is the area under the actual process path in the T-s diagram, so that the difference between these two areas clearly represents the aerodynamic work. In this way, the actual aerodynamic work can be represented for a compression process, as shown on the bottom left of Figure 2.16.

Considering an isentropic process, as added in the bottom-right part of this figure, then the pressure change work for the isentropic process is the same as the enthalpy change, given that $h_{2s}–h_1 = y_{12rev}$. Thus, the area denoted as y_{12rev} is the isentropic aerodynamic work that would be needed in an isentropic process with the same pressure rise. However, the increase in work needed for the real irreversible process is larger than the additional dissipation, j_{12}, and an additional contribution to the aerodynamic work is required. This can be represented as the white triangular area 1,2s,2,1 in Figure 2.16. The additional work needed in a real compression process is known as the 'preheat' effect to express the fact that the dissipation losses apparently cause the gas to be preheated as it is compressed. The later parts of the compression process then take part in a region where the temperature is higher and more work is needed for the same pressure rise, as shown in (2.80). The preheat effect causes the isentropic efficiency of a multistage compressor, to be less than that of its stages, as explained in Section 4.3.6.

2.7.3 The Polytropic Process

To simplify calculations, it is necessary to distinguish between the real process in the compressor and its approximate mathematical representation by a polytrope used for calculations as shown in Figure 2.17. The real process depends on the local acceleration or deceleration of the flow and on the rate of work input, dissipation or heat transfer on its way through the machine. The precise path of the flow process is much more complex than an assumption of a polytropic process with $pv^n = constant$. The real path is difficult to capture in experiments as usually the locations for measurement planes are limited, due both to technical difficulty and to the high cost of suitable instrumentation. Modern computational methods are able to provide more insight into the real compression path when analysed appropriately. The first publication to do this with CFD simulations of a radial stage was Moore et al. (1984).

Despite the simplification, the polytropic process is often used as a representation of a real compression process. The equation to describe the change between state 1 and state 2 can be written as

$$p_1 v_1^n = p_2 v_2^n, \quad v_1/v_2 = (p_2/p_1)^{1/n}. \tag{2.58}$$

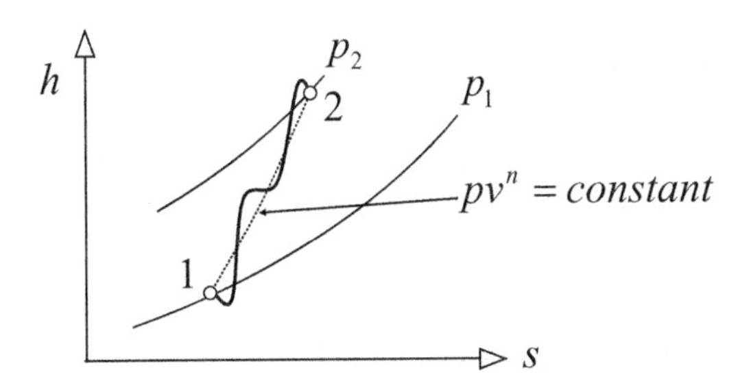

Figure 2.17 The real compression path and its approximation as a polytropic process.

Along the compression path, the polytropic exponent n, which expresses the compressible nature of the process, remains constant. For a value of $n = \infty$, the specific volume does not change with pressure and, with successively lower values of the exponent, a given pressure rise causes a larger change in specific volume showing that the gas volume can be compressed more easily.

For an infinitesimal change along the polytropic path (the dotted line in Figure 2.17)

$$pv^n = constant, \quad \Rightarrow pnv^{n-1}dv + v^n dp = 0, \quad npdv + vdp = 0. \tag{2.59}$$

For an ideal gas

$$pv = RT, \quad \Rightarrow pdv + vdp = RdT = (R/c_p)dh. \tag{2.60}$$

These equations lead to another relationship introducing the polytropic ratio

$$\upsilon = \frac{dh}{vdp} = \frac{(n-1)/n}{R/c_p} = \frac{(n-1)/n}{(\gamma-1)/\gamma}. \tag{2.61}$$

Equation (2.61) shows that a process with a constant polytropic exponent also has a constant polytropic ratio, $\upsilon = dh/vdp$. For an ideal gas, the isentropic exponent γ in these equations is the specific heat ratio.

Next consider an infinitesimal compression process from p to $p + dp$ with both dissipation and heat transfer. If this process were reversible and adiabatic, with $ds = dq = dj = 0$ the Gibbs equations show that it would require a work input of

$$dh = vdp. \tag{2.62}$$

For an irreversible adiabatic process in a compressor, more work is required, and this can be characterised by the small-scale polytropic efficiency, η_p, so that the required shaft work becomes

$$dw_s = \frac{1}{\eta_p} vdp. \tag{2.63}$$

The dissipation is the difference between these equations

$$dj = (Tds)_{irrev} = \left(\frac{1}{\eta_p} - 1\right) vdp = \left(\frac{1}{\eta_p} - 1\right) dy \tag{2.64}$$

so that in an adiabatic flow, the dissipation can be defined as a fraction of the aerodynamic work in terms of the polytropic efficiency.

Considering a reversible process with heat transfer alone, the heat transfer can also be defined as a fraction, ζ_q, of the aerodynamic work:

$$dq = (Tds)_{rev} = \zeta_q vdp = \zeta_q dy. \tag{2.65}$$

For the situation with combined heat transfer and work input, these equations can be substituted into (2.61). This leads to the following equation for the polytropic ratio:

$$v = \frac{dh}{vdp} = 1 + \frac{Tds}{vdp} = 1 + \frac{dq + dj}{dy_{12}} = \zeta_q + \frac{1}{\eta_p}. \tag{2.66}$$

This shows that the polytropic process is of particular value in the analysis of turbomachines in that both the heat transfer and the dissipation processes may be considered together through the fact that both can cause a change in entropy (Casey and Fesich, 2010). This is not possible in the isentropic analysis because the entropy is taken as constant in this ideal comparison process.

2.7.4 Special Features of Polytropic Processes

Some special features of these equations are worth discussing. First, consider an adiabatic process in which the heat transfer is zero. In this case, the preceding equations reduce to the standard well-known expressions for the polytropic efficiency of an adiabatic compressor:

$$\zeta_q = 0, \quad \eta_p = \frac{1}{v} = \frac{\gamma - 1}{\gamma}\frac{n}{n-1}, \qquad \frac{n-1}{n} = \frac{(\gamma - 1)}{\gamma\eta_p}. \tag{2.67}$$

This identifies that for all real adiabatic compression processes with $\eta_p < 1$, the polytropic exponent is larger than the isentropic exponent and the polytropic ratio is $v > 1$. For example, in air with $\gamma = 1.4$, taking for convenience a typical peak polytropic efficiency of $\eta_p = 6/7 = 0.857$, the polytropic exponent is $n = 1.5$.

Furthermore, for this adiabatic process and taking the polytropic efficiency to be unity, the polytropic equations describe a reversible isentropic adiabatic process. In the limit of no losses in an adiabatic process, the isentropic exponent is the same as the polytropic exponent:

$$\zeta_q = 0, \quad \eta_p = v = 1, \quad n = \gamma. \tag{2.68}$$

It is this ideal, reversible adiabatic polytropic process which is used to define the aerodynamic work when considering the isentropic efficiency in Section 4.3. As a result, in an adiabatic process with no losses, the polytropic and isentropic processes are the same.

Another issue of interest is the use of these equations to model extreme off-design operating points which have zero efficiency and a pressure ratio of unity. These conditions may be reached during the operation of a turbocharger with an engine: it is quite possible for the engine to force the compressor to operate at low pressure ratios, as explained in Section 18.5. In mismatched multistage compressors at a low pressure ratio, one of the stages may also be forced to operate at very low pressure ratios. In these cases, the impeller work input is dissipated or appears as kinetic energy, such that no pressure rise occurs. This would be a process at constant pressure and, as expected, for this process the polytropic exponent is zero:

$$\zeta_q = 0, \quad \eta_p = 0, \quad v = \pm\infty, \quad n = 0. \tag{2.69}$$

This corresponds to the point on the compressor stage characteristics with a pressure ratio of unity.

For diabatic processes the heat transfer coefficient is not zero and the effective polytropic exponent then depends on both heat transfer and energy dissipation, (2.66). Consider first an ideal process in which the polytropic efficiency is unity, but which experiences heat addition, that is, a reversible diabatic process. The heat addition leads to an increase in entropy. If the amount of heat addition is exactly the same as the actual dissipative entropy increase in a real process, the diabatic process with no losses has the same polytropic ratio and polytropic exponent as the adiabatic process with losses, and in terms of their end states they are indistinguishable. Thermodynamically we can visualise an irreversible adiabatic polytropic process as a reversible polytropic process with heat transfer. A reversible diabatic process can thus have the same polytropic exponent as an irreversible adiabatic process if the heat added exactly matches the heat produced by the dissipation in the real process. The special feature of the polytropic analysis which makes it so useful is that heat transfer and energy dissipation are interchangeable in this way.

Another interesting special case in terms of diabatic processes is one in which the heat removed exactly cancels the entropy increase produced by dissipation such that there is no change in specific entropy. The polytropic exponent is then the same as the isentropic exponent, with

$$\zeta_q = 1 - \frac{1}{\eta_p}, \quad \upsilon = 1, \quad n = \gamma. \tag{2.70}$$

This process would be isentropic, but it is neither adiabatic nor reversible.

An isothermal compression process can also be modelled with these equations. In this case, all the energy from the work input and the frictional dissipation is removed by heat transfer such that the temperature remains constant and

$$\zeta_q = -\frac{1}{\eta_p}, \quad \upsilon = 0, \quad n = 1. \tag{2.71}$$

A summary of these different cases is given in Table 2.1, and these are shown graphically in Figure 2.18. This figure provides an extension to diabatic flows as illustrated in a similar figure given by Cordes (1963) for adiabatic flows. The cases

Table 2.1 The polytropic process as a model for other processes.

Case	Exponent	Process	Equation	Polytropic ratio
a	$n = 1$	Isotherm	$pv =$ constant	$\upsilon = 0$
b	$1 < n < \gamma$			$0 < \upsilon < 1$
c	$n = \gamma$	Isentrope	$pv^\gamma =$ constant	$\upsilon = 1$
d	$\gamma < n < \infty$			$1 < \upsilon < \gamma/(\gamma - 1)$
e	$n = \pm\infty$	Isochor	$v =$ constant	$\upsilon = \gamma/(\gamma - 1)$
f	$-\infty < n < 0$			$\gamma/(\gamma - 1) < \upsilon < \infty$
g	$n = 0$	Isobar	$p =$ constant	$\upsilon = \infty$

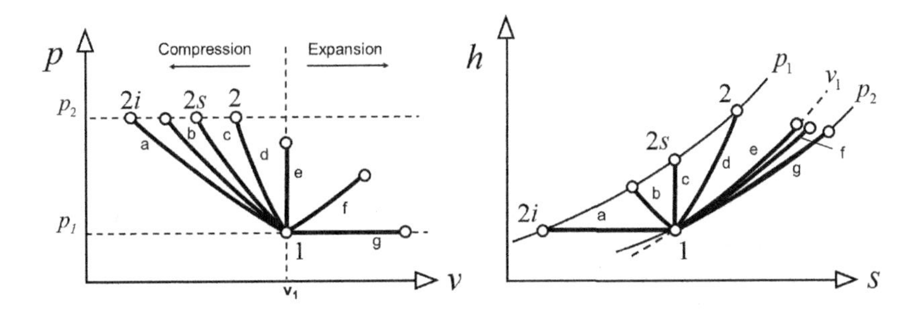

Figure 2.18 Polytropic processes with different exponents as described in Table 2.1.

denoted by (a)–(d) involve compression with a density increase, (e) has constant density and (f) and (g) are expansion processes with a density decrease. These latter cases are, of course, seldom relevant in compressors and occur only at extreme operating points near the low-pressure part of the characteristics. For completeness, it should be noted that the polytropic processes sketched in the *p-v* diagram of Figure 2.18 are described by (2.58), and the polytropic process as shown in the *h-s* diagram is described by

$$s_2 - s_1 = R\left(\frac{\gamma}{\gamma - 1} - \frac{n}{n-1}\right) \cdot \ln\left(\frac{h_2}{h_1}\right), \tag{2.72}$$

which is derived from (2.58) with the Gibbs equation.

2.8 The Aerodynamic Work

2.8.1 Flow Work, Aerodynamic Work and Displacement Work

Because of the potential for confusion between the flow work, $d(pv)$, the aerodynamic work, vdp, and the displacement work, pdv, these are discussed here in more detail with some interesting conclusions. As explained earlier, the $d(pv)$ terms in the steady flow energy equation are related to the difference between the flow work done by the surroundings on the open system to bring a small element of mass into the control volume at the inlet pressure p_1 and the work of the system on the surroundings to expel a similar mass from the control volume at outlet pressure p_2. In fact, these terms are sometimes known separately as the work of introduction and the work of expulsion; see Figure 2.15. Because they are always present during an energy transfer process in an open system, they can be combined with the change in internal energy to give a useful property, enthalpy, as defined in (2.23).

Furthermore, (2.43) and (2.44) show that the useful work in raising the pressure of the gas is given by the aerodynamic, or vdp, work, so some clarification is needed. It is useful to note that, by the chain rule, $\Delta(pv) = p\Delta v + v\Delta p$. The flow work terms, $d(pv)$, can be seen to be made up of the sum of the aerodynamic work, vdp, and the displacement work, pdv. Thus, the flow work can be shown to be

$$p_2v_2 - p_1v_1 = \int_1^2 d(pv) = \int_1^2 pdv + \int_1^2 vdp. \tag{2.73}$$

This rearrangement demonstrates that the flow work is the sum of two parts. The first part is known as the displacement work, needed to compress the volume of a small mass of fluid at pressure p, (2.18). The displacement work can be visualised as the area below the pressure volume curve in a p-v diagram, shown in Figure 2.7. The second part is the aerodynamic work, the pressure change work or the vdp work. The integral of vdp can be visualised as the area between the process compression curve and the vertical axis in a p-v diagram, shown in Figure 2.7. The aerodynamic work is sometimes known as the polytropic head because the integration needed to determine its value requires the specific volume to be defined in terms of the pressure, and this is often approximated as a polytropic process, $pv^n = constant$, with a polytropic exponent n. The aerodynamic work for an ideal liquid with constant density is $\Delta p/\rho$. In pumps, the aerodynamic work is often converted to an equivalent potential energy, using $\Delta p/\rho g$, and is denoted as the head rise in metres of water. It is from this analogy with the pumping of liquids that the aerodynamic work in compressors is sometimes also given the name *head*.

The first term on the right-hand side of (2.73) is negative in a compressor as the specific volume decreases from inlet to outlet, and the second is positive as the pressure increases. It is, therefore, useful to rearrange this equation to give positive values as follows:

$$\int_1^2 vdp = (p_2v_2 - p_1v_1) + \int_2^1 pdv. \tag{2.74}$$

It is necessary to know how the pressure changes with a change in volume to carry out the integrations in this equation. If the process between states 1 and 2 is taken as a polytropic process with $pv^n = constant$, then integration of the term under the integral of the displacement work term determines that

$$\int_2^1 pdv = \frac{1}{(n-1)} p_1v_1 \left[\left(\frac{p_2}{p_1} \right)^{\frac{n-1}{n}} - 1 \right]. \tag{2.75}$$

The flow work term for a polytropic process can be rearranged as follows:

$$p_2v_2 - p_1v_1 = p_1v_1 \left[\left(\frac{p_2}{p_1} \right)^{\frac{n-1}{n}} - 1 \right]. \tag{2.76}$$

The integration for the aerodynamic work term leads to

$$\int_1^2 vdp = \frac{n}{(n-1)} p_1v_1 \left[\left(\frac{p_2}{p_1} \right)^{\frac{n-1}{n}} - 1 \right]. \tag{2.77}$$

From the preceding equations, the fraction of the aerodynamic work which appears as displacement work and the fraction which appears as flow work is as follows:

$$\frac{\int_2^1 pdv}{\int_1^2 vdp} = \frac{1}{n}, \qquad \frac{p_2v_2 - p_1v_1}{\int_1^2 vdp} = \frac{n-1}{n}. \tag{2.78}$$

With an increase in the polytropic exponent, n, there is less change in volume for a given change in pressure, as shown in the left-hand part of Figure 2.18. At the limit, for an ideal incompressible fluid with constant density, the exponent n is effectively infinity and there is no change in the specific volume, as shown by line (e) in Figure 2.18. Similarly, the ratio of pressure change work to volume change work increases as the polytropic exponent increases. As a result, at the limit with an incompressible fluid, with $dv = 0$, all of the flow work appears as aerodynamic work, the vdp term. and the proportion of displacement work, the pdv term, is zero.

2.8.2 Effect of the Gas Properties on the Aerodynamic Work

For other processes, with different gases and polytropic exponents, different ratios can result. If we consider an ideal gas undergoing an isentropic compression process with pv^γ *constant*, we obtain

$$\frac{\int_2^1 pdv}{\int_1^2 vdp} = \frac{1}{\gamma} = \frac{c_v}{c_p}, \qquad \frac{p_2v_2 - p_1v_1}{\int_1^2 vdp} = \frac{\gamma-1}{\gamma} = \frac{c_p - c_v}{c_p} = \frac{R}{c_p}. \tag{2.79}$$

For an ideal gas with $\gamma = 1.4$, such as air, undergoing an isentropic pressure rise according to the isentropic process, $pv^\gamma = \text{constant}$, 71.4% of the aerodynamic work, or of the enthalpy change, appears as displacement work with an equivalent change in the internal energy of the gas, and 28.6% appears as flow work. Note that these ratios are not related to the pressure ratio, p_2/p_1, but only to the isentropic exponent and apply for all isentropic compression processes in an open system independent of the pressure ratio.

These ratios clearly vary for gases with different isentropic exponents. Gases with much lower ratios of the isentropic exponent, such as a compressible flow of steam which has a value of γ close to 1.111, the ratios would be 90% of the aerodynamic work appearing as displacement work and only 10% appearing as flow work. For high molecular weight refrigerant gases, the exponent is often close to 1.2. For machines operating with these gases, the effective density change is much higher than that with air at a similar pressure ratio. This becomes most obvious in the end stages of a steam turbine, where the flow volume expands far more than in a gas turbine operating with air at the same pressure ratio. As a consequence, a compressor designed for a given gas may require adaptation of the area of the flow channels when operating with a different gas.

The aerodynamic work for an ideal liquid is $\Delta p/\rho$. Given that the density of water is roughly a thousand times that of air, the specific work required to increase the pressure of air by a certain amount is roughly a thousand times that of water. It is for this reason that the steam turbine Rankine cycle was so much easier to implement than the gas turbine Joule or Brayton cycle in the early days of the development of fossil fuel power generation: the work absorbed in the compression process as a liquid in the Rankine cycle was so much smaller than that of the air in the Joule cycle that the losses in the compression process were relatively unimportant. New developments in gas turbines are utilising this by operating a closed Joule cycle on supercritical CO_2. This is a fluid state of carbon dioxide in which the gas is held above its critical pressure and temperature so that the density is similar to that of a liquid. This allows the aerodynamic work in the compressor to be reduced significantly.

Assuming that the gas being pressurised is an ideal gas with $pv = RT$, it follows from (2.77) that

$$y_{12} = \int_1^2 v dp = \frac{n}{(n-1)} RT_1 \left[\left(\frac{p_2}{p_1} \right)^{\frac{n-1}{n}} - 1 \right]. \tag{2.80}$$

The specific aerodynamic work for a compression process with a given pressure ratio is directly proportional to the inlet temperature of the gas. This underlines the importance of intercooling in multistage compressors, as the specific work required to generate the same pressure ratio is reduced if the inlet temperature of each stage is lower. The modification to this equation for real gases involves replacing the gas constant by the product of the gas constant and the real gas factor, as is discussed in Section 3.5.

The value of R in (2.80) is the specific gas constant for the gas being compressed, which is given by the universal gas constant divided by the molecular weight of the gas. From this, it is clear that it requires less work to achieve a given pressure ratio with a high molecular weight gas, such as a refrigerant, than with a low molecular weight gas, such as hydrogen or helium. With low molecular weight gases, a given amount of work produces a lower pressure ratio, so more energy and more compressor stages are needed for the compression.

In possible future gas transmission pipelines in a decarbonised energy system using hydrogen rather than methane, more work is required to pressurise the gas and more compressor stages are needed. In a recent analysis of the design of a new particle accelerator aiming at higher energy particle collisions to study possible unknown particle physics, the cooling requirements of the superconducting magnets near to absolute 0 K have been analysed (Podeur et al., 2019). To reduce the power requirements of the compression, one promising solution is to use a mixture of helium and neon, also called nelium, as the process gas. By adding neon, acting as a ballast gas, the use of a multistage turbocompressor becomes economically more viable, especially when looking at the required number of stages for a given pressure ratio.

2.8.3 The Effect of the Efficiency on the Aerodynamic Work

From (2.67), the polytropic exponent for an ideal gas varies with the efficiency and the isentropic exponent as follows:

$$n = \frac{\gamma \eta_p}{1 - \gamma(1 - \eta_p)}. \tag{2.81}$$

Using the example of a high-efficiency compression process in air, with $\gamma = 1.4$ (which is 7/5) and with a polytropic efficiency of $\eta_p = 6/7 = 0.857$, the polytropic exponent is $n = 3/2 = 1.5$, and one third of the aerodynamic work appears as flow work and two thirds as displacement work. As the efficiency drops, the value of the polytropic exponent increases so that, for example, with $\eta_p = 3/7 = 0.429$, then the polytropic exponent increases to $n = 3$ and the proportion of the aerodynamic work which appears as flow work rises to two thirds and the proportion as displacement work drops to one third. An efficiency of $\eta_p = 2/7 = 0.286$ leads to a polytropic exponent of infinity so that all of the aerodynamic work appears as flow work and none as displacement work. In fact, at this efficiency the density does not change with pressure, and the flow is effectively incompressible, as in line (e) in Figure 2.18. Below this efficiency, the exponent becomes negative, and so the pressure rise is accompanied by a volume expansion with an increase in specific volume. The important conclusion from this is that in multistage compressors operating at extreme off-design conditions, a very low efficiency at low pressure ratios can rapidly lead to a high volumetric mismatch between the stages.

2.9 The Compressor Stage as the Sum of Its Components

Considering that the typical purpose of a compressor stage is to raise the pressure of a fluid from inlet flange, 1, to outlet flange, 3, and assuming that the changes in kinetic energy and potential energy at the flanges are negligible, the shaft work of the machine can be written

$$w_{s12} = y_{13} + j_{13}. \tag{2.82}$$

The terms of this equation can now be considered as the sum of the terms for the individual components, the impeller from state 1 to state 2, and the diffuser from state 2 to state 3.

If the impeller is considered alone – from state 1 to 2 – then (2.41) applies, with an increase in velocity between stage inlet, state 1, and impeller outlet, state 2. This identifies that the shaft work is used both to overcome the dissipation in the impeller and to increase the pressure and kinetic energy across the impeller. If this equation is applied downstream of the impeller across the diffuser alone – from state 2 to state 3 – there is a decelerating flow and no shaft work input, and so it is the reduction of kinetic energy across the diffuser which overcomes the frictional dissipation and provides the aerodynamic work to increase the outlet pressure

$$½\left(c_2^2 - c_3^2\right) = j_{23} + y_{23}. \tag{2.83}$$

In this way, it can be seen that the first and second laws can be applied both to individual components and to the whole stage.

There are two additional points to consider given that the aerodynamic work and the dissipation are not state variables but depend on the path of integration. The first is that the polytropic paths from state 1 to 2 and from state 2 to 3 do not necessarily have the same polytropic exponent as the whole path from state 1 to 3. If the polytropic exponent from state 1 to state 3 is used to calculate the aerodynamic work, there is only an approximate equality between these terms.

$$\begin{aligned} y_{13} &\approx y_{12} + y_{23} \\ j_{13} &\approx j_{12} + j_{13}. \end{aligned} \tag{2.84}$$

If state 2 lies on the polytropic path between states 1 and 3, as is shown in Figure 2.14, there is no approximation in the polytropic exponents. If the integration from state 1 to 3 is made in two steps – from 1 to 2 and from 2 to 3 – then there is again no approximation, so additional information at state 2, the impeller outlet or diffuser inlet, is useful. In the discussion in Section 2.2.5, the degree of reaction is defined in terms of the enthalpy change from the inlet total condition to the impeller outlet static condition. One advantage of this procedure is that the static state at impeller outlet near to design often lies close to the polytropic path from inlet total to outlet total conditions and minimises the error in (2.84).

Huntington (1985) developed a more accurate numerical procedure for this integration by splitting the polytropic compression path between two pressure levels into a large number of subpaths with smaller steps in the pressure. For the small steps, he could use the definition of the polytropic efficiency to find the outlet conditions and calculate the polytropic head as the sum of each part; see Section 3.6.3. In test measurements of compressors, there is often only data at the inlet and outlet flanges. It is then assumed that there is a single value of the polytropic exponent from inlet to outlet and the performance of the components is indeterminate. In some radial compressor tests, a static pressure measurement near the impeller outlet allows more precision in the integration by including state 2 in the integration path from 1 to 3.

A second issue is that in order to be exact, the determination of the aerodynamic work requires an integration of *vdp* along the actual flow path as shown in Figure 2.12, and not simply an assumption of an equivalent polytropic process. This exact integration should become possible with data available from CFD simulations. In modern simulations with full data as to the conditions in the whole flow domain, it is possible to carry out the integration correctly. Issues to be addressed in such integrations are the mixing and increase in entropy introduced across a mixing plane at the impeller outlet and the availability of an appropriate algorithm for the determination of the local mean values. An algorithm for CFD calculations is set out in the Verein Deutscher Ingenieure (VDI) standard on this subject, VDI 4675 (VDI, 2012) based on an approach by Kreitmeier, but this has not been widely adopted. Approximations still arise, however, as in a compressor a form of averaging across the flow channel is required to determine the mean 1D control volume values, which always results in a loss of information. Based on one example taken from simulations by the authors, the difference in the aerodynamic work between a correct integration and the assumption of a polytropic path is less than 1%. As a result, the value from extra effort on CFD postprocessing may not be very high especially given that the determination of the appropriate mean values for the integration of *vdp* is never completely free of approximations for compressors.

3 Equations of State

3.1 Overview

3.1.1 Introduction

Compressors operate on gases to increase the gas pressure and density, whereas pumps increase the pressure of a liquid at constant density. To understand the compression process properly, an equation of state is required to determine the relationships between the intensive thermodynamic variables (p, T and v), typically to calculate the specific volume, enthalpy and entropy for given values of the pressure and temperature. Equations of state define the relations between these properties when in thermodynamic equilibrium. The simplest equations of state are those for a perfect gas and for an incompressible fluid. Centrifugal compressors are used with such a wide range of fluids that complex real gas models are often needed to characterise the behaviour of the fluid. In this chapter, an introduction is provided to the cubic equations of state with the Van der Waals equation and more details are provided of one specific cubic equation of state, the Aungier–Redlich–Kwong equation. The objective of the chapter is to introduce some of the issues related to equations of state for real gases so that compressor engineers can move on to the many other books and references with much more detail.

The real gas equations not only change the relationship between the intensive properties, but also require special care in determination of the integration of a polytropic process to determine the aerodynamic work. Different methods of carrying out this integration are discussed.

3.1.2 Learning Objectives

- Know the equations of state for perfect gases and for ideal incompressible liquids.
- Explain the difference between the thermal and calorific equations of state.
- Study the extension of the ideal equation of state to take into account the attractive force between molecules and the volume of the molecules, leading to the Van der Waals equation.
- Understand the concept of corresponding states for real gases.
- Know the most practical real gas equations, in the cubic form and in the virial form.
- Be aware of the need for corrections terms in the estimation of aerodynamic work with real gases.

3.2 Equations of State for Perfect Fluids

3.2.1 Terminology

The terminology related to the equations of state is, in increasing complexity, as follows:

(a) An *ideal gas* follows the ideal gas law, $pv = RT$, such that the real gas compressibility factor $Z = pv/RT = 1$.
(b) A *calorically perfect gas* also follows the ideal gas law, $pv = RT$, so is also an ideal gas with a real gas factor of $Z = 1$, but has constant specific heats, c_p and c_v.
(c) An *elementary model of a real gas* can be obtained with the ideal gas law, combined with a constant real gas factor, $pv = ZRT$ with $Z \neq 1$ and constant specific heats.
(d) A *thermally perfect gas* (also known as a semiperfect gas) also follows the ideal gas law, $pv = RT$, but has variable specific heats, c_p and c_v, that are functions of temperature alone $c_p = c_p(T)$ and $c_v = c_v(T)$.
(e) A *real gas* has neither a constant real gas factor nor specific heats that are constant or a function of only temperature and therefore requires more complex equations.
(f) An *incompressible fluid* is a fluid in which the volume of a given fluid mass is constant and the internal energy is solely a function of temperature.

For most of the simplified considerations examined in this book, it is assumed that the equation of state can be represented either by the thermal and caloric equations of state for a perfect gas or by those for an incompressible fluid. The extension to real gases is given as separate sections where relevant. Lou et al. (2014) point out that even in compressor applications with air as the medium it is sometimes important to make use of real gas equations.

It is important to remember that the equation of state defines the relations which exist between properties in a state of thermodynamic equilibrium. It should not be confused with a process equation, such as those for a polytropic or isentropic process, which define the compression path which the fluid takes between specific end states. Both the process equation and the equations of state hold at each point on this compression path.

3.2.2 Perfect Gas

The thermal equation of state of a gas defines a functional relationship among the state variables of pressure, p, specific volume, v, and temperature, T, in the state of equilibrium. In 1662, Robert Boyle (1627–1691) discovered that the volume of a simple gas held at constant temperature was inversely proportional to the pressure (Boyle's law). This law can also be derived from the kinetic theory of gases, following Jeans (1940) (James Jeans, 1877–1946). If a closed vessel containing a fixed number of molecules is reduced in volume, more molecules will strike a given area of the sides of the container per unit time, causing a greater pressure. In 1787, Jacques Charles (1746–1823) found that the volume of a simple gas at constant pressure was directly proportional to its absolute temperature (Charles's law). This effect is caused by the kinetic energy of the molecules in the gas, which increases with temperature, causing an increase in volume at constant pressure. The combination of Boyle's and Charles's laws gives the simplest and most well-known expression for the equation of state of an ideal gas:

$$pv = RT, \tag{3.1}$$

where R is a constant of proportionality, known as the specific gas constant. For a gas with a molecular weight of M_w, R is determined from the universal gas constant R_m as

$$R = R_m/M_w, \quad R_m = 8.31447\ (kJ/kmol K). \tag{3.2}$$

There are no gases which fully obey the ideal gas law, but it is a very good approximation to the real behaviour of gases at low pressure. Deviations from this law become significant at high pressures and low temperatures, especially when the gas is close to the critical point or the saturated vapour line. Gases with a low boiling point temperature (such as He, H_2, Ne and even N_2, O_2, Ar, CO and CH_4) have deviations of less than 5% up to 50 bar and 2% up to 10 bar.

The caloric equation of state for a perfect gas holds that

$$h = c_p T, \quad u = c_v T, \quad c_p - c_v = R, \quad \gamma = c_p/c_v = \text{constant}. \tag{3.3}$$

Note that γ is used for the ratio of specific heats to distinguish this from κ, which is used for the isentropic exponent. For an ideal gas, these parameters are the same, $\kappa = \gamma$, but for a real gas they differ. Broadly, this means that any thermodynamic equation in the text containing the symbol γ relates to a perfect gas, or one with constant real gas factor.

Chapter 2 identifies the importance of enthalpy as a means of keeping track of work input and entropy as a means of monitoring the losses in a turbomachine. Based on these parameters, expressions are needed which show the change in other fluid properties as a result of the changes in entropy and enthalpy. For the compressible gas flows that can be represented by the thermal and caloric equations of state for a perfect gas, integration of the Gibbs equation leads to the following equations:

$$\begin{aligned} s_2 - s_1 &= c_p \ln(T_2/T_1) - R\ln(p_2/p_1) \\ s_2 - s_1 &= c_v \ln(T_2/T_1) - R\ln(\rho_2/\rho_1) \end{aligned} \tag{3.4}$$

These can be expressed in such a way as to determine temperature, pressure and density from the enthalpy and the entropy, as follows:

$$\begin{aligned} T_2/T_1 &= (h_2/h_1) \\ p_2/p_1 &= (h_2/h_1)^{\gamma/(\gamma-1)}\ e^{-(s_2-s_1)/R} \\ \rho_2/\rho_1 &= (h_2/h_1)^{1/(\gamma-1)}\ e^{-(s_2-s_1)/R}. \end{aligned} \tag{3.5}$$

For a fixed enthalpy rise, the pressure rise falls with increasing losses. Note that with no losses ($s_2 = s_1$), these equations become the well-known isentropic relationships for a perfect gas:

$$\begin{aligned} p_2/p_1 &= (T_2/T_1)^{\gamma/(\gamma-1)} \\ \rho_2/\rho_1 &= (T_2/T_1)^{1/(\gamma-1)} \\ p_2/p_1 &= (\rho_2/\rho_1)^{\gamma}. \end{aligned} \tag{3.6}$$

The Gibbs equation indicates that for an ideal gas lines of constant pressure (with $dp = 0$) in an *h*-*s* diagram have a slope given by

$$\left.\frac{dh}{ds}\right|_p = T \tag{3.7}$$

so that the slope of an isobar in the *h*-*s* diagram increases with temperature. For an ideal gas, this slope increases with increasing enthalpy, as shown in Figure 3.1, leading to isobars which are curved lines. This figure gives the impression that the isobars diverge, but (3.4) indicates that for a given temperature two isobars of constant pressure in the *T*-*s* diagram are shifted horizontally relative to each other by $s_2 - s_1 = -R\ \ln(p_2/p_1)$, as shown in Figure 3.1. This horizontal shift with increasing pressure causes the vertical gap between two isobars to increase, and this shows that for a given pressure ratio at a higher temperature, more enthalpy change is required.

In a mixture of ideal gases, the momentum transfer between each molecular species and the container walls is not affected by the presence of other molecules. The pressure can then be determined by Dalton's law as the sum of all partial pressures. The perfect gas law can be applied to each component of a gas mixture as if it were alone.

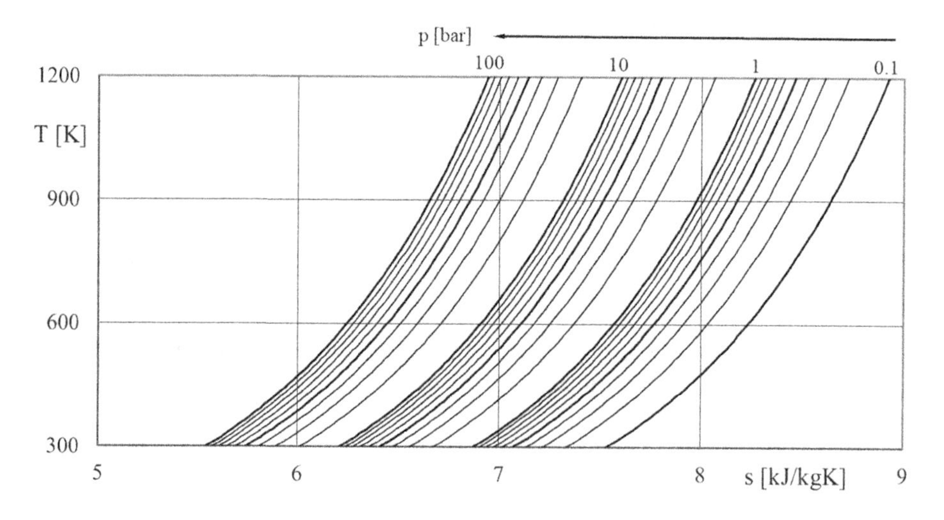

Figure 3.1 Temperature entropy diagram of a perfect gas with $\gamma = 1.4$.

3.2.3 Idealised Incompressible Fluid

For pumps and hydraulic turbines, and for ventilators operating at low speeds with little change in density, the equation of state for an incompressible fluid is needed. This is included here for comparison with that applied to compressors. For an idealised incompressible fluid, the volume of a given fluid mass (or its density) is constant, regardless of temperature or pressure, and the internal energy is solely a function of temperature and not of pressure. The second law of thermodynamics in the Gibbs form can be then written as $Tds = du$. For any substance

$$du = \left.\frac{\partial u}{\partial T}\right|_v dT + \left.\frac{\partial u}{\partial p}\right|_T dp = c_v dT + \left.\frac{\partial u}{\partial p}\right|_T dv \tag{3.8}$$

and for an incompressible fluid where the internal energy is not a function of pressure, $du = c_v dT$, and by combining these equations

$$ds = c_v(dT/T). \tag{3.9}$$

Together with the Gibbs equation ($dh = Tds + vdp$), this also yields

$$dh = c_v dT + vdp \tag{3.10}$$

and on integration we obtain

$$\begin{aligned} s_2 - s_1 &= c_v \ln (T_2/T_1) \\ h_2 - h_1 &= c_v(T_2 - T_1) + v(p_2 - p_1). \end{aligned} \tag{3.11}$$

In general, the temperature rise in a hydraulic machine is very small, so that the entropy change equation can also be approximated by

$$s_2 - s_1 = c_v \ln [1 + (T_2 - T_1)/T_1] \approx c_v(T_2 - T_1)/T_1 \tag{3.12}$$

and the equations for the other fluid properties in terms of entropy and enthalpy become

$$\begin{aligned} \rho_2 &= \rho_1 = \rho \\ T_2 - T_1 &= T_1(s_2 - s_1)/c_v \\ p_2 - p_1 &= \rho(h_2 - h_1) - \rho T_1(s_2 - s_1). \end{aligned} \tag{3.13}$$

These equations highlight several interesting and subtle differences between the thermodynamics of incompressible turbomachines and those of compressible thermal turbomachines operating on ideal gases. First, in hydraulic machines the losses causing an increase in entropy lead to an increase in the temperature across the machine, as shown in (3.13). In thermal turbomachines, it is the change in enthalpy (from the work done on the fluid without external heat transfer) that is related to the total temperature change (from the caloric equation of state, $h = c_p T$). In hydraulic machines with no losses, there would be no temperature increase. Typically, the temperature rise in a hydraulic machine is a fraction of a degree, and the temperature not only increases across a pump but also across a water turbine. This fact is not immediately obvious to engineers who deal only with gas or steam turbines where the

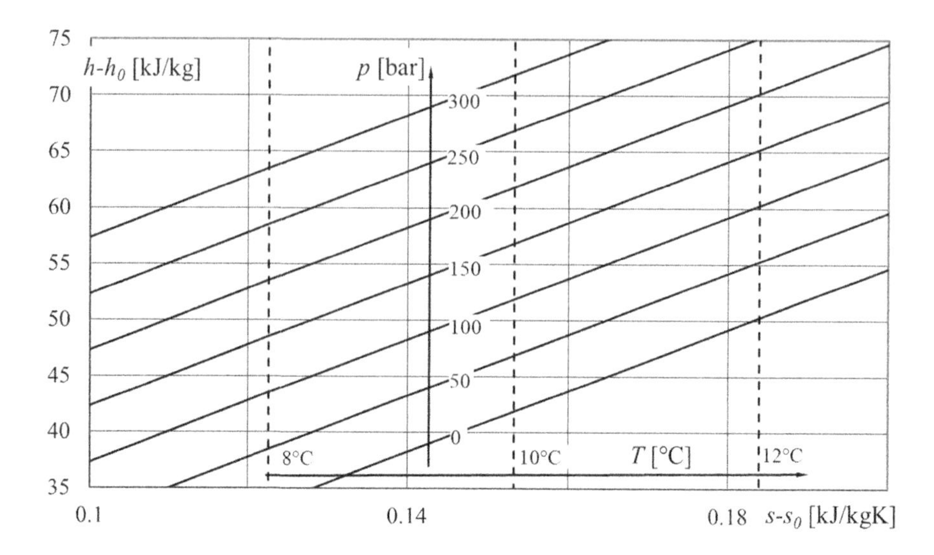

Figure 3.2 *h-s* diagram for an ideal liquid.

gas always cools on expansion through the turbine. This small temperature change in hydraulic machines represents the losses and is actually used as a method of efficiency measurement in large-scale pumps and water turbines; see Gülich (2008).

A second difference is that for a truly incompressible fluid, the entropy is a function of the temperature only, so that lines of constant temperature are vertical lines in the *h-s* diagram, and not horizontal lines as with a perfect gas. The stagnation and static conditions, which lie on an isentropic line, are then at the same temperature for a liquid. Again, this is not immediately obvious to engineers concerned with compressible flows, where the stagnation temperature is always higher than the static temperature.

The third difference is that the preceding equations show that in liquids, the isobars have the equation $h_2 - h_1 = T_1(s_1 - s_2)$, so for the small temperature changes that occur, the constant pressure lines in the *h-s* diagram are straight lines with a slope given by the temperature, and not curved lines as with an ideal gas. In a liquid, the slope of the constant pressure lines in an *h-s* diagram are also proportional to the local temperature, but the temperature increases with entropy and not with the enthalpy, so that the slope at a certain entropy is constant for all pressure levels. The lines of constant pressure thus have a similar form but are displaced vertically in the *h-s* diagram, as shown in Figure 3.2, and not displaced horizontally as with an ideal gas.

3.3 Equations of State for Real Gases

3.3.1 Real Gases with a Constant Compressibility Factor *Z*

The simplest way to include some features of a real gas while retaining the form of the perfect gas equations is to make use of a constant compressibility factor, *Z*, such that,

$$pv = ZRT, \quad h = c_p T. \tag{3.14}$$

A value of the real gas factor which is less than unity, $Z < 1$, indicates that for a given pressure and temperature the volume of the gas is reduced compared to that of a corresponding ideal gas. The thermally perfect gas is a special case with $Z = 1$. The form of the ideal gas equations required for real gases with a constant compressibility factor can be derived by integration of the Gibbs equation ($Tds = dh - vdp$) for a general gas and leads to the following equations:

$$\begin{aligned} s_2 - s_1 &= c_p \ln(T_2/T_1) - ZR \ln(p_2/p_1) \\ s_2 - s_1 &= c_v \ln(T_2/T_1) - ZR \ln(\rho_2/\rho_1). \end{aligned} \tag{3.15}$$

These equations can be expressed in the form to determine pressure and density from the enthalpy and the entropy, as follows:

$$\begin{aligned} p_2/p_1 &= (h_2/h_1)^{\gamma/(\gamma-1)} \; e^{-(s_2-s_1)/ZR} \\ \rho_2/\rho_1 &= (h_2/h_1)^{1/(\gamma-1)} \; e^{-(s_2-s_1)/ZR}. \end{aligned} \tag{3.16}$$

The relationship to the isentropic exponent becomes

$$\begin{aligned} c_p &= c_v + ZR \\ ZR/c_p &= (c_p - c_v)/c_p = 1 - 1/\gamma = (\gamma - 1)/\gamma \\ \gamma &= \frac{1}{1 - (ZR/c_p)}. \end{aligned} \tag{3.17}$$

Note that the speed of sound becomes

$$a = \sqrt{\gamma ZRT}. \tag{3.18}$$

In all of these equations, the gas constant in the ideal gas relationships is simply replaced by the product of the gas compressibility factor and the gas constant. However, with a real gas the compressibility factor may not be constant over the full range of the compression process. Aungier (2000) suggests that the best results with this model for a compression process from state 1 to 2 with variable compressibility factor are obtained with the following approximations:

$$\begin{aligned} &Z = \sqrt{Z_1 Z_{2s}}, \quad c_p = (h_{2s} - h_1)/(T_{2s} - T_1), \quad c_v = (u_{2s} - u_1)/(T_{2s} - T_1) \\ &\kappa = \ln(p_{2s}/p_1)/\ln(v_{2s}/v_1), \end{aligned} \tag{3.19}$$

where $2s$ is the state with the outlet pressure p_2 and inlet entropy s_1, and κ is the isentropic exponent and not the ratio of specific heats. The integration of vdp to calculate the aerodynamic work then leads to the following:

$$y_{12} = \int_1^2 vdp = \frac{n}{(n-1)} ZRT_1 \left[\left(\frac{p_2}{p_1}\right)^{\frac{n-1}{n}} - 1 \right], \tag{3.20}$$

so that the work required by the compression process is proportional to the real gas factor, Z.

3.3.2 Principle of Corresponding States

In Section 3.3.1, the treatment of a real gas with a nearly constant compressibility factor, Z, was considered. Generally, however, the compressibility factors for different gases are not constant, but the variations are found to be qualitatively very similar. When plotted in a certain way, different gases appear to deviate from ideal behaviour to the same extent. This is referred to as the law of corresponding states. For this generalised presentation, the compressibility factor Z is plotted against a dimensionless reduced pressure p_r for different values of the dimensionless reduced temperature T_r defined as

$$p_r = p/p_c, \quad T_r = T/T_c, \quad v_r = v/v_c, \tag{3.21}$$

where p_c, T_c and v_c are the pressure, temperature and volume at the critical point of the gas. The critical temperature is the temperature above which it is not possible to liquify a given gas, and the critical pressure is the minimum pressure required to liquify a given gas at its critical temperature. The liquid–vapour critical point in a pressure–temperature phase diagram is at the high-temperature extreme of the liquid–gas phase boundary. Above this point, there exists a state of matter that is continuously connected and distinct gas and liquid phases do not exist. At temperatures which are greater than the critical point, there is no change of phase as the gas is compressed from the gaseous state to the liquid state. This results in a generalised compressibility chart of the form $Z = f(p_r, T_r)$. Figure 3.3 shows experimental data for 10 different gases on a chart of this type adapted from Su (1946). The solid lines denoting reduced

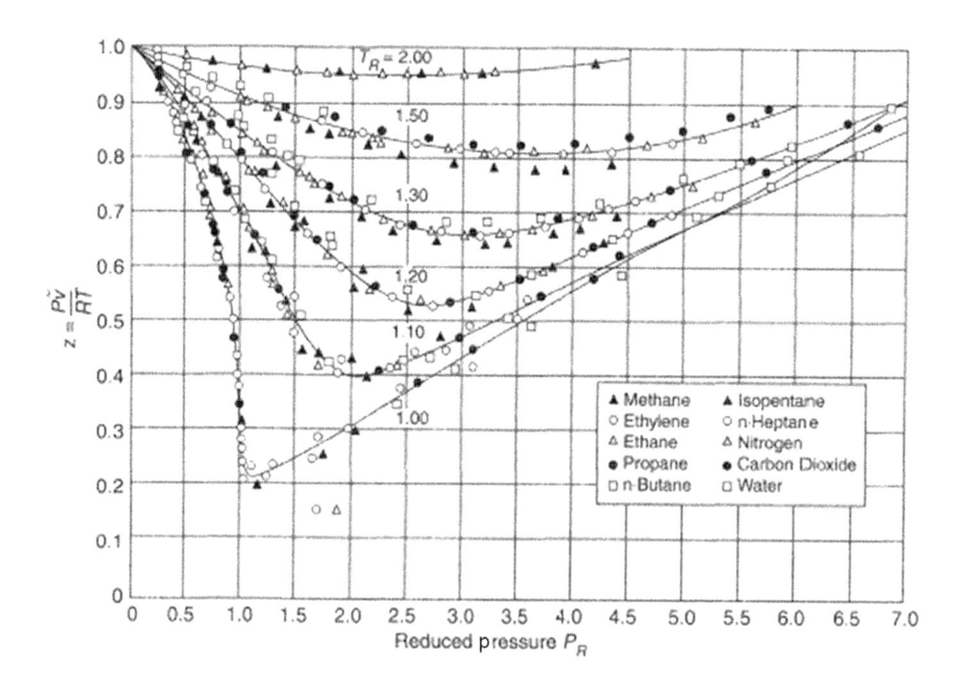

Figure 3.3 Generalised compressibility chart for various gases. (adapted from Su, 1946)

isotherms represent the best curves fitted to the data. The law of corresponding states indicates that many substances at equal reduced pressures and temperatures have very similar reduced volumes and deviate from ideal behaviour to the same amount. This relationship is approximately true for many substances, but becomes increasingly inaccurate for large values of the reduced pressure, p_r. Figure 3.3 identifies that at very low pressure ($p_r \ll 1$), the gases behave as an ideal gas regardless of temperature, at high temperatures ($T_r > 2$), gases behave almost as ideal gases and that the largest deviation from ideal gas behaviour is in the vicinity of the critical point.

3.3.3 Van der Waals Equation

While looking for a way to link the behaviour of liquids to that of gases, the Dutch physicist Johannes van der Waals (1837–1923) made several breakthroughs for which he was ultimately awarded the 1910 Nobel Prize for physics. He developed an explanation for some of the limitations of the ideal gas law and suggested an equation that was able to fit the behaviour of real gases over a much wider range of pressures. Van der Waals realised that two of the basic assumptions of the molecular theory of gases were questionable. Kinetic theory assumes that gas particles occupy a negligible fraction of the total volume of the gas and also that there is no force of attraction between gas molecules.

The first assumption is valid at pressures close to normal ambient pressures, but at much higher pressures this is no longer true. As a result, real gases are not quite as compressible at high pressures as an ideal gas, as the molecules require some space, known as the covolume. The volume of a real gas is therefore larger than predicted by the ideal gas equation, and so van der Waals proposed to subtract a term for the excluded volume of the molecules, b, in the ideal gas equation. He corrected for the second assumption of the kinetic theory by including a term, a/v^2, that accounts for the attraction of the molecules. The attraction between the molecules reduces the momentum transferred to the wall by the molecules, which in turn results in a lower pressure compared with an ideal gas:

$$p = \frac{RT}{(v-b)} - \frac{a}{v^2}. \tag{3.22}$$

Other cubic equations, discussed later, have a similar cubic form, but the interesting features of these can be usefully demonstrated with the van der Waals equation. Equation (3.22) can be written as

$$v^3 - \left(b + \frac{RT}{p}\right)v^2 + \left(\frac{a}{p}\right)v - \frac{ab}{p} = 0. \tag{3.23}$$

At the critical point, there is an inflection (Klein and Nellis, 2012), such that

$$\left(\frac{\partial p}{\partial V}\right)_T = \left(\frac{\partial^2 p}{\partial V^2}\right)_T = 0. \tag{3.24}$$

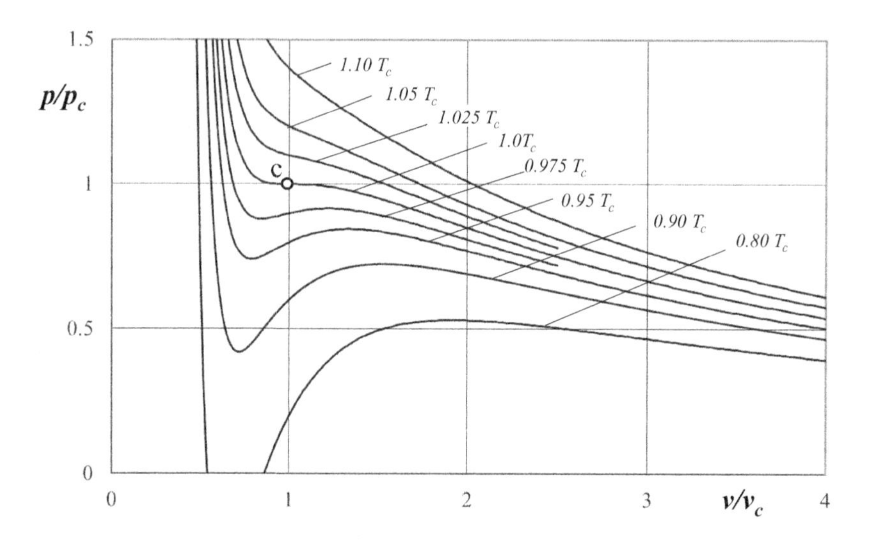

Figure 3.4 The p-v diagram of a Van der Waals gas.

Solving the preceding condition for the van der Waals equation, the critical point can be derived as

$$T_c = 8a/27b, \quad v_c = 3b, \quad p_c = a/27b^2. \tag{3.25}$$

In terms of the reduced state variables, the van der Waals equation becomes

$$p_r = \frac{8T_r}{(3v_r - 1)} - \frac{3}{v_r^2}, \tag{3.26}$$

where the reduced states are defined as $p_r = p/p_c$, $T_r = T/T_c$ and $v_r = v/v_c$. Equation (3.26) is plotted in Figure 3.4 showing the reduced pressure versus the reduced specific volume for different values of the reduced temperature. This relatively simple equation scales the shape of the isotherms of the gas around the properties at the critical point. It also predicts a sort of phase transition between liquid and vapour. The isotherms in the two-phase region are not realistic but can be delineated by assuming that the areas above and below the isotherms are equal. The van der Waals equation may not be the best match to any real gas, but it demonstrates that the cubic equation of state includes a critical point and that the coefficients can be expressed in terms of the critical properties.

3.3.4 Cubic Equations of State

The van der Waals gas is clearly an improvement on the ideal gas equation, but more recent developments show that this can be developed into a more general cubic equation with coefficients that can be adapted to further reduce the deviations from the ideal gas law. Several authors have published generalised forms of a cubic equation and have demonstrated that many different well-known cubic equations of state can be related to the generalised form. Many of these cubic equations of state have found widespread use in turbomachinery applications. An example of a modern generalised cubic equation is

$$p = \frac{RT}{v - (b - c)} - \frac{a(T)}{v^2 + \delta v + \varepsilon}. \tag{3.27}$$

This equation of state is simple to apply, involving few parameters, and does not require complex computational techniques.

In this equation, the term b–c is the covolume of the molecules, the second term accounts for the repulsive force of the molecules, whereby a is a function of the temperature, and the terms with the constants δ and ε represent the attraction between molecules. Note that at high specific volumes these equations gradually become the same as the ideal gas law. The van der Waals equation is obtained from the generalised form with $c = \delta = \varepsilon = 0$. Redlich and Kwong (1949) introduced a temperature dependency to the attraction parameter, a. and, in the general form, their equation is obtained with $c = \varepsilon = 0$ and $\delta = b$. Soave (1972) modified the Redlich–Kwong equation by introducing a term called the acentric factor, ω, to the attraction parameter, a, to quantify for the nonspherical shape of molecules. Peng and Robinson (1976) also proposed a cubic equation, which in the generalised form has coefficients with $c = 0$. $\delta = 2b$ and $\varepsilon = -b^2$. This maps natural gas systems with higher accuracy, especially for liquid states. Aungier (1995c) also modified the Redlich–Kwong equation, taking account of the acentric factor, so that it provides much better accuracy near the critical point. In the generalised form, the Aungier–Redlich–Kwong equation has the coefficients $\varepsilon = 0$ and $\delta = b$, and c is a more complex function of the critical parameters, as given later in this chapter. The Peng–Robinson and Soave–Redlich–Kwong equations are widely used in the petroleum industry, and the Aungier–Redlich–Kwong equation is popular in turbomachinery design and analysis software.

3.3.5 Virial Equations of State

A completely different approach to the modelling of the equations of state for a real gas is to use a virial equation of state, which is a series expansion of the following, or of an even more complex, form:

$$pv = RT\left(1 + \frac{B}{v} + \frac{C}{v^2} + \frac{D}{v^3} \cdots\right), \quad Z = 1 + \frac{B}{v} + \frac{C}{v^2} + \frac{D}{v^3} \cdots \tag{3.28}$$

All of the virial coefficients (B, C, D, etc.) are functions of temperature, and successive terms generally become less important. Polishuk (2009) has shown that the virial form with three constants, B, C and D can also be reformulated as a generalised cubic equation, (3.27):

$$p = \frac{RT}{(v-b)} - \frac{a(T)}{(v+c)(v+d)}$$
$$B = b - \frac{a}{RT}, \quad C = b^2 + \frac{a(c+d)}{RT}, \quad D = b^3 - \frac{a(c^2+cd+d^2)}{RT} \tag{3.29}$$

so that the simplest virial equations of state can also be considered to be close to cubic equations of state.

One general application of this virial form, for instance the Benedict, Webb and Rubin (BWR) equation, with an extension by Starling (the BWRS equation), has 11 coefficients, Starling (1973). This is popular for hydrocarbon gases as it has the ability to cover both liquids and gases. A modified version, known as MBWR, has been developed by Younglove and Ely (1987) with 32 terms. This is the basis of a popular software for gas properties from the National Institute of Standards and Technology (NIST) which is an industry standard and well-regarded reference for gas properties in the design and analysis of turbomachinery.

3.3.6 Multiple Variable Equations of State

To obtain relations of the gas properties in a wider range of conditions, multiple variable equations of state have been developed. The Lee–Kesler equation from Lee and Kesler (1975), with an extension by Plöcker and Knapp (1978), has 12 coefficients. The Span–Wagner equation has a different background and has 51 terms, Span and Wagner (2003). International agreement has been obtained on a standard set of equations for the properties of water and steam by the International Association for the Properties an Water and Steam (IAPWS) (Wagner et al., 2000). The advantage of these complex equations is that they can be adapted to provide very good matches to experimental data on individual gases and mixtures in the single-phase regions, and even near the critical point. Their disadvantage is the complexity and the difficulty of finding a solution, which usually involves applying Newton–Raphson iteration techniques. In some cases, the equations have also been developed into a form which allows backward calculations between the properties. Lüdtke (2004) and Kumar et al. (1999) give recommendations for which equation of state should be used for certain gases.

3.3.7 Real Gas Mixtures

For intensive properties (u, h, s, c_v and c_p), the contribution of each component in the mixture is added according to its mole fraction, x_i:

$$h_m = \sum_i x_i h_i, \quad s_m = \sum_i x_i s_i, \quad \cdots$$
$$x_i = \frac{N_i}{N_m}, \tag{3.30}$$

where N_i is the amount of a constituent expressed in moles. This is valid for real and ideal gas mixtures. Mixtures of real gases can be treated as with cubic equations of state as if they were pure substances with pseudocritical properties, p_c, T_c, v_c, derived by averaging the critical constants in a similar way. The pseudocritical properties may also be calculated to reduce the error in the equation of state if the behaviour of the mixture is well known.

Gas mixtures are important in air compressors with humidity, and in gas compressors from natural gas wells in which the gas mixture composition (methane, butane propane, etc.) may change over time as the well is depleted.

3.3.8 Software Packages for Real Gases

The science of chemical thermodynamics is constantly evolving. New and more advanced equations and coefficients are often developed for single gases, or even for technically relevant thermodynamic regions (close to the critical point, for example). The range of possible gases for use in compressors is also expanding. For example, at the time of writing there are numerous developments involving new refrigerant gases which have less ozone depletion potential and reduced global warming potential. This is an onerous requirement for modern software systems: to include modelling provision for all the possible gases and gas mixtures that might occur in compression systems.

The incorporation of real gas properties into compressor design software via look-up tables of values rather than as equations of state has some significant advantages. First, this procedure only has to be programmed once for all possible gases and mixtures. Secondly, interpolation within tables can be much quicker than complex iterations in the equations of state. But it requires skill to set up the table, making sure that the points are appropriately distributed to capture the gradients of properties in the region of interest and care to ensure the compressor design tool does not go out of range of the table.

3.4 The Aungier–Redlich–Kwong Cubic Equation of State

3.4.1 Aungier–Redlich–Kwong

In this section, an example of a modern cubic equation of state, the Aungier–Redlich–Kwong equation is given in more detail. In this model, the thermal equation of state takes the form

$$p = \frac{RT}{(v - b + c)} - \frac{\alpha(T)}{v(v + b)} = \frac{RT}{(v - b + c)} - \frac{\alpha_0 T_r^{-n}}{v(v + b)}, \tag{3.31}$$

where

$$n = 0.4486 + 1.1735\omega + 0.4754\omega^2, \quad \alpha_0 = 0.42747 R^2 T_c^2 / p_c$$

$$a = \alpha(T) = \alpha_0 (T/T_c)^{-n}, \quad b = 0.08664 RT_c/p_c, \quad c = \frac{RT_c}{p_c + \dfrac{\alpha_0}{v_c(v_c + b)}} + b - v_c. \tag{3.32}$$

3.4.2 Departure Functions for the Caloric Equation of State

In addition to the thermal equation of state, an additional equation for the caloric equation of state is needed. This is determined on the basis of departure functions (Poling et. al., 2000), which define the deviation from the properties of a perfect gas. Consider the internal energy:

$$u = (T, v)$$
$$du = \left(\frac{\partial u}{\partial T}\right)_v dT + \left(\frac{\partial u}{\partial v}\right)_T dv. \tag{3.33}$$

The last term is converted into a form which can be determined by integration from the equation of state with the help of the Maxwell equations, and the integration then yields

$$du = c_v dT + \left[T\left(\frac{\partial p}{\partial T}\right)_v - p\right] dv$$
$$u(T, v) = u^0 + \int c_v^0(T) dT + \int_\infty^v \left[T\left(\frac{\partial p}{\partial T}\right)_v - p\right] dv. \tag{3.34}$$

It is assumed in this process that the first term is integrated at a pressure level at which the specific heat is independent of pressure (that is, at a low pressure or nominal zero pressure). The internal energy of a real gas then becomes the sum of three terms: the reference internal energy of the perfect gas; the temperature dependence of the internal energy of the perfect gas; and a so-called departure term, related to the pressure dependence of the specific heat of the real gas. The departure term can be obtained by integration of the thermal equation of state using derivatives from the cubic equation of state. The ideal enthalpy can then also be determined using the definition of enthalpy as $h = u + pv$.

A similar procedure can be derived for the departure functions for the enthalpy, the entropy and the specific heats. With the Aungier–Redlich–Kwong equation of state, the analytic form of the departure functions for each of these becomes

$$h = h^0 - h_{dep} \text{ where } h_{dep} = -pv + RT - \frac{1}{b}\left(T\frac{\partial a}{\partial T} - a\right)\ln\left(\frac{v+b}{v}\right)$$

$$c_p = c_p^0 - c_{p,dep} \text{ where } c_{p,dep} = -p\left(\frac{\partial v}{\partial T}\right)_p + R + \frac{\partial a}{\partial T}\frac{(1+n)}{v}\ln\left(\frac{v+b}{v}\right) - \frac{a(1+n)}{v(v+b)}\left(\frac{\partial v}{\partial T}\right)_p$$

$$s = s^0 - s_{dep} \text{ where } s_{dep} = -R\ln\left(\frac{v-b+c}{v_0}\right) - \frac{1}{b}\frac{\partial a}{\partial T}\ln\left(\frac{v+b}{v}\right). \tag{3.35}$$

3.4.3 Temperature Dependence of the Specific Heat

The preceding equations need to be combined with additional equations taking into account the temperature dependence of the specific heat. For the specific heat, the temperature dependence is often specified empirically as a single interval polynomial in T, as follows:

$$c_p^0 = f(T), \qquad \text{for } T_{\min} < T < T_{\max}: \quad c_p^0(T) = a_1 + a_2 T + a_3 T^2 + \ldots \tag{3.36}$$

The ideal enthalpy is obtained by integrating the definition of the specific heat with temperature:

$$\begin{aligned} c_p^0(T) &= \left(\frac{\partial h^0}{\partial T}\right)_p \\ h^0 = h_{ideal} &= \int c_p^0 dT \\ h^0(T) &= a_1 T + a_2 T^2/2 + a_3 T^3/3 + \ldots \end{aligned} \tag{3.37}$$

The entropy is obtained by integration of the Gibbs equation:

$$\begin{aligned} Tds &= dh - vdp \\ ds &= \frac{dh}{T} - \frac{v}{T} dp, \\ ds^0 &= \frac{dh^0}{T} - R\frac{dp}{p} \end{aligned} \tag{3.38}$$

as follows:

$$s^0 = s_{ref} + \int c_p^0 \frac{dT}{T} - R \int \frac{dp}{p}$$

$$s^0 = s_{ref} + \int_{T_{ref}}^{T} c_p^0 \frac{dT}{T} - R \ln\left(\frac{p}{p_{ref}}\right) \quad \text{and}$$

$$\int_{T_{ref}}^{T} c_p^0 \frac{dT}{T} = a_1 \ln\left(\frac{T}{T_{ref}}\right) + a_2 (T - T_{ref}) + a_3\left(T^2 - T_{ref}^2\right)/2 + a_4\left(T^3 - T_{ref}^3\right)/3 + \cdots \tag{3.39}$$

3.5 Isentropic and Polytropic Processes with Real Gases

3.5.1 Isentropic Process

From Chapter 2, we know that the aerodynamic work in a compression process is the integral of the *vdp* work and for a real gas, with a constant compressibility factor, this is given by

$$y_{12} = \int_1^2 vdp = \frac{n}{(n-1)} ZRT_1 \left[\left(\frac{p_2}{p_1}\right)^{\frac{n-1}{n}} - 1\right]. \tag{3.40}$$

For an isentropic process with an ideal gas, $Z = 1$ and $n = \gamma$. The simple isentropic exponent of a perfect gas, which is identical to the ratio of the specific heat capacities, is not valid for real gases, but small isentropic changes of state of real gases are often approximated by the ideal gas equations given in (3.6). With larger changes of state, this approximation may be inaccurate even when the value of the isentropic exponent corresponds to the correct local value of c_p/c_v of the real gas under examination. Equation (3.17) provides a small correction to include the compressibility factor, Z. Kouremenos and Antonopolous (1987) suggests that better agreement can be obtained when three different isentropic exponents are used for real gases κ_{Tv}, κ_{pv} and κ_{pT} as follows:

$$Tv^{(\kappa_{Tv}-1)} = \text{constant}, \quad pv^{\kappa_{pv}} = \text{constant}, \quad p^{\left(1-\kappa_{pT}\right)}T^{\kappa_{pT}} = \text{constant}. \tag{3.41}$$

Replacement of the classical exponent $\gamma = c_p/c_v$ by these exponents from (3.6) in (3.41) improves the accuracy of these equations when applied to real gases. Isentropic changes of state of real gases may, therefore, be calculated directly from the preceding equations with κ replaced by the new exponents accordingly. Kouremenos and Antonopoulos also show that these exponents depend on the local state variables as follows:

$$\kappa_{Tv} = 1 + \frac{v}{c_v}\left(\frac{\partial p}{\partial T}\right)_v \quad \kappa_{pv} = -\frac{v\,cp}{p\,cv}\left(\frac{\partial p}{\partial v}\right)_T \quad \kappa_{pT} = \frac{1}{1-(p/c_p)(\partial v/\partial T)_p}. \tag{3.42}$$

Mean values between the end states may be used for a better match with large changes in pressure.

3.5.2 Polytropic Process

Beinecke and Lüdtke (1983) and Lüdtke (2004) show that, in a similar way, different polytropic exponents are needed for compressor performance calculations. These can be written in a similar notation giving

$$pv^{n_{pv}} = \text{constant}, \quad p^{\left(\frac{1-n_{pT}}{n_{pT}}\right)}T = p^{m_{pT}}T = \text{constant} \tag{3.43}$$

and with a polytropic efficiency of η_p, these give the following equations for the respective polytropic exponents:

$$n_{pv} = \frac{\kappa_{pv}}{\left\{1+\kappa_{pv}\left(\frac{\kappa_{pT}}{\kappa_{pT}-1}\right)\left(1-\frac{1}{\eta_p}\right)\right\}} \qquad m_{pT} = \frac{ZR}{c_p}\left(\frac{1}{\eta_p}-1\right)+\left(\frac{\kappa_{pT}-1}{\kappa_{pT}}\right). \tag{3.44}$$

The coefficient m_{pT} is used to calculate the end temperature of a polytropic process for a given pressure ratio, and the exponent n_{pv} is used in the integration of the aerodynamic vdp work to give the polytropic head. Once again, the mean values of the parameters at the suction and end states should be used for better agreement with measured values.

3.6 The Aerodynamic Work with Real Gases

3.6.1 Schultz Method for the Polytropic Head

The aerodynamic work is given by the integration of vdp on the compression path, and as a consequence it is not a state variable. To be consistent with other publications, the aerodynamic work in this section is called the polytropic head. The end states of the process are given directly by the gas equations, but for the integration, additional assumptions are needed. One approach to calculate the polytropic process for real gases in compressors has been given by Schultz (1962), and this has formed the mainstay of many international standards for compressor performance calculations with real gases over many years. The average value of the polytropic exponent can be calculated from the end states of a compression process using (3.43) as

$$n_{pv} = \frac{\ln (p_2/p_1)}{\ln (v_2/v_1)}. \tag{3.45}$$

Schultz (1962) recognised that the polytropic process with a real gas could affect the accuracy of the head predicted by (3.40) and developed a correction factor to account for the path differences by examining the error for real gases undergoing an isentropic compression process. The correction factor was developed for an isentropic compression process from state 1 to 2s, which was assumed to be approximately the same as that for a polytropic compression process:

$$y_{12} = \int_1^2 vdp = f_s \frac{n_{pv}}{(n_{pv} - 1)} Z_1 R T_1 \left[\left(\frac{p_2}{p_1} \right)^{\frac{n_{pv}-1}{n_{pv}}} - 1 \right]. \tag{3.46}$$

The correction factor f_s is given by

$$f_s = \frac{h_{2s} - h_1}{\left(\frac{\kappa_{pv}}{\kappa_{pv} - 1} \right) (p_2 v_{2s} - p_1 v_1)}. \tag{3.47}$$

3.6.2 Mallen and Saville Method for the Polytropic Head

Mallen and Saville (1977) suggested an improved method. Mallen and Saville recognised that the polytropic path represented by pv^n = constant is a more or less intuitive extension of the isentropic path pv^γ = constant, which may not be valid for real gases, and suggested replacing the polytropic path by a process path in which

$$T\frac{ds}{dT} = C = \text{constant}. \tag{3.48}$$

They also showed that for an ideal gas with constant specific heats, this is equivalent to a constant polytropic exponent. Integration of (3.48) gives the following equations for the polytropic work calculated from the end states of the process:

$$s_2 - s_1 = C \ln (T_2/T_1), \quad q = \int_1^2 T ds = C(T_2 - T_1)$$
$$y_{12} = (h_{t2} - h_{t1}) - q = (h_{t2} - h_{t1}) - \frac{(s_2 - s_1)(T_2 - T_1)}{\ln (T_2/T_1)}. \tag{3.49}$$

3.6.3 Huntington Method for the Polytropic Head

More recently, Huntington (1985) suggested an improved procedure after noting that the simplifications in the Schulz and Mallen and Saville methods caused computational errors. Firstly, he developed an accurate numerical reference procedure in that he split the polytropic compression path between two pressure levels into a large number of subpaths with smaller steps in the pressure. For the small steps, he could use the definition of the polytropic efficiency to find the outlet conditions, whereby for each subpath an iteration is required to find the final temperature of the subpath. By this method, he found that the Schultz method typically had a lower polytropic head than the reference method by 1.7% and that the Mallen and Saville method was typically 1.1% higher than the reference.

Following this he developed a method, in which he accounted for the fact that in a real gas the effect of the isentropic exponent on the isentropic work is not quite the same as the effect of the polytropic exponent on the polytropic work. He suggested a correction in which an empirical relationship for the variation of the compressibility factor, Z, along the polytropic path, was postulated as follows:

$$Z = a + b(p_2/p_1) + c \ln (p_2/p_1). \tag{3.50}$$

This method can also be couched in terms of a Schulz correction factor estimated by the following relations:

$$f_s = \frac{\dfrac{h_2 - h_1}{\left(n_{pv}/\left(n_{pv} - 1\right)\right)(p_2 v_2 - p_1 v_1)}}{1 + \dfrac{(s_2 - s_1)/R}{a \ln (p_2/p_1) + b(p_2/p_1 - 1) + (c/2)\{\ln (p_2/p_1)\}^2}}, \tag{3.51}$$

where a, b and c are given by

$$a = Z_1 - b, \quad b = \frac{(Z_1 + Z_2 - 2Z_3)}{\left\{\sqrt{(p_2/p_1)} - 1\right\}^2}, \quad c = \{Z_2 - a - b(p_2/p_1)\}/\ln (p_2/p_1). \tag{3.52}$$

The states 1 and 2 are the end states of the process, and state 3 is at an intermediate pressure and temperature given by

$$\begin{aligned} p_3 &= \sqrt{p_1 p_2}, \quad T_3 = \sqrt{T_1 T_2} \\ T_3' &= T_3 \exp\left\{\left(s_3' - s_3/p_1\right)/c_p\right\} \\ s_3' &= s_1 + (s_2 - s_1)d \\ d &= \frac{(a/2)\ln(p_2/p_1) + b\left(\sqrt{p_1 p_2} - 1\right) + (c/8)\{\ln(p_2/p_1)\}^2}{a\ln(p_2/p_1) + b\left(\sqrt{p_1 p_2} - 1\right) + (c/2)\{\ln(p_2/p_1)\}^2}. \end{aligned} \tag{3.53}$$

Iterative calculations are needed to update the symbols with a prime, including the value b.

3.6.4 International Standards

The importance of these equations is that they give an estimate of the aerodynamic work, or the polytropic head, for a real gas. This is needed in the calculation of the efficiency, as shown in Section 4.8. The American Society of Mechanical Engineers (ASME) 'Performance Test Code on Compressors and Exhausters', ASME PTC 10 (ASME, 1997), has served worldwide as the standard for determining centrifugal compressor thermodynamic performance during factory acceptance and field performance testing. The methodology applied in PTC 10 is still based upon the work of Schultz (1962). Recent work in this area, as outlined in this chapter and also by Sandberg and Colby (2013), has revealed that the application of PTC 10 can result in significant errors in the estimation of compressor polytropic head and efficiency. A revised form of calculation procedure has not yet become standard.

4 Efficiency Definitions for Compressors

4.1 Overview

4.1.1 Introduction

Two examples should suffice to whet the reader's appetite for the importance of clarity in efficiency definition in radial compressors. First, the commonly used isentropic efficiency – which is sometimes called adiabatic efficiency – compares the actual work transfer to that which would take place in an ideal isentropic adiabatic process with no losses and no heat transfer. Unfortunately, the isentropic efficiency does not actually represent the real quality of a machine at all well. For example, consider a two-stage turbocharger with a pressure ratio of 2 in both stages. If each stage achieves the same isentropic efficiency of 80%, then on combining them to a two-stage compressor with a pressure ratio of 4, the isentropic efficiency is then lower than that of the individual stages (78.1%). How strange and disorientating. The so-called polytropic efficiency overcomes this problem and, in this case, if both stages have 80% polytropic efficiency, the two-stage compressor would have the same polytropic efficiency as its individual stages.

Second, a radial compressor impeller may have, at the same time, a total–total polytropic impeller efficiency of over 90%, a static–static isentropic efficiency of well below 60% and an impeller wheel efficiency of 30%. You can guess which definition a sales engineer would prefer to use to sell his product. It turns out that one can misuse efficiencies just as one can misuse statistics.

In this chapter, the background to the systematic definition of the isentropic, polytropic and isothermal efficiencies, with and without kinetic energy, is considered. This chapter concerns only the different definitions of the efficiencies, and Chapter 10 gives more detail on the source of losses and expected efficiency levels.

4.1.2 Learning Objectives

- Be aware of the many different forms of efficiency definition and the need to decide which one is appropriate for use in different situations.
- Understand the different reversible reference processes used for definition of isentropic, polytropic and isothermal efficiencies.
- Understand the implications of the kinetic energy of the fluid for total–total, total–static and static–static efficiencies.

- Understand why the aerodynamic work is integrated along the path of static states and not along the path of total states in the definition of total–total polytropic efficiency and why this is acceptable for the total–total isentropic efficiency.
- Be able to distinguish between the different terms; reversible work, aerodynamic work and the polytropic head.

4.2 Compressor Efficiency

4.2.1 Issues with the Definition of Efficiency

The efficiency is the key parameter on which the assessment of the process in a compressor is based. It is used to compare the quality of different compressors, or to identify the variation in performance of a particular machine with a change in operating conditions or the deterioration over time. The efficiency is important as it is directly related to the power consumption, the environmental impact and the overall operating costs of a compressor.

Compressor efficiency is best defined as the ratio of the minimum effort that would be required to compress the flow in an ideal reversible compression process between two pressure levels relative to the effort actually expended to achieve this pressure rise. It is expressed in such a way that the efficiency has the value of unity (or 100%) for a perfect machine. This statement looks very straightforward but is actually highly complex. It is necessary to consider slowly and methodically all the issues that may cause difficulties and require clarification, and this discussion is started in this section and continued throughout the chapter.

The ideal thermodynamic process in the reversible machine which is being used as a basis to define the minimum work has to be defined and, in practice, this reference process could be isentropic, polytropic or isothermal. The isentropic adiabatic process might naturally be considered to be the most useful definition of the reference process for this perfect machine as it has no losses and no heat transfer. This leads to the definition of the isentropic efficiency. But as shown in this chapter, it is sometimes more appropriate to consider a reversible machine with a reference process that is the same as the polytropic process which actually occurs in the real machine. This leads to the definition of the polytropic efficiency, which in compressors is larger than the isentropic efficiency. In intercooled compressors, a reversible isothermal process may be more useful as reference, and this leads to the definition of the isothermal efficiency. Because of the divergence of the constant pressure lines in an h-s diagram towards high temperature, an isothermal process defines the minimum possible work between two pressure levels. The reversible isothermal process leads to the lowest possible work for a perfect machine, and so the isothermal efficiency is lower than the isentropic and polytropic efficiencies

The effort in the compression process is best considered to be the sum of the useful aerodynamic work together with any change in kinetic energy of the fluid. The aerodynamic work is the integral of vdp along the change of static conditions in the reversible process from the inlet to outlet conditions. In pumps and water turbines, the change in potential energy also needs to be included. Chapter 2 shows that the aerodynamic work in a thermodynamic process is defined in terms of an integration along the process path of vdp, with static conditions, and not as an integration of $v_t dp_t$, along a process path of total conditions. This is an important distinction to be examined in more detail in this chapter.

Connected with the difference between total and static states is the fact that clarity is needed about the value of any kinetic energy present at the inlet and outlet planes. Is the kinetic energy in these planes negligible, and if not, is it useful or not? In some cases, the changes in kinetic energy are sufficiently small that they can be neglected, in some they must be included and in others including the outlet kinetic energy can lead to a distortion of the quality of the compression process. The locations of the planes used for the definition of the inlet and outlet of the process also need to be clearly defined. These planes may be the inlet and outlet of a single rotor, a single component, a single stage, multiple stages, or the whole machine from flange to flange. In the special case of a stationary component with no work input, such as a diffuser, there is no shaft work so the definition of efficiency for such a component needs to be able to include this possibility. Another issue, not discussed in detail here, is how to determine the appropriate 1D mean value of the thermodynamic properties at the inlet and outlet planes. For this, the reader is referred to Cumpsty and Horlock (2006), who recommend area averaging of the pressures and mass-averaging of the velocity components and enthalpy for the applications considered here.

4.2.2 The Ideal Reversible Reference Process

The choice of the reversible process used as an ideal reference is a matter of convention. It depends on the application of the machine and results in different efficiency definitions. Common practice makes use of isentropic processes, but polytropic or isothermal processes can also be used to define efficiencies, as shown schematically in Figure 4.1. The isentropic efficiency is the easiest to understand and considers the relevant reference process to be an ideal reversible process at constant entropy from state 1 to 2s between the inlet and outlet pressure levels. In this reference process, there is no heat transfer, so the isentropic efficiency is also known as the adiabatic efficiency.

The polytropic efficiency considers the ideal reference process to be a reversible process whose static states coincide with the actual change of state of the gas in the compressor from state 1 to 2. This ideal reversible process has no losses so that heat has to be added within the process to bring the change of state along this reversible path into line with the change in entropy due to the dissipation in the real process. In this way, the reversible reference polytropic path is actually not adiabatic.

The isothermal efficiency considers the ideal reference process to be a reversible process between the inlet and outlet pressure levels but which have the same

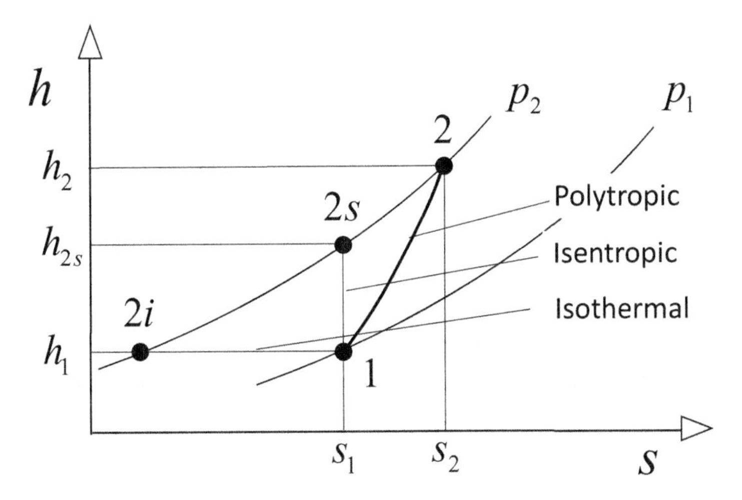

Figure 4.1 Reference reversible processes process for different efficiency definitions.

temperature of the gas as at the inlet, from state 1 to $2i$. This is important because the lowest possible work between the two pressure levels would be in an isothermal process. The ideal isothermal process is also not adiabatic. Heat has to be removed along the reference isothermal compression path to account for the reduction in entropy.

4.2.3 The Minimum Work of Compression

The minimum shaft work required to compress the flow in a compression process along the flow path is given by the reversible steady flow shaft work, which is the sum of the aerodynamic work and the change in kinetic energy as defined in (2.43). The reversible work along the flow path from state 1 to state $2rev$ is then defined as

$$w_{s12,rev} = y_{12,rev} + \tfrac{1}{2}\left(c_{2rev}{}^2 - c_1{}^2\right) = \int_1^{2rev} vdp + \int_1^{2rev} cdc. \tag{4.1}$$

The first term in these expressions is the aerodynamic work, and the second term is the change in kinetic energy. It is important to note that in this definition, the aerodynamic work is defined by the integration of vdp along the path of static states and the kinetic energy terms at inlet and outlet are added to this to give the total reversible work. The bookkeeping of the change in kinetic energy is done separately from the bookkeeping of the aerodynamic work. The internal work does not include losses external to the flow path, such as the disc friction losses or leakage losses, and these need to be considered separately, as described in Section 4.6. Furthermore, the actual effort expended is taken to be work done by the shaft,

w_{s12} on the fluid. Therefore, the general form of efficiency for compressors can be written as

$$\eta_c = \frac{w_{s12,\,rev}}{w_{s12}}. \tag{4.2}$$

On this basis, several different definitions of compressor efficiency can all be expressed as

$$\eta_c = \frac{\int_1^{2rev} vdp + \frac{1}{2}\left(c_{2rev}{}^2 - c_1{}^2\right)}{w_{s12}}. \tag{4.3}$$

where the integrations are carried out along the chosen reversible compression path.

For the case of the polytropic efficiency, the reversible reference path coincides with the static states of the actual compression path, giving

$$\eta_p = \frac{\int_1^2 vdp + \frac{1}{2}\left(c_2{}^2 - c_1{}^2\right)}{w_{s12}}, \tag{4.4}$$

where state 2 is the actual state of the gas at the outlet condition. Furthermore, for this process the shaft work is the sum of the reversible work and the dissipation energy that occurs in the real process and can be written as

$$w_{s12} = y_{12} + \frac{1}{2}\left(c_2{}^2 - c_1{}^2\right) + j_{12} = \int_1^2 vdp + \frac{1}{2}\left(c_2{}^2 - c_1{}^2\right) + \int_1^2 Tds. \tag{4.5}$$

On this basis, the polytropic efficiency is defined as

$$\eta_p = \frac{\int_1^2 vdp + \frac{1}{2}\left(c_2{}^2 - c_1{}^2\right)}{w_{s12}} = 1 - \frac{\int_1^2 Tds}{w_{s12}} = 1 - \frac{j_{12}}{w_{s12}} \tag{4.6}$$

as already given in (2.44), but there it was derived without considering the kinetic energy terms. An ideal reversible machine with no dissipated energy would have an efficiency of unity. For the polytropic efficiency, the deviation from unity is directly related to the dissipation incurred along the real process path as a fraction of the shaft work. The common engineering use of the value 1-η as a convenient statement of the losses is then more or less self-evident.

The integration of *Tds* along the actual compression path is the only correct measure of the internal dissipation or lost work that has occurred in the process. By the same token, from (4.6), the polytropic efficiency might be considered to be the true efficiency of a compressor. However, this definition requires detailed knowledge of the thermodynamic properties along the actual compression path. It is shown in this chapter that both the isothermal and isentropic efficiencies have the advantage that the reference reversible compression path is fully defined by the end states alone. This detailed knowledge of the true compression path is generally not available, and the integration of *vdp* along it is then not possible. In practice, the polytropic efficiency has to be defined by reference processes based on a good approximation to the real compression path.

4.3 Isentropic Efficiency

4.3.1 The Static–Static Isentropic Efficiency

In the first instance, the isentropic efficiency is considered here in terms of static states assuming the kinetic energy is negligible. The extension to include total states and the changes in kinetic energy is given in Sections 4.3.2 and 4.3.3. If the kinetic energy is negligible, there is no relevant difference between the total and static conditions and the shaft work is then the same as the change in the static enthalpy. The isentropic efficiency defines an isentropic (constant entropy) compression path as the reference process to determine the reversible work. This isentropic path lies between the initial state and the final pressure that is achieved in the actual process, as represented by a vertical line in the *h-s* diagram shown in Figure 4.1. The final condition of the reference process, 2*s*. is the state at the same outlet pressure as the real process, p_2, but with the entropy of the initial state, s_1.

For a small isentropic step with $ds = 0$ along the compression path, as shown in Figure 4.2, the Gibbs equation reduces to $(vdp)_s = dh_s$, demonstrating that the reversible work along the isentropic path is equal to the isentropic enthalpy change in the process. The small-scale isentropic efficiency is therefore simply defined as the ratio of the isentropic to the actual enthalpy rise. This can be written differentially as

$$\eta_s = \frac{(vdp)_s}{dh} = \frac{dh_s}{dh}. \tag{4.7}$$

The overall static–static isentropic efficiency of the machine is defined as the ratio of the overall isentropic enthalpy rise to the actual enthalpy rise in the compressor based on static conditions:

$$\eta_s = \frac{\int_1^{2s}(vdp)_s}{\int_1^2 dh} = \frac{\int_1^{2s} dh_s}{\int_1^2 dh} = \frac{h_{2s} - h_1}{h_2 - h_1}. \tag{4.8}$$

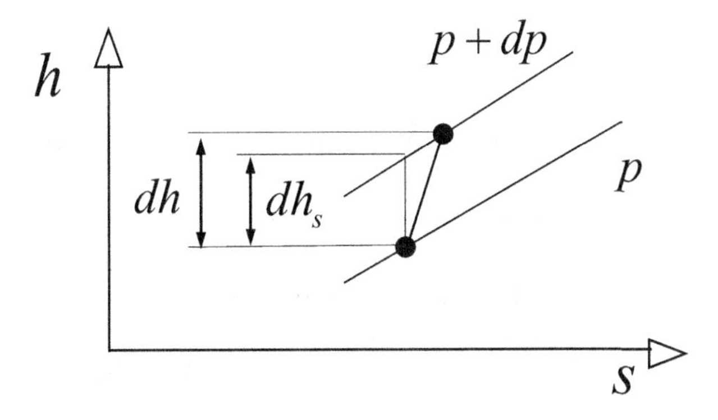

Figure 4.2 Reference process for the small-scale isentropic efficiency.

The reference isentropic process is both reversible and adiabatic. The denominator in the equation is the change in static enthalpy, but if the change in kinetic energy is negligible, then this is the same as the change in the total enthalpy, which is the proper definition of the shaft work. The isentropic efficiency is also referred to as the adiabatic efficiency, although this terminology could be misleading as an adiabatic process is not necessarily isentropic.

Isentropic efficiency is easy to understand and can be straightforwardly applied to real gases as well as ideal gases. Furthermore, the reversible work of the reference isentropic process is a function of the inlet temperature and the pressure ratio; it is not affected by the actual performance of the machine. This means that knowledge of the end state of the real process is not required in order to calculate the reversible aerodynamic work. In connection with thermodynamic cycle calculations, the isentropic efficiency leads to simple expressions, and is therefore generally used in relation to processes in single-stage turbochargers, refrigeration cycles and gas turbine cycles, as shown in Sections 18.9.1 and 18.10.1.

Making use of the isentropic relations for a perfect gas from Section 2.7.1, the static–static isentropic stage efficiency can be written as

$$\eta_s^{ss} = \frac{T_{2s} - T_1}{T_2 - T_1} = \frac{(T_{2s}/T_1) - 1}{(T_2/T_1) - 1} = \frac{(p_{2s}/p_1)^{\frac{\gamma-1}{\gamma}} - 1}{(T_2/T_1) - 1} = \frac{(p_2/p_1)^{\frac{\gamma-1}{\gamma}} - 1}{(T_2/T_1) - 1}. \tag{4.9}$$

The difference in enthalpy between states $2s$ and 2, shown in Figure 4.2, at the pressure level p_2, can be obtained by integration of the Gibbs equation for a constant pressure process, $dp = 0$, $(dh)_p = Tds$, between states $2s$ and 2, which shows that

$$h_2 - h_{2s} = \int_{2s}^{2} dh_p = \int_{2s}^{2} Tds. \tag{4.10}$$

This indicates that the measure of the detrimental effect of real irreversibilities on the isentropic efficiency is the integral of Tds from 2 to $2s$. Thus, an entropy-based isentropic efficiency can be defined as

$$\eta_s^{ss} = \frac{h_{2s} - h_1}{h_2 - h_1} = \frac{(h_2 - h_1) - (h_2 - h_{2s})}{h_2 - h_1} = 1 - \frac{\int_{2s}^{2} Tds}{h_2 - h_1}. \tag{4.11}$$

Equation (4.11) shows that the effective lost work in the definition of isentropic efficiency is related to integration of the dissipation along an isobar from $2s$ to 2. This is in contrast to the integration of the dissipation along the actual compression path from state 1 to state 2, which is the true measure of the losses in the compressor as is used in the definition of polytropic efficiency described in the next section. This points to a major disadvantage of the isentropic efficiency and is discussed in Section 4.3.6.

Following Denton (1993) and Greitzer et al. (2004), a useful engineering approximation to (4.11), with only a small error when the efficiency is high, may be obtained by assuming that the temperature is nearly constant along the path from $2s$ to 2 such that

$$\eta_s \approx 1 - \frac{T_2(s_2 - s_1)}{h_2 - h_1}. \tag{4.12}$$

This indicates that the isentropic efficiency is related to both the entropy change in the process as well as the temperature at the compressor outlet. In practice, the isentropic efficiency of the compressor is normally presented in the form of total–total and total–static isentropic efficiencies, where these are defined in terms of the thermodynamic state of the gas (total or static) that is used to represent the inlet and the outlet conditions.

4.3.2 The Total–Total Isentropic Efficiency

The total–total isentropic efficiency extends the definition of static–static isentropic efficiency, given in the preceding section, to consider changes in the kinetic energy in the flow and is visualised in Figure 4.3. The total energy available at the inlet of the compressor is given by the total or stagnation conditions. If the outlet kinetic energy can be used in a downstream component, the total energy at the outlet is also of interest. This is the case at the outlet of an intermediate stage in a multistage compressor. It also occurs when a further increase in static pressure can be achieved by downstream stationary components. In such instances, the outlet total condition may be specified as the useful end state of the process. The actual work required in the machine is then compared with the work in an ideal machine which compresses the flow isentropically from the inlet total condition to the same total pressure as the actual process.

Following the procedure given by (4.3) for the efficiency, in which the changes in the kinetic energy are added to the reversible aerodynamic work to give the reversible work, this results in the total–total isentropic stage efficiency defined as

static–static stage isentropic efficiency

$$\eta_s^{ss} = \frac{(h_{2s} - h_1)}{(h_2 - h_1)} = \frac{a}{b}$$

total–total stage isentropic efficiency

$$\eta_s^{tt} = \frac{(h_{t2s} - h_{t1})}{w_{s12}} = \frac{(h_{2s} - h_1) + \tfrac{1}{2}(c_2^2 - c_1^2)}{(h_2 - h_1) + \tfrac{1}{2}(c_2^2 - c_1^2)} = \frac{c}{d}$$

total–static stage isentropic efficiency

$$\eta_s^{ts} = \frac{(h_{t2s} - h_{t1}) - \tfrac{1}{2}c_2^2}{w_{s12}} = \frac{(h_{t2s} - h_{t1}) - \tfrac{1}{2}c_2^2}{(h_{t2} - h_{t1}) + \tfrac{1}{2}(c_2^2 - c_1^2)} = \frac{e}{d}$$

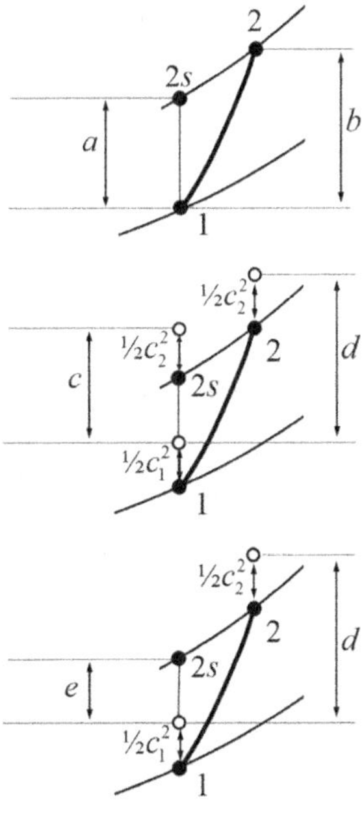

Figure 4.3 Reference processes for the static–static, total–total and total–static efficiencies.

$$\eta_s^{tt} = \frac{\int_1^{2s}(vdp)_s + \int_1^{2s}(cdc)_s}{\int_{t1}^{t2} dh_t} = \frac{\int_1^{2s} dh_s + \frac{1}{2}\left(c_{2s}^2 - c_1^2\right)}{\int_{t1}^{t2} dh_t} = \frac{\int_{t1}^{t2s} dh_{ts}}{\int_{t1}^{t2} dh_t} = \frac{h_{t2s} - h_{t1}}{h_{t2} - h_{t1}}. \tag{4.13}$$

An important aspect of the total–total isentropic efficiency is that the integration of the aerodynamic work vdp along the isentropic path of static states and subsequently adding the change in kinetic energy leads to the same result for the reversible work as the integration of $v_t dp_t$ along the isentropic path of total states because of the following identity from the Gibbs equation $(v_t dp_t)_s = dh_{ts}$:

$$w_{rev} = \int_1^{2s} dh_s + \frac{1}{2}\left(c_{2s}^2 - c_1^2\right) = \int_{t1}^{t2s} dh_{ts} = \int_{t1}^{t2s} (v_t dp_t)_s. \tag{4.14}$$

This identity only holds for an isentropic process and cannot be used for other processes. It does not hold for a polytropic or isothermal process, but is often mistakenly used in the efficiency definition for these processes.

Equation (4.13) can be expressed as

$$\eta_s^{tt} = \frac{h_{t2s} - h_{t1}}{h_{t2} - h_{t1}} = \frac{(h_{2s} - h_1) + \frac{1}{2}\left(c_{2s}^2 - c_1^2\right)}{(h_2 - h_1) + \frac{1}{2}\left(c_2^2 - c_1^2\right)}. \tag{4.15}$$

In practical applications, it is not easily possible to determine the isentropic outlet velocity at the virtual state 2*s* which would occur in a reversible isentropic process, and so c_{2s} is taken to be the same as c_2. Where the inlet and outlet kinetic energy are the same, or where they are negligible in comparison with the stage enthalpy rise, (4.15) reverts to (4.8), and then the total–total efficiency is the same as the static–static efficiency. For a perfect gas, the total–total isentropic stage efficiency can then be written with the help of the isentropic relations as

$$\eta_s^{tt} = \frac{T_{t2s} - T_{t1}}{T_{t2} - T_{t1}} = \frac{(T_{t2s}/T_{t1}) - 1}{(T_{t2}/T_{t1}) - 1} = \frac{(p_{t2s}/p_{t1})^{\frac{\gamma-1}{\gamma}} - 1}{(T_{t2}/T_{t1}) - 1} = \frac{(p_{t2}/p_{t1})^{\frac{\gamma-1}{\gamma}} - 1}{(T_{t2}/T_{t1}) - 1}. \tag{4.16}$$

4.3.3 The Total–Static Isentropic Efficiency

In many applications of ventilation fans and single-stage centrifugal compressors, the kinetic energy at the stage outlet is dissipated further downstream. Here, the resulting losses can be included in the efficiency of the compressor by subtracting the outlet kinetic energy from the reversible work. Following the same procedure given by (4.3), this leads to the total–static isentropic efficiency as

$$\eta_s^{ts} = \frac{\left[\int_1^{2S}(vdp)_s + \frac{1}{2}\left(c_{2s}^2 - c_1^2\right)\right]}{\int_{t1}^{t2} dh_t} = \frac{\int_{t1}^{t2s} dh_s - \frac{1}{2}c_1^2}{\Delta h_t} = \frac{(h_{2s} - h_1) - \frac{1}{2}c_1^2}{h_{t2} - h_{t1}} = \frac{h_{2s} - h_{t1}}{h_{t2} - h_{t1}}. \tag{4.17}$$

In other words, the total–static isentropic efficiency compares the actual work required in the real machine with the work in an isentropic machine which compresses the flow from the inlet total condition to the same outlet static pressure. The integration of the aerodynamic work is still considered to be along the reversible path of static conditions. For a perfect gas, the total–static isentropic stage efficiency can be written as

$$\eta_s^{ts} = \frac{T_{2s} - T_{t1}}{T_{t2} - T_{t1}} = \frac{(T_{t2s}/T_{t1}) - 1}{(T_{t2}/T_{t1}) - 1} = \frac{(p_{2s}/p_{t1})^{\frac{\gamma-1}{\gamma}} - 1}{(T_{t2}/T_{t1}) - 1} = \frac{(p_2/p_{t1})^{\frac{\gamma-1}{\gamma}} - 1}{(T_{t2}/T_{t1}) - 1}. \tag{4.18}$$

4.3.4 Comparison of Total–Total and Total–Static Isentropic Efficiencies

Equations (4.13) and (4.17) can also be used to analyse the development of the isentropic efficiencies through a compressor. By placing the end point, state 2, at any location along the compression path, such as the impeller outlet or the diffuser outlet, the isentropic efficiency from the inlet up to that point is obtained. The difference between the total–total and total–static isentropic efficiencies at any plane in the compressor, ξ, is the ratio of the isentropic kinetic energy available at that plane to the overall shaft work as

$$\xi = \eta_s^{tt} - \eta_s^{ts} = \frac{h_{t2s} - h_{2s}}{h_{t2} - h_{t1}} = \frac{\frac{1}{2}c_{2s}^2}{h_{t2} - h_{t1}} = \frac{c_{2s}^2}{2\lambda u_2^2}, \quad \lambda = \frac{h_{t2} - h_{t1}}{u_2^2}, \tag{4.19}$$

where λ is the nondimensional work coefficient. The total–total isentropic efficiency at a specific location in the compressor shows the theoretical upper limit for the total–total stage isentropic efficiency that can be achieved at the outlet of the downstream components if no further losses are incurred. The total–static isentropic efficiency at that location shows the theoretical lower limit for the total–total stage isentropic efficiency if none of the kinetic energy at that location is converted into pressure rise and there are no further losses. In this way, the ratio of the local kinetic energy at a specific location to the shaft work, ξ, is a measure of the maximum potential for improvement in isentropic efficiency that can be gained downstream of that plane.

The difference between these two efficiencies, ξ, is the ratio of the kinetic energy at that plane to the shaft work input. The isentropic kinetic energy, $\frac{1}{2}c_{2s}^2$, is the maximum available kinetic energy that can be transformed to static pressure rise in downstream components in an isentropic process (without incurring any losses). In practical applications, it is assumed that the isentropic velocity, c_{2s}, is the same as the actual local velocity, c_2. Considering state 2 to be at the impeller outlet, the kinetic energy at this location is commonly accounted by the so-called kinematic degree of reaction as described in Section 2.2.5, defined as the ratio of the total–static to the total–total enthalpy rise across the impeller:

$$r_k = \frac{h_2 - h_{t1}}{h_{t2} - h_{t1}} = \frac{h_{t2} - h_{t1} - \frac{1}{2}c_2^2}{h_{t2} - h_{t1}} = 1 - \xi_2. \tag{4.20}$$

For typical stages with a degree of reaction of close to 0.6, the difference in the total–total and the total–static efficiencies at the impeller outlet is then 0.4. The total–total efficiency is then a highly misleading measure of the impeller efficiency as it can be increased simply by reducing the degree of reaction of the impeller through increasing the kinetic energy at impeller outlet, but as its value is often above 90% many authors like to use it.

4.3.5 Evolution of the Isentropic Efficiency in the Compressor

The evolution of the total–total and total–static isentropic efficiencies at different locations downstream of a typical centrifugal compressor impeller operating at its design point is shown schematically in Figure 4.4. The total–total isentropic efficiency at the impeller outlet is the maximum isentropic efficiency that can be achieved by the compressor if the impeller exit flow is brought to rest isentropically. Its value is obviously less than unity due to the losses already incurred in the impeller and will further reduce as more losses are incurred in the downstream stationary components. The total–static isentropic efficiency at the impeller outlet gives the minimum isentropic efficiency that can be achieved by the compressor stage if no further static pressure rise is achieved downstream of the impeller, whereby all the dynamic head at the impeller outlet is lost in the machine. This is the most pessimistic measure of compressor efficiency. As the flow proceeds through the compressor to the diffuser and the volute outlet, the total pressure decreases due to losses, while the static pressure rises depending on the pressure recovery in the diffuser and volute sections. Therefore, the value of ξ reduces, meaning that there is less potential for further gains in efficiency. In other words, the impact of downstream components on efficiency reduces as the flow decelerates downstream of the impeller outlet. The value of ξ, given in (4.19), identifies the sensitivity of the stage efficiency to the losses in the downstream components. At the diffuser inlet, the typical ratio of the local kinetic energy to the stage work is 0.4. At the inlet to a return channel, it may be 0.08, which indicates that in terms of

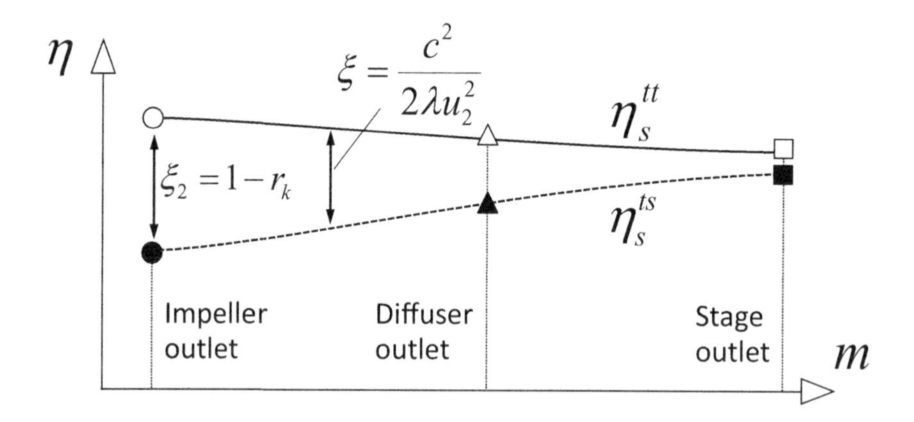

Figure 4.4 Evolution of the isentropic efficiency downstream of an impeller.

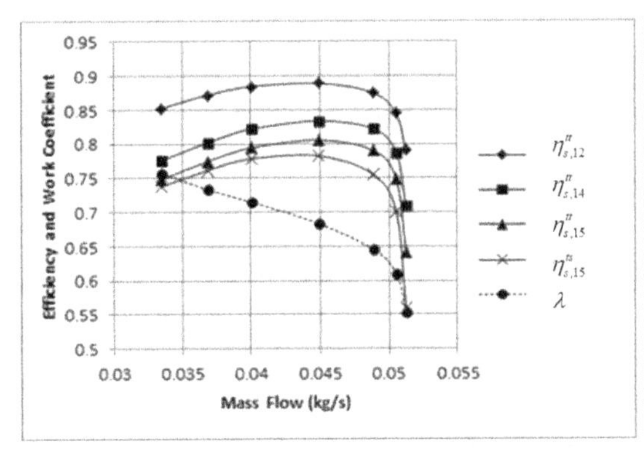

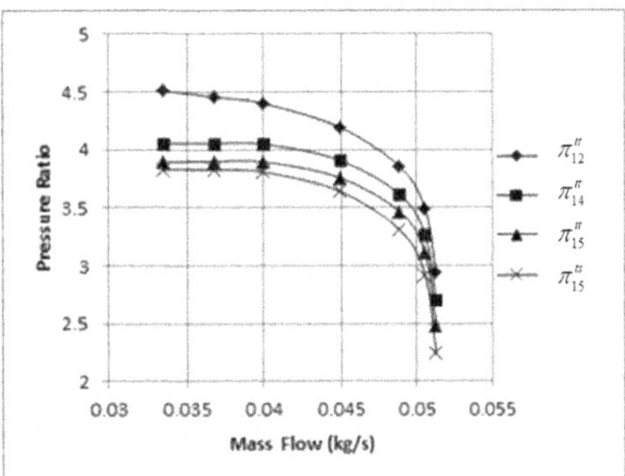

Figure 4.5 Calculated performance of a stage and impeller.

efficiency an improvement in the loss coefficient of the diffuser will be about five times more effective than a similar improvement to the return channel.

The trend in the variation of the total–total efficiency and the associated total–total pressure ratio through a turbocharger stage from a CFD calculation is shown in Figure 4.5. The total–total efficiency and pressure ratio drops from the impeller outlet, 2, via the diffuser outlet, 3, the volute outlet, 4, to the flange, 5, as the kinetic energy at impeller outlet is not completely converted into pressure rise.

4.3.6 Divergence of Constant Pressure Lines

Despite its simplicity, a major disadvantage of isentropic efficiency is that it does not accurately reflect the aerodynamic quality of the machine. In practice, the isentropic efficiency falls as the pressure ratio increases due to a purely thermodynamic effect. The isentropic efficiency of a compressor with two stages, each having the same isentropic efficiency, is lower than that of the individual stages due to the displacement of the constant pressure lines. The vertical gap between the enthalpy of two isobars in an *h-s* diagram increases with the entropy such that a larger enthalpy change is required to produce the same pressure ratio at a higher temperature. The isentropic efficiency considers the enthalpy rise from state 1 to state 2*s* as the reversible work for comparison with the actual work input in its definition. The actual increase in the isentropic enthalpy from state 1 to the outlet pressure of the two-stage machine is then less than that of the two stages considered individually. This can be seen in the following equations and in Figure 4.6.

$$\begin{aligned}
\eta_{s(1)} &= \frac{\Delta h_{1s}}{\Delta h_1} = \eta_{s(2)} = \frac{\Delta h_{2s}}{\Delta h_2} = \frac{\Delta h_{1s} + \Delta h_{2s}}{\Delta h_1 + \Delta h_2} \\
\eta_{s(1+2)} &= \frac{\Delta h_{1s} + \Delta h_{2ss}}{\Delta h_1 + \Delta h_2} = \frac{\Delta h_{1s} + \Delta h_{2s} + (\Delta h_{2ss} - \Delta h_{2s})}{\Delta h_1 + \Delta h_2} \\
\Delta h_{2s} &> \Delta h_{2ss}, \quad \Rightarrow \quad \eta_{s(1+2)} < \eta_{s(1)}, \eta_{s(2)}.
\end{aligned} \tag{4.21}$$

The drop in isentropic efficiency for the two-stage machine is related to integration of the dissipation energy along an isobar from 2*s* to 2, as opposed to the integration along

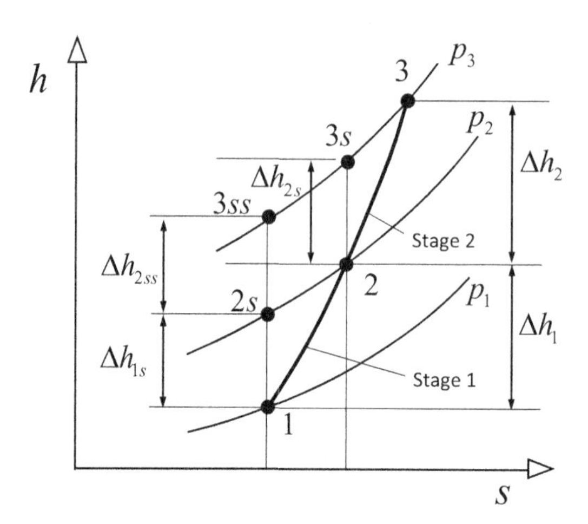

Figure 4.6 The isentropic efficiency of a two-stage compressor.

the actual compression path from state 1 to 2, as shown in (4.11). The average temperature increases from 2*s* to 2 with increasing the pressure ratio, so the second stage with the same dissipation losses has a lower efficiency! The increase in the work required in the second stage accounts for the fact that the inlet to the second stage is preheated by the first stage, and as a consequence requires more work for the same pressure ratio, as shown in Section 2.7.2 and in Figure 2.16. This problem can be avoided by using polytropic efficiency.

4.4 Polytropic Efficiency

4.4.1 The Polytropic Process

The key difference between the polytropic and the isentropic efficiency is that the reference compression process in the definition of the reversible work is taken as a reversible compression path from the inlet state 1 to outlet state 2. Of course, the real process from state 1 to state 2 is not reversible, but in the determination of the reversible work the changes of state that are used are those that occur in the real process. It is shown in this section that the intermediate states can be achieved in a reversible diabatic process from state 1 to state 2 if the reversible heat added in the reversible process is the same as the dissipation occurring at each step along this path.

Looking at a differential compression step shown in Figure 4.2, the small-scale polytropic efficiency is defined as the ratio of the reversible enthalpy rise, vdp, to the actual enthalpy rise, vdp/dh. This leads to the definition of the small-scale polytropic efficiency as

$$\eta_p = \frac{vdp}{dh} = \frac{dh - Tds}{dh} = 1 - \frac{Tds}{dh}. \tag{4.22}$$

In a real compression process, this small-scale efficiency varies along the actual compression path, as can be seen in the h-s diagram of the compression path of a typical compressor in Figure 2.17. To calculate the aerodynamic work along this path, the integration of the aerodynamic work, vdp, must be carried out as

$$\eta_p = \frac{\int_1^2 vdp}{\int_1^2 dh} = \frac{\int_1^2 vdp}{h_2 - h_1} = \frac{y_{12}}{h_2 - h_1}. \tag{4.23}$$

This integration requires a detailed knowledge of the thermodynamic properties along the actual path, which is not available in practice. Instead, the aerodynamic work in this process can be calculated along a so-called polytropic path, hypothesised as a path between the actual initial and final states, along which the small-scale efficiency remains constant. It has been shown in Section 2.5.7 that, for an ideal gas, this corresponds to a process described with pv^n = constant. The exponent n is the polytropic exponent and from Section 2.5.8 is related to the polytropic efficiency in an adiabatic process as

$$\eta_p = \frac{n/(n-1)}{\gamma/(\gamma-1)}, \quad \frac{n-1}{n} = \frac{(\gamma-1)}{\gamma\eta_p}, \quad n = \frac{\gamma\eta_p}{\gamma\eta_p - (\gamma-1)}. \tag{4.24}$$

The equation, pv^n = constant does not suffice to define the polytropic process for real gases, and the calculation of the aerodynamic work and the polytropic efficiency becomes more complex, as outlined in Section 3.6. This is the main weakness of the polytropic efficiency compared with the isentropic efficiency.

The assumption of a constant efficiency along the compression path is consistent with the view that two compressor stages with the same efficiencies should result in the same overall efficiency if they are combined to give a two-stage compressor. The overall reversible work in a polytropic process is referred to as 'polytropic head' or aerodynamic work and is denoted in (4.23) by y_{12}. The polytropic process is discussed in detail in Section 2.7.3 for ideal gases and in Section 3.5.1 for real gases.

The polytropic efficiency is also referred to as the 'small-scale isentropic efficiency'. For a small compression step along a polytropic path, the Gibbs equation shows that the isentropic enthalpy rise, dh_s, is equal to the polytropic work, vdp, and therefore the polytropic and isentropic efficiencies of a small compression step are identical. This can be understood by considering that a small compression step from p to $p + dp$ along the polytropic path can be replaced by an isentropic work input, dh_s, followed by an isobaric reversible heat addition that generates the same entropy increase as the irreversible loss generates in the real process. The work done during the heat addition is zero, and so the reversible (polytropic) work from p to $p + dp$ is equal to the isentropic work, dh_s. Thus, for a small compression step, the isentropic and polytropic efficiencies are identical.

The overall polytropic work is then calculated by integrating the small-scale isentropic enthalpy rises along the polytropic path. Given that the overall isentropic enthalpy rise is obtained by summation of the small-scale isentropic enthalpy rises along an isentropic path from p_1 to p_{2s}, it is always smaller than the polytropic head.

The polytropic efficiency is, therefore, always larger than the isentropic efficiency for a compressor.

The choice of the thermodynamic conditions, either total or static conditions, to define the end states of the polytropic process results in different definitions for polytropic efficiency. These are discussed in Sections 4.4.4–4.4.6.

4.4.2 Preheat Effect

Following (4.11), the static–static isentropic efficiency can be related to the static–static polytropic efficiency as

$$\eta_s^{ss} = \frac{h_{2s} - h_1}{h_2 - h_1} = \frac{(h_{2s} - h_1)}{\int_1^2 vdp} \frac{\int_1^2 vdp}{h_2 - h_1} = \frac{(h_{2s} - h_1)}{\int_1^2 vdp} \eta_p^{ss} = P_h \eta_p^{ss}, \tag{4.25}$$

where P_h is known as the preheat factor, which directly relates the polytropic efficiency to the isentropic efficiency of the stage. The physical background to the preheat effect is given in the discussion in Section 2.7.2. From the Gibbs equation, the preheat factor can be rearranged as

$$P_h = \frac{(h_2 - h_1) - (h_2 - h_{2s})}{\int_1^2 vdp} = \frac{\int_1^2 vdp - \int_1^2 Tds - \int_{2s}^2 Tds}{\int_1^2 vdp} = 1 - \frac{\int_{2s}^2 Tds - \int_1^2 Tds}{\int_1^2 vdp}. \tag{4.26}$$

Considering that the entropy changes from 1 to 2 and from 2*s* to 2 are the same, the integral of *Tds* from state 2*s* to state 2 is always larger than that from 1 to 2 as the integral is carried out at a higher average temperature. This indicates that the preheat factor in a compressor is always less than unity and that the isentropic efficiency is always smaller than the polytropic efficiency. The difference between the isentropic and polytropic efficiencies with increasing pressure ratio, as shown in Figure 4.7, is given by

$$\eta_s = \frac{\left(\pi^{(\gamma-1)/\gamma} - 1\right)}{\left(\pi^{(\gamma-1)/\eta_p\gamma} - 1\right)}. \tag{4.27}$$

4.4.3 Polytropic Efficiency in Terms of Entropy Changes

The Gibbs equation for a thermally perfect gas can be shown to be

$$ds = c_p\left(\frac{dT}{T}\right) - R\left(\frac{dp}{p}\right). \tag{4.28}$$

For a small-scale process at constant pressure, that is with $dp = 0$, the Gibbs equation leads to $(ds)_p = c_p(dT/T)$ and for a process at constant temperature, with $dT = 0$,

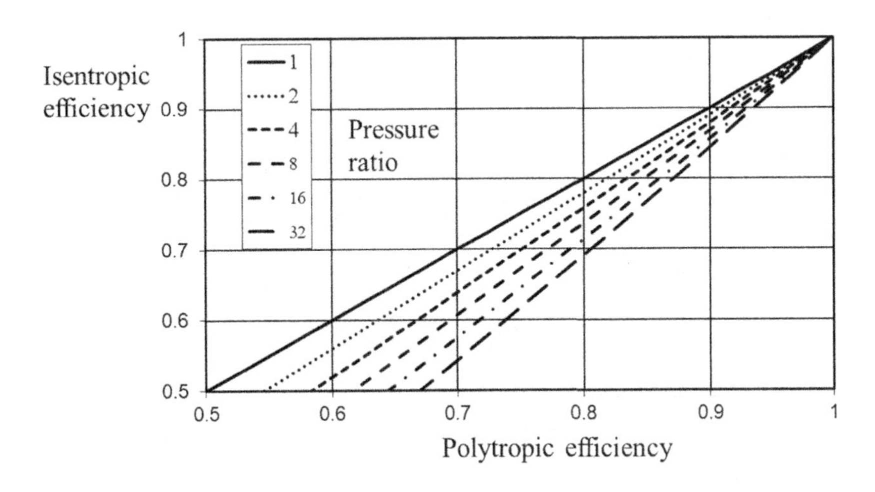

Figure 4.7 Isentropic and polytropic efficiency for compressors with $\gamma = 1.4$.

it leads to $(ds)_T = -R(dp/p)$. Hence, the small-scale polytropic efficiency in (4.22) can be derived to be

$$\eta_p = \left(\frac{R\dfrac{dp}{p}}{c_p \dfrac{dT}{T}} \right) = -\frac{(ds)_T}{(ds)_p}, \tag{4.29}$$

which, on integration, for a large-scale process leads to the entropy-based polytropic efficiency

$$\eta_p = -\frac{(\Delta s)_T}{(\Delta s)_p} = \frac{s_{1h} - s_2}{s_{1h} - s_1}. \tag{4.30}$$

The denominator is the change in entropy along a line of constant pressure (the inlet pressure) between the inlet and outlet temperature, and the abscissa is the change in entropy along a line of constant temperature between the inlet and outlet pressure. This formulation allows the polytropic efficiency to be represented in an *h-s* diagram as the ratio of two horizontal lengths (*c*/*d* in Figure 4.8) on the entropy scale (abscissa) just as the isentropic efficiency can be identified as the ratio of two vertical lengths (*a*/*b*) on the enthalpy scale (ordinate) (Barbarin and Mikirtichan, 1982; Casey, 2007).

In practice, the polytropic efficiency may, like the isentropic efficiency, take different forms depending on the choice of the thermodynamic conditions (total or static conditions) to define the end states of the process. These forms are discussed in the following.

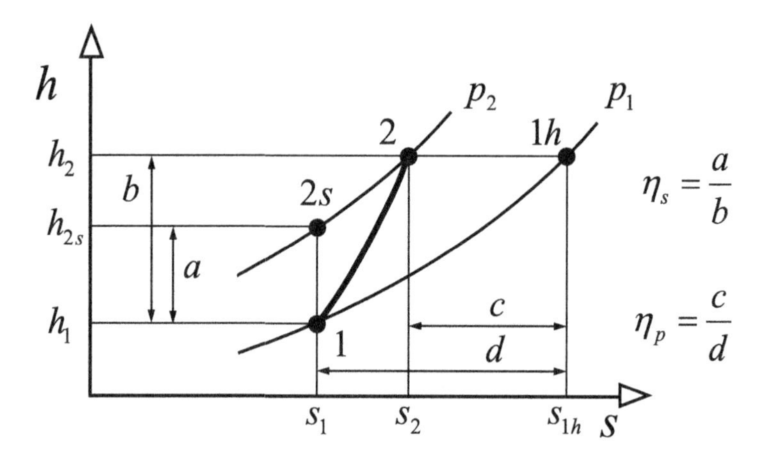

Figure 4.8 The isentropic and polytropic efficiencies in an h-s diagram.

4.4.4 Static–Static Polytropic Efficiency

If it is only the quality of the static enthalpy rise in the machine that is of interest, a so-called static-to-static polytropic efficiency is used. It is defined as the ratio of the polytropic enthalpy rise along a polytropic path between inlet static and outlet static conditions to the actual enthalpy rise along the same path; see Figure 4.9. The differential and integral forms of the static-to-static polytropic efficiency can be written as

$$\eta_p^{ss} = \frac{vdp}{dh} \tag{4.31}$$

$$\eta_p^{ss} = \frac{\int_1^2 vdp}{\int_1^2 dh} = \frac{\int_1^2 vdp}{h_2 - h_1} = \frac{y_{12}}{w_{s12} - \tfrac{1}{2}(c_2^2 - c_1^2)}. \tag{4.32}$$

where $\tfrac{1}{2}(c_2^2 - c_1^2)$ is the change in kinetic energy from state 1 to state 2. If the change in kinetic energy across the machine is negligible, the static–static polytropic efficiency is the true efficiency of the machine.

For ideal gases, where the polytropic process can be represented by the relation $pv^n = constant$, the aerodynamic work in (4.32) can be derived from integration of vdp,

$$y_{12} = \int_1^2 vdp = \frac{n}{(n-1)} p_1 v_1 \left[\left(\frac{p_2}{p_1} \right)^{\frac{n-1}{n}} - 1 \right] = \frac{n}{(n-1)} RT_1 \left[\left(\frac{p_2}{p_1} \right)^{\frac{n-1}{n}} - 1 \right]. \tag{4.33}$$

For a perfect gas, the static–static polytropic efficiency can be derived by direct integration of (4.31) as

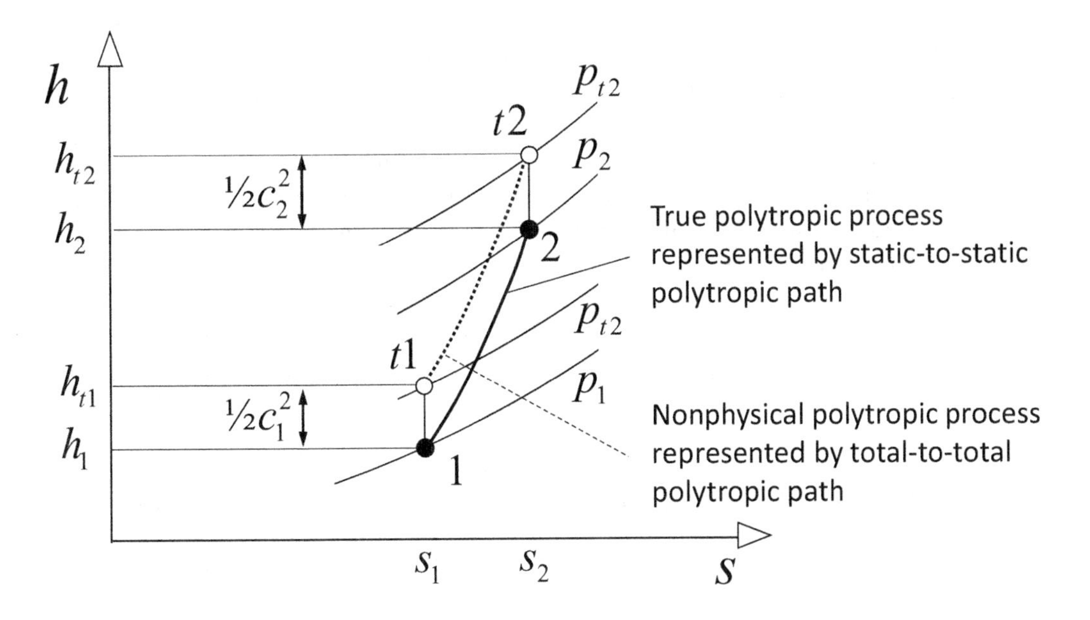

Figure 4.9 Polytropic efficiency defined on two process paths.

$$\eta_p^{ss} = \left(\frac{vdp}{dh}\right) = \left(\frac{RTdp}{c_p pdT}\right) \quad \Rightarrow \quad \left(\frac{dp}{p}\right) = \eta_p^{ss}\left(\frac{c_p}{R}\right)\left(\frac{dT}{T}\right) = \eta_p^{ss}\left(\frac{\gamma}{\gamma-1}\right)\left(\frac{dT}{T}\right)$$

$$\ln\left(\frac{p_2}{p_1}\right) = \eta_p^{ss}\left(\frac{\gamma}{\gamma-1}\right)\ln\left(\frac{T_2}{T_1}\right) \quad \Rightarrow \quad \eta_p^{ss} = \frac{\gamma-1}{\gamma}\frac{\ln(p_2/p_1)}{\ln(T_2/T_1)}. \tag{4.34}$$

These equations are generally used to determine the aerodynamic work and the efficiency from measurements of temperature and pressure.

4.4.5 Total–Total Polytropic Efficiency

As with the isentropic efficiency, a polytropic efficiency between the total conditions at the inlet and outlet can be defined. This definition assumes that the kinetic energy at the outlet is useful and can be included in the useful output of the compressor with the aerodynamic work that is achieved. From (4.3), the reversible work is the sum of the aerodynamic work and the change in kinetic energy, so the definition of the total–total polytropic efficiency is then given as

$$\eta_p^{tt} = \frac{y_{12} + ½(c_2^2 - c_1^2)}{w_{s12}} = \frac{\int_1^2 vdp + ½(c_2^2 - c_1^2)}{w_{s12}}. \tag{4.35}$$

This can be considered to be the true definition of the total–total polytropic efficiency as it integrates the aerodynamic work with the static states and considers the kinetic energy separately.

The total–total polytropic efficiency is, however, commonly defined in a less cumbersome but inexact way. In a similar approach as given in (4.14) for the isentropic efficiency, this definition posits a polytropic path traversing all the total conditions between the inlet total conditions, $t1$, and outlet total conditions, $t2$. The polytropic efficiency is then taken as the ratio of the aerodynamic work along this polytropic path of total conditions to the actual enthalpy rise; see Figure 4.9. This (incorrect) definition of the total–total polytropic efficiency can be written in differential and integral forms as

$$\eta_{pt}^{tt} = \frac{v_t dp_t}{dh_t} \tag{4.36}$$

$$\eta_{pt}^{tt} = \frac{\int_{t1}^{t2} v_t dp_t}{h_{t2} - h_{t1}} = \frac{y_{12}^{tt}}{w_{s12}}. \tag{4.37}$$

In line with the derivations of (2.80) and (4.34), the total–total aerodynamic work can then be written as

$$y_{12}^{tt} = \frac{n}{(n-1)} p_{t1} v_{t1} \left[\left(\frac{p_{t2}}{p_{t1}} \right)^{\frac{n-1}{n}} - 1 \right] = \frac{n}{(n-1)} R T_{t1} \left[\left(\frac{p_{t2}}{p_{t1}} \right)^{\frac{n-1}{n}} - 1 \right] \tag{4.38}$$

and efficiency as

$$\eta_p^{tt} = \frac{\gamma - 1}{\gamma} \frac{\ln (p_{t2}/p_{t1})}{\ln (T_{t2}/T_{t1})}. \tag{4.39}$$

The total–total polytropic efficiency defined in this way is, of course, not the true polytropic efficiency. The definition is based on the reversible work calculated along a path of total conditions from $t1$ to $t2$ rather than the path of static conditions from 1 to 2 with separate consideration of the kinetic energy, as shown in Figure 4.9. The difference between the two equations for the polytropic efficiency arises because of the following inequality:

$$\int_1^2 v dp + ½\left(c_2^2 - c_1^2\right) \neq \int_{1t}^{2t} v_t dp_t \tag{4.40}$$

such that

$$w_{shaft} = \int_1^2 v dp + ½\left(c_2^2 - c_1^2\right) + \int_1^2 T ds \neq \int_{1t}^{2t} v_t dp_t + \int_{1t}^{2t} T_t ds. \tag{4.41}$$

The difference between the two formulations for the reversible work is the difference in dissipation energy along the total and the static paths, which is larger for the total path as the total temperature is larger than the static temperature. This difference reduces to zero for an isentropic process with $ds = 0$. This implies that the total–total polytropic efficiency, as defined in (4.39), is always slightly less than the true polytropic efficiency of the compressor, (4.35). In practice, the difference is small, of the order of 1% or less depending on the level of kinetic energy change in the machine compared to the work input. This difference may be smaller than the difference due to the fact that the integration of vdp is generally carried out along a virtual polytropic path rather than the path of real states between the inlet and the outlet.

4.4.6 Total–Static Polytropic Efficiency

As with isentropic efficiency, a total–static polytropic efficiency can be defined to measure the efficiency of the compression from the inlet total condition to the outlet static condition to consider those cases where the outlet kinetic energy is not a useful output from the compressor. Different definitions for total–static polytropic efficiency have been used in the literature, a number of which are discussed here.

Considering that the exit kinetic energy is lost, the useful total–static polytropic work from $t1$ to 2 can be defined as the reversible work minus the kinetic energy at the outlet. Following the same procedure as given in (4.3) yields

$$\eta_p^{ts} = \frac{\left[\int_1^2 vdp + ½(c_2^2 - c_1^2)\right] - ½c_2^2}{w_{s12}} = \frac{\int_1^2 vdp - ½c_1^2}{h_{t2} - h_{t1}}. \tag{4.42}$$

This is analogous to the definition of the total–static isentropic efficiency, where the isentropic work required to compress the flow from state $t1$ to state 2 is compared with the actual shaft work consumed. If this were to be used for the definition of the total–static efficiency of an impeller it would lead to values near to 0.5 (50%), given that the kinetic energy at impeller outlet is about 40% of the work input.

It is therefore more useful to define the total–static polytropic efficiency based on the actual shaft work minus the kinetic energy at the outlet. This represents the fraction of the actual work expended to compress the flow from the inlet total to the outlet static conditions:

$$\eta_p^{ts} = \frac{\left[\int_1^2 vdp + ½(c_2^2 - c_1^2)\right] - ½c_2^2}{w_{s12} - ½c_2^2} = \frac{\int_1^2 vdp - ½c_2^2}{h_{t2} - h_{t1} - ½c_2^2}. \tag{4.43}$$

A slightly different approach may be adopted, as presented by Mehldahl (1941) and used by Dalbert et al. (1988), whereby a polytropic path is assumed between inlet total and outlet static conditions, and the total–static polytropic work is calculated by integration of vdp along this path from $t1$ to 2. This is usually an acceptable approximation as the inlet and outlet kinetic energy in the inlet and outlet planes are small. The total–static polytropic efficiency is then derived as

$$\eta_p^{ts} = \frac{\int_{t1}^2 vdp}{h_{t2} - h_{t1} - ½c_2^2}. \tag{4.44}$$

A popular definition for total–static polytropic efficiency is similar to that given in (4.39) except that the total pressure at the outlet is replaced with the static pressure:

$$\eta_p^{ts} = \frac{\gamma - 1}{\gamma} \frac{\ln(p_2/p_{t1})}{\ln(T_{t2}/T_{t1})}. \tag{4.45}$$

The popularity of this equation is due to the fact that it is directly related to the measurable thermodynamic properties at the inlet and the outlet and so obviates the need for any knowledge of the exit dynamic head. However, it cannot be derived from direct integration of vdp from static states along a polytropic path to give the aerodynamic work and, unlike (4.38), has a less physical basis for its definition. It is an acceptable approximation for situations where the exit kinetic energy is small.

4.4.7 Stage and Component Polytropic Efficiencies

In Section 4.4.4, the static–static polytropic efficiency of a compression process was defined as the ratio of the polytropic work along the polytropic path to the static enthalpy change between the initial and the final states. By dropping the suffix p (for polytropic) for convenience, the static–static efficiencies for the stage, impeller and the diffuser can be written as

$$\eta_{stage}^{ss} = \frac{\int_1^3 vdp}{h_3 - h_1} = \frac{\int_1^3 vdp}{\Delta h_t - ½(c_3{}^2 - c_1{}^2)} \tag{4.46}$$

$$\eta_{imp}^{ss} = \frac{\int_1^2 vdp}{h_2 - h_1} = \frac{\int_1^2 vdp}{\Delta h_t - ½(c_2{}^2 - c_1{}^2)} \tag{4.47}$$

$$\eta_{diff}^{ss} = \frac{\int_2^3 vdp}{h_3 - h_2} = \frac{\int_2^3 vdp}{½(c_2{}^2 - c_3{}^2)}, \tag{4.48}$$

where 1, 2 and 3 denote the stage inlet, impeller outlet and the stage outlet. In this extension of the earlier derivations, the polytropic efficiency for the diffuser is defined in terms of the aerodynamic work achieved as a fraction of the change in kinetic energy across the diffuser. If it is assumed that the polytropic head for the stage is the sum of the polytropic head in each of its components, which is only exactly the case if state 2 lies on the polytropic path from state 1 to 3, then

$$\eta_{stage}^{ss}\left(1 - \frac{½c_3^2}{\Delta h_t} + \frac{½c_1^2}{\Delta h_t}\right) = \eta_{imp}^{ss}\left(1 - \frac{½c_2^2}{\Delta h_t} + \frac{½c_1^2}{\Delta h_t}\right) + \eta_{diff}^{ss}\left(\frac{½c_2^2}{\Delta h_t} - \frac{½c_3^2}{\Delta h_t}\right) \tag{4.49}$$

or

$$(1 - \xi_3 + \xi_1)\eta_{stage}^{ss} = (1 - \xi_2 + \xi_1)\eta_{imp}^{ss} + (\xi_2 - \xi_3)\eta_{diff}^{ss}, \tag{4.50}$$

where ξ_1, ξ_2, and ξ_3, are the ratios of the local kinetic energy to the shaft work at planes 1, 2 and 3. It is possible to extend the analysis to include additional locations within the stage, for example a location between the diffuser and the downstream volute or return channel could be added. This would lead to the following extension to (4.50), where state 1 is the stage inlet, state 2 the impeller outlet, state 3 the diffuser outlet and state 4 the stage outlet:

$$(1 - \xi_4 + \xi_1)\eta_{14}^{ss} = (1 - \xi_2 + \xi_1)\eta_{12}^{ss} + (\xi_2 - \xi_3)\eta_{23}^{ss} + (\xi_3 - \xi_4)\eta_{34}^{ss}. \tag{4.51}$$

With knowledge of which contribution each component makes to the overall efficiency, it is relatively easy to determine whether it is worthwhile expending development effort on a particular part of the machine. In common practice, the total–total efficiency is used to measure the overall performance of the stage giving $\xi_1 = \xi_3 = 0$ in (4.50). Substituting the degree of reaction for ξ_2 from (4.19) into 4.50 and putting $\xi_1 = \xi_3 = 0$ for inlet and outlet total conditions, the total–total efficiency of the stage can be written in terms of the degree of reaction, impeller total–static and diffuser static–total efficiencies:

$$\eta_{stage}^{tt} = r_k \eta_{imp}^{ts} + (1 - r_k)\eta_{diff}^{st}. \tag{4.52}$$

An exact derivation leading to the same expression is provided by Mehldahl (1941).

This equation allows the performance of the components of the impeller and the diffuser to be investigated separately once the stage total–total polytropic efficiency and the static pressure at impeller outlet are known. A procedure for this is given in Section 20.4.4. The first derivation of these equations in this form goes back to the work of Mehldahl (1941) and subsequent relevant publications are the papers of Hausenblas (1965), Ambuehl and Bachmann (1980) and Dalbert et. al. (1988).

An example showing the stage efficiency and the separate impeller and diffuser efficiencies for a backswept industrial process compressor stage tested with a range of different vaned diffusers with different vane inlet angles is given in Figure 4.10, adapted from the data given by Dalbert et al. (1988). This diagram does not include a scale, as the original data was published in a normalised form, but absolute values are not necessary to show how the impeller efficiency and the diffuser efficiency combine with the degree of reaction to determine the stage efficiency. In this example, at a tip speed Mach number of 0.8 the impeller has a wide range of high efficiency from 80% to 120% of the design flow. Each diffuser tested has a much narrower operating range covering only part of the full operating range of the impeller. The degree of reaction remains nearly constant across the operating range and does not change when the diffuser changes.

In this case, the research stage was tested with many diffusers, but in many test campaigns, this procedure is not possible. The best stage efficiency and widest operating range is obtained at the tip speed Mach number of 0.8 with the diffuser setting angle of 64°. If the information of the impeller and the diffuser efficiency were not available, then the stage test data would not immediately identify why the efficiency drops at high flow on this characteristic. The separation of the stage efficiency into separate diffuser and impeller efficiencies identifies in this case that the impeller efficiency remains high over the operating range of this diffuser and that the shape of diffuser efficiency characteristic essentially determines the stage efficiency and hence the shape of the pressure rise characteristic. The shift in the stage

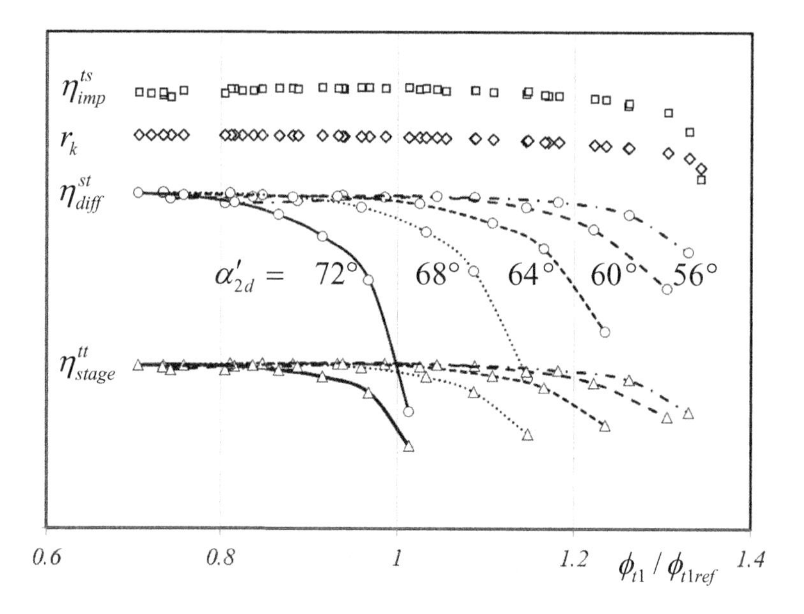

Figure 4.10 Impeller efficiency, degree of reaction and the diffuser efficiency combine to define the stage efficiency with different diffuser angle settings. (data provided by courtesy of MAN Energy Solutions)

characteristic with diffuser setting angle is determined by the throat area of the diffuser vanes; see Section 12.9.

If, as in Figure 4.10, the impeller efficiency remains fairly constant over a wide range, the diffuser efficiency must drop to very low values to lead to a stage efficiency of zero, as at zero stage efficiency the diffuser efficiency is given by

$$\eta^{st}_{diff} = -\frac{r_k}{(1 - r_k)} \eta^{ts}_{imp}. \tag{4.53}$$

By examining a virtual case where the diffuser has no loss, $\eta^{st}_{diff} = 1$, it is possible to show that the total–total efficiency of the impeller becomes

$$\eta^{tt}_{imp} = r_k \eta^{ts}_{imp} + (1 - r_k) = 1 - r_k \left(1 - \eta^{ts}_{imp}\right). \tag{4.54}$$

In this form, it is easy to see that for a given total–static efficiency of the impeller a decrease in the degree of reaction (by increasing the kinetic energy at the impeller outlet) leads to an increase in the total–total impeller efficiency. This feature makes the use of the total–total impeller efficiency rather awkward as a comparative measure of the quality of a stage, as the associated losses are only really specified if the degree of reaction is also specified.

A further advantage of the three efficiency definitions given in (4.46)–(4.48) is that at peak efficiency, as shown in Figure 2.14, the static state at impeller outlet lies close to the polytropic path of the stage compression process so that the numerical values of the three efficiencies are similar.

4.5 The Impeller Wheel Efficiency

The static enthalpy rise in the impeller includes a substantial portion which is often considered to be isentropic as it is due to the lossless centrifugal effect from the change in the blade speed from inlet to outlet. Vavra (1970a) suggested a definition of the impeller efficiency, called the wheel efficiency, to assess the efficiency of the diffusion process in the impeller without the centrifugal effect. Because the work done by the centrifugal effect was effectively loss-free, he argued that the wheel efficiency without the centrifugal effect was a better measure of the quality of the rotor performance than the impeller static–static efficiency. The wheel efficiency is not often used but has been examined by Moore et al. (1984) and by Larosiliere et al. (1997) in the interrogation of their CFD simulations.

Vavra defined the wheel efficiency in terms of an isentropic efficiency as

$$\eta_s^w = \frac{h_{2s} - h_u}{h_2 - h_u} = \frac{h_{2s} - h_1 - \tfrac{1}{2}(u_2^2 - u_1^2)}{h_2 - h_1 - \tfrac{1}{2}(u_2^2 - u_1^2)}, \tag{4.55}$$

where h_u is the enthalpy at state u as shown in Figure 2.18. Vavra evaluated the inlet blade speed as that at the casing, but Moore et al. (1984) and Larosiliere et al. (1997) used the mean wheel speed at inlet. In terms of a polytropic analysis, the Vavra wheel efficiency can be defined as

$$\eta_p^w = \frac{y_{12} - \tfrac{1}{2}(u_2^2 - u_1^2)}{h_2 - h_1 - \tfrac{1}{2}(u_2^2 - u_1^2)}. \tag{4.56}$$

From the Euler equation, the value of the divisor in these equations can be seen to be the change in the relative kinetic energy in the rotor, $\tfrac{1}{2}(w_1^2 - w_2^2)$, and in this way the wheel efficiency relates only to the effectiveness of the diffusion process in the relative flow of the impeller. Of course, the losses include not only those due to the diffusion but also all of the other sources, such as tip clearance, secondary flows and viscous losses. Vavra (1970a) comments that the relatively low value of the wheel efficiency of radial stages compared to the higher values possible in axial compressors (usually with no centrifugal effect) suggests that some improvement in aerodynamic design should be possible.

Rearrangement of (4.56) together with the definition of the static–static polytropic impeller efficiency and in combination with other dimensionless parameters leads to

$$\eta_p^w = \frac{\eta_p^{ss}\lambda r_k - \tfrac{1}{2}\left(1 - (r_1/r_2)^2\right)}{\tfrac{1}{2}(w_1/u_2)^2\left(1 - (w_2/w_1)^2\right)}, \tag{4.57}$$

where λ is the work coefficient and r_k is the degree of reaction. This shows that, for a given diffusion ratio in the impeller, the wheel efficiency falls with a higher inlet radius ratio as the centrifugal effect is reduced. Larosiliere et al. (1997) examined a stage with a small inlet casing radius of $r_1/r_2 = 0.486$ and with a design de Haller number of $w_2/w_1 = 0.714$. The peak total–total isentropic efficiency of the impeller was found to be nearly 94%, but the peak wheel efficiency without the centrifugal effect was only 20%. Moore et al. (1984) examined a stage with a higher inlet casing radius of $r_1/r_2 = 0.646$ and a design de Haller number of $w_2/w_1 = 0.72$. The peak total–total isentropic efficiency of the impeller was also found to be over 93% but the peak wheel efficiency without the centrifugal effect was 60%.

The large difference between the wheel efficiency of Larosiliere et al. and that of Moore et al., despite the total–total efficiency and the de Haller number being nearly the same, indicates that the wheel efficiency is strongly affected by the impeller radius ratio as given in (4.57). It turns out that the wheel efficiency is not a very useful parameter to identify the aerodynamic quality of the diffusion process in the impeller, as it is not really true that the enthalpy rise due to the centrifugal effect is loss-free. The tip clearance losses and the friction losses within the impeller can just as well be attributed to the centrifugal effect as to the change in static enthalpy due to the relative flow deceleration.

4.6 External Losses and Sideloads

In compressors with sideloads, such as the economiser flows in a multistage refrigeration compressor, each compressor section has a different mass flow. The overall polytropic efficiency can then be calculated with the mass average of the changes in the enthalpy and aerodynamic work through the machine

$$\eta_p = \frac{\sum \dot{m} y_{12}}{\sum \dot{m} \Delta h_{t12}}, \tag{4.58}$$

where the different mass flows in the different compressor sections are taken into account. This approach was suggested by Traupel (2000) for the case of feed heating in steam turbines.

In addition to the fluid dynamic sources of loss in the flow path outlined in the previous efficiency definitions, parasitic losses and mechanical losses are present, such as the bearing losses, windage in the bearings, gearbox losses and the disc friction losses of the impeller backplate and the shroud if present. If the mechanical power absorbed by the bearings and gears is P_m and the power required internally by the compressor is P, then the mechanical efficiency is defined as

$$\eta_m = \frac{P}{P + P_m}. \tag{4.59}$$

Consideration of the parasitic losses is given in Section 10.6.

4.7 Efficiency in Diabatic Processes

4.7.1 Diabatic Compression Processes

The efficiency of a diabatic compression process, that is, one with heat transfer, is also defined as the ratio of the reversible work to the actual work expended to achieve the pressure rise. For a small compression step with heat transfer, the Gibbs equation, the first law and the second law of thermodynamics can be written respectively as

$$
\begin{aligned}
dh &= vdp + Tds \\
\delta w_{act} + \delta q_{act} &= dh + dke \\
Tds &= \delta q_{act} + Tds_{irrev}.
\end{aligned}
\tag{4.60}
$$

where δw_{act} and δq_{act} are the actual work and heat transfer per unit mass, *dke* is the change in kinetic energy and Tds_{irrev} is the entropy production caused by the irreversibilities such as friction and heat transfer though a finite temperature difference. The preceding equations can then be combined to give

$$\delta w_{act} = vdp + dke + Tds_{irrev} = \delta w_{rev} + Tds_{irrev}. \tag{4.61}$$

This demonstrates that, just as with adiabatic processes, the reversible work in diabatic processes is calculated by the integration of *vdp* and *dke* along the reference reversible compression path. Heat transfer to the compressor does not appear explicitly in this equation. The amount of heat transfer does, however, influence the integration path from state 1 to state 2 and so impacts on both the reversible work and the dissipation. The overall actual work expended is still the shaft work and therefore the compressor efficiency defined in (4.3) is valid for both diabatic and adiabatic processes. For a diabatic process, the overall enthalpy change is a function of both the shaft work and the overall heat transfer. For the case with heat transfer, the general form of the compressor efficiency can therefore be written as

$$\eta_c = \frac{\int_1^2 vdp + \frac{1}{2}(c_2^2 - c_1^2)}{w_{s12}} = \frac{\int_1^2 vdp + \frac{1}{2}(c_2^2 - c_1^2)}{(h_{t1} - h_{t2}) - q_{12}}. \tag{4.62}$$

As in the case of adiabatic flows, by choosing different ideal reference processes for the calculation of the reversible work, different efficiency definitions may be derived for diabatic flows. However, a reversible diabatic flow (with no dissipation losses) is not isentropic. Thus, there is no justification to use an isentropic process as a reference for the ideal work required by a perfect diabatic compressor (Casey and Fesich, 2010). The isentropic work between the initial and final pressures is essentially independent of any heat transfer to the system. The application of polytropic and isothermal efficiencies as a measure of the quality of compression in diabatic processes is discussed in the following sections

4.7.2 The Effect of Heat Transfer in Turbocharger Compressors

Sirakov and Casey (2013) point out that in the testing of turbocharger compressors in hot gas stands in close proximity to the turbine, there is a large heat exchange from the turbine to the compressor. The efficiency of the compressor is usually determined from the pressure ratio and temperature ratio using (4.16), but the measured temperature rise includes the heat transferred from the turbine as well as the work done on the flow. The neglect of this heat transfer causes the apparent work input to be higher than in reality. Conventional performance maps then underestimate the efficiency of the compressor stage by as much as 20 points at low speeds. A correction procedure for this effect based on (4.62) can be used to convert performance maps obtained with heat transfer to performance maps for adiabatic conditions. The effect of the heat transfer on the apparent work input, and hence on the efficiency, is given by a nondimensional parameter. This is defined as the difference between the apparent work input and the actual work input expressed nondimensionally with the square of the blade tip-speed

$$\lambda_{act} - \lambda = \frac{q}{u_2^2} = \frac{\dot{Q}}{\rho_{t1}\phi_{t1}u_2^3D_2^2} = \frac{\dot{Q}}{\rho_{t1}a_{t1}^3D_2^2}\frac{1}{\phi_{t1}M_{u2}^3} = k_c\frac{1}{\phi_{t1}M_{u2}^3}, \tag{4.63}$$

where k_c is a dimensionless coefficient that depends on the rate of heat transfer per unit area into the compressor. This coefficient can be determined by experiment and, depending on the design of the turbocharger housing, typical values are found to vary widely between $0.002 < k_c < 0.004$. When tested in hot gas stands driven by the hot turbocharger turbine, the apparent efficiencies on the low-speed characteristics are affected strongly by the heat transfer in different turbocharger configurations. This makes comparison of the aerodynamic quality of compressors measured in such test stands very difficult. Equation (4.63) allows this effect to be estimated and the true efficiency of the performance map to be calculated.

4.7.3 Diabatic Polytropic Efficiency

If the compression path between the initial and final states in a diabatic process is approximated by a polytropic path, and the kinetic energy terms are negligible, then the static–static polytropic efficiency is simply defined as the ratio of the reversible work to the actual shaft work:

$$\eta_p^{ss} = \frac{\int_1^2 vdp}{w_{s12}} = \frac{\int_1^2 vdp}{(h_{t1} - h_{t2}) - q_{12}}. \tag{4.64}$$

The aerodynamic work is still determined using (4.38), derived for adiabatic processes. The differential form of the polytropic efficiency can be derived by considering that for a polytropic process, the polytropic ratio, which is defined as the ratio of the actual enthalpy rise to the polytropic enthalpy rise, remains constant. The polytropic ratio has the properties of the polytropic efficiency in an adiabatic process and can be written as

$$\upsilon = \frac{dh_t}{v_t dp_t} = \frac{(n-1)/n}{(\gamma-1)/\gamma}, \tag{4.65}$$

where n and γ are the polytropic and isentropic exponents. The reversible work and the actual work in the process are given by $\delta w_{rev} = v_t dp_t$ and $\delta w_{act} = dh_t - \delta q$ such that the polytropic efficiency of a diabatic process can be written as

$$\eta_p^{tt} = \frac{v_t dp_t}{dh_t - \delta q} = \frac{1}{(dh_t/v_t dp_t) - (\delta q/v_t dp_t)} = \frac{1}{\upsilon - \zeta_q}, \tag{4.66}$$

where the heat transfer ratio, ζ_q, is defined as the ratio of the heat transfer to polytropic work and is assumed to stay constant along the compression path.

4.7.4 Isothermal Efficiency

In intercooled multistage compressors, where heat is removed between intermediate compression stages (or during the compression, using external cooling jackets), the required aerodynamic work is lowered by reducing the temperature at the inlet of the individual stages. The minimum theoretical compression work is achieved by a reversible isothermal process which can be viewed as a compression from p_1 to p_{2i} using an infinite number of intercoolers which remove all the energy from the work input by heat transfer (see Figure 4.1). The end state in an isothermal process can then be denoted as $2i$ in the h-s diagram, which has the temperature of the initial state and the pressure of the final state of the actual process. In real gases, the enthalpy varies with both pressure and temperature and therefore does not remain constant in an isothermal process. The static–static isothermal efficiency is defined as the ratio of the reversible work, integrated along the isothermal compression path from state 1 to state $2i$, to the compressor shaft work and can be written as

$$\eta_i^{ss} = \frac{\int_1^{2i} v dp}{w_{s12}} = \frac{\int_1^{2i} (dh - Tds)}{(h_{t2} - h_{t1}) - q_{12}} = \frac{(h_{2i} - h_1) - T_1(s_{2i} - s_1)}{(h_{t2} - h_{t1}) - q_{12}}. \quad (4.67)$$

For ideal gases, the enthalpy is a function of temperature only and therefore remains constant along the isothermal path. Furthermore, the entropy at state $2i$ can be determined from the Gibbs equation, leading to the conventional equation for the isothermal efficiency for an ideal gas:

$$\eta_t^{tt} = \frac{R \ln (p_2/p_1)}{w_{s12}}. \quad (4.68)$$

Intercooling of compressors provides a reduction in the compression shaft work. The efficiency given in (4.68) is, however, purely theoretical since it requires an infinite number of intercoolers and takes no account of the pressure losses in the heat exchangers. As a result of including the pressure losses in the heat exchangers, a finite number of intercoolers is determined as the theoretical optimum (Simon, 1987); Vadasz and Weiner, 1992). In a two-stage compression process with intercooling with no pressure losses, the optimum intermediate pressure is given by $p = \sqrt{(p_1 \cdot p_2)}$ (Haywood, 1980). The optimum location of the intercoolers in terms of the pressure ratio for the individual stages with multiple intercoolers is given by Vadasz and Weiner (1992). In reality, the costs of the heat exchangers play a role as the effectiveness is determined by the surface area, so economic considerations also need to be taken into account, which leads to a smaller number of intercoolers than the thermodynamic optimum.

4.8 Efficiency Definitions for Real Gases

The basic equations for the efficiency definitions given in this chapter are valid for both real gases and ideal gases. For the isentropic and the isothermal efficiency, the intermediate states are fully defined, so an integration of vdp with real gas equations to obtain the aerodynamic work is possible. With real gases, a difficulty occurs with the polytropic efficiency as the integration of vdp to calculate the aerodynamic work requires knowledge of the intermediate states of the process, and these have to be estimated. There is a large literature on the difficulty of calculating the polytropic head by integrating vdp for real gases, and on the inaccuracies that this causes. This is summarised in Section 3.6. Provided that the aerodynamic work is calculated with the equations for real gases given there, the efficiency definitions in this chapter are valid.

The entropy definition of polytropic efficiency, (4.30), is sometimes considered to be accurate for real gases; see, for example, Casey (2007). However, this is not correct for real gases as its derivation makes use of ideal gas relationships. Its accuracy for real gases can be increased by splitting the overall pressure rise into several steps with one or more virtual intermediate pressures (Dubbel, 2001). This is shown in Figure 4.11 with a single intermediate pressure. Using a large number of intermediate pressures, the following equation can be used for the entropy-based polytropic efficiency for real gases:

$$\eta_p = \frac{\sum(\Delta s)_h}{(\Delta s)_p} = \frac{\sum(\Delta s)_h}{\Delta s + \sum(\Delta s)_h}. \tag{4.69}$$

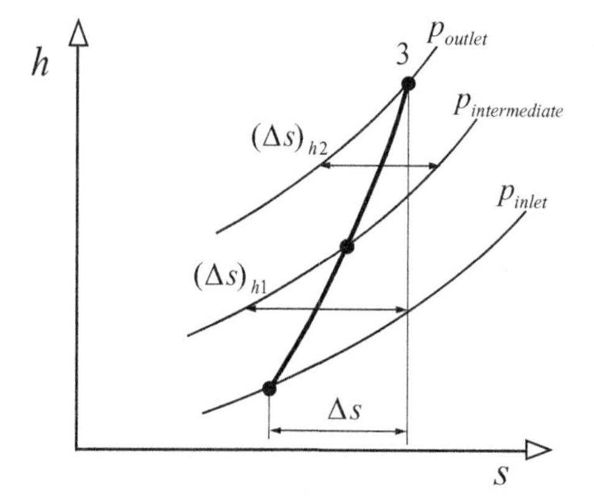

Figure 4.11 The entropy changes related to the polytropic efficiency with an intermediate pressure.

5 Fluid Mechanics

5.1 Overview

5.1.1 Introduction

This chapter summarises the basic laws of fluid mechanics relevant for turbomachinery fluid flows. Some developments of these laws are given that are of special importance for explaining the motion of the fluids and the structure of the flow patterns in radial turbocompressors. Other basic aspects of the fluid dynamics are covered in Chapter 6 on compressible flows of gases; Chapter 7 on the aerodynamic loading limits of aerofoils, cascades and diffusers; Chapter 8 on nondimensional similarity parameters; Chapter 9 on the specific speed; and Chapter 10 on performance and losses. Because the design of compressors is based on many different aspects of fluid mechanics, it is important for these topics to be covered in some detail so that designers feel very comfortable with the equations and physical descriptions of the flow patterns in these chapters before embarking on any particular design.

This chapter begins with a section highlighting the important fluid dynamic principles that are fundamental to understanding the motion of fluids. These include the continuity and the momentum equations in various forms. These equations are then used to delineate the effect of the fluid motion on pressure gradients on the flow. The simple radial equilibrium equation for a circumferentially averaged flow is introduced in Section 5.3.4, as a first step towards a more complete analysis of the meridional average flow in Section 15.4. Special features of the flow in radial compressors due to the radial motion are considered, such as the effects of the Coriolis and centrifugal forces and the relative eddy which gives rise to the slip factor of a radial impeller. A short overview of boundary layer flows of relevance to radial compressors is provided. The flow in radial compressor impellers is strongly affected by secondary flows and tip clearance flows and an outline is provided of the current understanding of the physics related to these. The phenomenon of jet-wake flow in compressors is described.

5.1.2 Learning Objectives

- Know the different forms of the continuity equation with respect to the meridional velocity and with the absolute velocity and how these relate to area changes in turbocompressors.

- Know the significance of blockage in increasing the meridional velocity in compressor flow channels.
- Know the momentum equation in its various forms and how it leads to pressure gradients in turbomachinery flows.
- Understand the background to the Coriolis and centrifugal forces in impellers.
- Know the effect of the relative eddy on the generation of the slip velocity.
- Have an understanding of the types of boundary layers that can be encountered in compressors and the importance of the Reynolds number.
- Have an understanding of the nature of tip clearance flows, secondary flows and flow separations in impellers and their role in the jet-wake flow pattern.

5.2 The Laws of Fluid Mechanics

5.2.1 Internal Flows in Compressors

Turbocompressor internal flows are among the most complex to be encountered in fluid dynamic practice. The geometry of the blade passages is highly complex, with stationary vanes, rotating blades, curved passages and obstructions in the flow channels. Further geometrical complexity is caused by shrouds and sealing devices, tip clearance gaps and nonaxisymmetric inlet and outlet casings. These flows include a wide range of different flow physics.

The flows may be incompressible, subsonic, transonic or supersonic, and in the high-speed flows strong shocks with boundary layer interaction may occur. The geometry and the flows are generally three dimensional. Strong vortices often dominate the flow structure due to secondary flow vortices in the blade channels and tip clearance vortices in the flow leaking over the blade tips. The free stream turbulence is high, so that the boundary layers are usually transitional or turbulent. The turbulent regimes are complex due to the appreciable pressure gradients, rotation, curvature, high turbulence levels, separation and interaction of wakes and boundary layers. The flows are unsteady in both the rotating and absolute frames of reference. The flow is decelerating while it moves into a region of higher pressure with a high risk of reverse flow and flow separation.

This book touches on these issues in this chapter and the next chapters. More mathematical and very good physical and mathematical descriptions of many of these issues can be found in Greitzer et al. (2004).

5.2.2 Fundamentals of Fluid Motion

As mentioned in Section 2.3.6, there are two different mathematical approaches to analyse the flow in fluid dynamics. The first is known as the Lagrangian approach in which the motion of a small particle of a fluid through space and time is followed as it crosses the region of interest. The mathematics of such an analysis are seldom used as they are far too laborious for simple understanding of a turbomachine, as the particle undergoes an unsteady process and deforms its shape and volume during its passage through the machine. The analysis most commonly used is called the Eulerian approach, in which the state of the fluid in a fixed region called a control volume is analysed. The fate of an individual fluid particle is less interesting than the mean behaviour of the changing particles that pass the boundaries of the fixed region. A control volume with fixed boundaries is defined and analysed using the conservation equations across the control volume. For this control volume analysis, the laws for the conservation of mass and momentum are applied at the fixed boundaries of the control volume. The conservation of energy and conservation of the moment of momentum are also applied in this way with the derivation of the steady flow energy equation and the Euler turbine equation.

When applied to the flow in compressors, the basic laws of fluid mechanics for an inviscid fluid in a suitably defined control volume provide information on the physics of flow acceleration, flow deceleration and flow turning within the flow channels of the compressor. Flow deceleration in the compressor blades rows has a large effect on the performance, as in a viscous flow this may lead to flow separation as discussed in Section 5.5. Flow turning is related to the shape of the blades and the flow passage causing pressure gradients in the flow, and these determine the change in direction of the flow between the inlet and outlet of a blade row. This then determines the change in swirl velocity of the blade row and, in an impeller, the amount of work via the Euler turbine equation.

In modern analysis methods of turbomachinery, numerical calculations of turbulent flows with computational fluid dynamics (CFD) solve the conservation equations for a viscous fluid on a fine grid of small volume elements representing the flow domain of the compressor. The results can be used to assess the detailed flow patterns, the velocity and the pressure fields, the forces and the losses in the unsteady viscous and turbulent flow within the complex three-dimensional geometry of the flow passages of a radial compressor. This will be discussed in detail in Chapter 16, but the basic equations for such analyses, the Navier–Stokes equations, are introduced here.

5.2.3 Continuity Equation

The continuity equation is the equation of the conservation of mass applied across a control volume. It is used here to indicate that in a steady flow across a control volume, mass is conserved such that with one inlet and outlet the following applies:

$$A_1\rho_1\left(\vec{c}_1\cdot\vec{n}_1\right)+A_2\rho_2\left(\vec{c}_2\cdot\vec{n}_2\right)=0, \tag{5.1}$$

where 1 is the inlet, 2 is the outlet, c is the component of velocity normal to the surface area A and n is the normal vector to the surface considered. In turbomachinery design and analysis, this equation is applied in two ways. Firstly, it can be applied to the meridional velocity associated with the meridional streamlines, which are approximately normal to the annulus area. Secondly, it can be applied it to the velocity normal to the area representing the gap between blades.

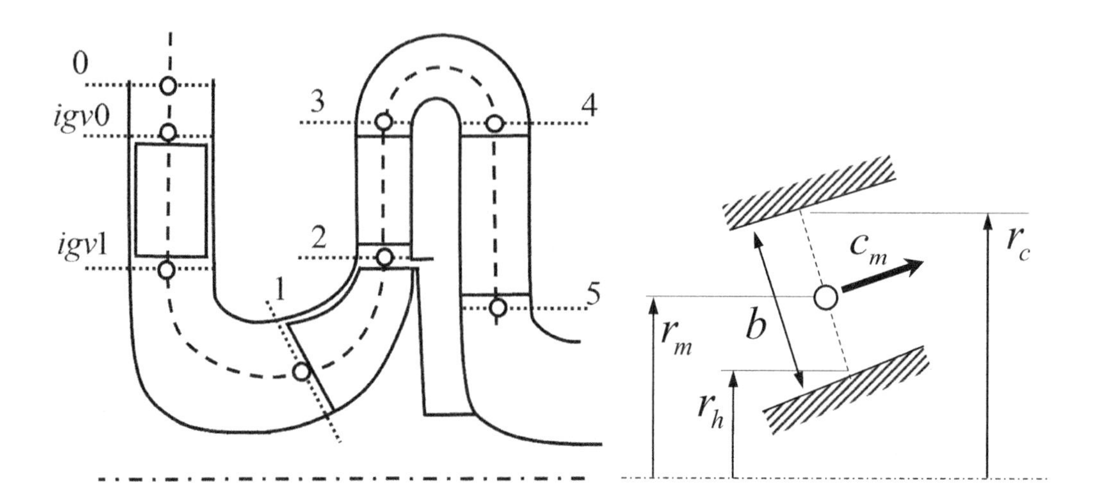

Figure 5.1 Continuity equation applied to the annular area.

Considering the meridional velocity first, as shown in Figure 5.1 and assuming that the flow at all meridional positions through the machine has the same mass-flow rate as at the inlet, then conservation of mass gives

$$\dot{m} = \rho_1 k_1 A_{a1} c_{m1} = \rho_2 k_2 A_{a2} c_{m2} = \rho_3 k_3 A_{a3} c_{m3}, \tag{5.2}$$

where 1, 2, 3 and so on are different locations in the flow, ρ is the local density, A_a is the annular cross-sectional area at the location concerned, c_m is the meridional velocity and k is an empirical local blockage factor and is often referred to as the blockage, B, where $B = 1 - k$. With no blockage, $B = 0$ and the blockage factor is $k = 1$. The geometrical annulus area in this equation is considered the area normal to the mean flow direction and is calculated as

$$A_a = 2\pi b r_m, \quad r_m = (r_c + r_h)/2, \tag{5.3}$$

where b is the channel width and r_m is the mean radius.

Viscous effects give rise to low velocities near to the walls leading to blockage, an effective reduction in the flow area as a consequence of the boundary layer thickness. This affects the mean flow velocities in the flow channel, which may cause a change in the work input and can have large effects on the mass-flow capacity of the compressor. The blockage factor accounts for the fact that the effective area available for the flow is less than the geometrical cross-sectional area due to the blockage of the boundary layers as discussed in Section 5.5 and any blockage due to the presence of tip clearance vortices in the flow, Section 5.7.

The blockage factor typically has a value between 1 and 0.8 and its effect on the calculation is shown in Figure 5.2. The diagram on the left shows the velocity profile in a straight flow channel without consideration of boundary layers. The central diagram represents the real velocity profile, which due to the boundary layer on the end-walls then has a higher velocity in the middle of the channel. The right-hand

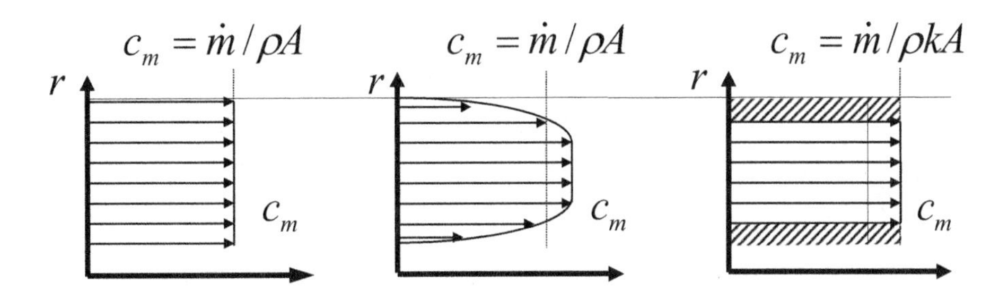

Figure 5.2 Continuity equation for the meridional flow in an axial annulus with blockage.

picture shows that including an end-wall blockage factor with the geometrical area in the continuity equation increases the meridional velocity at the centre of the channel to that of the real flow. Blockage may also arise from wakes downstream of the blade rows or from reduced flow velocities in the tip leakage vortices.

If the end-wall blockage is neglected and if the designer chooses to keep the meridional velocity of the flow constant through the machine, then (5.2) shows that as the density increases, the annulus flow area reduces. The reduction of the annulus flow area as the density increases can be seen in all of the multistage compressor meridional views shown in Chapter 1 as the outlet width of the impellers decrease steadily through the machine. There is, however, no fundamental reason why the meridional velocity should stay constant through an impeller, and in Section 11.6.4 it is shown that typically the meridional velocity decreases by about 50% across an impeller and is lower in impellers designed for low flow coefficients.

The continuity equation can also be applied to the flow area normal to the streamlines between the blades and, in this case, gives

$$\dot{m} = A_1\rho_1 c_1 = A_2\rho_2 c_2, \quad A_2 > A_1, \quad \rho_2 c_2 < \rho_1 c_1, \tag{5.4}$$

where 1 and 2 are locations in the flow; ρ is the local density; and A is the normal flow area between the blades perpendicular to the flow velocity, c, as shown in Figure 5.3. If the area between the blades increases, then the product of density and velocity decreases. In compressor blades, a deceleration in the flow is achieved by increasing the available flow area between the blades in this way, highlighted by the increasing diameter of the inscribed circles in Figure 5.3. The deceleration is then accompanied by an increase in static pressure (Bernoulli equation), provided that the physical limits on the deceleration causing flow separation are not exceeded, as given in Section 7.5. In turbines, the area between the blades decreases in the flow direction to provide an acceleration of the flow.

5.2.4 Momentum Equation

The momentum equation of fluid dynamics is the statement of Newton's three laws of motion for a control volume. Newton's first law states that every object in a state of uniform

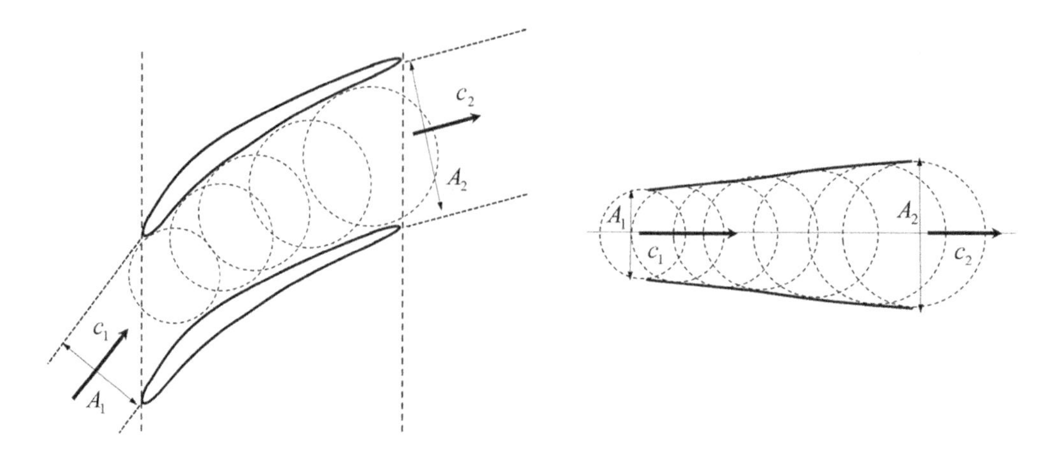

Figure 5.3 Continuity equation applied to the area between two adjacent compressor blades and the equivalent area change in a 1D analogy using a straight diffusing channel.

motion tends to remain in that state of motion unless an external force is applied to it. This is sometimes known as the law of inertia. In terms of a control volume, it would state that if there is no change in the momentum of fluid between the inlet and the outlet of a control volume, then no forces are acting. Flow continues to move in a straight path at constant velocity unless forces are acting to make it change direction or to accelerate or decelerate.

Newton's second law provides a relationship between an object's mass, m; its acceleration, a; and the applied force F such that $F = ma$. The force applied leads to a change in momentum. Acceleration and force are vectors, so that this law states that the direction of the force vector is also the same as the direction of the change in momentum. When applied to a turbomachinery control volume, this equation reduces to the statement that the forces acting on the control volume cause a change in the momentum of the fluid between inlet and outlet. Given that this is a vector equation, the force components in three directions can be identified from the changes in the respective components of the momentum (axial, circumferential, radial).

Newton's third law is that for every action there is an equal and opposite reaction. The force imposed by the blade on the flow is accompanied by an equal and opposite reaction which defines the force imposed by the flow on the blade.

In a fluid, the application of a pressure force changes the velocity of the fluid, and the rate of change in velocity is proportional to the force applied. The conservation of momentum can be applied to a control volume, and according to Newton's second law of motion the sum of all the forces acting on a body equals the change in momentum:

$$\sum \vec{F}_i = \sum_i \left[\dot{m}\left(\vec{c} \cdot \vec{n}\right)\right]_i = \sum \left[\rho\, \vec{c}\, \left(\vec{c} \cdot \vec{n}\right) A\right]_i . \tag{5.5}$$

From the perspective of the fluid, the forces acting are of two types. They are either surface forces ($F = pA$) acting on the boundaries of the system, such as shear and pressure forces, or they are body forces which act throughout the control volume, such as gravity, Coriolis and centrifugal forces.

If we apply (5.6) in the meridional direction, we obtain

$$F_{s,m} + p_1 A_{a1} - p_2 A_{a2} = -(\rho_1 A_{a1} c_{m1}) c_{m1} + (\rho_2 A_{a2} c_{m2}) c_{m2}$$
$$F_{s,m} + p_1 A_{a1} - p_2 A_{a2} = \dot{m}(c_{m2} - c_{m1}) \tag{5.6}$$

along the direction of the applied force. In the meridional direction, the change in meridional velocity is related to the resultant force acting on the control volume, both through pressure on the boundaries and body forces. In the axial direction, the axial thrust can be seen to be related to the inlet and outlet pressure across the control volume and the change in axial velocity.

If the momentum equation is applied to forces acting in the circumferential direction, then

$$F_{s,u} = -(\rho_1 A_{a1} c_{m1}) c_{u1} + (\rho_2 A_{a2} c_{m2}) c_{u2}$$
$$F_{s,u} = \dot{m}(c_{u2} - c_{u1}). \tag{5.7}$$

Once again, the fluid experiences these forces as pressure forces, but this time acting normal to the meridional flow direction. If there is no pressure force acting, then there is no change in the circumferential component of velocity, a fact already known from the Euler turbine equation.

5.2.5 Moment of Momentum Equation

Newton's laws can also be applied to the angular momentum across a control volume, in which case the increment in angular momentum across the control volume is equal to the moment of the forces about the axis. This leads to the Euler turbine equation, as described in Section 2.2.2.

5.2.6 Navier–Stokes and Euler Equations

The complete governing equations in fluid dynamics are described by the so-called Navier–Stokes equations. These are partial differential equations including the conservation of mass and momentum, plus the conservation of energy. They are valid for the three-dimensional unsteady motion of a compressible viscous fluid and include terms for the friction of a fluid and heat transfer processes (Greitzer et al., 2004). These equations are defined and described in Chapter 16, and the so-called Euler equations are obtained if the viscosity and heat transfer in the Navier–Stokes equations are neglected.

5.2.7 Average Values of Complex Flows

An engineer finds it useful to characterise a complex turbomachinery flow field with a few parameters such as mean velocities, pressures and temperatures, pressure ratio or efficiency. The changes in these average parameters can be used to provide useful insights into the flow behaviour, to define performance and to make engineering decisions. There are various averaging schemes which may be used to compute these parameters for a three-dimensional flow pattern with a complex distribution of velocity and total pressure. No single 1D description is adequate which simultaneously matches all the significant parameters of a nonuniform flow (Pianko and Wazelt, 1983). Some of the different types of averaging that can be considered are area averages, mass averages or momentum averages. In turbomachinery applications with only small nonuniformities, the difference between these are rather small, and it is usual to carry out area-averaging of pressures, and mass-averaging of velocity components and enthalpy, following Cumpsty and Horlock (2006). In more complex flows with large separations, other procedures may be preferable (VDI 4675, 2012).

5.3 Pressure Gradients in Fluid Flows

5.3.1 Euler Equation of Motion along a Streamline

The Euler equation for steady flow along a streamline is a relation between the velocity, pressure and density of an inviscid moving fluid. It is based on Newton's second law of motion. If the momentum equation is applied along a stream tube, it leads to

$$
\begin{gathered}
pdA - \left(p + \frac{\partial p}{\partial s}ds\right)dA = \rho c dA\left(c + \frac{\partial c}{\partial s}ds\right) - \rho c dA c \\
-\frac{\partial p}{\partial s} = \rho c \frac{\partial c}{\partial s} \quad \Rightarrow \quad \frac{\partial p}{\partial s} + \rho c \frac{\partial c}{\partial s} = 0.
\end{gathered}
\tag{5.8}
$$

This useful form of the momentum equation indicates that if there is a pressure increase along the direction of the stream tube, then the flow must inevitably decelerate, and that the changes in velocity are related to the pressure gradient. This means that in a blade row with an increasing pressure in the streamwise direction, we are always concerned with flow deceleration. For this reason, an important parameter related to the pressure rise in an impeller is the amount of deceleration of the flow, which in impellers is often expressed as the ratio of the outlet relative velocity to the inlet relative velocity, w_2/w_1, and is known as the de Haller number, after de Haller (1953), as discussed in Section 7.5.4.

If this momentum equation is integrated along the stream-tube for incompressible flow with constant density, it leads to the Bernoulli equation,

$$
p + ½c^2 = \text{constant} = p_t, \tag{5.9}
$$

which is a statement that in an inviscid flow the total pressure is conserved along streamlines.

5.3.2 Euler Equation of Motion Normal to a Streamline

The momentum equation can also be applied normal to a stream-tube, as shown in Figure 5.4. The centrifugal force acting on a fluid particle moving with a

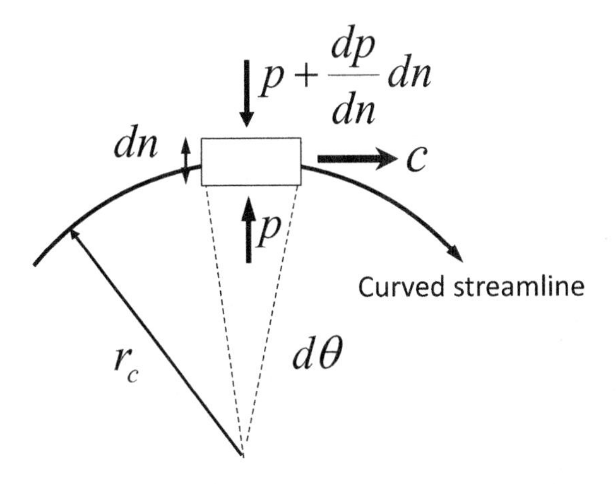

Figure 5.4 Momentum equation applied normal to an element of a stream-tube.

velocity c along a curved streamline with a radius of curvature r_c is given by $\rho c^2/r_c$. This is balanced by the pressure force acting across the particle so that

$$d\theta dn\rho r_c \frac{c^2}{r_c} = \left[p + \frac{\partial p}{\partial n}dn\right](r_c + dn)d\theta - pr_c d\theta - 2\left(pdn\frac{d\theta}{2}\right) = \frac{\partial p}{\partial n}r_c d\theta dn$$
$$\Rightarrow \quad \frac{dp}{dn} = \rho\frac{c^2}{r_c}. \tag{5.10}$$

This equation shows that if there is a pressure increase normal to a stream–tube, then the flow must travel in a curve with a radius of curvature, r_c, to balance this pressure gradient. The turning of the flow in compressor blades and in the meridional channel is always linked to a pressure gradient normal to the flow direction. The pressure gradient forces the flow to change direction towards the local centre of curvature of the streamline. If the streamlines are not curved, then there is no pressure force acting normal to the streamline.

Equation (5.10) shows that in relation to the flow in any curved part of a turbocompressor flow, the pressure further from the centre of curvature of the curved region is larger than that closer to the centre of curvature. Further, if the flow has a constant total pressure, then this equation together with the Bernoulli equation, (5.9), indicates that there is generally a decrease in velocity to be expected with an increase in the distance from the centre of curvature. This is fundamental to how a wing of an aeroplane produces lift and to how a turbomachinery blade generates forces to cause a flow deflection.

This equation also applies to curvature in the meridional plane, where the equation for the pressure gradient normal to a meridional stream-tube is then given by

$$\frac{dp}{dn} = \rho\frac{c_m^2}{r_c}, \tag{5.11}$$

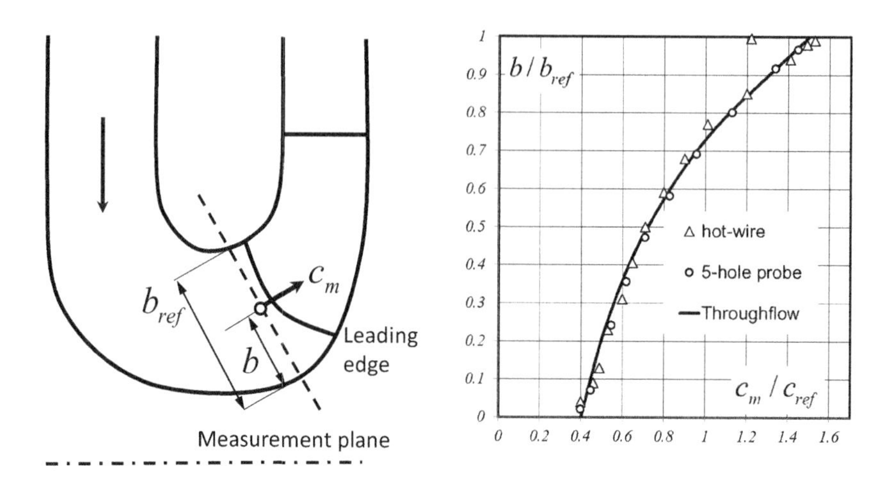

Figure 5.5 Comparison between the calculated and measured inlet velocity profile of a model of a process radial compressor inlet. (data courtesy of MAN Energy Solutions)

as shown in the example of the inlet to a radial impeller in Figure 5.5. This shows a comparison of a throughflow simulation with the measured meridional velocity in a model of an inlet bend to a radial compressor impeller. Here the curvature causes the pressure on the outside of the bend to be larger than that on the inside of the bend, leading to a large velocity gradient across the inlet of the impeller (Casey and Roth, 1984). This is an extreme example with a short impeller and a high inlet curvature, but a similar effect takes place in all radial impeller inlets.

5.3.3 Swirling Flow in an Annulus

If we consider (5.10) in relation to a swirling flow in an annulus, then this shows that in the presence of swirl the centrifugal force is kept in balance by the radial pressure gradient such that the pressure increases with radius:

$$\frac{dp}{dr} = \rho \frac{c_u^2}{r}. \tag{5.12}$$

In an axial compressor, there is generally a swirling flow between the blade rows so that the pressure at the casing is higher than the pressure on the hub. In fact, from the Euler turbine equation the swirl component of velocity in an axial compressor is larger downstream of the rotor than downstream of the stator so that different radial pressure gradients occur at these two different positions. This equation is also the fluid mechanical basis of the centrifugal effect in radial flow turbocompressors. At the inlet to an impeller, the flow usually has no swirl and on its passage through the impeller the swirl increases so the radial pressure gradient increases towards the outlet.

5.3.4 The Simple Radial Equilibrium Equation for an Axial Flow

The pressure gradient forces described earlier in this chapter do not act individually on the flow but are all simultaneously present. Together with additional forces from the blade and frictional forces, they form the basis of the so-called radial equilibrium equation for the circumferentially averaged flow in the meridional plane. The pressure gradient forces are in equilibrium with the velocity gradients (accelerations) of the mean flow, and these determine the distribution of the fluid in the meridional plane. This will be described in detail in Chapter 15, where the streamline curvature throughflow method is introduced.

As a preliminary to this, a simplified form of this equation known as the simple radial equilibrium equation is described in this section. A rotationally symmetric inviscid swirling flow in the concentric annulus between two axial blade rows is considered. It is also assumed that there is no radial component of velocity, $c_r = 0$, and that the streamlines have no curvature in the meridional plane. Equation (5.12) determines the value of the local radial pressure gradient. The variation of the local total enthalpy along a radial plane in the flow leads to an equation for the radial gradient of the enthalpy at this position as

$$\begin{aligned} h_t &= h + ½c^2 = h + ½c_u^2 + ½c_z^2, \quad (c_r = 0) \\ \frac{dh}{dr} &= \frac{dh_t}{dr} - c_u\frac{dc_u}{dr} - c_z\frac{dc_z}{dr}. \end{aligned} \tag{5.13}$$

This can be combined with the Gibbs equation to give another expression which includes the radial pressure gradient:

$$dh = vdp + Tds, \qquad \Rightarrow \frac{dh}{dr} = \frac{1}{\rho}\frac{dp}{dr} + T\frac{ds}{dr}. \tag{5.14}$$

Combining these equations gives

$$c_z\frac{dc_z}{dr} = \frac{dh_t}{dr} - T\frac{ds}{dr} - \frac{c_u}{r}\frac{d(rc_u)}{dr}. \tag{5.15}$$

If this flow has no gradient of entropy and has constant total enthalpy with radius, the simple radial equilibrium equation is obtained as

$$c_z\frac{dc_z}{dr} = -\frac{c_u}{r}\frac{d(rc_u)}{dr}. \tag{5.16}$$

This shows that the presence of swirl in an inviscid rotational flow in an annulus gives rise to a radial gradient in the axial velocity across the annulus. This equation was once used as the basis for the preliminary design of axial compressor blading. The designer specified the velocity triangles at a mean section of the blade, and this equation enabled the variation of the velocity triangles across the blade at other radial sections to be determined. For this purpose, the radial variation of the swirl was defined and the variation of axial velocity across the span was calculated from this. For the case of a free-vortex design, in which $rc_u = constant$ across the annulus, there is no gradient in the axial velocity. Other distributions of swirl across the radius in which the axial velocity is not constant may be more advantageous; see Horlock (1958), McKenzie (1967) and Lewis (1996).

The more general form of the radial equilibrium equation is outlined in Chapter 15 and considers enthalpy and entropy gradients, streamline curvature, radial flows, forces from the blades and dissipation forces from the viscosity of the flow. It also allows a general arbitrary orientation of the calculating planes and meridional annulus and is then suitable for design calculations of the mean two-dimensional flow through a radial compressor stage. The key difference between the simple radial equilibrium equation and the more general equation is the inclusion of the streamline curvature terms, and it is these terms which give their name to the more general method, known as the streamline curvature method.

5.3.5 Radial Equilibrium for the Inlet of a Radial Impeller

Neglecting blade and viscous forces in the meridional section of a radial impeller inlet, then the balance of forces normal to the meridional streamline leads to the following pressure gradient across the span:

$$\frac{1}{\rho}\frac{\partial p}{\partial q} = \frac{c_u^2}{r}\cos\varepsilon - \frac{c_m^2}{r_c}, \tag{5.17}$$

where ε is the meridional lean angle of the stream surface. At the outlet of the impeller, where the meridional radius of curvature is large and the meridional flow is nearly radial, $\varepsilon = 90°$, only small spanwise pressure gradients are expected. At the inlet and upstream of the impeller, where the swirl is negligible, and there is a constant total pressure across the span, this leads to

$$\begin{gathered} p = p_t - \tfrac{1}{2}\rho c_m^2, \quad \frac{dp}{dn} = -\rho c_m \frac{dc_m}{dn}, \quad \rho c_m \frac{dc_m}{dn} = \rho \frac{c_m^2}{r_c} \\ \Rightarrow \frac{dc_m}{dn} = \frac{c_m}{r_c}, \quad \frac{\Delta c_m}{c_m} \approx \frac{r_{casing} - r_{hub}}{r_c}. \end{gathered} \tag{5.18}$$

From this, higher velocities are present at the casing related to the lower pressure due to the curvature. For gas turbine and turbocharger compressor stages with a long axial length (typically $L/D_2 = 0.35$) and a low radius of curvature, this effect leads typically to a 15% difference in meridional velocity across the leading edge. In process compressor stages with high curvature and short axial length, the difference is substantially larger and may be above 50%, as shown in Figure 5.5. The high curvature of the inlet portion of impellers for multistage compressors necessitates careful calculations of the distribution of meridional velocity across the span for correct design of the blade leading edge angle.

5.3.6 Pressure Gradients in the Circumferential Direction

Along the mean meridional streamline between two adjacent blades, the circumferential component of the blade force causing a change in swirl is derived in Chapter 15 and is given by

$$\frac{c_m}{r}\frac{\partial(rc_u)}{\partial m} = F_u. \tag{5.19}$$

An element of fluid between two blades also experiences the circumferential force per unit mass from the gradient of the pressure force acting on the fluid by the blades, as follows

$$-\frac{\partial p}{\partial \theta} = \rho r F_u. \tag{5.20}$$

The combination of these two equations gives an equation for the circumferential pressure gradient that is consistent with the distributed force representing the blades in the meridional plane:

$$-\frac{\partial p}{\partial \theta} = (\rho c_m)\frac{\partial(rc_u)}{\partial m}. \tag{5.21}$$

The swirl gradient term depends on both the blade angles and the deviation of the flow from the blade direction. On the assumption that the gradient in pressure is linear between adjacent blades, then the difference between the static pressure on the suction and the pressure surface of the blade is obtained as

$$\left(p_{ps} - p_{ss}\right) \approx \rho c_m \Delta\theta \ \frac{\partial(rc_u)}{\partial m}. \tag{5.22}$$

If the blade-to-blade pressure difference is small, and if in this region the variations in properties in the relative system can be considered to be incompressible and loss-free, then from the Bernoulli equation the following expression for the velocity difference between the suction and pressure surfaces is obtained:

$$\Delta p_{ps-ss} = p_{ps} - p_{ss} \approx \frac{\rho}{2}\left(w_{ss}^2 - w_{ps}^2\right), \tag{5.23}$$

where the relative velocities are relative to the respective blade row (absolute velocities in a stator). The velocity term can be expanded, and on replacing the sum of suction surface and pressure surface velocities by the midpassage velocity, $2w = w_{ps} + w_{ss}$, the following equation is obtained:

$$\Delta p_{ps-ss} \approx \frac{\rho}{2}\left(w_{ss}^2 - w_{ps}^2\right) = \frac{\rho}{2}\left(w_{ss} + w_{ps}\right)\left(w_{ss} - w_{ps}\right) = \rho w\left(w_{ss} - w_{ps}\right). \tag{5.24}$$

The difference between the suction and pressure surface velocities can be estimated as

$$\left(w_{ss} - w_{ps}\right) \approx \Delta p_{ps-ss}/(\rho w) = \Delta\theta(\rho c_m)/(\rho w)\frac{\partial(rc_u)}{\partial m} = \Delta\theta \frac{c_m}{w}\frac{\partial(rc_u)}{\partial m}. \tag{5.25}$$

Thus, the final result for the approximate suction and pressure surface velocities is given by

$$w_{ss} = w + (\Delta\theta/2)\frac{c_m}{w}\frac{\partial(rc_u)}{\partial m}, \quad w_{ps} = w - (\Delta\theta/2)\frac{c_m}{w}\frac{\partial(rc_u)}{\partial m}. \tag{5.26}$$

In this equation, $\Delta\theta$ is the angular spacing, or pitch, between the suction and pressure surface of the blades.

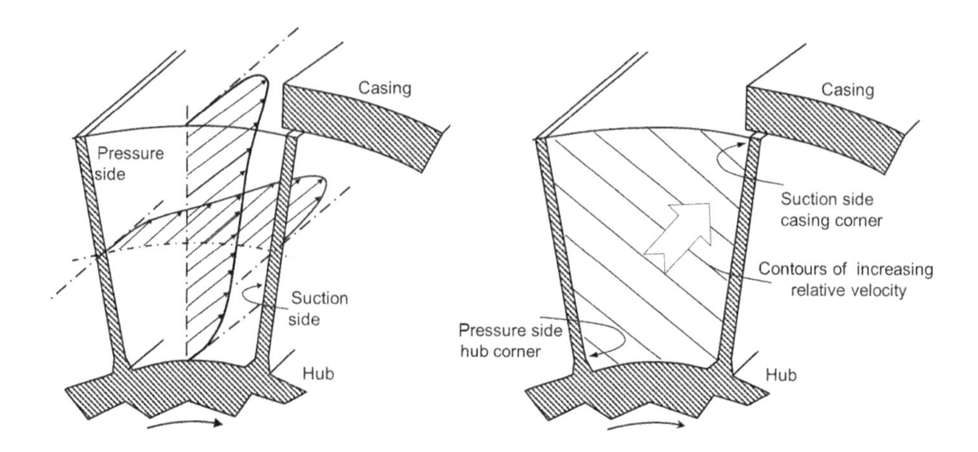

Figure 5.6 Sketch of velocity distribution and idealised velocity contours within the inducer section of an impeller.

This equation shows that in an inviscid flow, the velocity levels increase in the circumferential direction from the pressure surface towards the suction surface. Combined with (5.18), which shows that the velocity levels increase towards the casing, the isolines of constant velocity in the inducer of a radial impeller channel are orientated from the suction side/casing corner with the highest velocities towards the pressure side/casing corner with the lowest velocities (Hirsch et al., 1996), as sketched in Figure 5.6.

5.4 Coriolis and Centrifugal Forces in Impellers

5.4.1 Coriolis and Centrifugal Forces

In the analysis of a radial compressor impeller, it is generally convenient to consider the flow in the relative coordinate system. If we neglect gravitational forces, then in the absolute coordinate system there are two forces which control the fluid motion, the pressure forces and the viscous forces. However, as a consequence of viewing the motion in a relative coordinate system, the two real forces have to be supplemented by two fictitious forces due to the rotation of the frame of reference – the centrifugal force and the Coriolis force (Greitzer et al., 2004). The Coriolis forces are named after the French engineer and physicist who first discovered their importance, Gaspard Gustave de Coriolis (1792–1843). The balance of forces in an inertial coordinate system (which is called the absolute system here) equates the pressure gradient forces and the viscous forces directly to the acceleration. Coriolis found that in the rotating system the apparent centrifugal and Coriolis forces must be added; otherwise, the flow will not appear to move according to Newton's laws of motion for a stationary observer. In a rotating frame of reference, the centrifugal force has a magnitude of $\omega^2 r$ and acts along the radial direction. The Coriolis force has a magnitude of $2\omega w$ and acts normal to both the relative velocity direction and to the rotational axis.

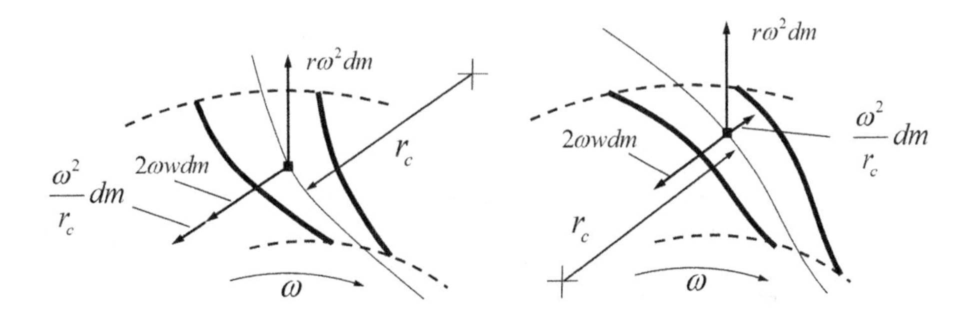

Figure 5.7 Forces acting on a fluid element in the radial part of an impeller: (left) with little backsweep; (right) with higher backsweep.

Balje (1978) provides a sketch of the Coriolis and centrifugal forces acting on an element of fluid in the radial part of an impeller, which is given in the notation of this volume in Figure 5.7. In a purely radial flow with a radial component of velocity of w_r the Coriolis force has a magnitude of $2\omega w_r$. This acts in a direction normal to both the rotational axis and the radial direction, and thus leads to an apparent force in the circumferential direction acting against the sense of the rotation. This effect can be seen if a ball is thrown radially outwards by an observer rotating on a carousel as the ball appears to fly sidewards in a curved path as it moves outwards. In an axial compressor and in the inducer section of a radial impeller, with only a small radial component of velocity and with the flow aligned in the direction of the rotational axis, the Coriolis force is negligible, but this increases towards the outlet of a radial impeller as the radial flow component increases. In a swirling flow with a swirl component of velocity of velocity of w_u the Coriolis force has a magnitude of $2\omega w_u$, which is directed radially inwards. The actions of the two components of the Coriolis force are shown together in Figure 5.7.

5.4.2 Effect of the Coriolis Force on Pressure Gradients

When considering the relative flow in the impeller, (5.12) becomes

$$c_u = w_u + \omega r, \quad \frac{dp}{dr} = \rho\left(\frac{w_u^2}{r} + 2\omega w_u + \omega^2 r\right), \tag{5.27}$$

where the second and third terms take into account the Coriolis and the centrifugal forces respectively. In the rotating frame of reference, the pressure gradients only appear to be in accordance with the Euler equations if these fictitious forces are taken into account. Greitzer et al. (2004) comment that in the view of an observer in the rotating system, the radial pressure gradient expected from the Euler equations appears to be reduced by the Coriolis and centrifugal forces.

One of the major effects of the Coriolis force in the radial section of a radial impeller is that it gives rise to a pressure gradient in the circumferential direction with a high pressure on the pressure surface, the leading surface of the blade, as follows:

$$\frac{dp}{d\theta} = -2\rho\omega r w_r. \tag{5.28}$$

In this case, the pressure gradient is normal to the relative flow direction and the pressure gradient opposes the backwards Coriolis force. This pressure gradient occurs even if the blades themselves are straight with no curvature. At the impeller inlet with nearly no radial component of velocity, the blade-to-blade pressure gradient in the flow channel is determined mainly by the curvature of the blades. Towards the outlet of an impeller with nearly straight blades and no axial flow component of velocity, the circumferential pressure gradient is determined almost entirely by the Coriolis force. Balje (1970) and Moore (1973) noted that the strong Coriolis effect in a radial channel naturally caused a low radial velocity on the pressure surface. They identified that this could lead to a form of separation on the blade surface if the mean throughflow velocity was less than half the blade-to-blade velocity difference. Moore termed this form of separation, which is not related to viscous effects, an eddy. This suggests a maximum loading limit for radial impellers, or a minimum number of blades to avoid this type of separation, and is discussed in Section 7.5.13.

An important insight into the blade loading of radial compressors from this is the fact that the designer has nearly no influence over the blade-to-blade loading in the radial part of the impeller, as this is determined mainly by Coriolis forces. In this region, it would be possible to curve the blades to provide a force to counteract the Coriolis force, as explained by Van den Braembussche (2019) and as shown in Figure 5.8. But other mechanical considerations usually cause modern impellers to have only a small blade curvature near the trailing edge, as explained in Section 11.8.1, so the loading due to curvature is small and the Coriolis forces tend to dominate in this region.

Close to the impeller outlet, the blade can no longer support a circumferential pressure gradient, and this leads to the generation of a slip velocity near the impeller outlet; see Section 5.4.4.

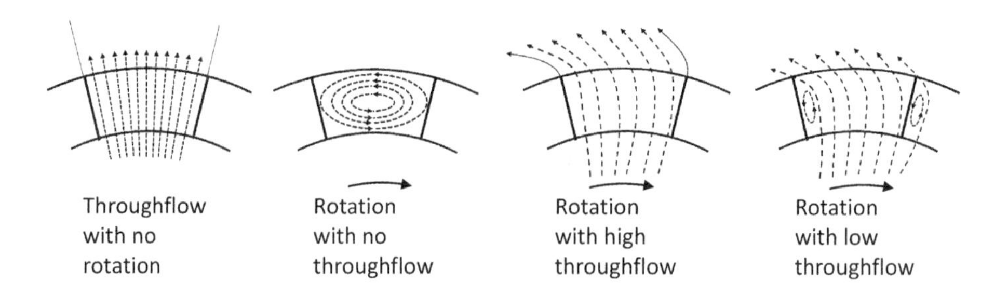

Figure 5.8 The combination of throughflow and rotation in a radial vaned impeller in the relative system.

5.4.3 The Relative Eddy

A sketch of the effect of the Coriolis forces in a radial vaned impeller from the view of an observer in the rotational frame is shown in Figure 5.8. The flow within the impeller blades can be considered to be a combination of a throughflow which has a constant velocity across the channel between the blades and a circular eddy motion between the blades in the opposite direction to the impeller rotation. The left-hand sketches show the flow to be expected within a blade passage with a purely radial flow with no blade rotation and the flow with no throughflow during rotation, the shutoff condition. The eddylike flow within the blades with only rotation is known as the relative eddy and is such that at the shutoff condition the flow at impeller outlet slips backwards relative to the impeller. Combination of these two different flow patterns leads to a flow with a higher velocity on the suction surface and in which the impeller outlet flow does not follow the direction of the blades at outlet as shown in the right-hand sketches. This feature of the flow is an important aspect of the determination of the work achieved by the impeller and may lead to a form of flow separation on the blade pressure surface if there are too few blades.

Calculations of such inviscid flow patterns were provided by Stodola (1905), Kucharski (1918) and Busemann (1928) and are reported in Traupel (1962). A modern calculation of these inviscid effects at different flow rates for a backswept impeller using the inviscid 2D calculation method of Hetzer, Epple and Delgado (2010) is given in Figure 5.9. Although the inviscid flow model cannot represent the viscous effects, it emphasises that the underlying structure of the impeller flow is a combination of a throughflow and a vortex flow such that the outflow deviates due from the blade direction at all operating points due to the relative eddy. In Figure 5.9,

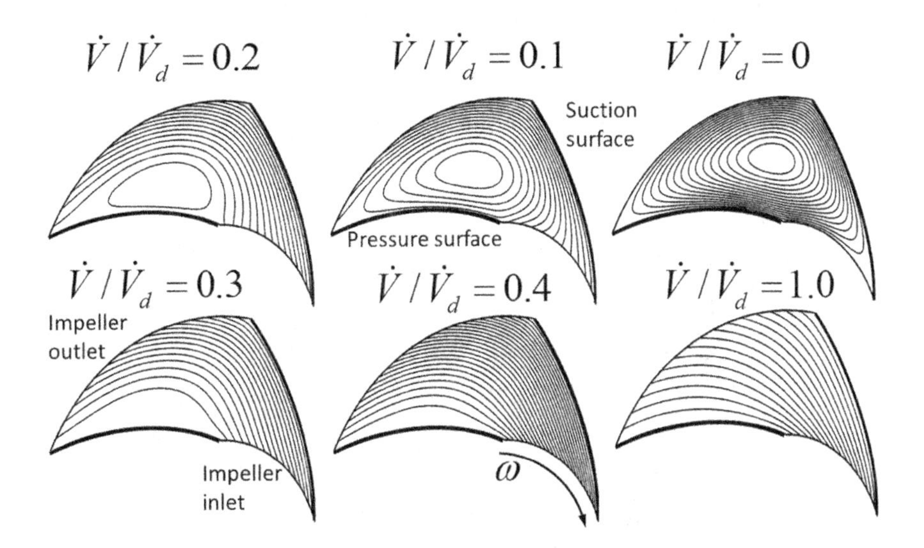

Figure 5.9 An inviscid 2D calculation of the relative eddy for a backswept impeller. (images courtesy of Philipp Epple using the software of Hetzer, Epple and Delgado, 2010)

the flow reduces in a clockwise direction from bottom-right image at the design flow rate to the top-right image with no throughflow.

5.4.4 Slip Velocity in Radial Compressors

Even under ideal frictionless conditions, the flow leaving a blade row is not perfectly guided by the blades and the deviation of the flow from the blade direction at impeller outlet leads to a reduction in the work produced compared to ideal blades. Ideal in this sense would be the aerodynamicist's ideal of an infinite number of very slippery blades with no friction and with zero thickness in which the flow would precisely follow the blade direction at outlet.

The deviation between the exit flow angle and the blade metal angle is known as the deviation angle in axial compressors. In radial impellers, this effect is best modelled by the use of a so-called slip velocity, as shown in the velocity triangle in Figure 5.10. The slip is not primarily related to the losses in the impeller but may be increased due to flow separation on the blade suction surface. There is usually no intrinsic advantage in minimising the deviation angle or slip velocity since the blade outlet angle can nearly always be changed to give the required gas outlet angle. The equation relating the deviation between the impeller outlet flow angle and the blade angle and the slip velocity for a radial impeller can be derived from the velocity triangle as

$$\delta = \beta_2' - \tan^{-1}\left(\tan\beta_2' - \frac{c_s/u_2}{c_m/u_2}\right). \tag{5.29}$$

For an impeller with a backsweep of $\beta_2' = -30°$, a slip factor of $c_s/u_2 = 0.1$ and a flow coefficient at outlet of $c_{m2}/u_2 = 0.3$, the deviation angle is $\delta = 12.3°$, leading to an underturning of the flow, and the outlet flow angle is $\beta_2 = -42.3°$. Despite the high

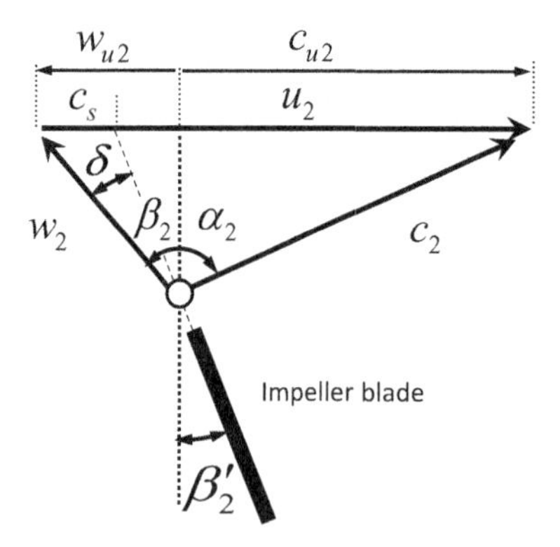

Figure 5.10 Outlet velocity triangle of an impeller showing the slip velocity, c_s.

solidity of the blades, the underturning or deviation caused mainly by the relative eddy is surprisingly high.

There are several different ways of defining the amount of slip. Traupel (2000), together with much of the European literature, considers the slip as a factor reducing the work (*Minderleistungsfaktor* in German) and defines the slip factor as the ratio of the work with and without slip leading to

$$\sigma' = \frac{(u_2 c_{u2})_{with\ slip}}{(u_2 c_{u2})_{without\ slip}} = \frac{(c_{u2})_{with\ slip}}{(c_{u2})_{without\ slip}} = 1 - \frac{c_s}{(c_{u2})_{without\ slip}} = 1 - \frac{c_s}{u_2 + c_{m2}\tan\beta_2'}, \tag{5.30}$$

where the outlet blade angle is negative for a backswept impeller in the notation used in this book. Given that the slip is generated in the relative frame, it seems inappropriate to compare this with the absolute swirl component of velocity. Wiesner (1967), in line with much other American literature on this subject, defines a slip factor relative to the blade speed at impeller outlet:

$$\sigma'' = 1 - \frac{c_s}{u_2}. \tag{5.31}$$

For a radial bladed impeller with no backsweep, $\beta_2' = 0°$, and both definitions lead to the same value, but for backswept blades they differ. The typical values of the Wiesner slip factor are close to 0.9, where the only information of any relevance is the deviation from a value of unity. A change of 10% in the slip velocity leads to a 1% change in the slip factor, as the part of the velocity triangle of interest in the work transfer is the swirl component in the absolute flow at outlet, c_{u2}. To overcome this unnecessary confusion, the slip factor is defined here directly as a nondimensional slip velocity, or simply the slip, as follows:

$$\sigma = c_s/u_2. \tag{5.32}$$

The relatively simple concept of slip has been analysed theoretically by many early researchers and was first described by Stodola (1905). The Stodola analysis considers the relative eddy within the blade channel, as shown in Figure 5.8, to lead to an additional circumferential component of relative velocity at the impeller outlet, w_s, which is against the direction of motion. The slip velocity is assumed to be the mean velocity of the relative eddy at the impeller outlet and is given by

$$c_s = w_s = \omega D_d/2, \tag{5.33}$$

where ω is the rotational speed of the impeller and D_d is the diameter of the blade discharge circle, whose diameter is the distance between the suction side trailing edge and the pressure surface of an adjacent blade, as shown in Figure 5.11. From the geometry of the blades near the impeller outlet, the blade discharge circle is given approximately by

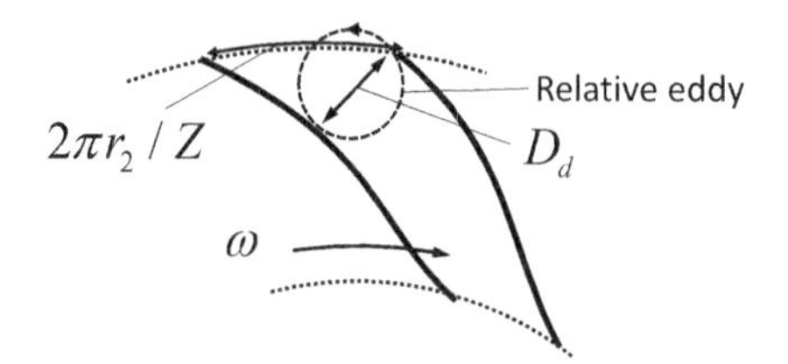

Figure 5.11 The relative eddy at impeller outlet, following Stodola (1905).

$$D_d = \frac{2\pi r_2}{Z} \cos\left|\beta_2'\right|, \tag{5.34}$$

where Z is the number of blades at the impeller outlet and β_2' is the blade outlet angle (negative in the notation of this book). This leads to an equation for the nondimensional slip velocity given by

$$\sigma = \frac{c_s}{u_2} = \frac{\pi \cos\left|\beta_2'\right|}{Z}. \tag{5.35}$$

Several researchers have suggested that a separation on the blade suction surface increases the slip factor by reducing the effective diameter of the blade discharge circle (Stanitz and Prian, 1951; Eckert and Schnell, 1961; Stuart et al., 2019). A more complex theoretical analysis of blade-to-blade inviscid flow, usually with nonarbitrary blade shapes, was provided by Busemann (1928) and more recently by Hetzer et al. (2010), but neither of these methods take into account the viscous and 3D effects. A pragmatic approach to estimating the value of the slip velocity from performance measurements is given in Section 11.6, where it is shown that it is given by the value of the work coefficient extrapolated to shutoff or zero flow.

These days the slip velocity is best calculated using modern 3D CFD methods, but for practical purposes, a good rule of thumb to estimate its magnitude during preliminary design studies is the Wiesner correlation. This is based on an experimental and theoretical analysis of Wiesner (1967), with a simple expression for the nondimensional slip velocity of

$$\frac{c_s}{u_2} = \frac{\sqrt{\cos\left|\beta_2'\right|}}{Z^{0.7}}. \tag{5.36}$$

In this, Z is the blade number and β_2' is the blade outlet angle. It is interesting to note that as the backsweep angle increases, fewer blades are used, so in fact both the nominator and denominator decrease, and the slip velocity generally lies between 0.09 and 0.12 of the impeller outlet blade speed for most impellers independent of backsweep. Nevertheless, better correlations are always useful in preliminary and throughflow design. Many publications provide more recent attempts at improving the

Wiesner correlation, such as those of Paeng and Chung (2001), Backström (2006) and Qiu et al (2011).

The difficulty of developing a sound correlation is explained by Cumpsty (2006) and others, who make it clear that there are four contributions to the slip velocity. The most important is the relative eddy in the flow, as discussed earlier. The second is that the trailing edge cannot support a pressure difference as there is a condition of zero loading at the trailing edge, known as the Kutta–Joukovsky condition. This condition is imposed on the flow near the trailing edge, so that the flow cannot possibly exactly follow the blade direction at this point. The third is that any difference in suction surface and pressure surface velocities near the trailing edge causes a deviation in the flow angle due to the difference in stream-tube thickness on the suction and pressure surfaces leading to a deviation from the suction side. Finally, the suction surface boundary layer flow is expected to be thicker than that on the pressure surface due to the higher diffusion levels.

5.5 Boundary Layers and End-Wall Flows

5.5.1 The Boundary Layers

Due to the friction between the wall and the flow, there is no slip at the wall so that the fluid on the wall has zero relative velocity. Above this, there is a layer where the fluid velocity becomes steadily faster away from the wall until it reaches the velocity of the freestream flow, u_e. This layer was called the boundary layer by the German aerodynamicist Ludwig Prandtl (1875–1953). The velocity profile normal to the wall is called the boundary layer profile. The characteristic nondimensional similarity parameter of relevance to boundary layers in fluid flows is known as the Reynolds number, $Re_L = u_e L/\nu$. It is found experimentally that in flows over bodies at high Reynolds number ($Re_L > 100{,}000$) the boundary layer is thin relative to the length of the body, L.

In classical boundary layer theory for isolated aerofoils (Prandtl, 1938), the flow is split into two separate regions: the freestream flow with low viscous effects, and the boundary layer close to the surface with large viscous forces. The boundary layer itself affects the freestream flow as it displaces the main flow outwards, leading to boundary layer blockage. The pressure distribution on the surface experienced by the boundary layer is dependent on this blockage, and this in turn affects the freestream pressure gradients. In some situations, the boundary layer separates from the surface of the body, and in these cases the interaction between the two regions is much stronger. The boundary layer plays a critical role in determining whether the flow will separate and how much lift will be generated around a lifting surface, but the equations of inviscid flow are a good approximation in the bulk of the flow field external to the boundary layer.

In the development of the flow in ducts and channels, the boundary layers on the walls may grow in the flow direction and may merge in the centre of the channel, giving rise to a fully developed viscous flow. It is the channel width, or its hydraulic diameter, D_h, which characterises this type of flow, and the Reynolds number is then defined as $Re_D = u_e D_h / \nu$. In most turbomachinery applications, there are two types of boundary layers to consider, the layers on the blades which begin at the leading edges and grow towards the trailing edges, and the end-wall boundary layers which are present upstream and downstream of the blade rows as well as between the blade rows. The boundary layers on the blade surfaces do not usually merge but are joined downstream of the blade in the form of a wake.

In multistage axial compressors, the end-wall boundary layers do not continue to grow through the machine but develop a repeating pattern from stage to stage (Smith, 1970). In the narrow diffusers of radial compressors with a small outlet width ratio, the end-wall boundary layers may merge and lead to a flow that is closer to a fully developed duct flow (Senoo et al., 1977). The end-wall boundary layers in radial impellers are usually strongly influenced by secondary flows and by the tip clearance flows as outlined in Sections 5.6 and 5.7. The development of the boundary layers within blade rows causes the flow passage to be partly blocked by the slow-moving fluid near the walls, and there is usually a strong interaction between the main flow and the boundary layers, especially at off-design operating points where the flow may separate.

In radial impellers with curvature of the blades and of the meridional flowpath, together with Coriolis forces, the flow patterns are more complex, and the boundary layers are swept across the flow channel by secondary flows and may form a jet-wake flow pattern, as described in Section 5.8. The utility of simple boundary layer methods based on a one-dimensional shear flow is known to be limited in such flows, but such tools are still very useful in preliminary design methods, so it is deemed worthwhile to describe some aspects of the physics of boundary layers, as a useful guide to the nature of the flow over compressor blade rows.

5.5.2 Boundary Layer Thickness

The boundary layer on the blade surfaces of a blade-to-blade section of a compressor develops in a two-dimensional way from the leading-edge stagnation point and grows in the meridional direction along both surfaces of the blade. As more flow becomes retarded on the surface by friction, the thickness of the layer increases towards the trailing edge with the following order of magnitude:

$$\frac{\delta}{L} \approx \sqrt{\frac{\nu}{u_e L}} = \frac{1}{\sqrt{Re_L}}. \tag{5.37}$$

In most turbomachinery applications, the viscosity is small and the Reynolds number is large (typically $Re_L \approx 10^6$), so the boundary layer can be seen to be thin relative to the length of the blade, about $\delta / L \approx 1/1000$. The thin boundary layer may have gradients of static pressure in the meridional direction, but in boundary layer theory it is generally considered to be so thin that it cannot support a gradient of static pressure away from the wall. The flow outside of the boundary layer thus imposes the local freestream static pressure through the thin boundary layer. Increasing the Reynolds number of a compressor generally reduces the thickness of the boundary layers so a small increase in flow is noted due to the lower blockage of the boundary layers. The losses of the boundary layer also decrease with an increase in the Reynolds number, and this increases the efficiency.

There is some difficulty to define the actual boundary layer thickness, δ, as the velocity profile merges steadily with the main flow and there is always some departure from the mainstream value of the velocity at any finite distance from the surface. In practice, a point is reached beyond which the influence of viscosity is imperceptible. The normal convention is to define the thickness as the distance to the wall where the velocity reaches 99% of the freestream velocity, u_e. This approach is not really possible in the curved flows in turbomachinery, where the curvature leads to complex gradients in the main flow velocities so the edge of the boundary layer is not clearly delineated. The fact that the real flow may also be highly three dimensional with tip clearance jets, skewed boundary layers with secondary flows and flow separation leads to further difficulty.

5.5.3 Displacement Thickness and Boundary Layer Blockage

Comparing the mass-flow rate of a viscous and inviscid flow through the same channel, the mass flow with the viscous fluid is reduced due to the friction in the boundary layer. To evaluate the deficit in mass flow of the boundary layer in viscous flows, the displacement thickness, δ^*, is defined as follows:

$$\delta^* = \int_0^\delta \left(1 - \frac{\rho u}{\rho_e u_e}\right) dy. \tag{5.38}$$

This represents the distance that the flow channel wall has to be displaced outwards so that the mass flow with a nonviscous flow, with the freestream velocity, u_e, all through the boundary layer, is the same as with the viscous flow. This leads to the effective flow channel being slightly blocked by the boundary layers on its wall, which is important for determining the mass flow and velocity level, which in turn affects the static pressure. Zonal methods of calculation of the 2D blade-to-blade flow in a cascade together with the boundary layer include this interaction between the two zones and are important tools in the development of axial blade rows (Drela and Youngren, 1991). The displacement thickness is also related to the blockage of the boundary layer as mentioned in Section 5.2.3. in connection with the continuity equation and is discussed further in Section 7.4 on diffusers. The growth of this blockage as the compressor approaches the stability limit is a crucial aspect in surge and stall in axial compressors.

Eckardt (1976) was able to determine the mean quantitative effect of the blockage from measurements in a radial impeller near the outlet by comparing the area-average meridional velocity with the mass-averaged meridional velocity, as follows,

$$k = 1 - B = \bar{c}_m^{area} / \bar{c}_m^{mass}, \tag{5.39}$$

to determine that the flow blockage in his measurements, $B = 1 - k$. A similar definition of blockage was used by Dring (1984). Eckardt found that the blockage at impeller outlet was approximately 18%. The background to this very large blockage of the flow in some radial impellers near the outlet is discussed in Section 5.8 on the jet-wake flow pattern.

The problem of defining the edge of the boundary layer and separating the two zones in complex radial compressor flows has not yet been successfully solved. Johnson and Moore (1980) examined gradients in the rotary stagnation pressure, which is conserved in inviscid incompressible flow in a rotor, to identify the location of the boundary layers. Another useful method to determine the actual cutoff position between the boundary layer and the freestream in a three-dimensional axial compressor flow has been suggested in a work by Khalid et al. (1999). The first step in their blockage calculation process involves identifying the main flow direction and subsequently to delineate regions with high-velocity gradients compared to this, indicating regions with a deficit in the mass flow. The procedure to determine the exact cutoff between the main flow and the deficit regions is complex and relies on a lot of engineering judgement, as the flow also includes tip clearance vortices with large changes in flow direction. This system is still a research area for radial compressors but has been used by Garrison and Cooper (2009) to calculate the blockage near the exit of centrifugal impellers. The process is made difficult in modern radial impellers as the flow field often contains large regions with low momentum but with relatively moderate gradients between the low- and high-momentum regions.

5.5.4 Mass-Flow Thickness, Momentum Thickness and Energy Thickness

A further boundary layer thickness, defined by Head (1960), is the mass-flow thickness denoted by $\Delta = \delta - \delta^*$, and in the physics of turbulent boundary layers, this appears to have a key role in the growth of the boundary layer through the entrainment of the freestream flow into the boundary layer via turbulent eddies. The physics of this was first delineated by Coles (1956) when he postulated that the outer part of a turbulent boundary layer could be viewed as a continuously growing turbulent wake modified by the presence of a wall. The assumption is that the rate at which freestream fluid is incorporated into the boundary layers is determined by the velocity defect in the outer part of the layer. For boundary layers proceeding to separation in an adverse pressure gradient, the entrainment is reduced, and for boundary layers in an accelerating flow, it is increased.

The boundary layer not only has a deficit in mass flow, but also in momentum flux. A deficit thickness known as the moment thickness, θ, is defined as follows:

$$\theta = \int_0^\delta \left(1 - \frac{u}{u_e}\right) \frac{\rho u}{\rho_e u_e} dy. \tag{5.40}$$

This, added together with the displacement thickness, represents the distance from the wall that the flow channel has to be displaced outwards so that the mass flow with a nonviscous flow at the freestream velocity, u_e, has the same momentum flux as with the viscous flow. This is related to the drag force of the boundary layer on the surface.

In a similar vein, the boundary layer also has a deficit in kinetic energy flux. An equivalent deficit thickness known as the kinetic energy thickness, ε, is defined as follows:

$$\varepsilon = \int_0^\delta \left(1 - \frac{u^2}{u_e^2}\right) \frac{\rho u}{\rho_e u_e} dy. \tag{5.41}$$

This represents the distance from the wall that the flow channel has to be displaced so that the mass flow with a nonviscous flow at the freestream velocity, u_e, has the same kinetic energy flux as with the viscous flow. This is related to the losses generated by the frictional energy dissipation of the boundary layer on the surface.

These three parameters describe the fullness of the boundary layer profile, and ratios of these are used to define shape factors to characterise this:

$$H = \delta^*/\theta, \quad H_\varepsilon = \varepsilon/\theta, \quad H^* = (\delta - \delta^*)/\theta. \tag{5.42}$$

These parameters take on an important role in the theory of integral boundary layer methods, which solve the Von Karman integral momentum equation for the boundary layer (Schlichting and Gersten, 2006):

$$\frac{d\theta}{dx} + (2 + H)\frac{1}{u_e}\frac{du_e}{dx}\theta = \frac{c_f}{2}. \tag{5.43}$$

This equation allows the development of the momentum thickness of the boundary layer to be predicted for a given externally imposed velocity gradient in the free-stream. It needs to be solved together with empirical relationships for the variation of the shape factor H and the skin friction coefficient c_f. No detail will be given here, but the further development of the method of Head (1960), in Head and Patel (1968), leading up to the work of Green et al. (1977), Melnik et al. (1986) and Drela et al. (1986) are suggested for further reading. The method of Drela and Giles has now become a standard tool for estimation of boundary layer growth in the two-dimensional blade-to-blade plane of all types of turbomachinery, Drela and Youngren (1991), as it includes models for the compressible freestream flow, transition, reverse flow and blunt trailing edges. A method of calculating the end-wall boundary layer thickness of axial diffusers using this approach in a throughflow calculation can be found in Hirschmann et al. (2013).

5.5.5 Boundary Layer Structure in the Meridional Direction

In this section, the structure of the boundary layer on the blade surface in the meridional direction along an axial compressor blade is considered as that is where the majority of the research has been carried out. At the typical Reynolds number levels in compressors, $>2 \times 10^5$, the forward few per cent of the length of a compressor blade row the boundary flow may be laminar. In this short region, the flow is dominated by molecular viscosity and slides over the wall in smooth layers with increasing velocity farther from the surface and is steady in time. Downstream of this region, the flow becomes turbulent, and towards the trailing edge the turbulent boundary layer might separate.

5.5.6 Boundary Layer Transition

There are several modes in which this transition to turbulence might take place, and this description of them is based on that of Langtry and Menter (2005), as shown in Figure 5.12. The early research on transition in flows with low turbulence levels ($<1\%$) was based on inviscid stability theory. This suggested that all boundary layer flows were only unstable if there was an inflexion point in the velocity profile. It was later predicted physically by Prandtl and then proven mathematically by Tollmien that a laminar boundary layer can be destabilised by the presence of viscous instability waves, often referred to as Tollmien–Schlichting waves (Schlichting and Gersten, 2006).

In turbomachinery flows, the freestream has a higher turbulence level ($>1\%$) and contains many disturbances travelling with the main flow, for example the wakes from upstream components, and these interact with the laminar boundary layer to make it unstable. It quickly undergoes transition over a short transition length to a turbulent flow by a process called 'bypass' transition, as this bypasses any natural tendency for the laminar layer to become unstable from the growth of Tollmien–Schlichting waves. Bypass transition causes the external disturbances to propagate into the boundary layer, creating low-frequency perturbations that break down to turbulence (Mayle, 1991). It leads to a very rapid growth of disturbances and to transition in any device with adverse pressure gradients in the laminar region. Bypass transition can also occur

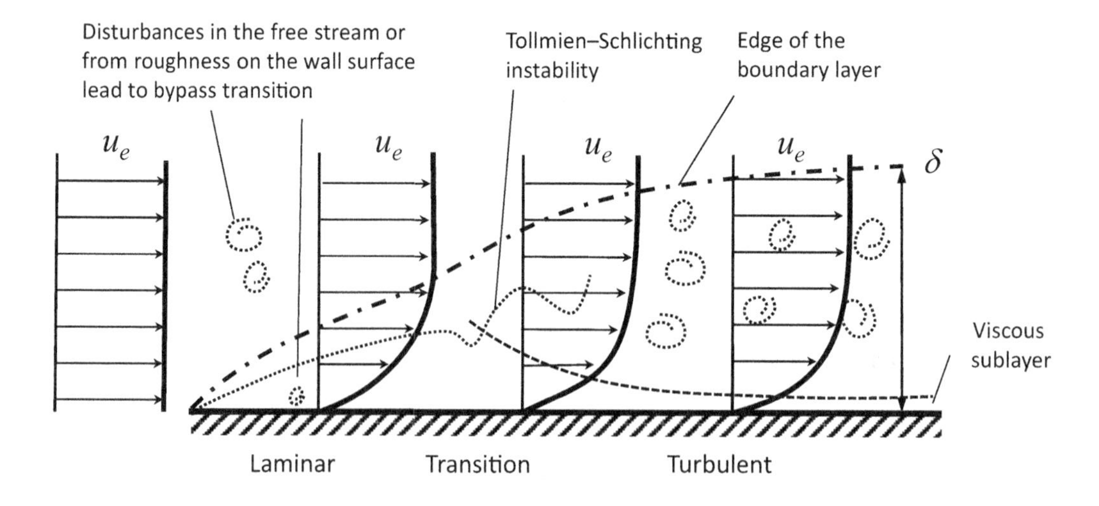

Figure 5.12 Laminar-to-turbulent boundary layer transition.

due to surface roughness where the disturbances originate from the perturbations at the wall instead of from the freestream turbulence.

An important instance of bypass transition arises in turbomachinery flows where the blade rows are subjected to periodically passing turbulent wakes. It has been found experimentally that the wakes are so disruptive to the laminar boundary layer that turbulent spots often form in the location where the wake impinges on the surface. It is for this reason that this mode of transition is usually differentiated from bypass transition and is often referred to as wake-induced transition. This leads to a situation where the bypass transition is also intermittent as the wake passes downstream and has been the subject of extensive study in axial compressors and turbines (Halstead et al., 1997). The process of transition varies with time as the wakes pass over the blade with the position of the start of the fully turbulent boundary layer moving with the wake disturbances along the blade chord with time. After the passage of the disturbances, the laminar layer retains the velocity profile of the turbulent layer, and is known as a calmed region (Halstead et al., 1997).

The importance of the leading-edge shape on the transition to turbulence and the loss production in axial compressors has been well known since the work of Tain and Cumpsty (2000); the change in transition due to the leading edge shape may also influence the separation of the end-wall flow (Auchoybur and Miller, 2018). There is no reason to believe that this is not the same for radial compressors.

5.5.7 Boundary Layer Separation

The fluid near the wall in a boundary layer is slowed down by wall friction. In the presence of an adverse pressure gradient, the momentum of the fluid particles near the wall will be further reduced. The momentum may become too small to overcome the pressure rise, and the forward motion of fluid particles near the wall eventually becoming arrested or even becoming directed towards the backwards direction. At some point (or line), the viscous layer may then depart or break away from the solid surface and the boundary layer is said to separate. In this way, the distribution of the pressure along the surface of a compressor blade has a large effect on the development of the boundary layer. In particular, if the region of increasing pressure is sufficiently long and the adverse pressure gradient sufficiently steep, then the boundary layer profiles change so radically that the shear stress at the wall reduces to zero and the flow separates from the surface, leaving a region of reverse flow next to the surface. The velocity profiles at different positions along the surface then develop as shown in Figure 5.13. When separation occurs, flow instabilities can occur with a large effect on the compressor characteristics.

The turbulent boundary layer has higher velocities close to the surface than does a laminar layer and its momentum in a region close to the surface is higher. It is then less likely to separate in a region of increasing pressure than a laminar boundary layer. Because of the risk of separation, the boundary layer on the suction surface of compressor blades sets some particularly difficult aerodynamic design challenges. The fundamental difficulty is that the fluid is being asked to travel uphill against an

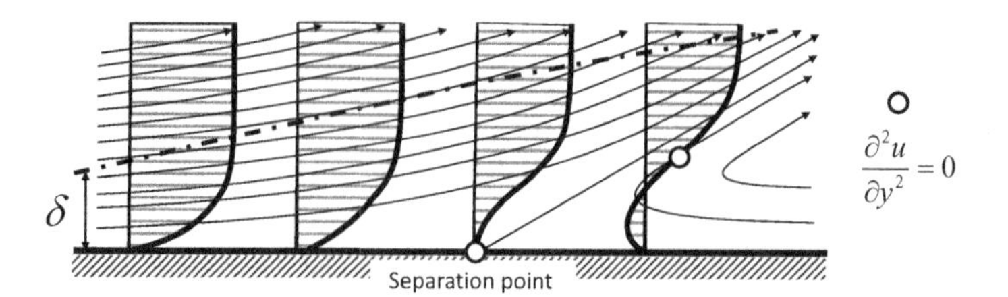

Figure 5.13 Turbulent boundary layer velocity profile in an adverse pressure gradient leading to separation and reverse flow.

increasing pressure. This is the opposite case to a turbine, where the fluid travels in its natural direction from a higher to a lower pressure. The viscous effects near the wall limit the pressure rise that can be produced. Attempts to exceed this limiting pressure rise lead to local flow instability resulting from separations. Many different empirical parameters are described in Section 7.5 to try to define this limit.

For a compressor, a bad design with separation may have serious consequences. The flow separation is frequently quite sensitive to small changes in the shape of the body, particularly if the pressure distribution can be strongly affected by the change of shape of the body, such as a change in the thickness and the blade angle distribution through erosion. In axial compressor design, the predicted boundary layer parameters are controlled by the use of so-called supercritical or controlled diffusion blade profiles (Behlke, 1986). These profile shapes have more curvature near the leading edge and are flatter towards the trailing edge. They minimise the risk of separation and the losses associated with the separation by controlling the blade boundary layer growth along the blade chord in such a way that the adverse pressure gradient decreases towards the trailing edge, a strategy proposed by Stratford (1959). Stratford suggested that if the boundary layer was taken close to separation, and maintained in that condition, the profile loss should be small because the surface shear stress and hence the blade drag would be small. However, that argument neglects the dissipation in the boundary layer, which continues, and even increases, as the shear stress falls to zero, as discussed in Section 10.3.3.

An interesting insight from Walreavens and Cumpsty (1995) is that the growth of the boundary layer in an adverse pressure gradient is not strongly affected by the skin friction. The skin friction is negligible if the flow is close to separation and the momentum integral equation reduces to

$$u_e \frac{d\theta}{dx} = -(2+H)\frac{du_e}{dx}\theta. \tag{5.44}$$

The growth of the momentum thickness is then proportional to the local value of the momentum thickness and the velocity gradient. In fact, it is higher than this, because the shape factor will also increase towards the trailing edge. The use of velocity gradient parameters as a means of assessing the loading of compressor blades, such as the Lieblein diffusion factor or the de Haller number, follow on automatically from this; see Section 7.5.

5.5.8 Laminar Separation Bubble

In some situations, the laminar boundary at the leading edge may separate due to a strong adverse pressure gradient before the onset of transition, for example if the boundary layer is still laminar when the flow enters a suction surface shock wave, or if the shape of the leading edge includes a sharp discontinuity in curvature, leading to a so-called spike in the velocity distribution. The separated boundary layer may remain separated but, in some circumstances, the separated layer may undergo transition and the higher entrainment caused by the turbulence may cause the flow to reattach to the surface in a laminar separation bubble (Horton, 1967; Gaster, 1969). The mean flow structure of a laminar separation bubble is shown in Figure 5.14.

A very useful study of the leading-edge separation bubbles on axial compressor blades can be found in Walreavens and Cumpsty (1995). Subsequent work in this area (Tain and Cumpsty, 2000) showed the extent to which the laminar flow accelerates around the leading edge is very dependent on the precise shape of the leading edge. They found that the flow may become sonic in the spike at the blend point between a circular leading edge with the suction surface of the blade and recommended using elliptical shapes for leading edges to weaken this peak in the velocity distribution. Subsequent work by Goodhand and Miller (2011) shows that a blade with a continuous curvature at the leading-edge blend point helps to eliminate the separation bubble completely, thus improving the blade performance. CFD simulations of radial impellers by Bousquet et al. (2014) and Itou et al. (2017) have identified the onset of separation at the leading edge forming a 3D separation bubble which reattaches further downstream on the suction surface, especially at part load operating points prior to the onset of rotating stall.

The length of the separation bubble depends on the transition process within the shear layer and the freestream turbulence level. Traditionally, separation bubbles have been classified as long or short based on their effect on the pressure distribution around an aerofoil (Mayle, 1991). Short bubbles reattach shortly after separation and only have a local effect on the pressure distribution. Long bubbles can completely alter the pressure distribution around an aerofoil. Any change in bubble length can

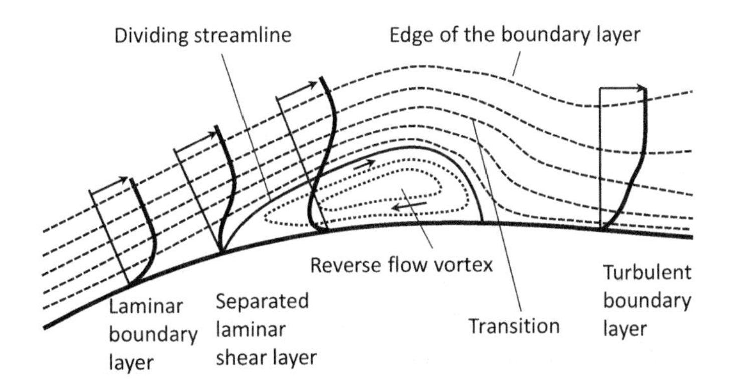

Figure 5.14 The mean flow structure of a laminar separation bubble.

result in a dramatic loss of lift and can cause an aerofoil to stall if the bubble fails to reattach; see Chapter 7.

5.5.9 Boundary Layer Structure Normal to the Blade Surface

The velocity profiles of the turbulent and laminar layers are different and give rise to the different properties of the two layers. The outer layers of the turbulent boundary layer, being composed of many small eddies, are very effective in mixing the mainstream flow with the flow in the boundary layer, a process known as entrainment. This disperses the high momentum freestream fluid deep into the boundary layer and makes the velocity near the surface higher than is found in a laminar layer, as shown in Figure 5.15. In a laminar boundary layer, the Newtonian viscous stresses which arise due to molecular exchange between the layers dominate the flow. In a turbulent boundary layer, the mechanism for shear stresses arises from the high momentum exchange between adjacent layers caused by the turbulence. The different velocity profiles are characterised by their shape factor; $H = 2.5$ is typical of laminar flows, while $H = 1.35$ is typical of turbulent flows with no pressure gradient, and this increases to $H = 4.0$ when the flow is close to separation in an adverse pressure gradient.

A turbulent flow can be considered as the combination of a mean and a fluctuating velocity, $u + u'$, where the fluctuating velocities are typically in the order of 10% or less of the local mean velocities. For this flow, the appropriate mean equations of motion are known as the Reynolds-averaged Navier–Stokes equations (RANS) (Schlichting and Gersten, 2006. Additional terms are needed in the time mean momentum equation to account for the transport of the momentum by the turbulent fluctuations. These terms appear like a viscous effect but are caused by turbulent stresses and are known as the Reynolds stresses:

$$\tau_{turb} = \rho \overline{u'_x u'_y}. \tag{5.45}$$

Over most of the turbulent boundary layer, the Reynolds stresses are much larger (up to three orders of magnitude) than the viscous stresses and dominate the flow except for a thin region close to the wall.

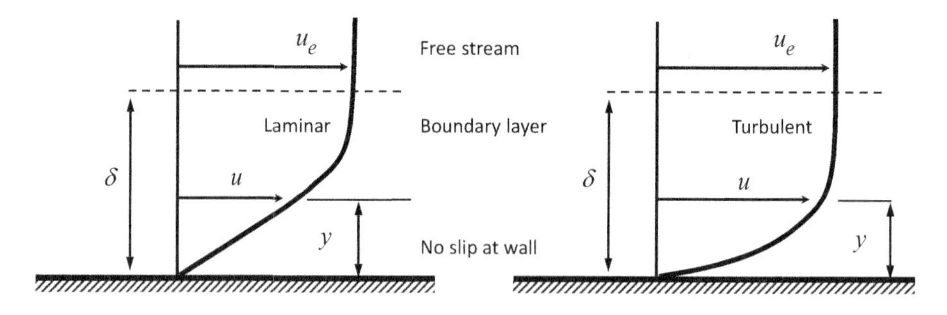

Figure 5.15 Boundary layer velocity profiles on a flat plate (left laminar, right turbulent).

Understanding of turbulence and of Reynolds stresses is still incomplete, and the extremely difficult nature of this subject caused one expert, Bradshaw (1994), to refer to turbulence as 'the invention of the Devil on the seventh day of creation, when the Good Lord wasn't looking'. The physicist Heisenberg, when asked what questions he would like to ask God, is said to have replied, 'I would ask God two questions: "Why quantum mechanics, and why turbulence?" I think he will have an answer for the former.' Fortunately, over the last few decades, some practical and very effective turbulence models have been developed to estimate the Reynolds stresses for use with the RANS equations, and these are discussed in Section 16.6.

The higher velocity near the surface of a turbulent boundary layer has two important effects. First, the velocity gradient at the surface is higher. The frictional shear stress near the wall is thus much greater for a turbulent boundary layer than for a laminar one, and higher losses occur. In turbine profile design, turbulent boundary layers are avoided, if possible, by maintaining accelerated flow along the turbine blade surfaces to retain a laminar flow with lower losses until close to the trailing edge, where the flow may decelerate. This is not possible in compressors, except in small regions near the leading edge, as the flow decelerates over most of the suction surface and the laminar layer is not retained. The major effect of having a turbulent velocity profile is that with a higher momentum close to the wall, there is less tendency for a turbulent layer to separate in an adverse pressure gradient and cause a stall.

5.5.10 Law of the Wall

Experiments and boundary layer analysis show that the turbulent boundary layer does not extend completely to the wall; close to the wall there is no space for turbulent eddies with momentum exchange, so that a viscous sublayer is retained underneath the turbulent layer. The viscous sublayer is laminar close to the wall, where the flow is completely dominated by viscous shear stresses. Above this laminar sublayer is a buffer zone where the flow is still strongly dominated by the viscosity. Above the viscous sublayer, the turbulent region can also be considered to have two regions: the so-called log-law region, where the turbulence stresses are still affected by the viscous sublayer, and the outer region, where the flow has the nature of a turbulent wake and is dominated by turbulent mixing processes (Figure 5.16).

For the inner region of the boundary layer, it is convenient to consider that the viscous shear stress dominates the flow. Further outwards, the turbulent shear stresses dominate the viscous stress. In the inner region, the flow is dominated by the wall shear stress, τ_w; the kinematic viscosity, ν; the density, ρ; and the distance from the wall, y. Dimensional analysis shows that a key nondimensional parameter that determines the nature of this near-wall flow is the so-called friction velocity, $u_\tau = \sqrt{(\tau_w/\rho)}$, so the near-wall boundary layer velocity profile can be plotted in universal boundary layer coordinates as

$$u^+ = \frac{u_x}{u_\tau} = f\left(\frac{yu_\tau}{\nu}\right) = f(y^+). \tag{5.46}$$

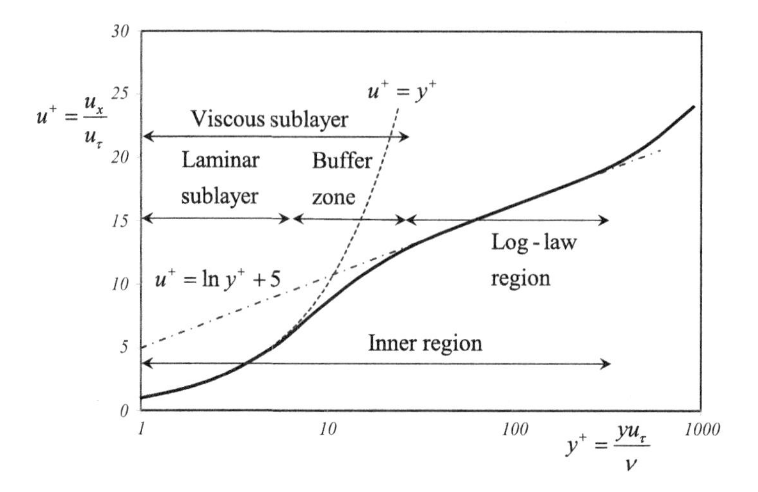

Figure 5.16 Turbulent boundary layer velocity profile for a flat plate.

The different regions of the universal velocity profile of the turbulent boundary layer are shown in Figure 5.16. The laminar sublayer, in which $u^+ = y^+$, extends out to about $y^+ = 5$. The buffer zone extends from $y^+ = 5$ to about $y^+ = 30$. The fully turbulent flow begins about $y^+ = 30$ and exhibits a velocity profile given by

$$u^+ = \frac{1}{\kappa} \ln (y^+) + B, \tag{5.47}$$

where $\kappa = 0.41$ is the von Karman constant and the constant B lies between about 4.9 and 5.5 for a smooth surface. This turbulent profile extends out into the outer region, where influences of the Reynolds number and pressure gradient can be seen. The log-law of (5.47) is really only valid for a zero-pressure gradient boundary layer but is generally applied to other wall-bounded flows with adverse and positive pressure gradients, as the outer part of the boundary is more strongly influenced by the pressure gradients than the inner region. Clearly this equation will become incorrect for flows near separation or reattachment points, since the wall shear stress and hence the friction velocity is zero. The application of more generalised wall functions to complex 3D turbulent boundary layer flows is an active area of research.

This universal velocity profile gives details of how the flow varies down to the wall. In simulations, it is clear that if we wish to capture the details of this near-wall flow, then the most reliable way is to use a very fine grid. This can be very expensive in computation effort, particularly in 3D flows, and slow convergence can also be a problem. Some turbulence models used in CFD simulations are valid for the turbulent flow region but fail in the viscous sublayer close to the wall. Various so-called low Reynolds number versions of turbulence models have been proposed incorporating modifications which remove this limitation. Alternatively, the standard models can be used in the interior of the flow and coupled to the wall functions which are used to resolve just the wall region. This is the traditional industrial solution for such simulations.

In turbomachinery blade surface boundary layers, as with the fully developed pipe flow, we need to distinguish between the effects due to laminar and turbulent flow, so there is also a lower critical Reynolds number for flat plates below which the flow is effectively laminar. In addition, the effects due to roughness at higher Reynolds numbers lead to an upper critical Reynolds number above which the flow is fully rough and the friction factor is not affected by the Reynolds number. Between these extremes, there is a transition region where the friction factor depends on the roughness and the Reynolds number, such that it can also become hydraulically smooth if the roughness is low. A basic model would be that for a flow over a flat plate, but this needs to be modified for describing turbomachine flows. Pressure gradients, secondary flows, vortices or other flow structures such as laminar separation bubbles make the flow through turbomachines far more complex than that over a flat plate.

Successful prediction of frictional drag for external flows, or pressure drop and losses for internal flows, depends on the fidelity of the prediction of the local wall shear. Losses caused by the wall shear in the boundary layer are discussed in Section 10.3.3.

5.5.11 End-Wall Boundary Layer

The fluid mechanics of the end-wall boundary layer region are critical in developing performance prediction methods and in spite of well over 70 years of research on the topic, the flow in this region is still not very well understood. In axial compressor throughflow methods, pitchwise averaged boundary layer equations have been used which have avoided dealing with the complex endwall flow structure, for example Wright and Miller (1991). In radial compressors, there has been little success in developing general predictive end-wall flow procedures along such lines. The end-wall flow includes the primary end-wall boundary layer, the secondary flow and the leakage jet due to the tip clearance. Detailed measurements and CFD simulations have provided new insights into this, but, because of the complexity of the flow, much empiricism remains.

5.5.12 Turbulent Boundary Layers on Rough and Smooth Surfaces

A rough surface finish leads to a higher wall friction and can cause a considerable reduction in the efficiency of compressors (Kawakubo et al., 2008). In most applications, the machined surfaces are generally made as smooth as necessary to avoid this problem. In practice, machined surfaces do not need to be completely smooth, as the surface roughness has no hydraulic effects once it is less than the thickness of the laminar sublayer of the turbulent boundary layer as defined by the following equation,

$$\frac{\rho u_e k_s}{\mu} < 100, \tag{5.48}$$

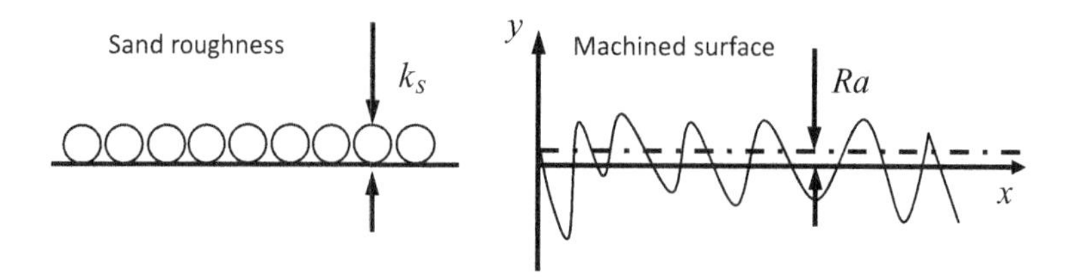

Figure 5.17 Different types of surface roughness.

where k_s is the equivalent sand roughness of the surface. A roughness level of this size and below gives rise to a hydraulically smooth surface and allows fluids to flow with a minimum dissipation loss. The definition of the equivalent sand roughness of a surface is an approximate way of relating the true roughness patterns in machined objects to fundamental studies of surfaces roughened by the application of sand grains (Nikuradse, 1950).

There are several ways to characterise the roughness of machined surfaces. The most widely used is the arithmetic average roughness or the centre-line average roughness, Ra. This is the arithmetic average of deviations about the centre line of a given roughness profile, as shown in Figure 5.17, and defined as

$$Ra = \frac{1}{l}\int_0^l |y| dx. \tag{5.49}$$

Surfaces of different roughness can also be classified with ISO class number specification, ISO 1302 (2002), indicating the roughness of a particular surface compared to standard surfaces. Roughness classes of most relevance in compressor technology are between N9 and N7. N9 indicates that $Ra = 6.3$ μm, which would be the roughness of a sandblasted metal surface of an impeller following a welding process, and N7 indicates that $Ra = 1.6$ μm, which would represent a typical smooth milled surface of a milled impeller or diffuser.

The conversion of a particular arithmetic average Ra value to an equivalent sand grain k_s value is generally done by defining an equivalence factor between the two. The equivalence factor is not unique and depends on the structure of the machined surface, such as the orientation of any finishing marks and the skewness of the roughness profile (Gülich, 2008). It can vary within wide boundaries as the local geometrical properties of a machined surface differ from that of a surface with closely spaced sand grains, and the flow near the roughness elements is different. This is an active area of research and there are still several different ways of defining the equivalent sand roughness of a machine surface based on a value of the average roughness (Koch and Smith, 1976; Casey, 1985).

The law of the wall can also be modified to consider the roughness of the wall surfaces. Hama (1954) observed that the main effect of surface roughness on the mean turbulent velocity profile was to generate a downward shift in the log-law velocity, Δu^+, leading to a rise in momentum deficit compared to the smooth-wall case.

He noted, however, that shape of the mean velocity profile in the overlap and outer layer was unaffected by the roughness. For this reason, the log-law for rough walls can be expressed as

$$u^+ = \frac{1}{\kappa}\ln(y^+) + B - \Delta u^+, \tag{5.50}$$

where κ is the von Karman constant, and B is the smooth-wall intercept. This is the way in which rough surfaces are taken into account in CFD simulations.

5.5.13 Friction Factor for Fully Developed Pipe Flow with Roughness

The important issues with surface roughness can best be visualised by considering the variation of the friction factor for fully developed pipe flow, as given experimentally in the so-called Moody diagram (Moody, 1944), of which a modern example is given in Figure 5.18. The friction factor, f_D, is given in this diagram as a function of two nondimensional parameters, the Reynolds number based on mean velocity and the hydraulic diameter D_h, and the relative roughness, k_s/D_h, which is the equivalent sand roughness, k_s, divided by the hydraulic diameter:

$$f_D = f\left(\frac{\rho c D_h}{\mu}, \frac{k_s}{D_h}\right). \tag{5.51}$$

Different regimes can be identified in the Moody diagram. A laminar region occurs below what is known as the lower critical Reynolds number, and above this a transition region occurs between the laminar and the turbulent region. In the turbulent region, the friction factor is related to roughness and Reynolds number, and in this region a hydraulically smooth turbulent flow, with the lowest possible losses, is possible when the roughness effects are negligible. At very high Reynolds number, above what is known as the upper critical Reynolds number, a region occurs where the friction factor is determined solely by the roughness.

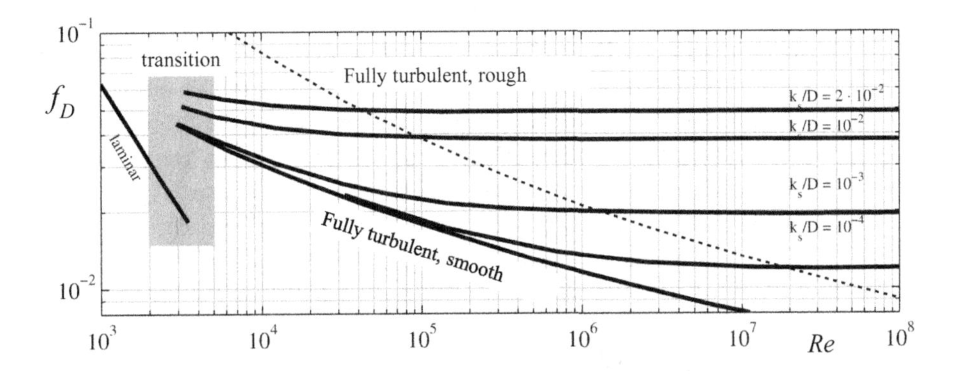

Figure 5.18 Moody diagram for the friction factor of a pipe flow.

The Moody diagram and the associated Darcy–Weisbach friction factor, f_D, are a powerful tool for calculating the total pressure drop in a fully developed pipe flow using the Darcy–Weisbach equation as

$$\Delta p_t = f_D \frac{L}{D_h} \frac{\rho c^2}{2}. \tag{5.52}$$

This equation identifies that the total pressure losses increase with the square of the mean pipe velocity, linearly with the length of the pipe and inversely with the hydraulic diameter. In practical terms, short, wide flow channels with low velocities have the lowest pressure drop. The carryover of this to radial stages is that we can expect poor efficiencies if the flow channels are narrow. It is worth noting that for geometrically similar pipes, in which L/D_h is constant, then a change in the Reynolds number, the size or the roughness will cause a change in the friction factor, and this will cause a change in the total pressure losses as follows:

$$\delta(\Delta p_t) = \frac{L}{D_h} \frac{\rho c^2}{2} \delta f_D. \tag{5.53}$$

In this way, the change in the pressure losses can be related to a change in the friction factor rather than to a change in the Reynolds number. An equation of this form is the basis of several methods for estimating the effect of Reynolds number, size and roughness on losses and performance, as discussed in Section 10.3.4.

Given the Reynolds number, *Re*, and the relative roughness parameter, k_s/D_h, the friction factor in the turbulent flow region of the Moody diagram can be calculated from the Colebrook–White formula as follows:

$$\frac{1}{\sqrt{f_D}} = -2 \log_{10} \left(\frac{k_s}{3.7 D_h} + \frac{2.51}{Re \sqrt{f_D}} \right). \tag{5.54}$$

An important feature of this equation is the important effect of roughness at high Reynolds numbers. If we consider an extremely high Reynolds number, then the term including the Reynolds number becomes zero and the second term of the equation in the bracket tends to zero. The bracket reduces to just the first term and is known as the von Karman equation. This expresses the physics of fully turbulent flow where the skin friction factor is simply a function of the roughness. Several other more convenient intrinsic equations for the friction factor of pipe flows have been published which give similar results with less effort and are within 1 or 2% of the friction factor predicted by the Colebrook–White equation. It is worth keeping in mind that the Colebrook–White formula itself may be up to 5% in error as compared to experimental data, so this error is of no practical significance. The equation of Haaland (1983) is apparently more accurate over a wider range,

$$\frac{1}{\sqrt{f_D}} = -\frac{1.8}{n} \log_{10} \left(\left[\frac{k_s}{3.75 D_h} \right]^{1.11n} + \left[\frac{6.97}{Re} \right]^n \right), \tag{5.55}$$

where $n = 0.9$.

Fully developed pipe flow is not such a realistic model of the flow in compressor blading, where the suction surface and pressure surface boundary layers seldom grow sufficiently to merge together on the two sides of the flow channel. However, there is some evidence that the end-wall boundary layers in the diffusers of narrow low-flow coefficient stages may merge in the diffuser flow channels (Senoo et al., 1977). As a first estimate, a more appropriate representation of the effects of friction on turbomachinery pressure losses would be the viscous flow on a flat plate. A key difference between the fully developed pipe flow and the flat plate flow is the choice of the representative Reynolds number. In the flat plate flow, the length scale in the Reynolds number needs to be defined with the streamwise coordinate, as opposed to the hydraulic diameter for the fully developed pipe flow.

5.6 Secondary Flows

5.6.1 Generation of Secondary Flows

If a flow with a nonuniform velocity profile, such as in a boundary layer, is made to follow a curved path, a cross-flow is generated with velocity components normal to the primary flow direction, as shown in Figure 11.17. This is known as a secondary flow as it has a different direction to the primary flow, but the secondary flow velocity components can be a substantial fraction of the primary velocity. The determination of the secondary flow velocities requires a clear view of the primary flow, so that in radial compressors with complex 3D pressure gradients and other cross-flows due to tip clearance effects it is not always so straightforward to determine the secondary flow velocities.

The boundary layers are generally relatively thin so that the external freestream static pressure extends through the boundary layer to the wall. In the freestream, the inviscid balance of forces determines the strength of any pressure gradients, so that the pressure gradients are in equilibrium with the centrifugal forces from the curvature of the streamlines in the blade-to-blade plane and in the meridional plane, and with any Coriolis forces. In the thin boundary layers close to the surface, the velocities are lower and the total pressure reduces towards the wall, but the transverse pressure gradient imposed on these is broadly the same as that of the bulk flow and leads to cross-flows in the boundary layers that are perpendicular to the main flow direction. In a stationary coordinate system, secondary flows with a streamwise component of vorticity are produced when a primary flow with nonuniform stagnation pressure, such as a boundary layer or a wake, is subjected to pressure gradients perpendicular to the main flow direction, Figure 5.19.

The Euler equation normal to a streamline, (5.10), can be used to explain secondary flows, which are flows with a velocity component normal to the main throughflow direction. Broadly speaking, the pressure gradient normal to the bulk flow direction in this equation is determined by the mean flow velocities in the area largely unaffected by viscosity. But in the thin boundary layers close to the surface, the velocities are

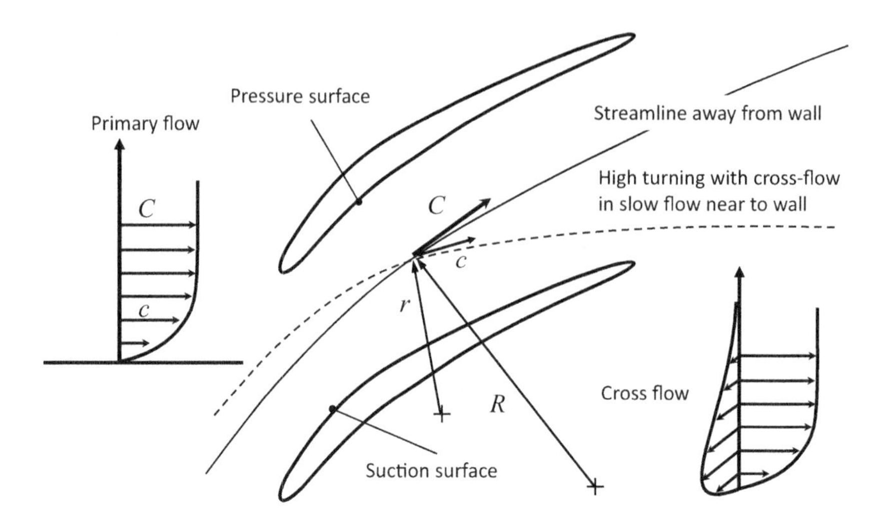

Figure 5.19 Generation of secondary flows in a compressor blade row.

lower but the pressure gradient imposed on these is the same as that of the bulk flow. The pressure gradient determined by the mean flow, causes the slow-moving fluid in the boundary layers to turn with a smaller radius of curvature to counteract the misbalance in the pressure gradient, as shown in Figure 5.19 and as determined by

$$\frac{1}{\rho}\frac{dp}{dn}=\frac{C^2}{R}=\frac{c^2}{r}, \quad c<C \ \Rightarrow\ r<R. \tag{5.56}$$

The end-wall boundary layer has a lower velocity than, but experiences roughly the same cross-stream pressure gradient as, the freestream. The streamline radius of curvature near the end-wall is thus smaller than in the freestream, leading to cross-passage motion and the accumulation of low stagnation pressure fluid near the suction surface hub corner. In this way, the boundary layer fluid near the end-wall migrates to the blade suction surface corner. Such flows are generated when a primary flow with a nonuniform total pressure, as in a boundary layer, is subjected to accelerations that are perpendicular to the main flow direction and are called secondary flows. If the blade loading is high enough, this low stagnation pressure fluid is not able to negotiate the pressure rise in the blade passage and hub-corner stall occurs, increasing passage blockage, lowering the static pressure rise capability of the compressor blade row and increasing the entropy rise from flow mixing downstream.

The cross-flow generated in this case is from the pressure to the suction surface and is accompanied by a flow in the return direction elsewhere in the flow channel, leading to the formation of a passage vortex, as shown in Figure 5.20. The passage vortex causes an over turning of the flow towards the suction surface adjacent to the end-wall, known as secondary deviation, and an underturning outside of the end-wall boundary layer and leads to a skewed boundary layer profile.

In axial turbomachinery, the secondary flows are weaker in compressors than in turbines due to the lower curvature of the flow path leading to weaker

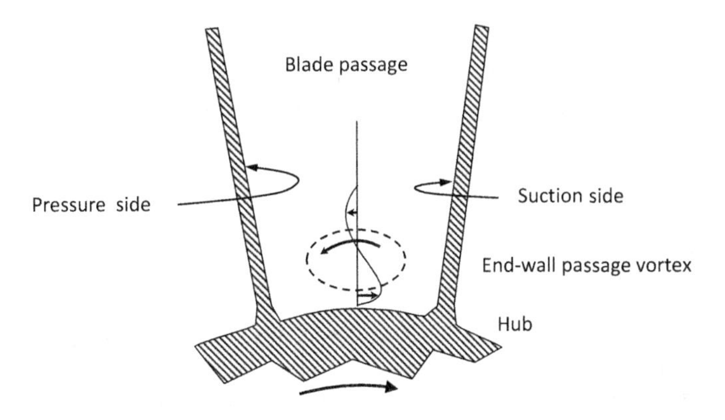

Figure 5.20 Generation of an end-wall passage vortex.

blade-to-blade pressure gradients. Also, the leading-edge thickness of compressors is less, and this leads to a smaller vortex in the stagnation region on the end-wall near the leading edge. In addition, the end-wall boundary layers in compressors are thicker. However, because of the adverse pressure gradient, the boundary layers are much more prone to separate in a compressor, and this is more likely to occur wherever boundary layer flow accumulates on the suction surface due to secondary flow.

5.6.2 Secondary Flows in Radial Compressors

The details of inviscid secondary flows and the generation of streamwise vorticity for flows in rotating passages was first derived by Squire and Winter (1951) and by Smith (1955). Hawthorne (1974) derived a simpler relationship which was only published as an internal report in Cambridge University. This was used by Johnson and Moore in their seminal description of the secondary flows in impellers in the late 1970s and 1980s. The Hawthorne equation for the rate of change of the secondary circulation, Ω_s, along a streamline is derived in detail in Johnson (1978) and also given by Hirsch et al. (1996) and Greitzer et al. (2004) as

$$\frac{\partial}{\partial s}\left[\frac{\Omega_s}{w}\right] = \frac{2}{\rho w^2}\left[\frac{1}{R_n}\frac{\partial p^*}{\partial b} + \frac{\omega}{w}\frac{\partial p^*}{\partial z}\right], \tag{5.57}$$

where p^* is the rotary stagnation pressure introduced in (2.52). The first term has its origin in the flow turning and the second term has its origin in the Coriolis forces, and R_n is the radius of curvature normal to the streamline. The curvature of an impeller flow passage is complex, so it is useful to consider the blade-to-blade curvature separately from the hub-to-shroud curvature and to include these terms separately, as given by van den Braembussche (2006),

$$\frac{\partial}{\partial s}\left[\frac{\Omega_s}{w}\right] = \frac{2}{\rho w^2}\left[\frac{1}{R_m}\frac{\partial p^*}{\partial b} + \frac{1}{R_b}\frac{\partial p^*}{\partial n} + \frac{\omega}{w}\frac{\partial p^*}{\partial z}\right], \tag{5.58}$$

where R_m is the radius of curvature of a streamline in the meridional plane and R_b is the radius of curvature in the blade-to-blade plane.

In these equations, p^* is the rotary stagnation pressure, that is, the pressure associated with the rothalpy, I, and the rotary stagnation temperature, T^*, which for an ideal gas would be as follows:

$$\frac{p^*}{p} = \left(\frac{T^*}{T}\right)^{\gamma/(\gamma-1)} = \left(\frac{I}{c_p T}\right)^{\gamma/(\gamma-1)} = \left(1 + \frac{w^2 - u^2}{2c_p T}\right)^{\gamma/(\gamma-1)}. \tag{5.59}$$

The rotary stagnation temperature is conserved, just like the rothalpy, in a steady adiabatic flow in a rotor. The rotary stagnation pressure is also conserved in a rotor if there are no losses. The rotary stagnation pressure is the pressure achieved when the relative flow is brought to rest without losses on the axis of a rotating machine.

Combination of (5.59) and (2.48) shows that, in an incompressible flow, the generation of secondary vorticity is determined by velocity gradients as follows (Hirsch et al., 1996; van den Braembussche, 2006):

$$\frac{\partial}{\partial s}\left[\frac{\Omega_s}{w}\right] = \frac{2}{w}\left[\frac{1}{R_m}\frac{\partial w}{\partial b} + \frac{1}{R_b}\frac{\partial w}{\partial m} + \frac{2\omega}{w}\frac{\partial w}{\partial z}\right]. \tag{5.60}$$

This equation describes the development of the secondary vorticity due to different types of secondary flow, as explained in more detail later in this section. The three terms in this equation are related to the generation of vortical flows known as blade surface vortices, end-wall passage vortices and Coriolis vortices as shown in Figure 5.21.

The first term of the right-hand side of (5.60) describes the vorticity related to the blade surface vortices, generated by the meridional curvature, where R_m is the radius of curvature in the meridional plane, and b is the coordinate in the blade-to-blade

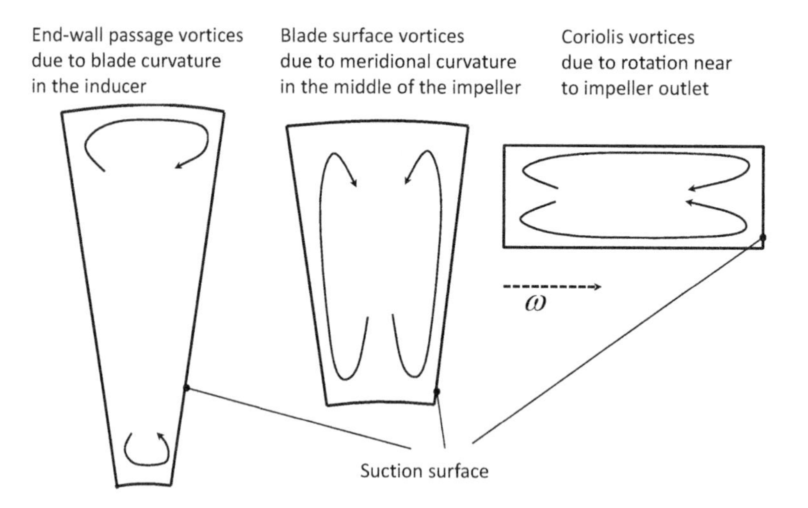

Figure 5.21 Idealisation of secondary flow vortices at different meridional positions from inlet to outlet in a radial impeller flow passage.

direction. The gradient of relative velocity in the blade-to-blade direction close to the blade surfaces in combination with the pressure gradient due to the meridional curvature causes a secondary flow to develop on both pressure and suction surfaces towards the casing. This motion starts at the leading edge, as the streamlines are already curved at this position. It will become stronger as the bend from the axial-to-radial direction of the channel becomes sharper and the blade boundary layer gets thicker. It will then vanish progressively as the curvature reduces in the radial part of the channel, as shown in Figure 11.16. The meridional radius of curvature is smaller at the casing than near the hub, so stronger secondary flows are found near the casing. The boundary layer develops more rapidly on the suction surface with higher diffusion and may grow rapidly downstream of a suction surface shock, so stronger blade surface secondary flows are expected along the suction side. The pressure gradient from the meridional curvature forces the blade surface boundary layers along both of the blade surfaces towards the casing. In two-dimensional impellers with no meridional curvature, this source of secondary flow does not exist, as the meridional radius of curvature is effectively infinite. In process compressor impellers with highly curved meridional channels, the blade surface vorticity dominates.

The second term on the right-hand side of (5.60) represents what are conventionally known as the end-wall passage vortices, generated by the flow turning with a radius of curvature of R_b in the blade-to-blade plane and the velocity gradient with respect to the meridional coordinate. The strength of these will depend on the blade shape rather than the meridional channel shape. In old-fashioned impellers, with a short inducer with high turning to the axial direction, this will cause very strong secondary flows in the end-wall boundary layers towards the suction surface of the blade both on the hub and the casing, causing a collection of the slow moving end-wall boundary layers on the suction surface. In the radial region of the impeller with no further change in the blade angle, no additional blade passage vortices are developed. On the casing, the passage vortex opposes the direction of the tip clearance jet and develops into a region of slow-moving fluid in blade suction surface/casing corner. The passage vortex also starts from the leading edge because the hub and shroud boundary layer already exist. Moreover, the relative velocity being higher near the shroud than at the hub, the gradients will be stronger at the shroud, where stronger secondary flows can be expected. In more modern impellers in which the inducer does not turn the flow to the axial direction, these secondary flows are weaker and may extend over a longer section of the blade. In blades that have a large backsweep, which increases towards the blade trailing edge, the passage vortices may even act in the opposite direction as the trailing edge is approached as the blade curvature is in the opposite direction.

The last term of the RHS of (5.60), originating from the Coriolis forces, only has an effect if there is a velocity gradient in the axial direction. This will be the case for the end-wall boundary layers in the radial and nearly radial parts of the impeller. This causes the secondary flows in the end-wall boundary layers to move towards the suction surface. They will either contribute to the passage vortices or oppose them depending on the blade curvature, as sketched in Figure 5.7. The three contributions in (5.60) act together and determine the ultimate location of any low momentum fluid from the boundary layers. An interesting study on the relative strength of the different contributions in a radial impeller was made by Johnson (1978), where he concluded that the influence of the axial to radial bend and the Coriolis forces are equally important and that the blade passage vortices are of least importance. Johnson and Moore (1980) showed that the boundary layer fluid is transported by the secondary flows to the location of the shroud suction side corner, but its exact location depends on the design and the operating point.

The secondary flows convect the fluid with low total pressure, but large additional losses are not thought to be generated by the small secondary flow velocities themselves (Denton, 1993). In radial impellers, the secondary flows may not be so important for loss generation but are very important for redistributing the slow-moving boundary layer fluid to the suction surface/casing corner. This, together with the high diffusion of the flow on the suction surface, causes growth in the boundary layers and may be responsible for low forward momentum, leading to flow separation in a strong adverse pressure gradient. The mixing of the separated flow in and downstream of the impeller may then be a cause for further losses. The motion of the secondary flows leads to an accumulation of low energy boundary layer fluid with high losses towards the suction surface casing region and a high velocity region around the pressure surface hub region. The ultimate position of the low energy fluid will depend on the balance of the different driving forces in (5.60) and the strength of any tip leakage flows; see Section 5.7.

In a recent publication on the use of splitters in turbine stators, Clark et al. (2017) have pointed out, following Squire and Winter (1951), that the amount of secondary flow vorticity is mainly dependent on the transverse pressure gradients. However, the kinetic energy losses that arise from this can be reduced by splitting the vorticity into separate smaller vortices where the kinetic energy of the of the vorticity is then smaller. This may be an additional unsung benefit of the use of splitter vanes in radial impellers.

5.7 Tip Clearance Flows

5.7.1 Structure of the Leakage Flow Pattern

Tip leakage flow over unshrouded blades in axial compressors has received considerable attention as it accounts for a reduction in performance (Koch and Smith, 1976). When the size of the tip clearance is increased, the efficiency drops and the stall margin deteriorates drastically. The work of Bindon (1989) is a careful study of the complex leakage flow pattern over turbine blade tips, and a similar study on compressor blades was completed by Storer (1991). Accurate measurement of the velocity field near the end-wall and inside the tip clearance of rotating blade rows is very difficult so that CFD has been widely applied to understand the physics of tip clearance flow. The description of tip leakage flow given here draws on such studies and is based partly on a lecture given by Miller and Denton (2018).

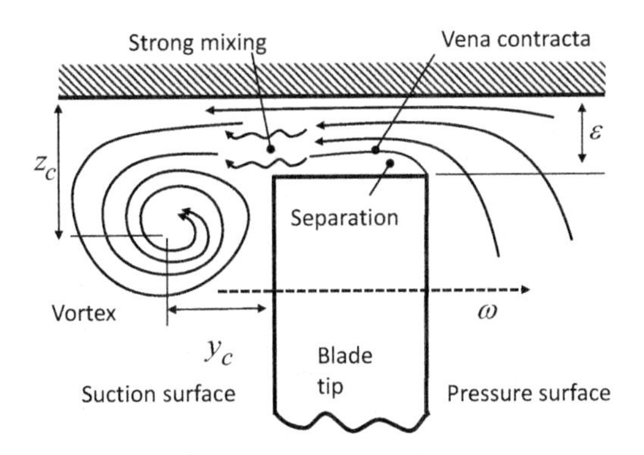

Figure 5.22 Sketch of idealised flow pattern over a compressor rotor tip gap.

The driving mechanism for a leakage flow across the blade tip gap at the end of unshrouded blade rows is the difference in pressure between the pressure surface and the suction surface causing a cross-flow in the end-wall region of the blade tip. This tip clearance flow generally opposes the end-wall secondary flows in a compressor flow channel. The flow leaking over the tip from the pressure side to the suction side will usually separate from the corner between the pressure surface and the blade tip and lead to a separation bubble on the blade end, as sketched in Figure 5.22. The separated flow forms a vena contracta in which the flow is confined to an area less than the tip clearance gap, and it is this area, above all else, that determines the amount of leakage flow. The leakage jet then becomes turbulent and expands to fill the whole clearance gap. In compressor blade rows with thin blades, this mixing process is not usually complete within the clearance gap, whereas in turbine blade rows with thicker blades the flow may reattach on the end of the blade before leaving the tip gap. In a compressor impeller, the leakage flow is reinforced by the movement of the casing relative to the blade. The idealised structure of the leakage flow in an axial compressor rotor can be visualised as in Figure 5.22.

Before passing over the tip gap, the leakage flow entrains any fluid in the pressure surface boundary layer near the blade tip. After passing over the tip of the blade, the tip clearance flow mixes with the flow already on the end-wall, entrains boundary layer fluid from the suction surface and rolls up to form a clear vortex structure beginning at the leading edge and containing any entrained boundary layer fluid from the end-wall and the pressure and suction surfaces. This vortex grows as the flow passes downstream, and more leakage flow enters the vortex core. At very low tip clearance gaps, there is little shed vorticity and often no strong evidence of a tip clearance vortex. The presence of boundary layers on the adjacent surfaces and on the blades can cause additional vortices, separations and shear flows which interact with the tip clearance flow, such as scraping vortices, interactions with the casing wall boundary layer, end-wall separation, rolling up of blade boundary layers into tip clearance vortex, free shear layer in the tip gap and tip separation vortex (Kang and Hirsch, 1993). The relative motion of the end-walls is also important such that in compressor rotors, the end-wall motion is in a direction that tends to increase the flow through the tip gap and in turbine rotors ittends to decrease the flow through the tip gap.

A simplified inviscid analysis of the trajectory of the tip leakage vortex in an axial compressor blade row with a constant chordwise distribution of blade loading and no secondary flows is given by Chen et al. (1991). This shows that the depth of the centre of the vortex from the casing end-wall, z_c, remains at a constant radial position as the vortex develops and is proportional to the tip clearance gap width, ε. The theory predicts that the position of the vortex core from the blade suction surface, y_c, is also proportional to the tip clearance gap width and increases linearly as the vortex develops downstream. This approach has been extended by Aknouche (2003) to a centrifugal pump impeller.

In a real viscous flow in a radial impeller, the flow is more complex. The combined flow pattern of the different vortices can be simulated with CFD, and there are many published papers showing the computed trajectory of the secondary flow vortices and the leakage flow vortices and their relationship to the flow distortion at the impeller outlet. The tip vortex entrains boundary layer flow from the suction surface and contains fluid with low total pressure at its core, giving it a wakelike structure along its entire length. The core of the vortex has a throughflow velocity substantially different from the freestream velocity and is of course a strongly swirling flow. The tip clearance vortex may also interact with any suction surface shock waves and grow by a mechanism of vortex breakdown (Hazby, 2010). In a radial compressor, the blade chords are long compared to the spacing, so that as the vortex passes downstream it may also be ingested by the tip clearance gap of the next adjacent blade or interact with a splitter blade leading edge.

An example from an impeller CFD simulation is given in Figure 5.23. The Mach number contours in the blade-to-blade view of this simulation near the tip section cut through the tip vortex and identify its location as a trough in the velocity level which shows its trajectory across the flow channel. In this case, it would appear that there are two vortices in each flow channel, one travelling across the flow channel to the pressure side of the next blade. This grows in strength as it passes through the shock

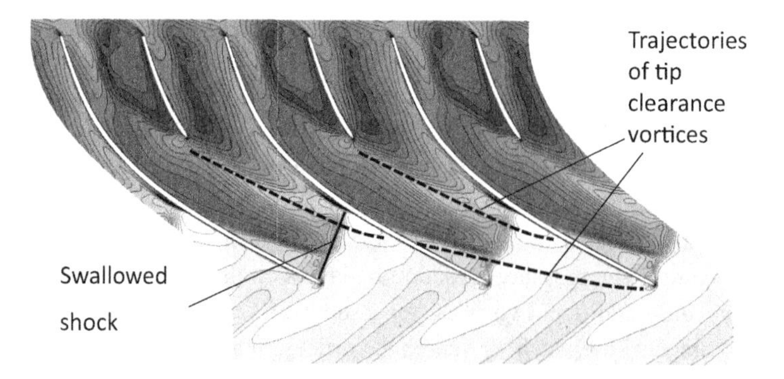

Figure 5.23 Trajectories of tip clearance vortices near the casing in a simulation of a transonic impeller with a splitter blade

wave in the flow channel and then passes over the adjacent blade emerging as a second vortex, which can be seen reaching to the splitter vane in the adjacent channel.

Due to the high inlet blade angle, of close to $\beta_{c1} = -60°$, the leakage flow over the blade tip has a strong forward component of velocity within the jet especially near the leading edge. The exact trajectory of the vortices cannot be generalised as it depends on the operating point, the number of blades, the shape of the blades, the location of the splitter blade leading edge and the clearance level. The strength of the tip clearance jet becomes stronger with a more circumferential orientation at flow rates below design, and at very low flow rates the vortex may even be in the upstream meridional direction. Thanks to modern CFD simulations, the flow patterns in open impellers are well known and clearly understood, and are discussed in more detail in Section 5.8.

In multistage radial compressors, the impellers are generally fitted with shroud cover plates so that the leakage gap does not exist, but there is then a leakage over the shroud cover plate from impeller outlet to impeller inlet, and usually some seals on the cover plate with low clearance to reduce the strength of this. Experimental data obtained on a high flow coefficient impeller published by Dalbert et al. (1999) indicated that for an open impeller with a small clearance gap and the same impeller equipped with a shroud cover plate, the performance, efficiency and operating range were indistinguishable within the accuracy of the industrial measurements. There is, however, evidence from low-speed axial compressor research tests that a very small clearance gap can be more efficient than no gap at all, as the leakage jet blows away the slow-moving boundary layer on the suction surface corner and may in this way impede a separation (Cumpsty, 2004).

5.7.2 Effect of Blade-to-Blade Loading

Many publications are available which highlight the sensitivity of efficiency to the tip clearance levels in radial compressors. An increase in the size of the tip gap not only reduces the efficiency, it also reduces the flow capacity due to blockage of the annulus by the tip clearance vortex. The blockage may also reduce the work and as a consequence there is less pressure rise, which may also reduce the surge margin of the compressor. Few experimental and theoretical studies have been made of the fine detail of the tip clearance flows in radial impellers. Fortunately, there is substantial literature available on the nature of tip clearance flow in axial compressors and turbines, so that the fundamentals of the fluid physics of tip clearance flow are well understood. The work on axial turbines, which generally have a large blade thickness, is less relevant in compressors, as in turbines the leakage flow over the blade tip causes a separation bubble on the blade end adjacent to the pressure surface and reattaches on the blade tip surface before the end of the clearance gap on the suction surface. In compressors with thinner blades, the jet of the tip clearance flow travels through the gap unimpeded. The work on tip clearance in axial compressors outlined by Rains (1954), Storer and Cumpsty (1994), Chen et al. (1991) and Denton (1993) provides the basis of the physical description of the tip clearance flow given here.

The main factor in the tip clearance flow is the pressure difference between the pressure and suction surface of the blade, which drives a leakage flow through the gap over the blade tip. The amount of leakage flow is clearly also influenced by the area of the tip gap so that any losses associated with the leakage flow tend to increase as the gap becomes a larger proportion of the blade span. No work is done on the flow leaking over a blade tip, and so one effect of a clearance is a reduction in the work input. The loss of work occurs because the blade load decreases towards the tip and is not related to any losses caused by the leakage flow. In addition, the blockage caused by the leakage jet increases as the clearance gap increases so that there may be a reduction in the flow capacity of the impeller.

In a compressor, the flow leaking over the tip from the pressure side to the suction side will usually separate from the corner between the pressure surface and the blade tip. The separated flow forms a vena contracta in which the flow is confined to an area less than the tip clearance gap. Typically, a vena contracta has a discharge coefficient of 0.61, but Storer and Cumpsty (1994) found that a value of 0.8 matched their data better. Following this, the jet forms a very clearly defined shear layer which remains separated, and any mixing takes place after the jet has left the clearance gap. The process is completed in a distance of the order of five tip clearances, so that in axial and radial compressors where the blade thickness is less than this, it will not be complete when the leakage flow emerges from the clearance space onto the suction side of the blade (Miller and Denton, 2018). The loss generated by the leakage jet is discussed in Chapter 10.

Clearly this theoretical outline of the physics is essentially inviscid in nature and quite approximate. In reality, the tip clearance flow does not occur as a single phenomenon; it is also linked to other 3D secondary flows on the end-walls, as outlined in other sections of this chapter. But the inviscid nature of the tip clearance flow mechanisms suggests that numerical CFD simulations of such flows in radial compressors will capture all the essential features of this flow.

5.8 Jet-Wake Flow in Impellers

5.8.1 Early Descriptions of the Jet-Wake Flow

The flow patterns in radial compressor impellers depend on their geometrical design and the operating point and can be strongly influenced by flow separations, secondary flows and tip clearance flows as described in the preceding sections. In the late 1960s, many impellers used in turbochargers and gas turbines were rather old-fashioned types and were very strongly influenced by flow separations even at the design point. These impeller types consisted of an axial inducer with a blade section of high turning which brought the relative flow to the axial direction in a short axial distance, followed by purely radial blades with no backsweep in a highly curved meridional channel turning the flow to the radial direction. Such impellers do not differ greatly from the original impellers of Elling, Ohain and Whittle as used in the first functioning gas turbine and jet engines (see Section 1.6). The geometrical description of such an impeller used in the Rolls-Royce Ghost jet engine is given by Moore (1976). An impeller of this type was studied by Eckardt (1975) as impeller 0 in his research described in Section 5.8.2, and the elliptic camber line shape and meridional flow channel of another similar impeller is given in Krain (1981). An example of such an impeller from Elling dating back to 1904 is shown in Figure 1.17. At this time, industrial impellers for process gas applications tended to already have backsweep and often had no inducer with the leading edge in the bend from the axial to radial direction.

The loss production and the performance characteristics of centrifugal compressors with impeller blades with no backsweep are strongly dependent on the separation of the flow inside the impellers, an aspect which could not be analysed by their designers at the time, as no CFD tools for the analysis of this were available. The radial part was designed with purely radial blades to minimise bending stresses in the blade roots, and the axial inducer with radial blade elements was needed to guide the inlet flow into the impeller and transfer it to the radial part. In many cases, the two parts were even manufactured separately, the axial part being composed of radial blade elements, and the radial part as a simple flat section, and the two parts were joined together for use.

Dean and his coworkers named the flow pattern in such impellers the jet-wake flow (Figure 5.24). They highlighted the formation of a strong jet-wake pattern at the outlet of such impellers, in which a region of low-energy fluid accumulated in the shroud suction side corner of the blade channel accompanied by a high-velocity fluid in the hub pressure side part of the flow passage (Dean and Senoo, 1960; Dean, 1971). The jet is found on the pressure side of the blade and shows the highest radial velocity. This region has the stagnation pressure of a nearly loss-free flow in the relative frame

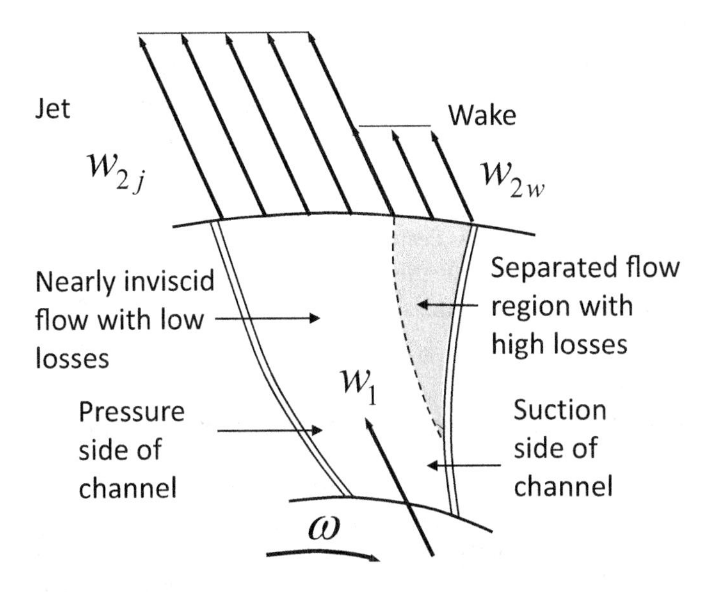

Figure 5.24 Sketch of the analytical model of jet-wake flow suggested by Dean and Senoo (1960).

of reference. The wake region is found on the suction side of the blade and shows a much lower radial velocity associated with a very high tangential velocity in the absolute frame. In the original model, it was also assumed that these regions could have slightly different flow directions in the relative frame.

It was realised that the separation into a jet and wake was a source of loss in the impeller and that the strong mixing of the two regions might also give rise to additional losses in the downstream diffuser. At this time, both Moore (1973) and Rothe and Johnston (1979) made large steps in our understanding of separated flow in straight rotating flow channels. Moore (1973) measured the existence of a separated flow on the suction surface of a rotating channel. Rothe and Johnston (1979) experimentally studied a series of low-speed rotating straight diffusing channels with an area ratio of 2, in which a separation was forced by a backward-facing step on the diffuser wall. The effect of curvature and rotation on the turbulence properties in the separated layer caused the separated shear layer to become stabilised such that it did not undergo transition and did not reattach. The separated region grew by the incorporation of secondary flows from the end-walls of the channel and led to a 2D separation zone similar to that in Figure 5.24. At this time, Hawthorne (1974) completed his analytical work on secondary flows, showing how curvature and Coriolis forces contribute to the development of pronounced cross-flows in radial impellers, as discussed previously, and help to feed the separated wake region with low-momentum fluid, but a complete understanding was not possible until the advent of practical CFD simulations.

The poor level of understanding of this flow structure at that time can be seen in the publications of Traupel (1975) and Balje (1978). Traupel (1975) made theoretical studies of the flow pattern and surmised that the stratification effect in a curved flow was due to the turbulence: the pressure gradients across the flow channel transported those elements of the flow with high local velocity to the pressure side and those elements with a lower velocity to the suction side. This is essentially the same mechanism as in a secondary flow (Figure 5.19), except the local differences in velocity due to the turbulence are the mechanism for the energy stratification into a jet and a wake rather than the differences in velocity level in the freestream and the boundary layer. Balje (1978) argued that the boundary layer on the concave side of the blade-to-blade flow channel is destabilised and in contrast the flow on the convex side is stabilised. He argued that this reduces the boundary layer growth on the pressure side and increases it on the suction side, leading to a stratification effect in the flow channel.

5.8.2 Breakthrough in Understanding through Laser Measurements

The breakthrough in the visualisation of this jet-wake flow pattern in real radial impellers was made in a series of German research projects at the German Aerospace Centre (DFVLR), the forerunner of the Institute of Propulsion Technology of the DLR. The projects from this group have continued to the present day, and the publications of the DLR group are emphasised in this section, even though publications from many other sources can be found to support the arguments given.

The DLR projects began by using the laser two-focus velocimetry method of Schodl (1980) in a radial compressor test rig equipped with windows, to investigate the internal flow patterns in impellers, as highlighted in the publications of Eckardt (1975), Eckardt (1976), Eckardt et al. (1977), Eckardt (1979) and many subsequent publications by his successors at the DLR research laboratory, for example, Krain (1987), Krain and Hoffman (1989), Krain et al. (1995) and Krain et al. (2007). Many of these research projects were financed by the Research Association for Combustion Engines or *Forschungsvereinigung Verbrennungskraftmaschinen* (FVV) in Germany, and an overview of the first 40 years of important research on radial compressors financed by this group is given in Hederer (2011).

Following the elucidation of the flow patterns through laser velocimetry, the publications of Johnson and Moore then used theoretical considerations and CFD simulations to clarify the flow field (Johnson, 1978; Moore and Moore, 1981; Johnson and Moore, 1983). These publications showed clearly that it is the strong decelerations of the relative flow on the casing streamline which lead to the formation of the jet-wake flow pattern in the blade-to-blade plane, and that this flow pattern is highly three-dimensional due to the transport of low-momentum flow from secondary flows and tip clearance flows.

As a short overview of the initial research work by this group, the meridional geometry of the first three impellers tested by Eckardt at the DLR, known as impeller 0, A and B, are shown in Figure 5.25, together with the measured meridional velocity profile measured at impeller outlet near the peak efficiency point of the stages. These stages were designed before the advent of practical CFD simulations. All three stages were constrained to have the same meridional channel on the casing wall to facilitate the use of the same windows for the laser velocimetry.

The outlet meridional velocity profile for the impeller 0 of the DLR series is shown on the bottom-right of Figure 5.25. This impeller has radial vanes with no backsweep following the inducer so that a high diffusion of the flow is present in the impeller. The impeller outlet velocity profile shows a strong stratification of the flow reminiscent of the jet-wake model of Dean (1971). The internal flow measurements halfway through the impeller show that this flow pattern is caused by a strong separation from the blade suction side on the inducer on the casing.

The middle right of Figure 5.25 shows the meridional velocity profile of a modified version of impeller 0, known as impeller A. This has the same inlet blade geometry but has a short section of backsweep near the trailing edge, leading to a backsweep angle of –30° near the outlet. The hub contour is modified to adjust the impeller area ratio to account for the different pressure ratio with backsweep, but the casing contour is similar to impeller 0. In this impeller, the flow is first turned to the radial direction before a backsweep is introduced, so that the early part of the impeller also has a strong deceleration of the flow leading to a suction surface flow separation on the casing. The distortion of the flow at the outlet still resembles the jet-wake model, but the depth of the wake region is much reduced and the wake is farther from the suction surface.

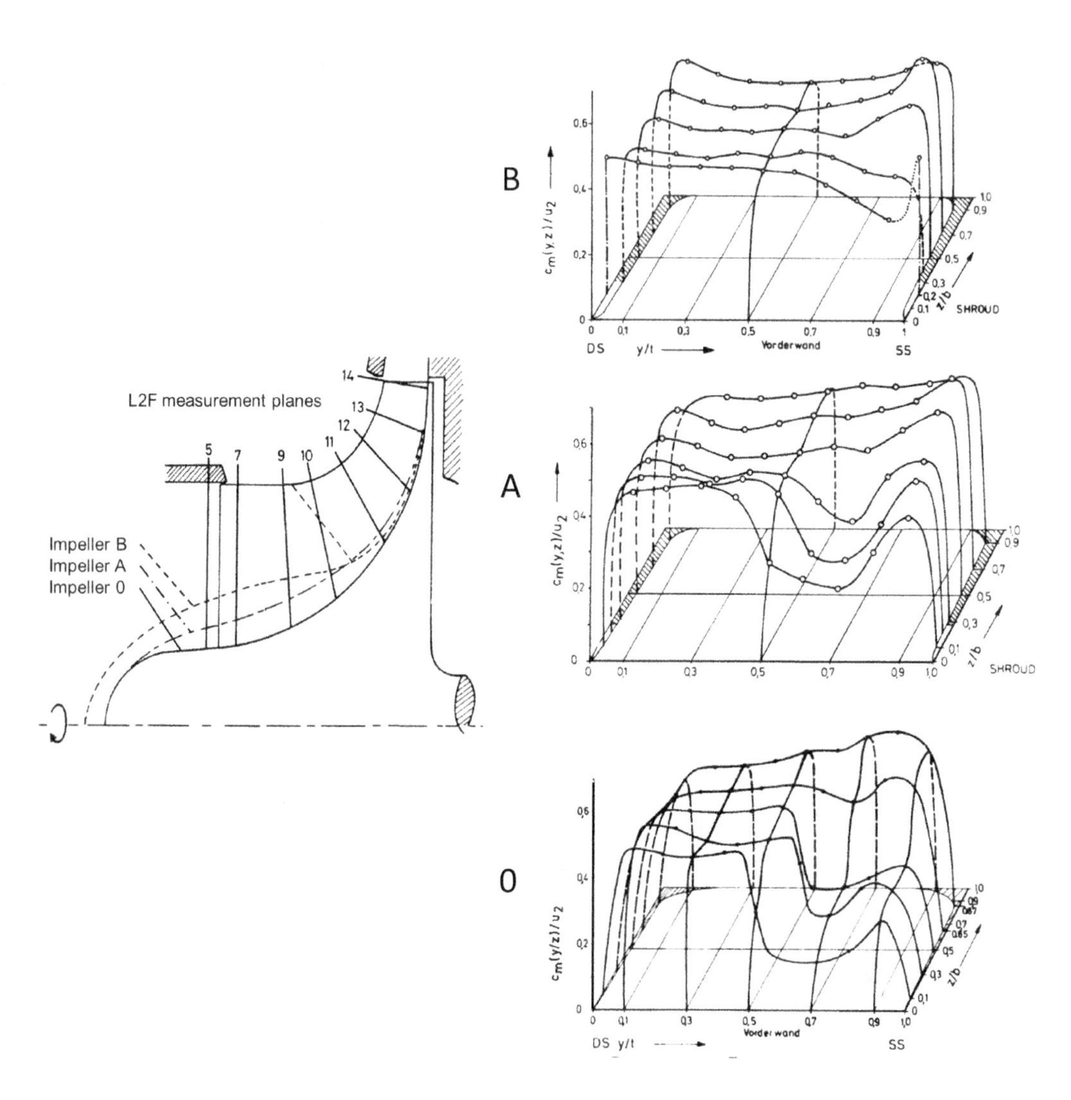

Figure 5.25 Meridional flow channel of the impellers 0, A and B studied by Eckardt and the measured outlet meridional velocity profiles of the impellers. (image adapted from the work of Dietrich Eckardt and courtesy of the FVV Research Association for Combustion Engines)

On the top-right of Figure 5.25, the meridional velocity distribution of impeller B is shown. In this impeller there was no inducer as the impeller was designed to represent a typical process compressor stage with the leading edge in the bend from axial to radial direction. This stage had a backsweep of 40°. The distribution of blade angles from inlet to outlet was more uniform and more typical of modern impeller designs. In this case, there was no evidence of any suction surface flow separation, and the jet-wake flow pattern does not occur: there is simply a region of low meridional velocity near the suction surface casing wall at the impeller outlet.

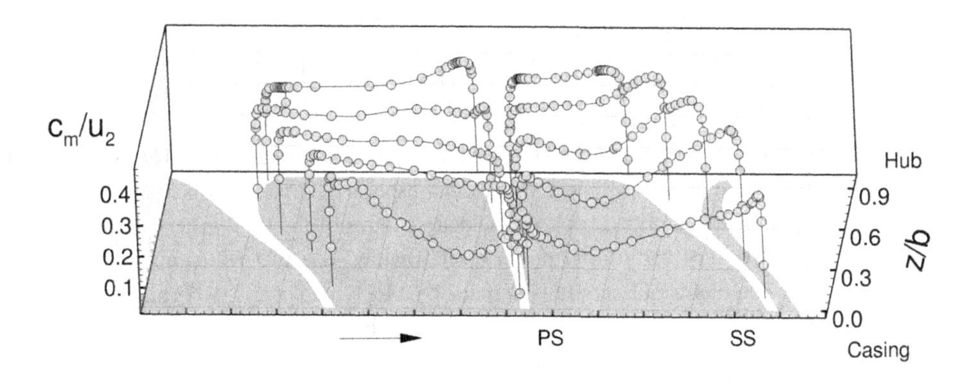

Figure 5.26 Distorted meridional velocity distribution at an impeller outlet designed with CFD using the TRACE CFD code; PS = pressure side, SS = suction side. (image by David Wittrock courtesy of DLR Institute of Propulsion Technology)

As an example of more recent work by this DLR research group, Figure 5.26 shows the meridional profile calculated from a CFD solution at impeller outlet of a more modern design (Elfert et al., 2017), in which a CFD optimisation design system was used to complete the design of a backswept impeller with splitter blades with a lower overall diffusion (Voss et al., 2014). This impeller, known as SRV5, also exhibits a more uniform distribution of blade angles from impeller inlet to outlet, as impeller B, but the blade surface is not defined by straight line generators, so the blades are curved from the hub to the casing as shown in the photograph of the same impeller in Figure 11.28. In addition, the splitter blade is designed separately to the main blade and is not simply a cutoff version of the main blades. In this case, there is hardly any evidence of a jet-wake flow in the blade-to-blade direction. The meridional velocity profile shows a hub-strong tendency across the whole span, and the region of lowest meridional velocity is closer to the middle of the casing between the main blades than it is to the suction surface corner. The highest meridional velocities occur on the suction side of the channels near the hub, contrary to that expected by the jet-wake model, but high velocities are also present on the pressure side of the channels.

Similar results showing the limited validity of the jet-wake model as a description of the flow in a well-designed impeller with lower diffusion levels were published by Rohne and Banzhaf (1991) and by other publications of the DLR group such as that of Krain et al (2007). Rohne and Banzhaf (1991) measured two impellers with splitter blades of modern design with the L2F method to identify that the main meridional distortion in the outlet flow is a meridional velocity which decreases from the hub to the casing, with a fairly constant distribution of absolute swirl velocity across the span. Krain et al. (2007) describe L2F measurements of an impeller called SRV4 designed with the aim of achieving a more homogeneous flow distribution at impeller outlet. This work shows that it is possible to avoid the distinct jet-wake flow at impeller outlet arising from flow separation and also to reduce the hub to casing flow inhomogeneity.

5.8.3 A Modern Synthesis

Following the early studies of Eckardt, numerous publications have appeared over the last 30 years examining the flow development in impellers in fine detail. In many of these, the flowfield is examined both through measurement data and CFD simulations. It is not possible to review all these studies, but aspects of several publications well known to the authors are described here to elucidate features of the flow pattern at impeller outlet.

Two very early CFD publications making use of CFD simulations to analyse the measurements of Eckardt were made by McDonald et al. (1971) and Casey et al. (1992). Casey et al. (1992) made simulations with the Dawes BTOB code, which used the Baldwin–Lomax turbulence model described in Section 16.6.4, to study the Eckardt impeller B, with 40° of backsweep. This impeller did not turn the flow to the radial direction before the backsweep was applied but like most modern impellers used a blade angle distribution with a smooth linear change in blade angle from inlet to outlet, which leads to no strong evidence of a separated jet-wake flow (see Figure 5.27). The distribution of the meridional flow at the impeller outlet from the Dawes BTOB simulations is in reasonable agreement with the L2F measurements of Eckardt and show the reverse flow very close to the casing wall that is not visible in the measurements. There is no clear wake region, and the region of low meridional velocity near the suction surface casing wall is related to the tip clearance flow and the accumulation of the secondary flows rather than to a flow separation. These and other results of this period showed the enormous potential of CFD to aid the understanding of the flow in impellers.

The pattern of the hub-strong radial velocity in modern impellers is dominated by the tip clearance flow and with a large clearance there is local flow reversal near the casing. The results of Jaatinen-Värri et al. (2013) show that increasing tip clearance considerably increases the reversed flow into the impeller at the outlet and that the

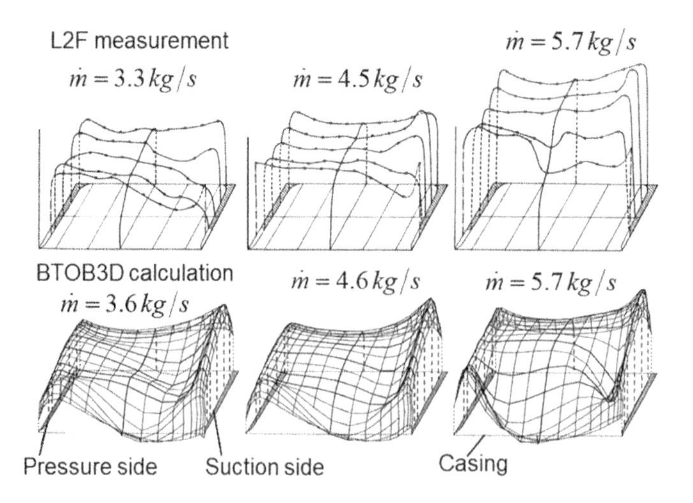

Figure 5.27 Distorted meridional flow velocity of Eckardt impeller B with 40° backsweep as measured by Eckardt (1977) and simulated with the Dawes BTOB3D code. (image adapted from those of Casey et al., 1992, using data of MAN Energy Solutions)

reversed flow then partly mixes into the flow in the same blade passage where it entered and the rest migrates over the blade, mixing with the tip clearance flow.

An interesting feature of the simulations and the measurements in Figures 5.26 and 5.27 is the small ridges of high meridional velocity adjacent to the blade suction and pressure surface from the casing to the hub near the trailing edge. The analysis of Hazby et al. (2013) of the measurements of Schleer et al. (2008) was the first to identify this ridge of high velocity as an effect of slip. Due to the relative eddy, the flow is not completely turned to the blade direction at outlet but is underturned at the trailing edge. This causes a high velocity on the pressure surface as the flow accelerates around the sharp trailing edge corner. In many publications, this small local region of high velocity adjacent to the blade surfaces is incorrectly confused with the jet of the jet-wake model.

The picture that emerges from the measurements and the simulations in modern impellers is that the flow structure does not comply with the classical jet-wake flow pattern shown in Figure 5.24. The main feature shown in simulations and measurements in modern impellers is that the radial velocity on the casing stream surface is much less than that on the hub, as in Figures 5.26 and 5.27. The key features in the measured flow field at impeller outlet cannot be linked directly to the impeller flow fields from the measurements alone. The understanding of the flow features at impeller outlet can best be appreciated by using the CFD calculations to trace back the origins of the outlet flow patterns with streamlines in the relative frame of reference to identify their source from details of the flow within the impeller.

The predicted contours of absolute tangential velocity at 2.5% downstream of an impeller trailing edge for the SRV5 impeller described by Voss et al. (2014) are shown in Figure 5.28 The flow streamlines in the relative frame of reference are also shown in

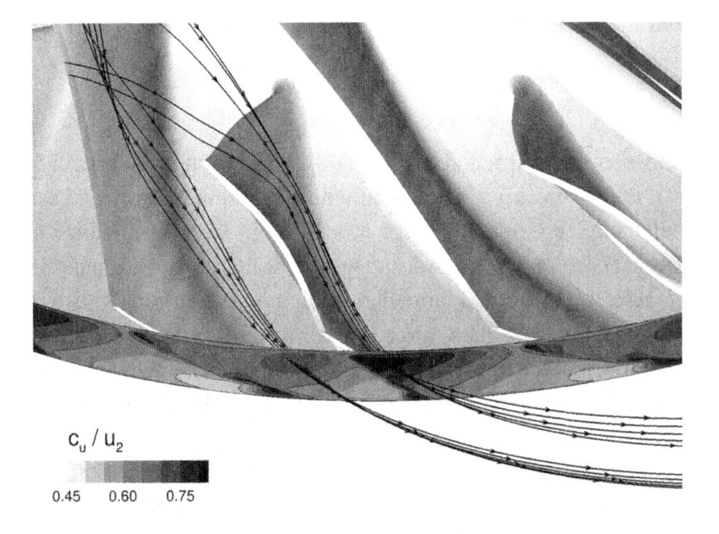

Figure 5.28 Streamlines from the leakage flow over the impeller tip, CFD using the TRACE CFD code. (image by David Wittrock courtesy of DLR Institute of Propulsion Technology)

the figure to demonstrate the origins of the observed flow features at the impeller outlet. The CFD calculations have been carried out with the DLR code TRACE. Several distinct regions in the contours of absolute tangential velocity can be identified. The same flow features can be observed in many modern impeller simulations, for example, those of Guo et al. (2008), Hazby et al. (2017) and Elfert et al. (2017). A small region of high tangential velocity is associated with the blade wake immediately downstream of the impeller cutoff trailing edge on the hub of both the main blade and the splitter blade. Not surprisingly, the flow patterns appear to be different at the exit of the two passages on either side of the splitter vane, and this is a common feature of most simulations which arises from the unequal channel flows as a result of the velocity gradient from suction to pressure surfaces at the inlet to the splitter blade.

The relative flow streamlines released in the region of moderately high tangential velocity midway between the suction surface of the main blade and the splitter blade and traced backward inside the impeller show that the leakage flow from the front part of the main blade has reached the outlet of the impeller at this location and operating condition. The splitter blade tip leakage flow joins with leakage over the rear part of main blade and creates a strong low-momentum region close to the suction side of the splitter blade, which also entrains the secondary flows from the suction side of the splitter blade. This might be considered to be a wake, in the sense of Dean and Senoo (1960), but is in fact the core of the tip leakage vortex. The resultant blockage of the vortex pushes the main flow towards the hub region, resulting in a higher meridional velocity in this region and consequently to a lower absolute tangential velocity near the hub wall. The low tangential velocity in the first splitter passage on the suction side of the main blade has almost disappeared in the second splitter passage. As discussed in detail by Hazby et al. (2013) this region is primarily formed from the flow close to the suction side near to the tip of the main blade that turns rapidly in the relative frame of reference at the impeller exit.

5.8.4 Conclusions

There is no clear jet-wake structure due to a flow separation in Figure 5.28, nor in any of the other of the publications mentioned with flow simulations or measurements in modern impellers, except towards surge at off-design when a suction surface separation may form. These research results suggest that the jet-wake phenomenon identified in early research on radial impellers with no backsweep with a strong flow stratification in the blade-to-blade direction is mostly attributable to the flow separation on the suction surface near the casing. This is caused by the large deceleration of the relative velocity in impellers of that period with purely radial blades at outlet and highly loaded inducers which turned the relative flow to the axial direction. Impellers with more backsweep and with a more uniform blade angle distribution through the impeller do not show the classical jet-wake phenomenon with low flow velocities on the blade suction surface due to flow separation. In modern impellers, the effect of the tip clearance flow and the secondary flows dominate the flow pattern at impeller outlet, causing a lower meridional velocity on the casing than the hub. The complex

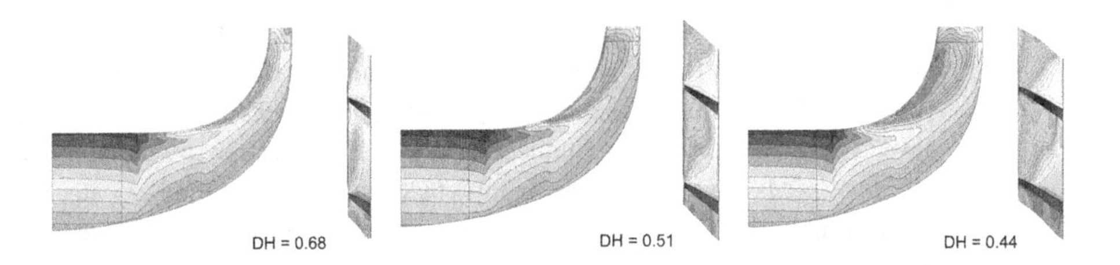

Figure 5.29 Flow separation at different levels of de Haller number in a radial impeller.

flow patterns just downstream of the impeller are affected by a mixture of the secondary flows, the tip clearance flows, the relative eddy and the trailing edge blade wake and depend on the number of blades and the operating point.

The designer has little control over the details of this flow pattern. The designer does, however, have control over the possible flow separation in the impeller by the choice of the curvature of the meridional flow channel, the blade angle distribution and the overall level of deceleration as expressed by the de Haller number $DH = w_2/w_1$ on the casing streamline of the impeller. This is demonstrated in the simulations shown in Figure 5.29. In these simulations, an impeller designed for a very moderate diffusion level, or a high de Haller number, is steadily widened at the impeller outlet to increase the level of diffusion in the impeller. The blade angle distribution along the hub and the casing, the inlet area, the hub contour, the flow rate and rotational speed remain the same. The way in which widening of the impeller outlet increases the amount of diffusion in the relative flow in the impeller is described in Section 11.6.5. Figure 5.29 shows contours of the area-averaged relative Mach number in the impeller in a meridional view together with contours of the relative Mach number at the impeller outlet. The de Haller number given in the figure has been determined from the meridional throughflow calculation at the design point of each impeller, using the method of Chapter 15 and not from the CFD simulations shown.

The narrowest impeller shown on the left has a de Haller number of $DH = 0.68$, which is typical of an impeller designed for a work coefficient of 0.6 for use with a vaneless diffuser. There is little evidence of any separation along the casing streamline, although there is a weak meridional throughflow on the casing due to the effect of the clearance. This impeller exhibits a hub-strong relative Mach number distribution at the outlet with a fairly uniform velocity distribution in the circumferential direction at the outlet plane, similar to that of Figure 5.26. Decreasing the de Haller number to $DH = 0.51$ by widening the impeller at outlet causes a region of low meridional velocity to form at the casing, and the outlet velocity distribution begins to show a jet-wake tendency with a larger distortion in the circumferential direction. The diffusion is most severe at the shroud, and with a wider impeller as shown on the right, the flow breaks down on the suction surface on the casing, exacerbated by the clearance flow. In the widest impeller shown, with a de Haller number of $DH = 0.44$ on the casing streamline, the flow separation is extremely strong, and there is a clear jet and a wake in the outlet plane reminiscent of impeller 0 of Eckardt and the jet-wake model of

Dean and Senoo (1960). These results confirm that well-designed impellers with a moderate diffusion tend to exhibit a hub-to-casing distortion in the outlet meridional velocity profile and that impellers with a high diffusion exhibit a casing separation and have a jet-wake distortion in the blade-to-blade direction at the impeller outlet. Moreover, it is clear from these results that the de Haller number calculated by an inviscid throughflow method provides the designer with excellent guidance as to the risk of separation, so that it can be avoided by the use of throughflow calculations in the preliminary design phase of the impeller.

Despite the advent of well-designed impellers using throughflow simulations and modern CFD analysis in which the very pronounced jet-wake flow, as depicted in Figure 5.24, has practically disappeared, this has not caused the demise of loss models based on the jet-wake model, as described in Section 10.5. But the other simple 1D loss models also described there are also unable to capture the details of this highly complex flow pattern and its loss generation processes. Fortunately, modern CFD simulations within the impeller can identify the flow patterns due to clearance and secondary flows and provide an estimate of the losses generated.

6 Gas Dynamics

6.1 Overview

6.1.1 Introduction

This chapter summarises the laws of gas dynamics, that is, the fluid dynamics of compressible flows in which substantial changes in density occur. Such flows are relevant to understanding the high-speed gas flows in channels of variable area and in turbocompressor blade rows. A one-dimensional approach is used to describe the important phenomena, such as the nature of choking, normal and oblique shock waves and expansion waves. This 1D approach enables important engineering estimates to be made and gives powerful insights as a complement to more complex numerical simulations.

Special emphasis is given to the nature of the transonic flow at inlet to a radial compressor inducer, and how this is affected by the blade shape and the operating conditions. The compressible flow equations are considered for flows of ideal and real gases.

6.1.2 Learning Objectives

- Understand the different nature of subsonic flow, transonic flow and supersonic flow.
- Be familiar with the 1D theory of gas dynamics and how it affects compressible flow in turbocompressor flow channels.
- Know the meaning and significance of corrected flow per unit area or the mass-flow function.
- Know the important features of choking, expansion waves, normal and oblique shock waves and unique incidence in compressor blade rows.
- Know the difference between the compressible flow of real and ideal gases.

6.2 Gas Dynamics of Ideal Gases

6.2.1 Background

This section considers the effect of compressibility on the flow. The real flow in a compressor blade row may be highly three dimensional, unsteady, viscous and turbulent, but useful understanding can be obtained by first considering an inviscid steady one-dimensional flow in a straight duct of varying area. The nondimensional parameter of interest in such flows is the Mach number, $M = c/a$, which is the ratio of the local gas velocity, c, to the local speed of sound, a. A subsonic flow has a Mach number $M < 1$ and a supersonic flow has a Mach number $M > 1$. Further discussion of the Mach number as a similarity parameter is given in Chapter 8. It can be shown that the speed of sound is the speed of small isentropic sound waves through the gas, $a = \sqrt{(\partial p/\partial \rho)_s}$ so that in an ideal gas the speed of sound is given by $a = \sqrt{(\gamma RT)}$. A sonic Mach number then implies that no signals can move upstream in the flow. In most compressor fluids, the speed of sound of the gas is of the same order as the mechanical blade speed, and large density changes may occur.

There are three main issues related to compressibility in compressors that need to be properly understood. The first is that the compressible flow does not have constant density, so that for a fixed mass flow the volume of the gas changes through the machine. Application of the continuity equation in hydraulic machines is much simpler than in thermal turbomachines as with a constant density the velocity varies in inverse proportion to the area change of the flow channels. In compressible flows, this is not the case, and both fluid dynamics and thermodynamics are used to deal with the change in density. It is found that the sensitivity of the changes in velocity to an area change are highest when operating close to a Mach number of $M = 1$. Also, in a supersonic flow with $M > 1$, an increase in area further increases the gas velocity, whereas in a subsonic flow with $M < 1$ an increase in area causes a deceleration of the flow.

The second issue is the phenomenon of choking, in which the local Mach number in the throat of the flow passage concerned reaches unity and no further increase in the nondimensional mass-flow function is possible. The third issue is the phenomenon of shock waves, which are a discontinuous change in properties in a supersonic compressible flow, and these become important at the design point in compressors with high subsonic, transonic and supersonic relative inlet flow, and at off-design points with low pressure ratio.

Excellent introductions to compressible flow with more extensive details are given in many fluid dynamic textbooks, and the relevance of such flows in turbomachinery are given by Greitzer et al. (2004) and Bölcs (2005) and in other references given there.

6.2.2 One-Dimensional Flow in Variable Area Ducts

An ideal compressible fluid flowing from a large container, or from the ambient environment, is considered. This has a fixed total temperature and total pressure at the inlet as it flows into a duct of varying area. The steady flow energy equation can be applied to obtain that the local temperature along the duct varies with the local Mach number as

$$h_t = h + \tfrac{1}{2}c^2, \quad T_t = T + \frac{c^2}{2c_p},$$
$$\frac{T_t}{T} = 1 + \frac{c^2}{2c_pT} = 1 + \frac{(\gamma - 1)c^2}{2\gamma RT} = 1 + \tfrac{1}{2}(\gamma - 1)M^2. \tag{6.1}$$

This equation identifies the variation of temperature with the Mach number along the duct. If the flow is considered to be isentropic, the isentropic relations for an ideal gas can be used to find relations for the variation of pressure and density with Mach number:

$$\frac{p_t}{p} = \left(\frac{T_t}{T}\right)^{\frac{\gamma}{\gamma-1}} = \left(1 + ½(\gamma - 1)M^2\right)^{\frac{\gamma}{\gamma-1}}, \quad \frac{\rho_t}{\rho} = \left(\frac{p_t}{p}\right)^{\frac{1}{\gamma}} = \left(1 + ½(\gamma - 1)M^2\right)^{\frac{1}{\gamma-1}}. \quad (6.2)$$

These equations identify that the variation of pressure and density in an isentropic flow along the duct at points with different cross-sectional areas depends on the local Mach number. If the Mach number is known, then all the actual static properties can be calculated, including the local speed of sound:

$$\frac{a_t}{a} = \sqrt{\frac{T_t}{T}} = \left(1 + ½(\gamma - 1)M^2\right)^{\frac{1}{2}}. \quad (6.3)$$

6.2.3 The Continuity Equation in Compressible Flow

The continuity equation for one-dimensional compressible flow is given in Section 5.2 and can be written in differential form as

$$A\rho c = \text{constant}, \quad \frac{dA}{A} + \frac{d\rho}{\rho} + \frac{dc}{c} = 0. \quad (6.4)$$

This shows that for an incompressible flow (with $d\rho = 0$), the velocity decreases as the duct cross-sectional area becomes larger. But for compressible flow, the density varies with the Mach number as given by (6.2), and it is not immediately clear how the velocity varies with the change in area. The steady flow energy equation in a duct can be expressed in differential form and combined with the differential form of the Gibbs equation for isentropic flow, together with the equation for the speed of sound, to obtain

$$\begin{aligned} &h_t = h + ½c^2 = \text{constant}, \quad \Rightarrow dh + cdc = 0 \\ &Tds = dh - dp/\rho \quad \Rightarrow dh = dp/\rho \quad \Rightarrow dp = -\rho cdc \\ &a^2 = dp/d\rho, \quad \Rightarrow a^2 d\rho = -\rho cdc, \Rightarrow d\rho/\rho = -\left(c^2/a^2\right)dc/c. \end{aligned} \quad (6.5)$$

So that an expression for change in the velocity for a change in area in a compressible flow is obtained as

$$\frac{dA}{A} = -\frac{dc}{c}\left(1 - M^2\right). \quad (6.6)$$

If the flow is subsonic (with $M < 1$), the flow must accelerate if the area is reduced as in a nozzle and decelerates as the area is increased as in a diffuser. If the flow is supersonic (with $M > 1$), the flow decelerates if the area is reduced and accelerates as the area is increased. When the Mach number is unity, (with $M = 1$), the flow velocity is the same as the local speed of sound and the duct has a throat ($dA = 0$). Under such conditions, the duct is said to be choked as no further increase in mass flow is possible with the given inlet conditions and any change to the area or conditions downstream of the throat can no longer influence the flow upstream of the throat.

To achieve a Mach number greater than unity, the channel must first decrease in area until $M = 1$ is attained at the throat and then, beyond the throat, increase in area to continue to accelerate the flow above $M > 1$, in a form known as a Laval nozzle. As the area is reduced and the Mach number is increased in the subsonic region, the change in velocity for a given change in area becomes progressively larger as $M = 1$ is approached. Similarly, if the flow is supersonic and close to $M = 1$, the change in velocity for a given change in area is also large. In subsonic flow or supersonic flow close to $M = 1$, known as transonic flow, then a small decrease in area of 1% can lead to a large change in velocity, as shown for the following examples:

$$\begin{aligned}
&\frac{dc}{c} = -\frac{dA}{A}\frac{1}{\left(1 - M^2\right)}, \quad \frac{dA}{A} = -1\%, \\
&M = 0.0, \quad \frac{dc}{c} = 1.0\%, \quad M = 0.5, \quad \frac{dc}{c} = 1.3\%, \\
&M = 0.9, \quad \frac{dc}{c} = 5.3\%, \quad M = 0.95, \quad \frac{dc}{c} = 10.2\%, \\
&M = 1.05, \quad \frac{dc}{c} = -9.8\%, \quad M = 1.1, \quad \frac{dc}{c} = -4.8\%.
\end{aligned} \tag{6.7}$$

From this it can be seen that there is a high sensitivity of velocity levels to a small change in effective flow area in transonic compressors with flow Mach numbers close to unity. The presence of boundary layers and losses in transonic blade rows necessitate a more detailed calculation where these effects are considered.

The topology of a typical compressor flow channel, both in the impeller blades and in the diffuser vanes, is similar to a Laval nozzle. There is a region with a decreasing area near the leading edge of the blades which is followed by a minimum area called the throat. Subsequently the area between the blades increases towards the trailing edge. When a compressor blade row is operating towards choke, this effect is amplified by the negative incidence on the blades, leading to a clear decrease in effective area to the throat.

6.2.4 Corrected Flow per Unit Area

Another important parameter which characterises compressible flow in a duct on a one-dimensional basis is the corrected flow per unit area, or corrected mass-flow function, through the duct. If the continuity equation is written for any point along the duct, then the the corrected flow per unit area can be derived from (6.2) and (6.3) as follows:

$$\dot{m} = A\rho c, \quad \dot{m}_{cor} = \frac{\dot{m}}{A\rho_t a_t} = \frac{\rho c}{\rho_t a_t} = M\frac{\rho a}{\rho_t a_t} = \frac{M}{\left(1 + ½(\gamma - 1)M^2\right)^{\frac{(\gamma+1)}{2(\gamma-1)}}}. \tag{6.8}$$

The corrected mass flow increases with subsonic Mach number up to $M = 1$ and decreases at supersonic Mach numbers. The maximum value of the corrected flow per unit area at $M = 1$ is often taken as as a reference, giving

$$\frac{\dot{m}_{\max}}{A^{*}\rho_t a_t} = \left(\frac{2}{\gamma+1}\right)^{\frac{\gamma+1}{2(\gamma-1)}}, \tag{6.9}$$

where A^{*} is the area of the duct with a local Mach number of $M = 1$. Using this area as a reference the variation of area through the channel becomes

$$\frac{A}{A^{*}} = \frac{1}{M}\left(\left(\frac{2}{\gamma+1}\right)\left(1+\frac{(\gamma-1)}{2}M^2\right)\right)^{\frac{(\gamma+1)}{2(\gamma-1)}}. \tag{6.10}$$

This provides the value of the area, A, for any Mach number, M, in terms of the area A^{*} at the location with $M = 1$. The area decreases through the duct with subsonic flow up to $M = 1$ and increases for supersonic flow. The smallest area with the maximum corrected flow per unit area is known as the throat. The variation of corrected mass flow with Mach number is shown in Figure 6.1, for an ideal gas with $\gamma = 1.2$, 1.4 and 1.6. The maximum corrected mass flow increases slightly with a decrease in the isentropic exponent so that a change in the isentropic exponent of the gas affects the maximum mass flow. This causes a small shift in the operating characteristic of the compressor operating in a choked condition with different gases; see Section 8.7.4.

A practical application of the corrected flow is that, if the local area of the channel compared to that at the throat is known, then the Mach number for isentropic flow at this position can be determined. From (6.2) and (6.3), the local temperature, pressure, density and speed of sound relative to stagnation conditions are then known. Another application of these equations is to determine the isentropic Mach number

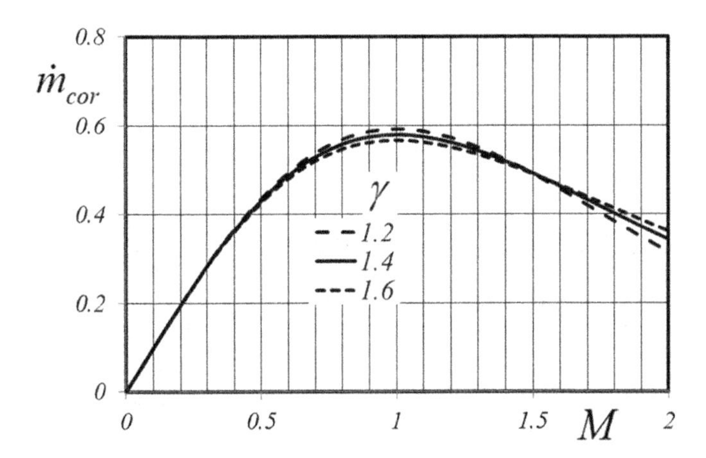

Figure 6.1 Corrected mass-flow function in an isentropic converging diverging nozzle as a function of the Mach number and the isentropic exponent.

at a position in the flow where the static pressure is measured and the inlet total pressure is known. This is the Mach number that would apply at this position if there were no losses in the flow:

$$M_{isentropic} = \sqrt{\frac{2}{\gamma - 1}\left[\left(\frac{p_t}{p}\right)^{\frac{\gamma-1}{\gamma}} - 1\right]}. \tag{6.11}$$

A further application is in a flow situation where the local mass flow, total temperature and flow area are known, so that the total pressure can be determined from the measured static pressure.

6.2.5 Choked Flow

The designer of compressors has a particular interest in the maximum mass flow that occurs when the Mach number reaches unity in the throat of the flow channels. For an isentropic one-dimensional choked flow, (6.1)and (6.2) show that conditions at the throat, denoted by an asterisk and called the critical conditions, are given by

$$\frac{T^*}{T_t} = \frac{2}{\gamma + 1}, \quad \frac{\rho^*}{\rho_t} = \left(\frac{2}{\gamma + 1}\right)^{\frac{1}{\gamma-1}}, \quad \frac{a^*}{a_t} = \left(\frac{2}{\gamma + 1}\right)^{\frac{1}{2}}, \quad \frac{\rho^* a^*}{\rho_t a_t} = \left(\frac{2}{\gamma + 1}\right)^{\frac{\gamma+1}{2(\lambda-1)}}. \tag{6.12}$$

From this, an expression can be obtained for the choking mass flow at the throat with a Mach number of unity and a throat area of A^* as

$$\dot{m}_{ch} = A^* \rho^* a^*, \quad \frac{\dot{m}_{ch}}{A^*} = \rho^* a^* = \rho_t a_t \left(\frac{\rho^* a^*}{\rho_t a_t}\right) = \rho_t a_t \left(\frac{2}{\gamma + 1}\right)^{\frac{\gamma+1}{2(\gamma-1)}}. \tag{6.13}$$

And from this we can define the maximum value of the nondimensional mass-flow function as

$$\frac{\dot{m}_{ch}}{A^* \rho_t a_t} = \left(\frac{2}{\gamma + 1}\right)^{\frac{(\gamma+1)}{2(\gamma-1)}}. \tag{6.14}$$

Note that for air with $\gamma = 1.4$, the numerical value of the mass-flow function at choke and the temperature and pressure ratio up to the throat are given by

$$\frac{T_t}{T^*} = 1 + \frac{\gamma - 1}{2} = \frac{\gamma + 1}{2}, \quad \frac{p_t}{p^*} = \left(\frac{\gamma + 1}{2}\right)^{\gamma/(\gamma-1)}, \quad \frac{\rho_t}{\rho^*} = \left(\frac{\gamma + 1}{2}\right)^{1/(\gamma-1)}$$
$$\frac{T^*}{T_t} = 0.8333, \quad \frac{p^*}{p_t} = 0.5283, \quad \frac{\rho^*}{\rho_t} = 0.6339, \quad \frac{\dot{m}_{ch}}{A^* \rho_t a_t} = \frac{\rho^* a^*}{\rho_t a_t} = 0.5787. \tag{6.15}$$

A duct or an orifice with a pressure ratio in air less than 0.5283 (or an expansion ratio of 1.892) between inlet and outlet will be choked.

In most of the available technical literature, the nondimensional mass-flow function is not described in these terms, but the speed of sound is replaced by its expression in an ideal gas, $\sqrt{(\gamma RT_{t1})}$, and the total density is replaced by the density of an ideal gas, p_{t1}/RT_{t1}. In some situations, R is replaced by c_p, or either γ or R may be omitted, leaving the mass-flow function simply as $m\sqrt{T_{t1}}/p_{t1}$, which is no longer nondimensional (but has rather wonderful units: kg $\sqrt{\text{K}}$/s bar). In this chapter, the preceding form is retained, as it keeps its validity for a real gas and has the advantage that it is clearly nondimensional just by inspection, being the ratio of the mass flow to a reference mass flow given by the product of inlet density, inlet speed of sound and a reference area. It then becomes immediately clear that any processes which increase the inlet density or the inlet speed of sound cause a higher mass flow at choke.

6.2.6 Variation of Pressure in a Nozzle at Different Back Pressures

The area ratio and Mach number given by (6.10), together with the gas properties given by (6.2), describe the operation of a convergent-divergent flow channel at its design point with $M = 1$ at the throat. For this steady isentropic flow, the local area determines the Mach number and fixes the ratio of static to total pressure at all points in the channel. In many turbomachinery situations, however, a turbomachinery flow channel is forced by other components to operate under different conditions. For example, in a radial compressor operating with a vaned diffuser, the pressure downstream of the diffuser may be lower than that required by the isentropic flow equations so that a more complex Mach number and pressure distribution occurs in the diffuser flow channels. This issue is best examined by considering the flow field in a one-dimensional convergent-divergent channel with varying back pressure.

The effect of the back pressure on the flow field in a converging-diverging channel is shown in Figure 6.2 from Greitzer et al. (2004), for an ideal gas with $\gamma = 1.4$. The top picture on the left gives a sketch of the channel and the downstream chamber. The middle picture on the left shows the pressure distribution in the channel at different back pressure, and on the bottom-left there is the Mach number distribution through the channel. At pressure ratios that are higher than that required to reach $M = 1$ at the throat, the flow is subsonic everywhere. The pressure first decreases and then increases and the velocity and Mach number first increase and then decrease, as shown in lines 1 and 2. As the back pressure is reduced, the corrected mass flow per unit area increases, but the flow remains subsonic. At a pressure ratio of 0.5283 at the throat, a Mach number of 1 is attained at the throat and the corrected mass flow per unit area reaches its maximum choked value. For pressure ratios just above this value, the flow remains subsonic everywhere and the Mach number decreases to the exit of the duct, as depicted by line 3. For all back pressures below this value, the subsonic flow upstream of the throat is unchanged as no information can reach this domain once the throat has attained $M = 1$.

A very low back pressure that is sufficient to cause continuous supersonic flow to the exit of the channel could be considered to be the design condition for this channel. Then the Mach number continues to increase downstream of the throat in line with the

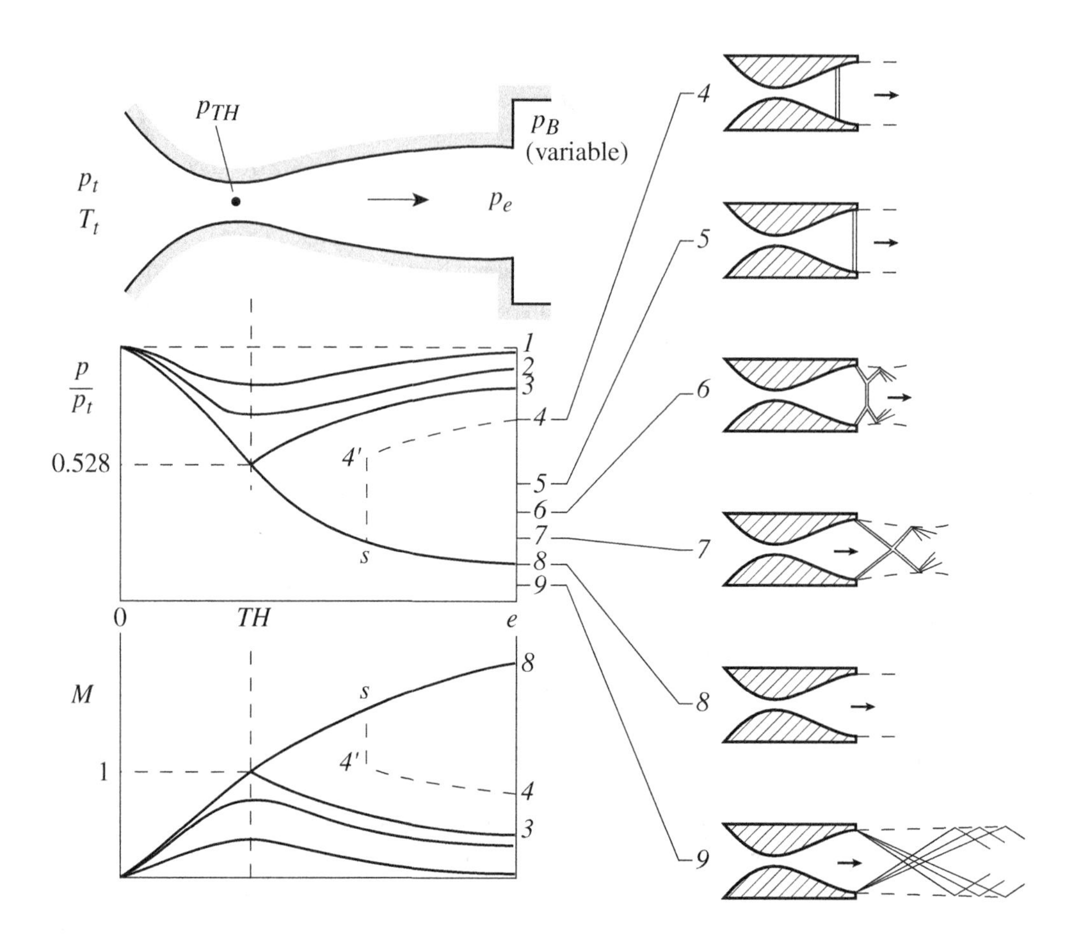

Figure 6.2 The effect of the back pressure in a converging diverging nozzle. (image courtesy of Ed Greitzer and Cambridge University Press)

area change, as shown in line 8. The isentropic flow in the channel is now perfectly adapted to the pressure in the downstream chamber and has an outlet pressure consistent with the back pressure, with a supersonic jet into the downstream chamber. Lines 3 and 8 are the only solutions to the flow field for an isentropic flow in the channel with a Mach number of $M = 1$ at the throat.

In the situation with a back pressure between line 3 and 8, there are no solutions for the flow field that are consistent with an isentropic flow, so that a flow field with isentropic flow is not possible and the flow becomes discontinuous with loss production in shock waves. As the back pressure ratio falls below line 3, the outlet flow remains subsonic with a pressure determined by the back pressure, but the flow immediately downstream of the throat becomes supersonic. Between the supersonic and subsonic regions, there is an abrupt normal shock in the channel with a step change from a supersonic to a subsonic Mach number and a step change in the pressure level in the flow. The nature of the normal shock wave is described in Section 6.3.1.

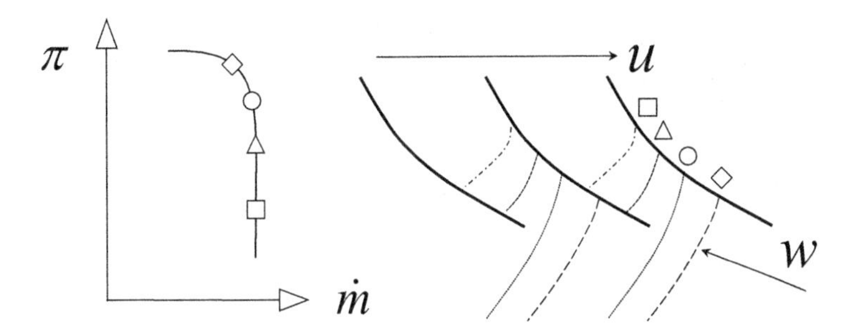

Figure 6.3 A sketch of the shock position changing with pressure ratio in the inlet of a radial compressor impeller flow channel.

As the back pressure is lowered further, the shock moves farther downstream, with a larger step change in the Mach number and pressure ratio, indicating a stronger shock. Further decreases in the back pressure draw the shock backwards in the channel until at line 5 it stands at the exit of the divergent channel, and downstream of the throat the channel flow is completely supersonic. Clearly no additional changes can take place within the channel as the flow is supersonic downstream of the throat and the adaptation of the outlet flow to the chamber back pressure occurs downstream of the duct. For back pressure ratios between line 5 and 8, the flow in the duct is expanded to a lower pressure than the actual back pressure and is said to be overexpanded. The adjustment to the actual back pressure then takes place is a series of normal and inclined shock waves, as sketched for lines 6 and 7 in Figure 6.2. For lower back pressures, below that of line 8, the channel cannot expand the flow to reach the pressure in the chamber and is said to be underexpanded. The adjustment to the actual back pressure then takes place through a series of inclined expansion waves, as shown for line 9.

All of the pressure ratio conditions below that of line 3 in Figure 6.2 have the same inlet mass flow to the nozzle, so that the pressure ratio at choke is not fundamentally related to the mass flow but is determined by the back pressure and the associated strength of the shocks in the flow channel. It is as if the losses across the shock and its position in the channel automatically adjust to balance the pressure in the flow with that of the back pressure. This effect determines the position of the shock in the inlet region of a supersonic radial compressor, and a sketch of this effect in a radial impeller operating at several points on its characteristic is shown in Figure 6.3.

6.3 Shock and Expansion Waves

6.3.1 Normal Shock Waves

In a compressible flow, the speed of motion of an infinitesimal small pressure wave is known as the speed of sound. Large disturbances of finite amplitude can occur, and these are known as shock waves, as shown in Figure 6.2. Consider a one-dimensional flow across a shock wave, in a frame of reference relative to the

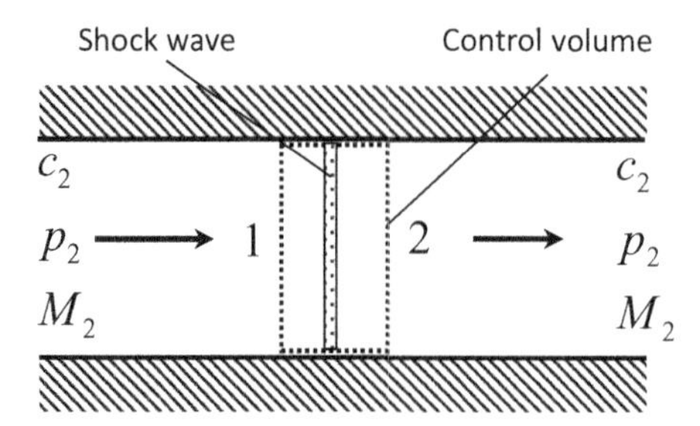

Figure 6.4 Control volume for a normal shock wave fixed relative to the shock.

shock as shown in Figure 6.4; then the relevant conservation equations for the conditions across the shock are the continuity equation, the momentum equation, the energy equation and the caloric and thermal equations of state for an ideal gas, as follows:

$$
\begin{aligned}
&\rho_1 c_1 = \rho_2 c_2 \\
&p_1 + \rho_1 c_1^2 = p_2 + \rho_2 c_2^2 \\
&h_1 + {}^1\!/\!_2 c_1^2 = h_2 + {}^1\!/\!_2 c_2^2 = h_t \\
&h = c_p T, \quad pv = RT.
\end{aligned}
\tag{6.16}
$$

Rearrangement of these equations, as shown in many reference books on this subject such as Zucker and Biblarz (2002), leads to the following equations relating the conditions across the shock:

$$
\begin{aligned}
&M_2^2 = \frac{(\gamma - 1)M_1^2 + 2}{2\gamma M_1^2 - (\gamma - 1)} && M_2 < M_1 \\
&\frac{p_2}{p_1} = \frac{1 + \gamma M_1^2}{1 + \gamma M_2^2} && p_2 > p_1 \\
&\frac{T_2}{T_1} = \frac{2 + (\gamma - 1)M_1^2}{2 + (\gamma - 1)M_2^2} && T_2 > T_1 \\
&T_{t2} = T_{t1} \\
&\frac{p_{t2}}{p_{t1}} = \left(\frac{1 + \gamma M_1^2}{1 + \gamma M_2^2}\right)\left(\frac{1 + [(\gamma - 1)/2)]M_2^2}{1 + [(\gamma - 1)/2)]M_1^2}\right)^{\gamma/(\gamma-1)}.
\end{aligned}
\tag{6.17}
$$

The subsonic Mach number downstream of the shock is accompanied by a rise in the static pressure and the static temperature. The conservation of total enthalpy determines that, for an ideal gas, there is no change in total temperature across the shock. These equations show that there is a loss in total pressure across the shock due to the generation of entropy. The entropy generated is given by the Gibbs relationships as

$$s_2 - s_1 = c_p \ln(T_2/T_1) - R\ln(p_2/p_1) = -R\ln(p_{t2}/p_{t1}). \tag{6.18}$$

Further analysis shows that for moderate shock strengths the stagnation pressure ratio increases with the third power of the upstream normal shock strength, and the entropy rise is given by

$$\frac{s_2 - s_1}{R} = \frac{2}{3}\frac{\gamma}{(\gamma+1)^2}\left(M_1^2 - 1\right)^3. \tag{6.19}$$

This implies that normal shock waves have a high efficiency in producing a pressure rise without any blades. Up to an inlet Mach number of 1.5, the efficiency of the pressure rise across a normal shock is greater than 0.9, which accounts for why compressors with supersonic relative inlet flow at the tip can still achieve high efficiency. In fact, a shock Mach number of 1.3 is generally considered the sensible upper limit for radial compressors because otherwise the increase in pressure across the shock may interact with the boundary layer, causing boundary layer separation with higher losses.

6.3.2 Oblique Shock Waves

The discussion of overexpanded converging-diverging ducts in Section 6.2.6 described inclined, or oblique, shocks. Oblique shocks in turbocompressors generally occur near to the leading edge in supersonic flows. To analyse this situation, a section of an oblique shock is analysed to examine the flow relative to the normal component of the inlet velocity, as shown in Figure 6.5. It is assumed that the tangential component of velocity, c_{t1}, does not change across the shock, and that the normal component, c_{n1}, behaves like a normal shock. With a shock wave angle of ε, this gives rise to a deflection of the flow across the shock wave of δ, where

$$\tan\delta = \frac{2\cot\varepsilon\left(M_1^2\sin^2\varepsilon - 1\right)}{2 + M_1^2(\gamma + 1 - 2\sin^2\varepsilon)}. \tag{6.20}$$

Equation (6.20) gives the shock deflection for a specific Mach number to the shock and the shock wave angle. With an oblique shock, the flow direction always turns towards the shock front. There is no deflection if the shock is a normal shock with $\varepsilon = 90°$, and there is also no deflection if $M_1 \sin\varepsilon = 1$, or $\sin\varepsilon = 1/M_1$. The latter

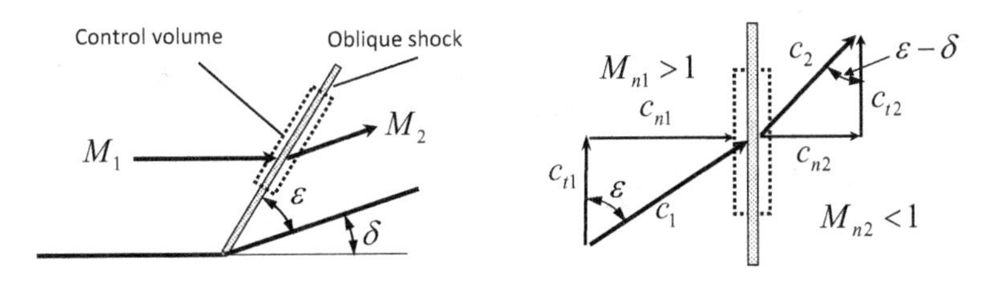

Figure 6.5 Analysis of an oblique shock with shock wave angle of ε.

condition is known as a Mach wave and represents an infinitesimal pressure wave travelling with the speed of sound. These weak waves can combine in supersonic flow to become a shock wave if sufficient Mach waves are present at any location.

In a supersonic flow, there is a maximum deflection, which cannot be exceeded and is dependent on the Mach number. For a flow where the deflection is larger than this, an oblique shock cannot exist. The maximum deflection for each Mach number is usually given in a diagram in most compressible flow textbooks. There are two types of oblique shocks: strong shocks with a high shock wave angle which are followed by subsonic flow behind the shock, and weak shocks with a low shock wave angle which retain supersonic flow downstream of the shock. The deceleration across the shock is lower for a weak shock than for a strong shock, and the pressure ratio across the shock is also smaller.

The efficiency of the pressure increase across an oblique shop is higher than that of a normal shock, so that the design strategy for inlets of gas turbine engines in supersonic flight is to attempt to decelerate the flow by a series of weak oblique shocks rather than a single strong normal shock.

6.3.3 Prandtl–Meyer Expansion Waves

The discussion of underexpanded converging-diverging channels in Section 6.2.6 also brought attention to expansion waves in the flow as a means of increasing the speed and decreasing the pressure of a supersonic flow. These are shown in Figure 6.2 at the bottom-right emanating from the sharp corner of the channel outlet. A supersonic flow can be accelerated by a series of expansion waves, each of which increase the Mach number and deflect the flow away from the expansion wave, as shown in Figure 6.6. Any disturbance upstream of the wave travels faster than a disturbance in the downstream region, and the wave spreads out. If a supersonic flow in the direction of the surface meets a convex corner with a small change of direction, the flow sets up an expansion wave at the corner which accelerates and deflects it. In the case of a large corner, the expansion wave becomes a fan of such waves and this is known as

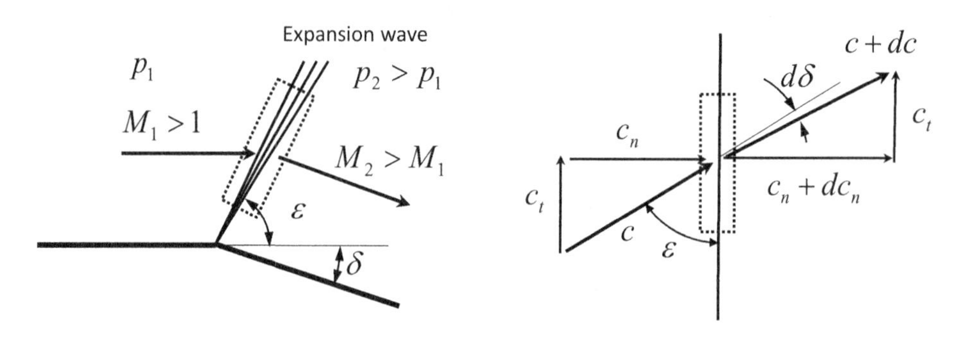

Figure 6.6 Expansion wave.

Prandtl–Meyer flow. The expansion waves make an angle of $\sin^{-1}(1/M)$ to the flow and are known as characteristic lines.

If a supersonic flow passes over a convex surface, the expansion waves cause the flow to accelerate, and over a concave surface the expansion waves decelerate the flow and build to an oblique shock farther from the surface, as shown in Figure 6.7. The relevance of this in compressors is that in a blade row inlet if the blade suction surface presents a convex surface to the inlet supersonic flow, then the flow will accelerate to a higher Mach number as it enters the flow passage. To avoid this, the blades at the tip section of the inducer in a radial impeller are generally very straight.

6.4 Shock Structure in Transonic Compressors

6.4.1 Detached Curved Shock on a Leading Edge of a Blade Row

The stagnation streamline onto a leading edge deflects by 90° at the stagnation point. For situations where the geometrical shape of the boundary forces a supersonic flow to deflect by more than the maximum possible deflection, the oblique shock equations provide no solution and the shock becomes a mixture of a weak and a strong shock, as shown in Figure 6.8. Near the stagnation streamline, the shock moves forward to become a strong detached normal shock followed by a subsonic region. The shock becomes weaker farther away from the leading edge, and the flow no longer becomes subsonic as it passes through the weak shock.

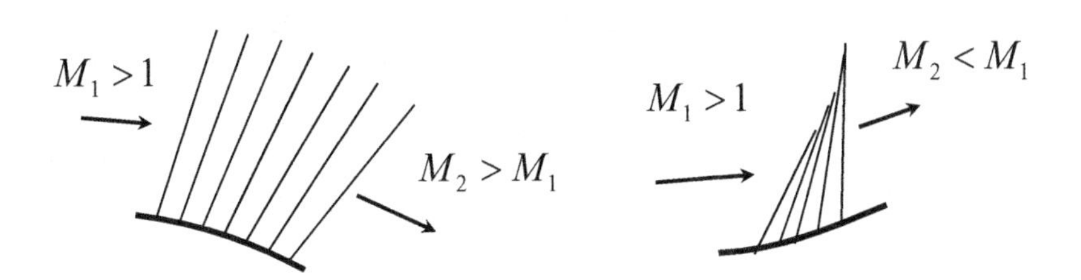

Figure 6.7 Expansion waves in a supersonic flow over a convex surface (left) and a concave surface (right).

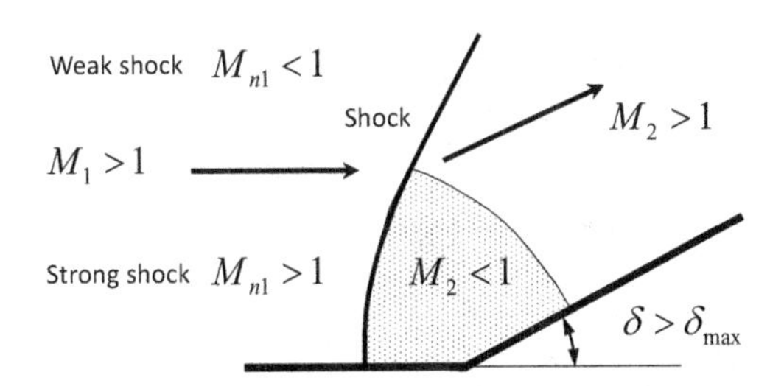

Figure 6.8 Curved shock upstream of a wedge with a deflection greater than the maximum.

Close to the leading edge of the cascade of blades in a transonic radial compressor, this leads to a flow pattern as shown in Figure 6.9. The shock is just upstream of the leading edge of one blade but reaches to the suction side of the adjacent blade. Because there is subsonic flow downstream of the shock, downstream disturbances are sensed upstream. Increasing the back pressure increases the relative inlet flow angle, with a reduced axial velocity, and moves the shock forward to become stronger. This leads to a higher preshock Mach number due to the Prandtl–Meyer expansion on the suction surface downstream of the leading edge. The stagnation streamline moves closer to the suction surface and increases the blade loading at the leading edge, which together with tip clearance effects is a possible cause of stall. Decreasing the back pressure moves the bow shock backwards into the channel.

6.4.2 Unique Incidence

If the blade has a sharp leading edge, then decreasing the back pressure causes the shock from the suction surface to attach to the leading edge, the shock is swallowed and this mode of operation is called a started inlet, as shown in Figure 6.10. The first

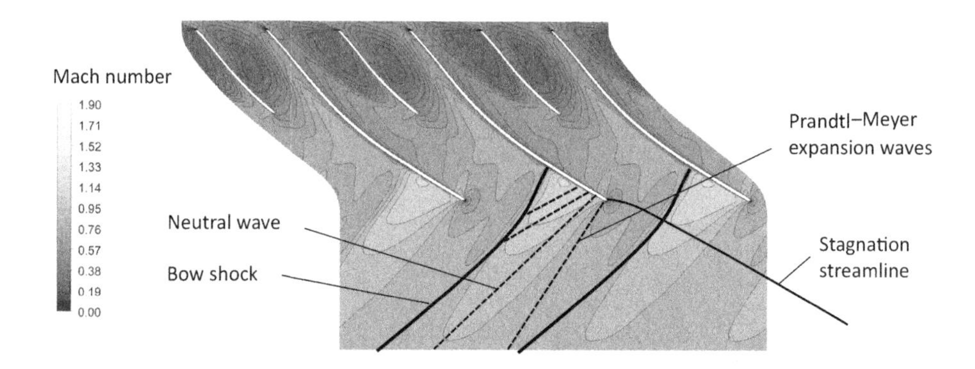

Figure 6.9 Curved bow shock upstream of a compressor leading edge in a CFD simulation.

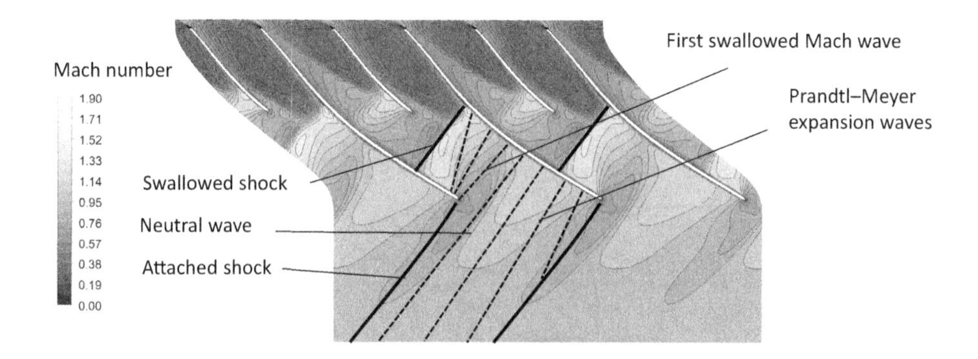

Figure 6.10 Operation with an attached shock at the unique incidence condition.

swallowed Mach wave leaves the suction surface and intersects the leading edge of the adjacent blade, and as there are no upstream running waves downstream of this, no downstream disturbances can be sensed upstream. The blade is choked and said to be operating at the unique incidence condition. This cannot happen if the passage chokes first in a downstream throat such that this mass flow cannot be attained.

In the unique incidence condition, a neutral wave leaves the suction surface and runs to the inlet, passing between the shock waves from adjacent leading edges. The inlet flow angle is the same as the local flow angle at where this wave leaves the suction surface. For a specific inlet Mach number M_1, the inlet flow angle β_1 is determined by the uncovered part of the suction surface. There is then a unique relationship between M_1, and β_1 leading to a unique incidence for each Mach number. In a real blade row, the effects of blade thickness and of boundary layer displacement thickness need to be considered.

6.4.3 Variation of Shock Structure with Operating Point

The shock structure at the tip of a transonic compressor impeller from CFD simulations at different operating points is given in Figure 6.11. The predicted impeller total–total efficiency versus mass-flow characteristic is shown in the map,

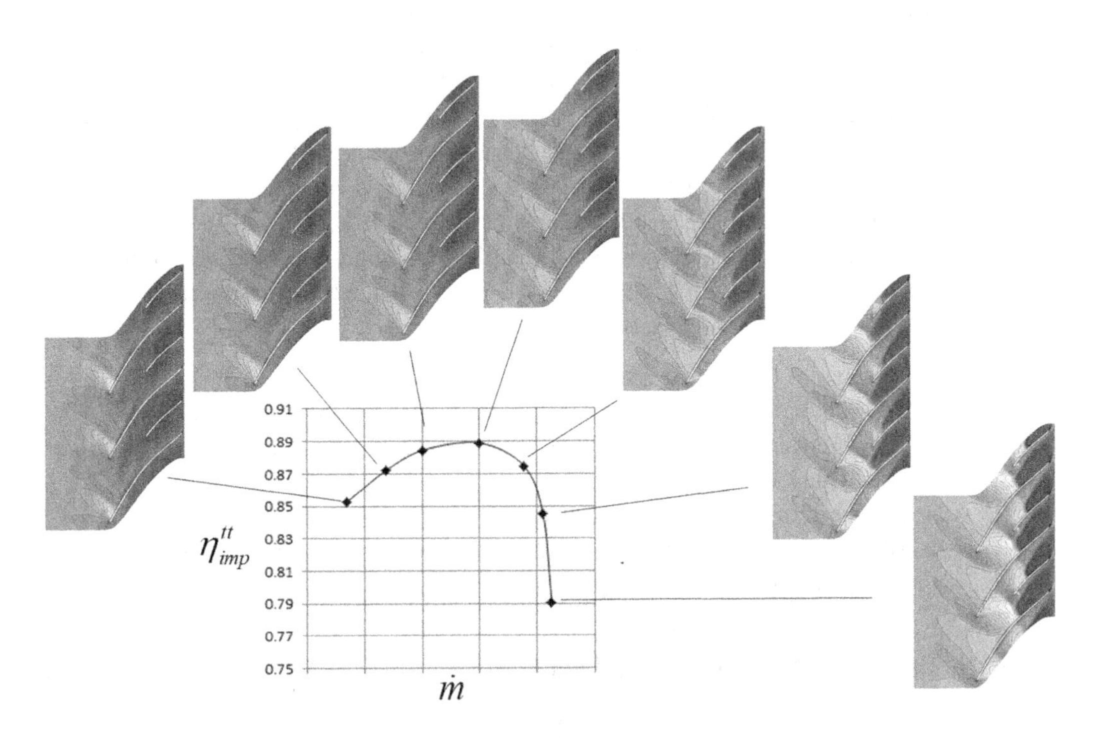

Figure 6.11 Shock structure at 90% span of a transonic radial compressor impeller for different operating points.

and the Mach number distribution at the 90% span section of the impeller for several operating points is given. The points are shown in the efficiency map, and the Mach number distribution for points from choke to surge are placed counterclockwise around the map. At low pressure ratio with high flow, there is an oblique shock attached to the leading edge with the main passage shock deep inside the passage. The tip section is clearly choked for the operating points with a swallowed shock in the flow channel, and operation is at the condition of unique incidence. However, the flow continues to increase over these operating points as the hub sections are not choked and can still swallow more flow. With increasing back pressure and lower flow, the passage shock moves upstream and peak efficiency tends to occur just before the shock is expelled (Freeman and Cumpsty, 1992). At the peak total–total efficiency of the impeller, operation is with the shock from the suction surface of the blade just expelled upstream of the adjacent blade. At higher back pressure and lower flow towards the stall point, the shock is expelled farther from the leading edge.

6.5 Gas Dynamics of Real Gases

6.5.1 Background

Most of the derivation of the appropriate equations for the compressible flow of real gases as given in this section is reported in the work of Baltadjiev et al. (2015). In the 1D compressible flow analysis given earlier, and in most courses on gas dynamics, the equations are derived for the case of an ideal gas with constant specific heats, in which the ratio of the specific heats, $\gamma = c_p/c_v$, is the same as the isentropic exponent, κ, and as such are entirely interchangeable, as used in the previous equations. The equations all then include terms related to the ratio of the specific heats.

This section examines some aspects of these equations for the case that the ideal gas relations no longer hold, that is, for a real gas with nonconstant specific heats. This has the following consequences

- The ratio of the specific heats is not a constant.
- The isentropic exponent is not the same as the ratio of the specific heats.
- The difference between the specific heats at constant volume and constant pressure is not equal to the gas constant.
- The speed of sound is not defined by the ratio of the specific heats.
- The isentropic exponent rather than the ratio of specific heats must be used in all the relevant compressible flow equations.

6.5.2 Ratio of Specific Heats in a Real Gas

In a real gas, the specific heat at constant volume can be determined from the specific heat at constant pressure, as shown in the following equations (Edminster, 1961; Stephan and Mayinger, 2000):

$$
\begin{aligned}
c_p &= T\left(\frac{\partial p}{\partial T}\right)_s\left(\frac{\partial v}{\partial T}\right)_p \\
c_v &= -T\left(\frac{\partial p}{\partial T}\right)_V\left(\frac{\partial v}{\partial T}\right)_s.
\end{aligned} \tag{6.21}
$$

The difference between the specific heats is given by the following equation:

$$c_p - c_v = T\left(\frac{\partial v}{\partial T}\right)_p\left(\frac{\partial p}{\partial T}\right)_v. \tag{6.22}$$

Thus, if the specific heat at constant pressure is known as a function of temperature, then the ratio of the specific heats can be calculated from this.

6.5.3 Isentropic Exponents

In a real gas, as outlined, the isentropic exponent for the pressure volume relationship between two end states of the gas satisfies the following equation:

$$pv^{n_s} = \text{constant}. \tag{6.23}$$

In the notation of Section 3.5, $n_s = \kappa_{pv}$. This can be rewritten in differential form as

$$\frac{dp}{p} + n_s\frac{dv}{v} = 0 \Rightarrow n_s = -\frac{v}{p}\left(\frac{\partial p}{\partial v}\right)_s = -\frac{v}{p}\left(\frac{\partial p}{\partial T}\right)_s\left(\frac{\partial T}{\partial v}\right)_s. \tag{6.24}$$

Combining this with the preceding,

$$n_s = -\frac{v\,c_P}{p\,T}\left(\frac{\partial T}{\partial v}\right)_p\left[-\frac{T}{c_v}\left(\frac{\partial p}{\partial T}\right)_v\right] = \frac{c_P}{c_v}\frac{V}{p}\left(\frac{\partial p}{\partial v}\right)_T. \tag{6.25}$$

A different exponent is required for the isentropic pressure exponent, which is the exponent for the pressure–temperature relationship between two end states of the gas to satisfy the following equation:

$$p^{m_s}/T = \text{constant}. \tag{6.26}$$

In the notation of Section 3.5, $m_s = (1 - \kappa_{pt})/(\kappa_{pt})$. In a similar way, the following can be shown:

$$m_s = \frac{p}{T}\left(\frac{\partial T}{\partial p}\right)_s. \tag{6.27}$$

6.5.4 Speed of Sound in a Real Gas

For a real gas in an isentropic pressure-volume process, the following equation for the speed of sound is obtained:

$$a = \sqrt{n_s ZRT} = \sqrt{n_s pv}. \tag{6.28}$$

6.5.5 Isentropic Flow Process in a Real Gas

For an isentropic process, the Gibbs equation reduces to $dh = vdp$. Integrating both sides of this equation from total to static conditions yields

$$h_t - h = \int_p^{p_t} vdp. \tag{6.29}$$

Using the equation for the isentropic process and assuming that the value of the isentropic exponent remains the same between total and static states, the integration leads to

$$h_t - h = \int_p^{p_t} v_t \left(\frac{p_t}{p}\right)^{\frac{1}{n_z}} dp = v_t p_t^{\frac{1}{n_s}} \int_p^{p_t} p^{\frac{1}{n_s}} dp = v_t p_t \left[1 - \left(\frac{p}{p_t}\right)^{\frac{n_s-1}{n_s}}\right] \frac{n_s}{n_s - 1}. \tag{6.30}$$

Note that an approximation has been made here that the isentropic exponent is constant along the whole process.

Consider a one-dimensional isentropic and adiabatic flow of a real gas through a stream-tube of cross-sectional area A from a reservoir with fixed total pressure and temperature. For this flow process, with no work and heat transfer, the total enthalpy remains constant so that

$$h_t - h = \frac{(c)^2}{2} = \frac{(aM)^2}{2} = \frac{M^2}{2} n_s pv = v_t p_t \left[1 - \left(\frac{p}{p_t}\right)^{\frac{n_s-1}{n_s}}\right] \frac{n_s}{n_s - 1}. \tag{6.31}$$

This can be rearranged to give the following:

$$\frac{n_s - 1}{2} M^2 = \frac{p_t v_t}{pv} \left[1 - \left(\frac{p}{p_t}\right)^{\frac{n_s-1}{n_s}}\right] = \left(\frac{p_t}{p}\right)^{\frac{n_s-1}{n_s}} \left[1 - \left(\frac{p}{p_t}\right)^{\frac{n_s-1}{n_s}}\right] = \left(\frac{p_t}{p}\right)^{\frac{n_s-1}{n_s}} - 1. \tag{6.32}$$

So, for a real gas, the final form of the pressure ratio, density ratio and temperature equations for an isentropic flow process are given by

$$\frac{p_t}{p} = \left(1 + \frac{n_s - 1}{2} M^2\right)^{\frac{n_s}{n_s-1}}, \quad \frac{\rho_t}{\rho} = \left(1 + \frac{n_s - 1}{2} M^2\right)^{\frac{1}{n_s-1}}$$
$$\frac{T_t}{T} = \left(1 + \frac{n_s - 1}{2} M^2\right)^{\frac{n_s m_s}{n_s-1}}, \quad \frac{Z_t}{Z} = \frac{p_t}{p} \frac{V_t}{V} \frac{T}{T_t} = \left(1 + \frac{n_s - 1}{2} M^2\right)^{1 - \frac{m_s n_s}{n_s-1}}. \tag{6.33}$$

These equations are the real gas equivalent of the well-known isentropic gas dynamic equations given in (6.2). In order to calculate the compressible flow of a real gas, we may use the same equations as used in an ideal gas, but the ratio of the specific heats, γ, needs to be replaced by the true value of the isentropic volume exponent, n_s, of the real gas.

6.5.6 Corrected Mass Flow per Unit Area

Usually of interest is the equation for the mass flow per unit area in the duct or the corrected mass flow, which is defined in (6.8) as

$$\dot{m}_{cor} == \frac{\dot{m}}{A\rho_t a_t} = \frac{A\rho a}{A\rho_t a_t} M. \tag{6.34}$$

Here the total density and the speed of sound at the reservoir conditions have been used to make the mass flow nondimensional and, for an ideal gas, this is the same as the more usual use of the total pressure and total temperature to derive a term of the form $m\sqrt{T}/p$. The expression for the density ratio has already been given, and the expression for the ratio of the speeds of sound in this equation can be derived as

$$\frac{a}{a_t} = \frac{\sqrt{n_s ZRT}}{\sqrt{n_s Z_t RT_t}} = \sqrt{\frac{ZT}{Z_t T_t}} = \frac{\sqrt{pv}}{\sqrt{p_t v_t}}. \tag{6.35}$$

The combination of these gives the corrected mass flow as

$$\dot{m}_{cor} = M\sqrt{\frac{pv}{p_t v_t}}\frac{v_t}{v} = M\sqrt{\frac{p}{p_t}\frac{v_t}{v}} = M\left(1 + \frac{n_s - 1}{2}M^2\right)^{\frac{-n_s-1}{2(n_s-1)}}. \tag{6.36}$$

This shows that any equations related to choking are also the same as those required for an ideal gas but with the ratio of specific heats replaced by the isentropic volume exponent.

7 Aerodynamic Loading

7.1 Overview

7.1.1 Introduction

This chapter highlights the important fluid dynamic principles of the flows around aerofoils, cascades of blades and the diffusing flow in radial compressor flow channels both in impellers and vaned diffusers. The content is fundamental to understanding what is meant by high aerodynamic loading in compressors. In practice, this term is used with several separate meanings. The primary meaning is that there is a large difference in pressure between the two surfaces of a blade profile. This pressure difference is crucial in leading to the mechanical forces and the torque needed to drive an impeller and the turning of the flow in the diffuser. The pressure difference between the adjacent blades may be high even if the work coefficient or pressure rise of the stage is moderate, simply by dint of having few blades. The second meaning is that the stage is designed for a particularly high work coefficient, whereby for radial compressors anything above $\lambda = \Delta h_t/u_2^2 = 0.7$ is considered to be high. The third meaning is related to the amount of diffusion in the flow between the inlet and the outlet of a cascade or a diffuser vane. In 2D axial compressor blade sections, these different forms of aerodynamic loading are directly related to each other, but this is not so in radial compressor geometries due to the change in radius and the curvature of the meridional flow channel.

The important differences between isolated aerofoils, cascades and diffuser flow channels with respect to aerodynamic loading are explained. The nature of decelerating flows in two-dimensional channel diffusers is explained together with their interesting flow regimes, which limit the amount of diffusion that is possible. This introduces the reader to the physics of diffusers with a view to help the understanding of the performance of the different zones of radial compressor vaned diffusers, discussed in Section 12.6. A final section summarises the different parameters used to define the fluid dynamic loading limits in compressors blade rows related to diffusion limits and blade loading.

7.1.2 Learning Objectives

- Explain the differences between isolated aerofoils, cascades and flow channels and their relevance to radial compressor flow.

- Know the effect of incidence and blade number on the pressure distribution around blade cascades.
- Have knowledge of the physics and operation of two-dimensional diffusers with flow deceleration.
- Know the different blade-to-blade loading limits and limits to the pressure rise capability that are suitable for use in compressors.

7.2 Isolated Aerofoils

Before considering the nature of the flow in cascades of blades, this section briefly looks at the flow over an isolated aerofoil, the objective being to provide some general information about the forces acting, the effects of incidence and compressibility and the nature of blade stall. Aerodynamic body shapes with small drag are slender with a sharp trailing edge and their main axis is approximately parallel to the relative flow direction. The body shape can also be curved in such a way that it produces a large lift force component perpendicular to the flow relative to the moving body. It is this lift force which defines the aerodynamic nature of turbomachinery flows. Such bodies producing lift are called aerofoils, aerofoil sections, wing sections, profile sections or simply profiles. In compressors and other turbomachines, they are generally called blades when they rotate or vanes when they are stationary.

A common misconception is that the lift force on a profile is caused because the airflow moving over the top of the curved aerofoil has a longer distance to travel and needs to go faster to join up again with the air travelling along the lower surface. The faster flow is associated with a lower pressure through the Bernoulli equation, (5.9), thus producing a lift force. The flow does travel faster over the upper surface, but what actually causes the lift force is the curvature of the flow and not the distance travelled. A curved blade shape, or a straight blade moving through air inclined to the flow, causes the streamlines to curve and introduces pressure changes normal to the flow direction; see Section 5.3.2. The consequence is that the pressure on the convex surface of the aerofoil is lower than the distant pressure, while on the concave surface it is higher than the distant pressure. The two surfaces are then said to be the suction surface and the pressure surface. It is this pressure difference across the profile generated from the flow curvature that causes a force in the direction towards the convex surface, approximately perpendicular to the oncoming flow, that defines the blade-to-blade loading. In Section 12.3, the swirling flow in a radial vaneless diffuser is described. In this special case, the flow without blades does not have a straight path, but tends to be a logarithmic spiral with a fixed flow angle. When blades are added to make a vaned diffuser, it is the concave surface that becomes the suction surface.

The design of the best aerodynamic profiles for wings and for impellers is mainly related to adjusting the amount of curvature and its distribution along the chord to control pressure differences across the profile. The design of the profile is carried out such that the pressure distribution is adequate to generate the lift force needed and may be forward-loaded or rear-loaded depending on the design strategy. When the aerofoil is operating off-design, it is the incidence of the flow at the leading edge that becomes the limiting factor as it leads to a high loading at the leading edge. The pioneers of manned flight experimented on the appropriate curvature needed to provide the correct amount of lift (Tobin, 2004). Lilienthal observed the camber of bird's wings and tried to copy this, but overdid the curvature of the wings of his kites, using a camber angle of over 30°. These wings provided high lift at low incidence but became overloaded at incidence, which caused flow separation with an abrupt stall. In the end, when one of his kites stalled it dropped from the sky and caused his premature death. The Wright brothers were also inspired by bird flight and decided from their successful wind tunnel and gliding experiments to use a shallower curvature with a profile camber of just over 15°. These wings required more surface area to provide the lift but gave a more docile stall at incidence. These were subsequently adapted for use in the first successful powered flight of the Wright flyer in 1903, a biplane with two lifting wings. A survey of early research on the stalling characteristics of wing sections is provided by Young (1951).

More modern aircraft at higher speeds have an even lower camber. A clear demonstration of the effect of curvature on increasing the lift coefficient of a wing is the change in the camber shape of a modern aeroplane wing during landing. Ailerons and flaps at the rear and various slats and drooped nose configurations at the front of the wing are used to increase the curvature in the flow, causing a larger lift coefficient which enables a lower flight speed for the same lift force, thus allowing a safe landing at low speed. In the design of compressors, it is not usually possible to use blades with a variable camber, so that the designer is forced to consider the issues of blade loading and the effect of incidence on stall.

7.2.1 Lift and Drag

In classical fluid mechanics textbooks on aerofoil sections, such as Abbott and von Doenhoff (1949) and Riegels (1961), it is shown that the lift force, L, on an isolated aerofoil with no losses is perpendicular to the oncoming flow. In the presence of friction losses, there is an additional drag force, D, in the direction of the flow. These can be expressed nondimensionally in terms of a lift and drag coefficient:

$$C_L = \frac{L/A}{½\rho_\infty c_\infty^2}, \quad C_D = \frac{D/A}{½\rho_\infty c_\infty^2}, \quad \tan\varepsilon = \frac{C_D}{C_L}. \tag{7.1}$$

In this equation, A is the plan area of the profile (that is, span times chord); $\tan\varepsilon$ is the drag to lift ratio and the terms with suffix ∞ are the upstream velocity and the upstream density, such that the denominator is the dynamic pressure far upstream of the profile.

The performance of an isolated aerofoil can be presented in two ways: as a plot of the lift and drag coefficient as a function of the angle of attack, or as a polar diagram of the lift versus drag coefficient with a parameter variation of the angle of attack, as shown in Figure 7.1. The lift increases with the angle of attack and ultimately reaches a limit due to stall of the aerofoil. The drag coefficient also increases with the angle of attack. In the polar diagram, the tangent from the origin to the lift-drag curve

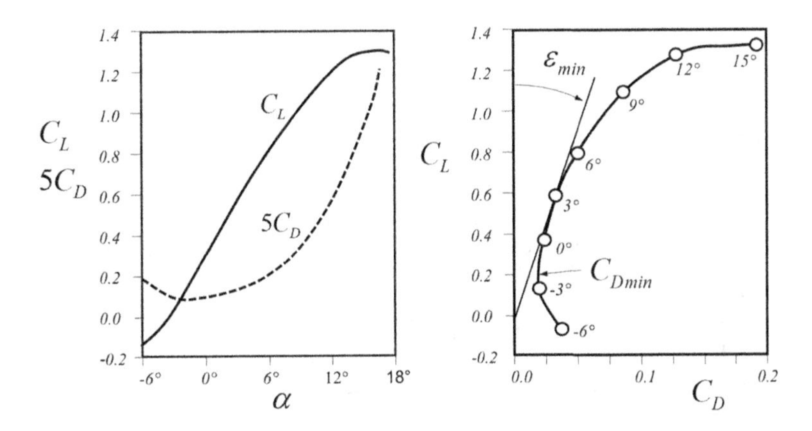

Figure 7.1 Sketch of the lift and drag coefficient and polar diagram for an aerofoil profile.

determines the angle of attack with the maximum lift-to-drag ratio, and for an isolated aerofoil this is known as the angle of glide. The angle of glide can be considered to be the optimum for an isolated aerofoil.

7.2.2 Aerofoil Profile Shape

The aerofoil shape can be described in terms of a thickness distribution and a camber line shape, and a similar approach is also used for more complex three-dimensional profiles in radial compressor blades and vanes, as described in Chapter 14. The chord line in a wing profile is a straight line joining the leading and trailing edges, and the camber line is described as the distance from the chord line. The difference in the slope angle of the camber line between the leading edge and the trailing edge is known as the camber angle. The thickness is defined as the distance between the suction and pressure surface normal to the camber line. Points on the aerofoil surface are at half the thickness set perpendicularly to both sides of the camber line. Standard profile types are described as tables of values, or as equations, for the camber and half-thickness distribution as a function of the fractional distance along the chord, x/c:

$$y_c/c = f(x/c), \quad y_t/c = f(x/c). \tag{7.2}$$

Combination of the coordinates of the camber and thickness distributions at specified distances along the chord line gives rise to the coordinates of the upper and lower surface as

$$\begin{aligned} x_u &= x - y_t \sin\theta, \quad & y_u &= y_c + y_t \cos\theta \\ x_l &= x + y_t \sin\theta, \quad & y_l &= y_c - y_t \cos\theta, \end{aligned} \tag{7.3}$$

where $\tan\theta$ is the slope of the camber line. The leading edge is the starting point of the camber line at the front and the trailing edge is the final point at the rear. This method of construction causes the cambered profile sections to project slightly forward of the leading-edge point, as shown in Figure 7.2.

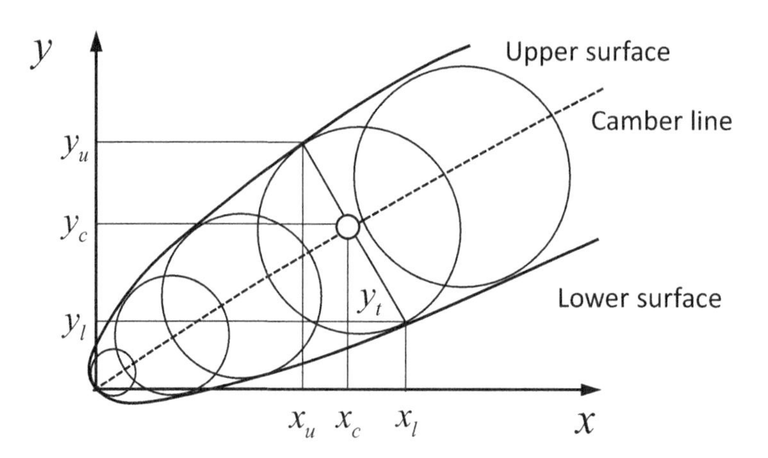

Figure 7.2 Superposition of blade thickness on camber line profile.

7.2.3 Leading and Trailing Edges

Profile leading edges have historically been rounded with a circular arc, or with an ellipse, joining up to the blade surfaces at a tangent point. The sharp change in curvature at the tangent point leads to a spike in the local surface pressure distribution followed by a trough; see Section 7.2.5. In compressor profiles, the rapid deceleration of the flow downstream of the spike may lead to additional losses if the spike is too large, as discussed in Section 10.4.2. More complex shapes with smooth curvature can provide lower losses due to the removal of the spike, but provided the velocity ratio between the trough and the peak of the spike is kept above 0.9, then Goodhand and Miller (2011) have shown that the spike is not detrimental at the design point in axial compressor profiles.

The trailing edges may also be round or simply truncated at constant radius as in most radial impellers. In some standard thickness distributions for axial compressor aerofoil shapes, notably the commonly used NACA 65 series, the trailing edge has no thickness, so a mechanically acceptable aerofoil is obtained by adding a small additional thickness near the trailing edge. Gülich (2008) gives examples where the impeller trailing edge is not simply truncated but is sharpened by underfiling from the suction side or overfiling on the pressure side, which changes the effective blade outlet angle. A similar example is given by Medic et al. (2014), where it is noted that replacing a constant radius truncated trailing edge with a rounded trailing edge, while maintaining the same maximum radius, caused a noticeably higher total pressure ratio as it reduced the effective backsweep angle.

7.2.4 Incidence and Angle of Attack

The inclination angle between the oncoming flow with a freestream velocity of c_∞ and the x-axis along the aerofoil chord is known as the aerofoil angle of attack. The angle

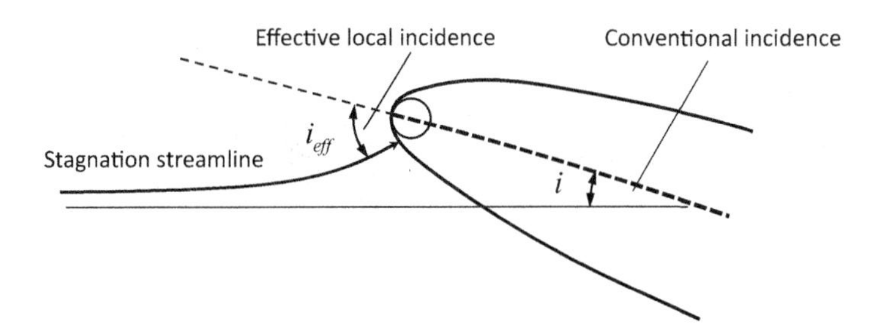

Figure 7.3 Incidence angle and effective incidence angle.

between the tangent to the camber line at the leading edge and the direction of the freestream velocity of c_∞ is known as the incidence angle. For an uncambered aerofoil, their values are the same. In a situation between two closely spaced blade rows where the freestream flow direction is difficult to determine, an additional incidence is sometimes used (Tain and Cumpsty, 2000; Galloway et al., 2018). This should not be confused with the others. It is the effective local incidence angle of the flow as it approaches the stagnation point on the aerofoil surface to the camber line of the aerofoil, as shown in Figure 7.3. This considers the upward flow of air as it begins to travel around the leading edge of an aerofoil, also known as the upwash.

7.2.5 Effect of Incidence

The aerodynamic loading of an aerofoil can be represented as a graph of the static pressure coefficient along the suction and pressure surfaces plotted against chord. Alternatively, it may be presented as the isentropic Mach number along those surfaces or the equivalent velocities. Figure 7.4 shows how the aerofoil suction surface pressures vary with the angle of incidence for an uncambered isolated aerofoil in an incompressible flow, on the 12% thick RAE 101 (Royal Aircraft Establishment) symmetrical profile using data from Brebner and Bagley (1956). The increase in the adverse pressure gradient with incidence is clear. Similar effects take place in a cascade of blades as the inlet flow angle changes. At zero incidence, the oncoming flow approaches the leading edge such that the stagnation streamline is directly onto the nose of the aerofoil, although it may not exactly be at the end of the camber line. The surface pressures are smooth with more acceleration at the front of the suction surface than on the pressure surface, leading to a positive lift force. On the pressure surface, the Mach number distribution is relatively flat. After attaining the peak velocity, and minimum pressure, on the suction surface, there is a region where the flow decelerates and the pressure recovers towards the trailing edge. This region has an adverse pressure gradient and may be at risk of flow separation. It is the diffusion on the blade suction surface that is important in determining the performance of the blade section at the design point. At off-design towards surge, this is where the flow is most likely to separate and determines the maximum incidence before stall.

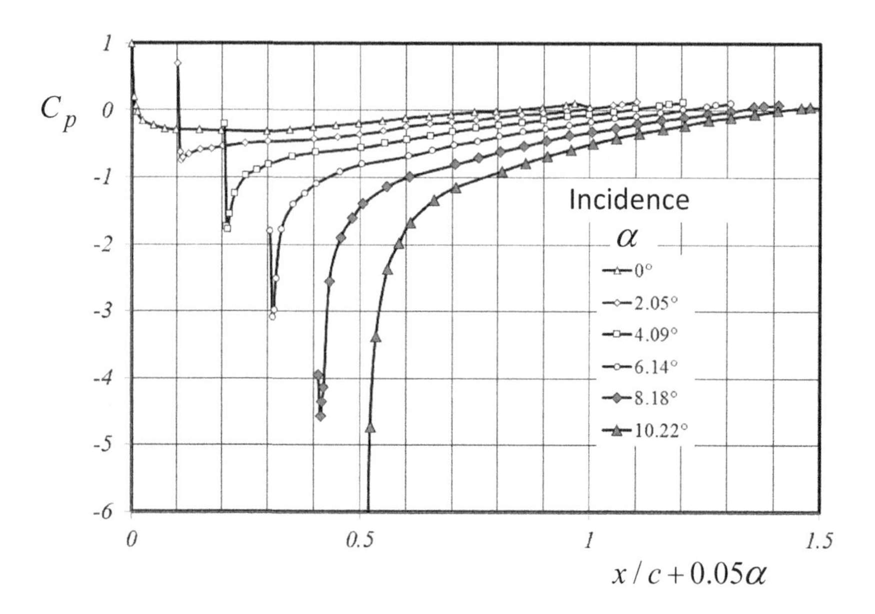

Figure 7.4 Pressure rise coefficient of an uncambered RAE 101 aerofoil at different incidence angles and a Reynolds number of $Re = 1.6 \times 10^6$, with data from Brebner and Bagley (1956).

At positive incidence, the flow approaches the underside of the profile more steeply such that the stagnation streamline impinges on the blade pressure surface, as shown in Figure 7.3. The suction surface flow moves over the sharply curved leading edge and accelerates to a suction peak, or leading-edge spike, on the suction surface. The deceleration and adverse pressure gradient on the suction surface increases with positive incidence, and the local deceleration downstream of the leading-edge spike may lead to suction surface separation, leading to stall and higher losses. This becomes more extreme as the incidence angle rises and increases the leading-edge loading and the lift coefficient up to the point where the aerofoil stalls. In a cascade of blades, the increase in incidence leads to a higher turning in the blade row, but the outlet flow direction remains nearly constant. On the pressure surface, the flow accelerates less, and the pressure increases with incidence.

The effect of the local pressure field surrounding the aerofoil is to curve the incoming streamlines, including the stagnation streamline. The streamline leading to the stagnation point curves sharply near the leading edge, and locally the effective incidence of the stagnation streamline is much higher than the actual angle of incidence (see Figure 7.3). The term *incidence* generally refers to the difference between the blade inlet angle and angle of the upstream flow. The effect of positive incidence is to cause a lower pressure along the whole of the suction surface with a large suction pressure peak which increases with incidence. On the pressure surface, the pressure increases with incidence so both surfaces contribute to the increase in lift. In this way, positive incidence is defined as incidence which increases the aerodynamic loading.

At negative incidence, the flow approaches the upper side of the wing profile more steeply such that the stagnation streamline impinges on the blade suction surface and the pressure surface flow accelerates over the nose to form a suction peak, but now this is on the pressure side. In the inlet region of the aerofoil, the loading has reversed but the lift may still be present as the remainder of the profile makes a positive contribution. The highest velocity and the highest deceleration is now on the pressure surface, and this may be a cause of pressure surface flow separation downstream of the suction peak.

The streamline at the trailing-edge leaves the blade at the relatively sharp edge at all incidence levels. It cannot shift to the suction or pressure surface as the trailing edge is usually so sharp that the flow must separate if it tries to round the trailing edge. A body with a sharp trailing edge creates a circulation of the flow of sufficient strength to change the lift coefficient sufficiently to keep the rear stagnation point at the trailing edge, known as the Kutta condition. The flow downstream of a thick trailing edge is always unsteady with vortex shedding from both surfaces, and this also effectively imposes the flow direction, as there is no tendency for the stagnation point to shift on to the pressure or suction surface at a thick trailing edge (Denton, 2010).

The lift increases with the angle of attack in an almost linear manner at low angles, but at high angles the lift stops increasing due to boundary layer separation on the suction surface, as discussed in Section 5.5.7, and the aerofoil is said to be stalled. The sharp deceleration just downstream of the leading edge on the suction surface may cause the laminar boundary layer to separate, creating a separation bubble on the suction side of blade. A low-speed aerofoil with 12% thickness stalls abruptly at an incidence of between 12 and 15°.

Gault (1957) categorised three types of stall found on isolated aerofoils on the basis of the thickness of the aerofoils near the leading edge. Trailing edge stall occurs on profiles with a thickness of over 4% of chord at 1.25% of the chord length. In these cases, a turbulent boundary layer separation near the trailing edge of a thick aerofoil moves forward with increasing angle of attack and leads to a steady decrease in the lift-to-drag ratio as the incidence increases. Leading-edge stall occurs on profiles with a thickness of less than 4% but more than 2% of chord at 1.25% of the chord length. In these cases, an abrupt separation of the boundary layer on a less thick aerofoil near the leading edge at incidence causes a rapid drop in the lift-to-drag ratio, as shown in Figure 7.1. Thin aerofoil stall occurs on profiles with a thickness of less than 2% of chord at 1.25% of the chord length. In these cases, the separation takes place abruptly at the leading edge, leading to a narrower range of incidence with thin leading edges.

This basic categorisation of the stalling behaviour of different aerofoils exhibits some overlap between the types and is also a function of the Reynolds number and the shape of the leading edge contour, but shows that early research identified that stall is strongly affected by the shape of the leading edge thickness and its influence on the boundary layer development. A similar behaviour can also be identified in considerations of the stability of compressor blade rows, which may show a steady decrease in pressure rise towards surge or may show an abrupt drop in pressure rise caused by stall at incidence.

7.2.6 Effect of Mach Number

The preceding considerations are related to incompressible flow. If the incoming flow increases to higher Mach numbers, then the density at positions of high local velocity falls, so that the volume flow rate within a stream-tube increases. To maintain the local mass flow rate in the stream-tube, the continuity equation determines that the local velocity must increase beyond its incompressible value, and the velocity distribution on a blade contour therefore changes with the inlet Mach number, increasing in a manner shown qualitatively in Figure 7.5. If the flow locally exceeds the speed of sound, it does not afterwards become subsonic gradually, but usually through a shock normal to the profile contour, where the flow velocity suddenly drops from supersonic to subsonic values, as described in Section 6.3.1.

In the design of aerofoil wing shapes, considerable effort has been expended to avoid the additional drag associated with the high Mach numbers and their effect on the boundary layers. Passenger aeroplanes fly at around a Mach number of 0.8 to avoid these issues. Aeroplane wings are swept backwards and this mitigates the shock strength by reducing the component of Mach number normal to the wing. Sweep is also used at the leading edge of radial compressor impellers to weaken the effect of high Mach numbers, as discussed in Section 11.9.3.

In order to reduce the effects of a high Mach number, several design strategies have been adopted for the aerofoil sections. An important development was the use of profiles with a so-called rooftop pressure distribution. These have a shape that leads to a relatively flat pressure distribution on the forward upper surface so that the flow is not accelerated in this region, and this delays the advent of supersonic flow. A popular aerofoil series that adopts this strategy is the NACA 6-series (NACA is the forerunner of NASA), whereby the type known as NACA 65 has often found application as an axial compressor profile. The ‘6’ refers to the series, and the ‘5’ refers to extent of the rooftop style loading extending to 50% of the chord of the isolated aerofoil from the leading edge. When this profile shape is used in a cascade, this feature of the pressure distribution is lost.

Special aerofoil profile shapes, known as supercritical profiles, were also developed which allowed a small area of supersonic flow to exist without a downstream shock, creating a shock-free supersonic flow region. In order to obtain a high lift without

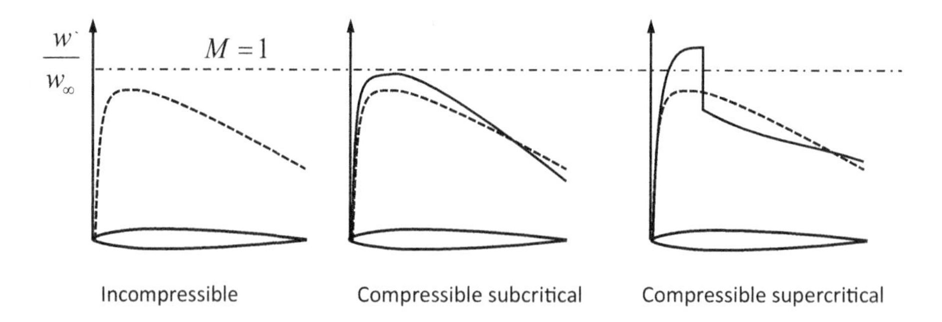

Figure 7.5 Effect of compressibility on the velocity distribution.

supersonic flow on the suction surface, rear-loaded profiles were introduced. These have a lower surface that is highly cambered towards the rear, and this generates lift in the rear part of the aerofoil by increased pressure on the pressure surface. This has been used as the basis of the profiles for the Airbus 300 aircraft and many civil transport aircraft. Finally, if supersonic flow cannot be avoided, supersonic aerofoil profiles are used, and these are usually thin and sharp at the leading edge. This strategy is used in radial compressor impellers near the leading edge at the casing where the relative flow is often supersonic. Both sweep and supercritical aerofoil profile designs have been used in compressor blades, for example Behlke (1986).

7.3 Profiles in Cascade

7.3.1 Loss Coefficient and Deviation

In turbomachines, the individual aerofoil sections are not isolated but are placed equidistant next to each other around the circumference. Both the inducer of the impeller and the diffuser vanes can usefully be considered as a 2D cascade of aerofoils with flow curvature to change the flow direction between inlet and outlet of the cascade; but in compressor cascades, it is sometimes more useful to think in terms of diffusion. The two effects are related, and Figure 5.3 shows how the turning of the profile sections produces an increase in stream-tube area through the cascade leading to a diffusion of the flow. The ideal pressure rise coefficient in an axial cascade with incompressible flow with no change in the stream-tube width is given by

$$C_p^{id} = \frac{p_2 - p_1}{p_{t1} - p_1} = 1 - \left(\frac{A_1}{A_2}\right)^2 = 1 - \left(\frac{\cos\beta_1}{\cos\beta_2}\right)^2. \tag{7.4}$$

The aerodynamic theory of cascades (Scholz, 1977; Gostelow, 1984) considers the performance of a linear array of 2D blade profiles in which, due to the close proximity to each other, the flow over each profile is greatly affected by its near neighbours. The cascade is effectively an unrolled approximation of a cylindrical section through the blades around the circumference. As a result of being in a cascade, the relevant geometrical parameters of the blades include the inlet and outlet camber angles and profile shape, as in an aerofoil, and a new parameter known as the space–chord ratio, s/c, or its inverse, known as the solidity, $\sigma = c/s$. In radial compressors, the solidity is high, $\sigma > 1$, as the chord is usually much larger than the spacing between the blades.

The basic properties used to describe the performance of such a cascade are not the lift and drag coefficients but rather the loss coefficient and the deflection or turning angle of the flow by the cascade, $\Delta\alpha$, or the deviation angle of the flow from the direction of the blade camber angle at its trailing edge, δ, defined as

$$\zeta = \frac{\Delta p_t}{½\rho c_1^2}, \quad \Delta\alpha = \alpha_2 - \alpha_1, \quad \delta = \alpha_2' - \alpha_2. \tag{7.5}$$

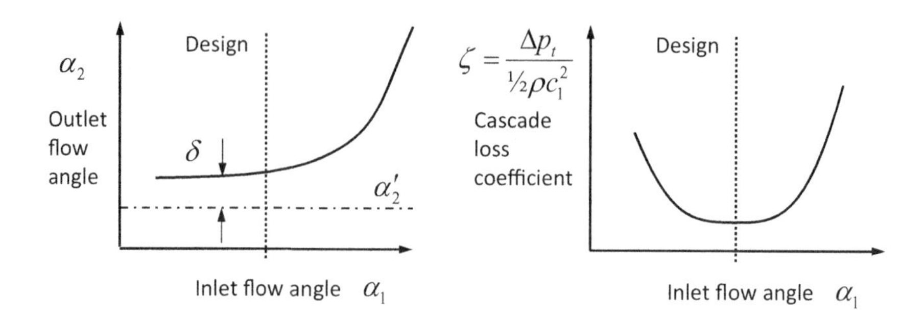

Figure 7.6 Outlet flow angle and profile loss coefficient as a function of inlet flow angle.

In compressor cascades, the loss coefficient is typically defined in terms the total pressure loss relative to the inlet dynamic head to the cascade, as shown here. Each of these parameters is a function of the inlet flow angle of the cascade, α_1. The deviation is the angle between the tangent to the camber line at the trailing edge and the direction of the freestream velocity downstream of the cascade. An example of the typical variation of the flow outlet angle and the profile loss coefficient for an incompressible flow in a 2D cascade is given in Figure 7.6. The deviation remains almost constant until the suction surface boundary layer starts to grow in thickness and eventually separates at positive incidence conditions, which causes the deviation to increase significantly.

The losses increase with both positive and negative incidence, and the diagram showing losses as a function of inlet flow angle is often known as the loss bucket. The incidence range of low speed blades is a strong function of leading-edge thickness. Thick turbine blades with a low inlet Mach number can have an incidence range of the order $\pm$ 30°, while thin compressor blades at a higher subsonic inlet Mach number may have an incidence range of only $\pm$ 5°.

Unlike an isolated aerofoil, the cascade causes a change in the direction of the flow. Because of this, the lift and drag in a 2D cascade are not defined relative to the oncoming flow but are defined normal and parallel to the direction of the mean flow direction which is given by

$$\tan \alpha_\infty = (\tan \alpha_1 + \tan \alpha_2)/2. \tag{7.6}$$

It is possible to relate the loss coefficient and the flow turning to the lift coefficient and drag coefficient of the individual profiles, and the equations for this are developed by Horlock (1958) and lead to

$$\begin{aligned} C_D &= \frac{\zeta}{\sigma} \frac{\cos^3 \alpha_\infty}{\cos^2 \alpha_1} \\ C_L &= \frac{2}{\sigma}(\tan \alpha_1 - \tan \alpha_2) \cos \alpha_\infty - \frac{\zeta}{\sigma} \frac{\cos^2 \alpha_\infty \sin \alpha_\infty}{\cos^2 \alpha_1} \\ &= \frac{2}{\sigma}(\tan \alpha_1 - \tan \alpha_2) \cos \alpha_\infty - C_D \tan \alpha_\infty. \end{aligned} \tag{7.7}$$

The lift coefficient in a compressor cascade is often more than around 20 times larger than the drag coefficient, so the last term in the equation for the lift coefficient is usually negligible, and this allows the following rearrangement of this equation

$$C_L\sigma = 2(\tan\alpha_1 - \tan\alpha_2)\cos\alpha_\infty = \frac{2}{c_x}\cos\alpha_\infty\Delta c_y. \tag{7.8}$$

where c_x is the velocity normal to the cascade and Δc_y is the change in the tangential velocity along the plane of the cascade. This equation shows that it is the product of the lift coefficient and the solidity that determines the turning of the flow in a cascade. To achieve the same turning, a low solidity cascade needs a higher lift coefficient than a high solidity cascade. An isolated aerofoil has zero solidity and so produces no turning of the flow. Two of the important loading criteria for compressor blades are considered in Section 7.5, the Lieblein diffusion factor and the Zweifel number; both include the term $\Delta w_u/\sigma$ in their formulation, which is the change in the circumferential component of velocity divided by the solidity, so that (7.8) shows the direct link between these loading coefficients and the lift coefficient of a blade in terms of a 2D cascade.

The use of linear cascade theory is rather cumbersome in practice even for axial compressors as the mean-flow direction, α_∞, becomes a function both of the lift coefficient and the inlet flow angle. Compressor engineers seldom mention or make use of the lift coefficient of the profiles. Linear cascade theory then becomes a rather highfalutin theory of no real practical significance. In addition, in a radial machine neither the solidity nor the throughflow velocity normal to the cascade remains constant through the cascade. A linear 2D cascade may be a reasonably good approximation to an axial cascade if the flow remains two dimensional, but the real limit on performance for cascades or blade rows is probably set in most cases by the flow in the corners (Goodhand and Miller, 2012). In radial stages, however, there is a mixed-flow cascade (axial to radial) in the impeller and a purely radial cascade in the diffuser, both of which have a change of the meridional velocity and solidity through the blade row. In addition, in radial impellers there are strong secondary flows causing a component of velocity tangential to the cascade plane. For these reasons, cascade theory is seldom used in radial machines, but nevertheless Scholz (1977) derives various cascade theory equations for these applications.

7.3.2 Profile Types

Common standard aerofoil profiles used in low-speed axial compressors are the American NACA 65 series and the broadly equivalent British C-series of profiles, as described in Cumpsty (2004). In old-fashioned blades, the camber line is often taken to be a circular arc. If the suction and pressure surfaces are also taken as circular arcs, this leads to a double circular arc (DCA) blade type, which was often used at higher Mach numbers because of its thin leading edge. More modern subsonic profile types in axial compressors have the camber line tending to become straight towards the trailing edge. There are no such standard profiles for radial compressors, although some two-dimensional impellers and diffusers make use of circular arc blading; see Section 14.5.3.

Considerable research has been carried out on the most suitable profile form in axial compressor blade sections (Hobbs and Weingold, 1984; Behlke, 1986; Calvert and Ginder, 1999). This has led to the development of the so-called controlled diffusion type of aerofoil. Controlled diffusion profiles have more turning and diffusion near the leading edge where the suction surface boundary layer is thin and becomes straighter towards the trailing edge. This leads to less deviation and relieves the boundary layer diffusion towards the trailing edge. The discussions of Calvert and Ginder (1999) provide an overview of their design strategy for transonic axial fan blades at different span-wise positions and of the compromises that need to be made there. At the tip of a transonic fan with a high Mach number the blades are thin and nearly straight to avoid acceleration in the supersonic inlet flow. After the flow has reached the suction surface shock, more turning is then applied downstream of the shock so the blades tend to be straighter in the inlet region with more turning in the rear sections of the cascade. The blade sections on the hub are of the controlled diffusion type, with more turning near the leading edge and becoming straighter towards the trailing edge.

7.3.3 Alpha-Mach Diagram

The overall performance of blade sections at higher relative inlet Mach numbers is illustrated in Figure 7.7 in the form of an alpha-Mach diagram, adapted from the research of Raw and Weir (1980) and Goodhand and Miller (2011). This shows the location of the stall and choke incidence on a graph of the flow inlet angle (or blade incidence) plotted against relative inlet Mach number. At very low Mach numbers, the limit of operation at negative incidence is better represented by a negative incidence stall condition rather than choke, and this is also shown on the diagram. Interestingly, recent research, discussed later in this chapter, has also shown that even at higher Mach

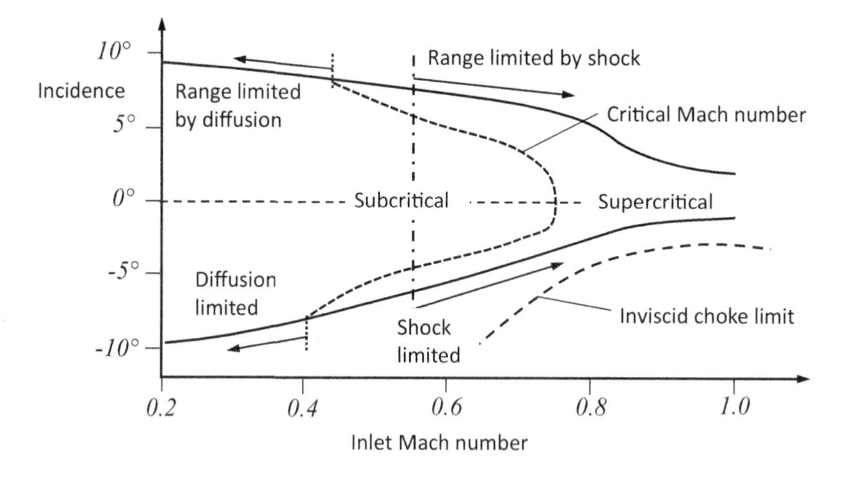

Figure 7.7 Sketch of the effect of incidence on range in an alpha-Mach diagram. (image adapted from the work of Raw and Weir (1980) and Goodhand and Miller (2011))

numbers the negative stall incidence is more related to flow separation than choke. This diagram clearly shows the reduction in the operating range of blade sections at higher Mach numbers, which makes the design of high-speed compressors significantly more difficult than those with lower Mach numbers. A similar form of diagram is derived in Section 17.4 to show the reduction of range with increased Mach number which determines the width of the characteristic of radial compressor stages.

The alpha-Mach diagram shows how the performance of the profile changes as the Mach number increases and indicates that the critical Mach number with the first sign of any flow with $M > 1$ is affected by the incidence. First, the minimum loss of the sections increases with Mach number, and the flow incidence at the minimum loss shifts to a higher incidence. Second, the range of the inlet angle where the losses remain relatively low reduces with increasing Mach number, indicating a reduction in the operating range of the sections. The stalling incidence and the choking incidence are closer together as the Mach number increases. At low inlet Mach numbers, the effect is similar to that shown in Figure 7.5, whereby the peak suction surface velocity at incidence increases, leading to a higher deceleration and an earlier stall. In typical axial compressor profiles that have a typical high subsonic inlet Mach number at their design point, several factors play a role in the further reduction of incidence range as the Mach number increases.

Goodhand and Miller (2011) show that the positive incidence range is determined by the formation of a shock at about 15% chord on the suction surface, which terminates the supersonic patch, as shown in Figure 7.5 for an isolated aerofoil. The shock separates the suction surface laminar boundary layer, resulting in three separate additional sources of loss: first, mixing in the reattaching laminar separation bubble; second, mixing in a trailing edge separation, caused because the thickened reattached boundary layer fails to negotiate the required diffusion on the suction surface; and third, an increase in the mixing downstream of the trailing edge due to the thickened wake. The limit of negative incidence is determined by the formation of a shock which terminates a small (~1 mm long), supersonic patch around the leading edge. The shock separates the pressure surface boundary layer. The mechanism which causes the rise in loss is mixing within the pressure surface separation bubble, not irreversibilities within the shock. The fluid dynamic mechanism usually considered responsible for the limit on incidence at high flows is commonly referred to as choke. Goodman and Miller show, however, that at the limit of negative incidence the flow is not yet choked at the throat.

An interesting result was published by Carter (1961). He tested a sharp leading edge and was surprised to find that it operated with low loss over a surprisingly wide operating range. Goodhand and Miller (2011) determined that the best leading edge turned out to be a near-parabolic shape that merged smoothly onto the aerofoil surface with no spike over the whole operating incidence range. Goodhand suggests that Carter had probably stumbled across this shape in his experiments.

7.3.4 The Effect of Solidity on Loading

The effect of blade number on the profile velocity distribution in an axial cascade can be visualised as in Figure 7.8. Doubling the number of blades reduces the pitch

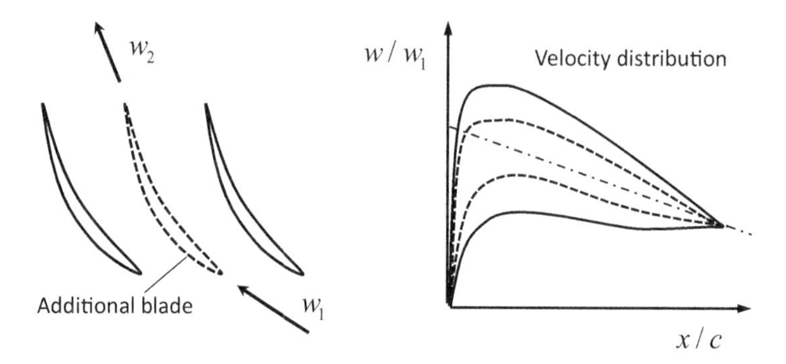

Figure 7.8 The approximate effect of doubling the blade number on the velocity distribution in an axial cascade.

between the blades, and this reduces the difference in velocity between the suction and pressure surface by one half. The peak velocity on the suction surface is reduced, and the minimum velocity on the pressure surface is increased. The diffusion on the suction surface is reduced. These effects combined with the increase in the wetted surface area of the blades, and the additional blockage cause by a larger number of blades, causes a change in the losses of the cascade. The design of compressor blade profiles is therefore usually a compromise between reducing profile loss at design incidence, by increasing pitch to chord, and increasing operating incidence range by reducing pitch to chord. In a radial compressor with splitter blades, the increase in solidity in the rear of the impeller reduces the blade-to-blade loading in this region.

7.3.5 The Compressor as a Diffusing Flow Channel

A useful approximation to the behaviour of a blade row channel in an impeller or a diffuser in a radial machine is to consider the flow path as a diffusing channel, as already shown in Figure 5.3. The physics of straight two-dimensional diffuser flow channels is described in Section 7.4, and this is a much better guide to the understanding of the breakdown of the flow in radial compressor flow channels than the theory of cascades. The 2D diffuser approach, however, does not usually consider the curvature of the channel nor the effect of incidence at the leading edge, so a mixed approach is sometimes more useful. One approach called TEIS (two elements in series) has been described by Japikse and Baines (1994) and is a good description of the blade row performance. At the leading edge and up to the throat, the flow is most strongly influenced by the area change in the oncoming stagnation streamlines as the incidence changes. Up to the throat, there can be acceleration or deceleration of the flow depending on the flow inlet angle, as shown in Figure 7.9. The leading-edge blade angle can only lead to a single operating point with zero incidence somewhere between the surge and choke conditions, probably at the design conditions. Downstream of the throat, the channel curvature and diffusion

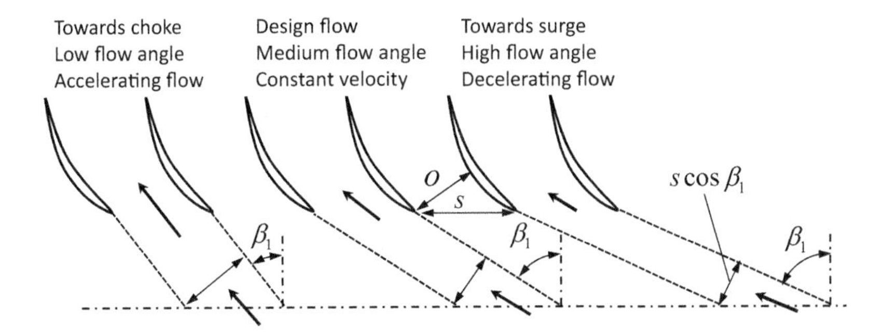

Figure 7.9 Blade-to-blade view of stagnation streamlines in a compressor cascade with different inlet flow angles.

effects dominate. This approach is used in Section 12.6 to discuss the operation of vaned diffusers.

An example of the pressure distribution on the mean line in a radial impeller with a splitter blade at different operation points is shown in Figure 7.10. It can be seen that the largest changes in the pressure distribution as the incidence changes is in the forward part of the flow channel and up to the throat. At high incidence, the suction peak on the leading edge of the main blade and the splitter blade can be seen. The effect of incidence towards choke is to reverse the loading on the leading edge of the main blade.

7.3.6 Cascade Loading Limits

The preceding description identifies that there are three separate issues that can cause excessive aerodynamic loading in a compressor cascade. First, there is the overall diffusion of the flow: the cascade may be poorly designed to have an overall deceleration and turning that is too great, and then the flow is likely to stall within the channel, either on the blade or most probably in the end-wall region or the corners. This is usually assessed by considering the ratio of the outlet to inlet relative velocity, known as the de Haller number, $DH = w_2/w_1$, as described in Section 7.5.4. Second, the blade-to-blade loading may be too high such that local diffusion levels on the blade profile downstream of the suction peak are too high. This occurs if there are too few blades for the amount of turning such that the suction surface has a high velocity at the suction peak and the deceleration downstream of the suction peak is too great and the flow may separate. This is usually assessed by considering the Lieblein diffusion factor, which considers the amount of deceleration from the suction peak velocity, $DF = (w_{max} - w_2)/w_1$ as outlined in Section 7.5.10. Third, at off-design conditions, the incidence may become too high, either as a positive incidence or a negative incidence causing separation from the suction peak at the leading edge. These issues are discussed further in Section 7.5.

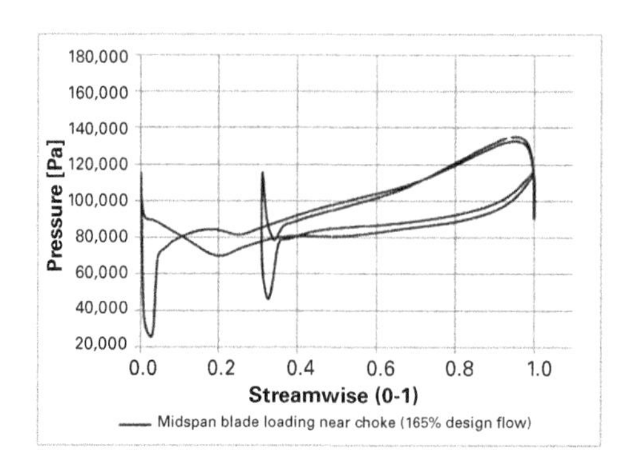

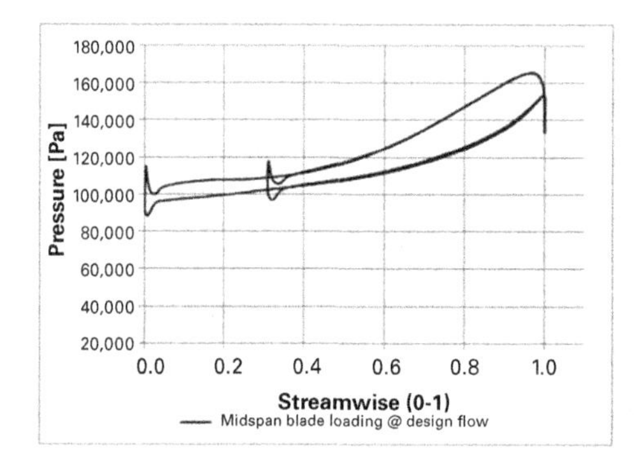

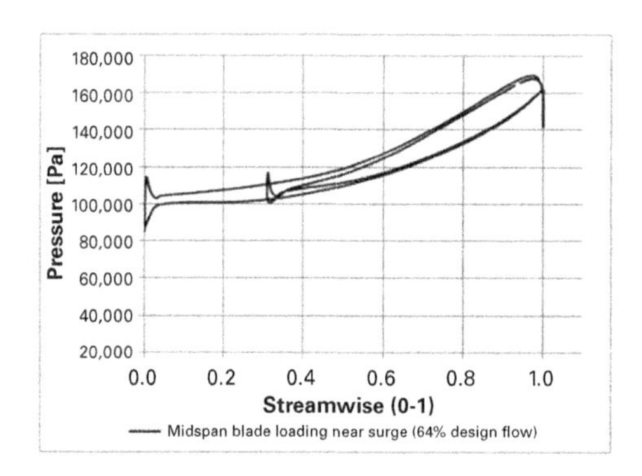

Figure 7.10 Pressure distribution in a radial impeller with splitter blades on the midspan mean line near choke (164% flow), near design (100% flow) and near surge (64% flow).

7.4 Diffusers

7.4.1 Functionality and Applications

In a radial turbocompressor, the flow decelerates both in the rotor and in the stator in order to increase the static pressure as determined by the Bernoulli equation, (5.9). Section 2.2.5 shows that the breakdown of the pressure rise for a typical radial

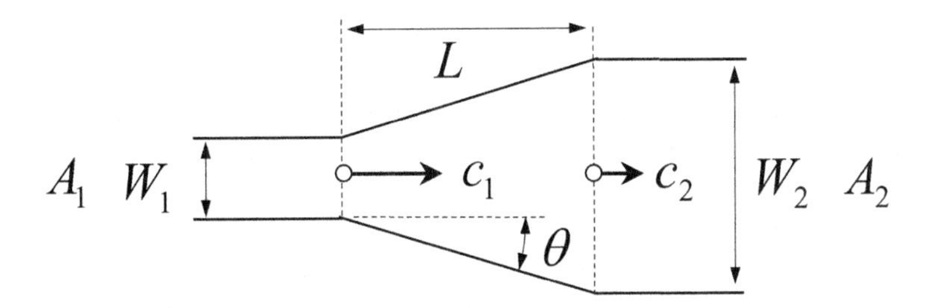

Figure 7.11 Geometrical parameters of a planar diffuser.

turbocompressor stage is approximately 35% due to the deceleration across the stator, 15% due to the deceleration of the relative flow in the rotor and the remaining 50% due to the centrifugal effect. The relative flow velocity in the impeller and the absolute flow in the diffuser decrease through each component, so an understanding of the limitations of diffusion as a means of raising the pressure is important for both components. In this section, the background to diffuser performance as given by the substantial literature on two-dimensional planar diffusers is considered. This is used in Section 12.6 to identify the expected performance of the different zones in radial compressor vaned diffusers.

A diffuser is a channel for the purpose of decelerating the flow and causing an increase in static pressure. In subsonic flow, this is achieved with an enlargement of the cross-sectional area in the direction of the flow. The diffuser design can be generally classified into planar, rectangular and axisymmetric diffusers, the latter of which can in turn be subdivided into conical and annular diffusers. In addition, there are also diffusers with an abrupt change in the cross-sectional area, known as Carnot or step diffusers. Miller (1990) gives an excellent general description of the behaviour of diffusers, and Japikse and Baines (1998) give a comprehensive overview on the design and performance of such diffusers for turbomachinery.

The basic design of planar diffusers is defined by three nondimensional geometrical parameters as shown in Figure 7.11. These are the ratio of the outlet to inlet area or width ratio, $A_2/A_1 = W_2/W_1$, where W is the width of the flow channel, the dimensionless length to inlet width ratio, L/W_1, where L is the length of the diffuser, and the half divergence angle, θ, where $\tan\theta = (W_2 - W_1)/2L$. In engineering practice, the full divergence angle, 2θ, is used to categorise the planar diffuser opening angle. Similar geometrical parameters can also be defined for rectangular, conical and annular diffusers, as shown by Sovran and Klomp (1967).

7.4.2 Performance Coefficients of Diffusers

For evaluating the performance of diffuser flows, dimensionless coefficients are necessary. In general, these are defined in terms of pressure differences in an incompressible flow, as most generic test data have been carried out at low subsonic velocities. The most important coefficient for diffuser design is the static pressure recovery coefficient:

$$C_p = \frac{p_2 - p_1}{p_{t1} - p_1}. \tag{7.9}$$

This gives the static pressure rise between the diffuser inlet and the diffuser exit relative to the dynamic pressure at the inlet. The pressure recovery coefficient represents the amount of the available kinetic energy at the diffuser inlet, which is converted into static pressure rise at the diffuser exit. Due to the continuity equation, some kinetic energy must be left at the exit, so that even if there are no losses, the value of C_p is always smaller than unity.

The total pressure difference across the diffuser due to losses caused by dissipative processes (friction, separation, mixing, etc.) is also made nondimensional with the dynamic pressure at the inlet, and this leads to the total pressure loss coefficient, ζ,

$$\zeta = \frac{p_{t1} - p_{t2}}{p_{t1} - p_1}. \tag{7.10}$$

At a particular location, the ratio of the local kinetic energy to that at the inlet represents the fraction of kinetic energy which has not yet been converted into static pressure rise. This gives rise at the diffuser outlet to a kinetic energy loss coefficient (or leaving loss coefficient), ξ,

$$\xi = \frac{p_{t2} - p_2}{p_{t1} - p_1}. \tag{7.11}$$

All three coefficients together account for the total energy at diffuser inlet, and so they add up to unity:

$$C_p + \zeta + \xi = 1. \tag{7.12}$$

Assuming a frictionless incompressible flow, an ideal pressure rise coefficient, C_p^{id}, can be derived for a diffuser from the Bernoulli and continuity equation:

$$\begin{gathered} p_{t2} = p_{t1}, \quad p_2 + \tfrac{1}{2}\rho c_2^2 = p_1 + \tfrac{1}{2}\rho c_1^2, \quad c_1 A_1 = c_2 A_2, \\ C_p^{id} = \frac{p_2 - p_1}{p_{t1} - p_1} = \frac{\tfrac{1}{2}\rho c_1^2 - \tfrac{1}{2}\rho c_2^2}{\tfrac{1}{2}\rho c_1^2} = 1 - \frac{c_2^2}{c_1^2} = 1 - \left(\frac{A_1}{A_2}\right)^2. \end{gathered} \tag{7.13}$$

This coefficient describes the maximum possible C_p value in an ideal flow where there are no dissipative losses and with incompressible flow. It depends only on the area ratio and reaches the value of unity when the area ratio becomes infinite and there are no leaving losses. As a reference value for this parameter, a diffuser with an area ratio of 2 can achieve an ideal pressure recovery of $C_p^{id} = 0.75$. In a real diffuser, there are dissipation losses, ζ, and boundary layer blockage on the walls at inlet and at outlet, ε_1 and ε_2, and we can estimate their effect on the static pressure rise coefficient as follows:

$$\begin{gathered} p_{t2} = p_2 + \tfrac{1}{2}\rho c_2^2 = p_1 + \tfrac{1}{2}\rho c_1^2 - \zeta\tfrac{1}{2}\rho c_1^2 = p_1 + \tfrac{1}{2}\rho c_1^2(1 - \zeta) \\ C_p = \frac{p_2 - p_1}{p_{t1} - p_1} = \frac{(1-\zeta)(\tfrac{1}{2}\rho c_1^2) - \tfrac{1}{2}\rho c_2^2}{\tfrac{1}{2}\rho c_1^2} = 1 - \zeta - \frac{c_2^2}{c_1^2}, \quad k_1 c_1 A_1 = k_2 c_2 A_2 \\ C_p = 1 - \zeta - \left(\frac{k_1 A_1}{k_2 A_2}\right)^2 = 1 - \zeta - \left(\frac{1-\varepsilon_1}{1-\varepsilon_2}\right)^2 \left(\frac{A_1}{A_2}\right)^2. \end{gathered} \tag{7.14}$$

This equation shows that an increase in losses and an increase in boundary layer blockage across the diffuser, $\varepsilon_2 > \varepsilon_1$, both lead to a reduction in the achievable pressure rise coefficient.

Another diffuser parameter that is found in the literature is the diffuser effectiveness or diffuser efficiency, η_{diff}, defined as

$$\eta_{diff} = \frac{C_p}{C_p^i}, \tag{7.15}$$

which relates the actual diffuser pressure recovery to the isentropic pressure recovery in a compressible flow, C_p^i. This parameter has the advantage that the effect of the Mach number, as described in Section 7.4.3, is similar on both the real and the isentropic C_p values, so the effect of the Mach number is decoupled from the performance parameters (Gao et al., 2017).

7.4.3 Diffusers in Compressible Flow

For compressible flow, which is typical in radial compressors, the compressibility of the fluid can have a beneficial effect on the static pressure rise, leading to a higher static pressure across the diffuser than would be obtained in an incompressible flow. If the incoming flow has a higher Mach number, then the density at positions of low local velocity increases, so that the volume flow rate there decreases. The local velocity decreases beyond its incompressible value and the pressure recovery increases. This effect is best described by the equation for the ideal pressure recovery coefficient in a compressible flow rearranged by dividing by the inlet total pressure as follows:

$$C_p^{comp} = \frac{p_2 - p_1}{p_{t1} - p_1} = \frac{(p_2/p_{t11}) - (p_1/p_{t1})}{1 - (p_1/p_{t1})}. \tag{7.16}$$

From Section 6.2.2, the isentropic relation for the pressure ratio term in the denominator is given by

$$\frac{p_{t1}}{p_1} = \left(1 + ½(\gamma - 1)M_1^2\right)^{\frac{\gamma}{\gamma-1}} \tag{7.17}$$

and the exit Mach number is given from the continuity equation by

$$M_2 = M_1 \frac{A_1}{A_2} \frac{\left(1 + ½(\gamma - 1)M_1^2\right)^{\frac{\gamma+1}{2(\gamma-1)}}}{\left(1 + ½(\gamma - 1)M_2^2\right)^{\frac{\gamma+1}{2(\gamma-1)}}}. \tag{7.18}$$

The compressible pressure recovery coefficient is then given by

$$C_p^{comp} = \frac{\left(1 + ½(\gamma - 1)M_2^2\right)^{\frac{\gamma}{1-\gamma}} - \left(1 + ½(\gamma - 1)M_1^2\right)^{\frac{\gamma}{1-\gamma}}}{\left[1 - \left(1 + ½(\gamma - 1)M_1^2\right)^{\frac{\gamma}{1-\gamma}}\right]\left(1 + ½(\gamma - 1)M_2^2\right)^{\frac{\gamma}{1-\gamma}}}. \tag{7.19}$$

The iterative solution of these equations for a given area ratio leads to Figure 7.12 showing that the ideal pressure rise coefficient increases for increasing inlet Mach number.

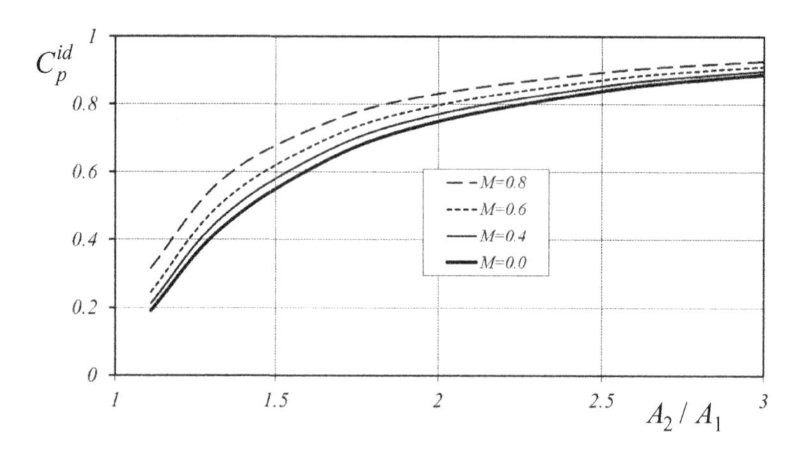

Figure 7.12 Effect of Mach number on the isentropic static pressure rise coefficient.

For a representative area ratio of 2, at an inlet Mach number of 0.8, the ideal pressure recovery coefficient is 11% higher than with incompressible flow at $C_p = 0.831$. All the preceding coefficients are related to a one-dimensional (1D) flow. For a three-dimensional (3D), flow the specific flow variables have to be averaged to provide a representative mean value.

7.4.4 Flow Regimes in Planar Diffusers

In experimental work on planar diffusers, Reneau et al (1967) and Kline at al. (1959) showed in a series of important publications that there are four different flow regimes depending on the area ratio and the length-to-width ratio. They developed a typical diffuser chart which plots the diffuser area ratio against the length-to-width ratio in a log–log plot. In this diagram, the lines of constant opening angle are straight parallel diagonal lines as can be seen from

$$\tan\theta = \frac{W_2 - W_1}{2L} = \frac{W_1}{2L}\left(\frac{W_2}{W_1} - 1\right),$$

$$\text{In}(\tan 2\theta) = -\text{In}\left(\tfrac{L}{W_1}\right) + \text{In}\left(\tfrac{W_2}{W_1} - 1\right). \tag{7.20}$$

A long diffuser with small area ratio and small opening angle is in the bottom-right, and short diffusers with high opening angle are in the top-left of the diagram. For a given nondimensional length, increasing the area ratio (or the divergence angle) of the diffuser causes four flow regimes to be passed through in the following order, as shown in Figures 7.13 and 7.14:

(a) No appreciable stall with attached flow at low divergence angles
(b) Transitory stall with separation varying in position and size with time and generating a fluctuating pressure recovery

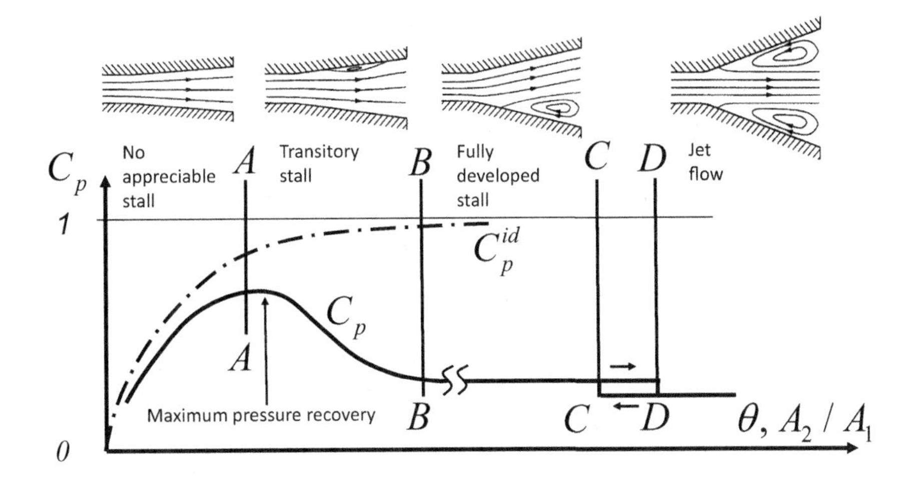

Figure 7.13 Effect of diffuser flow regimes on pressure recovery.

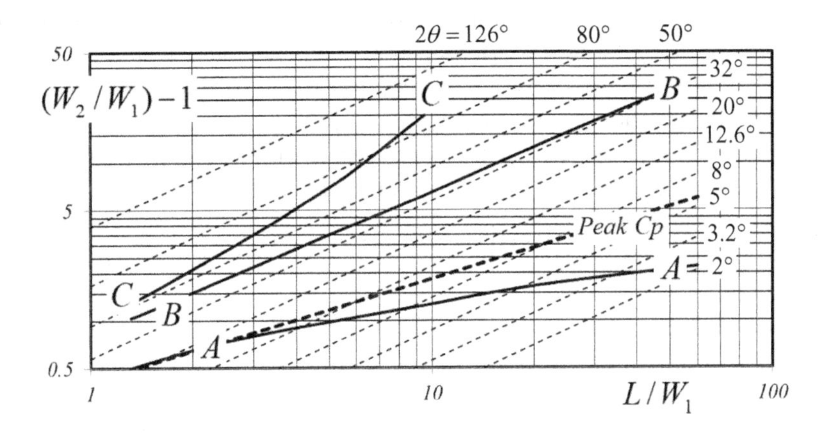

Figure 7.14 Flow regime chart for two-dimensional diffusers.

(c) Fully developed stall with a large, steady separation on one wall with very poor performance

(d) Jet flow with fully developed stall on both walls with no pressure recovery and with a hysteresis zone if the area ratio is subsequently reduced.

Kline et al. noted that maximum pressure recovery lies just within the region with transitory stall, so that all diffusers at their optimum performance level are close to stall. This is a key aspect to the design of diffusers for compressors. Any attempt to achieve higher performance by more deceleration of the flow may end up being disastrous as it may cause the diffuser to completely stall and, as a result, achieve a very low performance. The region of most interest in diffusers for radial compressors is in the lower half of Figure 7.14, close to the peak pressure recovery with an area ratio close to 2 and a nondimensional length of 5, and the diagram shows that at this point the opening angle is between 6 and 10°. It is not the case that a constant opening angle corresponds to the peak pressure recovery; shorter diffusers need a larger opening angle for optimum performance.

Kline gave a name to the regions above and below the maximum pressure recovery as he recognised different reasons for the lower performance in the regions with high and low area ratios. Below the line A-A, he named this 'inefficient diffusion', as the loss in pressure recovery was mainly caused by too much friction and a high leaving loss, which would be an increase in losses ζ and a high value of in A_1/A_2 in (7.14). In the region above line A-A, he named this 'insufficient diffusion', as he realised that the loss in pressure recovery was caused by too much blockage, as the boundary layers separated on the walls, and this would be an increase in blockage, with $\varepsilon_2 > \varepsilon_1$ in (7.14).

7.4.5 Reneau Diffuser Performance Charts

Reneau et al. (1967) investigated performance charts for two-dimensional diffusers with a wide range of geometries. Sovran and Klomp (1967) extended these charts to include conical and annular diffusers. Runstadtler et al. (1975) extended this to a range of diffusers at high inlet Mach number. Many additional plots can be found in the publications of the Engineering Sciences Data Unit (ESDU), such as ESDU 74015 (2007). Japikse and Baines (1998) and Miller (1990) also provide a range of relevant Reneau charts. A typical Reneau chart adapted from that of Sovran and Klomp and showing the main area of interest is given in Figure 7.15.

Contours of constant pressure recovery, C_p, are included in Figure 7.15. For the case with an area ratio of 2, the peak pressure recovery is 0.65, which is 85% of the ideal recovery, for this case with an inlet blockage of 3%. There are in fact two optima in this diagram. There is a separate locus of optimum points for a diffuser of limited

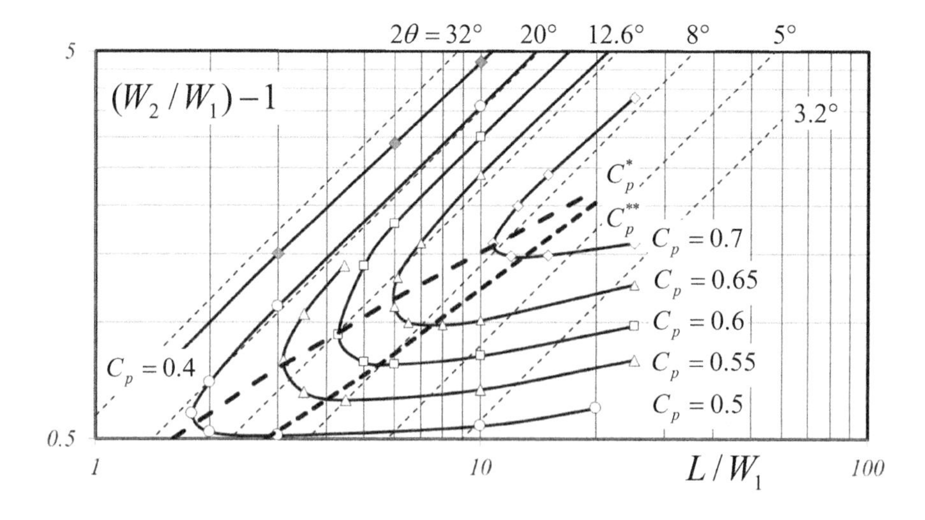

Figure 7.15 Reneau performance chart for two-dimensional diffusers with 3% inlet blockage.

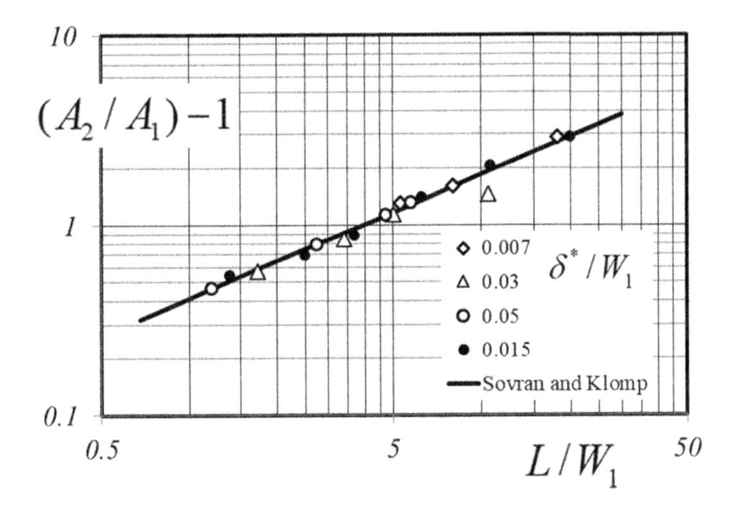

Figure 7.16 Optimum geometry of two-dimensional diffusers with different inlet blockage. (image uses data from the work of Sovran and Klomp, 1967)

length, C_p^*, and for a diffuser of limited area ratio, C_p^{**}. Of interest in this chart is the fact that the performance drops off with large opening angles. The optimum limited length diffuser has an opening angle given approximately by

$$2\theta_{opt} \approx \frac{33}{\sqrt{L/W_1}}. \tag{7.21}$$

If a large area ratio is needed, a practical way to avoid this issue is to subdivide the diffuser with intermediate walls so that each part of the diffuser then has a longer length to inlet width ratio and a smaller opening angle. The data in this diagram are used in Section 12.6 to compare the performance of different zones of a vaned diffuser with the expected performance of a 2D diffuser.

Reneau et al. (1967) included diagrams for different inlet blockage to the diffuser and found that the peak performance falls with increasing inlet blockage, as would be expected from (7.14). One of the major contributions made in the work of Sovran and Klomp (1967) was to identify that although the performance of the diffusers varies strongly with area ratio, length-to-width ratio and blockage, the optimum geometry for best performance was roughly the same for all types of diffuser (including conical diffusers and annular diffusers) and independent of the inlet blockage. This is shown in Figure 7.16. As an example, for the case of an area ratio of 2, a nondimensional length of between 4 and 5 is suggested. Although the prediction of the performance of diffusers is difficult and complex, the selection of the best area ratio for optimum performance of straight two-dimensional diffusers appears to be almost trivial. The equation for the line fitted to the data of Sovran and Klomp (1967) shown in Figure 7.16 is as follows:

$$\frac{A_2}{A_1} = 1 + 0.4109\left(\frac{L}{W_1}\right)^{0.6527}. \tag{7.22}$$

This provides a good guideline for the area ratio of a wide range of diffusers of a given length to achieve optimum performance. These optimum diffusers are close to transitory stall to achieve the maximum pressure recovery. When flow stability or uniform outlet flow distribution is important, or when the inlet flow conditions are poor, it is advisable to operate with an area ratio that is less than that given by this equation (Miller, 1990).

The maximum pressure recovery of optimum diffusers was considered by Stratford and Tubbs (1965) on the basis of boundary layer theory. They develop the following relation for the maximum recovery coefficient of optimum diffusers as

$$C_p^* = 1 - \frac{F}{\left(\frac{L}{\delta^*} + E\right)^{0.434}} = 1 - \frac{F}{\left(\frac{L}{W_1}\frac{W_1}{\delta^*} + E\right)^{0.434}}, \tag{7.23}$$

where F and E are constants depending on the Reynolds number, L is the diffuser length, W_1 is the inlet width and δ^* is the boundary layer displacement thickness at the inlet. Thompson (1979) analysed many diffuser charts given by Runstadler et al. (1975) and suggests for typical Reynolds numbers in turbomachinery that $E = 0$ and $F = 5.4$ so that an approximate estimate of the pressure recovery coefficient for the optimum diffusers with different levels of inlet blockage is given by

$$C_p^* = 1 - 5.4\left(\frac{\delta^*}{W_1}\frac{W_1}{L}\right)^{0.434}. \tag{7.24}$$

The performance of diffusers is not only related to the preceding geometrical parameters and the blockage; other parameters also play a weaker role. Another geometrical parameter that also influences the pressure recovery is the addition of a tailpipe downstream of the diffuser which tends to reduce the separation. The width of the diffuser, W_1, is not the only geometrical parameter of interest; the height of the flow channel, h_1, also has an effect, albeit small. The optimum ratio of W_1/h_1 is close to unity, and only small additional losses are experienced if $0.8 < W_1/h_1 < 2.0$. The Reynolds number determines the state of laminar or turbulent flow and therefore affects the tendency for separation of the flow. In typical compressor diffusers with a high Reynolds number, the flow is fully turbulent and therefore the impact of Reynolds number is often negligible.

As mentioned in Section 7.4.3, the static pressure rise is also affected by the Mach number, as an increase in this leads to a larger pressure rise coefficient. Although the ideal pressure recovery increases through this effect, the pressure gradients in the flow also increase and make it more prone to separation, which actually decreases the pressure recovery and increases the losses. In general, a higher turbulence results in a better momentum exchange of the high-energy core flow with the low-energy flow in the boundary layer and therefore might prevent separation. This mixing mechanism also contributes to the losses, but these might be much less than the benefit of avoiding any separation. It is wise to point out that not only the turbulence intensity, but also the turbulence length scale, is important.

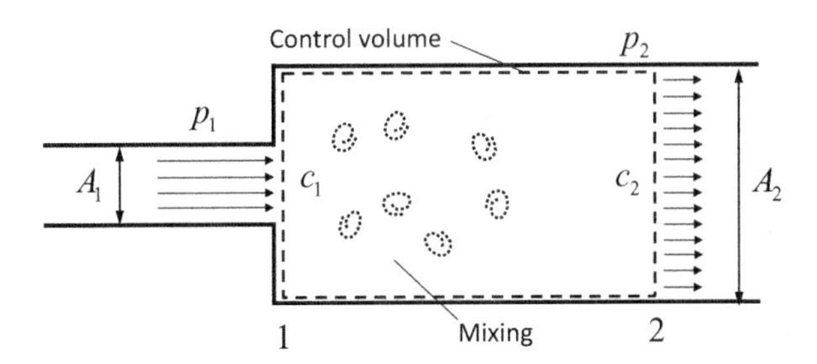

Figure 7.17 A step diffuser with complete mixing and no losses.

7.4.6 Step Diffusers

The ideal pressure recovery of a step diffuser, as shown in Figure 7.17, can be estimated from the momentum and continuity equations, as given by Greitzer et al. (2004). If we consider that there are no losses and that the flow is fully mixed out at the outlet plane, then the ideal pressure recovery is given by

$$A_2(p_2 - p_1) = \rho A_1 c_1^2 - \rho A_2 c_2^2, \quad c_1 A_1 = c_2 A_2,$$
$$C_p^{id} = \frac{p_2 - p_1}{\frac{1}{2}\rho c_1^2} = 2\frac{A_1}{A_2}\left(1 - \frac{A_1}{A_2}\right). \tag{7.25}$$

With an area ratio of 2, this gives an ideal pressure recovery of 0.5, so that even a step diffuser can recover 50% of the energy in the flow. This will of course be slightly less with dissipation losses. In cases where a diffuser has a fixed outlet passage, such that the diffuser area ratio exceeds that for maximum pressure recovery, it is better to reduce the diffuser area ratio and have a sudden expansion at the outlet to the required passage area. Miller (1990) gives examples showing that this strategy can significantly reduce the loss coefficient.

7.4.7 Diffusers in Series

It is important to realise that the pressure recovery of a diffuser depends on the available inlet kinetic energy. For example, if we consider two diffusers in series, both with an area ratio of 2, as shown schematically in Figure 7.18, we can determine that the effective pressure recovery of the combination is as follows:

$$C_{p13} = \frac{p_3 - p_1}{\frac{1}{2}\rho c_1^2} = \frac{p_2 - p_1}{\frac{1}{2}\rho c_1^2} + \frac{c_2^2}{c_1^2}\frac{p_3 - p_2}{\frac{1}{2}\rho c_2^2}$$
$$C_{p13} = C_{p12} + \frac{A_1^2}{A_2^2} C_{p23} \approx C_{p12} + \frac{1}{4} C_{p23}. \tag{7.26}$$

The overall performance is much more strongly affected by the first diffuser so that most design effort should always be put into the design of the first diffusing component with the highest inlet velocity. But this also shows that a second diffuser can often

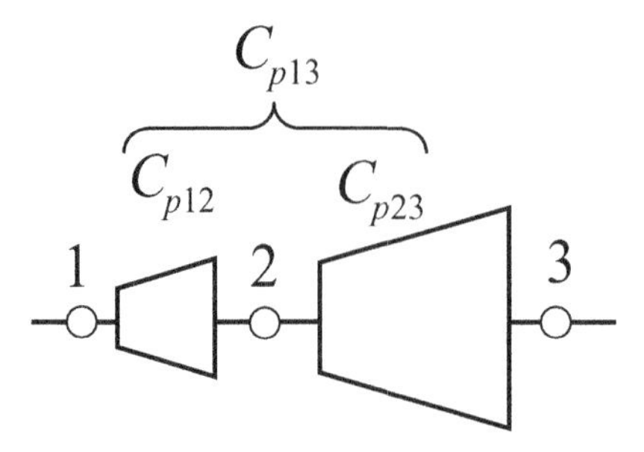

Figure 7.18 Two diffusers in series.

be a useful addition to increase the performance, and this strategy is sometimes used in radial turbocompressors with a decelerating conical diffuser downstream of the volute, as shown in Figures 1.8 and 1.15. It may also be satisfactory to use a step diffuser downstream of a diffuser flow channel to increase the pressure recovery.

7.5 Blade Loading Parameters

7.5.1 Limits to Operation of Blade Rows

The fundamental fluid dynamic aspects of loss generation, pressure rise capability, and choking in a turbocompressor provide limits to the operational effectiveness of the blade rows. The designer tries to do the best to reduce any unavoidable losses at the design point by choosing an optimum blade-to-blade loading and avoiding too much diffusion in the flow channels. At off-design, the designer tries to maximise the operating range of the blade row between choke and stall, but this is always limited by the change in performance with incidence. Skin friction and turbulent dissipation will always be present and increase with the wetted surface area if the number of blades is increased to decrease the blade-to-blade loading. At the same time, more blades will reduce the throat area between the blades and limit the maximum flow through the blade row. Fewer blades may decrease the frictional dissipation but may at the same time increase the blade-to-blade loading on the blade row above a critical level and lead to premature stall. Any attempt to reduce the friction losses by increasing the extent of laminar flow on the blade surface before transition may also make the flow at more risk of separation in adverse pressure gradients.

The designer tries to avoid extreme values of blade-to-blade loading limits at the design point, but due to the change in the velocity triangles at off-design conditions the limits may be reached in these operating points. Different sorts of loading limits may be relevant in different situations. The blade-to-blade loading of the design may be high due to having few blades. The machine may be operating at high flow rates close to choke with transonic flow such that a wide range to choke is not possible. The deceleration of the flow within the blade rows may be high such that the machine is forced to operate close to its peak pressure rise capability with nearly separated flows. It is often necessary for a compressor to operate over a range of flows such that the increase in incidence at lower flows then causes stall and limits the performance.

Many aspects of the geometry play a role in the aerodynamic limits. There is not a lot that can be done to reduce the tip clearance losses of an impeller, other than making the clearance gaps as small as practically possible, or by applying a shroud with its own clearance gap from the outlet to the inlet with leakage flow and parasitic disc friction on the shroud outer wall. The secondary flows are to an extent controllable by the blade and channel shape, and the designer can make some efforts to improve the design by the use of suitable blade shapes with lean to counteract these, as discussed in Section 11.8.3. The risk of separation is also controllable by the amount of deceleration that takes place in the flow and by minimising the preshock Mach number of any suction surface normal shock waves.

In these, and any many other aspects, the designer is forced to seek the best compromise taking into account the requirements of the application, strength of materials, manufacturing constraints, cost of components, assembly, operability, noise as well as the aerodynamic design. With the quest for improved levels of performance, it is tempting for designers to rely increasingly on CFD computations to estimate the risk of flow separation and increased losses from higher blade loadings. However, these methods are not in themselves entirely adequate to ensure that higher blade loading levels are acceptable due to limitations in the turbulence modelling for separated flows. A good use of the CFD methods is to compare the blade loading levels of a new design to those of similar designs for which test data are available. In this process, the many well-established classical blade loading parameters for compressors should be checked (for example, ideal pressure recovery coefficient, de Haller number, Lieblein diffusion factor, Zweifel number) and compared to values that have proved acceptable in earlier tests of similar machinery with similar clearance levels, so that the experience of nominally similar designs can be used.

It is wise to tread carefully as it is asking for trouble to ignore the enormous wealth of experience encapsulated in the simple blade loading criteria developed by the early giants of the business. To help in this process, it is useful to apply simple one-dimensional loading coefficients that can be adjusted to take account of experience with other earlier designs as an aid to the trade-offs that are inevitable. The remainder of this chapter describes some of these blade loading parameters, most of which have been developed for axial machines, and discusses their applicability to radial compressors.

7.5.2 Incidence

The most common aerodynamic limit of relevance to the off-design performance of radial turbocompressor impeller blades and diffuser vanes is the incidence of the flow on to the blade rows. The incidence needs be kept low at the design point for good efficiency and choke margin and will naturally increase as the flow is reduced. The velocity triangle at any point in a compressor can be expressed as

$$\tan\alpha = \frac{u}{c_m} + \tan\beta = \frac{1}{\phi} + \tan\beta. \tag{7.27}$$

If the flow coefficient varies by a small amount, $\Delta\phi$, the change in the absolute and relative flow angle can be expressed by differentiation as

$$\frac{1}{\cos^2\alpha}\Delta\alpha = -\frac{1}{\phi^2}\Delta\phi + \frac{1}{\cos^2\beta}\Delta\beta. \tag{7.28}$$

Considering the conditions at a stator inlet at the design point (denoted with superscript d), where the relative flow angle, β_2, can be considered to be approximately constant with a change in flow, giving $\Delta\beta = 0$, this leads to

$$\frac{\Delta\alpha_2}{\Delta\phi_2/\phi_2^d} = -\frac{\cos^2\alpha_2^d}{\phi_2^d}. \tag{7.29}$$

This is the change in incidence angle of the stator for a fractional change in the flow, or the incidence sensitivity. Under the assumption that the absolute flow angle is constant, a similar equation can be derived for the change in the rotor incidence with a fractional change in the flow, to be

$$\frac{\Delta\beta_1}{\Delta\phi_1/\phi_1^d} = -\frac{\cos^2\beta_1^d}{\phi_1^d}. \tag{7.30}$$

Using the typical values suggested in Chapter 11 for the impeller inlet velocity triangle with $\beta_1 = -60°$ and $\phi_1 = 0.577$, this leads to a sensitivity of a 2.5° increase in incidence angle of the impeller casing section for a 10% reduction in the flow. A similar value arises for the diffuser inlet with $\alpha_2 = 70°$ and $\phi_2 = 0.25$, leading to a sensitivity of a 2.7° increase in incidence of the diffuser for a 10% reduction in the flow. From this, it appears that both the impeller and the diffuser inlet have approximately the same incidence sensitivity to a change in the flow. Taking a typical multistage axial compressor stage, such as the mean line of the rotor reported by Smith (1970) with $\beta_1 = -50°$ and $\phi_1 = 0.6$, the incidence sensitivity to a 10% change in flow is 3.9°, that is, 60% higher than that of a typical radial stage. On this basis, an incidence of 10° is reached with a 40% reduction in flow in a typical radial stage but with only a 25% reduction in flow in an axial stage. This is one of the reasons why axial compressor stages have a narrower operating range than radial stages. The other main reason is related to the slope of their work input characteristics, as discussed in Section 11.6.3.

Many correlations can be found for the limiting incidence of inducers, such as that from Baines (2005), but none of them appear to be very precise or universally valid. Broadly, these state the incidence at surge decreases as the Mach number increases, but the large scatter shows that many other parameters play a role. Note that the mean incidence at surge at low Mach numbers is around 10°–15° as for an isolated aerofoil. As the incidence increases, then the pressure rise in the inlet region of the blade up to the throat also increases and some other researchers, such as Kenny (1972), Herbert (1980) and Elder and Gill (1985), prefer to use the increase in the inlet pressure rise in the semivaneless space, which is related to the incidence, as a criterion for the maximum off-design loading.

7.5.3 Stage Work Coefficient: $\lambda = \Delta h_t/u_2^2$

In some literature, the stage work coefficient is known as the stage loading coefficient as it is used as a measure of the stage loading. Increasing the compressor stage work coefficient at the design point reduces the number of stages needed for a given pressure rise or reduces the blade tip speed that is needed, with associated reduction in weight and cost. In the past, a great deal of effort has been expended in axial compressor research in trying to design blade rows with higher work coefficient, and this approach is known to be severely limited in axial compressors due to the flow separation associated with high deceleration in the blade rows. Based on the midspan of an axial compressor stage, a work coefficient of 0.2–0.4 is considered to be a conventional design, and a design with 0.6 is considered an upper limit with an extreme loss of efficiency (Dickens and Day, 2011; Hall et al., 2012).

Radial compressors are fortunate that most of the pressure rise is brought about by the centrifugal effect so that a stage work coefficient of $\lambda = 0.75$ or even higher is possible without exceeding the deceleration limits in the flow. Note that in a radial compressor this nondimensional parameter is defined relative to the impeller tip speed, so the enthalpy increase is in fact typically three times that of an axial compressor, where the work coefficient is based on a midspan speed, which is the same at inlet and outlet of the stage. In radial compressors, the upper limit on the stage work coefficient is mainly due to the fact that a higher loaded stage has a flatter work characteristic (Section 11.6.3) and less range to surge (Section 17.4). The designer can make the most important decision for a radial compressor without postulating details of the geometry, but by just selecting an outlet velocity triangle with a typical stage work coefficient at the design point, in the range of 0.6–0.65. Extreme values close to 0.8 can also be achieved at the cost of operating range and efficiency.

7.5.4 De Haller Criterion, $DH = w_2/w_1$

De Haller (1953) was the first to put forward a criterion to consider the development of the end-wall boundary layers on the hub and casing walls of 2D blade cascades. In tests on axial compressor blade profiles in a simple 2D cascade tunnel, de Haller determined that it was not possible to decelerate the flow by more than 72–75% below the inlet velocity. Attempts to decrease the velocity ratio below $DH = 0.72$ by increasing the outlet area of a cascade with more turning caused separated flow in the end-wall region of the cascade (end-wall stall) and large end-wall blockage. The flow in the channel effectively became blocked by this separation so that no further deceleration below this value was possible. This is similar to the limits discussed in Section 7.4 for 2D diffusers.

In a 2D axial cascades, each blade section experiences a similar deceleration level. In an axial compressor, the rotor hub section needs to achieve more or less the same pressure rise as the tip section, which has a higher inlet relative velocity. As a consequence, end-wall stall in axial compressor rotors is generally a phenomenon related to the hub section, which has a higher deceleration to achieve this pressure rise. In a radial compressor rotor, the tables are turned: the hub section has a higher radius change than the tip section, and so the tip experiences a higher deceleration than the hub as the flow slows to the mean velocity level at impeller outlet. In mixed flow impellers, both sections are similarly loaded.

From subsequent work by Koch (1981), described in Section 7.5.7, it is now known that the limiting de Haller number of $DH = 0.72$ applies to axial blade rows with a space/chord ratio close to unity (similar to those tested by de Haller), and that higher levels of deceleration can be achieved with more closely spaced blades as the length to inlet width ratio of the flow channel becomes larger, as in a 2D diffuser. A limiting value of the deceleration ratio on the casing streamline in radial compressors with inducers at the design point is $DH = w_2/w_1 = 0.6$, whereby the inlet velocity is taken as that on the casing and the outlet velocity is the mean level at impeller outlet. It is shown in Section 5.8 that more deceleration than this tends to cause the formation of the jet-wake flow pattern in impellers due to separation on the casing streamline. Tognola and Paranjpe (1960) suggested that higher values of the de Haller number (with less diffusion) might be more appropriate for backswept impellers without an inducer, as used in multistage process compressors. This is consistent with the work of Koch, as such impellers have a lower solidity.

7.5.5 Mach Number Ratio in the Isentropic Core Flow: $MR_2 = M_{rel1t}/M_{rel2j}$

On the basis of the jet-wake model of the impeller flow, Dean suggested a design parameter for the limiting diffusion based on the deceleration from the relative velocity at the inlet to the relative velocity in the isentropic core flow at the outlet, that is, in the jet as shown in Figure 5.25. This model is developed in detail by Japikse (1985), where the idea has been adapted to be a limiting Mach number ratio. The difficulty of using this parameter is that it requires an impeller flow model based on the notion of a two-zone jet and wake, and this may not be truly applicable to modern impellers, as discussed in Section 5.8.

7.5.6 Ideal Static Pressure Rise Coefficient: $C_p^{id} = (p_2 - p_1)/(\frac{1}{2}\rho c_1^2)$

Section 7.4 on diffuser flows identifies in (7.13) the ideal pressure recovery for incompressible flow as a function of the area ratio of a diffuser flow channel. This is simply a measure of the static pressure rise through a diffuser made nondimensional by the inlet relative dynamic head. If the limiting velocity ratio given by de Haller for an axial cascade is taken as $c_2/c_1 = 0.75$, then the equivalent ideal pressure recovery is close to $C_p^{id} = 0.44$, and this is close to the typical design value for an axial compressor. This implies that calculating with no losses and no blockage, the maximum ideal pressure recovery coefficient associated with the de Haller limit is 0.44. The value is known as the ideal value as it is that which would occur without separation and losses, based entirely on design flow angles and velocity levels.

A similar criterion can be used for the design of separate parts of the blade row. For example, Yoshinaka (1977) suggests using limits for the maximum pressure rise in the inlet region of an impeller blade row from the leading edge to the throat based on limit values of between 0.25 and 0.4 given for pipe and cambered diffusers by Kenny (1972).

7.5.7 Maximum Static Pressure Recovery Coefficient of Koch (1981)

Koch (1981) extended the idea of de Haller that stall takes place in the end-wall by correlating the maximum pressure rise coefficient of an axial compressor stage with a geometrical parameter derived from classical diffuser studies, as outlined in Section 7.4. He pursued the notion that the static pressure recovery of an axial compressor stage behaves essentially like that of a planar 2D diffuser. Based on extensive test data in a low-speed multistage test compressor at General Electric, Koch postulated a maximum value for the static pressure recovery of a whole stage based on the equivalent channel diffuser geometry. This representative stage was the third stage in a machine of four stages, so the measurements that are the basis for the Koch static pressure recovery coefficient include the effects of the momentum deficit in the inlet end-wall boundary layer coming into the blade.

The geometry parameter used by Koch is the arc length of the cambered aerofoil divided by the cascade trailing edge staggered spacing, L/g_2, which is a length to outlet width ratio as shown in Figure 7.19. This is similar to the conventional type of correlation used for diffuser flows, except that the outlet width is used rather than the inlet width. The justification for using the outlet width instead of the more usual inlet width is that in a compressor cascade it is the effective exit flow area that remains roughly constant over the range of operation while the effective flow area at inlet changes with the inlet flow angle and incidence. The correlation for the limiting stage pressure rise of an axial stage takes the following form:

$$C_p^{Koch} = \frac{\Delta p_{13}}{\frac{1}{2}\rho\left(w_1^2 + c_2^2\right)} = f(L/g_2). \tag{7.31}$$

The stage static pressure rise is made nondimensional with the average kinetic energy at the rotor and stator inlet. The stage average value of L/g_2 ratio is calculated by using the blade-row inlet dynamic head of rotor and stator as the weighting factor. Corrections are applied for the effect of Reynolds number and tip clearance. In addition, a form parameter of the velocity triangles called the effective dynamic pressure factor related to the inlet flow coefficient of the stage and the stage reaction was found to be important. This form parameter demonstrated that there is a strong link between the local flow coefficient and stagger leading to a rise in the maximum pressure rise in blade rows with high stagger and low flow coefficient. The underlying mechanism responsible for this effect is the reenergisation of the end-wall boundary layer due to the frame of reference change between blade rows, an effect known as the recovery ratio, which becomes larger as the flow coefficient is reduced or the stagger is increased (Smith, 1958).

The basic correlation is shown in Figure 7.19, and, as with the de Haller number, Koch emphasises that it is a criterion for the separation of the end-wall boundary layers. Although there is considerable scatter between the actual pressure rise achieved in multistage compressors and that predicted by Koch's method, no compressors achieve a higher pressure rise coefficient than the correlation of Sovran and Klomp (1967) for diffusers with 9% blockage. It is of interest that the pressure coefficient of

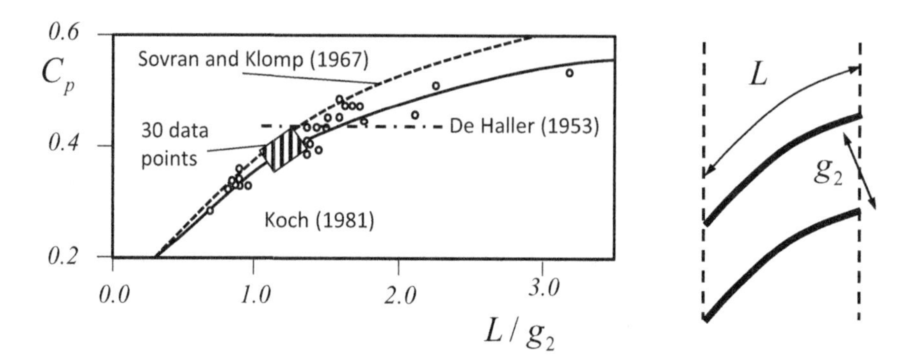

Figure 7.19 Koch correlation for maximum pressure rise of axial compressor stages. (image uses data from Casey (1994a) and Koch (1981))

stages at L/g_2 close to 1, similar to the cascades of de Haller, achieve a static pressure rise coefficient of around 0.44, which is in line with the de Haller criterion. Low aspect ratio axial compressors with a higher solidity (low space to chord ratio) can achieve a higher pressure rise coefficient, as the flow channels are better diffusers with higher nondimensional length-to-width ratio.

There is, of course, no one-to-one correspondence between axial compressor technology and radial compressor technology with a large centrifugal effect and three-dimensionally curved channels, but it is encouraging that the Koch correlation suggests a sound reason that radial compressor impellers may be designed with a lower de Haller number of 0.6 than axial ones with a typical limit of 0.7; they are simply more effective diffusers with a longer blade length compared to the width of their blade channels. On this basis, stages with an L/g_2 of 3 or larger, typical of radial compressors, can achieve a pressure rise coefficient of 0.6, corresponding to a de Haller number of 0.63, which is close to the design value for typical radial impellers.

Auchoybar and Miller (2018) discovered a further effect that reenergises the end wall flows in axial stages and leads to an increase in the pressure rise capability by reducing the rotor blade exit flow angle, β_2, in the end-wall region.

7.5.8 Equivalent Diffuser Opening Angle

A similar procedure to that of Koch (1981) for determining the maximum pressure rise capability of an axial compressor cascade based on an analysis of axial compressor blade rows as a diffuser was developed by Schweizer and Garberoglio (1984). They calculated an equivalent diffuser opening angle of the cascade based on the geometrical parameters of the blade row expressed in terms of an equivalent conical diffuser with an opening angle of $2\theta_{eq}$. A total of 121 cascade configurations were examined in a cascade tunnel, and these data were used to identify the maximum pressure rise capability of axial compressor blade rows at design conditions as

$$C_p^{id} = 1 - \left(\frac{A_1}{A_2}\right)^2, \quad \theta_{eq} = \tan^{-1}\left(\frac{(A_2/\pi)^{1/2} - (A_1/\pi)^{1/2}}{L}\right), \quad \Delta\beta = \beta_2 - \beta_1$$

$$C_p^{\max} = 0.087858 - 0.23355C_p^{id} + \frac{2.7847C_p^{id}}{\theta_{eq}^{1/2}} - \frac{2.0348C_p^{id}}{\theta_{eq}^{3/2}} - 0.019203C_p^{id}(\Delta\beta)^{1/2}, \tag{7.32}$$

where $\Delta\beta$ is the turning angle of the cascade in degrees. The data were corrected to a Reynolds number of 10^6 and a standard roughness. Corrections for the effect of the inlet boundary layer blockage, tip clearance and Mach number are given by Schweizer and Garberoglio. The correlation leads to a maximum pressure rise of the cascade with a half opening angle of $\theta_{eq} = 7°$. Many radial compressor designers also consider the equivalent opening angle of the diffuser channels as a criterion for the design of vaned diffusers, as is discussed in Chapter 12. A value of 2θ between 10–12° for the covered passage is a common design philosophy for the channel part of a vaned diffuser.

7.5.9 Zweifel Tangential Force Coefficient

The Zweifel loading coefficient was first developed by Zweifel (1945) as a criterion for the determination of the blade number in axial turbine cascades with very high turning. He realised that for turbine cascades, there is an optimum space–chord ratio that gives a minimum overall loss. More blades lead to higher frictional losses, and fewer blades may cause higher losses due to the higher blade-to-blade loading and the large deceleration of the flow on the suction surface downstream of the covered passage close to the trailing edge. Zweifel put forward a reference value for the tangential blade loading based on constant pressure on the suction surface at the inlet value, and constant pressure on the pressure surface at the outlet value. He suggested that the optimum number of blades would be obtained when the actual tangential force was 80% of this ideal value. More recent work with blade design using modern CFD methods shows that blades can achieve a higher value than this. A criterion for hub stall in axial compressors has been derived from this by Lei et al. (2008) based on the difference between the Zweifel number at midspan and that on a 2D section close to the hub streamline. This criterion has not been used in radial compressors, so no further information is given here

7.5.10 The Lieblein Diffusion Factor

This concept was introduced by Lieblein in the 1950s based on extensive tests on axial compressor cascades by NACA, the forerunner of NASA. As opposed to the de Haller number, which is related to the diffusion on the end-walls, this concerns itself with the diffusion along a blade surface, along a stream section. There have been several variants of this in various publications. Lieblein (1965) first defined the local diffusion factor as the ratio of the amount of deceleration on the suction surface from the maximum velocity to the outlet velocity relative to the blade cascade inlet velocity as

$$DF = \frac{w_{max} - w_2}{w_1}. \tag{7.33}$$

Lieblein suggested that losses due to separation increased dramatically above a limit value of 0.5 for this parameter.

In the early days of axial compressor aerodynamics, no CFD simulations were available, so it was not easy to predict the maximum suction surface velocity. Lieblein estimated the maximum surface velocity of typical cascades with circular are camber lines and 10% thickness chord ratio to be as follows:

$$w_{max} \approx w_1 + \frac{\Delta w_u}{2\sigma}. \tag{7.34}$$

He then suggested an alternative definition for the diffusion factor of a rotor blade that evolved from the local diffusion factor as follows:

$$DF_r = 1 - \frac{w_2}{w_1} + \frac{\Delta w_u}{2 w_1 \sigma}. \tag{7.35}$$

He suggested values of 0.45 at the design condition and considered that 0.65 would indicate the maximum value of this parameter before stall. Note that if the designer specifies the design value of the de Haller number together with the Lieblein diffusion factor, then this provides an estimate for the space–chord ratio and number of blades required. It is in this way that this parameter is used in the preliminary design of axial compressors.

An interesting experiment was reported by Greitzer et al. (1979) related to this. They tested an axial compressor, first with the number of blades as designed, and secondly again after removing blades such that the Lieblein diffusion factor criterion was exceeded. They observed that with the correct number of blades the stall occurred as end-wall stall, but if the number of blades was reduced stall occurred earlier by blade stall.

7.5.11 Modified Diffusion Factor for Radial Impellers

In the application of the Lieblein diffusion factor to radial impellers, there are two issues that cause difficulty. First, the hub-to-shroud curvature in a radial impeller causes an acceleration of the flow on the casing streamline at the inlet, leading to an additional component of diffusion in the impeller, so a correction for this effect is needed when drawing on the axial compressor experience of Lieblein. Secondly, the Lieblein diffusion factor includes the space/chord ratio, which is constant along an axial blade section, but not constant in a radial impeller as owing to the increase in radius the spacing increases towards the trailing edge of the impeller. The diffusion factor suggested by Rodgers (1978) was as follows:

$$DF_r = 1 - \frac{w_2}{w_{1rms}} + \frac{\pi r_2}{ZLw_{1rms}}\left[c_{u2} - \frac{r_{1rms}}{r_2}c_{u1}\right] + \left[\frac{0.02}{r_2/r_{1c} - 1}\right], \tag{7.36}$$

where w_2 is the outlet relative velocity, w_{1rms} is the inlet relative velocity at the rms radius, L is the length of the blade, Z is the blade number, r_{1c} is inlet casing radius, r_{1rms} is inlet rms radius and r_2 is the outlet radius. In his paper, Rodgers discusses the limitations of this equation and recognises that the relative velocity ratio from inlet to outlet is the most important parameter for radial impellers and suggests a de Haller number of 0.6 is appropriate for centrifugal stages. Recently, Dick et al. (2011) adapted the Rodgers diffusion factor for use with radial fans and found that it was useful for estimating the fall in efficiency with an increase in the diffusion factor. For a high efficiency in radial fans, the Rodgers diffusion factor should be close to 0.4, and 10 points in efficiency may be lost if the diffusion factor rises to 0.6.

Coppage et al. (1956) also defined an additional radial impeller diffusion factor as follows:

$$DF_r = 1 - \frac{w_2}{w_{1c}} + \frac{0.75\lambda \dfrac{w_2}{w_{1c}}}{\left[\left(\dfrac{Z}{\pi}\right)\left(1 - \dfrac{r_{1c}}{r_2}\right) + 2\dfrac{r_{1c}}{r_2}\right]}, \tag{7.37}$$

where w_2 is the outlet relative velocity, w_{1c} is the inlet relative velocity at the casing, λ is the work coefficient, u_2 is the impeller tip speed, Z is the blade number, r_{1c} is inlet casing radius and r_2 is the outlet radius. No details are provided by Coppage, but this was used to suggest the magnitude of an additional source of loss for impellers with high values of the diffusion factor.

7.5.12 Aungier Diffusion Factor and Blade Loading Parameter

Aungier (2000) also derives two additional loading parameters for the casing streamline based on a diffusion factor approach for radial impellers of

$$\begin{aligned} DF_r &= \frac{w_{\max}}{w_2}, \quad w_{\max} = \frac{w_{1c} + w_2 + \Delta w}{2w_2}, \quad \frac{\Delta w}{2u_2} = \frac{\pi D_2}{Z}\frac{\lambda}{c}, \\ BL &= \frac{2\Delta w}{w_{1c} + w_2}, \end{aligned} \tag{7.38}$$

where the c is the camber line length of the blade, which is typically close to $0.5D_2$. The limit value of these loading parameters suggested by Aungier (1995a) are $BL < 0.9$ and $DF_r < 1.7$. These are used by Podeur et al. (2019) to determine the blade numbers in industrial-style compressor impellers.

7.5.13 Limits to Blade-to-Blade and Hub-to-Casing Loading

As mentioned in Section 5.8.1, a form of reverse flow, known as an eddy after Moore (1973), can occur if the mean channel velocity is less than half of the blade-to-blade velocity difference determined by the blade curvature and the Coriolis effects. The reverse flow that occurs is on the pressure surface and is not related to boundary layer separation. This immediately suggests a criterion for the blade-to-blade loading to ensure that the pressure surface velocity level is not negative in the design, which would remove the eddy:

$$C_{btob} = \frac{w_s - w_p}{w_m} < 2. \tag{7.39}$$

A similar effect can occur in the hub-to-shroud loading if the curvature of the inlet of the impeller is too high within an impeller with a high blade span, so that following (5.21) the hub velocity becomes negative. A limit on the hub to casing loading parameter allows this to be avoided:

$$C_{htos} = \frac{c_{m,casing} - c_{m,hub}}{c_{m,mean}} < 2. \tag{7.40}$$

A more stringent blade-to-blade loading criterion was suggested by Morris and Kenny (1968) in order to reduce the strength of the secondary flows on the end-walls. Based on experiments on secondary flows in pipe bends, they suggested an upper limit to the blade loading of

$$C_{btob} = \frac{w_s - w_p}{w_m} < 0.7. \tag{7.41}$$

Typically, in radial impellers this can be achieved in the region with splitters, as the increase in the number of blades reduces the blade-to-blade loading in this region. Clearly, though, in stages with few blades this will not be attained. The reduction of the number of blades, however, reduces the wetted surface area and the number of trailing edges with wake losses and may lead to higher efficiencies. A more stringent blade-to-blade loading criterion is often used in the inducer section to avoid high loading in the inlet which has to cope with the incidence effects.

A similar criterion was also applied by Came and Robinson (1998) to suggest a similar limit to the hub to casing loading to reduce the secondary flows caused by the transverse pressure gradient due to meridional curvature, as follows:

$$C_{htos} = \frac{c_{m,casing} - c_{m,hub}}{c_{m,mean}} < 0.7. \tag{7.42}$$

8 Similarity

8.1 Overview

8.1.1 Introduction

With the help of similarity concepts and the associated nondimensional parameters, the preliminary design of a new machine can be based on features of an existing machine, even one which may have been designed for a different fluid, other flow conditions or a different rotational speed. Its performance can also be estimated from that of a similar machine, even though it may be larger or smaller (see Figure 8.1, which is an example of a series of scaled machines taken from Jenny, 1993). The principle of similarity and the associated nondimensional parameters provide an invaluable aid to the design and testing of all turbomachinery, and to the proper understanding of their performance characteristics. This chapter considers the relevance and importance of similarity and nondimensional performance parameters in turbocompressors. The key nondimensional parameters of relevance to radial compressors, such as flow coefficient, work coefficient, pressure coefficient and the blade speed Mach number, are explained. A good grasp of these is an excellent basis for rationalising compressor performance over a range of applications. For completeness, it aims to be quite comprehensive in the collection of the wide range of nondimensional parameters and definitions in regular use in the field, and Chapter 9 considers two other parameters known as the specific speed and specific diameter.

8.1.2 Learning Objectives

- Be aware of the concepts of similarity in turbocompressors, such as geometric similarity, fluid dynamic similarity and thermodynamic similarity.
- Know the key nondimensional parameters of relevance to radial compressors, such as flow coefficient, work coefficient and pressure coefficient.
- Understand the importance of the blade speed Mach number on thermodynamic similarity.
- Be aware of the difference between performance maps and characteristic curves.

Figure 8.1 A family of nearly similar turbochargers scaled to different sizes. (image courtesy of ABB Turbocharging)

8.2 Similarity of Fluid Flows

8.2.1 Nondimensional Parameters

There are three conditions required for two machines to be considered to be operating at similar nondimensional conditions. The first is geometrical similarity, in which the geometrical parameters of the two machines differ only through scaling in size with a certain scaling factor. The second is fluid dynamic similarity, in which the kinematic motion of the fluid and the dynamic forces acting on the blades have to be similar in nondimensional terms. The third is thermodynamic similarity, in which the change in gas conditions (ratios of temperatures, pressure and densities) have to be similar in the two machines.

The first two of these conditions are well understood as they both apply to all turbomachines, including hydraulic pumps and low-speed fans, and are thoroughly explained in the relevant textbooks. The condition of thermodynamic similarity is less clearly understood and is often included as one of the necessary conditions for fluid dynamic similarity, which it is in high-speed compressors. Here, however, it is treated as a separate condition for similarity. It applies only to thermal turbomachinery where the changes in pressure and temperature have an effect on the fluid properties, particularly the density, and lead to a change in volume flow through the machine. Thermodynamic similarity becomes progressively more important in the interpretation of the performance of radial flow turbocompressors as the pressure ratio increases.

Nondimensional parameters are often used in the presentation of turbomachinery characteristics because they allow the performance to be defined in a way that is nearly independent of the fluid, the absolute dimensions of the machine and the inlet conditions. The dimensionless parameters provide a convenient way to convert the measured overall performance (efficiency, pressure rise and flow rate) of a certain machine at a particular speed to other conditions, such as a different speed, different fluid, different inlet conditions or a geometrically similar machine of a different size. The dimensionless parameters also allow the local aerodynamic features of different machines, or of a single machine at different operating points, to be compared.

The simple scaling laws for incompressible low-speed machines such as pumps and fans are derived from dimensional analysis and the dimensionless parameters in Section 8.4 and lead to the so-called fan laws or pump affinity laws:

$$\frac{\dot{V}_1}{\dot{V}_2} = \frac{n_1}{n_2}, \quad \frac{\Delta p_1}{\Delta p_2} = \frac{n_1^2}{n_2^2}. \tag{8.1}$$

If a pump is operated at similar fluid dynamic operating conditions, such as at peak efficiency, and cavitation and viscosity do not affect the performance, then the volume flow changes linearly with the rotational speed, and the pressure rise or enthalpy rise is proportional to the square of the speed. The characteristic performance map of a simple pump at different speeds then comprises a series of curves of similar shape. Apart from the small effect of the Reynolds number with a change in speed, and any changes in the physics of the flow through the onset of cavitation, the different speed lines of the pump characteristic can be scaled to form a unique single nondimensional head coefficient versus flow coefficient characteristic curve (Gülich, 2008), as sketched in Figure 8.2 and as defined in Section 8.3.2.

In this way, it is possible to generalise test results for a pump or a fan measured at different speeds, or even at a single speed, in an economical way to generate a dimensionless performance curve, which in turn allows other, nonmeasured, speed lines to be produced. The best efficiency point on the characteristic then occurs for a single value of the head coefficient and the flow coefficient. These values are determined by the geometry of the machine, whereby the peak efficiency is generally close to the flow condition with smooth zero-incidence flow on the blading. The rise in pressure and the flow at the best point at different speeds are then related by (8.1). With variable speed, these form a similar set of fan curves in which the locus of dynamically similar points is such that the flow rate is proportional to rotational speed and the pressure rise proportional to the square of speed. As the pressure losses in the associated system usually also vary as the square of the flow rate, these characteristics

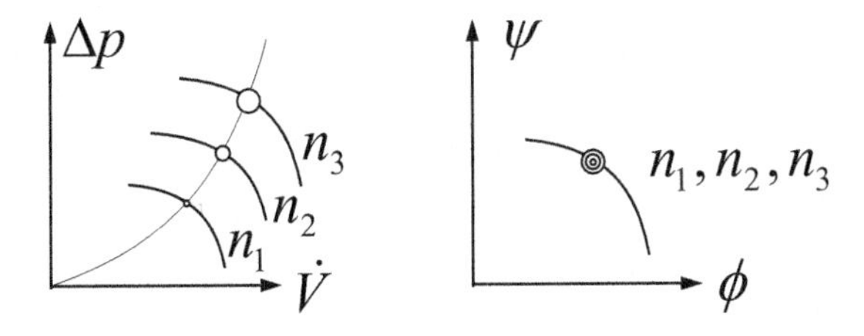

Figure 8.2 Sketch of the performance map (left) and characteristic curve (right) of a radial pump at different speeds.

allow the fan to be selected to match with the pressure loss law of the system such that, over a wide range of speeds, the fan will operate close to its peak efficiency point. The actual operating point of the fan is determined by the intersection of the fan curves with the system resistance curve, and if this resistance curve also has a parabolic form, then the operating conditions at each speed always lie on the same nondimensional operating point. Chapter 17 provides more details.

The scaling laws for compressible flow in thermal turbomachinery are more complicated than those for incompressible flow in hydraulic machines. Effects due to gravity, surface tension cavitation and heat conductivity can be ignored in compressors, so that the dimensionless parameters which are most useful depend on the application, but for an adiabatic flow the most useful nondimensional parameters used are based on four classical similarity parameters, following Cordes (1963), as outlined in the following section.

8.2.2 Newton or Euler Number

The effect of kinematic similarity is expressed by the dimensionless ratio of a pressure force acting on the system to the inertial force determined by a dynamic pressure acting over a certain area. In turbomachinery applications, the dynamic pressure can be expressed as a pressure rise or drop, and, via the Gibbs equation, as the product of the enthalpy change and the density. The inertial force can be related to the kinetic energy available in the rotational motion of the machine or to the kinetic energy in the fluid flow, which is usually expressed as the dynamic pressure defined in terms of the speed of rotation or the gas velocities:

$$C_p = \frac{\text{pressure force}}{\text{inertia force}} = \frac{F}{\frac{1}{2}\rho c^2 A} \sim \frac{\Delta p}{\frac{1}{2}\rho c^2} \sim \frac{\Delta h}{u^2}. \tag{8.2}$$

A wide range of dimensionless work coefficients, pressure rise coefficients, total pressure loss coefficients and head rise coefficients are used to characterise the flow on this basis.

8.2.3 Strouhal Number

A further condition for kinematic similarity is similarity of motion. This implies a similarity in the relative lengths and time intervals in the flows being compared. If the size of two machines is in a fixed ratio, then the velocities at corresponding positions are also in a fixed ratio, but only if the corresponding time intervals of the motion are also in the same fixed ratio. The dimensionless number that expresses this aspect of kinematic similarity is the Strouhal number, named after Vincenc Strouhal (1850–1922), which relates velocity, size and time intervals (frequency):

$$St = \frac{\text{velocity}}{\text{frequency} \times \text{scale}} = \frac{c}{nL} \sim \frac{c_m}{\omega r} \sim \frac{c_m}{u}. \tag{8.3}$$

In turbomachinery applications, this parameter usually appears as the local flow coefficient, c_m/u, and expresses the ratio of the throughflow component of the velocity to a reference blade speed (blade radius times rotational frequency). Physically, it relates the distance travelled by the flow in the meridional plane to the rotational translation of the impeller in the same time interval. A low local flow coefficient implies that the throughflow velocity is low relative to the blade speed and that more impeller revolutions are needed for a particle in the flow to pass through the machine.

The Strouhal number can also be used to express unsteady phenomena in the compressor. Sometimes the inverse is used in the form of a reduced frequency; see Greitzer et al. (2004). For example, unsteady effects are of importance in turbocharger compressors because the cylinders of the engine cause time-varying pressure and flow. The pulsations also need to be dynamically similar if two flows are considered to be similar. Usually the characteristics of turbochargers are measured on a test rig where the flow is steady, but then used in engines with unsteady disturbances from the cylinders, so a nondimensional parameter that can account for the unsteady effects is of relevance. The significance of the unsteady effects is usually determined by the reduced frequency (which is the inverse of the Strouhal number given previously):

$$\beta = \omega_d L/u, \tag{8.4}$$

where L is a suitable reference length, u a suitable reference velocity and ω_d is the frequency of an unsteady disturbance.

If we express the time-varying disturbance from the cylinders as a function such as $e^{j\omega_d t}$, then $1/\omega_d$ is a characteristic timescale for the disturbance. Small values of β imply that the fluid particles experience little change due to unsteadiness, so that the following distinctions can be made (Greitzer et al., 2004). For $\beta \ll 1$, unsteady effects are small and the flow is said to be quasisteady; for $\beta \gg 1$, the unsteady effects dominate; and for $\beta \sim 1$, both the unsteady and the quasisteady effects may be important.

A characteristic timescale for the flow in a compressor is the time in which a fluid element passes through the machine. For a radial wheel with impeller diameter D_2 and tip-speed u_2, a characteristic timescale is then related to D_2/u_2. The ratio of this to the timescale of the disturbance can then be used to identify the importance of the unsteady effects in radial compressors. This term is often used to examine the rotational frequency of instabilities in the compressor relative to the rotor rotational frequency:

$$\beta = \frac{D_2/u_2}{1/\omega_d} = \omega_d D_2/u_2 = \frac{\omega_d}{\pi n} = \frac{\omega_d}{\omega}. \tag{8.5}$$

In a turbocharger, the frequency of unsteady effects from the engine is related to the number of cylinders times the motor speed, and this is usually much lower than the compressor rotational speed so that the unsteady effects do not dominate and it is usually adequate to assume that the flow in the compressor is quasisteady. Nevertheless, short-term disturbances, such as valve opening and pulse effects, may occur on much shorter scales than the impeller rotational speed, and in these cases the stage cannot be considered to be quasisteady.

8.2.4 Reynolds Number

Dynamic similarity is determined by the dimensionless ratio of the inertia force to the viscous forces in the flow. This dimensionless parameter was first established by Osborne Reynolds (1842–1912) and is named the Reynolds number after his ground-breaking publication, Reynolds (1895):

$$Re = \frac{\text{inertia force}}{\text{viscous force}} = \frac{\tfrac{1}{2}\rho c^2 A}{\mu c L} \sim \frac{\rho c^2 L^2}{\mu c L} = \frac{\rho c L}{\mu}. \tag{8.6}$$

In most turbomachinery applications, the Reynolds number is large, and viscous effects are effectively confined to thin layers near the walls so that there is relatively little influence on performance with a change in the Reynolds number. As the Reynolds number becomes smaller due to small size or higher viscosity, increased frictional losses occur and performance is reduced.

8.2.5 Mach Number

As is shown in (8.7), thermodynamic similarity is categorised by the ratio of the flow velocity to the speed of sound in the form of a Mach number, first named by Stodola (1905) after Ernst Mach (1838–1916):

$$M = \frac{\text{velocity}}{\text{speed of sound}} = \frac{c}{a} = \frac{c}{\sqrt{\gamma R T}} \sim \frac{u}{\sqrt{\gamma R T_t}}. \tag{8.7}$$

Different Mach numbers can be used either to characterise the local velocity of the fluid relative to the local speed of sound, or to characterise the blade speed of the machine in relation to the speed of sound in the fluid medium at specified conditions, often the inlet total state.

8.3 Geometric Similarity

8.3.1 Scaling

The condition of geometric similarity simply implies that all dimensions of the machine scale with any reference length. In radial compressors and turbines, the impeller diameter is usually used as the reference scale. Thus, two geometrically similar machines have similar ratios of inlet to outlet diameter, have the same blade numbers giving the same pitch/chord ratio and have the same 'trim' (the blade channel height scales with the diameter); and all blade angles, which can also be interpreted as the ratio of two lengths, remain constant. Some of the most relevant geometrical parameters for a radial impeller are illustrated in Figure 8.3. The impeller tip clearance and the blade thickness may often be different from the inlet to the outlet of the impeller.

Geometrical similarity may be difficult to maintain in practice. The geometry of nominally similar machines may actually differ from machine to machine due to manufacturing tolerances or the manufacturing method, or even due to a change in the casing design with size. Some features tend to remain the same size in absolute

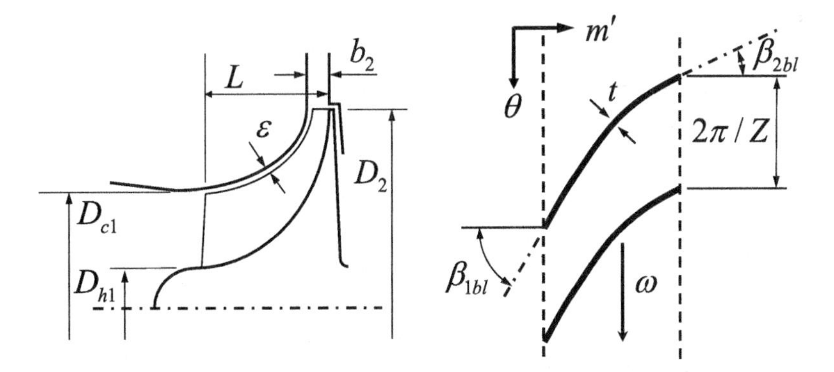

Figure 8.3 Key geometrical parameters of an open radial compressor impeller.

dimensions, such as the surface roughness or the fillet radii, or small clearance gaps and leakage flow paths, and so these parameters may in relative terms become progressively larger when a compressor is scaled down in size. Mechanical or manufacturing limitations may also cause the relative thickness of blades to increase in very small machines. Some geometrical features may change with the operating point, due to structural and thermal loads. For example, the clearance gaps of an open wheel depend on the stress-related centrifugal growth with increasing circumferential speed and on the temperature and pressure related strains of the impeller and casing.

8.3.2 Surface Roughness

The surface roughness is a product of the method of manufacture, and this usually remains the same when an impeller is scaled to a different diameter. The roughness may change over the life of the machine due to erosion. The similarity parameter of relevance for the roughness is the relative roughness, Ra/c, where Ra is the arithmetic average of the surface roughness profile and c is the chord length, as discussed in Section 5.5.12. The relative roughness reduces for larger impellers, and the surfaces become closer to being hydraulically smooth if the design is scaled to a larger diameter. This change has an effect on the boundary layers. It is usually considered to be a secondary effect in large machines and is taken into account in a similar way to the Reynolds number; see Section 8.4.4. In small machines, great care with the manufacturing method needs to be taken to attain a hydraulically smooth surface and, in some cases, low friction coatings may be appropriate.

8.4 Fluid Dynamic Similarity

8.4.1 Flow Coefficients

The condition of fluid dynamic similarity for two turbomachines requires similar nondimensional velocity fields, which implies fixed values of nondimensional flow, work and head coefficients as described in the next sections. The flow coefficient is derived from the Strouhal number but can also be interpreted as a dimensionless mass-flow ratio, where the actual mass flow is related to that which would occur if the total flow were to pass through the flow channels with annulus area A, at a velocity equivalent to the reference blade speed, giving

$$\phi = \frac{\dot{m}}{\rho \cdot A \cdot u} = \frac{\dot{V}}{A \cdot u} = \frac{c_m}{u}. \tag{8.8}$$

This local flow coefficient is often used to characterise the flow at a particular location within an impeller, where the local meridional velocity and the local blade speed are generally used. This requires a detailed knowledge of the internal conditions in the stage. For example, at the outlet of a radial impeller, the blade speed based on the impeller exit radius is used as the reference velocity, and the reference area is taken as the cylinder at the impeller outlet with an area of $A = \pi D_2 b_2$, giving

$$\phi_2 = \frac{c_{m2}}{u_2}. \tag{8.9}$$

A similar flow coefficient can be defined for the inlet of the impeller (with subscript 1), and at other positions in the stage. This form of flow coefficient is directly related to the local velocity triangles of the stage, as shown in Chapter 2. This flow coefficient is often used to emphasise the kinematic similarity of the flow at different speeds. A low flow coefficient implies that the wheel speed is much higher than the local through-flow velocity, and the impeller blades in this case have a high stagger angle and are more aligned to the circumferential direction. A higher flow coefficient implies that the blades are at lower stagger and are more aligned to the meridional direction.

In radial machines, this flow coefficient is not directly related to the absolute volume flow. Two radial compressor stages with the same outlet diameter and the same value of this local flow coefficient at outlet may not have the same flow capacity, since the internal flow channels may have different areas, as the result of, for example, different outlet widths (different 'trims'). Additional nondimensional coefficients are needed to show whether two stages have different flow capacities even with similar velocity triangles.

Two other flow coefficients related to the flow capacity are in regular use which can be determined from information known externally to the machine. The first is the global flow coefficient, or swallowing capacity, which relates the actual mass flow to the mass flow which would occur if the total flow were to pass through a virtual area A with the velocity of the reference blade speed and the density of the inlet total conditions. It requires no detailed information about the local density or the local flow areas, as it is based on the inlet total conditions and a virtual area. Two definitions of the virtual area are in common use: $A = D_2^2$ and $A = \pi D_2^2/4$. Since $\pi/4$ has a value quite close to 1, care must be taken to be sure which definition has been used.

If $A = D_2^2$ is used as the reference area, then

$$\phi_{t1} = \frac{\dot{m}}{\rho_{t1} A u_2} = \frac{\dot{m}}{\rho_{t1} D_2^2 u_2} = \frac{\dot{V}_{t1}}{D_2^2 u_2}. \tag{8.10}$$

This parameter therefore characterises the volumetric swallowing capacity of different machines in terms of the impeller diameter and tip-speed. Hence this flow coefficient is used to emphasise the relative flow capacity of different designs with the same nominal impeller size and tip speed.

Where $A = \pi D_2^2/4$ is used as the reference area, the flow coefficient becomes

$$\phi_M = \frac{4\dot{m}}{\pi \rho_{t1} D_2^2 u_2} = \frac{4\dot{V}_{t1}}{\pi D_2^2 u_2}. \tag{8.11}$$

Other forms are also possible in which the speed is defined in terms of the rotational speed (in rpm). These are less useful in compressor technology, as it is the blade speed (or the closely related tip-speed Mach number) which determines the pressure ratio so that a flow coefficient based on blade speed is more appropriate.

For a radial compressor impeller, the mass flow can be expressed as the product of annulus area, density and velocity at some point within the stage with a flow area of A. The global flow coefficient (8.10) can be rewritten as

$$\phi_{t1} = \frac{\dot{m}}{\rho_{t1} D_2^2 u_2} = \frac{\rho A c_m}{\rho_{t1} D_2^2 u_2} = \frac{\rho}{\rho_{t1}} \frac{A}{D_2^2} \frac{c_m}{u_2}. \tag{8.12}$$

where the first term is a density ratio, the second an area ratio depending on the geometry of the stage and the third a velocity ratio, or local flow coefficient, related to the velocity triangles. Considering conditions at the inducer at impeller inlet, then for a variety of different impeller designs

$$\frac{\rho_1}{\rho_{t1}} \approx 1.0, \quad 0.02 < \frac{A_1}{D_2^2} < 0.4, \quad 0.2 < \frac{c_{m1}}{u_2} < 0.4 \tag{8.13}$$

and thus, the global flow coefficient, which is the product of these terms, typically varies in the range $0.004 > \phi_{t1} > 0.16$. Best efficiencies are found in a much narrower range of $0.06 > \phi_{t1} > 0.10$. Higher values above $\phi_{t1} = 0.16$ tend to imply axial machines and lower values mixed or radial flow machines, so this parameter provides a useful distinction between the flow capacity of the different machine types and acts as a shape parameter. The global flow coefficient is also an excellent guide to the level of efficiency which can be achieved from a well-executed design; see Section 10.7.

The global flow coefficient is the appropriate parameter expressing similarity of flow at all points in the stage for an incompressible flow, as the ratio of meridional velocity to blade speed (and hence the velocity triangles) at any point in the flow field is directly proportional to the global flow coefficient. Consequently, a given stage geometry has the peak efficiency in incompressible flow at the same flow coefficient over a range of speeds, as shown for a pump in Figure 8.2. This is not the case in a compressible flow, however, because operation at different pressure ratios leads to a change to the density ratio in (8.12) and hence to slightly different flow volumes and velocity fields within the stage for the same value of the inlet flow coefficient. For a compressible gas, exact kinematic similarity occurs only if the values of the global flow coefficient (dynamic similarity) and the tip-speed Mach number (thermodynamic similarity) are the same, as explained in Section 8.5.1.

As an alternative, a similar flow coefficient can be defined by assuming that the flow in the reference area A has the velocity of the speed of sound and density at inlet total conditions, leading to

$$\Phi_{t1} = \frac{\dot{m}}{\rho_{t1} A a_{t1}} = \frac{\dot{m}}{\rho_{t1} A u_2} \frac{u_2}{a_{t1}} = \phi_{t1} M_{u2}. \tag{8.14}$$

This flow coefficient is the product of the global flow coefficient defined earlier and the tip-speed Mach number. This form is useful as a means to emphasise the compressibility effects and their variation with speed and with inlet conditions. All operating conditions covered by a pair of values of tip-speed Mach number, M_{u2}, and flow coefficient, ϕ_{t1}, are automatically covered when the product, $M_{u2}\,\phi_{t1}$, is used. This parameter is known as the corrected flow per unit area or the corrected mass-flow function; see Section 6.2.4.

The flow coefficient defined in (8.14) can best be understood when it is realised that it identifies the dependence of the flow capacity of a particular machine on the maximum possible flow capacity at the given inlet flow conditions. From one-dimensional gas dynamics (see Section 6.2.5), the maximum mass flow through an orifice of area A^* occurs when the flow at the orifice chokes by attaining sonic velocity, as follows:

$$\frac{\dot{m}_{\max}}{\rho_{t1} A^* a_{t1}} = \left(\frac{2}{\gamma+1}\right)^{\frac{\gamma+1}{2(\gamma-1)}}. \tag{8.15}$$

Combining this with (8.14) leads to

$$\Phi_{t1} = \frac{\dot{m}}{\rho_{t1} A a_{t1}} = \frac{\dot{m}}{\dot{m}_{\max}} \left[\frac{\gamma+1}{2}\right]^{(\gamma+1)/2(\gamma-1)}. \tag{8.16}$$

This shows that the corrected flow per unit area is directly proportional to the actual mass flow divided by the choking mass-flow rate through the reference area. Just as (8.15) takes into account the effect of the different inlet conditions on the mass flow through an orifice, (8.16) takes into account the similar effect of the different inlet conditions on the flow in the compressor. An increase in inlet total temperature or a decrease in inlet total pressure moves an operating point with a given mass flow closer to the choking condition.

This corrected mass-flow function is not often used directly in this form in industrial practice, especially if the gas being compressed is air at standard conditions and the machine being considered has a certain size. If a specific machine is being considered at different inlet conditions using the same gas, then an alternative is the corrected mass-flow rate or the corrected volume flow rate together with the corrected speed:

$$\dot{m}_{corr} = \dot{m}\sqrt{\frac{T_{t1}\, p_{ref}}{T_{ref}\, p_{t1}}}, \quad \dot{V}_{corr} = \dot{V}\sqrt{\frac{T_{ref}}{T_{t1}}}, \quad N_{corr} = N\sqrt{\frac{T_{t1}}{T_{ref}}}. \tag{8.17}$$

The reference temperature and stagnation pressure are often taken as those at the international standard atmospheric conditions at sea level, corresponding to a pressure of 101,325 Pa and a temperature of 15°C (288.15 K). Across the many companies involved with compressors, other standards may be used, so care and attention are needed. In Chapter 6, the authors recommend using a slightly different formulation for the corrected mass flow involving the inlet speed of sound and the inlet density as this is valid for real gases.

In some situations, the axial Mach number of the flow in the inlet channel of a compressor may be used as a measure of the flow rate, and this is essentially another measure of the corrected flow as

$$M_{ax} = \frac{c_{ax}}{\sqrt{\gamma RT}} = \frac{\dot{V}}{A\sqrt{\gamma RT}}. \tag{8.18}$$

8.4.2 Work Coefficients

In order to characterise the dynamic effects of the impeller rotational speed on the total enthalpy rise across a compressor a parameter based on the total enthalpy rise across the machine, similar to a Newton or Euler number, but known as the nondimensional work input coefficient, is used

$$\lambda = \frac{(h_{t2} - h_{t1})}{u_2^2} = \frac{\Delta h_t}{u_2^2}. \tag{8.19}$$

In typical radial compressors and pumps, this parameter typically lies between 0.55 and 0.80. In some of the turbomachinery literature, and more often in pumps and fans than compressors (see Gülich, 2008), the enthalpy rise or head rise is made nondimensional with $\frac{1}{2}u_2^2$ giving rise to numerical values that are twice as large as those from (8.19). This might be another source of confusion when using data from different sources, but as it generally leads to values greater than unity for radial compressors, it is quickly identified. From the Euler equation for the rotor, the total enthalpy rise across the machine is given by

$$\Delta h_{Euler} = u_2 c_{u2} - u_1 c_{u1}, \tag{8.20}$$

and so the work coefficient based on the Euler work input becomes

$$\lambda_{Euler} = \frac{u_2 c_{u2} - u_1 c_{u1}}{u_2^2} = \frac{c_{u2}}{u_2} - \left(\frac{u_1}{u_2}\right)\left(\frac{c_{u1}}{u_2}\right) \tag{8.21}$$

and the work coefficient, like the local flow coefficient, is thus the ratio of various velocity components in the local velocity triangles. For a typical radial compressor or pump impeller with no swirl at the inlet, this reduces to the simple expression for the Euler work coefficient:

$$\lambda_{Euler} = \frac{c_{u2}}{u_2}. \tag{8.22}$$

It should be noted that the actual work input absorbed by a compressor impeller is higher than the value predicted by the Euler equation due to the effect of parasitic work, which can be expressed by an additional power input factor, C_{PIF}, defined as follows:

$$C_{PIF} = \Delta h_t / \Delta h_{Euler}, \tag{8.23}$$

where $C_{PIF} > 1$. Parasitic work is usually taken to include recirculation of fluid via the tip clearance (such that it is compressed again and again) and disc friction due to drag on the external rotating surface of the impeller, both of which decrease the efficiency and increase the enthalpy rise across the stage to a level greater than the Euler value; see Section 10.6.

The work coefficient and the flow coefficient can be usefully combined to define the power required by the compressor:

$$P = \dot{m}\Delta h_t = \frac{\dot{m}}{\rho_{t1} u_2 D_2^2} \frac{\Delta h_t}{u_2^2} \rho_{t1} u_2^3 D_2^2 = \rho_{t1} \phi \lambda u_2^3 D_2^2. \tag{8.24}$$

This immediately identifies that the power of a compressor operating at dynamically similar conditions, that is, with similar work and flow coefficients, scales with the third power of the impeller tip speed and with the second power of the impeller diameter and is proportional to the gas inlet density, the work coefficient and the flow coefficient.

8.4.3 Pressure Rise or Head Rise Coefficients

In any real compressor, the work input to the impeller produces both a useful rise in pressure and overcomes the dissipation losses in the flow (friction on walls, mixing, secondary flows, leakage and shock losses). In this way, not all of the work input appears as useful pressure rise, and a different parameter is needed to characterise this. The pressure rise or head coefficient is used and is defined in a similar way as the work coefficient. There are two common forms, the polytropic head coefficient or the isentropic head coefficient, depending on which process is being used to define the efficiency.

The isentropic head rise coefficient is based on a reference process at constant entropy between the inlet and outlet pressures, which requires an ideal specific work input of Δh_{ts}, as follows:

$$\psi_s = \Delta h_{ts} / u_2^2 = (\Delta h_{ts} / \Delta h_t)(\Delta h_t / u_2^2) = \eta_s \lambda, \tag{8.25}$$

where $\eta_s = \Delta h_{ts}/\Delta h_t$ is the isentropic efficiency of the compression process. Similarly, the polytropic head rise coefficient makes use of the definition of the aerodynamic work, y_{12}, and polytropic efficiency, η_p, and is defined as

$$\psi_p = \frac{\int v dp}{u_2^2} = \frac{y_{12}}{u_2^2} = \frac{y_{12}}{\Delta h_t} \frac{\Delta h_t}{u_2^2} = \eta_p \lambda. \tag{8.26}$$

In an incompressible flow, for a compressor or pump operating at its peak efficiency point, the pressure rise is proportional to the square of the tip speed and the flow coefficient varies linearly with the speed, so these two nondimensional parameters are the basis of the fan-law relationships already given in (8.1) and shown in Figure 8.2.

8.4.4 Reynolds Number

Real fluids impede any motion imposed on them through the property known as viscosity, which describes the fluid's own internal resistance to flow and is a measure of internal fluid friction. It is a direct measure of the resistance of a fluid to deformation by a shear stress. The internal friction causes any applied velocity of the motion to decrease through dissipation of the kinetic energy, and in steady flow in turbomachinery additional energy must be added to overcome this dissipation.

The usual symbol for dynamic viscosity is the Greek letter mu, μ. The physical unit of dynamic viscosity in the SI system is the pascal-second, Pa s, which is identical to kg/(ms). If a fluid with a viscosity of one Pa s is placed between two plates, and one plate is pushed sideways with a shear stress of one pascal, it moves a distance equal to the thickness of the layer between the plates in one second. In such a situation, the shear stress of the fluid is given by Newton's law of viscosity:

$$\tau = \mu \frac{dc}{dy}, \tag{8.27}$$

where the dynamic viscosity is the constant of proportionality between the shear stress and the velocity gradient. Fluids with this linear relationship are known as Newtonian. Other fluids with a viscosity which does not depend linearly on the shear stress are described as non-Newtonian.

Of interest is the ratio of viscous forces to the inertial forces, the latter characterized by the fluid density ρ. This ratio is called the kinematic viscosity (Greek letter nu, ν), defined as follows:

$$\nu = \frac{\mu}{\rho}, \tag{8.28}$$

where μ is the dynamic viscosity, Pa s; ρ is the density, kg/m$^{3;}$ and ν is the kinematic viscosity in m^2/s. Dynamic viscosity in gases arises principally from the molecular diffusion that transports momentum between adjacent layers of flow. The kinetic theory of gases allows accurate prediction of the behaviour of gaseous viscosity. For gases, viscosity is not strongly dependent on pressure but increases as temperature increases, that is, the dynamic viscosity of air depends largely on the temperature. At 20°C and 1 bar, the dynamic viscosity of air is 18×10^{-6} kg/(ms) and the kinematic viscosity is 15×10^{-6} m^2/s. Sutherland's law expresses the approximate relationship between the dynamic viscosity of an ideal gas with temperature:

$$\mu = \mu_{ref} \left(\frac{T}{T_{ref}} \right)^{3/2} \frac{T_{ref} + S}{T + S} = C_1 (T)^{3/2} \frac{1}{T + S}. \tag{8.29}$$

For air, $\mu_{ref} = 17.16 \times 10^{-6}$ kg/(ms), $T_{ref} = 273.15$ K, $S = 110.4$ K and $C_1 = 1.458 \times 10^{-6}$ kg/(ms/K).

In liquids, the additional forces between molecules become important. This leads to an additional contribution to the shear stress, which means that in liquids the viscosity falls as temperature increases, and it becomes even slightly dependent on pressure. The dynamic viscosities of liquids are typically several orders of magnitude higher than the dynamic viscosities of gases. The dynamic viscosity of water at 20°C and 1 bar is approximately 1×10^{-3} Pa s while the kinematic viscosity is about 1×10^{-6} m^2/s.

At the contact surface between a flowing fluid and a wall, the fluid velocity has the same velocity as the wall. This is known as the no-slip condition, so that on a stationary wall the velocity is zero next to the wall. The fluid motion farther away from the wall causes a velocity gradient normal to the wall. The viscosity then produces a wall shear stress

$$\tau_w = \mu \left(\frac{dc}{dy} \right)_{y=0}, \tag{8.30}$$

where y is the coordinate normal to the wall and c is the velocity parallel to the wall. The wall shear stress gives rise to a local differential frictional force on the wall. Integration of this along the wall determines the total frictional force on the body as

$$dF_f = \tau_w dA, \quad F_f = \int_A \tau_w dA. \tag{8.31}$$

This frictional force opposes the motion of the fluid and is ultimately responsible for the most of the energy losses in turbomachinery and other fluid dynamic components.

In many simple fluid dynamic applications (such as flat plates, circular pipes, rotating discs, etc.), it is possible to determine the general relationships between the shear stresses and the frictional forces. The shear stresses can be expressed in non-dimensional form as friction factors or as frictional drag coefficients:

$$c_f = \frac{\tau_w}{(1/2)\rho c^2}, \quad C_D = \frac{D}{(1/2)\rho c^2 A}. \tag{8.32}$$

In addition to the frictional drag, there may also be forces from the pressure field around the body, known as pressure drag, but these also scale with the velocity, density and area in the same way.

In radial compressor stages, several definitions of the Reynolds number are often used:

$$Re_{D_2} = u_2 D_2 / \nu, \quad Re_{b_2} = u_2 b_2 / \nu, \quad Re = w_1 c / \nu. \tag{8.33}$$

The last of these, the chord-based Reynolds number using the locally relevant relative inlet flow velocity, w_1, and flow-path length (chord, c) of each component, is clearly more closely related to the real state of the boundary layers than the other definitions given here. These other definitions are similar to the global flow coefficient and make use of information that is external to the machine. The Reynolds number combines three effects into a single relevant parameter, the effect of viscosity, the effect of flow velocity and the effect of size (Reynolds, 1895).

The normally small second-order effect of the Reynolds number is often considered using correlations. Some successful older correction methods rely on the Reynolds numbers based on tip speed and tip diameter, such as Gülich (2008), or for narrow low-flow coefficient stages on tip width (Casey, 1985), as these are the more easily accessible global geometrical and flow parameters. The chord-based Reynolds number was successfully used in a unified correlation for the effect of Reynolds number for axial and radial compressors, which also includes roughness effects (Casey and Robinson, 2011; Dietmann and Casey, 2013).

In the majority of radial turbocompressor applications, the Reynolds number is sufficiently large that the flow remains in the turbulent region, so that most experiments in which the Reynolds number is varied show that it only has a secondary effect. The friction factor decreases with the Reynolds number, and the extremely high efficiency of large pumps and large water turbines depends as much on the high Reynolds numbers of these devices rather than on specific features of their aerodynamic design. A low Reynolds number may occur due to low inlet pressure, for example when a compressor operates at altitude in a jet engine compared with operation at sea level. The Reynolds number may also be low if it is scaled down (for example, small machines compared with large machines) or if the rotational speed is very low (as in ventilators and fans), such that laminar flow may occur. Very high Reynolds numbers may also occur in compressors for industrial gases with large inlet pressures of 200–300 bar.

8.5 Thermodynamic Similarity

8.5.1 Tip-Speed Mach Number

Thermodynamic similarity is not an issue in incompressible flow machines as the density does not change. The key factor controlling thermodynamic similarity in compressible flow machines is the nondimensional tip-speed Mach number of the machine. Considering a compressor producing a total enthalpy change of $\Delta h_{t,13}$ from total conditions at state 1 at stage inlet to state 3 at stage outlet, then

$$T_{t3} = T_{t1} + \Delta h_{t,13}/c_p, \quad T_{t3}/T_{t1} = 1 + \Delta h_{t,13}/\left(c_p T_{t1}\right). \tag{8.34}$$

This implies that the condition for thermodynamic similarity in two machines (i.e., constant temperature ratio between states 1 and 3) is that $\Delta h_{t,13}/(c_p T_{t1})$ is the same for both. If the compression process is considered to take place along a polytropic compression path, $pv^n = constant$, with polytropic exponent n, then a simple expression for the density and temperature ratio can be derived as follows:

$$\rho_{t3}/\rho_{t1} = (T_{t3}/T_{t1})^{\frac{1}{n-1}} = \left(1 + \Delta h_{t,13}/\left(c_p T_{t1}\right)\right)^{\frac{1}{n-1}}, \tag{8.35}$$

which shows that the relevant nondimensional parameter for similar density ratios in two similar compressors is also $\Delta h_{t,13}/(c_p T_{t1})$.

For radial compressors operating with fluid dynamic similarity and having similar velocity triangles, the Euler turbomachinery equation states that the total enthalpy rise is proportional to the square of the tip speed. Therefore, thermodynamic similarity also requires similarity of

$$\Delta h_{t,13}/\left(c_p T_{t1}\right) \propto \lambda u_2^2/\left(c_p T_{t1}\right) = (\gamma - 1)\lambda u_2^2/(\gamma R T_{t1}) \propto M_{u2}^2, \tag{8.36}$$

where $M_{u2} = u_2/\sqrt{(\gamma R T_{t1})}$ is the tip-speed Mach number of the stage, sometimes known as the machine Mach number. Thus, it is clear that for similar thermodynamic behaviour (temperature ratio, pressure ratio, density ratio, etc.) of the flow in two machines, the tip-speed Mach number is the relevant dimensionless parameter. In addition, from (8.35) to (8.36), the isentropic and polytropic exponents also need to be the same in the two cases.

The tip-speed Mach number should not be confused with the local Mach number of the gas flowing in the machine, whose value depends on the local gas velocity and the local temperature $M = c/\sqrt{(\gamma R T)}$. A tip-speed Mach number of $M_{u2} = 1.0$ does not imply that sonic conditions are attained anywhere in the compressor. But as the gas velocity is related to the blade speed, there is usually a close correspondence between the local gas Mach number and the tip-speed Mach number, and the latter provides a good and easily derived index of the likely extent of any critical Mach number effects, see Section 11.4.3. Thermodynamic similarity is not relevant in low-speed fans, hydraulic turbines and pumps where the tip-speed Mach number is zero or very close to zero and, as a result, there is then no strong effect of the tip speed on the fluid properties in the machine, other than via the changes in Reynolds number.

The tip-speed Mach number itself is often used as the characterising parameter for the flow in industrial compressors which may operate on different gases (Casey and Marty, 1985; Lüdtke, 2004), as it is a convenient and physical way of combining the similar effects of a speed change, a change in inlet temperature or a change in gas properties or inlet conditions on the stage temperature ratio. For a specific compressor operating on a specific gas, as in a typical turbocharger compressor which always operates on dry air, the effect of the speed of the compressor on the pressure and density ratio is often characterised by a simpler ratio, the so-called speed parameter $N/\sqrt{T}$, which is no longer nondimensional:

$$M_{u2} = u_2/\sqrt{\gamma R T_{t1}} = \omega r_2/\sqrt{\gamma R T_{t1}} = (2\pi N/60) r_2/\sqrt{\gamma R T_{t1}} \propto N/\sqrt{T_{t1}}. \tag{8.37}$$

In turbochargers operating on air, this may sometimes be approximated to simply $u_2/\sqrt{T_{t1}}$. If a specific machine is being considered at different inlet conditions, then another alternative is the corrected speed, which still has the units of rpm or rev/sec, as given in (5.18):

$$N_{corr} = N\sqrt{\frac{T_{t1}}{T_{ref}}}, \quad n_{corr} = n\sqrt{\frac{T_{t1}}{T_{ref}}}, \tag{8.38}$$

where the reference temperature used is the same as that used for the corrected mass flow.

Relatively recently, Berdanier et al. (2014) highlighted the fact that compressors operating in humid air do not follow perfect gas laws. Equations (8.17) and (8.38) are then insufficiently accurate to correct the performance to reference conditions, and more appropriate equations need to use the speed of sound as N/a_{t1} rather than $N/\sqrt{T}$ to avoid errors in the order of 0.5–1% in mass flow and pressure rise. Better still is to use the tip-speed Mach number.

8.5.2 Polytropic, Isentropic and Incompressible Analysis

The key relationships between the nondimensional parameters in a compressible gas flow can be derived by considering a compressor running on an ideal gas at a particular tip-speed Mach number with known inlet conditions, at an operating point with a work input coefficient, λ, as follows:

$$\begin{aligned} T_{t2} - T_{t1} &= \Delta T_t = \Delta h_t / c_p = \lambda u_2^2 / c_p = (\gamma - 1)\lambda u_2^2 / (\gamma R) \\ \Delta T_t / T_{t1} &= (\gamma - 1)\lambda u_2^2 / (\gamma R T_{t1}) = (\gamma - 1)\lambda M_{u2}^2 \\ T_{t2}/T_{t1} &= 1 + (\gamma - 1)\lambda M_{u2}^2. \end{aligned} \tag{8.39}$$

In a polytropic analysis, we assume that the compression takes place along a polytropic compression path with a polytropic efficiency, η_p, with the relations

$$pv^n = \text{constant}, \quad n/(n-1) = \eta_p \gamma / (\gamma - 1), \tag{8.40}$$

so that the pressure ratio, π, is given by

$$\pi_{t21} = p_{t2}/p_{t1} = [T_{t2}/T_{t1}]^{n/(n-1)} = \left[1 + (\gamma - 1)\lambda M_{u2}^2\right]^{n/(n-1)}. \tag{8.41}$$

These equations show clearly that two stages with identical velocity triangles (i.e., with the same work input coefficient, λ) will produce the same pressure ratio only when they have similar tip-speed Mach numbers, operate on fluids with the same gas properties and have the same polytropic efficiency giving the same polytropic exponent. Thus, thermodynamic similarity is a necessary condition for similar velocity triangles so that dynamic and kinematic similarity can be achieved in a compressible flow and illustrates the importance of the tip-speed Mach number in defining this.

Similar equations can be derived in combination with the isentropic efficiency. These make use of the isentropic process between the state 1 and the virtual state, 2*s*, which is at the same entropy but with the same pressure as state 2. They are generally considered easier to follow as they do not utilise the polytropic process, and can be written as

$$\begin{aligned} &T_{t2} - T_{t1} = \Delta h_t / c_p = \lambda u_2^2 / c_p = (\gamma - 1)\lambda u_2^2 / (\gamma R) \\ &T_{t2s} - T_{t1} = \eta_s (\gamma - 1)\lambda u_2^2 / (\gamma R), \quad T_{t2s}/T_{t1} = 1 + \eta_s (\gamma - 1)\lambda M_{u2}^2 \\ &p_{t2}/p_{t1} = [T_{t2s}/T_{t1}]^{\gamma/(\gamma-1)} = \left[1 + (\gamma - 1)\lambda \eta_s M_{u2}^2\right]^{\gamma/(\gamma-1)} = \left[1 + (\gamma - 1)\psi_s M_{u2}^2\right]^{\gamma/(\gamma-1)}. \end{aligned} \tag{8.42}$$

In an isentropic analysis, the fact the pressure ratio is determined by the product of efficiency and work coefficient is immediately clear. In both a polytropic analysis and an isentropic analysis, the pressure ratio is determined by the tip-speed Mach number, the efficiency and the work coefficient.

Considering a low-speed machine where $M_{u2}^2 \ll 1$, the expression in (8.41) can be expanded binomially to give

$$p_{t2}/p_{t1} = 1 + \frac{n(\gamma - 1)\lambda M_{u2}^2}{n - 1} + \cdots \tag{8.43}$$

and on using (5.41) and neglecting the higher-order terms:

$$\begin{aligned} p_{t2}/p_{t1} &= 1 + \eta_p \lambda \gamma M_{u2}^2 = 1 + \psi_p \gamma M_{u2}^2 \\ p_{t2} - p_{t1} &= \psi_p \gamma p_{t1} u_2^2 / \gamma R T_{t1} = \psi_p \rho_{t1} u_2^2. \end{aligned} \tag{8.44}$$

This makes it clear that it is the head coefficient (the product of the efficiency and the work input coefficient) that determines the pressure rise and that the pressure rise is directly proportional to the dynamic head based on impeller blade speed. This is precisely what is expected from an incompressible analysis of a low-speed machine from the incompressible fan laws as expressed in (8.1).

Expansion of the result of (8.42) from the isentropic analysis for low Mach numbers leads to the same equation as the expansion of (8.44) for the polytropic analysis. This means that at low Mach numbers, the polytropic analysis and the isentropic analysis are identical, and in pumps, ventilators and low-speed machines there is no need to distinguish the different types of analysis or to distinguish between polytropic and isentropic efficiencies. For this incompressible analysis, an alternative form of (8.44) can be written, and as the polytropic and isentropic analyses are the same, the suffixes can be dropped and

$$\frac{\Delta p}{\rho u^2} = \psi = \eta \lambda = \frac{\eta \Delta h_t}{u^2}. \tag{8.45}$$

This shows that the stage pressure rise coefficient characteristic of an incompressible stage is determined by the product of the work coefficient and efficiency. Similarly, the form of the pressure rise stage characteristic when the flow varies is determined by the variation in these two parameters. This dependence on two parameters is also true in the isentropic and polytropic analysis given earlier, but there it is slightly hidden in the algebra. This has important consequences and helps in the understanding of the performance characteristics as explained in Section 8.6.1.

8.5.3 Normalised Nondimensional Parameters

In some situations, it is useful to express the performance of a compressor in terms of normalised nondimensional parameters in such a way that the relative speed, the relative volume flow and the relative power all take the value of unity at the rated conditions. This is a trivial task if the rated conditions have the same inlet conditions and gas as the operating point, as it is then simply the ratio of the speeds, the flows and the powers. In a real situation where the inlet conditions change, or even the gas properties change, a useful way of doing this is to define the speed ratio as a Mach number ratio where the actual Mach number of the machine is relative to that at its rated conditions, $M_{u2}/(M_{u2})_r$. The appropriate relative volume flow parameter is then $(\phi_{t1}M_{u2})/(\phi_{t1}M_{u2})_r$, which can be written as

$$\frac{\phi_{t1}M_{u2}}{(\phi_{t1}M_{u2})_r}=\frac{V_{t1}}{(V_{t1})_r}\frac{(a_{t1})_r}{a_{t1}}. \tag{8.46}$$

This considers changes in the volume flow through changes in speed, fluid and inlet conditions. In some cases, this may be plotted at the peak efficiency point, and in others with respect to the choking flow. From (8.24), the relative power parameter can be derived to be $(\phi_{t1}\lambda M_{u2}^3)/(\phi_{t1}\lambda M_{u2}^3)_r$, which can be written as

$$\frac{\phi_{t1}M_{u2}}{(\phi_{t1}M_{u2})_r}\frac{\lambda M_{u2}^2}{(\lambda M_{u2}^2)_r}=\frac{P}{P_r}\frac{(\rho_{t1}a_{t1}^3)_r}{\rho_{t1}a_{t1}^3}. \tag{8.47}$$

8.6 Applications of Similarity Parameters

8.6.1 Idealised Performance of Compressors

In an ideal machine, that is, one with no losses, the work coefficient alone would determine the shape of the performance characteristic when the flow varies, as shown in Figure 8.4. The work characteristic is determined from the outlet velocity triangles of the blade rows, and thus the essential feature of the steepness of the performance characteristic is also determined by the velocity triangles; see Section 11.6.3.

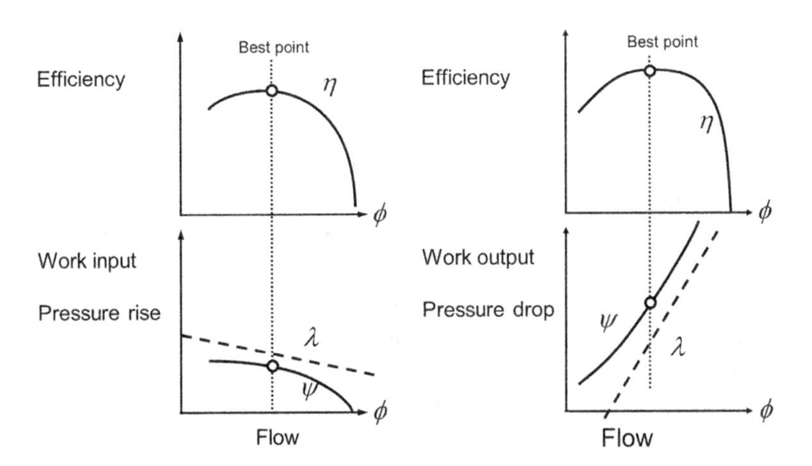

Figure 8.4 Idealised characteristic curves for a compressor and a turbine.

Moreover, as the outlet flow angles of the blade rows generally remain almost constant as the flow varies, the typical work coefficient versus flow coefficient characteristic is nearly linear. It has a similar negative slope at all flows, giving a nearly straight line for this characteristic, except at choke. At the peak efficiency point of a real machine, where, by definition, the efficiency does not change with flow, the slope of the pressure versus flow characteristic is then also determined entirely by the slope of the work versus the flow characteristic. This is especially relevant with regard to the use of backswept impellers.

The fact that the flow in a compressor decelerates and there are limitations to the amount of deceleration, as discussed in Section 7.5, leads to the work coefficient of a typical radial compressor being less than that of a radial turbine. In compressors, the energy dissipation results in a lower pressure rise than that in the ideal machine, so the pressure rise characteristic lies below the work characteristic. Away from the best or peak efficiency point, the efficiency falls off at both higher and lower flows, and it is this feature that leads to the typical curvature of pressure versus flow performance characteristics. At high flows, choking may occur, which limits the flow capacity such that the curves tend to become very steep in this region. In turbines, the pressure drop needs to be larger than in an ideal machine to overcome the dissipation so the pressure coefficient characteristic lies above that of the work coefficient as shown in Figure 8.4.

8.6.2 Similarity of Pressure Ratio

An important aspect of centrifugal compressor performance is the duty of the machine, as exemplified by the nondimensional parameter, the pressure ratio. Many industrial air applications require a stage pressure ratio of 1.5, or even less for simple blowers. To achieve higher pressure ratios in industrial applications, multistage machines are often used, rather than going to higher pressure ratios in a single stage, since intercooling can then be applied between the stages to reduce the power consumption of subsequent stages, for example in the main air compressor of an air liquefaction plant with a pressure ratio of around 8. Typical turbochargers for automotive diesel engines require a pressure ratio of around 2.5–4, and turbochargers for the diesel engines of large ships go up to a pressure ratio of 4–6. Small gas turbines in microturbines or in helicopter gas turbines require higher pressure ratios for simple-cycle engines, with 6 being commonplace and the latest engines demanding 10 or more from a single compressor stage.

The key equation defining the pressure ratio that is achievable is (8.41). In terms of nondimensional parameters, the pressure ratio is determined by the isentropic exponent of the gas, the impeller work coefficient, the square of the impeller tip-speed Mach number and the polytropic efficiency of the compression process. The pressure ratio when operating in air is determined essentially by the tip-speed Mach number, as most good impeller designs operate within a small range of work coefficients and have similar efficiency levels. The tip-speed Mach number is itself related through the

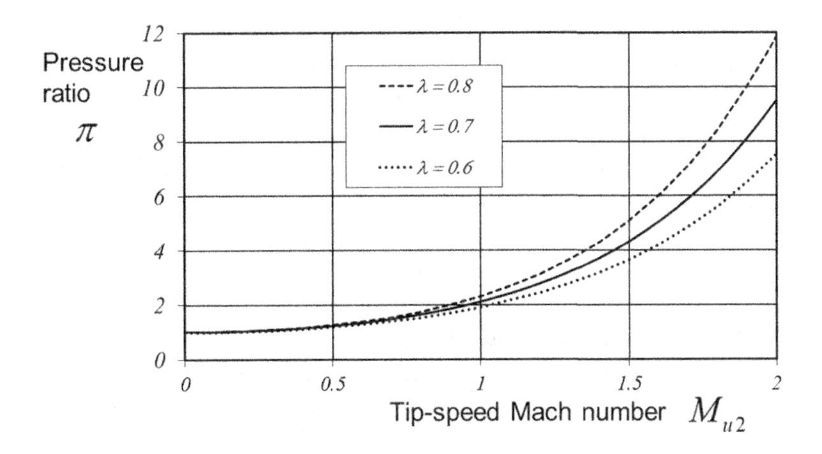

Figure 8.5 Pressure ratio in air as a function of tip-speed Mach number and work coefficient for a polytropic efficiency of η_p = 0.857.

velocity triangles to the local gas velocities, and this means that the flow Mach number levels of a single-stage design are not something chosen by the designer, but are determined by the specified duty. In axial compressor applications, lower stage pressure ratios and lower Mach numbers may be chosen given that additional stages may be added to reach the required pressure ratio, so the number of stages determines the Mach number level. Single-stage radial machines simply have to rotate faster to attain the required duty via a higher tip speed.

Figure 8.5 shows the variation in pressure ratio with a tip-speed Mach number for a typical range of radial impeller work coefficients from 0.6 to 0.8, for air with $\gamma = 1.4$ and a polytropic exponent of $n = 1.5$, which implies a polytropic efficiency of $\eta_p = 6/7 = 0.857$ (a fairly high level but achievable in the best large compressors).

Using the same numerical values in (8.41), again for air, with a tip-speed Mach number of 1.0 (i.e., an impeller tip-speed close to 340 m/s), a compressor with a work input coefficient of 0.65 and a polytropic efficiency of 85.7% and $n = 1.5$, leads to a pressure ratio of exactly 2, as $(1 + 0.4 \times 0.65 \times 1.0^2)^3 = 2.0$. These round numbers make this case fairly memorable: the tip speed of most turbocharger stages with pressure ratio near 2 is close to the speed of sound in air, at roughly 340 m/s, but may be a little higher on account of the lower efficiency. In cold weather, with a lower inlet temperature, a turbocharger compressor produces a higher pressure rise for a given rotational speed as the speed of sound decreases. Figure 8.5 also shows that a pressure ratio of around 4 requires a tip-speed Mach number close to 1.5, or a tip-speed in air of around 500 m/s. At a tip-speed around 250 m/s, or a Mach number of near 0.75, a pressure ratio of 1.5 is produced.

A compressor with the same work coefficient and efficiency but operating on hydrogen gas, which also has $\gamma = 1.4$, would also produce a pressure ratio of 2 at a tip-speed Mach number of 1.0. However, due to the low molecular weight of 2, the speed of sound in hydrogen is approximately 1300 m/s. As a consequence, this pressure ratio would be impossible to achieve in practice since the stresses in the impeller (proportional to the square of tip speed) would be excessive and beyond the capability of known materials. When operating at 340 m/s tip speed, near the maximum for a shrouded impeller, a hydrogen compressor would have a tip-speed Mach number of $Mu_2 = 340/1300 = 0.263$, the pressure ratio would be 1.055, as (8.41) gives $(1 + 0.4 \times 0.65 \times 0.263^2)^3 = 1.055$. Compression of low molecular weight gases to a certain pressure ratio requires many stages.

A refrigeration compressor operating on a gas with a high molecular weight such as R134a (with a molecular weight of 102, a speed of sound of 160 m/s and with $\gamma = 1.216$) would produce a slightly lower pressure ratio of 1.88 at a tip-speed Mach number of 1.0, as (8.41) gives $(1 + 0.216 \times 0.65 \times 1.0^2)^3 = 1.88$. But this would only require a tip speed of 160 m/s.

Clearly the consideration of thermodynamic similarity as outlined earlier in this chapter is of major importance not only when comparing stages operating at different speeds in air but also when comparing the compression of different gases. Although the pressure ratio is a nondimensional parameter, it is clearly not at all useful as an indicator of the severity of the aerodynamic duty when comparing compression of different gases, although engineers always working with a particular gas get an implicit feeling of the difficulty of a particular duty from the pressure ratio.

Of additional interest is the sensitivity of the pressure ratio to the polytropic efficiency, as shown for air in Figure 8.6, for a work coefficient of $\lambda = 0.7$. This figure illustrates that the pressure ratio becomes strongly dependent on the efficiency for a tip-speed Mach number above 1.3, which in this case would be a pressure ratio above 3 in air. This is one of the reasons that compressors designed for high pressure ratio in marine turbochargers and aeroengine radial compressors nearly always use a vaned diffuser to help to achieve a higher efficiency.

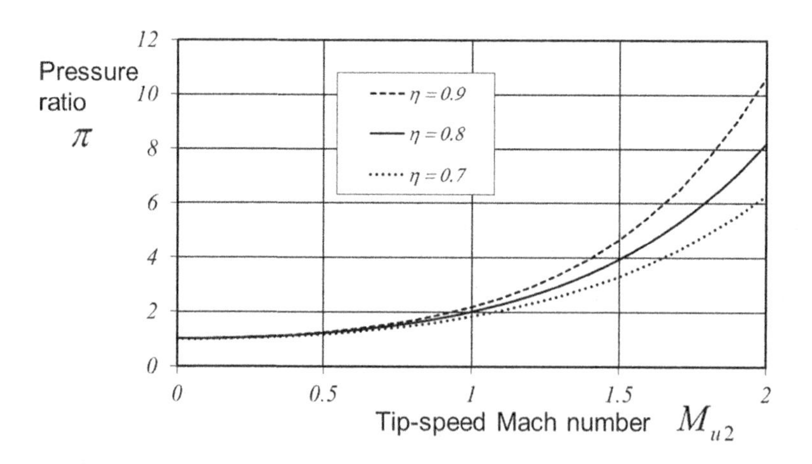

Figure 8.6 Pressure ratio as a function of tip-speed Mach number and polytropic efficiency for a stage with a work coefficient of $\lambda = 0.7$.

8.6.3 Compressor Performance Maps

The performance of a compressor stage depends on the geometry, flow rate, inlet temperature, inlet pressure, gas properties and speed of rotation. As is the case with most turbomachines, the clearest understanding can be obtained by the use of the nondimensional parameters introduced earlier. In many applications, the user is most interested in the efficiency, the pressure ratio and the volume flow, and these can be presented in the form of a compressor performance map. A compressor performance map gives the pressure ratio and efficiency as a function of the corrected volume flow for lines of constant corrected speed, as defined in (8.17), as follows:

$$\pi, \eta = f\left(\dot{V}_{corr}, n_{corr}\right). \tag{8.48}$$

In this diagram, each speed line represents a different corrected speed, and contours of constant efficiency can also be included to highlight the region of best efficiency. Such performance maps are dependent on the inlet conditions, the gas being compressed and the size of the compressor.

The compressor map can be defined for a single stage, as shown in Figure 8.7, but it is also used in multistage turbomachinery applications to define the performance of a whole machine. Important features of the performance maps of radial compressors are shown in Figure 8.7. The pressure ratio at the peak efficiency point increases approximately with the square of the speed, and the volume flow increases roughly linearly with speed (~ parabola) so that the pressure ratio is low at low-volume flows with low speed. As the volume flow decreases along a constant speed line (towards the left), the pressure ratio increases, but close to the peak pressure rise the compressor becomes unstable (surge or rotating stall). The surge line splits the map into an unstable region (on the left of the map) and a stable region (on the right). Compressor operation in the

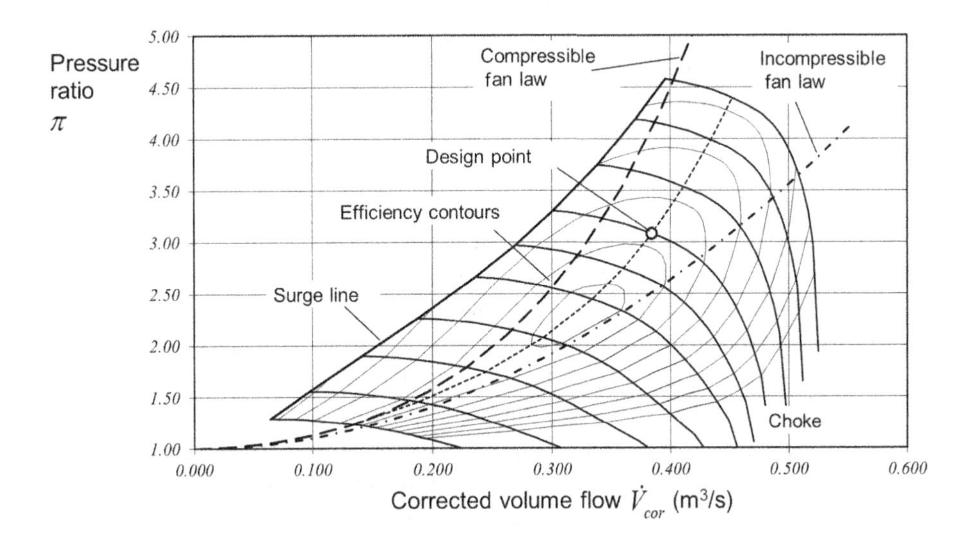

Figure 8.7 Performance map of a turbocharger stage with a vaneless diffuser.

unstable region is not allowed owing to high pressure and flow fluctuations, mechanical blade vibrations and noise. The maximum mass flow on the right of each speed line is limited by choking of the flow channels, which is indicated by vertical speed lines associated with very high gas velocities in the choked cross sections of the machine. The design point of a compressor should lie in the stable region close to peak efficiency, and with sufficient margin from the surge line and from choking, but in many applications, it is also necessary to operate at other points and ways of modifying the characteristic are needed; see Section 17.5. The operating range between the left-hand limit (surge) and the right-hand limit (choke) decreases as the speed increases, and a typical example is given in Section 17.4.

For practical applications, a wide margin between choking and surge is required, and the design of a single-stage compressor tries to achieve the maximum flow at high speeds and the minimum possible flow with an acceptable pressure ratio at lower and intermediate speeds. This inevitably requires a compromise as the physics of choking and the onset of instability limits what is possible. The design choices that lead to stable characteristics at low flow tend to reduce the maximum possible flow at high speeds, so the usable operating range is limited by both effects.

Figure 8.7 is not based on measurements but is calculated using the technique that is described in Section 18.3. It represents a typical turbocharger stage with a vaneless diffuser, a pressure ratio of $\pi = 3.1$ at the design point, a rotational speed of 91,765 rpm, an impeller diameter of 100 mm, a tip speed of 480 m/s and a tip-speed Mach number of 1.4. The stage shown is designed with a flow coefficient of $\phi_{t1} = 0.08$, a design work coefficient of $\lambda = 0.6$, and a backsweep angle of $\beta_2 = -45°$, which are close to the recommended values from Chapters 10 and 11. The polytropic efficiency at the design point is 0.834, which gives a pressure coefficient of $\psi = 0.5$ at the design point.

Contours of polytropic efficiency are plotted on the map. The peak efficiency occurs at about 88% of the design speed at a tip-speed Mach number of 1.24. At Mach numbers above design, the peak efficiency drops because of the additional shock losses associated with very high gas velocities in the inlet and the increased levels of diffusion in the impeller associated with high pressure ratios and associated density ratio. At lower Mach numbers, the peak efficiency also drops. This is essentially an effect of the reduced Reynolds number and mismatch of the impeller with the vaneless diffuser at low Mach numbers.

It is interesting to compare what happens if the pressure ratio and volume flow from the incompressible fan law given in (8.1) are plotted into the figure using the values of the pressure rise coefficient and flow coefficient from the peak efficiency point at low speed. This is shown in Figure 8.7 as the line marked 'Incompressible fan law', and along this line for a given flow the stage then produces a lower pressure ratio than the actual one. The main effect is that the binomial expansion given in (8.43) includes only the first term in the series, but as the tip-speed Mach number increases, the other terms become relevant and lead to a higher pressure. An obvious alternative would be to calculate this using (8.41), which is effectively a compressible form of the fan law. In this case, which is also shown in Figure 8.7 and marked 'Compressible fan law', the volume flow for a given pressure ratio is lower than the actual value. The main effect here is that as the tip-speed Mach number increases, the flow coefficient at the best point does not stay constant. It increases as the speed lines shift to higher flow, as is shown in Section 18.3.5.

8.6.4 Stage Characteristic Curves

For single-stage compressors, an alternative form of performance presentation can be made based on the inherent nondimensional parameters that have been introduced in this chapter. In this book, these are referred to as the compressor characteristic curves, characteristics and colloquially simply as chics. For a compressor stage, the functional relationship that describes the characteristic performance can be obtained by dimensional analysis (see Dixon and Hall, 2010; Dick, 2015) as

$$\psi_p, \lambda, \eta_p = f(\phi_{t1}, M_{u2}, Re, \gamma, L_1/D_2, L_2/D_2, \ldots), \tag{8.49}$$

where L_1/D_2, L_2/D_2 and so on are geometrical ratios representing the geometry of the stage, including the blade angles. Some of these geometrical terms are shown in Figure 8.3, such as the outlet width ratio, b_2/D_2; the axial length ratio, L/D_2; the inlet casing and hub diameter ratios, D_{1c}/D_2 and D_{1h}/D_2; and the tip clearance ratio, ε/b_2, together with the impeller inlet and outlet blade angles. The significance of the other terms in this equation has been explained previously. The assumption is that a given machine has the same performance when operating on different gases or with different inlet conditions, provided the nondimensional parameters have a similar value. In practical applications for a specific turbocompressor stage with a defined geometry at a certain size, (8.49) is often simplified to

$$\psi_p, \lambda, \eta_p = f(\phi_{t1}, M_{u2}). \tag{8.50}$$

The pressure rise coefficient may be omitted as it is the product of the efficiency and the work coefficient. For an incompressible flow the tip-speed Mach number may be omitted as it has no effect, and it has only a small effect at low Mach numbers in ventilators and fans. The effect of the secondary parameters, Re and γ, is neglected in the first instance, but is discussed in the sections that follow. In this conventional form, the efficiency, work coefficient and pressure coefficient are functions of the flow coefficient and tip-speed Mach number. The flow coefficient determines the point on the characteristic being considered and large changes in performance are apparent when the flow coefficient changes. In turbocompressors, the nondimensional characteristic curves at different Mach numbers may show large differences in the shape of the characteristics, becoming much narrower at high Mach numbers.

The characteristics are usually plotted as a diagram in the form of (8.50), that is. efficiency, work coefficient and pressure coefficient versus the inlet flow coefficient for different tip-speed Mach numbers. Such diagrams are then more or less independent of gas properties, inlet conditions and size, except for any effects of the Reynolds number. The same data for the radial compressor stage already plotted as a compressor map in Figure 8.7 are plotted in the form of stage characteristics in Figure 8.8.

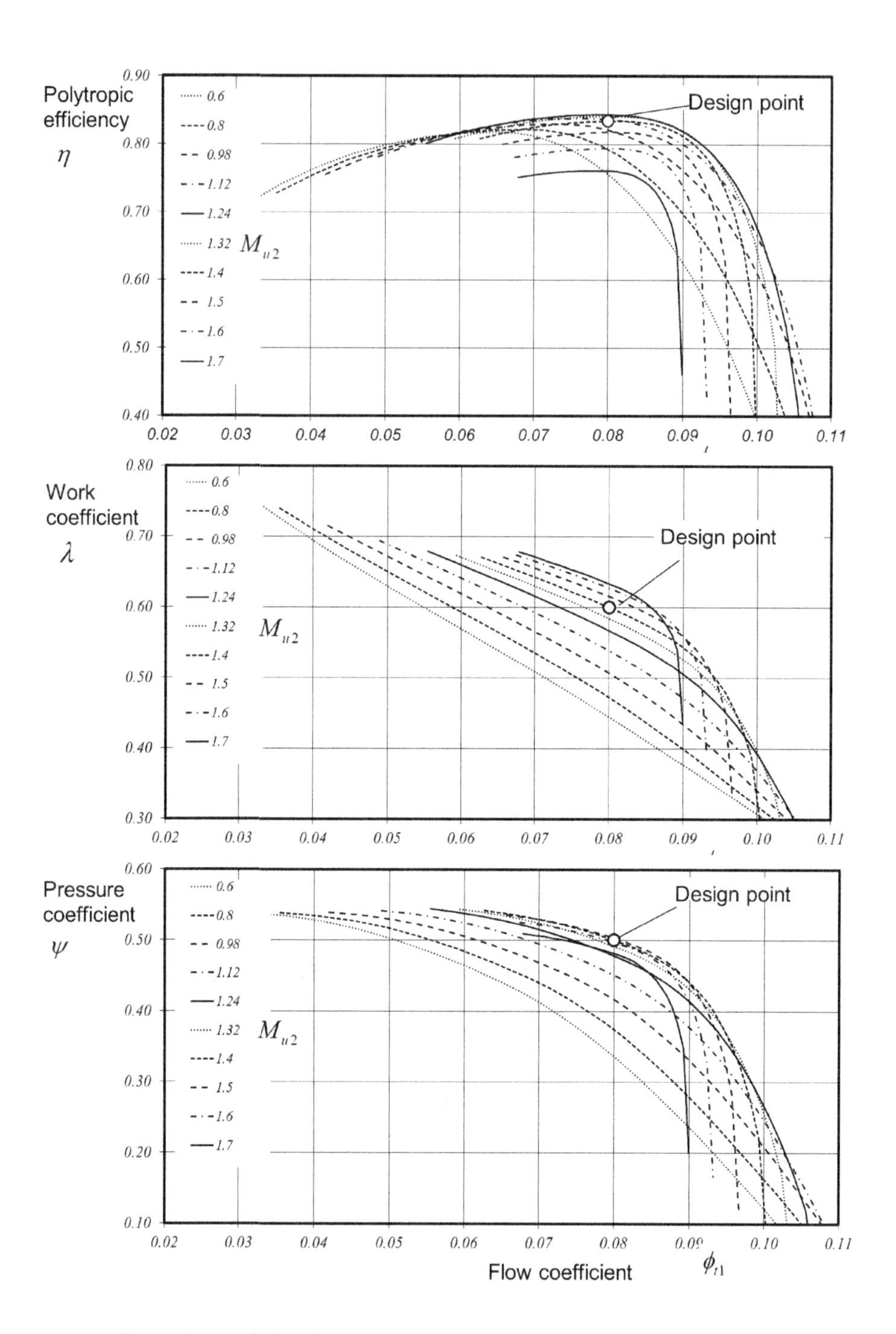

Figure 8.8 Stage characteristics of a turbocharger stage with a vaneless diffuser.

This approach to scaling the characteristics is only accurate where strict geometric similarity is maintained between the stages under consideration. As mentioned earlier, exact scaling of the components is, however, not always possible since certain dimensions cannt be assumed to scale in proportion to the impeller outlet diameter. Corrections may therefore be needed for changes both in the relative roughness of the wetted surfaces, Ra/D_2, and for the change in the clearance ratio, ε/b_2, where Ra is the arithmetical mean roughness and ε is the clearance between the impeller blades and casing of an open impeller.

In a compressible flow, the value of the work coefficient and the flow coefficient at the best efficiency point at different speeds are no longer constant, as shown in the pressure rise stage characteristics of Figure 8.8. These values become a function of the tip-speed Mach number as a result of the change in thermodynamic similarity at different speeds. This feature is not visible in the plot of the same data shown in the compressor performance map. The characteristic values do not vary much, and in fact, if they did vary a great deal this would completely negate the usefulness of such nondimensional parameters for the analysis of characteristics. In the case shown, the stage has a peak efficiency close to $\phi_{t1} = 0.08$ for nearly all tip speeds, with a pressure rise coefficient at this flow of close to $\psi = 0.5$ on the high-speed characteristics. These are typical values for turbocharger stages and are the same as the design values for this particular stage. Nevertheless, the values do vary slightly with the tip-speed Mach number, and it is this small variation which makes the character and anatomy of the compressor performance maps at high speeds different from those of pumps and low-speed ventilators.

The work coefficient changes with Mach number at a fixed flow coefficient close to the optimum flow, near $\phi_{t1} = 0.08$. Figure 8.8 illustrates that the work input coefficient of the impeller at this flow coefficient progressively increases as the Mach number rises. This is related to the increase in density at the outlet of the impeller with increasing speed and the effect that this in turn has on the impeller outlet velocity triangle; see Section 11.6.

An interesting effect is that of the Mach number on the work input coefficient at very high flows, where the impeller chokes. The work coefficient curves in Figure 8.8 for the lowest Mach number show that the work coefficient drops linearly as the flow increases, just as in a low-speed ventilator or pump, and choking does not occur. At higher Mach numbers, however, the work coefficient, λ, drops more rapidly towards the maximum flow end of the speed lines (as it does for all for backswept impellers). This is due to sonic conditions being reached in the throat of the impeller, such that no further increase in mass or volume flow is possible at the inlet. The inlet mass flow therefore no longer continues to increase as the exit pressure is reduced. The choking in the impeller inlet causes a fall in pressure and density across the associated shock waves. This in turn leads to an abrupt increase in volume flow at the impeller outlet and a corresponding decrease in λ at high inlet flow coefficients. The same data plotted as a function of the outlet flow coefficient would not show this drop but would continue to show a linear relationship, as discussed in Chapter 11.

Next the change in the efficiency with the tip-speed Mach number is considered. This stage achieves its peak polytropic efficiency at a tip-speed Mach number of approximately $M_{u2} = 1.24$. Higher Mach numbers reduce the peak efficiency because of the additional shock losses associated with very high gas velocities in the inlet and the increased levels of diffusion in the impeller. At lower Mach numbers, the peak efficiency also drops, and this is partly an effect of the reduced Reynolds number.

A further effect on the peak efficiency and, in particular, on its location on the characteristic curve is the matching of the impeller and diffuser. At the design Mach number, near the peak efficiency of the stage, the impeller and diffuser are designed such that the peak efficiency of both components occurs together. At other Mach numbers, changes in matching between the components occurs due to the change in density (and therefore in volume flow) at the impeller outlet. If, as before, the situation where the Mach number increases at a constant inlet flow coefficient is considered, then the density change causes an increase in the work coefficient and a decrease in the outlet flow coefficient, which in turn gives rise to an increase in the flow angle at diffuser inlet. At high Mach numbers, therefore, the diffuser then has a more highly swirling flow and wants to swallow more flow, and at low Mach numbers it has a more radial inlet flow and wants to swallow less. This effect means that compressor characteristics generally pass more flow at high speed and to be flatter than would be expected from the application of compressible fan laws.

Inspection of the efficiency characteristics in Figure 8.8 shows that the flow coefficient for peak efficiency moves from 0.06 to 0.08 as the Mach number increases. In a low-speed machine, the flow coefficient would remain at a fixed value. The effect of this has already been shown in the performance map of Figure 8.7 (the dashed line marked as 'Compressible fan law'). This locus lies close to peak efficiency at low speed but to the left of the peak efficiency or even close to the surge line at higher speeds. This type of mismatch between the components also leads to characteristics that are flatter than the fan-law parabola in multistage compressors (see Lüdtke, 2004). In fact, this effect due to compressibility helps the use of radial compressors in turbochargers as it causes the low-speed characteristic to reach to lower flows and the higher-speed characteristics to occur at higher flows than would otherwise be available, and actually aids matching with an engine. In this way, the effect of compressibility leads to a flow rate at high speeds that is 20–30% higher than would be achieved with an incompressible fluid, or alternatively a flow rate at lower speeds that is 20–30% lower. This effect of a change in matching between the components has been demonstrated here using the example of a stage with a vaneless diffuser. Vaneless diffusers are generally able to operate over a wider range than vaned diffusers at levels of impeller exit absolute gas angle where both are an option. For turbochargers in large diesel engines, it is usual to try to achieve a higher efficiency through the use of a vaned diffuser and, as shown in Section 18.3.5, this mismatching effect causing a shift to lower flows on the low-speed characteristics is much stronger with vaned diffusers.

A further aspect of the effect of Mach number is the clear trend demonstrated in Figure 8.8 towards a much narrower operating range as the Mach number increases, such that the low Mach number curve has an operating range between surge and choke that is much wider than at a higher Mach number. Details of this are discussed in Section 17.4. The reduced range at high Mach numbers is related to the decrease in the maximum flow coefficient due to the effect of choking and to the increased sensitivity of the losses to incidence as the Mach number rises. This is a general feature of all compressor blade rows and is well documented for both axial and radial compressors; see Section 7.3.3.

8.6.5 Similarity in Performance Testing

Despite the high quality of modern CFD simulations, it is common practice that ultimately a test program is used to judge the quality of a new compressor design, as described in Chapter 20. In addition, the tests allow fine modification or adjustment of components, such as an adjustment of the diffuser inlet angle to provide more precise control over the matching of the impeller with the diffuser. Compressor tests are of two basic types: flange-to-flange tests to determine the overall performance and internal component tests to determine the inner fluid behaviour. The test results, along with those from other test configurations, are used to set development objectives and to develop improved design tools and correlations.

For large industrial multistage compressors, acceptance tests are often made by the manufacturer together with the customer on the full-scale machine. Machines that are due to be delivered, or are already on site, are tested to prove that the specified operation and performance guarantees have been met. As it is not always possible to carry out such tests at full load with the actual process gas, tests with a model gas in dynamic similarity with the real gas are carried out (so-called Class 2 tests). The principles of the nondimensional parameters are then used to correct the measured performance relationship between pressure ratio, speed and volume flow to that expected with the real process gas.

Component tests can also be made on the original machine, but it is more usual to carry out these measurements as rig tests on geometrically similar, smaller machines. The advantage of model tests is that smaller components can be used, cutting down on manufacturing costs and the power and operating costs of the testing process. Use of production components for research testing runs the risk of delays to the delivery to the end user. More detailed measurements are normally more possible in a model test campaign than on a production machine, and ultimately design improvements can be made before the real machine is committed to manufacture. In some cases where the original machine is small, rig tests may be made on a scaled-up version of the original machine to facilitate the inclusion of more detailed instrumentation or simply to make use existing rig hardware (Medic et al., 2014; Diehl, 2019)

For geometrical similarity, the testing should clearly be carried out on a pure, linear scale model of the original machine. This model has a scale factor, k, such that all dimensions are scaled down by the same factor, $L = kL_m$. The blade angles are the same, areas scale according to $A = k^2 A_m$, and volumes scale according to $V = k^3 V_m$, so that the mass of the test rig, which partly determines the costs, also scales with k^3.

For thermodynamic similarity, the model machine should run with the same tip-speed Mach numbers as the actual machine. If the gas is the same and has the same inlet conditions, then the rotational speed must be increased to retain the blade tip speed, so that $u = u_m$. As a consequence, the gas velocities and the Mach numbers in the flow are the same. The power consumption of the model then scales with $P = k^2 P_m$, as does the volume flow rate and the mass-flow rate. The Reynolds number inevitably cannot be maintained, and this scales with the scale factor, such that $Re = kRe_m$, which is dealt with by a correction factor.

Interestingly, the use of thermodynamic similarity and geometric similarity lead to similarity of centrifugal forces and bending stresses, as described in Chapter 19. The blade natural frequencies of the model also retain the same relationship to the full-scale harmonics.

8.6.6 Scaling and Power Density of Small Machines

An important application of similarity parameters is in the scaling of machines from one size to another. There are some general trends for the changes in volume flow rate, enthalpy rise, torque and power that can be derived from the nondimensional parameters in the equations in this chapter, as follows:

$$\dot{V} \propto u_2 D_2^2, \quad \Delta h_t \propto u_2^2, \quad M \propto \rho u_2^2 D_2^3, \quad P \propto \rho u_2^3 D_2^2. \tag{8.51}$$

From (8.24), it can be seen that for a constant tip speed the power required by the compressor scales with the square of the impeller diameter. The power density is the power required for a given volume of machine and is of interest for such small machines. The volume of the compressor is proportional to the cube of the impeller diameter, $V_c = k_c D_2^3$, and the power density is given as

$$\frac{P_c}{V_c} = \frac{P_c}{k_c D_2^3} = \frac{\rho \phi \lambda u_2^3 D_2^2}{k_c D_2^3} = \frac{\rho \phi \lambda u_2^3}{k_c} \frac{1}{D_2}. \tag{8.52}$$

A new application of centrifugal compressors is the application of ultra-high-speed miniature centrifugal compressors with an impeller diameter less than 30 mm using high-speed electric motors to provide rotational speeds between 200,000 and 600,000 rpm. Casey et al. (2013) have examined the scaling rules with regard to such machines. Clearly from (8.52), a geometrically similar machine which is downsized but operates at similar nondimensional operating points (with the same flow coefficient and work coefficient) and with the same blade speed (to attain the same pressure rise) will have a higher power density as the size decreases. However, the actual volume flow and mass flow reduce as the size decreases, so several machines may be required for the same overall duty. As an example, if the angular velocity is doubled, reducing the size by one half, then four smaller machines would be required for the same duty. The machine volume and weight for the same duty is proportional to the cube of the diameter and reduces by a factor of eight, with the result that the four high-speed machines still have a weight and volume advantage over the lower-speed machines. This interesting and somewhat surprising result has also been pointed out by others working in this field (Epstein, 2004; Zwyssig et al., 2008).

The more challenging case of a microcompressor application is one in which the physical mass flow of the compressor remains the same as the machine is downsized. This is achieved by moving to a design with a higher flow coefficient, that is, geometrical similarity is not present, then

$$\frac{P_c}{V_c} = \frac{P_c}{k_c D_2^3} = \frac{\dot{m}\lambda u_2^2}{k_c D_2^3} = \frac{\dot{m}\lambda u_2^2}{k_c}\frac{1}{D_2^3}. \tag{8.53}$$

Here there is a much more rapid increase in the power density with reduction in size. A doubling of the angular velocity, reducing the diameter by one half, requires a single machine, but the power density goes up by a factor of eight with a massive weight and volume advantage. This is an important factor in trying to design machines with a higher flow coefficient.

In high-speed motor design, the torque required has a strong influence on the motor size. It is often of interest to design not for minimum power consumption but for minimum torque, as required torque is roughly proportional to the volume of the electrical machine (Rahman, 2004). If we consider the case with a fixed duty (that is, the mass flow and head rise are fixed, and with a fixed work coefficient the tip speed is also fixed), then we obtain

$$M = \frac{P_c}{\omega} = \frac{\dot{m}\lambda u_2^2}{2u_2/D_2} = \frac{\dot{m}\lambda u_2}{2}D_2. \tag{8.54}$$

This indicates that the torque required for a given application scales with the impeller diameter, giving lower torque for smaller machines, so from the point of view of the electrical machine there is an advantage to use smaller sizes. Note that these very general conclusions may be slightly affected by any changes in flow coefficient and efficiency, causing a different head coefficient as the size decreases. Some applications of these arguments to microcompressor designs are given by Casey et al. (2010) and Casey et al. (2013).

8.6.7 Corrected Flow

The volume flow rate of a compressor is a clearly defined term, and for a certain speed and size it remains essentially the same at the peak efficiency point independent of the inlet conditions. However, the mass-flow rate requires additional information for its meaning to be clear. Gases are compressible, which means that a given volume flow rate does not indicate the mass-flow rate, which depends on the inlet gas density. The compressor delivers the same volume, though not mass, at any other inlet density.

The corrected flow, (8.17), is meant to relate all compressors to a common set of conditions independent of the actual conditions that occur in the real application. These conditions are often taken as those at the international standard atmospheric conditions at sea level, corresponding to a pressure of 101,325 Pa and a temperature of 15 °C (288.15 K). The relative humidity (e.g., 30% or 0%) is also included in some definitions of standard conditions, and other standard definitions can be found. In countries using the SI metric system of units, the term 'normal cubic metre' (Nm^3) is very often used to denote gas volumes at some normalised or standard condition. In the USA, the units are standard cubic feet per minute (SCFM), at the standard conditions of 14.696 psia, 60° F and 0% relative humidity. Since there is no universally accepted set of normalised or standard conditions, these need to be specified. Confusion is assured.

8.7 Performance Corrections for Deviation from Similarity

8.7.1 Generalised Approach

In practice, it is not always possible to maintain similarity and a way of dealing with deviations from similarity is required. Dalbert et al. (1988) describe a general procedure for the correction of characteristics for a deviation from similarity as follows:

$$\eta_p = \eta_{pB}\left[\left(\frac{\eta_p}{\eta_{pB}}\right)_{M1}\left(\frac{\eta_p}{\eta_{pB}}\right)_{M2}\left(\frac{\eta_p}{\eta_{pB}}\right)_{M3}\cdots\right] + \left[(\Delta\eta_p)_{A1} + (\Delta\eta_p)_{A2} + (\Delta\eta_p)_{A3} + \cdots\right]$$

$$\phi_{t1} = \phi_{t1B}\left[\left(\frac{\phi_{t1}}{\phi_{t1B}}\right)_{M1}\left(\frac{\phi_{t1}}{\phi_{t1B}}\right)_{M1}\left(\frac{\phi_{t1}}{\phi_{t1B}}\right)_{M1}\cdots\right] + \left[(\Delta\phi_{t1})_{A1} + (\Delta\phi_{t1})_{A2} + (\Delta\phi_{t1})_{A3} + \cdots\right]$$

$$\lambda = \lambda_B\left[\left(\frac{\lambda}{\lambda_B}\right)_{M1}\left(\frac{\lambda}{\lambda_B}\right)_{M2}\left(\frac{\lambda}{\lambda_B}\right)_{M3}\cdots\right] + \left[(\Delta\lambda)_{A1} + (\Delta\lambda)_{A2} + (\Delta\lambda)_{A3} + \cdots\right], \quad (8.55)$$

where the values from the base characteristic with the functional relationship given by (8.50) have the suffix B. Corrections are applied for the other effects, which may be multiplicative corrections, $M1$, $M2$, $M3$ and so on, or additive corrections, $A1$, $A2$, $A3$, and so on. In most cases, only one or two of the different forms of correction factors are used, and the choice between a multiplicative or additive correction depends on an analysis of the physics related to the phenomenon being corrected.

The corrections may be necessary for small effects of Reynolds number and gas properties and for small changes in geometric similarity, such as a change in the tip clearance gap of an open impeller. The inclusion of a term expressing the shift in flow coefficient is useful. Many deviations from similarity lead to changes in performance due to a shift in the flow characteristic, along the lines of the corrections for the effects of blockage due to tip clearance leakage flow in axial compressor stages given by Smith (1970). It can be generally assumed that the corrected curves with a shift in the flow coefficient retain the same shape, that is, a correction term derived for the change at the peak efficiency point is applicable to all points on the characteristic as reinforced by the examples that follow.

These equations consider the primary importance of the flow coefficient, ϕ_{t1}, and the tip-speed Mach number, M_{u2}, on the performance, while relegating secondary fluid effects such as the Reynolds number to the role of corrections. This prediction technique provides a practical and accurate method of calculating the performance of radial compressor stages at both design and off-design conditions. Systematic performance measurements on several members of a standardised impeller family at different aerodynamic conditions are needed as a database for interpolation. The use of nondimensional parameters reduces the number of dependent variables required for the interpolation. A variety of different off-design problems can then be investigated using the same database. For example, the change in performance can be calculated for changes in flow rate, gas, inlet temperature, inlet pressure and rotational speed.

8.7.2 Corrections for Geometrical Changes on Performance

The functional equation (8.49) can be used to consider many effects on performance. The simplified equation given as (8.50) considers only the primary importance of the tip-speed Mach number, M_{u2}, and the flow coefficient, ϕ_{t1}, on the performance and the shape of the characteristics, and these form part of the performance maps and characteristic curves. The effect of Reynolds number, Re, and the isentropic exponent, γ, are neglected in this since these are secondary influences. Their effects can be included as a small correction to the basic characteristics, as described in this section. Small changes in the geometry may also be considered as a correction to the basic performance curves.

Consider as an example the change in efficiency due to a change in tip clearance between the blades and the casing. A Taylor expansion of the functional relationship in (8.49) gives

$$\eta = \eta_{ref} + \frac{\partial f}{\partial(\varepsilon/b_2)}\delta(\varepsilon/b_2) + \cdots, \tag{8.56}$$

where η_{ref} is the performance at a certain reference level of clearance. Neglecting the higher-order terms, a correction for the change in performance from the reference level can be determined. The actual correction can be determined using a component test for a specific geometry and aerodynamic conditions. With experience, this may then be applied to other geometries, perhaps with a further correction factor. The simplest linear correction formula is

$$\delta\eta = \eta - \eta_{ref} = K_{cl}\left[\left(\frac{\varepsilon}{b_2}\right) - \left(\frac{\varepsilon}{b_2}\right)_{ref}\right]. \tag{8.57}$$

The coefficient K_{cl} is sometimes known as the exchange rate for clearance effects on efficiency. For radial compressors, the value often used is $K_{cl} = -0.3$, indicating that a 1% increase in relative clearance, $\Delta\varepsilon/b_2$, produces a 0.3% loss in efficiency. This coefficient value has been examined in many publications, following on from the original suggestion by Pampreen (1973), and the sensitivity to a change in clearance varies strongly from stage to stage with no clear correlation with many stage design parameters (Lou et al., 2019). A recent study of tip clearance effects by Diehl (2019) shows that a correction for the efficiency is accompanied by a shift in the flow to lower flow coefficients and a decrease in the work input of the impeller. As the clearance increases, the blockage of the flow channel by the tip clearance vortex increases, so a work coefficient and a flow coefficient correction are also needed.

As a further example, the effects of variable inlet guide vanes (VIGV) or variable diffuser vanes on the efficiency, work coefficient and flow coefficient can be similarly dealt with. The performance maps as shown in Figure 8.7 can be repeated with different variable inlet guide vane setting angles, a_{igv}. As the effects of the guide vanes are known to be strong, an exact procedure would include a further parameter in the data for the base characteristics, giving rise to an additional dimension in the functional relationship:

$$\psi_p, \lambda, \eta_p = f\left(\phi_{t1}, M_{u2}, b_2/D_2, a_{igv}\right). \tag{8.58}$$

A totally comprehensive test campaign covering all stages available to a company, with all possible outlet width ratios or trims, at all Mach numbers and all VIGV settings, is impracticable, both in terms of cost and timescale and in terms of the large amount of data that would be generated requiring reduction. In addition, it would be difficult to deal with such data effectively within any compressor selection software. A simpler method is to interpret the change in performance due to VIGVs as a correction of the base characteristics. This is rather different from the correction for Reynolds number as the correction factors are significantly larger. This is discussed in the section on VIGV and variable diffuser control in Chapter 17, where Sections 17.5.4 and 17.5.5 show the approach.

There are other situations with a deviation from geometric similarity, as described by Dalbert et al. (1988) that can similarly be dealt with by correction factors to the base characteristics. These include small changes in the diffuser vane setting angle relative to that originally measured, a change in the inlet or outlet casing geometry, modification of labyrinth clearances, modification of the hub disc diameter to allow for a thicker shaft in a multistage compressor and cutback of the impeller blade trailing edges. A change in the outlet volute from that originally used for the initial measurements could also be accounted for in this way. The extent to which such correction factors (exchange rates) can be universally applied depends on the degree to which the effect has a fundamental effect on the performance leading to a drastic change in the shape of the map. Whether such a correction may be applied to a different compressor stage operating at a different flow condition is a matter of judgement and experience. Nonetheless, this procedure provides a logical way of allowing for the effects of deviation from exact geometric similarity.

8.7.3 Correction for the Effect of the Reynolds Number

An example of such a correction for a fluid dynamic effect is shown in Figure 8.9 for the change in the efficiency characteristic with the Reynolds number. In this case, the test data are for a low flow coefficient stage with $b_2/D_2 = 0.011$ from Casey (1985)

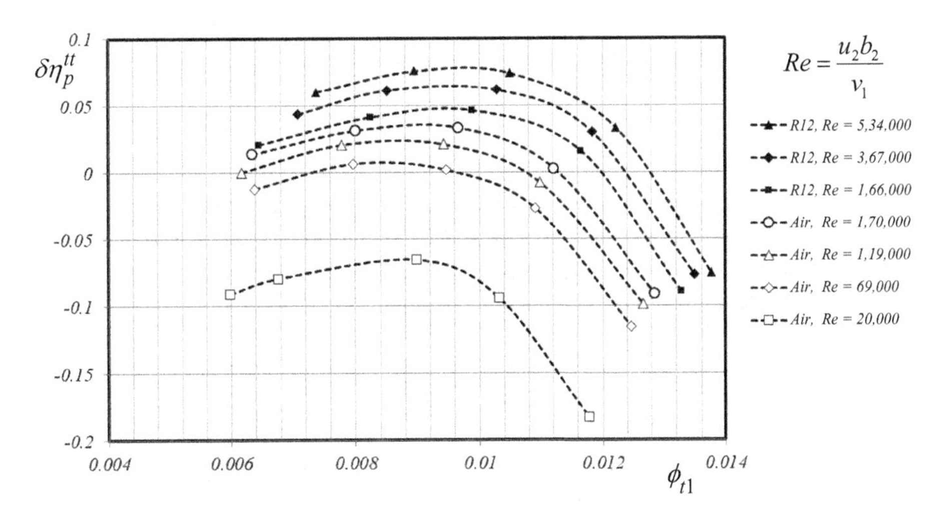

Figure 8.9 Reynolds number effect on efficiency in a low-flow coefficient stage. (image using data from Casey, 1985, with courtesy of MAN Energy Solutions)

showing a change in efficiency of 15 percentage points over the range of Reynolds number tested. Low flow coefficient stages are sensitive to the effect of Reynolds number as the frictional losses are high due to an increase in the ratio of the end-wall surface area to that of the blades as the flow coefficient decreases (see Chapter 10). Experiments on compressors across a range of Reynolds number do indeed show that the basic shape of the performance curve stays the same such that the corrections can be derived from the change in the peak efficiency point, and the same correction can be applied to all points on the characteristic. The performance curves adjust their position to correct for the effect of Reynolds number, and following Dietmann and Casey (2013), this can be corrected with an additive correction for the efficiency and a multiplicative correction for the flow coefficient.

8.7.4 Correction for the Effect of the Isentropic Exponent

The isentropic exponent is included in the gas dynamics equations determining the pressure and temperature ratio of the stage. This ratio influences the temperature and pressure rise that occur for a given Mach number. An additional condition for perfect fluid dynamic and thermodynamic similarity is that the isentropic exponent does not change. If a turbomachine operates on a variety of gases or gas mixtures with variable specific heat ratios or isentropic exponents, then its effect has to be considered separately from the effect of the Mach number but, as this effect is small, it is usually considered a correction to the performance. In this case, measurements show that the change in isentropic exponent at the same tip-speed Mach number and Reynolds number simply shifts the stage characteristic due to the change in the choke flow, which means a multiplicative correction of the flow coefficient is required.

From (8.15), for given inlet total conditions, the maximum flow through a channel with throat area A is given by

$$\Phi_{t1} = \phi_{t1} M_{u2} = \frac{\dot{m}_{\max}}{A \rho_t a_t} = \left(\frac{2}{\gamma + 1} \right)^{\frac{\gamma+1}{2(\gamma-1)}}. \tag{8.59}$$

Considering two different gases with different isentropic exponents, a and b, but the same throat area and tip-speed Mach number results in different maximum mass flows. The ratio of the two mass flows to the maximum flow coefficients for two fluids with isentropic exponents of a and b are given by

$$\frac{\dot{m}^a_{\max}}{\dot{m}^b_{\max}} = \frac{\phi^a_{t1}}{\phi^a_{t1}} = \frac{[2/(a+1)]^{\frac{a+1}{2(a-1)}}}{[2/(b+1)]^{\frac{b+1}{2(b-1)}}}. \tag{8.60}$$

This equation demonstrates, for example, that a change from air (with isentropic exponent $a = 1.4$) to CO_2 (with $b = 1.29$) gives an increase in the choking mass flow of 1.24%. A change from air (with $a = 1.4$) to argon (with $b = 1.67$) gives a decrease in the choking mass flow of 2.92%. A change from air to a refrigerant gas (with $b = 1.135$) gives an increase of 3.09%.

The next step is to consider the pressure ratio. The key equation defining the pressure ratio is (8.41). This can be written for a gas with isentropic exponent a, polytropic efficiency η_a and work coefficient λ_a, as follows

$$\pi_a = \left[1 + (a-1)\lambda_a M^2_{u2}\right]^{\eta_A a/(a-1)}. \tag{8.61}$$

Repeating this for another gas with a polytropic exponent of b, and assuming that both machines operate at the same tip-speed Mach number leads to

$$\pi_b = \left[1 + \frac{(a-1)}{(b-1)} \frac{\lambda_a}{\lambda_b} \pi_A^{(a-1)/\eta_A a}\right]^{\eta_B b/(b-1)}. \tag{8.62}$$

Although the nondimensional stage characteristics are similar as the stage is operating at the same Reynolds number and tip-speed Mach number, the pressure ratio with a change in the isentropic exponent is different, as shown in Section 8.6.2. In fact, the values of the flow coefficient of the two stages cannot be the same given that, from (8.59), the choking mass flow has changed. Thus, the efficiency of the stages and the work coefficient may be different as they operate at different flow coefficients with different gases. Assuming that the flow coefficient of the two stages has moved in line with the difference in the choking mass flow, then the shift in the stage characteristic for gas b from that of gas a can be calculated. The shape of the nondimensional characteristic is not changed but just shifted in flow by the change in the isentropic exponent. This is not entirely valid but nevertheless is a good approximation, as shown in Figure 8.10. This procedure is recommended by Roberts and Sjolander (2005) and Backström (2008) as the method to calculate the change in pressure ratio for a change in working fluid.

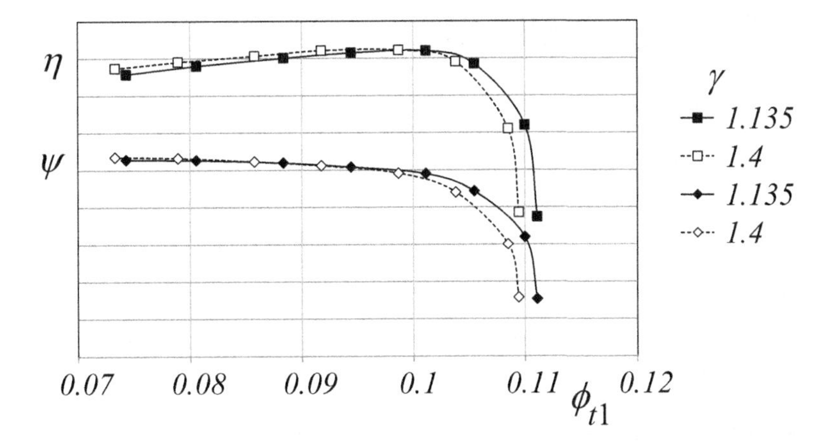

Figure 8.10 Change in the efficiency and pressure rise characteristic of an industrial vaneless stage at $M_{u2} = 1$ with a change in the isentropic exponent, γ, from 1.4 (air) to 1.135 (R12). (image using data courtesy of MAN Energy Solutions)

A similar problem is that of selecting a speed at which a compressor should be operated to give the same pressure ratio when operating on a different gas. The identification of the speed required to attain the same pressure ratio is required during the testing of a compressor operating on a substitute gas. This is given by solution of

$$\pi_a = \left[1 + (a-1)\lambda_a M_a^2\right]^{\eta_a a/(a-1)} = \pi_b = \left[1 + (b-1)\lambda_b M_b^2\right]^{\eta_b b/(b-1)}. \tag{8.63}$$

This results in the following equation:

$$M_b^2 = \frac{\left[1 + (a-1)\lambda_a M_a^2\right]^{\frac{\eta_a a(b-1)}{\eta_b B b(a-1)}} - 1}{(b-1)\lambda_b}. \tag{8.64}$$

This is essentially the equation given in the ASME performance test code, ASME PTC10 (1997), discussed in Chapter 20.

9 Specific Speed

9.1 Overview

9.1.1 Introduction

In many books and technical papers on radial turbocompressors, the nondimensional parameter known as specific speed is often used to categorise a particular type of design. The specific speed and specific diameter are two alternative dimensionless coefficients based on the same data which are used to define the flow coefficient and the head coefficients as described in Chapter 8. In their definition, both the flow coefficient and the head coefficient are included. The specific speed alone is often used to specify a particular type of design of radial compressor as, broadly speaking, an optimum specific speed can be defined, rather like an optimum flow coefficient. This optimum is often presented in the form of a Cordier diagram.

Both parameters are essentially incompressible in nature and are often used for hydraulic machines and in pump and ventilator design. In this chapter, the background to these parameters is described. This discussion gives clear guidance that the flow coefficient, work coefficient and tip-speed Mach number are more useful than specific speed and specific diameter and explains why the authors do not use specific speed in this book.

9.1.2 Learning Objectives

- Know the definitions of specific speed and specific diameter.
- Understand the background to the Cordier diagram.
- Be aware of the difficulty of using the concept of specific speed for compressor flows and the limitations of the related Cordier diagram.

9.2 Specific Speed and Specific Diameter

9.2.1 Background

The four key items of data needed to define a single-stage compressor application are the pressure rise, or isentropic enthalpy rise, Δh_s; the inlet volume flow rate, V_{t1}; the rotor casing diameter at impeller outlet, D_2; and the rotor angular velocity, ω. The first two represent the duty of the compressor, and the second pair define the machine design to meet this duty, which can be met by many combinations of size and speed. The isentropic enthalpy rise is used to define the duty in this chapter as it is directly related to the pressure rise but does not require any definition of efficiency in its specification; see Section 4.3.

To help with the selection of an optimum design, there are four alternative dimensionless coefficients which can be usefully defined on the basis of these dimensional parameters. The simplest and easiest to understand are the nondimensional isentropic enthalpy rise coefficient and global flow coefficient, which are defined in Section 8.2. These represent the nondimensional pressure rise and nondimensional inlet volume flow rate for a certain impeller size and impeller tip speed. Well-designed radial compressors have specific values of these parameters; for example, optimum values are considered in Sections 10.7.1 and 11.6 and are found to be close to $\phi_{t1} = 0.08 \pm 0.01$ and $\psi_s = 0.55 \pm 0.05$.

The specific speed and specific diameter are two alternative dimensionless coefficients based on the same data, which are often used for hydraulic machines. Both are essentially incompressible in nature and have been used extensively in pump and ventilator design (Bommes et al., 2003; Gülich, 2008). For hydraulic machines and low-speed fans, there is no change in density through the machine, so the inlet volume flow is the same as the flow rate everywhere in the machine. In compressible flow, this is not the case. In compressors designed for a high pressure ratio with a high tip-speed Mach number, there may be a considerable change in volume flow rate from inlet to outlet. A choice must then be made as to whether the specific speed is defined on the basis of the inlet volume flow, the outlet volume flow or the mean volume flow. Balje (1981) suggests that the largest value is taken whereas Rodgers (1980) uses the mean volume. Here the inlet volume flow is used.

9.2.2 Definitions

The specific speed and specific diameter are chosen to represent a form of nondimensional speed and a nondimensional size of the impeller for a certain pressure rise and volume flow rate and are defined as follows:

$$\omega_s = \omega \frac{\left(\dot{V}_{t1}\right)^{1/2}}{\left(\Delta h_s\right)^{3/4}}, \quad D_s = D_2 \frac{\left(\Delta h_s\right)^{1/4}}{\left(\dot{V}_{t1}\right)^{1/2}}. \tag{9.1}$$

The merit of the specific speed is that all single-stage centrifugal compressors have a value between certain limits, say 0.3 and 2.0. Lower and higher values would lead to impossible radial compressor designs. So its main use is to find out if a radial compressor can be built for the specified duty or whether another type of compressor is needed or other design specifications should be considered. When the rotational speed is specified due to the nature of the driver, this is very useful, but in many applications the choice of rotational speed is part of the design process and then specific speed is less useful. Just as there is an optimum flow coefficient, there is also an optimum specific speed for a radial compressor which is close to $\omega_s = 0.75$, according to Rodgers (1980), but note that his definition is based on the mean volume flow between inlet and outlet, and the numerical value would be higher if it was based on the inlet flow.

The definition in (9.1) is the same as that given by Balje (1981), but Balje gives the specific speed the symbol n_s, and this terminology is quite common. However, the angular velocity is expressed in this equation in the units of radians per second, so the numerical values of the specific speed need also to be thought of as radians per second. In fact, a more correct terminology for this would be the 'specific angular velocity', and for this reason it is given the symbol ω_s (and not n_s) in this book.

This nondimensional parameter depends only on three aspects of the turbomachinery duty: the volume flow, the rotational speed (in radians per second) and the isentropic head rise, because the effect of size is eliminated. Noting that

$$
\begin{aligned}
u_2 &= \omega r_2 = \omega D_2/2 \\
\dot{V}_{t1} &= \phi_{t1} u_2 D_2^2 = \phi_{t1} \omega D_2^3/2 \\
\Delta h_{t,s} &= \psi_s u_2^2 = \psi_s \omega^2 D_2^2/4,
\end{aligned}
\tag{9.2}
$$

these expressions can then be arranged to eliminate the diameter as follows:

$$
\omega_s = \omega \frac{\left(\dot{V}_{t1}\right)^{1/2}}{\left(\Delta h_{t,s}\right)^{3/4}} = \omega \frac{\left(\phi_{t1} \omega D_2^3/2\right)^{1/2}}{\left(\psi_s \omega^2 D_2^2/4\right)^{3/4}} = 2 \frac{\phi_{t1}^{1/2}}{\psi_s^{3/4}}.
\tag{9.3}
$$

The specific diameter is another nondimensional combination of the head and flow coefficients, as follows:

$$
D_s = D_2 \frac{\left(\Delta h_s\right)^{1/4}}{\left(\dot{V}_{t1}\right)^{1/2}} = D_2 \frac{\left(\psi_s \omega^2 D_2^2/4\right)^{1/4}}{\left(\phi_{t1} \omega D_2^3/2\right)^{1/2}} = \frac{\psi_s^{1/4}}{\phi_{t1}^{1/2}},
\tag{9.4}
$$

which is independent of rotational speed. A further rearrangement of (9.3) and (9.4) leads to the following relationships:

$$
\begin{aligned}
D_s \omega_s &= 2 \frac{\psi_s^{1/4}}{\phi_{t1}^{1/2}} \frac{\phi_{t1}^{1/2}}{\psi_s^{3/4}} = \frac{2}{\psi_s^{1/2}} \\
D_s^3 \omega_s &= 2 \frac{\psi_s^{3/4}}{\phi_{t1}^{3/2}} \frac{\phi_{t1}^{1/2}}{\psi_s^{3/4}} = \frac{2}{\phi_{t1}}.
\end{aligned}
\tag{9.5}
$$

Neither the flow coefficient nor the pressure coefficient is directly related to the specific speed or the specific diameter alone, and vice versa.

Equation 9.3 can be expressed in terms of the isentropic efficiency and the work coefficient to give

$$
\omega_s = 2 \frac{\phi_{t1}^{1/2}}{\left(\lambda \eta_s\right)^{3/4}}.
\tag{9.6}
$$

If the impeller has no backsweep, then the work coefficient will be close to $\lambda = 0.8$, and for a good design of a reasonably large impeller in the region of the optimum flow coefficient, the isentropic efficiency will be close to $\eta_s = 0.85$. For such impellers, (9.6) shows that specific speed is directly related to the flow coefficient and, as noted by Vavra (1970a), there is then no need to specify the specific diameter. This may be the background to the common misuse of the specific speed alone to denote the design of compressor stages without mention of the specific diameter; as in the period when its use became common, there were few backswept impellers. With backswept impellers of different work coefficients, it is necessary to specify both the specific speed and the specific diameter to pin down a particular design, and this is not often done.

9.2.3 Other Definitions of These Parameters

Unfortunately, there are several ways to define specific speed and specific diameter, so great care is needed in making use of these parameters. Some of the common forms are not even truly nondimensional as they make use of inconsistent units. In Europe, the pump industry commonly uses a definition of specific speed based on

$$n_q = N\frac{(\dot{V}_{t1})^{1/2}}{(\Delta H)^{3/4}} = 52.9\omega_s. \tag{9.7}$$

N is the rotational speed in rpm, the volume flow is in m^3/s and the pressure rise duty is defined by the pressure head in metres of water. In the USA, it is customary in pumps to use a similar equation based on very specific units,

$$N_s = N\frac{(\dot{V}_{t1})^{1/2}}{(\Delta H)^{3/4}} = (52.9*51.6)\omega_s = 2729.6\omega_s, \tag{9.8}$$

where the numerical value of the speed is in rpm, the volume is in US gallons per minute and the head is in feet of water. An additional possibility in imperial units is to take the speed in rpm, the volume flow in ft^3/s and the isentropic head rise in $(ft/s)^2$ leading to a value that is $N_s = 129\omega_s$.

All of these have the same physical significance as the nondimensional specific speed used in this book but simply express the numerical value in different units and so are not truly dimensionless. There is little chance of confusion of the proper nondimensional version with these dimensional parameters as the numerical factor produces values which do not overlap. The range of the dimensionless specific speed for centrifugal compressors is from 0.3 to 2, the pump specific speed then varies from 15 to 100, the American version from 800 to 5500 and the imperial specific speed from 40 to 250.

In German textbooks, a slightly different definition is often used, based on the notation originally defined by Cordier (1953). The relevant section in the important reference book for German engineers by Dubbel (2001) also uses this definition, and this is also used by Bommes et al. (2003) with respect to radial ventilators:

$$\sigma_M = n\frac{(\dot{V}_{t1})^{1/2}}{(\Delta h_{t,s})^{3/4}}(2\pi^2)^{\frac{1}{4}}. \tag{9.9}$$

In this definition, n is in revolutions per second and not in radians per second. This has many names in German (*Laufzahl*, *Schnelllaufzahl* or *Schnellläufigkeit*) but is called the speed number here to distinguish it from the specific speed.

The definition given by Dubbel (2001) for the specific diameter is again consistent with that of Cordier (1953) and Bommes et al. (2003), as follows:

$$\delta_M = D_2\frac{(\Delta h_{t,s})^{1/4}}{(\dot{V}_{t1})^{1/2}}\left(\frac{\pi^2}{8}\right)^{1/4}. \tag{9.10}$$

This parameter is here called the diameter number (as it is often known as the *Durchmesserzahl* in German) to distinguish it from the specific diameter. Unfortunately, it is also sometimes known as the specific diameter, and was defined as such by Cordier (1953), which can easily cause confusion. The conversion factors between the diameter number and the specific diameter, and the speed number and the specific speed, are as follows:

$$D_s = \frac{2^{3/4}}{\sqrt{\pi}}\delta_M = 0.94885\,\delta_M, \quad \omega_s = 2^{3/4}\sqrt{\pi}\,\sigma_M = 2.98009\sigma_M. \tag{9.11}$$

9.2.4 Physical Significance of Specific Speed and Diameter

Although in practice the specific speed is more often used than specific diameter, it is easier to visualise the physical meaning of the specific diameter, so this is considered first. Following unpublished notes of Gyarmathy (1996), the definition of specific diameter can be reformulated as

$$D_s^4 = \left(\frac{\Delta h_s}{\dot{V}^2}\right)D_2^4 = \frac{\Delta h_s}{(\dot{V}/D_2^2)^2}. \tag{9.12}$$

The numerator is the specific isentropic head rise of the machine and the denominator represents the kinetic energy of the throughflow – as a throughflow velocity squared. From this, it becomes clear that if the specific diameter is large, then the turbomachine produces a high head rise relative to the kinetic energy of the throughflow.

The specific speed can be rewritten as follows:

$$\omega_s^2 = \left(\frac{\dot{V}\omega^2}{(\Delta h_s)^{3/2}}\right) = \frac{\dot{V}/\omega}{(\sqrt{\Delta h_s}/\omega)^3} = \frac{\dot{V}/\omega}{(c_s/2\omega)^3}. \tag{9.13}$$

The numerator represents the volume induced in each rotation of the machine (actually in a rotation through an angle of 1 radian). In the denominator, the isentropic head rise has been replaced by the isentropic spouting velocity, which is a characteristic velocity related to the enthalpy rise, defined as $\Delta h_s = c_s^2/2$. The term, $c_s/2\omega$, represents a characteristic dimension related to the spouting velocity and the rotational speed – this can be visualised as how far the flow would travel in a rotation through an angle of 1 radian if it were travelling at the spouting velocity. The cube of this dimension represents a characteristic volume related to the head rise. Stages with a high specific speed have a high flow for a given head rise, which generally implies an axial machine. Stages with a low specific speed have a low flow relative to the head rise and are generally radial machines.

9.2.5 The Range of Parameter Variation

In practical radial and mixed flow compressors or pumps, the inlet flow can vary through an enormous range of flow coefficients, between $0.004 < \phi_{t1} < 0.2$. As a result, this flow variation accounts for over 700% variation in the value of the specific speed:

$$\left\{\frac{(\phi_{t1})_{\max}}{(\phi_{t1})_{\min}}\right\}^{\frac{1}{2}} = \left\{\frac{0.2}{0.004}\right\}^{\frac{1}{2}} = \{50\}^{\frac{1}{2}} \approx 7. \tag{9.14}$$

The typical variation in isentropic pressure coefficient of radial compressors or pumps is between $0.4 > \psi_s > 0.7$ and accounts for only a 50% variation in specific speed:

$$\left\{\frac{(\psi_s)_{\max}}{(\psi_s)_{\min}}\right\}^{\frac{3}{4}} = \left\{\frac{0.7}{0.4}\right\}^{\frac{3}{4}} \approx \{1.75\}^{\frac{3}{4}} \approx 1.52. \tag{9.15}$$

This identifies the fact that the specific speed is more closely related to the changes in the flow coefficient than any variation of the pressure coefficient.

In Figure 9.1, lines of constant pressure rise coefficient in the range from 0.3 to 0.7 are shown in a plot of specific speed versus flow coefficient. It can also be seen that a specific speed value of 0.75 based on the mean volume flow, considered by many to be the 'optimum specific speed' following Rodgers (1980), varies in inlet flow coefficient from 0.04 to 0.07 for a range of 0.5 to 0.6 in the pressure coefficient. For impellers with no backsweep, the pressure coefficient would be approximately 0.7, and then a specific speed of 0.75 corresponds to a flow coefficient of 0.08 and is then compatible with the optimum flow coefficient of $\phi_{t1} = 0.08 \pm 0.01$. There is no evidence in any experimental data on optimum radial compressors that a higher pressure rise coefficient suggests a higher flow coefficient for the design as indicated by the lines for constant pressure rise in Figure 9.1, somewhat diminishing the usefulness of specific speed as the key parameter guiding the selection of an optimum design.

The major variations in the specific speed of radial turbocompressors are caused by changes in the volume flow coefficient rather than the pressure rise coefficient, confirming that the use of the inlet flow coefficient is a much more useful parameter for defining the shape of a compressor. More importantly, the physical relationship

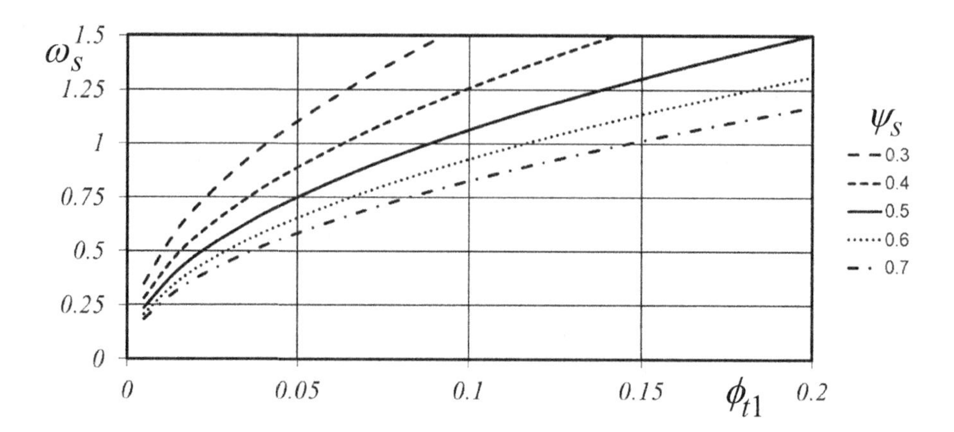

Figure 9.1 Specific speed as a function of pressure coefficient and flow coefficient.

between the flow coefficient and the work coefficient as given in the definition of the specific speed, (9.6), hides the fact that these two parameters are separate aspects of the design of a compressor. As shown in Chapters 10, 11 and 18, the duty of the stage, as defined by the flow coefficient, the work coefficient and the tip-speed Mach number, determines the geometry of the stage, the expected efficiency and its operating range and is much more useful.

9.3 The Cordier Diagram

9.3.1 Background

Well-designed axial machines have high specific speed and low specific diameter, whereas good radial machines have low specific speed and high specific diameter. This feature was noted by Cordier (1953); to distinguish different machine types, he produced an interesting plot, nowadays known as the Cordier diagram. The diagram from the original paper by Cordier is replotted in Figure 9.2, using the notation of this chapter but with the nondimensional specific speed, ω_s, as abscissa, following the convention of Balje (1981). In Figure 9.2, the specific diameter at the best operating point of a range of machines given by Cordier (1953) is plotted versus the specific speed.

Note that there is no condition of geometrical similarity between these different geometries; the condition of maximum efficiency is imposed so the plot shows only the best efficiency points of machines of different geometry. A change in specific speed thus implies a change in machine design and not a change in the operating point.

In broad terms, radial machines with a radial inlet and radial outlet are used when $\omega_s < 0.5$, radial machines with a more axial inlet and a radial outlet when $0.5 < \omega_s < 1.5$, mixed flow machines with a diagonal outlet when $1.0 < \omega_s < 2.0$ and axial machines when $\omega_s > 2.0$. The experimental data provided in Cordier's original publication are also reproduced in Figure 9.2. There is not sufficient experimental data in the axial region

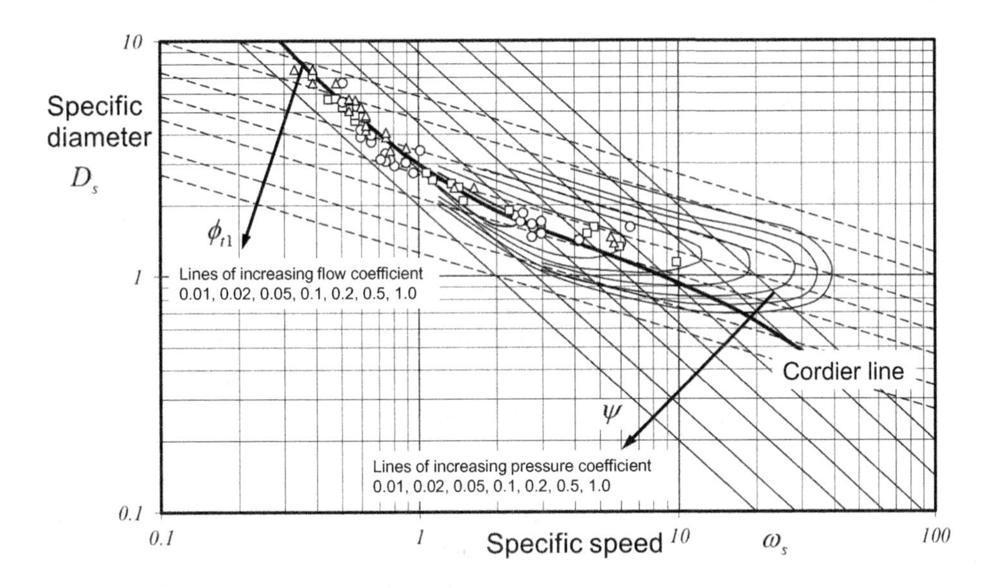

Figure 9.2 A Cordier diagram in the form recommended by Balje (1981).

to confirm the contours of efficiency given by Cordier; his original paper includes a short theoretical analysis of axial stage performance as a basis for the contours. There is, in fact, little published data on axial machines in this form, except from Epple et al. (2010), who derive equations for the Cordier line from first principles related to the geometry and the velocity triangles. For radial machines at low specific speed, the points lie so close together that Cordier did not even attempt to draw contour lines of efficiency.

The example of a Cordier diagram given in Figure 9.2 includes lines of constant flow and pressure coefficients, which are straight lines in the log–log presentation suggested by Balje (1981). The slopes of the lines are determined by the exponents in the preceding equations as can be seen from the differentiation of (9.5) leading to the following equations:

$$\ln D_s + \ln \omega_s = \ln \frac{2}{\psi_s^{1/2}}, \quad 3 \ln D_s + \ln \omega_s = \ln \frac{2}{\phi_{t1}}. \tag{9.16}$$

In the original Cordier diagram, and as depicted in Figure 9.2, the low specific speed machines (with a high specific diameter) are all radial machines and nearly all lie along a single band with a head coefficient close to 0.55 in a range between 0.4 and 0.7. At much higher specific speed, the data points have a lower head coefficient and are all axial machines with a higher flow coefficient. It can be seen from Figure 9.2 that the Cordier diagram is simply a mapping of the head and flow coefficients onto a more complex log–log plot based on the specific speed and specific diameter in which the lines of constant values have different slopes. The locus of the best machines for particular applications, known as the Cordier line, can also be plotted, and in this example the line given is taken from Balje (1981).

In some pump literature, the specific speed is called the type number to emphasise the fact that it characterises the geometry of the machine. Geometrical details of optimum impellers can be characterised as a function of the specific speed, and these values can be used for the preliminary design. Gülich (2008), for example, provides several diagrams and equations for the values of key geometrical parameters of centrifugal pumps as a function of specific speed (such as inlet to outlet radius ratio and outlet width to impeller diameter ratio). In general, it is argued that the specific speed represents a nondimensional shape factor giving a guide to the type of impeller that is most suitable for a given application. In fact, the flow coefficient has the same useful property and can be used to indicate different compressor types.

For a given duty, defined as isentropic head rise and inlet volume flow, a range of machine designs is possible based on different locations along the Cordier line. Each of these machines has a different speed and size. High specific speed implies a higher rotational speed with a smaller impeller for a given enthalpy change and flow rate, that is, an axial machine. Marcinowski (1959) commented that a very wide range of different machine types can be adapted to meet the same duty, and that in this process the diameter may vary by a factor of 8 and the angular velocity by a factor of 25. To demonstrate this, he illustrated machine types designed for the same duty at different speeds by different manufacturers, ranging from a small axial compressor with a high angular velocity to a large radial impeller at low speed (Casey et al., 2010). Of course, not all of these machines will be equally efficient for the purpose, and large differences can also be expected in the shape of the performance characteristics between axial and radial machines of different pressure rise coefficients and Mach number levels. For a low specific speed machine, increasing the speed moves the stage into the optimum specific speed region for a radial machine, and increasing the speed further moves the application into the mixed flow or even axial machine region.

9.3.2 Usefulness of the Cordier Diagram

At face value, the Cordier diagram looks well suited as a tool for preliminary design. It suggests that the specific speed determines the optimum specific diameter and head coefficient and ultimately the efficiency level and the shape of the machine. The following discussion shows, however, that this first impression is rather misleading and there is no benefit in the use of the rather obscure specific speed over the simpler flow coefficient.

To start with, there is no difficulty in classifying turbocompressors in terms of flow coefficients; see Figure 9.3 adapted from Strub (1984). Strub shows a range of turbines, pumps and compressors all classified in terms of typical head coefficients for specific flow coefficients. Just as a low specific speed on the Cordier line implies a high specific diameter, so a low flow coefficient of a radial compressor implies a high work coefficient compared to that of an axial or mixed flow device. The great advantage of the flow coefficient, however, is its simplicity and its direct linear relationship to the volume flow through the continuity equation, rather than involving

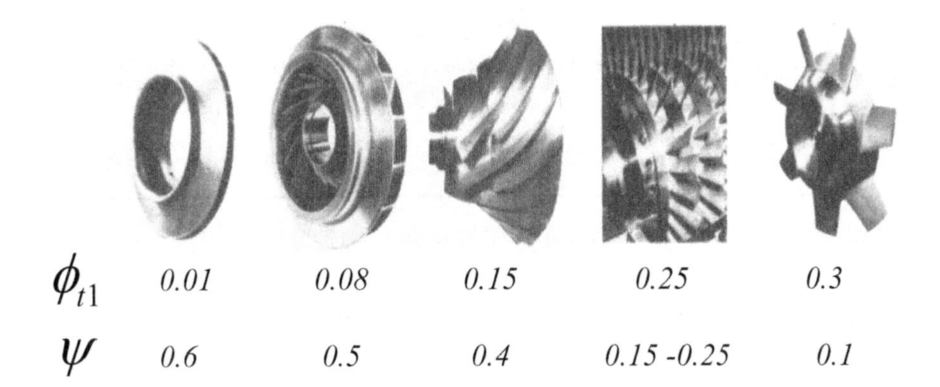

Figure 9.3 A range of different compressor rotors classified with typical head coefficients for specific flow coefficients. (image adapted from Strub, 1984, courtesy of Sulzer)

the square root of the volume flow. The flow coefficient is also linearly coupled to the work coefficient through the Euler equation, so it is of great value in predicting off-design performance and is used in Chapter 18 for this purpose. No nondimensional performance characteristics are ever plotted in terms of specific diameter versus specific speed, but always as head coefficient against flow coefficient.

In fact, there is not a unique Cordier line but a broad band of near-optimum machines, so both the specific speed and the specific diameter are needed to completely characterise the performance. This is similar for the flow coefficient and the head coefficient as both are needed to characterise axial machines in the Smith chart for axial turbines (see Smith, 1965). Unfortunately, in radial turbomachinery applications the specific speed alone is often used. It is argued that most good designs lie close to the Cordier line, and on this line the specific diameter is a function of specific speed, so only the specific speed is needed. The simplicity of representing the whole range of possible machines by a single Cordier line means that most users of the term 'specific speed' fail to recognise that the specific speed is not a unique measure of the performance.

Approximate equations for a range of Cordier lines for compressor stages which are highly loaded, medium loaded and lowly loaded in terms of their pressure rise coefficient were provided by Casey et al. (2010):

$$\begin{aligned} &\psi_s = \psi_m(1-A) + \psi_u A + (\psi_m - \psi_l)B \\ &A = 1/(1+e^{-t_1}); \quad t_1 = k_1(k_2 + \log_{10}\omega_s) \\ &B = e^{-t_2}; \quad t_2 = k_3(k_4 + \log_{10}\omega_s) \\ &k_1 = 4, \quad k_2 = -0.3, \quad k_3 = 5, \quad k_4 = 1. \end{aligned} \tag{9.17}$$

These equations blend the expected values of the pressure rise coefficient between the different specific speed zones using blending functions based on the specific speed. The values of the pressure coefficients for the different lines are given in Table 9.1.

Table 9.1 Coefficients in (9.17)

Specific speed	Pressure coefficient	Upper bound	Cordier line	Lower bound
Low ω_s	ψ_l	0.55	0.45	0.35
Medium ω_s	ψ_m	0.65	0.55	0.4
High ω_s	ψ_u	0.025	0.015	0.01

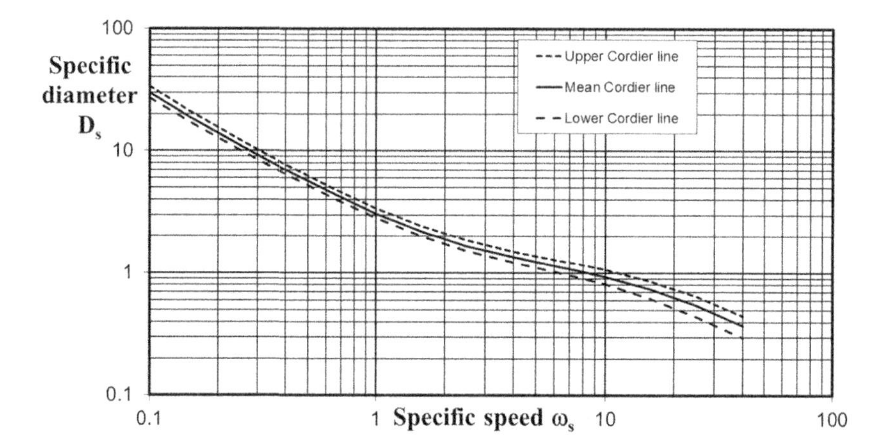

Figure 9.4 A range of Cordier lines in the Cordier diagram.

The lower Cordier line represents more highly loaded stages, in terms of ψ_s, and vice versa. In a numerical study of mixed flow impellers, Teichel (2018) suggests a higher value of $\psi_m = 0.9$ for the upper bound of medium specific speed stages designed using modern CFD methods. The values of the bounds at the low specific speed are lower than those at medium specific speed, as these stages have similar work coefficients but the lower specific speed stages have a lower efficiency. Based on an analysis of many stages defined by means of one-dimensional loss correlations, Mounier et al. (2018) suggest that the lower bound is more appropriate for small-scale compressor stages with a lower efficiency.

An important issue is the fact that the Cordier diagram is always plotted on a log–log scale. The Cordier lines from (9.17) are plotted in Figure 9.4 in a conventional Cordier diagram and again in Figure 9.5 as the head coefficient as a function of the flow coefficient, with a linear scale. The small band of the Cordier lines in Figure 9.4 appears to expand to an enormous range when plotted in this more useful way. The log–log scale and the glancing intercepts of the lines of constant head and flow in the Cordier diagram naturally cause all sensible designs to bunch close together, resulting in what looks like a unique line. The Cordier diagram with log–log axes is misleading in that it gives the impression that there is little difference between the stages close to this line. Once the same data are presented in the form of Figure 9.5, the range of pressure coefficients is in fact very wide. This knowledge is not new – a similar

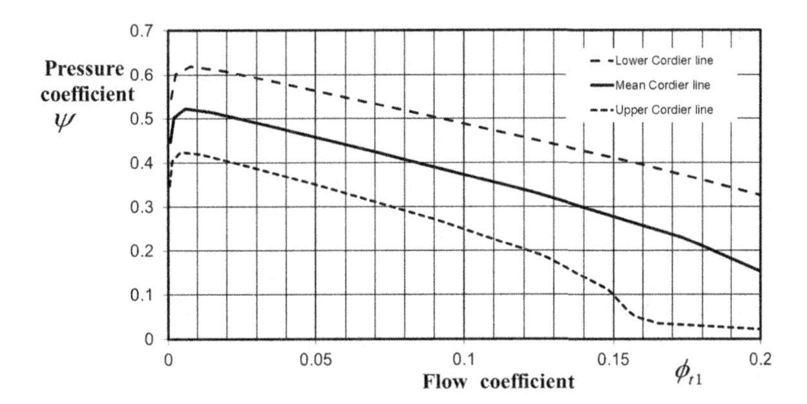

Figure 9.5 The Cordier lines from Figure 9.4 in a linear plot of pressure coefficient versus flow coefficient.

diagram can be found from the 1970s in early editions of Dixon and Hall (2010) and in Lewis (1996).

The Cordier diagram in Figure 9.2 with superimposed lines of constant head and flow coefficients shows that, along the Cordier line, the optimum pressure coefficient decreases as the flow coefficient increases. This feature is hidden if these lines are not included, as in Figure 9.4. Physically this is due to the fact that the inlet diameter has to increase relative to the outlet diameter as the flow coefficient increases. There is then a larger diameter ratio between inlet and outlet, that is, D_{1c}/D_2 increases. This implies there is a smaller centrifugal effect with a lower head rise as the design becomes more axial at higher flows (Casey et al., 2010). This effect is much clearer in a linear plot as in Figure 9.5 than in a Cordier diagram, and shows clearly that the flow coefficient is a more useful parameter for comparison of different centrifugal machines than specific speed.

If the rotational speed (in radians per second) is not known in advance, there is no significant benefit to using the specific speed parameter over the much simpler global flow coefficient for purely radial and mixed flow machines. Moreover, in most compressor applications a rule-of-thumb value for the acceptable tip speed is usually known, either from the required tip-speed Mach number or from mechanical considerations, so the flow coefficient is often much more useful as it includes the tip speed and not the rotational speed. Nevertheless, the availability of wide-ranging correlations based on the specific speed, for radial compressors from Rodgers (1991) and (1980), and many other sources for radial pumps, Gülich (2008), and for radial fans (Bommes et al., 2003) means that this rather complex parameter has wide acceptance among the centrifugal turbomachinery community.

Given that the location of the Cordier line is defined by the pressure rise coefficient, a combination of pressure coefficient and flow coefficient, or pressure coefficient and specific speed, appears to be more useful to the authors of this book than specific speed and specific diameter. Specific diameter has limited utility but still provides an interesting basis on which to compare radial machines with other types of turbomachinery, even including positive displacement machines with a substantially different pressure rise. This is used extensively by Balje (1981), where several examples can be found.

9.3.3 The Effect of Compressibility

The specific speed is basically an incompressible parameter and makes no reference to the speed of sound in the flow and the pressure ratio. Different volume flows are present at the inlet and outlet of a turbocompressor, and a convention is needed to define which to use, either the inlet flow (Balje, 1981) or the mean volume flow have been used (Rodgers, 1980). For an impeller at a given specific speed, rather different geometries are needed as the tip-speed Mach number changes because of the density change across the machine, as shown in Figure 11. To overcome this limitation of the specific speed, Mounier et al. (2018) include the impeller outlet width ratio, b_2/D_2, in their plots of the efficiency variation with specific speed of refrigeration impellers. The relationship between the outlet width ratio of the impeller, b_2/D_2, to the tip-speed Mach number is described in Section 11.6.5.

The specific speed is also unable to identify the risk of high entry Mach numbers and choking. In fact, this can be simply identified by the product of flow coefficient and tip-speed Mach number, as shown in Section 8.6.7. Balje (1962) derived equations for the impeller casing inlet relative Mach number as a function of specific speed and specific diameter and provides a Cordier diagram where limiting values of the relative inlet Mach number for different pressure ratios are included. Another dimensionless parameter, here called the speed of sound specific speed, can be defined to indicate the risk of high Mach numbers in the flow:

$$\omega_a = \omega \frac{(\dot{V}_{t1})^{1/2}}{(a_{t1})^{3/2}}, \tag{9.18}$$

where a_{t1} is the sonic velocity at inlet total conditions to the compressor. This parameter is defined and discussed by Pfleiderer and Petermann (1986), where it is called the speed of sound number (*Schallkennzahl* in German). The authors of this book have not seen this parameter in use in engineering practice, except in a single publication by Farkas (1977), and it is included here only for completeness. In combination with the definition of the flow coefficient and the tip-speed Mach number, the speed of sound specific speed can be written as

$$\omega_a = \omega \left(\frac{\phi_{t1} u_2 D_2^2}{a_{t1}^3} \right)^{1/2} = 2\omega r_2 \left(\frac{\phi_{t1} u_2}{a_{t1}^3} \right)^{1/2} = 2M_{u2}(\phi_{t1} M_{u2})^{1/2} = 2M_{u2}(\Phi_{t1})^{1/2}. \tag{9.19}$$

The speed of sound specific speed is seen to be directly related to the tip-speed Mach number and the corrected flow per unit area. The tip-speed Mach number and the corrected flow are generally used as a measure of the importance of compressibility for compressors, as these are much simpler to visualise and use.

10 Losses and Performance

10.1 Overview

10.1.1 Introduction

The previous chapters examined the aerodynamic and thermodynamic background required for understanding and assessing radial turbocompressor flow and performance. The next three chapters will show how this background affects the design of the individual components: Chapter 11 looks at impellers, Chapter 12 at diffusers and Chapter 13 at other stationary components. The objective of this chapter is to describe simple one-dimensional models that enable preliminary analysis of the expected performance of a radial turbocompressor stage at the design operating conditions on the basis of component losses, known as component stacking. Chapter 18 describes off-design performance predictions of single-stage compressors and of multistage compressors using stage stacking techniques.

The real flow in impellers is both unsteady and three dimensional (3D). Chapter 5 describes the complex flow patterns in some detail. The simple one-dimensional (1D) steady flow analysis given here facilitates a general understanding of the influence of the design parameters on the losses and performance and provides a basis for the preliminary design, The sources of loss in turbomachinery are viscous losses on the profiles and endwalls, losses across shock waves, clearance losses and leakage losses, and the physics of these loss mechanisms is described on the basis of a modern understanding of the loss generation processes. Historical correlations of different sources of loss often used in radial compressor analysis are also briefly described. A broad outline is given of the state-of-the-art performance levels that can be achieved and the associated design parameters. The background theory to mean-line calculation methods of radial impeller stages is explained.

10.1.2 Learning Objectives

- Be aware of the different ways of defining component losses.
- Understand the basic theory of loss mechanisms and the many different individual sources of loss in a radial compressor stage.
- Be aware of the very useful global correlations for efficiency showing that high levels of efficiency are found in stages designed for flow coefficients in the range $0.07 \leq \phi_{t1} \leq 0.10$.
- Be aware of parasitic losses and how they increase in relevance in radial compressor designs at low flow coefficients.

10.2 The Definition of Losses

10.2.1 Estimation of Losses

The accurate prediction of the components of loss in radial compressor stages is nontrivial as can be seen from the complex flow patterns at impeller outlet as discussed in Section 5.8. The generally recognised sources of loss in all turbomachinery are viscous losses on the profiles and end-walls, shock losses, clearance losses and leakage losses, but these cannot always be clearly separated from each other. In axial compressors, a rule of thumb going back to Howell (1945) is that the losses are roughly one-third profile losses, one-third end-wall losses and one-third tip clearance losses. This distribution of losses was difficult to confirm in the absence of computational methods, and it is quite probable that in the past some of the loss in the end-wall region was wrongly attributed to clearance flow (Miller and Denton, 2018). In radial machines, the viscous losses dominate, but there are also additional parasitic viscous losses due to disc friction on the impeller backplate (and on the outer surface of the shroud, if this is present). Shrouded impellers have no tip clearance losses but have leakage losses of the flow over the shroud.

Influential contributions to the understanding of the physics of losses were made by Denton and Cumpsty (1987) and Denton (1993). These authors emphasised that the losses can seldom be predicted with great accuracy and suggested that the error in loss prediction from simple models was not likely to be better than ±20%. At an efficiency level of 80% for a small turbocharger radial compressor stage, this error band would represent efficiencies between 76 and 84%, a difference that is greater than the difference between a very competitive and a very poor design. This error band has recently been confirmed by Velasquez (2017), who finds that errors of up to 8% in efficiency can be obtained from different simple loss models for radial machines.

The main contribution of Denton (1993) was to clarify the physics of the different sources of entropy production. Those that are of relevance to radial machines are due to boundary layers, wakes, shocks and mixing. The latter includes the mixing of a tip clearance flow with the main throughflow, the mixing of a wake downstream of a trailing edge and the mixing out of any nonuniform flow disturbances, such as a separation region or a tip clearance vortex. The physics of these loss mechanisms is described in the sections that follow. In principle, a good understanding of the physics of the loss sources enables the designer to achieve a high-quality design even if the predictive capability of simple models for individual loss sources is poor.

The ability to predict performance levels in centrifugal compressors with simple 1D methods lags well behind that of axial compressors. This is partly due to the fact that the geometry of axial machines is simpler, allowing the blade profile losses to be examined in 2D cascades, which are more amenable to parameterisation and correlation. Axial compressor cascades also benefit from a substantial experimental database, going back to the important work from the 1950s in the USA published as NASA SP36 (Bullock and Johnsen, 1965) and the published work of Howell's team in the Royal Aicraft Establishement (RAE) in the UK (Howell, 1945). The use of 2D blade-to-blade CFD models in axial compressor research, such as the code MISES of Drela and Youngren (1991), has also accelerated the understanding of the 2D profile losses in axial compressors (Goodhand and Miller, 2012).

Broadly speaking, four approaches are used for determining stage efficiency and losses. The most important and most exact method is measurement by testing with carefully calibrated instrumentation in a suitable test rig, either at full scale or as a scale model. Testing procedures in suitable test rigs are described in Chapter 20 and provide the precise global performance from flange to flange at different speeds, including the crucial information about the operating range from surge to choke at off-design conditions. In some cases, intermediate measuring planes may be used to provide more local information giving details of component performance within the compressor stage. Measurement planes at impeller outlet and diffuser outlet are particularly useful for separating the contributions of the impeller, the diffuser and the outlet components. The use of the similarity parameters given in Chapter 8 to relate the losses to changes in geometry or flow conditions allows performance testing to provide information either for other similar stages, or for the same stage at different conditions.

A second method of loss prediction involves 3D CFD simulations with RANS equations, and Chapter 16 gives a short overview. This approach provides detailed information about the local flow field and, if done correctly, is very reliable. However, in the experience of the authors, this method is still not able to predict efficiency consistently to better than $\pm 1\%$. This level of accuracy in radial compressors depends on modelling all the geometrical features of leakage flow paths, fillet geometry and clearance flows of the whole stage (Hazby et al., 2019). Steady 3D CFD simulations are notoriously weak when predicting compressor surge, and this is discussed in Chapter 17, but they are reasonably accurate when estimating choke flows. CFD may not provide predictions of efficiency with absolute accuracy but is generally viewed as a very good predictor of the trends in the performance due to changes in geometry or operating conditions.

A third approach is to use either a general 1D mean-line loss model, as described in this chapter, or a similar 2D throughflow compressor model with correlations, as described in Section 15.7. In this approach, the losses for each of the principal flow elements of the stage and for each individual source of loss is estimated from analytical loss models usually with empirical coefficients calibrated with test data.

A fourth approach, favoured by the authors for preliminary estimates, simply correlates global efficiency measurements on similar types of compressors with the most relevant parameters such as impeller diameter, clearance level and the key nondimensional parameters representing the duty of the machine, such as the flow coefficient, tip-speed Mach number, work coefficient and Reynolds number. This last method cannot distinguish between good and bad designs as it has no details of the geometry. It does, however, provide reliable target efficiencies that can be expected for designs of a given duty that are well executed based on historical precedent. This method is described in Section 10.7.

10.2.2 Loss Coefficient Definitions for a Component

In a mean-line method based on the summation of the losses in the individual components, it is necessary to specify the losses on a coherent basis. The only rational measure of loss in an adiabatic machine is the entropy creation (Zweifel, 1941; Denton, 1993), as used in Chapter 4 to define the efficiency. The correct metric for the definition of losses based on entropy increase is the specific dissipation energy, j_{12}, which is the integral of the product of temperature and entropy change, *Tds,* through the component, as discussed in Chapter 2. A dissipation loss coefficient for compressor components can be defined in terms of the dissipation losses as a fraction of the available inlet kinetic energy, as follows:

$$\chi_{diss} = j_{12} \Big/ ½w_1^2 = \int_1^2 Tds \Big/ ½w_1^2, \tag{10.1}$$

where state 1 is the inlet to the component, state 2 represents the outlet of the component, j_{12} is the specific dissipation between these locations, T the local static temperature and w_1 the inlet velocity relative to the component being considered.

This loss coefficient is not in regular use, and there are many different loss definitions used to quantify the magnitude of the losses for different components. There are several different publications available to compare their merits, for example Horlock (1958), Brown (1972), Whitfield (1978) and Denton (1993). These show that the definitions can all be expressed in terms of a loss coefficient based on (10.1), such that the dissipation loss coefficient can be calculated from the other definitions of loss. For a process where the intermediate states are known, or can be assumed to follow a polytropic compression process, the integration in (10.1) is straightforward, as shown in Section 2.7.3, and leads to a formally correct definition of the dissipation loss coefficient. In most real processes, however, the integration cannot be carried out since the intermediate steps in the process are not known. Thus, a simpler definition is often needed.

A simpler entropy loss coefficient was introduced by Denton (1993) in an attempt to generalise the wide variety of loss coefficients available that can be used to define the losses in turbomachinery blade rows. For decelerating flow in compressor blade rows, this entropy loss coefficient is defined relative to the kinetic energy and temperature of the flow at the inlet, as follows:

$$\chi_c = T_1(s_2 - s_1)/\left(½w_1^2\right). \tag{10.2}$$

This is essentially the same as the loss coefficient in (10.1) in the case where there is no change in temperature. For incompressible flows, the entropy loss is related to the classical incompressible loss in total pressure, which leads to the following pressure loss coefficient for compressor blade rows:

$$\omega_c = (p_{t1} - p_{t2})/\left(½\rho v_1^2\right). \tag{10.3}$$

This is the most common definition of loss coefficients, and its application to diffusers is described in Chapter 7. Another loss coefficient definition, which can be used for compressible and incompressible flows, is the stagnation pressure loss coefficient defined as

$$Y_c = (p_{t1} - p_{t2})/(p_{t1} - p_1). \tag{10.4}$$

This avoids the need of an assumption of constant density. A further loss coefficient is the energy or enthalpy loss coefficient, which is defined as

$$\varsigma_c = (h_2 - h_{2s})/(h_{t1} - h_1), \tag{10.5}$$

where h_{2s} is the outlet static state of the component if there were no losses. The use of a change in enthalpy as the basis to define a loss coefficient is somewhat confusing given that it is the change in entropy that determines the dissipation in the flow. Denton (1993) notes, however, that the difference in magnitude between the enthalpy and entropy definitions of loss coefficient is negligible. The fact that these loss coefficients are numerically nearly identical means that conventional loss coefficients (often determined in low-speed tests of components at low Mach number with incompressible flow) can be directly used as dissipation loss coefficients, using conversions for the effect of Mach number given by Brown (1972).

Denton (1993) shows that the entropy gain for an adiabatic flow through a stationary component is given by the drop in total pressure:

$$\Delta s_{ij} = -ZR \ln \left(p_{tj}/p_{ti}\right). \tag{10.6}$$

For a stator component i with a total pressure loss coefficient of

$$Y_i = \left(p_{tj} - p_{ti}\right)/(p_{ti} - p_i), \tag{10.7}$$

the entropy change is given by

$$\Delta s_{ij} = -ZR \ln \left[1 + Y_1(1 - p_i/p_{ti})\right]. \tag{10.8}$$

If the entropy loss coefficient is used, for both accelerating blade rows (as in compressor inlet guide vanes and also in turbine rotors and stators) and for decelerating blade rows (compressor rotors and stators and also turbine outlet guide vanes), the entropy can be calculated as follows:

$$\begin{aligned} s_2 &= s_1 + \chi_{comp} \tfrac{1}{2} w_1^2 / T_1 \\ s_2 &= s_1 + \chi_{turb} \tfrac{1}{2} w_2^2 / T_2, \end{aligned} \tag{10.9}$$

where 1 denotes the leading edge and 2 the trailing edge. Compressor loss coefficients are usually normalised with the inlet kinetic energy, whereas in turbines the exit kinetic energy is generally used.

10.2.3 Component Losses Defined by Efficiency

In addition to the preceding loss coefficients, it is often useful to define the efficiency of a particular component in relation to the change in entropy across it. If the small-scale static-to-static polytropic efficiency is specified (using the formulation based on entropy in Chapter 4), then the change in entropy across a component is

$$s_2 = s_1 + \left(1 - \eta_p\right)\left(s_{1h} - s_1\right), \tag{10.10}$$

where s_{1h}, is the entropy at the pressure of state 1 with the enthalpy of state 2. In adiabatic flows, this equation should be used in such a way that entropy always increases. As a result, for a decelerating flow the equation defines a compressor efficiency, and for an accelerating flow with a decrease in static enthalpy it defines a turbine efficiency, as follows:

$$s_2 = s_1 + \left(1 - 1/\eta_p\right)\left(s_{1h} - s_1\right). \tag{10.11}$$

This efficiency can be applied both to blade rows and to the annulus between them to determine the losses in different components and between different calculating planes. The value of the entropy, s_{1h}, at the pressure of state 1 with the enthalpy of state 2 is needed and is determined by the equation of state, which for an ideal gas is

$$s_{1h} = s_1 + c_p \ln\left(h_2/h_1\right) \tag{10.12}$$

and for incompressible fluids is

$$s_{1h} = s_1 + c_p \ln\left\{1 + \frac{(h_2 - h_1)}{c_p T_1}\right\}. \tag{10.13}$$

If there are changes in static enthalpy due to acceleration or deceleration, this formulation can also be used to predict the losses generated in the spaces between the blade rows. If there is no change in the static enthalpy, this algorithm produces no losses, such that in a constant area duct with constant static enthalpy a loss coefficient or dissipation coefficient approach has to be used.

The real advantage of this entropy-based formulation becomes apparent for the case of a real gas. Here the value of the entropy, s_{1h}, at the virtual state with the pressure of state 1 with the enthalpy of state 2, can be determined directly from the real gas equations or by interpolation in tables of properties:

$$s_{1h} = f(p_1, h_2). \tag{10.14}$$

This approach can also be used to advantage in the formulation of 1D stage-stacking codes for real gases as these usually involve stepping through the machine from inlet to outlet, as described in Section 18.6. The value of the pressure at the state 1 and the enthalpy at state 2 (via the Euler work equation) are generally both known during this process such that in the case of a real gas, the value of the entropy at the virtual state $1h$ can easily be determined.

The small-scale polytropic efficiency is generally used in such a way that a single value for the calculation is specified. In most preliminary design processes, such as that described in Section 10.8 and in throughflow methods, correlations for efficiency are used similar to those described in Section 10.7.

10.3 Viscous Loss Mechanisms

10.3.1 A Modern Understanding

While a number of important questions have yet to be answered, significant progress has been made in the understanding of loss mechanisms in turbomachinery flows in recent years. The contribution to the understanding of the physics of losses developed by Denton (1993) for axial machines is described in this section in relation to radial compressors. The approach of Denton has been further elucidated in a lecture by Miller and Denton (2018) and this presentation provides the structure of the account of loss mechanisms presented here.

Viscous dissipation losses in shear layers are ultimately responsible for the main part of the loss generation in radial compressors. The viscous losses occur both on the blade profile surfaces, where they are known as profile losses, and on the end-walls. In radial compressors with strong secondary flows, it is not generally useful to distinguish between the end-wall and the profile losses as the strong secondary flows cause the boundary layers to migrate around the flow channel, as discussed in Section 5.6. The most important aspect of the viscous losses is that they are a function of the state of the boundary layer and so are a function of Reynolds number and the relative roughness of the wetted surfaces. In the description given here, studies of the effect of Reynolds number on the performance of compressors is used to give additional insight into the magnitude of the viscous dissipation losses, following the approach of Casey and Robinson (2011) and Dietmann and Casey (2013).

10.3.2 Profile and End-Wall Loss

Profile loss in axial compressors is generally considered to be the loss generated on the blades themselves in regions not influenced by the end-wall flows. It is the loss that is usually measured in axial profile cascade tests, but it is not strictly confined to two-dimensional flow since the effects of changes in spanwise stream tube height can be important. It includes loss due to blade surface boundary layers, to separations on the blade surfaces and at the leading and trailing edges, to dissipation in the wake, to the effects of off-design incidence and to shock waves. The loss mechanisms of these sources of loss are considered to be similar in radial and axial compressors, allowing some carryover from axial compressor knowhow to radial compressors. Even though strong spanwise flows are present in radial impellers, the effect of incidence on the profile losses can be usefully described on the same basis as in axial compressors, especially with regard to the alpha-Mach diagram as discussed in Section 7.3.3.

The end-wall losses are the additional losses occurring near to the hub and casing. In radial impellers with a low flow coefficient, and in most flow channels in radial diffusers, the wetted surface of the end-wall is much larger than that of the blade profiles so that the viscous end-wall losses tend to dominate. The end-wall losses also include the annulus losses which occur on the end-walls upstream and downstream of the blades. In radial compressors, as in axial compressors, the end-wall flows are complicated by such features as secondary flows, skewing of the end-wall boundary layers, interaction with tip leakage flows and a change from a rotating to a stationary component, as discussed in Section 5.5. These features lead to extremely complex flow patterns and make the development of simple loss models very difficult.

10.3.3 Losses in Viscous Boundary Layers

The rate at which energy in a boundary layer is dissipated by the action of viscosity to generate entropy can be shown to be determined by an equation of the type

$$\dot{S} = \int_0^{y_e} \frac{\tau}{T} \frac{\partial u}{\partial y}, \tag{10.15}$$

where τ is shear stress in the direction of the flow, and the integration is taken across the boundary layer from the wall to the edge of the boundary layer, y_e (Denton, 1993; Greitzer et al., 2004). This may be interpreted as the viscous shear work being dissipated at temperature T. A turbulent boundary layer on the suction surface of a compressor cascade was examined in detail by Dawes (1990), where he integrated (10.15) in a CFD simulation to calculate that the laminar sublayer accounted for 50% of the loss generation, the buffer zone accounted for 40% and the outer layer accounted for only 10% of the loss generation. A similar result was obtained for a turbulent boundary layer examined by Moore and Moore (1983) showing that 90% of the boundary layer loss is generated deep in the boundary layer between the wall and about $y^+ = 30$. The outer part of the boundary layer is most affected by the streamwise pressure gradient, and this controls the onset of separation, but in the outer layer, there is little shear stress, and as a result little entropy is generated (Greitzer, 2004).

Denton (1993) suggested a practical method for estimating the dissipation losses in turbomachinery. He defined a dimensionless dissipation coefficient, c_d, as

$$c_d = \frac{T\dot{S}}{\rho u_e^3}, \tag{10.16}$$

where the velocity at the edge of the boundary layer is given as u_e. Most losses are generated in the inner part of the boundary layer, and as the velocity profile of the viscous sublayer is quite resilient to changes in the pressure gradient, a nearly constant value of the dissipation coefficient might be expected for boundary layers with different pressure gradients. Based on a curve fit of experimental data for flat-plate boundary layers, Schlichting and Gersten (2006) suggest a value for the dissipation coefficient in a zero-pressure gradient boundary layer with a weak effect of the Reynolds number based on the local momentum thickness, Re_θ, as follows:

$$c_d = 0.0056\,(Re_\theta)^{-1/6}. \tag{10.17}$$

Denton further suggested that over a wide range of applications, there is little change in Re_θ, so an additional approximation could be made for attached smooth-wall turbulent boundary layers by taking a constant value of $c_d = 0.02$.

The application of this technique to estimate the total entropy production in the boundary layers on the wetted surface areas relies on integrating the dissipation of the blade surface and end-wall boundary layers along their path length:

$$S = \int_0^l c_d \frac{\rho u_e^3}{T} dl, \tag{10.18}$$

where l is the length of the boundary layer. This expression gives the change in entropy, S, not the change in specific entropy, s, for which it needs to be divided by the mass-flow rate. This equation can be applied along different stream tubes in the blade surfaces and on the end-walls to estimate the total increase in entropy.

A very similar approach was suggested by Traupel (2000), and an application of this to predict the change in efficiency of radial compressor stages with a change in the Reynolds number and surface roughness was given by Dietmann and Casey (2013). The Traupel loss model suggests that the total viscous dissipation loss for a small section of a flow channel may be estimated as

$$dS = (\rho c_d/T) w^3 P dl, \tag{10.19}$$

where c_d is a mean dissipation coefficient on the walls, w the local velocity relative to the surface outside the boundary layer and dl the flow path length of a small element of the flow channel. P is the perimeter of the wetted area given by

$$P = 2\left(\frac{2\pi r_m \cos\beta}{Z} + b\right), \tag{10.20}$$

where r_m is the mean radius, Z is the number of blades, b is the width of the meridional flow channel and β is the blade angle.

From (10.19), the dissipation losses related to the entropy increase can then be calculated as

$$j_{12} = \int_0^l T ds = \int_0^l \frac{\rho c_d P w^3}{\dot{m}} dl = \int_0^l \frac{c_d P w^3}{A w_m} dl, \tag{10.21}$$

where A is the area normal to the flow path and w_m is the mean relative velocity in the flow channel with this area. The hydraulic diameter, D_h, of a flow channel is defined as $4A/P$ so that this expression may be written as

$$j_{12} = \int_0^l \frac{4 c_d w^3}{D_h w_m} dl. \tag{10.22}$$

This equation shows very clearly the importance of the hydraulic diameter on the dissipation losses in a radial compressor; high flow coefficient stages with wide flow channels have a large hydraulic diameter giving low losses, and low flow coefficient stages with narrow flow channels have a small hydraulic diameter giving high losses.

The exact values for the dissipation coefficients are determined by the state of the boundary layer in the flow channel. Traupel (2000) identified the following interesting relationships between the dissipation coefficient, c_d, and the friction coefficient, c_f, in channel flows: for accelerated flows, as in a turbine flow channel, $c_d < c_f$; for fully developed channel flow, $c_d = c_f$; and for decelerated flows, as in diffuser flow channels, $c_d > c_f$. This explains the apparent paradox that for an accelerated fluid, as in a turbine, the boundary layers are thin and, as a consequence, the wall friction factors are high, but the losses are nevertheless low. By contrast, in diffusing flows, as in compressors, the boundary layers are thick and the wall friction factors are low, but the losses are high. Denton (1993) and Greitzer et al. (2004) show a similar effect by explaining the variation of shear stress with velocity for boundary layers with different pressure gradients.

The Traupel approach was used by Wachter and Woehrl (1981) and Casey (1985) for radial compressor stages. Wachter and Woehrl (1981) used the friction factor for a pipe with a similar hydraulic diameter and Reynolds number as the compressor flow channels to estimate the friction coefficient, but increased this with an empirical factor depending on the amount of deceleration in the flow to give the dissipation coefficient. Dietmann and Casey (2013) provided an analytic approach to the determination of the integral in (10.22) based on the geometry of typical process compressor stages.

The specific entropy generation in (10.22) is proportional to the surface velocity cubed divided by the mean velocity level. In a flow channel with no large variation of velocity across the flow area, this reduces to a dependence on the square of the mean velocity. In blade rows, however, the suction surface velocity is higher than the pressure surface velocity, so the majority of loss generation takes place on those parts of the blade or end-wall near the suction surface. In particular, this means that the loss generation on the suction surface in axial compressors is several times (typically five times) greater than on the pressure surface (Miller and Denton, 2018).

The losses predicted by this approach also include effects due to changes in solidity, as shown by Denton (1993) for the case of axial turbines. A blade row with high space-chord ratio will have higher suction surface velocities due to a higher blade-to-blade loading and consequently a higher loss. A blade row with a lower space–chord ratio has a higher wetted surface area with more blade surface and as such also has a higher loss, leading to an optimum space–chord ratio between these extremes. Wachter and Woehrl (1981) used this technique to suggest the blade number for radial impellers.

The dissipation coefficient suggested in (10.17) only considers the effect of the local Reynolds number and is valid for a turbulent flow on a smooth flat plate. In reality, the dissipation coefficient depends strongly on whether the boundary layer is laminar or turbulent, and whether the surface is rough or smooth. This means that boundary layer transition and roughness play a very important part in determining the loss. Boundary layer transition in turbomachines is an extremely complex process, much more so than on aircraft wings, due to the lower Reynolds number, high freestream turbulence and unsteady flow arising from rotor–stator interaction; see Section 5.5.6. A very approximate boundary layer transition rule for Reynolds numbers within the transitional range in turbomachinery is to take the profile boundary layer as being laminar up to the point of maximum surface velocity and turbulent downstream of it. Typically, however, in radial machines the boundary layer is considered to be turbulent throughout. Chapter 16 points out that a similar approximation is also generally made in CFD simulations.

In turbomachinery blade surface boundary layers, as with fully developed pipe flow, the effects due to laminar and turbulent flow need to be distinguished to determine the appropriate dissipation coefficient. As with the pipe flow, there is a lower critical Reynolds number below which the flow is effectively laminar. In addition, the effects due to roughness at higher Reynolds numbers lead to an upper critical Reynolds number above which the dissipation is determined fully by the surface roughness, and a region where the dissipation depends on the roughness and the Reynolds number. The simple model for the dissipation of a flow over a flat plate, as in (10.17), needs to be modified for describing turbomachinery flows as roughness, streamwise pressure gradients, secondary flows, vortices or other flow structures such as laminar separation bubbles make the flow through turbocompressors far more complex than that over a flat plate and may increase the local dissipation rates.

10.3.4 Dissipation Coefficients in Radial Compressors

The studies of Casey and Robinson (2011), Dietmann and Casey (2013) and Dietmann et al. (2020) worked towards an approximate estimate for the dissipation coefficient in compressors as a function of Reynolds number and roughness. Many sources of test data on the effects of Reynolds number, size and roughness on the efficiency of compressors were analysed in these studies. The clearance losses and leakage losses are not largely affected by the viscous effects, and so these test data provide a method of determining both the proportion of losses related to frictional dissipation and how this proportion changes with Reynolds number. After extensive examination, Dietmann et al. (2020) followed the approach of Gülich (2008) for pumps, Traupel (2000) for turbines and Casey and Robinson (2011) for radial compressors to define equations for the representative dissipation coefficient of radial compressor impeller blades as a function of Reynolds number and roughness in the laminar and turbulent regions as

$$c_{d,lam} = \frac{4k_1}{Re^{0.5}}, \quad c_{d,turb} = \frac{0.544}{\left\{-\log_{10}\left(0.2\dfrac{R_a}{c} + \dfrac{12.5}{Re}\right)\right\}^{2.15}}, \quad Re = \frac{w_1 c}{\nu}. \tag{10.23}$$

The Reynolds number in these equations is based on the mean blade inlet relative velocity, the inlet kinematic viscosity and the blade camber line length. Laminar and turbulent dissipation coefficients are combined by using a blending function P to determine the dissipation coefficient in the transition region to give

$$t = k_2\left(\frac{c_{d,lam}}{c_{d,turb}} - 1\right), \quad P = \frac{1}{1+e^{-t}}, \quad c_d = Pc_{d,lam} + (1-P)c_{d,turb}. \tag{10.24}$$

P varies between 0 for turbulent and 1 for laminar flow and blends the two regions together. Dietmann et al. (2020) recommends that values of $k_1 = 2.656$ and $k_2 = 5$ are used.

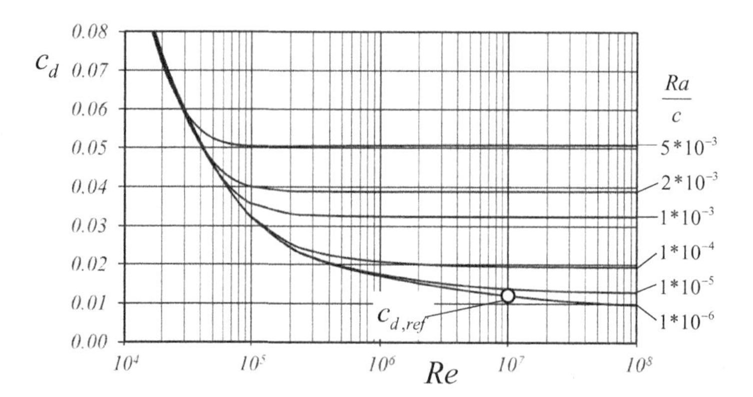

Figure 10.1 Dissipation coefficient for compressors using equations from Dietmann and Casey (2013).

The experimental data analysed with this system of equations were interrogated to determine how the sand grain roughness can be related to the actual surface roughness. This analysis suggested that with these equations the sand roughness should be taken to be the same as the centre line average roughness of the surface, $k_s = Ra$. These equations then lead to a lower critical Reynolds number of 200,000, below which the flow can be considered to be laminar; this is in line with many sets of experimental data on axial compressors, for example Schaeffler (1980) and Traupel (2000). Figure 10.1 shows the variation of this dissipation coefficient as a function of relative roughness and Reynolds number. For a radial stage with a diameter of around 300 mm operating in ambient air and with a tip-speed Mach number of 1, this gives a dissipation coefficient of 0.03, which is 50% higher than the constant value recommended by Denton (1993) for axial turbines but gives better predictions of the total dissipation losses in radial compressors.

10.3.5 Reynolds-Dependent and Reynolds-Independent Losses

There are many publications on scaling laws for the effect of size and Reynolds number on efficiency of turbomachinery going back well over 100 years. Dietmann (2015) and Diehl (2019) provide an overview of these publications, and this forms the basis for the short description given here. On the assumption that all the losses are viscous losses and can be scaled with the Reynolds number raised to a suitable power, an early formulation was to scale all the losses, as $1 - \eta$, as a function of the diameter or the Reynolds number as

$$\frac{1-\eta}{1-\eta_{ref}} = \left(\frac{D_{ref}}{D}\right)^m, \quad \frac{1-\eta}{1-\eta_{ref}} = \left(\frac{Re_{ref}}{Re}\right)^m, \tag{10.25}$$

where m is the Reynolds number exponent with a value close to 0.2, but varying from 0.1 to 0.4 in different literature sources for different types of turbomachinery. One of the important steps for improving the correlations was to acknowledge that not all the losses are caused by viscous dissipation and to split the losses into Reynolds-dependent and Reynolds-independent losses whereby the scaling equations changed to the form

$$\frac{1-\eta}{1-\eta_{ref}} = a + (1-a)\left(\frac{Re_{ref}}{Re}\right)^m, \tag{10.26}$$

where a is known as the Reynolds-independent loss fraction. Casey (1985) argues that this equation is incorrectly formulated as the Reynolds-independent loss fraction, a, cannot be a constant with a change in the Reynolds number, as is shown later in this section. Nevertheless, this approach established the idea that the losses are not all related to the viscous boundary layers, nor are they all affected by the Reynolds number.

The most rational efficiency definition for a compression process is the polytropic efficiency based on the dissipation losses or entropy rise, as defined in Section 4.4, as

$$\eta_p = \frac{\int_1^2 vdp + ½(c_2{}^2 - c_1{}^2)}{w_{s12}} = 1 - \frac{\int_1^2 Tds}{w_{s12}} = 1 - \frac{j_{12}}{w_{s12}}. \tag{10.27}$$

In this equation j_{12} is the dissipation loss related to the entropy increase, and w_{s12} is the shaft work. If we now consider separately the losses in each blade row of a multistage adiabatic compressor, an equation for inefficiency of the whole flow path of the machine can written as

$$1 - \eta_p = \frac{\sum_i j_i}{\sum_i \Delta h_{t,i}}, \tag{10.28}$$

where j_i is the dissipation loss in each individual blade row and its associated flow channel, i, and $\Delta h_{t,i}$ is the total enthalpy change of the rotor blade rows (the total enthalpy change is zero for stators).

The dissipation loss in each blade row can be written in terms of a nondimensional entropy loss coefficient and the local blade inlet relative velocity as in (10.1). The losses can be considered to be made up of many individual independent loss sources, only some of which depend on the Reynolds number, so the loss coefficient for any component in the compressor is the sum of a Reynolds-dependent loss and a loss which has no dependence on the Reynolds number:

$$\chi_i = \chi_{non\_Re} + \chi_{Re}. \tag{10.29}$$

The Reynolds-dependent loss coefficient can be defined from (10.1) and (10.22) as

$$\chi_{Re} = j_{12}/½w_1^2 = \frac{2}{w_1^2}\int \frac{4c_d w^3}{D_h w_m} dl. \tag{10.30}$$

This term can be integrated along the flow path in a throughflow calculation to determine the dissipation loss coefficient for a given blade row, as discussed in Section 15.7.2.

In combination with (10.16), (10.29) results in

$$1-\eta_p=\frac{\sum_i \frac{1}{2}w_i^2\left(\chi_{non\_Re}\right)}{\sum_i \Delta h_{t,i}}+\frac{\sum_i \frac{1}{2}w_i^2\left(\chi_{Re}\right)}{\sum_i \Delta h_{t,i}}=\alpha+\beta. \tag{10.31}$$

In this equation, the first term, α, represents the losses that are independent of the Reynolds number. The second term, β, represents the losses that are Reynolds dependent. The Reynolds-dependent losses are the profile losses and end-wall friction losses, but may include also disc friction losses where these are significant. The important contributions to the Reynolds-independent losses are the tip clearance losses, the leakage losses over shrouds and the shock losses. Both terms, α and β, are not physically the losses but are inefficiencies, that is, they are the losses divided by the work input.

For a given compressor design operating close to the peak efficiency point at a fixed speed, it may be taken that the first term in (10.31), α, is a constant, unless some aspect of the design, such as the nondimensional clearance level, is changed. The second term, β, changes with the Reynolds number, the size and the relative roughness together with the change in the associated dissipation coefficient. The expected variation in β with the Reynolds number for a given roughness is shown in Figure 10.2; β decreases with increasing Reynolds number until the upper critical Reynolds number is reached. The value of β at a high Reynolds number is then dependent only on the roughness. The variation of the Reynolds-dependent losses with the Reynolds number makes it clear that the Reynolds-independent loss fraction is not constant even though the Reynolds-independent losses do not change.

The changes in the Reynolds-dependent losses with the Reynolds number is related to the change in the dissipation coefficient so that with the introduction of reference conditions, at a certain Reynolds number and roughness to be discussed later, it follows that

$$\chi_{Re}=\chi_{ref}\frac{c_d}{c_{d,ref}}, \tag{10.32}$$

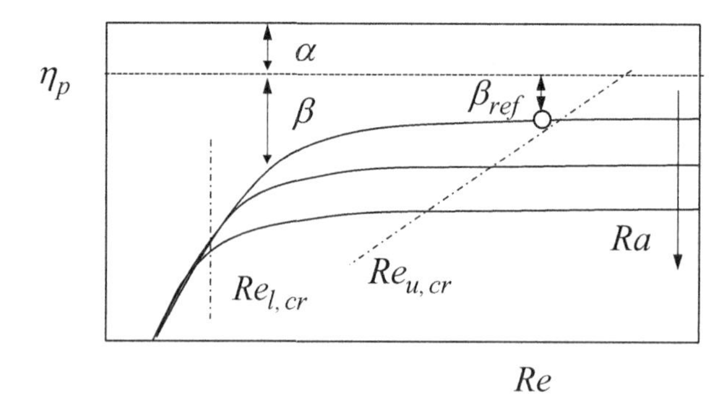

Figure 10.2 Expected variation of the Reynolds-dependent losses, β, and Reynolds-independent losses, α, with Reynolds number and roughness.

where c_d is the dissipation coefficient of an equivalent flow at the same Reynolds number and roughness. This states that the Reynolds-dependent losses scale proportionally to the change in the dissipation coefficient of the representative flow. Substituting this into (10.31) yields the following:

$$\beta = \frac{\sum_i \frac{1}{2} w_i^2 (\chi_{Re})_i}{\sum_i \Delta h_{t,i}} = \frac{\sum_i \frac{1}{2} w_i^2 \left(\chi_{ref} \frac{c_d}{c_{d,ref}}\right)_i}{\sum_i \Delta h_{t,i}}. \tag{10.33}$$

The summation across all components in (10.33) would allow this approach to be used for a multistage machine in that each individual component could have a different dissipation coefficient and its own influence on the losses. If sufficient information were available on the individual dissipation coefficients, it could be used to assess the Reynolds effects on a component-by-component basis, as is the case in the stage-stacking methods described in Section 18.6.

It is also possible to use (10.33) in a simpler way to estimate the sensitivity of the overall efficiency on the roughness or the Reynolds number. If the simplification is made that the change in dissipation coefficient in the first rotor row is representative for all blade rows, then this leads to an equation that can then be used for individual stages of the form

$$\beta = \frac{\sum_i \frac{1}{2} w_i^2 (\chi_{ref})_i}{\sum_i \Delta h_{t,i}} \frac{c_d}{c_{d,ref}} = \beta_{ref} \frac{c_d}{c_{d,ref}}, \tag{10.34}$$

where β_{ref} is the value of the Reynolds-dependent inefficiency (1-η) at the reference conditions for the first impeller. On this basis the effect on the inefficiency of a change in the friction factor due to a change in the Reynolds number, size or the relative roughness can be calculated for a typical stage as

$$1 - \eta_p = \alpha + \beta_{ref} \frac{c_d}{c_{d,ref}}. \tag{10.35}$$

The first term in this equation represents the Reynolds-independent inefficiency, and the second represents the Reynolds-dependent inefficiency. Note that α can be expected to remain constant, but that c_d, varies with the size, the Reynolds number or the roughness; β_{ref} depends on the geometry of the stage; and $c_{d,ref}$ depends on the reference conditions selected.

A change in the Reynolds number, roughness or size from condition 1 to condition 2 causes a change in the dissipation coefficient, and an expression is then obtained for the change in efficiency at the optimum point as

$$\Delta \eta_p = \eta_2 - \eta_1 = \beta_{ref} \frac{c_{d,2} - c_{d,1}}{c_{d,ref}} = -\beta_{ref} \frac{\Delta c_d}{c_{d,ref}}. \tag{10.36}$$

This equation is shown schematically in Figure 10.3, highlighting that the change in efficiency due to a change in the Reynolds number, roughness or size is proportional to a change in the associated dissipation coefficient. This approach is not directly coupled to any model for the individual losses, in that the coefficient β_{ref} is simply the

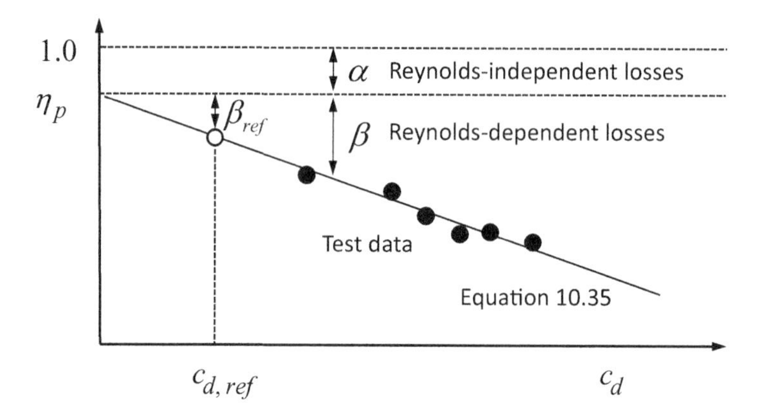

Figure 10.3 Expected variation of measured efficiency with dissipation coefficient.

inefficiency due to friction losses at the reference conditions, and can be determined by measurement or by simulations. The equation does not require information on the Reynolds-dependent losses and can be used to scale efficiency with changes in size, Reynolds number and roughness so long as these are taken into account in the dissipation coefficient, as proposed by Casey and Robinson (2011).

On the basis of (10.35), it is possible to establish the magnitude of the losses that are Reynolds independent in a particular compressor from numerical or experimental data. To this end, the performance of the compressor at different Reynolds numbers is needed, whereby for each different Reynolds number the appropriate dissipation factor can be determined from (10.24), and a figure similar to Figure 10.3 can be made with the experimental or simulated performance data for the efficiency. A straight line can be determined which is the best fit of (10.35) representing the expected linear variation of efficiency with the dissipation coefficient. The intercept of the straight line with the vertical axis represents a virtual situation where the dissipation coefficient is zero and the efficiency is only determined by the Reynolds number–independent losses. If the data being used were based on a simulation, the magnitude of Reynolds number–independent losses with no dissipation can be estimated by using a numerical simulation where all surfaces are treated as slip walls (Diehl, 2019). In the cases examined by Diehl (2019), such a calculation lies close to the linear extrapolation of simulations at higher Reynolds numbers to zero dissipation. In reality, the dissipation coefficient cannot reduce to zero. The reference value of the dissipation coefficient $c_d = 0.012$, as suggested by Casey and Robinson (2011), is the practical lower limit of the dissipation coefficient, which with the equations used here, is close to that of the dissipation factor for a hydraulically smooth surface at an extremely high-chord Reynolds number of 10 million (see Figure 10.1).

The work of Robinson et al. (2012) and Dietmann and Casey (2013) provides a useful estimate of the value of β_{ref} based on an analysis of extensive test data on specific machines at different levels of dissipation. This value can be determined from the test data by examining the slope of the line in a plot similar to Figure 10.3. In both of these publications, test data are presented for a change in size, Reynolds number and roughness and generally confirm the linear variation of the losses with the dissipation coefficient. Dietmann and Casey (2013) also describe how to estimate the value of β_{ref} by integrating (10.30) and also by interpreting the value from correlations to be described in Section 10.7. Dietmann (2015) also carried out CFD simulations of a range of different stages to determine the required slope, and this approach has also been used by Diehl (2019). For axial compressors, values of β_{ref} are lower than those of radial machines. In radial machines, the values of β_{ref} are found to decrease with an increase in the design flow coefficient of a stage, which is an effect of the hydraulic diameter of the flow channels identified in (10.22).

Dietmann et al. (2020) suggest the following correlation for the value of β_{ref}

$$\beta_{ref} = 0.05 + \frac{0.002}{\phi_{t1} + 0.0025}, \tag{10.37}$$

and on this basis, it is possible to estimate the expected value of the Reynolds-dependent losses of a radial compressor stage purely on the basis of its flow coefficient. The change in efficiency is also accompanied by a shift in the flow coefficient at the point of peak efficiency and a change in the pressure rise coefficient, whereby equations for this are given by Dietmann (2015). The correlation given by Dietmann (2015) is based on an analysis of all of the 30 test cases available to him. The level of agreement of the final correlation for the measured change in efficiency with the different test cases is generally within $\pm 25\%$, such that a change of 1% point in efficiency is calculated to within $1 \pm 0.25\%$ points. Given that the parameter β_{ref} varies by over 600% over the whole range of compressors considered, this scatter is actually quite small and is really a lot better than might be expected with such a model when applied to a wide range of compressor types, including axial and radial, single stage and multistage, incompressible and compressible, subsonic and transonic, and highly loaded and weakly loaded machines. The approach clearly requires further refinement, but if test data for a particular stage are available, the value of β_{ref} for a particular stage can be determined in the same way and may give better agreement than the use of the correlation above.

10.4 Other Aerodynamic Losses

10.4.1 Trailing-Edge Loss

The loss discussed in the preceding sections is the loss generated in the boundary layers of the profile and end-wall surfaces. Further entropy production occurs as the profile boundary layers mix out in the wake away from the wetted surfaces. For thin trailing edges, provided that the mixing takes place at constant area, a mixing calculation given by Denton (1993) gives a very simple result:

$$\varsigma = \Delta p_t \Big/ \tfrac{1}{2} w_2^2 = \frac{t C_p}{s \cos\beta_2} + \frac{2\theta_{te}}{s \cos\beta_2} + \left(\frac{\delta^* + t}{s \cos\beta_2}\right)^2, \tag{10.38}$$

where $2\theta_{te}$ is the total momentum thickness for both surfaces at the trailing edge, $s \cos \beta_2$ the width of the channel at the trailing edge, t the trailing-edge thickness, δ^* the trailing-edge displacement thickness of the wake and C_p the trailing-edge base pressure coefficient defined as

$$C_p = \frac{p_b - p_{ref}}{½ w_{ref}^2}. \tag{10.39}$$

The first term (10.38) is the loss due to the low base pressure acting on the trailing edge. Typical values of C_p are about –0.2, although the value varies greatly with the state of the boundary layer, the shape of the trailing edge and the ratio of trailing-edge thickness to boundary layer thickness. The second term is the boundary layer plus wake mixing loss. The last term is the mixing loss due to the trailing-edge blockage. The permissible trailing-edge thickness is usually limited by mechanical limitations, but it is clear from this that radial impellers with a cutoff trailing edge and radial diffusers with wedge shaped blades are disadvantaged by this loss mechanism. The effect of a separation is to increase the effective trailing-edge thickness by the displacement thickness of the separation, and this can also be modelled by (10.38) (Denton, 1993). If a separation occurs on the blade surface and reattaches before the trailing edge, that is, a separation bubble, then it is more difficult to estimate the effects on loss with a simple model as the dissipation in the separation bubble is not included.

10.4.2 Leading-Edge Loss

In the preceding discussion, the peak efficiency operating point with a near-zero incidence on to the leading edge has been considered. When operating off-design, the incidence changes. In a subsonic flow, the incidence affects the local surface velocity distribution and hence changes the entropy generation in the boundary layers through its effect on the velocity term in (10.22). The effects depend greatly on the detailed shape of the leading edge, and methods of predicting incidence loss which do not make allowance for this should be distrusted.

The losses increase at higher and lower off-design flows and a simple parabolic increase in losses with incidence are often assumed. In some publications, for example Pfleiderer and Petermann (1986), so-called shock losses are used, which has nothing to do with compressible flow shock waves (see Section 6.3). A simple mass and momentum balance for a leading edge with zero thickness (i.e., a flat plate which can exert no streamwise force on the fluid) shows that in incompressible flow the component of kinetic energy normal to the leading edge must be lost. In these methods, the difference between the circumferential velocity at blade inlet for zero incidence and the actual circumferential velocity at a given operating point are expressed as kinetic energy. In the shock loss models, a large fraction of this kinetic energy (between 50 and 70%) is assumed to be dissipated at the leading edge of a blade row to allow the flow to take up the direction determined by the blade geometry. However, there is no theoretical basis for such an assumption, as the incidence losses at the leading edge are related to the velocity profile around the leading edge and the associated boundary layer development, as described in Section 7.2.5. The mechanisms really responsible for the increase in loss at the leading edge with incidence may be either separation, increasing mixing loss, early boundary layer transition which increases the turbulent wetted area, a local shock in the flow or a thicker wake as a result of a separation bubble.

Leading-edge flow is especially important in transonic compressor blades with supersonic relative inlet flow. For supersonic inlet flow onto thin blades – such as are typical of transonic compressors – Freeman and Cumpsty (1992) show that the loss can again be predicted from a global mass, energy and momentum balance. This is possible because, for thin blades, the streamwise force exerted by the leading edge is a relatively small term in the momentum balance and can be estimated by making reasonable approximations. Lohmberg et al. (2003) show that this theoretical model also applies to transonic radial compressor inducers.

10.4.3 Mach Number Loss

An increase in losses and a decrease in operating range of a cascade of blades with an increase in the inlet Mach number has been described with regard to the alpha-Mach diagram in Section 7.3.3. Denton (1993) suggests that there are three main mechanisms for additional Mach number–related losses at the design point. In the first instance, the ratio of the peak suction surface velocity to the inlet velocity increases as the inlet Mach number increases, as described in Section 7.2.6 and as shown in Figure 7.5. This causes higher dissipation losses on the suction surface as the surface velocity increases. Further, if the inlet Mach number exceeds about 0.8, then a critical Mach number is reached where the suction surface locally attains supersonic flow. At higher Mach numbers, the flow may accelerate to form a normal shock in the passage with associated shock losses, as described in Section 6.3.1. Finally, if the Mach number before the shock is greater than 1.3, the pressure gradient associated with deceleration of the flow across the shock can cause separation of the flow on the suction surface with associated losses related to the mixing out of the separation and increased dissipation due to higher flow velocities caused by the blockage.

10.4.4 Tip Leakage Loss

The physical details of the tip clearance flow and the associated vortex and flow mixing are described in Section 5.7, where it is shown that the evolution of the tip clearance flow and leakage vortex is extremely complex. Although research into the losses caused by the tip clearance flow has been very active for several decades, the flow patterns are so complex that there is still no completely satisfactory simple loss model. The simple models all indicate that the efficiency deterioration is proportional to the amount of mass flow leaking through the clearance gap, and as such is then related to the area of the clearance gap that is available along the length of the blade. The empirical correction for the change in efficiency with clearance in centrifugal stages described in Section 8.7.2 supports this idea by suggesting that the loss is proportional to the size of the nondimensional tip clearance gap, and indicates that a 1% increase in relative clearance, $\Delta\varepsilon/b_2$, produces a 0.3% loss in efficiency for radial compressors. Many studies in centrifugal compressor impellers have identified a different sensitivity of the losses to changes in the size of the gap; a summary is provided by Diehl (2019). The different sensitivity may be due to the difficulty of estimating running clearances in a rotating impeller, and it may be that the clearance gap at the outlet is not always representative of the gap along the whole blade profile. The wide range of values for the sensitivity suggests that different loss mechanisms may also be involved in different stages.

Several loss mechanisms have been suggested to show how the losses increase with the size of the clearance gap related to the flow patterns. The simplest model for tip clearance loss is to model the complex process with momentum and continuity conservation equations as an inviscid process involving the mixing of two streams of different velocity, as shown in Figure 10.4. This was put forward by Senoo and Ishida (1986), also analysed by Denton (1993) and used by Storer and Cumpsty (1994) in studies of axial compressor tip clearance flow. Storer's model takes into account the angle at which the leakage jet enters the main flow. This angle is less than 90° because of the streamwise velocity of the leakage jet. The latter arises from the velocity of the leakage flow entering the clearance space and also because of any streamwise pressure gradient acting within the clearance space.

Another popular loss model in turbine flows considers the secondary kinetic energy in the flow to be a source of loss (Harvey et al., 1999). When applied to the tip clearance jet, this tends to underestimate the loss as it neglects any effect of the tip clearance flow in increasing dissipation in the primary flow through the action of the tip clearance vortex. One source of loss not included in the inviscid conservation analysis is the very high dissipation close to the blade tip in the shear layer formed between the leakage jet and the mainstream flow. Additional dissipation can occur because the tip clearance vortex collides with a splitter blade or interacts with a shock wave, or because the vortex passes over an adjacent blade tip gap. Zhang et al. (2020) have compared 1D models with CFD preditions in an attempt to identify the source of these additional losses.

In practice, the flow leaking across the blade tip also causes a distortion of the flow at the impeller outlet (Jaatinen-Värri et al., 2016), such that an additional

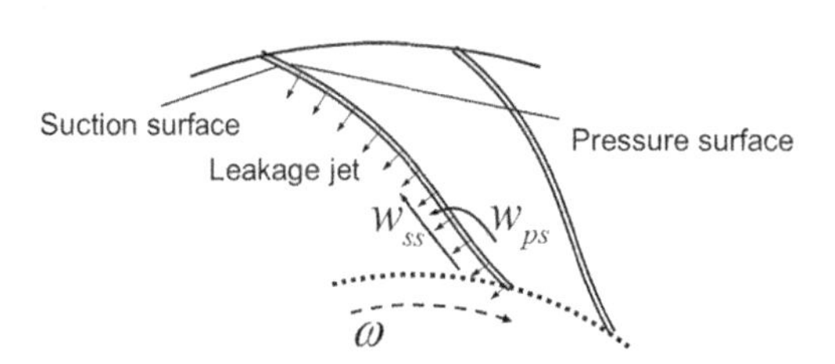

Figure 10.4 Tip leakage flow of an impeller viewed as a jet in a cross-flow.

source of loss is clearly related to the effect of this distorted flow pattern on the diffuser. Increasing the clearance leads to an increase in the diffuser inlet flow angle near the casing, causing a longer flow path along the casing wall in the diffuser with higher dissipation and possible reverse flow. A further issue is to what extent the tip clearance jet mixes out with the core flow and what proportion of this kinetic energy is lost. In addition, the tip-leakage flow results in a low streamwise velocity in the vortex core. The low velocity in the core of the vortex causes the vortex to increase in diameter as the flow moves into a region of higher pressure (Greitzer et al., 2004). This is responsible for flow blockage in the impeller flow channel, which increases the effective meridional velocity at outlet and causes a reduction in the work done.

The leakage flow pattern is so complex that there is still no final clarification of the best simple mixing loss model for clearance losses in centrifugal impellers. Fortunately for the designer, these flows and the losses associated with them are included in modern CFD simulations, allowing the designer to take their effects into account even if simple models are inadequate. Unfortunately, the greater understanding has not so far indicated any practical ways to reduce tip clearance losses for impellers at a given level of clearance. The target for the designer continues to be to achieve the tightest clearances possible.

10.4.5 Leakage Loss over Shrouded Impellers

For a shrouded impeller with a leakage flow path over the impeller, there is more mass flow passing through the impeller blade passages than if there was no leakage. The flow leaking over the shroud has already had work done on it by the impeller, and the dissipation of this energy in the leakage flow is an additional source of loss. In addition, when the leakage flow mixes with the main flow at the impeller inlet, it will have a higher enthalpy, a higher swirl velocity and a higher entropy than the main flow, and so additional entropy is generated during mixing of the two flows and possibly due to changes in the flow angle causing incidence losses at the impeller inlet. This is categorised as a parasitic loss and is discussed in Section 10.6.

Several studies have been published on the relative performance levels of specific open and closed impellers. It is generally considered that an open impeller has a lower performance due to the tip clearance losses between the moving blades and the casing. These increase if the clearance becomes a larger portion of the blade span, so that it can be expected that shrouded impellers perform better than open impellers if the design flow coefficient is low and the blade span is small. Harada (1985), however, found that an open impeller of a high flow coefficient stage with a small clearance has a higher performance towards surge compared to an identical shrouded one and that it also has a wider stable operating range. Dalbert et al. (1999) also confirmed this experimental result on a high flow coefficient stage, where CFD simulations of this stage identified a flow separation in this impeller with a shroud, but this was not present in the open impeller as it was cleaned up by the tip clearance flow.

10.5 Loss Correlations in Centrifugal Stages

10.5.1 Correlations

The key scientific questions are about the cause of the losses, not correlations. There is an important difference between understanding loss generation on the basis of physical models, as described in the preceding section, and developing a correlation based on available experimental data. A useful correlation can be developed based on test data, but it may not be adequate to predict performance data of stages not included in the development of the correlation. A physically based description of the process of loss generation based on a deeper understanding may be more effective over a wider range of cases. Nevertheless, loss prediction on a theoretical basis is so complex that many popular systems of empirical loss estimation for radial compressors have set up a pragmatic set of correlations for the individual loss sources from available experimental data, often with a rather flimsy scientific background.

The coefficients in these loss models must generally be adjusted to match experimental data to ensure that the method obtains reasonable agreement with a representative range of test cases. There are many publications available where authors have calibrated a particular scheme of loss modelling with their own data set of measurements on impellers and stages of different types, all showing reasonable agreement with measurements. These publications are too numerous to mention here, with well over 40 publications collected by the authors on this theme after 2000. In most of these publications, the loss correlations used were those developed by other authors before 2000. A recent publication by Velasquez (2017) finds that errors of up to 8 percentage points in efficiency can occur by the use of inappropriate correlations and so extensive calibration is needed to ensure that such methods are validated for particular stage types. After many years of experience with a simpler strategy, the authors of this book remain fairly unconvinced of the general validity of such approaches for centrifugal stages and prefer the global approach outlined in Section 10.7 for preliminary estimation of the efficiency of a well-executed stage. In any case, for more exact estimates of losses it is nowadays advisable to use 3D CFD simulations rather than correlations.

The most comprehensive method of loss prediction for radial compressor stages known to the authors is that devised by Herbert (1980), which embodies detailed loss models for IGV, impeller, vaneless space and vaned diffuser. Popular models that are often used and which are referred to consistently often in the literature are those of Galvas (1973) and Aungier (1995b). Oh et al. (1997) provide a summary of many similar methods with a recommendation for an optimum set of individual loss models. The actual equations used by these authors for the individual losses are not given here as they can be found in detail in these sources and in many secondary sources.

The full list of aerodynamic loss mechanisms in centrifugal stages considered in the correlations described by many authors includes the following:

- Inlet losses, for the frictional loss in the inlet duct, perhaps including IGV blades or an inlet silencer (especially in large turbochargers)

- Incidence loss on all blade rows, to account for the additional losses due to the difference in flow angle and the blade angle at the blade or vane inlet
- Shock loss, to account for losses associated with bow shocks and for shocks in the blade and vane passages, including the possibility of shock-induced separation
- Blade loading loss, to account for blade-to-blade pressure gradients that cause secondary flows
- Hub-to-shroud loading loss to account for the hub-to-shroud pressure gradients and their associated secondary flows
- Inlet distortion loss, to allow for the effects of a poor inlet flow distribution
- Skin friction loss, to account for surface friction
- Blade clearance loss, to account for the leakage flow of open impellers due to the passage of flow through the tip gap
- Recirculation loss, to account for reverse flow at the impeller tip at low flow rates, and any flow recirculation at the trailing edge
- Mixing loss, to account for the mixing of blade wakes and regions of separated flow
- Vaneless diffuser loss, for the frictional losses in the vaneless part of the diffuser
- Vaned diffuser loss, for the friction losses in the vaned diffuser
- Volute or return channel loss, for the losses occurring downstream of the diffuser
- Jet-wake loss models, considering any losses in stages which exhibit a strong jet and wake flow at impeller outlet

The preceding list of loss mechanisms includes many items that are related to friction losses on the wetted boundaries for the different components. In most methods, these are the dominant losses and are estimated using friction factors coupled with the flow path length and hydraulic diameter, or with dissipation factors somewhat similar to the methods of Traupel (2000) and Denton (1993) discussed in Section 10.3.3.

In most publications, the blade clearance losses in open impellers are estimated based on the flow area available for the leakage flow. Various correlations have been attempted to capture the additional losses due to high deceleration in the flow channels, making use of the diffusion factor, as described in Section 7.5.10. The secondary flow losses are determined on the basis of the hub-to-shroud velocity differences and the suction surface to pressure surface velocity differences. These are probably the weakest link in the correlation methods, and it is not clear that these empirical methods correctly predict the balance of losses between secondary flows and friction: more blades increase the frictional losses but weaken the blade-to-blade secondary flows.

Additional correlations are provided by many authors to determine key geometrical parameters such as the axial length of the impeller or the number of blades so that these are not determined from the loss correlations. Aungier (1995a) provides a correlation for impeller axial length as follows:

$$\frac{L}{D_2} = 0.014 + \frac{\pi}{4}\phi_{t1} + 0.023\frac{D_2}{D_{c1}}. \tag{10.40}$$

Came and Robinson (1998) suggest

$$\frac{L}{D_2} = \sqrt{\left[0.28(M_{ws1} + 2)\left(1 - \frac{D_{m1}}{D_2}\right)\left(\frac{D_{1c} - D_{1h}}{D_2}\right)\right]}. \tag{10.41}$$

Both of these correlations lead to values of L/D_2 between 0.3 and 0.4 for typical turbocharger and gas turbine centrifugal impellers. Industrial impellers are often much shorter because of the rotordynamic constraints on long shafts; see Section 19.13.

Eckert and Schnell (1961), Pfleiderer and Petermann (1986) and Rodgers (2000) provide correlations for impeller blade number. Just as in axial compressors, the blade number tends to be selected based on the value of a blade-to-blade loading coefficient, analogous to a Lieblein diffusion factor, and not from loss correlations themselves, as discussed in Chapter 7. The authors generally use the criterion based on a blade-to-blade loading coefficient as given in Section 7.5.13.

10.6 Parasitic Losses

10.6.1 Entropy Losses or Additional Work

The parasitic losses usually include three components. Disc friction loss accounts for the skin friction on the impeller backplate and shroud if present. Leakage loss accounts for any leakage from the seals over the shroud of a shrouded stage. Recirculation losses accounts for losses related to recirculation of the flow at inlet or outlet of the impeller, which generally occurs only at part-load conditions. These losses are not independent of each other. They depend on the fluid flow patterns in the impeller side spaces, which in themselves depend on the swirl and mass flow of the leakage entering these domains. The magnitude of the leakage flow in turn depends on the flow patterns in the side spaces, as this determines the pressure drop across the seals. The Reynolds number and the roughness also play a role. In shrouded impellers, the axial thrust is also strongly dependent on flow patterns and pressure drops in the impeller side spaces. More information on this is provided in Section 13.9.

It is important to distinguish between the loss of efficiency and extra work generated by parasitic losses. Traupel (2000) suggests that, as these parasitic losses are not part of the normal flow patterns in the gas path of the impeller at the design point, they should be calculated as additional sources of work that the impeller has to supply and then can be calculated as

$$\lambda = \lambda_{EULER} + \lambda_{DF} + \lambda_{leak} + \lambda_{rec}. \tag{10.42}$$

These additional contributions to the work should be considered as additional enthalpy change across the impeller and need to be added to the Euler work to determine conditions at the inlet of the diffuser. Two of these sources of loss have a similar property in that they involve more or less a fixed value of parasitic loss which is independent of the flow coefficient of the stage. The leakage flow over the seals depends on the pressure ratio across the impeller and not on the mass flow through the stage, so it becomes a more important fraction of the overall flow in low flow coefficient stages. The disc friction loss is a fixed additional power that needs to be supplied to overcome the friction on the backplate and, if present, the shroud. This also becomes relatively more important in low flow coefficient stages. These both have more impact at low flow coefficients, and this gives rise to a practical limit of the application of radial stages to flow coefficients above $\phi_{t1} = 0.005$. Below this value, the disc friction and leakage become proportionally very large, and an efficiency of 50% is hardly possible. At lower flows, positive displacement or rotary compressors are a more appropriate technology.

10.6.2 Disc Friction Loss

The torque needed to overcome friction on a rotating disc is related to the Reynolds number; the spacing between the rotating disc and the casing, s; the size of the disc; and on the leakage flow in the gap, as follows:

$$M_{DF} = (1/2)C_M \rho \omega^2 r^5, \quad C_M = f(Re, s/r, \dot{m}_{leak}) \tag{10.43}$$

The density in this equation is the density of the flow in the space between the disc and the stationary wall. These losses are proportional to the gas density at impeller discharge, ρ_2, which increases with increasing Mach numbers. But as the total power increases in a similar way with the Mach number, the disc friction losses generally remain a fixed proportion of the total power for designs at similar flow coefficients operating at different speeds. Low flow coefficient designs tend to have a comparatively small radius ratio (D_{1h}/D_2) with large discs and, consequently, comparatively large disc friction losses. Further details of the disc friction flows are given in Section 13.9.

Experiments by Dailey and Nece (1960) show that the moment coefficient for both surfaces of an enclosed rotating disc is given approximately by

$$C_M = \frac{0.102(s/r_2)^{0.1}}{Re^{0.2}} = \frac{0.102(s/r_2)^{0.1}}{(u_2 r_2/\nu)^{0.2}}. \tag{10.44}$$

Here, u_2 and ν are the blade speed and dynamic viscosity at the tip of the disc or at the impeller outlet. For the family of stages considered by Hazby et al. (2019), the Reynolds number based on diameter and tip speed is typically about 6.5×10^6 at the design point. With an axial spacing of $s/r_2 = 0.04$, this gives a moment coefficient of close to $C_M = 0.003$ for both surfaces. The magnitude of disc friction power is then given by

$$P_{DF} = \omega M_{DF} = \omega \frac{1}{2} C_M \rho u_2^2 D_2^3 = C_M \rho u_2^3 D_2^2. \tag{10.45}$$

The gas path power from the Euler equation is reduced at low flow coefficients, so the relative magnitude of disc friction power increases for lower flow coefficients as follows:

$$C_{DF} = 1 + \frac{P_{DF}}{P_{Euler}} = 1 + \frac{C_M \rho u_2^3 D_2^2}{\dot{m} \Delta h_t} = 1 + \frac{C_M}{\dfrac{\dot{m}}{\rho u_2 D_2^2} \dfrac{\Delta h_t}{u_2^2}} = 1 + \frac{C_M}{\lambda \phi_{t1}}, \tag{10.46}$$

where C_{DF} is known as the disc friction power input factor. These losses are proportional to the gas density at impeller discharge, which increases with increasing tip-speed Mach number. But as the total power increases in a similar way with the Mach number, the disc friction losses remain a fixed proportion of the total power for designs at similar flow coefficient operating at any speed.

Considering the general trend of variations, it is possible to use an experience-based correlation to relate the value of the disc friction power input factor directly to the flow coefficient as

$$C_{DF} = 1 + \frac{k_{DF}}{\phi_{t1}}. \tag{10.47}$$

For the family of the stages studied by Hazby et al. (2019), a value of $k_{DF} = 0.0017$ gives an excellent agreement with the numerically predicted data. The extra work due to windage manifested in the power input factor, C_{DF}, has a direct effect on the compressor efficiency. As the mass flow through the impeller remains constant, the impeller pressure rise and the work coefficient remain approximately the same, so the additional power required to support the friction on the backplate and the shroud leads to a loss in efficiency. Assuming that the power input factor is close to one, it is possible to derive that

$$\frac{\Delta\eta_{stage}}{\eta_{stage}} = \frac{1}{1 - C_{DF}} \approx -\frac{0.0017}{\phi_{t1}}. \tag{10.48}$$

Several authors have also identified that the mixing of the leakage flows with the core flow through the impeller may cause additional interference losses (Mischo et al., 2009; Guidotti et al., 2014; Hazby et al., 2019).

10.6.3 Leakage Loss over Shrouds

The main difference between an unshrouded impeller and a shrouded one is that the leakage jet in the latter is driven by the difference in pressure between upstream and downstream of the blade row rather than by the difference in pressure from the suction to the pressure surface. Seals are usually included in the spaces between the impeller and the stationary walls to reduce the leakage. The swirling flow of the leakage over the shroud has a strong influence on the pressure gradient in this region and on the axial thrust of the impeller and is considered in Section 13.9. The leakage loss as a parasitic loss is considered in Section 10.6.5.

10.6.4 Recirculation Loss

The nature of the inlet recirculation at part-load is described in Section 17.2.5. The recirculation loss can generally be considered to be negligible at the design point but may increase towards the surge line, and an empirical model for this has been suggested by Qiu et al. (2008) and further developed by Harley et al. (2014). At the design condition, the flow normally passes through the inducer without recirculation. At low mass flow, the casing streamlines experience a separation and the flow direction reverses at the blade tip due to the adverse pressure gradient. The recirculating flow leaves the impeller tip having already picked up a swirl component which adds to the work done by the impeller. The ring stall near the casing creates an aerodynamic blockage at the impeller inlet, driving the flow closer to the hub and reducing its incidence in this region. During this process, the recirculating fluid mixes with the incoming fluid, and this causes recirculation losses. The main flow entering closer to the hub gives rise to a higher centrifugal effect in the impeller so the stage can remain in stable operation even though it has higher losses (Ribaud, 1987; Schreiber, 2017).

10.6.5 The Effect of Parasitic Losses on Performance

The power that is input through the compressor shaft contributes mainly to increase the fluid pressure in the primary gas path and partly to overcome the parasitic losses. These are caused by the presence of a space between the impeller and the stationary walls and the windage on the impeller backface and shroud surfaces. In addition, extra power is required to repeatedly increase the pressure of the recirculating leakage over the shroud, representing a further loss in input power. This is proportional to the ratio of the shroud leakage flow to the total mass flow at inlet to the stage. The so-called power input factor, C_{PIF}, is defined as the ratio of the overall compressor input power to the primary gas path power that is expended to increase the fluid pressure and can be written as

$$C_{PIF} = \frac{P_{gas} + P_{back} + P_{shroud} + P_{leak}}{P_{gas}} = 1 + \frac{P_{back}}{P_{gas}} + \frac{P_{shroud}}{P_{gas}} + \frac{\dot{m}_{leak}}{\dot{m}_1}. \tag{10.49}$$

One use of a component-by-component mean-line analysis (see Section 10.8) is to estimate changes in performance due to adjustments in design or the configuration of the stage. This approach can be used is to estimate the change in performance when a simplified CFD analysis, carried out without leakage and disc friction losses, is modified to take account of the parasitic losses. This is useful in the preliminary design of a stage, since the CFD model at that point usually includes only a single passage for the vane and simplified housing geometry with no leakage flow paths. The simplified CFD methodology enables the grid to be developed easily and a rapid calculation consistent with the timescale of preliminary design. However, this simplified method captures too little of the real flow physics ultimately to give answers of sufficient fidelity. A more detailed CFD analysis, applying a fully featured geometrical model with few simplifications, gives more accurate predictions but requires more effort in grid generation and time to converge. A simple analysis to estimate the effect of the parasitic losses is given in this section.

Hazby et al. (2019) have developed a correlation for the terms in (10.49) based on a numerical CFD analysis of stages with and without leakage flow cavities and comparison with experimental data. The ratio of the shroud and backface powers to the gas path power is evaluated by calculating the overall torque on the corresponding surfaces in the CFD results. The power absorbed by friction on the backface and shroud surfaces remains more or less the same for a range of stages with different flow coefficients, meaning that their relative impact on the overall power increases at low flow coefficients. The contribution to loss from the impeller backface is shown to be larger than that from the shroud at all flow coefficients. The power loss due to leakage flow recirculation also increases as the flow coefficient is reduced, showing similar levels to the shroud friction losses, although this is dependent in detail on the individual arrangement of the seals and tip clearances.

The tip clearance losses of an open impeller are intimately connected with the flow through the blades and are considered an aerodynamic loss. Shrouded impellers are fitted with noncontacting sealing systems, such as labyrinth fins, to close off the regions of the cover discs between different pressure levels. Traupel (2000) provides an estimation of the leakage flow over shroud labyrinth seals as

$$\dot{m}_{leak} = C_c \pi \varepsilon D_{tip} C \sqrt{\frac{\Delta p}{ZRTn_{SL}}}. \tag{10.50}$$

The contraction coefficient C_c is 0.7; C is an empirical correction factor for different types of seal tooth geometry; ε is the clearance gap above the tip diameter of the seal, D_{tip}; and n_{sl} the number of sealing strips. Depending on the number and type of seals, and the clearance gap between the rotating and the stationary parts, the leakage flow over the shroud labyrinth seal can vary. For the stages examined numerically by Hazby et al. (2019), the shroud leakage flow is about 5% of the total mass flow for low flow coefficient stages with $\phi_{t1} = 0.01$, The hub leakage flow is approximately 6.5% leakage flow at a flow coefficient of $\phi_{t1} = 0.01$. Different numerical values of these coefficients may be appropriate for different designs of the sealing systems. For similar designs of seals, the proportion of shroud leakage to total mass flow increases as the stage flow coefficient is reduced approximately as follows:

$$\frac{\dot{m}_{leak}}{\dot{m}} = \frac{C_{leak}}{\phi_{t1}}, \quad C_{leak} = 0.0005. \tag{10.51}$$

The additional power required to repeatedly increase the pressure of the recirculating shroud leakage flow can also be expressed as a leakage flow power input factor, as follows:

$$C_{leak} = 1 + \frac{P_{leak}}{P} = 1 + \frac{C_{leak}}{\phi_{t1}} \approx 1 + \frac{0.0005}{\phi_{t1}}. \tag{10.52}$$

Because the leakage power input factor required increases as the flow coefficient is reduced additional seals, stepped in radius, or more effective tooth geometries for labyrinth seals are often used for low flow coefficient impellers.

The leakage flow predictions in (10.51) correspond to the peak efficiency operating point. The relative proportion of the leakage flow to the main flow increases as the surge point is approached and decreases towards choke. Furthermore, the flow over the impeller shroud is driven by the impeller static pressure at the outlet and is invariably in the direction from the impeller outlet to the inlet. In stages with return channels, such as those studied by Hazby et al. (2019), the direction of the hub leakage flow at peak efficiency condition is generally from stage outlet towards the impeller tip, although this may reverse for operating conditions towards choke. In a last stage compressor where the shaft leaves the housing, except of course in a vacuum application, the direction of the hub leakage flow will normally be from impeller tip towards the shaft.

The addition of a small leakage flow over the shroud does not significantly change the mass flow through the impeller, but the leakage replaces some of the fluid that had previously been drawn in from the inlet flange. As a result, the inlet mass flow reduces by

$$\Delta \dot{m}_1 = -\dot{m}_{leak} \tag{10.53}$$

and the inlet global flow coefficient reduces by the amount of leakage flow approximately as

$$\Delta \phi_{t1} = -\frac{\dot{m}_{leak}}{\rho_{t1} u_2 D_2^2} = -0.0005, \quad \frac{\Delta \phi_{t1}}{\phi_{t1}} = -\frac{0.0005}{\phi_{t1}}. \tag{10.54}$$

This is a relatively small effect at high flow coefficient, but becomes relatively larger as the flow coefficient decreases, increasing to 5% at a flow coefficient of $\phi_{t1} = 0.01$. Given that there is a leakage flow from the stage outlet to the diffuser inlet of a similar amount, the diffuser inlet mass flow remains approximately constant. In a process compressor stage with a return channel, the additional work input on the shroud contour heats the leakage flow over the shroud and increases the temperature of the flow at the impeller inlet. The leakage flow over the impeller backplate heats the flow at the diffuser inlet, and so both contributions to the parasitic losses ultimately appear as an increase in the impeller outlet temperature.

10.7 Global Estimate for Aerodynamic Efficiency at the Design Point

10.7.1 Efficiency as a Function of the Duty and Size

Rather than to sum the separate contributions of aerodynamic loss and parasitic loss derived from correlations from many sources and using empirical coefficients of somewhat doubtful individual validity, the authors of this book prefer a simpler approach for preliminary design purposes. This method uses correlations based only on global parameters representing the duty of the compressor. It is considered that this strikes the right balance in modelling the flow between the grossly oversimple and the unprofitably sophisticated during the preliminary design. There are several correlations of this type, such as those of Rodgers (1980), Casey and Marty (1985), Aungier (1995a) and Robinson et al. (2011) for compressors, together with those of Bommes et al. (2003) for centrifugal fans and Gülich (2008) for different types of centrifugal pumps. These methods estimate the expected stage total–total efficiency of a well-designed stage as a function of the design flow coefficient and the Mach number of

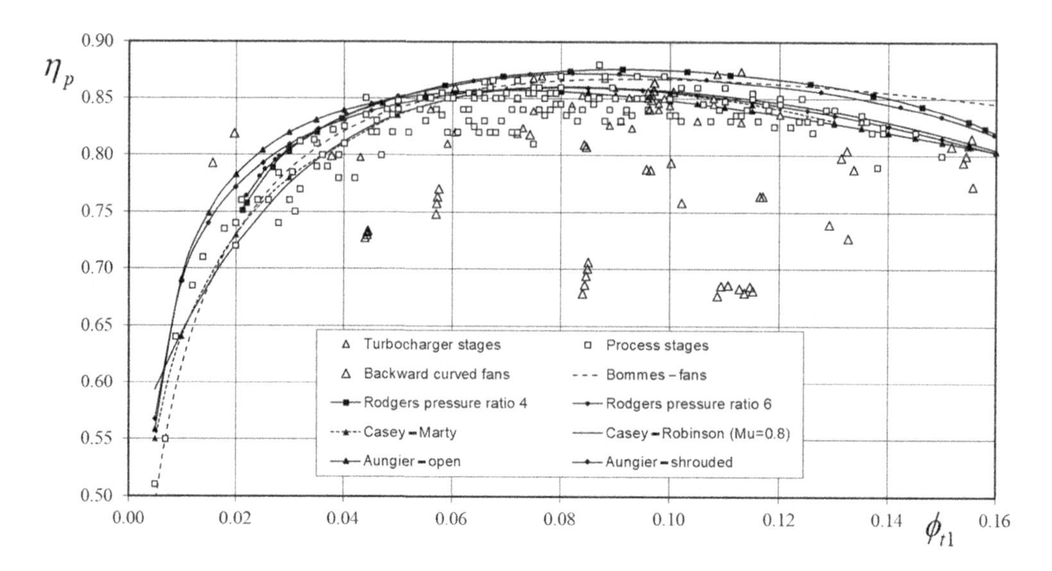

Figure 10.5 Polytropic efficiency versus flow coefficient.

typical large-scale machines with a diameter above 250 mm. Corrections for the effect of the work coefficient are needed, which are usually small. Corrections for the size, Reynolds number, roughness and clearance are needed for small-scale stages with impeller diameters less than 200 mm. The estimated efficiency of some of these different correlations is compared with test data in Figure 10.5.

Certain points are clear from the outset. First and foremost is that among compressors built with the same overall duty in terms of flow coefficient and tip-speed Mach number, there are different geometrical solutions, and some may have better performance than others. But since a general prediction method based on the duty alone is, by definition, required to ignore details of, for instance, rotor blade passage shape, it cannot discriminate between such cases. It can never give an absolute answer. Some empirical datum of expected performance for a standard geometry must be set, and the predictions need to agree with it. Relatively poor machines will fall below such a standard prediction and, it may be hoped, improved machines of the future will exceed it. Correlations of this type identify what Koch and Smith (1976) term the efficiency potential of the design, meaning that this efficiency can be expected at the design condition if the detailed design is carried out correctly using the best state-of-the-art design practices.

The Aungier (1995b) correlations provide equations for what he calls the target polytropic efficiency as a function of global parameters, and these are given in (10.56) and (10.57). This set of equations includes global effects due to open and closed impellers with vaneless and vaned diffusers. Aungier also provides a suggestion for the design value of work coefficient and the axial length of impellers to meet these efficiency targets but provides no information on losses due to high Mach numbers. The Casey and Marty (1985) correlation provides an estimate for the total–total polytropic stage efficiency of industrial shrouded stages (up to tip-speed Mach numbers of 1.0) and is also given by Cumpsty (2004). The Rodgers (1980) correlation provides the isentropic stage efficiency of gas turbine compressor stages with open impellers as a function of specific speed and pressure ratio, up to pressure ratios of 8, and has been published in several forms, one of which is given in Cumpsty (2004). The Robinson et al. (2011) correlation blends the separate correlations of Rodgers and that of Casey and Marty so that the trend of increased losses at high tip-speed Mach numbers is considered. These equations are given in (10.55). The comparison given in Figure 10.5 also includes correlations for the efficiency of radial and mixed flow fans from Bommes (2003), which were given in terms of specific speed in the original publication.

All of the compressor correlations represent more or less the state of the art at the time of their publication with medium-sized impellers ($200 > D_2 > 450$ mm), good vaned diffusing systems and close attention to minimising all clearance gaps. These may nowadays be slightly pessimistic as the use of CFD for compressor design has resulted in some improvement in performance levels, especially at high flow coefficients. It is noteworthy that the experimental data for radial fans tend to have lower efficiencies than the compressors, partly due to a lower technology base and also because of the lower Reynolds number at low speeds. Most companies involved in radial compressor design do not publish such data but will have in-house unpublished experimental experience for their particular type of compressor application and maintain their own correlations of this type, perhaps including other effects not shown here, such as impeller diameter, tip clearance and Reynolds number.

The curves given in Figure 10.5 are based on experimental testing and show that the best total–total polytropic stage efficiencies that can be reached are typically in the range from 86 to 88%, depending on the size of impeller, the quality of the diffuser and the running clearance levels that can be achieved. These high levels of efficiency are found in stages designed for flow coefficients in the range $0.07 \leq \phi_{t1} \leq 0.10$, and this range is known as the region of optimum flow coefficient. Low flow coefficient stages, with $\phi_{t1} < 0.05$, have inherently lower efficiency because of narrow channels with small hydraulic diameters and proportionally more friction losses on the end-walls than the blades, leading to higher overall frictional losses, as indicated by (10.22). In addition, parasitic losses (disc friction and leakage losses) increase as the flow coefficient decreases (as they become a larger part of the total energy input of the stage). High flow coefficient stages, with $\phi_{t1} > 0.10$, have lower efficiencies than those at the optimum flow coefficient because of the high gas velocities and high Mach numbers that occur particularly in the inducer at the casing. To accommodate higher flow rates in a radial wheel, it is necessary to increase the inlet eye diameter, which increases the blade speed at the casing in the inlet eye, and this naturally increases the relative velocity at inlet; see Section 11.4.3. It is not advisable to increase the eye diameter much above 73% of that of the impeller outlet diameter, as this typically results in high curvature on the casing wall with a risk of separation and a reduced centrifugal effect. This constrains the inlet area available, leading to high inlet velocities at higher flow coefficients with higher losses. For higher flow coefficients, mixed flow impellers are used which are less constrained in the ratio of eye diameter to impeller diameter; see Section 11.12.2.

There is a large scatter between the different correlations shown in Figure 10.5 of at least $\Delta\eta = \pm 0.02$. Part of this is due the level of the technology from the different applications in the diagram, including ventilators, fans, industrial compressors, turbocharger compressors and gas turbine compressors. Some is due to the different test methods and some is due to the effects of Reynolds number and size. For a more useful comparison between different stages for different applications, a normalisation of the conditions to a certain size and Reynolds number would be needed. Nevertheless, the trends are consistent in identifying a zone of higher efficiency in the middle of this range of flow coefficient.

Some test data for the peak efficiency from several manufacturers of process and turbocharger centrifugal compressor stages are also included in Figure 10.5, and these data show a scatter similar to that of the correlations. The correlations give information on the best compressors, but poorly designed machines may not reach these levels. Data are also included for backward-swept radial fans from Aldi et al. (2018). Not all the data in this diagram were available in digital form and have often been carefully digitised from the publications available, but they confirm the general trends and scatter. The low efficiency of some of the turbocharger stages is related to their small size, and the efficiency deficit is then due to the lower Reynolds number and high clearance. Few stages are available at high flow coefficients as such stages are generally avoided if possible, because of their lower efficiency, unless size constraints force them to be used. The correlations generally show the upper limit of the efficiencies that can be achieved.

The Casey and Robinson correlation (Robinson et al., 2011) takes account of recent experience of impellers designed with modern CFD methods, which leads to slightly higher peak efficiencies and a smaller drop of efficiency with higher flow coefficient when compared with earlier correlations. The full equations describing the polytropic efficiency by this method, $\eta_p = f(\phi_{t1}, M_{u2})$, as published by Rusch and Casey (2012), are as follows:

$$
\begin{aligned}
&\phi_{t1} < 0.08: \\
&\quad \eta_p = \eta_{\max}\left[1 - k_1(\phi_{\max} - \phi_{t1})^2 - k_2(\phi_{\max} - \phi_{t1})^4\right] + \delta\eta_p \\
&\phi_{t1} \geq 0.08: \\
&\quad \eta_p = \eta_{\max}\left[1 - k_3(\phi_{t1} - \phi_{\max})^2\right] + \delta\eta_p \\
&\qquad \eta_{\max} = 0.86, \quad \phi_{\max} = 0.08 \\
&\qquad k_1 = 27, \quad k_2 = 5000, \quad k_3 = 10 \\
&M_{u2} < 0.8: \quad \delta\eta_p = 0.0 \\
&M_{u2} \geq 0.8: \quad \delta\eta_p = -k_4 P - k_5 P^2 \\
&\qquad k_4 = 0.05, \quad k_5 = 3, \quad P = \phi_{t1}(M_{u2} - 0.8).
\end{aligned}
\tag{10.55}
$$

Aungier's correlation for shrouded stages are as follows, whereby VD stands for a vaned diffuser and VLD for a vaneless diffuser:

$$\begin{aligned} \phi_M &= \frac{4}{\pi}\frac{\dot{V}_{t1}}{u_2 D_2^2} = \frac{4}{\pi}\phi_{t1} \\ \lambda &= 0.62 - (\phi_M/0.4)^3 + 0.0014/\phi_M \\ \psi_{vd} &= 0.51 + \phi_M - 7.6\phi_M^2 - 0.00025/\phi_M, \quad \eta_{vd} = \lambda/\psi_{vd} \\ \eta_{vld} &= \eta_{vd} - 0.017/\left[0.04 + 5\phi_M + \eta_{VD}^3\right], \quad \psi_{vld} = \eta_{vld}\lambda. \end{aligned} \tag{10.56}$$

Aungier's equations for open impellers are

$$\begin{aligned} \lambda &= 0.68 - (\phi_M/0.37)^3 + 0.002/\phi_M \\ \psi_{vd} &= 0.59 + 0.7\phi_M - 7.5\phi_M^2 - 0.00025/\phi_M, \quad \eta_{vd} = \lambda/\psi_{vd} \\ \eta_{vld} &= \eta_{vd} - 0.017/\left[0.04 + 5\phi_M + \eta_{vd}^3\right], \quad \psi_{vld} = \eta_{vld}\lambda. \end{aligned} \tag{10.57}$$

Care is needed in the use of such correlations as it is clear that the efficiency of a compressor stage is not solely a function of the two parameters shown here; this can be seen in the test data shown in Figure 10.5. For very well-designed large machines, with a thorough experimental development and very low clearance, these curves will, in some cases, still be slightly pessimistic as higher peak efficiency may be achieved especially at high flow coefficients (Hazby et al., 2014). For small machines with poor relative clearances, the correlations will certainly be too optimistic. In effect, empirical corrections to these curves must be made for the effects of clearance, impeller diameter, Reynolds number, isentropic exponent, diffuser type, blade number, work coefficient and roughness. Such corrections are more straightforward to develop than loss correlations for individual loss sources, as they do not need to define the amount of the associated loss exactly, but just need to define the exchange rate for variation about a base level. This procedure is described in Section 8.7, where corrections for the effect of the isentropic exponent and the tip clearance are given, and in (10.36), where a correction for the effect of Reynolds number, size and roughness is given.

The preliminary 1D design process assumes a value of the stage efficiency derived from a correlation, and this guides the designer towards the optimum flow coefficient. This should not prejudice the primary objective of achieving the highest possible efficiency as the detail design progresses. The subsequent parts of the aerodynamic design process (throughflow and CFD) may ultimately lead to a higher efficiency than was first assumed. Some conservatism, however, is generally a good thing at the beginning of the design process. A design that overperforms on test in terms of the efficiency is not a disaster. If a design eventually overachieves in terms of pressure ratio, it is relatively straightforward to adjust to the precise targets with minor changes to the impeller and/or the diffuser if necessary. Conversely, if there is a shortfall on test, correction may require a more substantial rework of more components at higher cost and with protracted timescale.

10.7.2 Initial Sizing of the Compressor

The first step in any design procedure is the determination of the overall dimensions. Taking the design work coefficient according to Section 11.6, say $\lambda = 0.6$, as a starting point to ensure a sufficiently wide operating range and selecting the flow coefficient for a near-optimum efficiency level, say $\phi_{t1} = 0.09$ from Figure 10.5, then (5.36) gives the tip speed Mach number for an assumed polytropic efficiency as follows:

$$M_{u2}^2 = \frac{[\pi]^{\frac{(\gamma-1)}{\eta\gamma}} - 1}{(\gamma - 1)\lambda}. \tag{10.58}$$

The speed of sound is based on the inlet total conditions; hence the impeller tip speed can be calculated. Equation (10.58) applies to each individual stage being considered. The diameter of the impeller is then given as follows:

$$D_2 = \sqrt{\frac{\dot{m}}{\rho_{t1} u_2 \phi_{t1}}}. \tag{10.59}$$

The rotational speed is now known from the diameter and the tip speed. If there are constraints on the rotational speed or the impeller diameter, the process can be repeated for different values of the flow and work coefficients. The expected efficiency can then be corrected using (10.55) and the process repeated. The outcome may suggest a revision to the basic assumptions. For example, if the impeller Mach number is too high to give adequate operating range, or the tip speed is too high for mechanical reasons with a given material, the designer may consider adding stages. It is often useful to combine these equations in preliminary design systems to estimate the size and power requirements for a compressor design; see Section 11.6.8.

10.8 Mean-Line Calculation of the Flow Conditions through the Stage

A one-dimensional flow analysis using the thermodynamic and aerodynamic relationships in Chapters 2 through 4 provides a simple procedure for estimating the state of the fluid at various positions through the stage and gives an insight into the effect of many of the design parameters. The analysis assumes that the flow in the blade rows of the compressor is rotationally symmetric and steady. The flow is examined at specific planes within the machine, usually only on the mean line, and such an analysis is often called a 1D 'pitch-line' theory. This assumes that a representative mean value of the fluid properties can be found at each meridional position and gives a useful approximate indication of the state of the fluid for all types of radial machine provided appropriate geometrical parameters and empirical coefficients are used. The fundamental aspects of the pressure rise of a stage are determined by the velocity triangle at the impeller outlet, and the flow capacity by the velocity triangle at impeller inlet. This is discussed in more detail in Sections 11.4 and 11.6.

10.8.1 The Process of Compression

The notation for the stations in the compression process as used in this book are given in Figure 10.6 and as follows:

From the inlet flange

t0	Total inlet conditions at the inlet flange
0	Static inlet conditions at the inlet flange
igv0	Static conditions at the IGV inlet
igv1	Static conditions at the IGV outlet
1	Impeller inlet conditions
1*	Impeller throat conditions
2	Impeller outlet conditions
t2	Total outlet conditions at impeller outlet
d2	Diffuser inlet conditions (including leakage flows)
3	Diffuser vane inlet conditions
3*	Diffuser vane throat conditions
4	Diffuser vane outlet conditions
5	Diffuser outlet conditions and inlet conditions to downstream components

For a downstream volute

6	Volute outlet conditions
7	Conical diffuser outlet conditions
8	Outlet conditions at the outlet flange

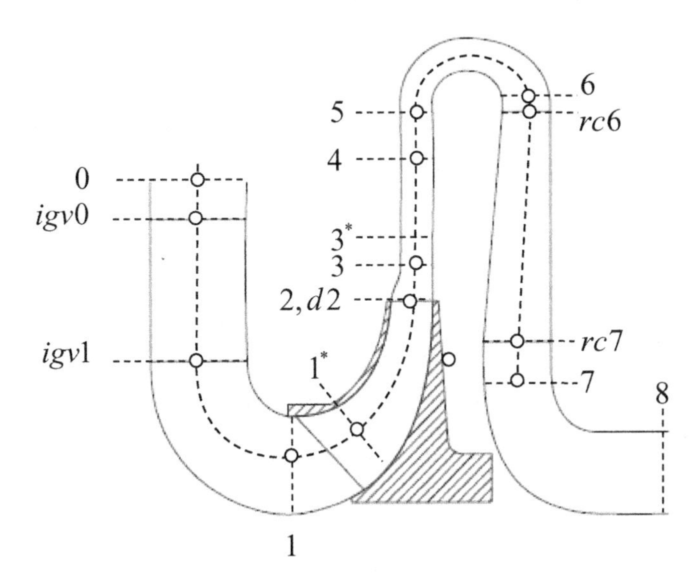

Figure 10.6 Notation for a radial compressor stage with a return channel

For a return channel

6	Crossover bend outlet condition
rc6	Return channel vane inlet conditions
rc7	Return channel vane outlet conditions
7	Return channel outlet conditions
8	Outlet conditions at the outlet bend

For an axial deswirl vane

6	Bend outlet conditions
ds6	Deswirl vane inlet conditions
ds7	Deswirl vane outlet conditions
7	Deswirl channel outlet conditions
8	Outlet conditions at the outlet

10.8.2 The Local Flow Conditions through the Stage

The mean-line calculation steps through the compressor from inlet to outlet, whereby at each step through a given component from i to $j = i + 1$, the conditions at the outlet, j, are calculated from those at its inlet, i, following a procedure essentially as described by Whitfield and Baines (1976). The step from the inlet to the outlet of component i considers the continuity equation, the Euler equation, the steady flow energy equation, the equation of state and empirical information on flow blockage, gas angles and the effect of irreversibility. The conditions at the stage inlet are assumed to be known as total conditions; at the inlet and outlet of each component, both static and total conditions are calculated. The general steps in the calculation based on an idealised real gas with constant real gas factor and specific heat, as discussed in Section 3.3.1, are described in this section. For this calculation it is assumed that the values of the following gas properties are constant through the stage.

Gas properties

Gas constant	R	
Real gas factor	Z	(10.60)
Isentropic exponent	γ	
Specific heat at constant pressure	$c_p = ZR\gamma/(\gamma - 1)$	

In the calculation of a specific operating point, it is assumed that the values of the following items are known at the inlet to a particular component.

Inlet conditions

Inlet total pressure	p_i	
Inlet total temperature	T_{ti}	
Inlet total enthalpy	$h_{ti} = c_p T_{ti}$	
Inlet entropy	$s_i - s_{ref} = c_p \ln\left(T_{ti}/T_{ref}\right) - ZR\ln\left(p_{ti}/p_{ref}\right)$	(10.61)

The operating point is defined in terms of speed and mass-flow rate:

$$
\begin{array}{lll}
\textit{Operating point} & & \\
\text{Inlet mass flow rate} & \dot{m}_i & \\
\text{Rotational speed} & N & \\
\text{Angular velocity of shaft} & \omega = 2\pi N/60 &
\end{array} \tag{10.62}
$$

and at each calculating plane, geometrical data are needed:

$$
\begin{array}{lll}
\textit{Geometrical data} & & \\
\text{Mean radius} & r_{mi} = (r_{hi} + r_{ci})/2 & \\
\text{Channel width} & b_i & \\
\text{Channel flow area} & A_i = 2\pi b_i r_{mi} &
\end{array} \tag{10.63}
$$

In some 1D analysis systems, the root-mean-square radius (rms), sometimes known as the Euler radius, is used instead of the mean geometric radius:

$$
\text{rms or Euler radius} \quad r_{rmsi} = \sqrt{\left(r_{h1}^2 + r_{ci}^2\right)/2} \tag{10.64}
$$

This radius has the property that in a uniform flow it splits the annulus into two regions with the same mass flow (Traupel, 2000) and may thus be considered to be more like a mass-averaged radius. The flow area, A_i, (10.63) is correct for annuli, cylinders and conical surfaces.

The inlet mass flow of the component must consider any leakage flows downstream of the previous component, such as hub and shroud leakage over an impeller casing. For a shrouded impeller, additional mass flow enters the inlet due to the leakage over the shroud. At impeller outlet. the mass low leaks to the inlet over the shroud and in a single-stage application from the outlet over the backplate. In a return channel design, the flow leaks from the outlet of the return channel up the backplate and enters the flow at diffuser inlet. In some applications, bleed flow is removed from the impeller outlet as a source for secondary air to seal bearing compartments and to balance the thrust load on the bearings.

Empirical information is needed to define the flow angles, the blockage and the losses. The absolute flow angle, α_i, is used downstream of a stator, and the relative flow angle, β_i, or the slip factor is used downstream of an impeller. There are many ways of categorising the losses; in the following equations, the entropy loss coefficient across the component is used. In an off-design calculation, these parameters may be defined in such a way that they are a function of the operating point, in terms of the incidence or the de Haller number, in order to calculate a performance map.

$$
\begin{array}{lll}
\textit{Empirical information} & & \\
\text{Flow blockage factor} & k_j & \\
\text{Flow angle or slip velocity} & \alpha_j,\ \beta_j,\ \sigma_j & \\
\text{Entropy loss coefficient} & \chi_i &
\end{array} \tag{10.65}
$$

For a one-dimensional flow, the meridional flow velocity, c_{mi}, and the swirl component of the velocity, c_{ui}, need to be known or calculated at the inlet plane of the component. Thermodynamic parameters are also needed, including the local speed of sound, and the total or static enthalpy, temperature, pressure and density. These can be calculated by means of an iteration assuming an initial estimate for the static conditions and using empirical data for the flow angle and the blockage at this location.

Initial guess of inlet static conditions

Static pressure at inlet $p_i = p_{ti}$

Static temperature at inlet $T_i = T_{ti}$

Iteration for inlet density

Static density $\rho_i = p_i/(ZRT_i)$

Meridional velocity component $c_{mi} = \dot{m}_i/(\rho_i A_i k_i)$

Swirl velocity component Upstream stator $c_{ui} = c_{mi} \tan \alpha_i$ or

Upstream rotor $c_{ui} = c_{mi} \tan \beta_i + \omega r_{mi}$

Absolute velocity $c_i = \sqrt{c_{mi}^2 + c_{ui}^2}$

Static enthalpy $h_i = h_{ti} - c_i^2/2$

Static pressure $p_i = p_{ti}(T_i/T_{ti})^{\gamma/(\gamma-1)}$

Repeat iteration for density until converged

(10.66)

The outlet conditions of component i and the inlet conditions of component j can be calculated in a similar way. Component i may be a rotor, an impeller, a stator or a vaneless region. Each lead to a slightly different equation:

Initial guess

Static pressure at outlet $p_j = p_{ti}$

Static temperature at outlet $T_j = T_{ti}$

Iteration for outlet density

Static density $\rho_j = p_j/(ZRT_j)$

Meridional velocity component $c_{mi} = \dot{m}_j/(\rho_j A_j k_j)$

Swirl velocity component Stator $c_{uj} = c_{mj} \tan \alpha_j$

Rotor $c_{ui} = c_{mi} \tan \beta_i + \omega r_{mi}$

Impeller $c_{ui} = c_{mi} \tan \beta_{bli} + \omega r_{mi}(1 - \sigma_i)$

Vaneless $c_{ui} = c_{ui}(r_{mi}/r_{mj})$

Absolute velocity $c_i = \sqrt{c_{mi}^2 + c_{ui}^2}$

Change in total enthalpy Stator or vaneless $\Delta h_{ij} = 0$

Rotor or impeller $\Delta h_{ij} = \omega r_{mj} c_{uj} - \omega r_{mi} c_{ui}$

Outlet total enthalpy $h_{tj} = h_{ti} + \Delta h_{ij}$

Static enthalpy $h_i = h_{ti} - c_i^2/2$

Entropy $s_j = s_i + \Delta s_{ij}$

Static pressure $p_j/p_i = (h_j/h_i)^{\gamma/(\gamma-1)} \; e^{-(s_j - s_i)/ZR}$

Static temperature $T_j = h_j/c_p$

Repeat iteration for density until converged

(10.67)

The prediction of stage performance outlined in the preceding equations can be used to estimate the change in performance for any variation in the configuration or the geometry. It is thus more complete than the global procedure described in Section 10.7. This does not necessarily imply that this approach is any more accurate for well-designed stages. Nevertheless, it allows the high losses associated with a poor design to be identified, and the effect of a change in configuration can be assessed, for example, the effect of an increase in the radial extent of the diffuser channel.

10.8.3 The Stage as the Sum of Two Components

A useful simplification of the procedure outlined in the preceding section for a typical stage without variable geometry is to consider the stage simply as the sum of two components, that is, the inlet and impeller as one component and all the subsequent stationary components as another. This essentially splits the performance into two parts which are linked by the velocity triangle at the impeller outlet. The stage total–total polytropic efficiency, η_{stage}, for this configuration may be estimated from the global flow coefficient, ϕ_{t1}, and the tip-speed Mach number, M_{u2}, using one of the correlations given in Section 10.7. It is then possible to estimate the impeller outlet velocity triangle with only a small amount of additional data.

If the compression in the impeller is considered as a polytropic process with a kinematic degree of reaction of r_k and a polytropic exponent of n_{imp} for the total-static compression process in the impeller, as used by Casey and Schlegel (2010), then the density ratio from inlet to outlet can be derived as

$$\frac{\rho_2}{\rho_{t1}} = \left[1 + (\gamma - 1) r_k \lambda M_{u2}^2\right]^{\frac{1}{n_{imp}-1}} \tag{10.68}$$

The flow coefficient at the impeller outlet can be related to the inlet flow coefficient using the continuity equation:

$$\phi_2 = \phi_{t1} \frac{1}{\pi} \frac{D_2}{b_2} \frac{\rho_{t1}}{\rho_2}. \tag{10.69}$$

Under the assumption that the flow has no swirl at the inlet to the impeller, then the Euler work due to the adiabatic work input of the impeller can be estimated approximately from the velocity triangles and the slip factor. The Euler work is related directly with the flow coefficient at impeller outlet, the assumed slip velocity – which may come from a correlation – and the impeller outlet blade angle, as follows:

$$\lambda_{Euler} = \frac{c_{u2}}{u_2} = 1 - \frac{c_s}{u_2} + \phi_2 \tan \beta_2'. \tag{10.70}$$

Note that the impeller outlet angle is defined in terms of the radial direction and with the notation used here is negative for a backswept impeller. A more detailed discussion of this equation is given in Section 11.6. The design value of the work coefficient includes both the Euler work and also the disc friction work such that

$$\lambda_d = \lambda_{Euler}\left(1 + \frac{k_{df}}{\phi_{t1}}\right), \tag{10.71}$$

where the k_{df} is a disc friction coefficient as outlined in (10.47). The degree of reaction can be calculated from the velocity triangle at outlet as

$$r_k = 1 - \frac{{\lambda_{Euler}}^2 + \phi_2^2}{2\lambda_d}. \tag{10.72}$$

In order to close this system of equations, it is still necessary to make additional assumptions. Firstly, the empirical values of the slip velocity and the disc friction coefficient are needed. An assumption is then needed that determines the impeller total–static efficiency. A good approximation is to take this to be is same as the stage total–total polytropic efficiency, as in Figure 2.14, and this allows the impeller polytropic exponent to be estimated:

$$\frac{n_{imp}}{n_{imp} - 1} = \frac{\eta_d \gamma}{\gamma - 1} = m, \quad n_{imp} = \frac{m}{m - 1}. \tag{10.73}$$

The justification for this is based on the typical compression process described in Section 2.9, where the static conditions at impeller outlet nearly lie on the polytropic process from inlet total conditions to outlet total conditions. Other assumptions are possible. Two pieces of additional information at the design point are needed, and these could be the impeller backsweep and the impeller outlet width ratio (b_2/D_2). With this information, the preceding equations can be solved iteratively in the order given, starting from an approximate value for the density ratio of (10.68).

The knowledge of the outlet flow coefficient and the work coefficient allows the exit velocity triangle to be defined so that the flow angles consistent with the design data can be determined. The flow angles are given by

$$\begin{aligned} \tan\alpha_2 &= \lambda_{Euler}/\phi_{2d} \\ \tan\beta_2 &= (\lambda_{Euler} - 1)/\phi_{2d}. \end{aligned} \tag{10.74}$$

Note that at the end of this process the diffuser static to total efficiency can also be calculated, and experience shows that this is similar to the impeller total to static efficiency:

$$\eta_{stage} = r_k \eta_{imp} + (1 - r_k)\eta_{diff}. \tag{10.75}$$

These equations are used here to relate the various 1D geometrical information to the velocity triangle at the design point. They are also used in Chapter 18 for the prediction of the performance map.

As a useful addition to these equations, the impeller total–total efficiency is given by

$$\eta_{12}^{tt} = r_k \eta_{imp} + (1 - r_k) = 1 - r_k(1 - \eta_{imp}) \tag{10.76}$$

and this efficiency may be used in some cases to calculate the conditions at impeller outlet if correlations for this are available, as for example from Rodgers (1980). Furthermore, it is possible to derive that the diffuser inlet Mach number, $M_{c2} = c_2/a_2$, is given by

$$M_{c2}^2 = (1 - r_k)2\lambda M_{u2}^2 / \sqrt{\left[1 + (\gamma - 1)r_k \lambda M_{u2}^2\right]}. \tag{10.77}$$

11 Impeller Design

11.1 Overview

11.1.1 Introduction

This chapter describes the essential aspects of impeller design taking into account the constraints from mechanical and aerodynamic considerations. As in earlier chapters, a one-dimensional (1D) steady flow analysis is used to obtain a general understanding of the effects of the impeller design parameters on the geometry. This analysis provides some clear preliminary design guidelines for obtaining values for specific flow parameters to obtain optimum performance. These preliminary guidelines are not simply tentative; they retain their validity through to the more complex 3D analyses. It is generally accepted that no amount of subsequent analysis with 3D CFD can correct for poor impeller design decisions made during the preliminary design.

The parameters for optimisation of the impeller inducer are described leading to a selection of the blade inlet angle and the casing diameter. The impeller blade inlet design attempts to achieve the maximum flow capacity with the lowest possible inlet relative Mach number. The inducer also determines the area of the throat between adjacent blade, and this determines the maximum flow capacity of the impeller. The effects of the impeller outlet velocity triangle on the work input and degree of reaction are explored. Examples are provided describing the type of outlet velocity triangle needed for different applications. The stability considerations that lead to the choice of backsweep at the impeller outlet are explained. Some generally applicable impeller design guidelines are suggested.

Sometimes it is necessary to adapt an impeller designed for one task to fulfil other requirements. Ways of doing this by means of trimming, flow cuts, overfiling and underfiling are described. A final section identifies some important differences between the velocity triangles used in radial flow compressor impellers and those used in the rotors of centrifugal pumps, axial compressors and radial turbines.

11.1.2 Learning Objectives

- Understand the importance of minimising the inlet relative Mach number in order to produce the most compact design and achieve the optimum blade inlet angle and inlet diameter.
- Be aware how the velocity triangle at inlet determines the flow capacity.

- Be aware how the velocity triangle at outlet determines the work coefficient and the slope of the pressure rise characteristic at the design point.
- Understand the impeller design guidelines provided.
- Be aware of how to adapt an impeller to different flow conditions by flow cuts and trims without a complete redesign.

11.2 Impeller Design

The impeller transfers energy to the fluid and is the only moving part of a compressor stage. There is an increase of enthalpy, pressure and absolute velocity across it. A typical high flow coefficient impeller is shown in Figure 1.10, and closed impellers with different flow coefficients are shown in Figure 1.11. The first axial part of a high flow coefficient impeller is often called the inducer, and the inlet of the impeller is often referred to as the eye, the pupil being the shaft at the hub. The inducer is in the axial part of the flow channel, so it has much in common with the rotor of an axial flow compressor. In transonic flow, the available design information from transonic axial fans provides a useful guide to the design of the inducer. Impellers for industrial multistage compressors often have no axial inducer, and the leading edge is in the bend to the radial part of the impeller. This leads to a shorter axial length of the stage and increases the shaft thickness, both of which allow for a more robust rotor dynamic design in multistage configurations. These days the outlet of the impeller generally has backswept blades leaning from the direction of rotation by between 25° and 55°.

The designer is concerned with the selection of the impeller blade geometry and channel shape to achieve the flow angles and velocity triangles that are considered to be optimal to achieve the duty required. In this chapter, the selection of the most appropriate velocity triangles is emphasised as this determines the blade angles that are needed. In axial flow turbines and compressors, an important aerodynamic aspect is the design of the optimum blade profile required to turn the flow in the circumferential direction. This is often discussed in terms of a combination of thickness and camber distribution, as in aerofoil design; see Section 7.2.2.

In axial machines, the profile design can be carried out on a two-dimensional basis in the blade-to-blade plane as the flow is nearly two dimensional in nature. In radial compressors, the blades are relatively thin and often of constant thickness along a meridional section (except for the rounding of the leading and trailing edges), so the thickness variation is not a major aerodynamic concern but is still important with regard to mechanical integrity and vibrations. The important aerodynamic aspects for the impeller geometry are the selection of, first, the curved meridional flow channel to turn the flow from the axial to the radial direction and, second, the inlet and outlet blade angles together with the blade camber angle distribution needed to turn the flow efficiently in the circumferential direction. The key nondimensional parameters in the design of radial compressors that determine the geometry are the global inlet flow coefficient, ϕ_{t1}; the impeller work coefficient, λ; the impeller outlet flow coefficient, ϕ_2; and the tip-speed Mach number, M_{u2}.

There are various mechanical and fluid dynamics constraints that impose limits on the design space. The blade tip speed, u_2, is limited because of the permissible rotor stresses, and this limitation is most severe for high flow coefficient shrouded wheels. The relative inlet Mach number at the tip of the impeller inlet, M_{w1}, is limited to a maximum value of about 1.3, because of compressible shock waves and the associated losses. The inlet hub diameter, D_{1h}, has to be large enough to permit the required number of blades with sufficient thickness and may not be too large that it acts as a constriction to the flow in the impeller inlet eye and reduces the centrifugal effect. For a multistage application, the required shaft diameter may also have an influence on D_{1h}. The inlet casing diameter, D_{1c}, should not exceed about $0.73{*}D_2$ to obtain an outer casing contour with acceptably small curvature in the meridional plane that retains some centrifugal effect on the casing streamline, but higher values are possible in mixed flow impellers. The ratio D_{1c}/D_2 is often known as the trim and is occasionally characterised by the square of that ratio, (11.59). From mechanical considerations, the span of the blade as determined by the ratio of D_{1c}/D_{1h} may require consideration. The impeller axial length, L_{imp}, affects the weight and inertia of the impeller and may not be too large to avoid rotor dynamic instability, and the ratio L_{imp}/D_2 determines the curvature of the meridional channel. The tip clearance gap, ε, should be kept small as a proportion of exit channel height in open impellers to avoid high losses. The level of diffusion in the impeller flow channels, as defined by the de Haller number, w_2/w_1, should not be too low to avoid flow separation. The outlet width ratio, b_2/D_2, cannot be smaller than certain limits to obtain good efficiencies. A certain amount of backsweep of the impeller blade at impeller outlet is needed to obtain a wide stable operating characteristic curve. The absolute flow angle into the diffuser, α_2, and the diffuser inlet Mach number, M_{c2}, are determined by the impeller and strongly affect the diffuser performance.

The influence of these parameters will be highlighted in this chapter.

11.3 Impeller Types

11.3.1 Open Impellers

In single-stage applications, the impeller is usually unshrouded and has a small clearance gap, between the tip of the rotating blades and the casing. As in all compressor applications, a large clearance gap between the rotor and the stator results in losses, and often in flow instability. The gap needs to be as small as possible and is typically between 0.3 and 0.5 mm for stages with an impeller diameter above 200 mm. Chapter 1 gives figures showing several examples of single-stage open impellers.

It is also possible to use open impellers for some multistage applications. The example given in Figure 11.1 shows a two-stage configuration on an overhung shaft with two open impellers for cooling of avionics. The rotor rotates in a counterclockwise direction and has a blade backsweep at outlet of around 40°; there is a lean of the tip section in the rotational direction with a rake angle of about 20°. To accommodate

Figure 11.1 A two-stage shaft configuration with open impellers.

the change in pressure and the reduction in volume flow rate between the stages, it would be possible to use an impeller of lower outlet width ratio and lower flow coefficient in the second stage. In this case, the change in volume flow is accommodated by making the second stage smaller than the first. Both stages have eight main blades and eight splitter blades. In this case, the blade does not have a clear *s*-shape, which is sometimes the case, but is more continuously curved in a single direction to the trailing edge.

If more than two stages are required on a single shaft, it is not generally possible to use open impellers: in multistage inline applications, it is difficult to maintain a small clearance between the impellers and the casing for all stages. Only one stage can be adjacent to the axial thrust bearing, which provides good control of its clearance. The changes in temperature during startup and operation cause the shaft and the casing to increase in length with changes in axial position of the rotor relative to the casing leading to a change in the clearance gaps. For this reason, multistage configurations mainly use shrouded impellers.

Figure 1.15 shows a multistage inline application in which the first stage is an open impeller. The advantage of such a configuration is related to the flow capacity of the open stage: as an overhung impeller, it can be designed with an axial inlet for a high flow coefficient. And as an open impeller, it can be larger and achieve higher tip speeds than the closed impellers on the same shaft, again increasing the flow capacity. This particular machine was built with a first-stage diameter of 1700 mm. The impeller has no splitter blades and no lean of the trailing edge but has *s*-shape blades. Figure 1.4 shows the shaft of a three-stage compressor with open impellers with splitters where the change in volume flow along the shaft is accommodated by using narrower impellers.

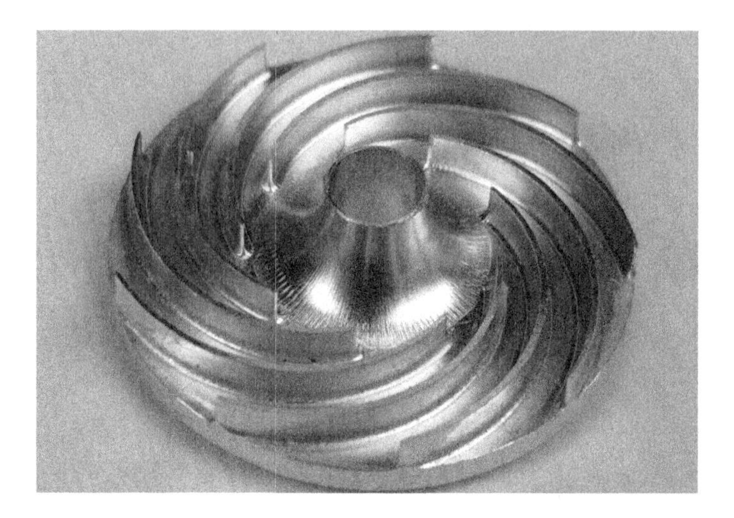

Figure 11.2 A low flow coefficient open impeller for a microcompressor. (image courtesy of Celeroton AG)

Open impellers are generally used for high flow coefficient stages and have strongly three-dimensional blade shapes as seen in the aforementioned figures. In most cases, the inlet casing diameter does not exceed about $0.73{*}D_2$. If a higher flow capacity is required, then a higher inlet diameter ratio is used, and in this case it is difficult to turn the flow completely to the radial direction and a mixed flow impeller is used, as discussed in Section 11.12.2.

In some special applications with low flow coefficients, two-dimensional open impellers may be used as shown in Figure 11.2. This impeller is from a two-stage microcompressor configuration (Casey et al., 2013). It has an impeller diameter of 21 mm and operates at 300,000 rpm. The high backsweep used for this low flow coefficient impeller, and the fact that the flow channel has a constant width through the impeller, indicates that its design follows the strategy of Casey et al. (1990) for impellers of this type, type C in Section 11.6.6.

11.3.2 Shrouded Impellers

In ventilators, two-dimensional impeller blade shapes with thin shrouds may be used (Bommes et al., 2003). In industrial compressors, the shroud is more robust, as can be seen from Figure 1.11. Axial or radial labyrinth seals are used to seal the flow over the shroud from the impeller outlet to inlet. A shrouded impeller achieves a lower maximum tip speed than an unshrouded one, due to the stresses caused by the mass of the shroud. Shrouded impellers, however, provide a more robust device that is not so prone to blade vibration problems, and so are highly suitable for industrial applications. The lower speeds and lower pressure ratio also open the way for the thermodynamic benefit of intercooling between the stages, as shown in Figure 1.15. The sealing systems of shrouded radial impellers for multistage compressors are usually arranged so that radial seal clearances allow axial movement of the rotor during warm-up and shutdown. The leakage flow over the shroud joins the flow at impeller inlet, and care is needed to ensure that this does not cause losses due to nonuniformity of the inlet flow through distortions in swirl and flow angle.

11.3.3 Impeller Tip Speeds

Mechanical aspects related to the design of impellers and the limits on tip speeds are discussed in Section 19.4. The following are typical values for the maximum tip speeds of impellers:

- Open impellers in cast aluminium: 200–300 m/s,
- Open impellers in forged aluminium: 450–560 m/s
- Open impellers in titanium alloys: 500–700 m/s
- Open impellers in steel: 350–500 m/s
- Shrouded impellers in steel: 280–340 m/s

11.4 Flow Conditions at the Impeller Inlet

11.4.1 Optimisation of the Inlet Velocity Triangle

A superficial inspection of the impellers of a range of radial compressor impellers with axial inducers shows that most impeller blades appear to be nearly indistinguishable from each other at inlet as the blade inlet angles are similar; see the aforementioned figures. This is because the simple one-dimensional analysis described here can be carried out to select the optimum inlet angle on the casing which leads to the most compact design giving the maximum flow per unit frontal area. Most impellers are designed to be as compact as possible to provide a smaller and cheaper machine, so impeller blade shapes are then superficially similar in the inducer inlet.

The basis of this methodology is to minimise the relative inlet velocity at the tip of the impeller inlet for a given mass flow. Because the highest inlet Mach number occurs at the casing, it is usual to consider the casing streamline as the primary aspect of the inlet velocity triangle in any 1D analysis of an impeller rather than the mean-line, which has a lower blade speed and a lower inlet relative Mach number. The casing streamline is also important with regard to the deceleration in the flow. The tip section with the highest inlet relative velocity has the maximum deceleration of the relative flow to the impeller outlet. In fact, due to the large radius change on the hub section, the relative flow along the hub streamline usually accelerates.

For centrifugal compressors with a high swallowing capacity and a pressure ratio above 4 in air, it is necessary to try to limit the inlet relative Mach number at the casing inlet, but impellers designed for other applications may require other design strategies. In the earlier technical literature, it is sometimes suggested that the use of preswirl can be used to lower the inlet relative flow Mach number. The design limit for the inlet relative Mach number has changed over the years. Shepherd (1956) suggests limiting the inlet Mach number to 0.7, or in extreme cases allowing it to rise to 0.85. The first edition of the book by Dixon, also in 1956 but now in its 6th edition as Dixon and Hall (2010), plots curves showing limits of 0.8 and 0.9 in the relative inlet Mach number. These suggested limits are outdated. Modern understanding of compressible flow in transonic compressors and the nature of choke suggests that a limit of between 1.2 and 1.3 is a more suitable maximum value for high flow capacity stages.

In transonic inducers, the entropy generation due to the shocks has to be controlled for high efficiency. The loss directly related to the entropy production across the shock is small, but it increases steeply with the upstream Mach number, discussed in Section 6.3.1, roughly as

$$\Delta s \propto \left(M^2 - 1\right)^3. \tag{11.1}$$

More importantly, the pressure rise at the shock can cause boundary layer separation and be a cause of further losses. According to Bölcs (2005), separation due to shock interaction in an axial compressor cascade occurs if the upstream Mach number of the shock exceeds approximately 1.3. The separated boundary layer itself causes additional losses due to the related secondary flow and mixing losses. Careful design of the impeller blade angles at inlet is required for higher relative supersonic inlet Mach numbers. If a shock is present it may be unstable and oscillate, causing flow instability in the impeller.

The typical inlet velocity triangle at the impeller inlet on the casing streamline is shown in Figure 11.3. This assumes that there are no inlet guide vanes to impart swirl to the inlet flow, so that $c_{u1} = 0$, and then $c_{m1} = c_1$. The use of inlet guide vanes to control the flow and pressure rise of the stage is considered in Sections 13.7 and 17.5.4. The main design consideration for the inlet velocity triangle is the impeller inlet casing diameter. If, for a given shaft speed and inlet volume flow, the inlet radius at the tip is made large, then the axial velocity is low but the blade speed at the tip is high. If the inlet radius at the impeller tip is low, the blade speed is low but the axial velocity is high, and between these extremes there is an inlet radius that minimises the inlet relative velocity, as shown in Figure 11.4. The simple analysis that follows was published by

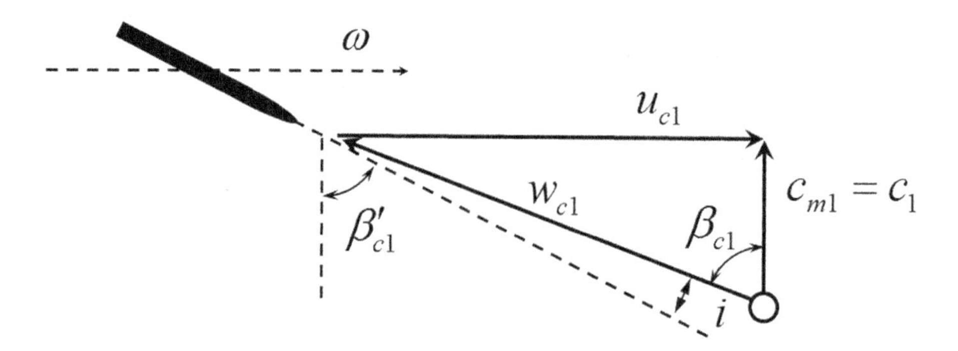

Figure 11.3 Velocity triangle of the impeller inlet at the casing with no inlet swirl.

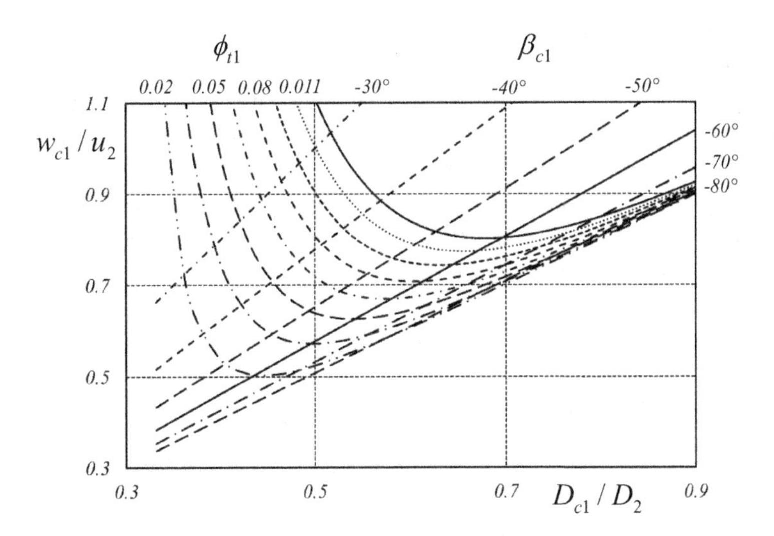

Figure 11.4 Variation of w_{c1}/u_2 with the casing diameter, D_{c1}/D_2, for different flow coefficients and with a hub diameter of $D_{h1}/D_2 = 0.3$.

Shepherd (1956) but has subsequently been republished in a different form by many authors. It leads to an inlet blade angle at the tip of the impeller of roughly $\beta_{c1} = -60°$ and accounts for the high similarity of all impeller inlet designs in the inducer region.

The simplest form of this analysis is to note that the relative inlet velocity at the tip of an impeller inducer with no swirl in the inlet duct is given by

$$w_{c1}^2 = c_{m1}^2 + u_{c1}^2. \tag{11.2}$$

On the assumption of a constant meridional velocity across the impeller leading edge, the meridional velocity can be determined from the continuity equation as

$$c_{m1} = \dot{m}/A_1\rho_1 = \dot{m}/\left(\pi\left(r_{c1}^2 - r_{h1}^2\right)\rho_1\right) \tag{11.3}$$

and making use of the definition of the global flow coefficient, ϕ_{t1}, leads to

$$c_{m1} = 4\phi_{t1}\rho_{t1}r_2^2u_2/\left(\pi\left(r_{c1}^2 - r_{h1}^2\right)\rho_1\right). \tag{11.4}$$

Thus, we can express this as a ratio to the impeller tip speed in dimensionless terms as

$$\frac{c_{m1}}{u_2} = \frac{4\phi_{t1}}{\pi}\frac{\rho_{t1}}{\rho_1}\frac{1}{\left(\nu_{c1}^2 - \nu_{h1}^2\right)}, \tag{11.5}$$

where $\nu_{c1} = D_{c1}/D_2$ is the inlet eye radius ratio and $\nu_{h1} = D_{h1}/D_2$ is the hub radius ratio relative to the impeller outlet diameter. In combination with (11.2), the relative inlet velocity becomes

$$\frac{w_{c1}^2}{u_2^2} = \left(\frac{4\phi_{t1}}{\pi}\frac{\rho_{t1}}{\rho_1}\right)^2\left(\nu_{c1}^2 - \nu_{h1}^2\right)^{-2} + \nu_{c1}^2. \tag{11.6}$$

Assuming that the inlet static density is the same as the inlet total density, and that the hub radius ratio is constrained by the shaft to be constant, the variation of the inlet casing relative velocity with the inlet eye radius ratio is shown in Figure 11.4. This figure illustrates that there is a minimum inlet velocity for a certain impeller eye diameter close to an inlet angle of $\beta_{c1} = -60°$.

The optimum inlet radius ratio to minimise the inlet relative velocity level can then be found by differentiation of (11.6) with respect to the tip radius ratio. On differentiation and equating this to zero to obtain the minimum, gives

$$\left(\nu_{c1}^2 - \nu_{h1}^2\right) = \sqrt[3]{2(4\phi_{t1}/\pi)^2}. \tag{11.7}$$

This can be expressed as

$$\frac{r_{c1}}{r_2} = \nu_{c1} = \sqrt{\nu_{h1}^2 + \sqrt[3]{2(4\phi_{t1}/\pi)^2}} \tag{11.8}$$

and from this, the optimum flow inlet angle (which for zero incidence would also be the blade casing inlet angle) can be determined to be

$$\tan\beta_{c1} = -\frac{\sqrt[3]{2(4\phi_{t1}/\pi)^2}\sqrt{\nu_{h1}^2 + \sqrt[3]{2(4\phi_{t1}/\pi)^2}}}{4\phi_{t1}/\pi}. \tag{11.9}$$

This rather complex looking equation has a very simple solution for the limiting case that the hub radius ratio is zero, which is that $\tan\beta_{c1} = -\sqrt{2}$, or $\beta_{c1} = -54.74°$ For a typical turbocharger impeller with a hub radius ratio of $\nu_{h1} = 0.275$ and a flow coefficient close to the optimum of 0.08, the equation leads to an inlet angle of $\beta_{c1} = -58°$. The optimum inlet angle at the casing thus becomes more negative as the hub diameter increases.

A rather more sophisticated analysis can be carried out considering the compressibility of the gas in the inlet, whereby the inlet static density is then not constant but decreases as the inlet meridional velocity increases (Whitfield and Baines, 1990; Dixon and Hall, 2010). Rusch and Casey (2013) have published some extensions to this analysis, and their derivation for zero inlet swirl using the nondimensional parameters in this book is given in this section. A similar analysis including the effects of inlet swirl has been published by Li et al. (2020). The mass-flow function of the stage is the mass flow relative to the flow which would pass through an area equivalent to the square of the impeller diameter with a gas velocity equal to the inlet total speed of sound and with the density at inlet total conditions:

$$\Phi = \frac{\dot{m}}{\rho_{t1}a_{t1}D_2^2} = \frac{\dot{m}}{\rho_{t1}u_2D_2^2}\frac{u_2}{a_{t1}} = \phi_{t1}M_{u2}. \tag{11.10}$$

If the mass flow at the impeller inlet is obtained from the continuity equation, the mass-flow function can be expressed as

$$\Phi = \frac{\rho_1 A_1 c_{1m}}{\rho_{t1}a_{t1}D_2^2}. \tag{11.11}$$

Following Dixon and Hall, the inlet area of the compressor eye is expressed using an impeller inlet shape factor k which represents the blockage of the inlet area caused by the hub:

$$A_1 = (\pi/4)\left(D_{c1}^2 - D_{h1}^2\right) = k(\pi/4)D_{c1}^2, \quad k = 1 - (D_{h1}/D_{c1})^2. \tag{11.12}$$

For uniform axial inflow, the velocity triangle yields

$$c_{m1} = w_{c1}\cos\beta_{c1}, \quad u_{c1} = w_{c1}\sin\beta_{c1} \tag{11.13}$$

with the angles and velocity components as shown in Figure 11.3. The axial velocity is assumed to be uniform over the inlet area. Assuming zero incidence angle, which is realistic at high transonic Mach numbers, the flow angle corresponds to the blade metal angle at the impeller inlet casing. With the preceding equations, the mass-flow function can now be expressed as follows:

$$\begin{aligned}\Phi &= k\frac{\pi}{4}\frac{\rho_1}{\rho_{t1}}\frac{w_{c1}^3}{a_{1t}u_2^2}\sin^2\beta_{c1}\cos\beta_{c1} \\ &= k\frac{\pi}{4}\frac{\rho_1}{\rho_{t1}}\frac{a_1^3}{a_{t1}^3}\frac{w_{c1}^3}{a_1^3}\frac{a_{t1}^2}{u_2^2}\sin^2\beta_{c1}\cos\beta_{c1}.\end{aligned} \tag{11.14}$$

The definitions of the relative and absolute inlet Mach numbers

$$M_{c1} = c_{m1}/a_1 = w_{c1}\cos\beta_{c1}/a_1 = M_{w1}\cos\beta_{c1}, \quad w_{c1} = M_{w1}a, \quad M_{c1} = M_{w1}\cos\beta_{c1} \tag{11.15}$$

are used to rewrite the mass-flow function as follows:

$$\Phi = k\frac{\pi}{4}\frac{\rho_1}{\rho_{t1}}\frac{a_1^3}{a_{t1}^3}\frac{M_{w1}^3}{M_{u2}^2}\sin^2\beta_{c1}\cos\beta_{c1}. \tag{11.16}$$

The static flow quantities at the impeller inlet are replaced by the total ones using the following expressions

$$\rho_1 = \rho_{t1}\left[1 + \frac{\gamma-1}{2}M_{c1}^2\right]^{-1/(\gamma-1)}, \quad a_1 = a_{t1}\left[1 + \frac{\gamma-1}{2}M_{c1}^2\right]^{-1/2} \tag{11.17}$$

to relate the relative Mach number at the impeller inlet casing and the relative flow angle to the mass-flow function. The final result is

$$\Phi = k\frac{\pi}{4}\frac{M_{w1}^3}{M_{u2}^2}\frac{\sin^2\beta_{c1}\cos\beta_{c1}}{\left[1 + \frac{\gamma-1}{2}M_{w1}^2\cos^2\beta_{c1}\right]^{1/(\gamma-1)+3/2}}. \tag{11.18}$$

This can be reformulated to yield a modified mass-flow function as

$$\Phi' = \Phi\frac{4M_{u2}^2}{k\pi} = \frac{M_{w1}^3\sin^2\beta_{c1}\cos\beta_{c1}}{\left[1 + \frac{\gamma-1}{2}M_{w1}^2\cos^2\beta_{c1}\right]^{1/(\gamma-1)+3/2}}, \tag{11.19}$$

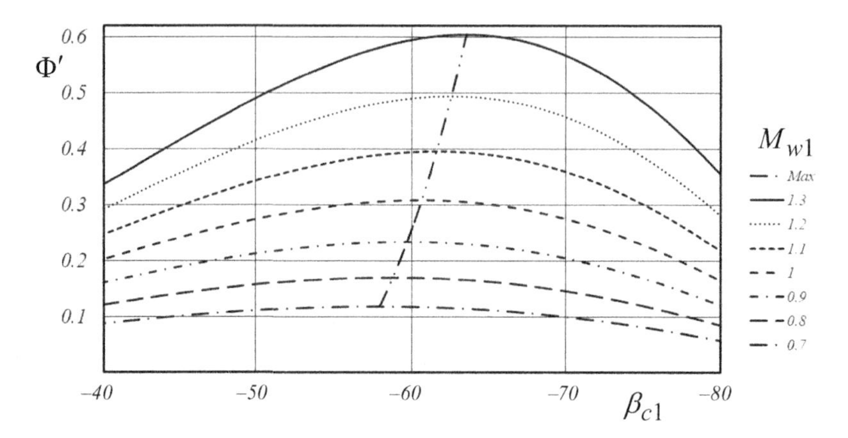

Figure 11.5 Modified mass-flow function for a centrifugal compressor with zero inlet swirl.

which is the same expression as derived by Dixon and Hall (2010) for their modified form of mass-flow function, and similar to that given by Whitfield and Baines (1990).

The modified mass-flow function is plotted over the relative inlet flow angle for a series of impeller inlet relative Mach numbers in Figure 11.5 for the isentropic exponent, $\gamma = 1.4$, representing air. The important feature of this curve is that for a given value of the modified inlet mass-flow function, there is a certain inlet angle leading to the minimum relative inlet Mach number. A curve connecting the maxima of the mass-flow functions for different relative Mach numbers at the impeller inlet on the casing is also included in Figure 11.5. This curve can be intersected by a given design mass-flow function value to yield the optimum design location with respect to inducer shock-related losses as it represents the lowest impeller inlet relative Mach number possible for this design duty. At low absolute inlet Mach number, the term on the bottom line of (11.19) becomes unity so that the maximum mass-flow function occurs at the peak value of the term $\sin^2\beta_{c1}\cos\beta_{c1}$, which is $\tan^2\beta_{c1} = 2$ or $\beta_{c1} = -54.74°$, as already derived in the previous incompressible analysis.

Compared to the incompressible analysis, the effect of compressibility moves the optimum inlet angle to a value closer to $\beta_{c1} = -60°$ for a typical turbocharger at a pressure ratio of 2.5 and to more negative values as the pressure ratio increases. Designers are well advised to use an inlet angle at the blade tip of around $\beta_{c1} = -60°$. The exact value of the optimum angle, given by Rusch and Casey (2012), is a function of the inlet relative Mach number at the tip, as follows:

$$\cos\beta_{c1,opt} = \frac{\sqrt{3 + \gamma M_{w1}^2 + 2M_{w1}} - \sqrt{3 + \gamma M_{w1}^2 - 2M_{w1}}}{2M_{w1}}. \tag{11.20}$$

This function leads to angles close to $\beta_{c1} = -60.6°$ at $M_{w1} = 1.0$, and becoming more negative with $\beta_{c1} = -65.4°$ at $M_{w1} = 1.5$ in air with $\gamma = 1.4$. For high molecular weight gases with $\gamma = 1.2$, the optimum angles are less negative by about 1°. Li et al. (2020) derive an equation for the optimum inlet angle for an impeller with swirl in the direction of rotation and show that for 30° of inlet swirl, the optimum flow angle is less negative by about 10°.

For a given relative Mach number, the dependence of the mass-flow function on the choice of inlet flow angle is weak near the optimum, so that other angles within ±5° from the optimum do not substantially change the mass-flow function. As a result, only a small increase in relative inlet Mach number generates an inlet angle that deviates by 5° from the optimum value. This suggests that there is no need for an impeller designer to spend a long time deliberating over this. For the design with the maximum modified inlet flow function, Rusch and Casey (2013) show that the inlet diameter ratio at the impeller eye is given by the following equation:

$$\frac{D_{1c}}{D_2} = \frac{M_{w1}}{M_{u2}} \sin\beta_{1c}\left[1 + \frac{\gamma - 1}{2} M_{w1}^2 \cos^2\beta_{1c}\right]^{-1/2}. \tag{11.21}$$

Impellers designed on this basis have a similar inlet diameter ratio, and this is an additional reason for the similarity of the inducer form of most impellers.

In some cases, considerations of a wide stable operating range may cause the designer to choose inlet blade angles that are more negative than the optimum to avoid high incidence at low flows, and considerations of choke may cause other values to be used to increase the throat area between blades to increase the maximum flow. The throat area is an aspect of the detailed design where other trade-offs may be appropriate, as discussed later, but these are only worthwhile when the correct inlet Mach number and inlet angles have already been selected to ensure a low relative inlet Mach number.

The choice of the inlet angle also determines the inlet eye diameter. In shrouded impellers for multistage compressors, it may be advisable to decrease the eye diameter below this optimal value to lower the resultant stresses in the shroud and to increase the centrifugal effect. A reduction of the actual eye diameter by 5% gives only a small increase in the inlet relative Mach number for typical process compressor stages with shrouds (Lüdtke, 2004). In the derivation of the preceding equations, any effects due to blockage by the blades, boundary layer blockage, incidence angle and nonuniform spanwise distribution of axial velocity have been neglected. The effects of flow nonuniformity and blockage can be included by changing the shape factor k if necessary.

Following this approach, Rusch and Casey (2013) have derived a preliminary design chart for impellers to maximise the flow capacity with a minimum inlet relative Mach number at the casing, M_{w1}. The upper part of Figure 11.6 shows such a design chart for the casing diameter. It shows lines of constant casing diameter ratio D_{c1}/D_2, calculated for air assuming that the hub blockage factor is $k = 0.9$. The design points of a variety of radial stages from many applications, shown as white circles, are included. Each design is unique, but the cluster of points identifies important trends. The bold solid line represents an inlet diameter ratio of $D_{c1}/D_2 = 0.73$, and few radial

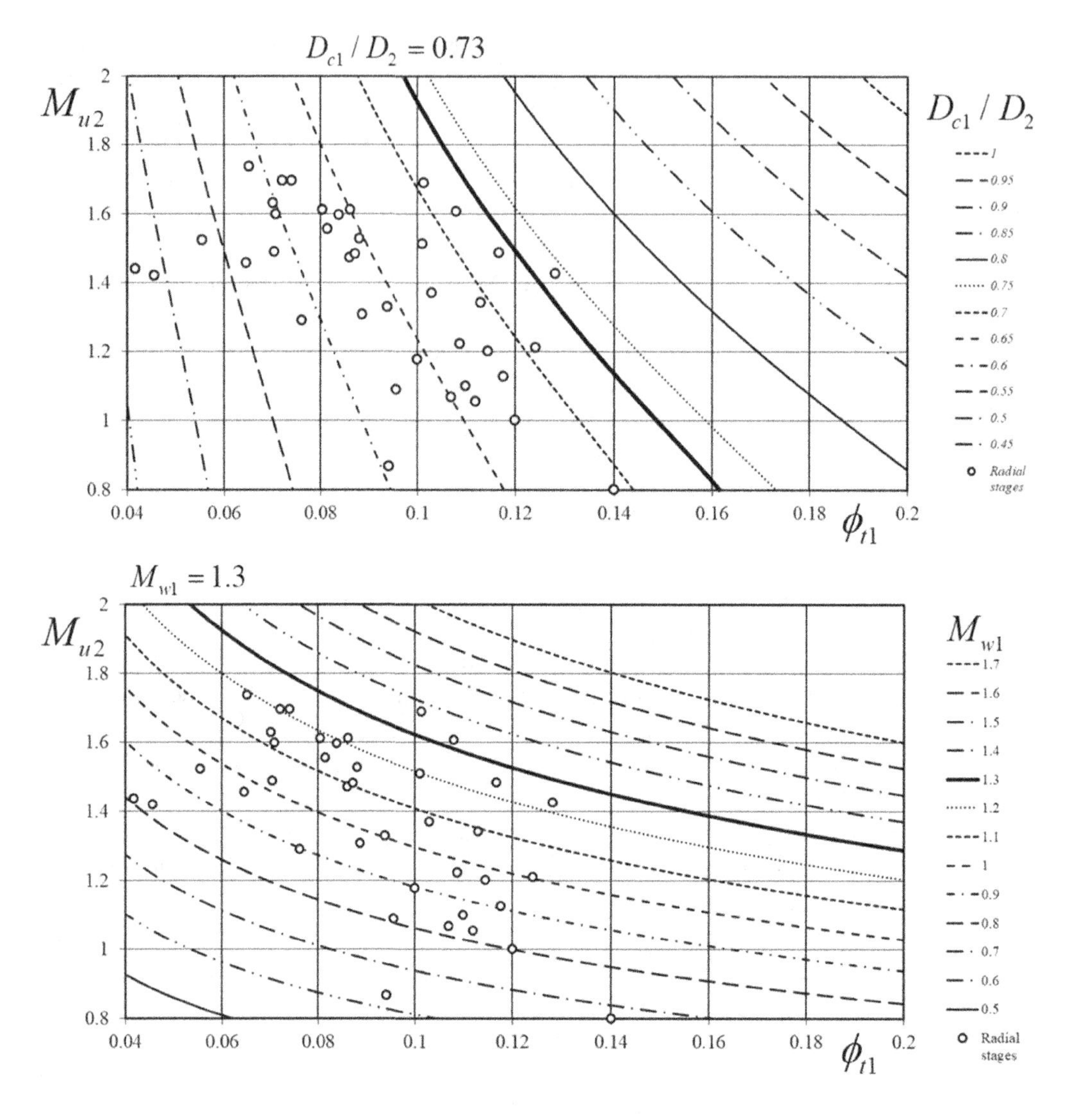

Figure 11.6 Design diagram for inlet diameter ratio, D_{c1}/D_2, and for the inlet relative Mach number at the casing for maximum swallowing capacity for a blockage factor of $k = 0.9$.

stages make use of a higher inlet diameter ratio than this. Values of the radius ratio that are too high reduce the centrifugal effect, and a practical upper limit of 0.73 for the inlet tip diameter ratio is suggested by the data, as shown by the thick line. For higher flow coefficients, a higher inducer radius ratio is needed, and this implies a move towards a mixed flow impeller, as discussed in Section 11.12.2.

The lower part of Figure 11.6 shows the inlet relative Mach number at the casing, M_{w1}, for the same conditions. The white circles are stages designed for high flow capacity, and it can be seen that values of the inlet relative Mach number significantly exceeding 1.25 seldom occur, and the thick solid line is $M_{w1} = 1.3$. This shows that an increase in tip-speed Mach number (or pressure ratio) needs to be accompanied by a decrease in the design flow coefficient to keep the inlet eye relative Mach number low. Stages designed for low flow coefficient are generally not critical in this respect. Figure 11.6 shows that at low tip-speed Mach numbers, the maximum swallowing capacity is determined by the limitation of the inlet eye diameter, and stages with high tip-speed Mach numbers are limited by the inlet relative Mach number at the casing. The contours of efficiency in the design diagrams of the original publication show that radial stages near or above these limits are not impossible but can be expected to be penalised in efficiency.

The optimum inlet angle for the impeller eye of a pump cannot be determined in this way. The risk of cavitation has to be considered, and the absolute pressure level at inlet has to be maximised, which leads to lower inlet angles closer to $\beta_{c1} = -75°$ than $\beta_{c1} = -60°$, as explained in Section 11.12.3. For low flow coefficient compressor stages, it is advisable to ensure that the friction losses are reduced. This means making the channel passages as wide as possible, equation 10.40, which also leads to relative inlet flow angles that are more negative, similar to those of pumps.

With an inlet flow angle at the casing of $\beta_{c1} = -60°$, the flow coefficient at the tip of the blade is given by

$$\frac{c_{m1}}{u_{c1}} = \frac{1}{\tan(-\beta_{c1})} = \frac{1}{\tan 60°} = \frac{1}{\sqrt{3}} \approx 0.577, \tag{11.22}$$

which is a useful rule of thumb. For a high flow coefficient impeller with $\nu_{c1} = D_{c1}/D_2 = 0.7$, the meridional flow coefficient based on the inlet meridional velocity relative to the blade speed at outlet becomes

$$\frac{c_{m1}}{u_2} = \frac{c_{m1}}{u_{c1}} \frac{u_{c1}}{u_2} = \frac{1}{\tan(-\beta_{c1})} \frac{r_{c1}}{r_2} = \frac{0.7}{\tan 60°} = \frac{0.7}{\sqrt{3}} \approx 0.4. \tag{11.23}$$

Conditions relating to the impeller outlet design, as given in Section 11.6, cause the typical value for the outlet flow coefficient of high flow coefficient radial impellers to be lower than 0.4, normally between $c_{m2}/u_2 = 0.2$ and 0.3. Consequently, the meridional flow velocity usually decreases by 50–75% through the impeller.

The preceding equations lead to a simple expression for the impeller inlet relative velocity at the casing relative to the impeller tip speed, as follows:

$$\frac{w_{c1}}{u_2} = \frac{w_{c1}}{u_{c1}} \frac{u_{c1}}{u_2} = \frac{1}{\sin(-\beta_{c1})} \frac{r_{c1}}{r_2} = \frac{1}{\sin 60°} \frac{r_{c1}}{r_2} = \frac{2}{\sqrt{3}} \frac{r_{c1}}{r_2}. \tag{11.24}$$

If the tip-speed Mach number and the inlet radius ratio are known, a quick estimate of the inlet relative tip Mach number can be made, a typical high flow coefficient impeller with $\nu_{c1} = r_{c1}/r_2 = 0.7$ leads to $w_{c1} \approx 0.8\ u_2$ and $M_{w1} \approx 0.8\ M_{u2}$. Although the machine Mach number is not directly related to the actual flow Mach numbers in the stage, there is a close correspondence in an impeller of a high flow coefficient stage optimised for maximum flow capacity.

11.4.2 Variation of the Flow Angle with Radius

The assumption of a constant axial velocity across the span together with the variation of blade speed with radius leads to a typical local inlet flow coefficient and flow angle at other radii of

$$\frac{1}{\tan(-\beta_1)} = \frac{c_m}{u} = \frac{c_m}{u_{c1}}\frac{u_{c1}}{u} = \frac{1}{\tan(-\beta_{c1})}\frac{r_{c1}}{r} \approx \frac{1}{\sqrt{3}}\frac{r_{c1}}{r}. \tag{11.25}$$

This equation allows the expected flow inlet angle and local flow coefficient across the span to be estimated. For a casing to hub diameter of $r_{c1}/r_{h1} = 2$, the hub inlet angle would be $\beta_{h1} = -40.9°$ for an impeller with a casing blade angle of $\beta_{c1} = -60°$. The blade inlet angle differs from this by the incidence selected, which is usually close to zero at the tip but may be higher on the hub, giving a less negative blade angle. A larger hub incidence is sometimes selected in order to introduce some curvature into the blade inducer and to adjust the first flap vibration frequencies of this component, or to increase the hub throat area, a common approach with turbocharger impellers. The flow on the hub streamlines is at low relative velocity and is generally accelerated through the impeller, so the hub can probably tolerate higher incidence. It is well known that turbine blade rows with accelerating flow generally have a higher tolerance against incidence than compressor blade rows. The incidence on the hub streamline should, however, be kept well below 10°, as a higher incidence of 15° is detrimental and causes a hub separation at design, as shown by Eisenlohr et al. (1998).

It can be shown that, for a blade with linear blade elements that are radially stacked at the inlet, the relationship between blade angle and radius is $\tan\beta/r$ = constant. A blade designed with no lean at the inducer inlet to avoid bending stresses in the inducer thus has a blade angle distribution that naturally matches the flow angle variation across the span fairly well. However, the assumption of a constant axial velocity across the span is actually not a good approximation if the impeller has curvature in the meridional direction; see Section 5.3.5, where it is shown that the variation of the meridional velocity across the span for a mean curvature of the meridional channel, r_{cm}, is given approximately by

$$\frac{\Delta c_m}{c_m} \approx \frac{r_c - r_h}{r_{cm}}. \tag{11.26}$$

The slower velocities on the hub streamline lead to a flow angle that is larger (more negative) on the hub streamline than would occur with the assumption of a constant axial velocity across the span. In an impeller with an axial inducer, this flow angle variation can be accommodated either by removing the restriction that the blade elements in the inlet section are precisely radial and allowing some small lean in the blade elements or by sweeping the leading edge slightly backwards at the tip. In a process compressor impeller without an inducer and with a leading edge in the curved bend to the radial section, the effect of the curvature in the inlet often causes the hub and casing blade angles to become very similar, as seen in Figure 11.7.

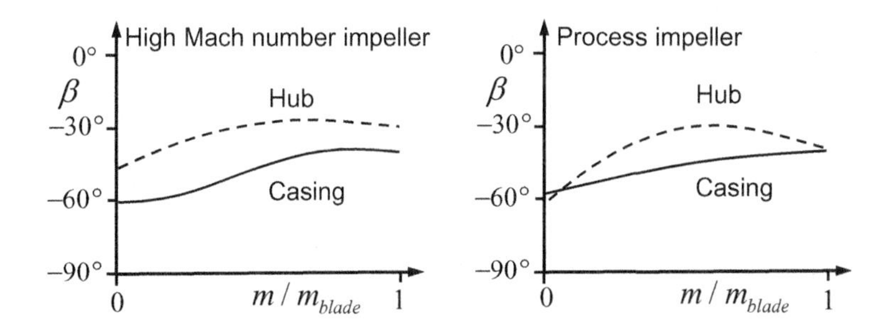

Figure 11.7 Examples of blade angle distribution on the hub and casing for a high Mach number impeller and a process impeller without an inducer.

11.4.3 Inlet Relative Mach Number

Rusch and Casey (2013) provide various equations for the geometry and Mach numbers for compact stages designed with a minimum relative Mach number at the impeller eye, and these are provided in the form of diagrams and equations that need to be solved iteratively. Further analysis of these equations has been carried out and an approximate empirical equation for the tip-speed Mach number has been found as a function of the inlet flow coefficient and the relative Mach number at the tip, as follows:

$$M_{u2} = \frac{M_{w1}(3.2\phi_{t1}/k)^{0.36}}{(3.2\phi_{t1}/k)^{0.36} + 0.15M_{w1}(0.45 + \phi_{t1}/k)}. \tag{11.27}$$

This is reasonably close to the original equation shown in Figure 11.6 for ideal gases with $\gamma = 1.4$, especially in the region of most interest. The advantage of this correlation over the original equations is that, for stages designed for the maximum compactness, it can be rearranged to provide a direct equation to calculate the relative Mach number at the tip as a function of the flow coefficient and the tip-speed Mach number, as follows:

$$M_{w1} = \frac{M_{u2}(3.2\phi_{t1}/k)^{0.36}}{1 - 0.15M_{u2}(0.45 + \phi_{t1}/k)}. \tag{11.28}$$

This is useful in preliminary design calculations.

This equation is plotted in Figure 11.8 and provides a very clear representation of the need to reduce the flow coefficient to avoid high inlet relative Mach numbers when designing for a high tip-speed Mach number. A rule of thumb for the maximum inlet relative Mach number of 1.3 is indicated in the diagram. The inlet shape factor, k, is smaller in stages with a higher hub diameter, such as typical stages downstream of an axial compressor in an axial-centrifugal combination or in a multistage process compressor. Figure 11.8 indicates that these designs naturally have a higher relative inlet Mach number at the casing due to the hub blockage.

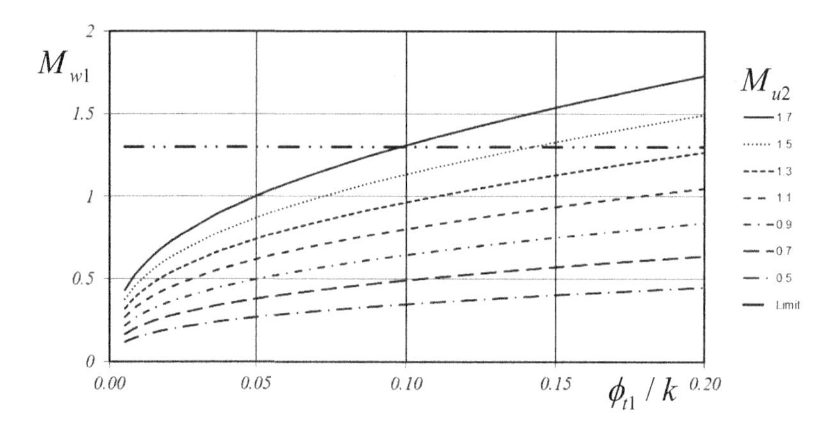

Figure 11.8 Inlet relative Mach number for compact stages as a function of the inlet flow coefficient, the tip-speed Mach number and the inlet shape factor *k*.

11.4.4 Splitter Blades

The analysis described in Section 11.4.1 assumes a constant meridional velocity across the inlet eye of the impeller. However, measurements show that the meridional flow tends not to be constant but to increase from the hub to the tip due to the curvature in the meridional plane, as explained in Section 5.3.5. This effect can be reduced by using a longer impeller with less curvature in the inducer. The minimisation of the relative velocity within the blade at inlet can be further enhanced by making use of splitter blades, whereby every second blade is cut back at the inlet to provide more area between the blades. This reduces the effect of blade blockage in the inlet region and so leads to a lower suction surface peak Mach number. It also increases the throat area between the blades to give more choke margin and allows a larger milling tool to be used in the manufacture, which cuts the cost of the impellers.

Splitters are invariably used in turbocharger impellers to reduce manufacturing costs. They are also normally used in high Mach number stages where the issues related to peak Mach number and choking are crucial. In some stages with a very high outlet pressure, double rows of splitters are deployed, giving four times as many blades at the outlet as at the inlet. This reduces blockage at inlet, and the increase in blade number at the outlet decreases the slip factor and increases the work coefficient. Splitters are not generally used in conventional low-speed compressors and pumps, which suggests that development experience at lower pressure ratios with lower Mach numbers in the flow shows that an impeller with full blades is more efficient than one with splitters. Splitters are often used in pumps with a high risk of cavitation, such as rocket fuel pumps, which usually also have a long inducer section.

Conventionally, a full blade that has had its upstream portion cut back from the leading edge and has been thinned and rounded at the leading edge is used as a splitter blade. In the development of more modern compressors, however, a completely different blade shape from that of the full blade can be selected to obtain a smoother flow around the splitter blade. The splitter blade may be thinner than the main blade as it is shorter and less prone to damaging vibrations. It can also incorporate blade leading-edge camber to improve the local incidence on to the blade (Came and Robinson, 1998). The motivation for modifying the splitter leading-edge camber is increased with fewer blades where the flow is less well guided by the main blades. The leading edge of the splitter is generally downstream of the throat of the main blades and forms its own throat; this is usually selected to be larger than that of the main blades by 10–15% to avoid choking at this location. If the splitter chokes first, it impedes the main blades in setting up the most efficient shock structure. Peak efficiency generally occurs at the position where the shock is just expelled from the covered passage, and choking by the splitter hinders this. Further discussion of this is given in Section 11.8.5 on the design of transonic inducers.

Other configurations of impellers with multiple splitter vanes are possible. Secondary splitter blades placed between the main blade and the splitter vane have been adopted in some designs, and two splitter vanes of different lengths have also been shown to provide an efficiency improvement in some cases (Malik and Zheng, 2019).

11.5 Flow Conditions at the Impeller Throat

Lown and Wiesner (1959) showed the importance of the prediction of the maximum flow capacity of an impeller by considering choking at the impeller throat using 1D methods of gas dynamics and provided evidence of the agreement between these simple analyses and test data on a range of compressor stages. Since then, other researchers have followed the same procedure.

The throat width of a particular section in the annulus is the minimum spacing between adjacent blades. It is possible to calculate the area of the impeller throat, A^*, if the annulus is subdivided into sections as shown in Figure 11.9 and the area of each element with a certain throat width is added. An approximation to the true area can be obtained from summation of the areas of each element with the following equation;

$$A^*_{imp} = \frac{LZ}{n}\sum_{i=1}^{n}\left(\frac{\pi D_i}{Z_{bl}}\cos\beta_1' - t_i\right), \tag{11.29}$$

where L is the length of the leading edge, Z is the number of blades, D_i is the diameter of the centre of each element, t_i is the thickness of the blade, β_1' is the blade inlet angle and there are n elements across the annulus. Different views of the impeller throat for a 3D impeller with an inducer are shown in Figure 11.9, but for clarity only four elements across the span are given.

If the impeller flow channel accelerates the flow up to the throat, it may reach a relative Mach number of unity at this position, and this then determines the maximum mass flow through the impeller. The analysis of choking in the impeller is slightly different to that given in Chapter 6 as the choking now takes place in the relative frame

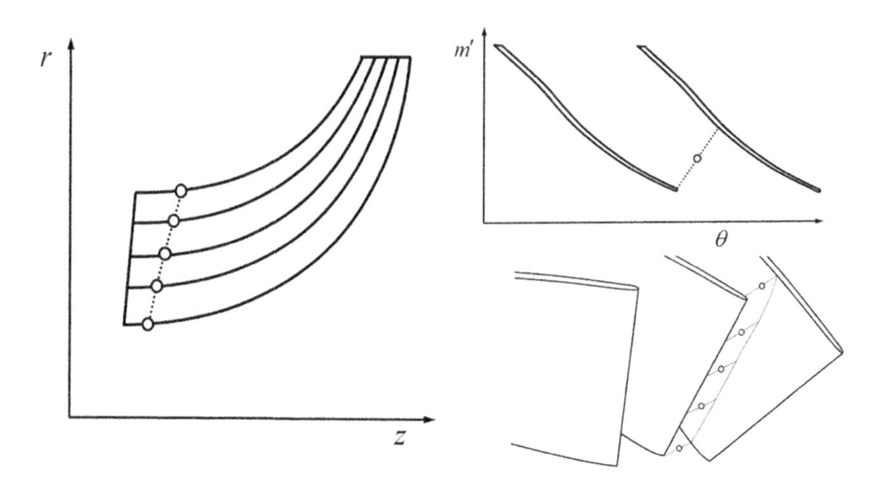

Figure 11.9 Sketch showing different views of the impeller throat in a 3D impeller with inducer (left: meridional view; top-right: blade-to-blade view; bottom-right: 3D view).

of reference. This means that the relative Mach number must be considered. When the throat chokes,

$$M_{rel} = \frac{w}{a} = \frac{w}{\sqrt{\gamma RT}} = 1, \quad w = \sqrt{\gamma RT}. \tag{11.30}$$

The temperature at the throat may be calculated from the rothalpy so that if there is no swirl component of the flow at inlet, $c_{u1} = 0$, then

$$c_p T_{t1} = c_p T_1 + \frac{\gamma R T_1}{2} - \frac{u_1^2}{2}, \quad \frac{T_1}{T_{t1}} = \left(\frac{2}{\gamma+1}\right)\left(1 + \frac{u_1^2}{2c_p T_{t1}}\right). \tag{11.31}$$

For isentropic flow from the inlet total conditions, the density and speed of sound at the throat are

$$\frac{\rho_1}{\rho_{t1}} = \left(\frac{T_1}{T_{t1}}\right)^{1/(\gamma-1)}, \quad \frac{a_1}{a_{t1}} = \left(\frac{T_1}{T_{t1}}\right)^{1/2}, \tag{11.32}$$

and from this the choking mass flow per unit area at the throat is given as

$$\dot{m}_{\max} = A^*_{imp}\rho_1 w_1 = A^*_{imp}\rho_{t1} a_{t1}\frac{\rho_1}{\rho_{t1}}\frac{a_1}{a_{t1}} = A^*_{imp}\rho_{t1}a_{t1}\left[\frac{2+(\gamma-1)u_1^2/a_{t1}^2}{\gamma+1}\right]^{\frac{(\gamma+1)}{2(\gamma-1)}}$$

$$\Phi_{\max} = \frac{\dot{m}_{\max}}{\rho_{t1}D_2^2 a_{t1}} = \frac{A^*_{imp}}{D_2^2}\left[\frac{2+(\gamma-1)u_1^2/a_{t1}^2}{\gamma+1}\right]^{\frac{(\gamma+1)}{2(\gamma-1)}} = \frac{A^*}{D_2^2}\left[\frac{2+(\gamma-1)(D_1/D_2)^2 M_{u2}^2}{\gamma+1}\right]^{\frac{(\gamma+1)}{2(\gamma-1)}}. \tag{11.33}$$

Alternatively, this can be written as

$$\frac{\phi_{t1,\,\max}}{A^*_{imp}/D_2^2} = \frac{1}{M_{u2}}\left[\frac{2+(\gamma-1)(D_1/D_2)^2 M_{u2}^2}{\gamma+1}\right]^{\frac{(\gamma+1)}{2(\gamma-1)}}. \tag{11.34}$$

It is difficult to represent the complex throat geometry with a 1D approximation, but the authors' experience is that the blade speed at the Euler diameter, or rms diameter, is most representative for choking in the 1D analysis. The maximum flow coefficient on a particular speed line with M_{u2} = constant is a function of the tip-speed Mach number and the area of the throat relative to that of the square of the impeller diameter. The throat has so far been represented as the physical area between the blades, but any boundary layer blockage can shift the effective location of the throat, and so reduce the throat area. Operating points towards choke typically involve flow acceleration from the inlet to the throat, so the throat blockage is usually small. High negative incidence towards choke can, however, reduce the available area in the flow passage through separation on the pressure side, and this may change the effective location of the throat to be downstream of the minimum geometrical area.

Equation (11.34) shows that choking in the impeller does not determine an absolute upper limit to the mass flow – the maximum mass flow changes if the blade rotational speed changes. This is not often properly documented in many texts where choke is referred to as a stonewall condition, as if there is a fixed limit. The fixed stonewall limit only applies if the rotational speed does not change. It is for this reason that the characteristic performance map of a compressor differs from that of a turbine. In a compressor, the rotational speed affects the choking mass flow of the impeller, and the speed lines in a compressor map are widely separated at choke. In a turbine, it is often the inlet nozzle vanes which choke so that no variation of the choking mass flow with the rotational speed is found; see Section 11.12.4.

Figure 11.10 shows the variation of the nondimensional choking parameter in (11.34) as a function of Mach number for a typical range of mean impeller inlet diameter ratios. The impeller choking parameter decreases strongly with an increase in tip-speed Mach number leading to a reduction of operating range between peak efficiency and choke. But the choking parameter reaches a plateau at around 1.5–2 depending on the diameter ratio, and the impeller choking parameter is then hardly affected by the tip-speed Mach number. This indicates that the range of operation from

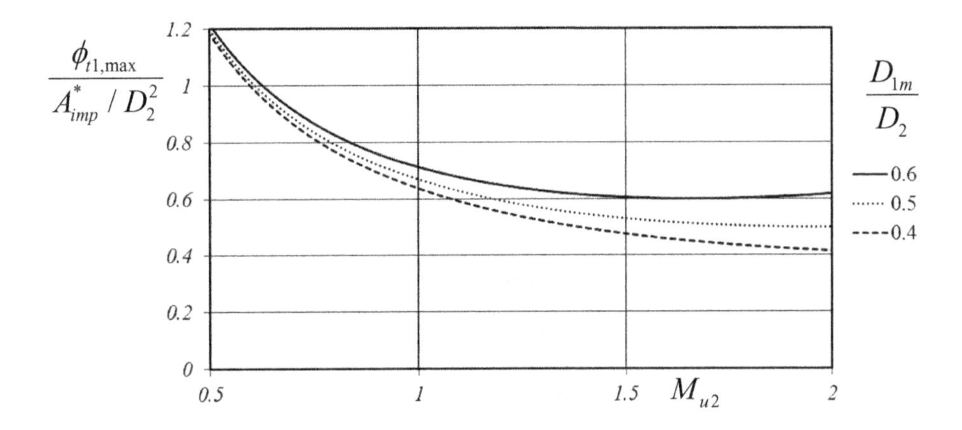

Figure 11.10 Variation of the impeller choking parameter (11.34) with tip-speed Mach number.

peak efficiency to choke at first becomes smaller with increasing Mach numbers but then remains relatively constant at high Mach number. Chapter 17 has more details of this behaviour from measurements on many compressor stages, and Chapter 18 uses this behaviour to predict the operating range to choke in a compressor map.

11.6 Flow Conditions at Impeller Outlet

11.6.1 Backsweep

The impeller shape at outlet is superficially similar for all modern impellers, as it is now commonplace to incorporate –25° to –55° of backsweep to enhance the operating stability of the stage. In fact, nearly all axial compressor rotors, turbine rotors and radial turbine impellers are also backswept at outlet, relative to the direction of rotation, so the term 'backswept' is something of a misnomer as it applies equally well to practically all turbomachinery rotors. But the technical literature only uses the term in the context of radial compressors; here it distinguishes modern designs from older designs which tended to have blades that were purely radial at impeller outlet; see Chapter 5 on the jet-wake phenomenon. Interestingly, centrifugal pumps and process compressors have been backswept for more than a century.

The background to the use of backsweep in radial compressors lies in the effect of the flow outlet angles on the work input and energy transfer when the flow rate changes. For an impeller for which the outlet flow direction is not the same as the blade outlet angle, with a slip velocity, c_s, and no swirl at the inlet, the Euler work input coefficient can be expressed from the velocity triangle as shown in Figure 11.11 as

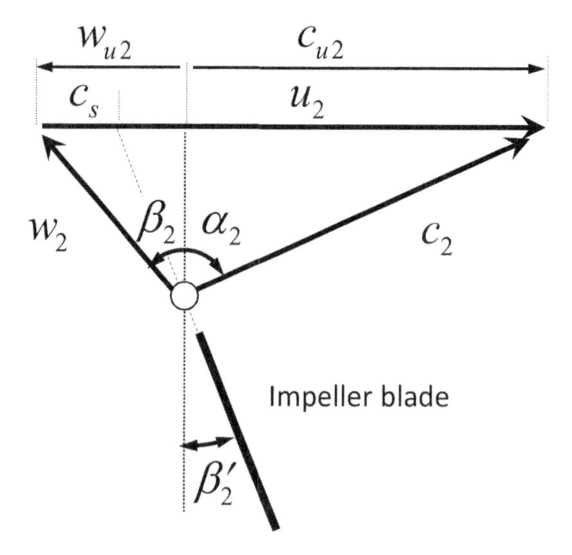

Figure 11.11 Impeller outlet velocity triangle with slip velocity.

$$\lambda_{Euler} = \frac{c_{u2}}{u_2} = 1 + \frac{w_{u2}}{u_2} = 1 - \frac{c_s}{u_2} + \phi_2 \tan\beta_2' = 1 - \sigma + \phi_2 \tan\beta_2'. \tag{11.35}$$

This equation shows that, under the assumption of constant slip velocity, the Euler work coefficient changes linearly with the flow coefficient at impeller outlet.

The relatively simple concept of slip was discussed in Section 5.4 and has been analysed by many researchers. For practical purposes, a useful rule of thumb for the magnitude of the slip velocity is the Wiesner correlation, given in Section 5.4.4, as

$$\frac{c_s}{u_2} = \frac{\sqrt{\cos|\beta_2'|}}{Z^{0.7}}, \tag{11.36}$$

but many other correlations are available. The slip depends on the impeller outlet blade angle, β_2', and the number of blades at impeller outlet, Z. As the backsweep angle increases, fewer blades are generally used, and as a result, both the nominator and denominator of (11.36) decrease together, and the slip velocity ratio generally lies between a fairly narrow range of $0.08 < c_s/u_2 < 0.13$. The slip velocity ratio remains within this range over a wide range of stages and operating points, as shown in an analysis of slip factor data for 250 cases given by Del Greco et al. (2007). Furthermore, the part of the velocity triangle of interest in the work transfer is the swirl in the absolute flow at the impeller outlet, c_{u2}. A large error of, say, 20% in the estimate of the slip velocity still causes an error of only 5% in the estimated work input coefficient. This helps to explain the longevity of the Wiesner correlation, which is widely used in preliminary design despite its imperfections.

If the parasitic losses were not present, then the Euler work coefficient at shutoff or zero flow would be simply given by

$$\lambda_{shutoff} = 1 - \sigma \tag{11.37}$$

such that the nondimensional slip velocity can be determined from the value of the Euler work coefficient extrapolated to zero flow. Taking the parasitic losses into account, using equations from Chapter 10, leads to an expression for the work coefficient as

$$\lambda = (1 + C_{PIF}/\phi_{t1})\lambda_{Euler} = (1 + C_{PIF}/\phi_{t1})(1 - \sigma + \phi_2 \tan\beta_2'). \tag{11.38}$$

In the notation of this book, a backswept impeller at outlet has a negative value for the impeller outlet angle. The last term of this equation is therefore negative, and the specific work input for a backswept impeller falls as the flow rate increases.

For an impeller with radial blades at outlet, the work input is constant with a change in the flow rate and for an impeller with a forward-swept blade the work input increases with flow rate. At high flow coefficient, the work coefficient characteristics are approximately straight lines, but at lower flow coefficients the additional power for the parasitic losses causes an increase in the work coefficient and the line deviates from a straight line to a more negative gradient. Equation (11.38) also indicates that an increase in the outlet flow coefficient due to a change in the channel blockage leads to a reduction in the work coefficient, which explains that an increase in tip clearance causing a higher blockage may reduce the work input.

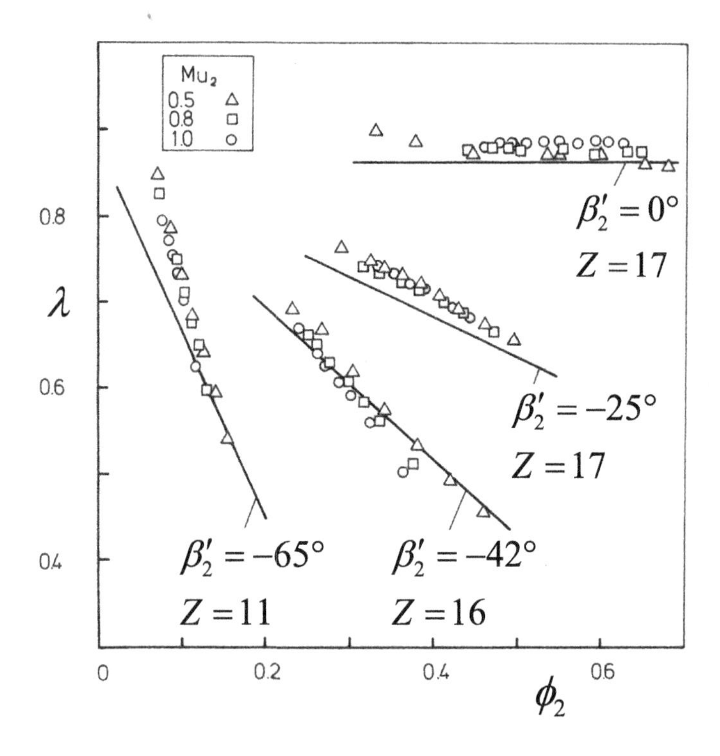

Figure 11.12 The effect of the blade angle and blade number on the work coefficient characteristic using (11.35). (image courtesy of the Institute of Refrigeration, www.ior.org.uk)

The extent to which such a simple model can be used to reproduce the energy transfer in radial compressors over a range of operation can be seen in Section 18.3, where (11.38) is used to predict measured performance characteristics of a variety of stages. A further comparison with this equation and measurement data for four different industrial impellers with different backsweep angles and blade numbers over a range of different tip-speed Mach numbers from Casey and Marty (1985) is shown in Figure 11.12.

11.6.2 The Influence of Backsweep on the Compressor Characteristic

If the efficiency remains constant with the flow rate, then the slope of the pressure rise characteristic with flow corresponds to that of the work input. To avoid instability, it is necessary for the stage to have a rising pressure as the flow is reduced, as discussed in Section 17.2.7, and backsweep is used to provide an inherently stable form of characteristic. The positive effect of backsweep on the stability and operating range of stage characteristics of radial compressor stages is illustrated in Figure 11.13.

The middle diagram of Figure 11.13 shows an impeller with radial blades at outlet. When operating near the peak efficiency point, the rate of change of efficiency with flow is low, and therefore the pressure rise characteristic is parallel to that of the work coefficient. The greater the deviation from the best point, the greater the losses and the greater the deterioration in the pressure rise characteristic. The peak efficiency point of the

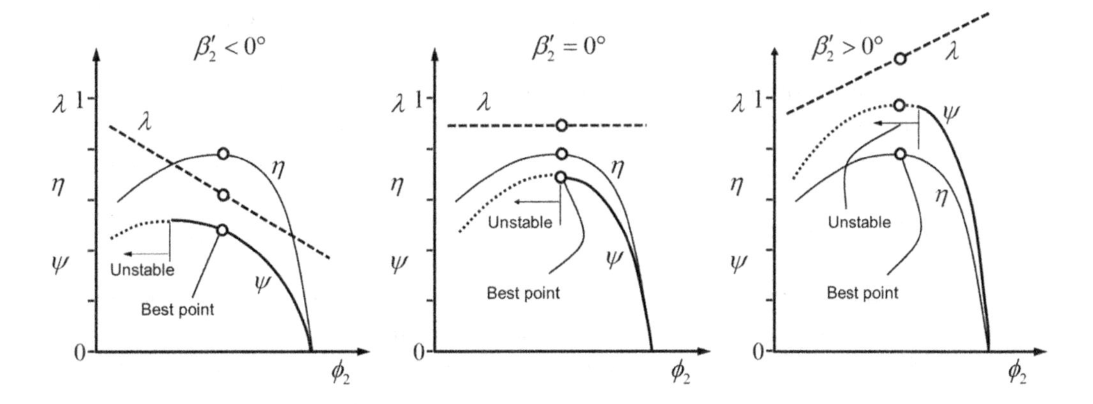

Figure 11.13 The effect of the design work coefficient on the compressor characteristic and the operating range (from left to right: backsweep, no backsweep, forward sweep).

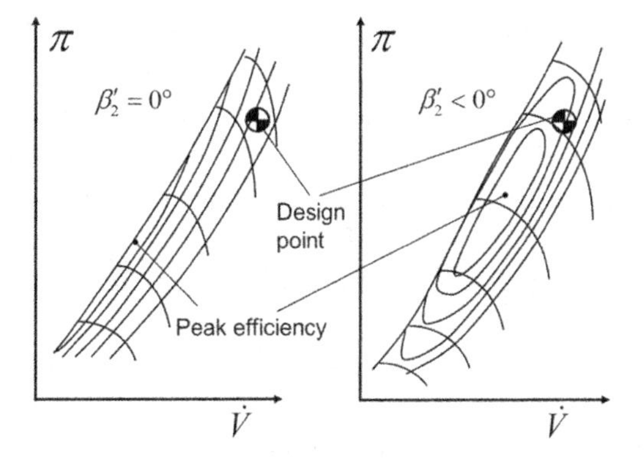

Figure 11.14 The effect of backsweep on the compressor map (left: no backsweep; right: with backsweep). (image adapted using data from Came et al, 1984)

characteristic for a radial bladed impeller is then exactly at the peak pressure rise. As explained in Section 17.2.7, this is the point where system instability often occurs, and then the peak efficiency is on the surge line. Operation with a safe margin to the surge line in a radial bladed impeller forces a move down the characteristic away from the peak efficiency point. For the forward-swept impeller on the right, the situation is worse: the point of peak efficiency actually lies to the left of the surge line, and it is not possible to operate stably at this point. The backswept impeller on the left has its peak efficiency at a point on the characteristic where the pressure rise continues to increase as the flow reduces, thus enhancing stability and providing a safe operating margin against surge at the best point.

Centrifugal impellers are generally designed with backsweep as this leads to a more negatively sloped characteristic. A reduction in backsweep is accompanied by a higher pressure rise but also by a flatter work characteristic, therefore giving a smaller operating range between surge and the peak efficiency operating point, as shown in Figure 11.14.

No backsweep causes the peak efficiency to occur on the surge line, which takes the safe operating line of the compressor well away from the peak efficiency (Came et al., 1984), although in order to achieve adequate range it may still be necessary for the design point to be towards choke operation. Backsweep provides not only enhanced stability, with the peak efficiency in the centre of the operating characteristic, but also brings higher peak efficiency through several other effects such that the region above 80% efficiency in the map is much larger.

The first positive effect of backsweep is that the absolute velocity at diffuser inlet, c_2, and the associated absolute Mach number are reduced. This makes the job of the diffuser easier, as less diffusion is needed in the diffuser and the diffuser losses are lower. Second, the associated decrease in kinetic energy at impeller outlet also increases the degree of reaction of the stage, thus making more pressure rise available by the centrifugal effect within the impeller, again increasing the efficiency. The improvement of efficiency of centrifugal stages as a result of increasing the degree of reaction has been demonstrated experimentally by Shibata et al. (2012). More backsweep also means that less deceleration of the relative flow is needed in the impeller so that there are lower impeller diffusion losses. An additional benefit of backsweep, described in Section 12.2, is that it causes an increase in the kinetic energy at diffuser inlet as the impeller moves off design towards lower flows. This enables the diffuser to continue to secure a good pressure recovery even if its nondimensional pressure recovery coefficient decreases as it moves off design towards surge.

The disadvantages of high backsweep are that it results in a lower work input, leading to higher blade speeds and higher stresses for a given pressure rise. In addition, there are even higher stresses due to bending forces on the roots of blades, which lean backwards. In some low-speed radial ventilator applications where surge, efficiency and stresses are not an issue, forward-swept blades are used to allow a higher work coefficient. The resultant lower blade speed leads to a markedly lower noise level. The aerodynamic noise of a rotating rotor is roughly proportional to the fifth or sixth power of the blade speed, although the lower efficiency is counterproductive in this regard.

An inherent characteristic of radial turbocompressors with backsweep is that, as system delivery pressure decreases, the flow capacity increases, giving the operating characteristic a negative slope. This advantage of backswept impellers is so overwhelming in its effect on the operating range that few radial compressors are now produced without backsweep. Typically, the range of work coefficient in modern radial impellers lies between 0.6 and 0.75 with backsweep. The lower value corresponds to stages which require more operating range and for typical process compressor stages and the higher value is used for aero engine applications. Higher values are used if absolutely necessary, but then some form of stability enhancement mechanism, such as a ported shroud, may be required to provide additional operating range; see Section 17.5.

11.6.3 Effect of Work Coefficient on the Steepness of the Characteristic

It is not the backsweep itself which determines the steepness of the work characteristic but rather the value of the work coefficient at the peak efficiency point and the slip factor. In this section, it is shown how the slope of the characteristic depends on the design work coefficient, and in Section 11.6.6 it is shown that it is possible to achieve the same work coefficient with different values of the backsweep.

At any point on the characteristic, the rate of change of the pressure rise coefficient with the flow coefficient is given by

$$\frac{\partial\psi}{\partial\phi} = \frac{\partial(\eta\lambda)}{\partial\phi} = \left(\lambda\frac{\partial(\eta)}{\partial\phi} + \eta\frac{\partial(\lambda)}{\partial\phi}\right). \tag{11.39}$$

Near the peak efficiency point, the first term in the bracket of this equation is zero as there is no change in efficiency with flow; close to design, the slope of the pressure rise coefficient is then determined entirely by the slope of the work coefficient with flow. Equation (11.35) can be rewritten at the design point (denoted with the subscript *d*) as

$$\lambda_d = 1 - \sigma_d + \phi_{2d}\tan\beta'_{2d}. \tag{11.40}$$

On combination with (11.35) and some reformulation to eliminate the outlet blade angle, and on the reasonable assumption that the slip factor remains unchanged with small deviations from design, this leads to

$$\frac{\lambda - \lambda_d}{\lambda_d} = \frac{\Delta\lambda}{\lambda_d} = \left(\frac{\lambda_d - 1 + \sigma_d}{\lambda_d}\right)\left(\frac{\phi_2 - \phi_{2d}}{\phi_{2d}}\right) = \left(\frac{\lambda_d - 1 + \sigma_d}{\lambda_d}\right)\left(\frac{\Delta\phi_2}{\phi_{2d}}\right). \tag{11.41}$$

This can be rearranged to give an expression for the steepness of the work characteristic as

$$\chi_d = \frac{(\Delta\lambda/\lambda^d)}{(\Delta\phi_2/\phi_{2d})} = \left(\frac{\lambda_d + \sigma_d - 1}{\lambda_d}\right). \tag{11.42}$$

The fractional change in work coefficient for a fractional change in the outlet flow coefficient in (11.42), χ_d, is a nondimensional expression for the steepness of the work characteristic at the design point. This demonstrates that the steepness is determined by the work input at the design point and the slip factor and not by the backsweep itself. For typical values of the work coefficient, the gradient is negative, such that an increase in flow causes a decrease in work coefficient. For a high value of work input coefficient of 0.88 and a slip velocity coefficient of 0.12, the slope is zero and the work input characteristic is horizontal. The pressure rise coefficient then reaches its maximum at the peak efficiency point and lower flows may not be possible because of system instabilities. For a typical value of work input coefficient of 0.6 and a slip coefficient of 0.1, the steepness is given by $\chi_d = -0.333$ and thus a decrease in flow of 1% produces a 0.333% increase in work coefficient. It is then possible to operate at lower flows than the peak efficiency as the pressure rise characteristic continues to rise as the flow is reduced.

Equation (11.42) indicates that for centrifugal stages designed with a high work coefficient, it is important to have a low slip factor to give a steeper work coefficient characteristic. For example, if an impeller with a work coefficient of 0.75 is designed with few blades leading to a large slip factor of 0.13, then the steepness is given by $\chi_d = -0.16$. Decreasing the slip factor to 0.08 by using more blades at impeller outlet increases the steepness substantially to a value of $\chi_d = -0.227$. High work impellers in aeroengine applications are often designed with many blades at outlet, and sometimes with multiple splitters, both to increase the work coefficient for a given backsweep and to increase the steepness of the work characteristic, as evidenced by the figures in Section 1.6.9.

An equation similar to (11.42) can also be derived for axial compressors (Horlock, 1958). Axial compressors generally have a much lower work coefficient than radial compressors due to the near absence of any centrifugal effect, although with an increasing hub radius some small centrifugal effect is present. Given a typical axial compressor with no slip and a work coefficient of 0.25 on the mean line the steepness is $\chi_d = -3$, which means that a 1% decrease in flow would increase the work coefficient by 3%. Axial compressor loading (in terms of the work coefficient) thus increases roughly 10 times more rapidly with a similar fractional decrease in flow than the loading of typical radial stages. This explains the fact that, for the same pressure ratio, multistage axial compressor characteristics are generally much steeper and narrower than those of radial compressors. These arguments also apply to pumps designed for low work coefficient at high flow coefficients, usually called high specific speed pumps, which generally have much steeper characteristics than pumps designed for lower flow coefficients with higher work coefficients, known as low specific speed pumps. See Gülich (2008) for examples and the discussion in Section 11.12.3.

A radial compressor design with a very low work coefficient of, say, 0.3 and a slip coefficient of 0.1 would have a steepness of $\chi_d = -1.5$, that is, the work characteristic becomes steeper and much more like that of an axial stage. It is clear that specifying a very high work coefficient leads to a narrow safe operating range between the peak pressure rise and the peak efficiency; see Section 17.4. Conversely, a low design work coefficient will cause the operating characteristic to become narrower due to its natural steepness such that the flow at choke point is close to that at the peak efficiency point. The slope of the work coefficient at the peak efficiency point thus becomes steeper with a decrease in the design work coefficient. Figure 11.15 shows the effect of different design work coefficients across the whole characteristic, as calculated with the method given in Section 18.3. These curves are for a process compressor impeller with a vaneless diffuser at a tip-speed Mach number of 0.9. At a low design work coefficient, the work versus flow characteristic becomes very steep and effectively reduces the operating range towards choke as the work and the pressure ratio drop very rapidly with increasing flow. At typical design values of

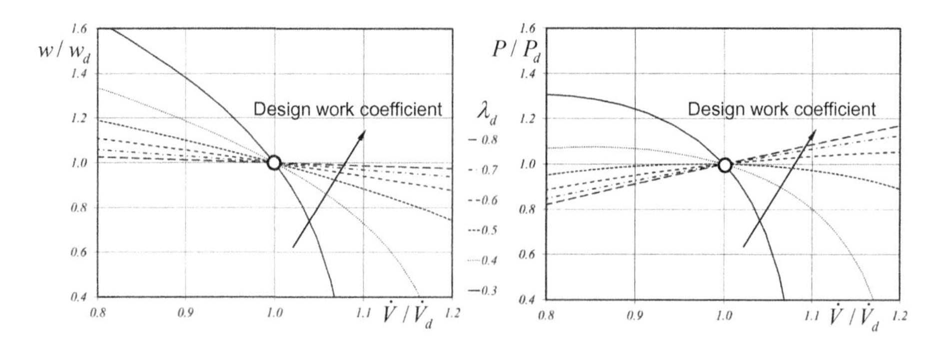

Figure 11.15 Variation in work input and power with flow relative to that at the design point for a range of designs with different design work coefficients (left: work; right: power).

the work coefficient, the power required increases with flow but decreases with flow for designs with low design work input coefficient.

11.6.4 The Link between the Inlet and Outlet Velocity Triangles

The preceding discussions have identified the main considerations for the selection of the velocity triangle at the inlet and outlet of the impeller. The link between the inlet and outlet velocity triangles is best illustrated by plotting these in the same diagram, as in Figure 11.16. This shows clearly that in its passage through the impeller the flow at the casing streamline with an increase in blade speed experiences a reduction in the meridional velocity, a reduction in the relative velocity, an increase in the absolute velocity and an increase in the swirl component of the absolute velocity.

In a strong adverse pressure gradient, the diffusion of the relative flow will cause the flow to separate, as discussed in Section 5.5.7. Figure 11.17 shows the surface

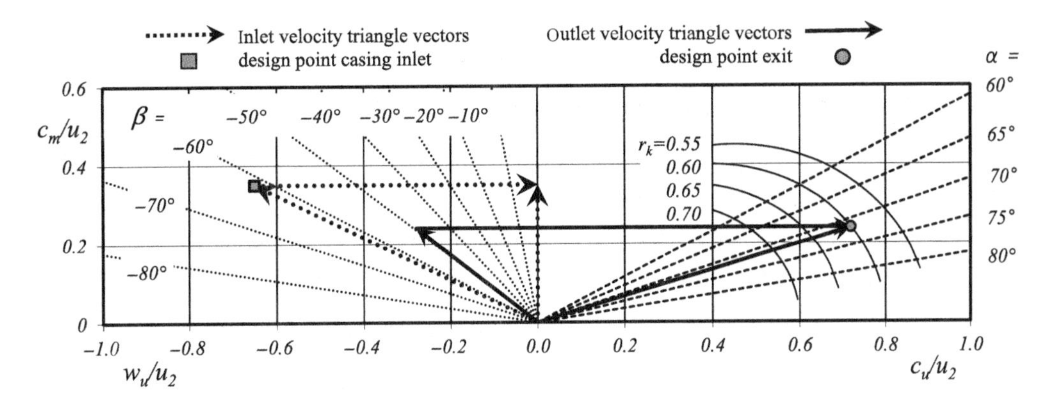

Figure 11.16 Inlet and outlet velocity triangles showing reduction in the relative velocity and meridional velocity through the impeller.

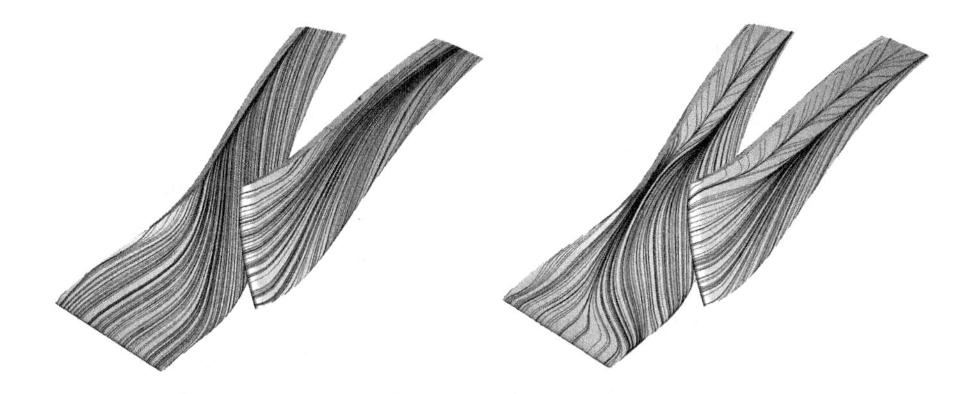

Figure 11.17 Surface streaklines in an impeller with and without flow separation on the casing based on a CFD simulation with ANSYS CFX.

streamlines from CFD simulations on the blades of an impeller that has not separated and one at a different operating point with a higher deceleration leading to a separation near the casing. In a radial compressor, the deceleration is best expressed as a limiting de Haller number for the flow on the casing streamline from inlet to outlet, $DH = w_2/w_{c1}$, whereby the limiting values to avoid high losses are typically taken to be between $0.5 < DH < 0.7$. Lower values down to about $DH = 0.45$ are possible but are generally only used in stages with an extremely high work coefficient and lower expectations for the efficiency. The limitation in that case is generally the absolute gas angle at diffuser inlet. Higher values of the de Haller number than $DH = 0.7$ would be considered to be lightly loaded in terms of diffusion.

11.6.5 The Outlet Width Ratio of the Impeller

The width at the impeller at outlet can be derived from the continuity equation on the assumption of no blockage as

$$b_2 = \dot{m}/c_{m2}\rho_2\pi D_2. \tag{11.43}$$

The density of the gas at impeller outlet is given by

$$\frac{\rho_2}{\rho_{t1}} = \left[1 + (\gamma - 1)r_k\lambda M_{u2}^2\right]^{\frac{1}{n_{imp}-1}}, \tag{11.44}$$

where r_k is the kinematic degree of reaction and n_{imp} is the polytropic exponent for the impeller compression process from inlet total to outlet static conditions. Further rearrangement gives

$$\frac{\phi_{t1}}{(b_2/D_2)} = \pi\phi_2\left[1 + (\gamma - 1)r_k\lambda M_{u2}^2\right]^{\frac{1}{n_{imp}-1}}. \tag{11.45}$$

Lettieri et al. (2014) define a useful term as $b_2{}^* = (b_2/D_2)/\phi_{t1}$, known as the nondimensional impeller exit width parameter. For a process stage operating in air at a tip-speed Mach number of $M_{u2} = 0.9$, with a polytropic efficiency of $\eta_p = 0.85$, a work coefficient of $\lambda = 0.64$ and an outlet flow coefficient of $\phi_2 = 0.25$, (11.45) leads to $b_2{}^* = (b_2/D_2)/\phi_{t1} = 1.0$. This shows that as a convenient rule of thumb, the outlet width ratio, b_2/D_2, of such an impeller is then given almost exactly by the value of the inlet flow coefficient, ϕ_{t1}. For impellers designed with a higher work coefficient or with a higher pressure ratio, the outlet width is less than this rule of thumb due to the decreased volume flow at impeller outlet. Stages with high flow coefficients also tend to have higher values of the outlet flow coefficient to reduce the span of the blades, and stages with low flow coefficients tend to have a lower outlet flow coefficient to increase the channel height. Lettieri et al. (2014) show that the optimum nondimensional impeller exit width parameter rises to $b_2{}^* = (b_2/D_2)/\phi_{t1} = 3$ for low flow coefficient impellers, so that $(b_2/D_2) = 3\phi_{t1}$ provides a good guideline for such impellers.

During the design of an impeller the outlet width ratio, b_2/D_2, relative to the global flow coefficient, ϕ_{t1}, is determined by (11.45) so that for a given global flow coefficient, the outlet width is determined by the choice of the outlet flow coefficient, ϕ_2. Some of the considerations leading to the choice of the outlet flow coefficient are given in Section 11.6.6. In terms of the parameters under the control of the designer, the outlet flow coefficient is given by

$$w_{c1} = c_{m1}/\cos\beta_{c1}, \quad w_2 = c_{m2}/\cos\beta_2$$

$$w_2/w_{c1} = \frac{c_{m2}/\cos\beta_2}{c_{m1}/\cos\beta_{c1}} \tag{11.46}$$

$$\frac{c_{m2}}{u_2} = \frac{c_{m1}}{u_2}\frac{w_2}{w_{c1}}\frac{\cos\beta_2}{\cos\beta_{c1}}.$$

This equation shows that the outlet flow coefficient is determined in part by the inlet flow coefficient, the de Haller number, w_2/w_{c1}, and the change in flow angle through the impeller. If the other parameters remain constant, a wider outlet width ratio leads to lower outlet flow coefficient and a higher deceleration with a lower de Haller number in the impeller.

If there are no large side-streams, the first-stage impeller in a multistage configuration has the highest flow coefficient and is always the widest stage. As a result, the first-stage impeller will usually be the most highly stressed impeller in a mechanical sense, and there is a strong incentive to reduce its outlet width ratio, b_2/D_2, by increasing its outlet flow coefficient, ϕ_2. Conversely, low flow coefficient stages will naturally have narrow impellers. Here there is a strong incentive to make the hydraulic diameter of the channels larger to reduce the friction losses so that such stages tend to have lower values of the outlet flow coefficient, ϕ_2. Care is needed with this if vaneless diffusers are used as this may cause instability if the appropriate pinch is not applied; see Section 12.2.4.

Equation (11.45) also shows that as the tip-speed Mach number is increased, the outlet width for a given flow coefficient and a fixed outlet flow coefficient will reduce. A design that is carried out for a low Mach number has a wider outlet flow channel and can be adapted for a high Mach number by altering the convergence of the flow channel without changing the blades; see Figure 11.18. Alternatively, the impeller may retain the same outlet width at higher Mach number but will then operate with a higher deceleration and a lower outlet flow coefficient. This gives a higher work input but for a vaneless diffuser would require readjustment of the pinch and for a vaned diffuser an increase in the diffuser inlet blade angle.

11.6.6 The Influence of Various Parameters on the Outlet Velocity Triangle

The outlet velocity triangle is determined by three aerodynamic parameters. If consideration of the range or of the required pressure ratio leads to a certain selection of work coefficient, $\lambda = c_{u2}/u_2$, and the required pressure ratio leads to a particular blade speed at the impeller outlet, u_2, then both the absolute swirl velocity, c_{u2}, and the relative swirl velocity, $w_{u2} = u_2 - c_{u2}$, are fully determined. One free parameter remains in the velocity triangle. This could be either the local flow coefficient at impeller outlet, $\phi_2 = c_{m2}/u_2$, the relative velocity, w_2/u_2, or the absolute flow angle into the diffuser,

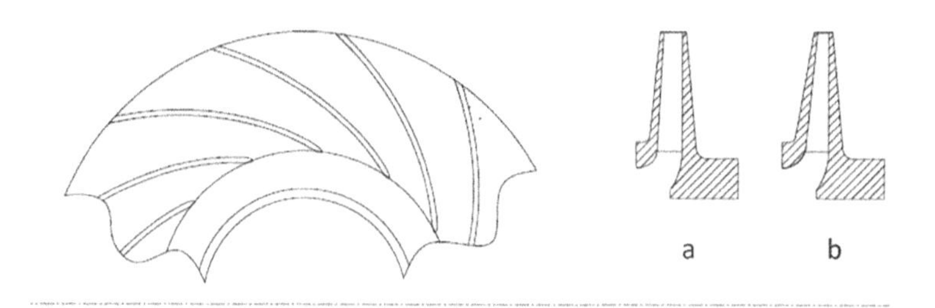

Figure 11.18 Increase in the convergence of the meridional channel at high tip-speed Mach numbers: (a) low tip-speed Mach number and (b) high tip-speed Mach number.

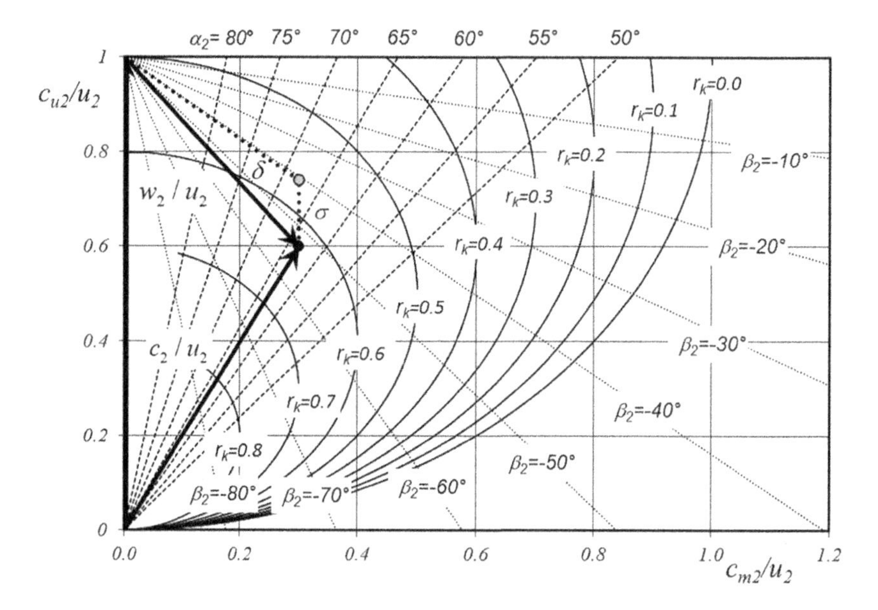

Figure 11.19 Work coefficient versus outlet flow coefficient showing the degree of reaction, flow angles and the velocity triangle for a design with $c_{m2}/u_2 = 0.3$ and $c_{u2}/u_2 = 0.6$.

α_2. Each of these may be used by designers to determine the outlet velocity triangle depending on the particular requirements of the design.

Figure 11.19 illustrates the effect of different velocity triangles at the impeller outlet on the absolute and relative flow angles as a function of the work coefficient and the outlet flow coefficient. This plot is made nondimensional with regard to the blade tip speed and shows that the designer has two free parameters for the outlet velocity triangle for a given tip speed. One of these is the nondimensional work coefficient which is plotted on the y-axis in this figure, c_{u2}/u_2. A higher value on this axis with a higher work input requires a lower relative velocity, w_2/u_2, giving a low de Haller number and a higher diffusion in the impeller, w_2/w_{c1}. The other free parameter plotted on the x-axis is the local flow coefficient at impeller outlet, c_{m2}/u_2. A higher value on this axis leads to a narrower impeller with a smaller outlet width ratio, b_2/D_2, and lower absolute flow angles into the diffuser, α_2. Together these values determine the relative flow angle and the absolute flow angle so that the impeller blade outlet angle and the diffuser blade inlet angle are also determined. The shape of the outlet velocity triangle of any specific impeller at design can be identified from a point on this diagram, shown as the black point in the diagram, by joining the point to the vertical axis at the coordinates (0,0) and (0,1); the outlet velocity triangle for a specific design with $c_{m2}/u_2 = 0.3$ and $c_{u2}/u_2 = 0.6$ is shown.

In the determination of the stage performance, the impeller effectively operates on the total to static enthalpy rise across the impeller, while the diffuser operates on the static to total enthalpy rise of the diffuser. The kinematic degree of reaction, r_k, is the ratio of the total to static enthalpy rise across the impeller to the total enthalpy rise across the stage. As described in Chapter 2, the degree of reaction is given by

$$r_k = \frac{h_2 - h_{t1}}{h_{t3} - h_{t1}} = \frac{h_{t3} - c_2^2/2 - h_{t1}}{h_{t3} - h_{t1}} = 1 - \frac{c_2^2}{2(h_{t3} - h_{t1})} = 1 - \frac{c_{u2}^2 + c_{m2}^2}{2\lambda u_2^2} = 1 - \frac{\lambda^2 + \phi_2^2}{2\lambda}. \tag{11.47}$$

This leads to

$$\phi_2^2 + \lambda^2 = 2\lambda(1 - r_k). \tag{11.48}$$

Lines of constant degree of reaction are plotted in Figure 11.19 for a range of reaction. As pointed out by Mehldahl (1941), the lines of constant reaction in this diagram are circles which all pass through the origin, (0,0), the centres of the circles lie along the ordinate with $c_{m2}/u_2 = 0$ and with $c_{u2}/u_2 = 1 - r_k$ and the radius of the circles is given by $1 - r_k$.

This rather elegant presentation was first published by Mehldahl. It is adapted here to show the components of the velocity triangles and the absolute and relative flow angles for a given style of design. Other useful forms of diagrams with different axes examining the parameter variation in the outlet velocity triangle have been published by Mehldahl (1941), Wiesner (1960), Casey (1994a), Ribi (1996), Lüdtke (2004) and Lou and Key (2019).

Other features of the velocity triangle can also be added to this diagram, for example, contours of constant absolute velocity c_2/u_2 are circles with their centre at the origin (0,0) and contours of constant relative velocity w_2/u_2 are circles with their centre at (0,1). This highlights the fact that in the region of most interest, a higher work coefficient gives rise to a higher absolute velocity at the diffuser inlet and to a lower relative outlet velocity, giving more deceleration and a lower de Haller number in the impeller. Interestingly, the blade angle at impeller outlet can also be added to the diagram as shown by the open point placed vertically above the black point at a distance corresponding to the slip factor. This also allows the deviation angle of the flow at impeller exit to be shown in the diagram.

An enlarged version of the most relevant part of Figure 11.19 is shown as Figure 11.20, where, due to the change in the scales of the axes, the circular segments have become elliptical. The values of c_{m2}/u_2 and c_{u2}/u_2 for four different impellers, A, B, C and D, are shown together with a rectangular band of typical process impellers from Lüdtke (2004). Individual impellers have a unique position in this diagram, but

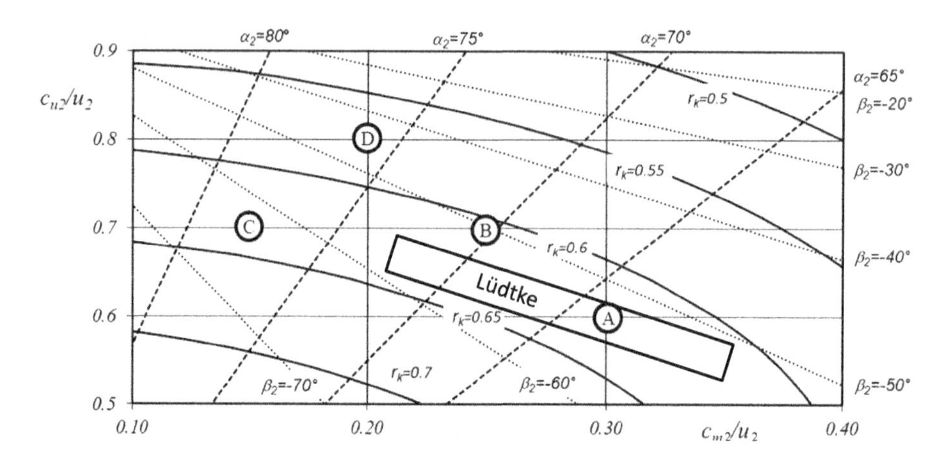

Figure 11.20 Work coefficient versus flow coefficient showing the degree of reaction and flow angles with the velocity triangle parameters for four design styles, A, B, C and D.

the outlet velocity triangles of most impellers cluster in the neighbourhood the points A, B, C and D. Impeller A is a stage suitable for a vaneless diffuser with a low absolute flow angle into the diffuser, α_2; impeller B is suitable for a vaned diffuser or a vaneless diffuser with pinch; C is a stage for a low flow coefficient with a vaned diffuser; and stage D is a high work stage with a vaned diffuser.

The fact that there are only two free parameters in this diagram means that changing one of these also changes others, and this interrelationship makes it difficult to do studies of the optimum velocity triangle. For example, if a work coefficient, $\lambda = c_{u2}/u_2 = 0.7$ is selected on the basis of the required pressure ratio and a flow coefficient of $\phi_2 = c_{m2}/u_2 = 0.25$ is chosen, then the velocity triangle is the same as stage B in this diagram. Reducing the outlet flow coefficient to obtain a wider impeller leads to a stage between B and C. Such a change also increases the relative outlet flow angle which requires either more back-sweep in the impeller or a lower slip with fewer blades. The absolute flow angle into the diffuser becomes larger such that the diffuser will need to be adjusted. At the same time, the level of diffusion in the impeller and the degree of reaction increases. It is not possible to determine with 1D arguments which of these changes will cause a change in efficiency or in the surge margin; the designer typically resorts to CFD in the detailed optimisation process.

Generally, the final selection of the velocity triangle at the impeller outlet is made on the basis of the amount of diffusion in the impeller. Adjustment of the outlet meridional velocity – by changing the impeller outlet width – allows the designer to achieve the desired diffusion. The simplest approach is to select a design value for the de Haller number along the casing contour, $DH = w_2/w_{c1}$, whereby values between $DH = 0.5$ and 0.7 are normally chosen. If an impeller with a reasonably high flow coefficient and with $r_{c1}/r_2 = 0.65$ is considered, the optimum inlet angle is close to $\beta_{c1} = -60°$, then (11.24) can be used to find an expression for the approximate relative velocity at the outlet as a function of the de Haller number as follows:

$$\frac{w_2}{u_2} = DH\frac{w_{c1}}{u_2} \approx DH\frac{2}{\sqrt{3}}\frac{r_{c1}}{r_2} \approx 0.75DH. \tag{11.49}$$

With modest diffusion and a de Haller number of $DH = 0.7$, this equation leads to numerical values of $w_2/u_2 = 0.525$; with a high diffusion giving $DH = 0.5$, the result is $w_2/u_2 = 0.375$. This also affects the relative and absolute flow angles at impeller tip, β_2 and α_2, as follows:

$$\frac{c_{m2}}{u_2} = \sqrt{\left[\left(\frac{w_2}{u_2}\right)^2 - \left(1 - \frac{c_{u2}}{u_2}\right)^2\right]}, \quad \tan\beta_2 = \frac{1 - \dfrac{c_{u2}}{u_2}}{\dfrac{c_{m2}}{u_2}}, \quad \tan\alpha_2 = \frac{\dfrac{c_{u2}}{u_2}}{\dfrac{c_{m2}}{u_2}}. \tag{11.50}$$

Looking at impeller A, (11.50) leads to outlet flow angles of $\beta_2 = -53.1°$ and $\alpha_2 = 63.4°$ for a work coefficient of $\lambda = c_{u2}/u_2 = 0.6$ and an outlet flow coefficient of $\phi_2 = c_{m2}/u_2 = 0.3$. This would be a typical outlet velocity triangle for an industrial process impeller or a turbocharger impeller with a vaneless diffuser designed for a wide flow range. Some industrial impellers with vaned diffusers may also operate close to this point. A vaneless diffuser ideally requires an absolute flow angle below 65° for successful stable operation and good range; see Section 12.3.2. This forces these designs into the lower-right part of this diagram so that stages with vaneless diffusers are limited in work input and require a high meridional velocity ratio at outlet. If a vaneless diffuser is used and a higher work is required then, without the use of pinch in the diffuser, the value of ϕ_2 would need to be increased towards 0.35 to avoid a very high flow angle into the diffuser. Lüdtke (2004) gives examples showing the variation of the outlet flow coefficient from $\phi_2 = c_{m2}/u_2 = 0.35$ down to 0.2 in process compressor stages as the global flow coefficient decreases from $\phi_{t1} = 0.1$ to 0.01, shown as the large rectangle labelled Lüdtke in Figure 11.20. The points to the right in this band of impellers is generally reserved for high flow coefficient process impellers.

Impeller B is a stage with a work coefficient of $\lambda = c_{u2}/u_2 = 0.7$ and an outlet flow coefficient of $\phi_2 = c_{m2}/u_2 = 025$: this is close to typical micro gas turbine–style impellers and turbocharger impellers for higher pressure ratios with vaned diffusers. The relative outlet flow angle is given by $\beta_2 = -50.2°$ and the absolute flow angle is $\alpha_2 = 70.4°$, so that this stage would normally be unsuitable for a vaneless diffuser and requires a vaned diffuser. A commonly used and memorable combination of parameters for impellers with vaned diffusers close to point B is $\lambda = c_{u2}/u_2 = 0.72$ and $\alpha_2 = 72°$. Alternatively, a vaneless diffuser can be used with a pinch, where the diffuser width is less than that of the impeller. The pinch needs to be sized to bring the absolute flow angle at diffuser inlet below 65°. Both impeller A and impeller B have a relative outlet angle close to $\beta_2 = -50°$, so that depending on the number of impeller blades and the associated slip factor, this means that a blade backsweep angle between −40° and −45° would be needed.

Impeller C is a stage in which the meridional flow coefficient has been deliberately reduced in order to increase the outlet width ratio of the impeller, b_2/D_2. This is a typical outlet velocity triangle for a very low flow coefficient impeller with $\phi_{t1} < 0.01$ and for nearly all centrifugal pumps. The associated hydraulic diameter of these stages is relatively large despite the low flow coefficient and, following (10.22), a high efficiency can be achieved despite the low flow coefficient. The work coefficient is still high with $\lambda = c_{u2}/u_2 = 0.70$, but the outlet flow coefficient is now $\phi_2 = c_{m2}/u_2 = 0.15$. The absolute flow angle is close to $\alpha_2 = 78.0°$, so that a vaned diffuser is essential. The relative flow angle is now $\beta_2 = -63.4°$ so that a high backsweep is needed. Despite the high level of backsweep, this stage can achieve a large work coefficient as evidenced by the many pumps designed with an outlet velocity triangle similar to impeller C (Gülich, 2008). Examples of compressor stages designed with outlet velocity triangles somewhat similar to stage C with high backsweep are given in the publications of Casey et al. (1990), Dalbert et al. (1999), and Casey et al. (2010). Lettieri et al. (2014) shows that stages of this type can achieve a wide operating range with a vaned diffuser.

Impeller D is a stage with a high work coefficient of $\lambda = c_{u2}/u_2 = 0.8$ and an outlet flow coefficient of $\phi_2 = c_{m2}/u_2 = 0.2$. This leads to an outlet relative flow angle of $\beta_2 = -45°$, and so the stage would probably have a backsweep of around −35°. The high absolute outlet flow angle would require a vaned diffuser, and to achieve the high work coefficient, the impeller would need a low slip factor with many blades at impeller outlet. Examples of modern stages for aeroengine applications with outlet velocity triangles close to point D are described by Braunscheidel et al. (2016).

A, B, C and D are examples of possible design styles, but many other designs are possible with velocity triangles anywhere in the neighbourhood of these specific examples. It is important to realise that the selection of the impeller outlet velocity triangle may limit its possible use with vaneless diffusers unless a pinch is used. To be most effective, impellers need to be specially designed for use with either vaneless or vaned diffusers.

11.6.7 Selection of Impeller Blade Number

There are no reliable guidelines on blade number, and values from experience or from manufacturing or mechanical limitations are usually used in the preliminary design. These may be refined based on the blade loading coefficients in the throughflow calculation, using the blade loading coefficient described in Section 7.5.13. In some cases, the hub stress and weight and inertia due to the number of blades may play a role (Rodgers, 2000), and in others noise from blade passing frequencies and instability may be important (Chen, 2017). More blades reduce the impeller throat area, which reduces the maximum flow and makes blade milling more difficult and expensive. Impellers for higher inlet relative Mach numbers, with $M_{w1} > 1.0$, invariably use splitter vanes to resolve the problem of the blade blockage at the leading edge and then have an even number of blades at outlet. Generally, the number of blades increases with the design work coefficient. Turbocharger impellers with a low work coefficient, such as stage A in Figure 11.20, may have as few as 10 blades at outlet (with five main blades and five splitter blades), and aeroengine designs for a high work coefficient design, such as stage D, may have up to 32 blades at outlet (with 16 main blades and 16 splitter blades). Designs similar to stage C with highly backswept blades may have 10 full blades and designs similar to stage B between 16 and 20 full blades.

11.6.8 One-Dimensional Impeller Preliminary Design in a Nutshell

There are a number of ways for the designer to determine the impeller geometry and velocity triangles for a given flow and pressure ratio. A good starting point is to select the work coefficient $\lambda = c_{u2}/u_2$, based on the conflicting considerations of operating range and pressure ratio. A value of $0.68 < \lambda < 0.72$ helps to minimise the impeller tip speed for a given pressure ratio, leading to lower stress levels, as in stages B and C in Figure 11.20. More extreme designs at higher pressure ratio may use a value of 0.8, as with stage D, but these may be limited in operating range, as outlined in Section 17.4. These impellers require a vaned diffuser to be most effective. A lower value, say $\lambda = 0.6$, results in a steeper work characteristic with a wider operating range and may be more effective with a vaneless diffuser, such as stage A in Figure 11.20. The work coefficient and an estimate of the expected efficiency allows the tip speed Mach number for a given pressure ratio to be determined from (10.58). This leads to an estimate of the blade tip speed for initial considerations of the mechanical aspects and may lead to a revision of the work coefficient that has initially been selected.

In the second instance, it is sensible to select a value of the global flow coefficient within the range of optimum flow coefficient described in Section 10.7.1, leading to a high levels of efficiency, that is, in the range $0.07 \leq \phi_{t1} \leq 0.10$. In this way, the initial estimate of the optimum diameter of the impeller, D_2, can be made from (10.59). If size and weight are at a premium, a smaller impeller diameter with a higher flow coefficient may sometimes be used. For impellers on a single-shaft multistage machine, it may not be possible to select the optimum diameter as this would require smaller impellers along the shaft to the rear of the machine. This may be done to a certain extent, but the flow coefficient of the rear stages is generally reduced to give similar diameter for all stages.

For a 3D impeller being designed for a given flow coefficient, ϕ_{t1}, and tip-speed Mach number, M_{u2}, (11.28) gives the relative Mach number at the casing of the impeller eye, M_{w1}, for the most compact design. Equation (11.20) then provides the optimum flow angle at impeller inlet on the casing, β_{c1}, and (11.21) gives the impeller inlet eye diameter ratio, D_{c1}/D_2, for a given hub diameter. Equation (11.25) provides a first approximation of the inlet blade angle across the span, but this needs to be revised taking into account the curvature of the flow during the throughflow design. The relative velocity at the inlet on the casing, w_{c1}/u_2, is given by (11.24) so that in a 1D sense the inlet geometry and inlet velocity triangle is then determined. Similar equations can be derived for 2D impellers.

From the inlet relative velocity on the casing streamline, the impeller relative velocity at outlet is determined by the de Haller number $w_2 = DH^*w_{c1}$. Together with the work coefficient, the de Haller number determines the impeller outlet flow coefficient, ϕ_2, and the impeller outlet width, b_2. In this way, a compact design following standard guidelines for the minimum inlet relative velocity and the acceptable diffusion in the impeller will be produced. All design engineers following these standard guidelines will then generate a similar impeller for this duty in terms of the velocity triangles. This similarity in designs for a given duty becomes important in

Chapter 18, when the prediction of a performance map is considered and it is shown that the duty itself is a good indicator of the shape of the performance map.

In some preliminary design systems, the rotational speed in rpm, N, the mass-flow rate and the total pressure ratio, π_t, are taken as given, for example if these are constrained by matching with the turbine as part of a turbocharger design or if the radial compressor is the last stage on an axial-centrifugal machine in a turboshaft gas turbine. If the de Haller number and the backsweep are then defined, the required impeller diameter and velocity triangle at impeller outlet can be determined iteratively from this information. An initial guess for the impeller diameter needs to be made, and the inlet velocity triangle can then be based on Section 11.6. The relative velocity at the impeller outlet is determined from the de Haller number, $w_2 = DH^* w_{c1}$. If, in addition, the outlet blade angle is chosen and the slip factor is calculated from the blade number using a suitable correlation, then the trigonometry of the velocity triangle at impeller outlet gives

$$\cos\beta_2 = \frac{c_{m2}}{w_2} = \frac{w_{u2} - c_s}{w_2 \tan\beta_2'} = \frac{w_2 \sin\beta_2 - u_2\sigma}{w_2 \tan\beta_2'}. \tag{11.51}$$

This equation can be solved for the relative outlet flow angle to provide

$$\cos\beta_2 = \frac{-k_1 k_2 + \sqrt{1 + k_1^2 - k_2^2}}{1 + k_1^2}, \quad \text{where } k_1 = \tan\beta_2', \quad k_2 = u_2\sigma/w_2. \tag{11.52}$$

Similarly, from the velocity triangle and the Euler equation it is possible to derive

$$u_2 = \frac{-w_2 \sin\beta_2 + \sqrt{(w_2 \sin\beta_2)^2 - 4(u_1 c_{u1} + \Delta h_t / C_{PIF})}}{2}. \tag{11.53}$$

From the blade tip speed and the rotational speed, the impeller diameter can be determined. For this equation, the rise in total enthalpy for a given total pressure ratio has to be determined from

$$\Delta h_t / h_{t1} = \pi_t^{(n-1)/n} - 1, \tag{11.54}$$

and this requires the polytropic exponent, n. This exponent can be guessed initially but must be revised as soon as the impeller diameter, the tip-speed Mach number and the flow coefficient have been determined. The exponent is determined using the correlations for efficiency from Section 10.7. As a result, the procedure requires an iteration. A commonly used software based on similar equations to these is Vista CCD from PCA Engineers Limited. A free, simplified version of this software is available for download to use on a smartphone from www.pcaeng.co.uk/apps.

11.6.9 Effect of the Impeller Outlet Velocity Triangle on the Diffuser

A vaneless diffuser can encounter flow instability and thus may cause rotating stall and surge if the absolute flow angle at the design point is above 65°; see Section 12.3.2. This implies that stages with a vaneless diffuser are unlikely to be successful if they are designed much above the line given as $\alpha_2 = 65°$ in Figure 11.20. The design point itself may be stable for stages designed above this line, but as the flow is reduced the outlet angle increases, and at some point the diffuser will become unstable. In such cases, the diffuser inlet needs to be equipped with a pinch in order to increase the meridional velocity and decrease the absolute inlet flow angle to match the impeller with the diffuser.

The design of the impeller outlet strongly influences the diffuser. An impeller that is optimised for low-speed operation typically has a wider outlet width compared to an impeller that is designed to operate at high speeds; see Section 11.6.5. At high speed, a wider tip results in a large absolute flow angle, increasing the losses in the vaneless diffuser. The tip width of an impeller, optimised for high-speed operation, would be too narrow for low-speed operation as it would reduce the diffusion inside the impeller with high outlet flow angles, which could also increase losses due to operation with an undersized volute.

In the case of a stage with a vaned diffuser, the diffuser throat is generally designed to match the impeller throat at the design speed; see Section 12.9. At lower speeds, the impeller delivers a higher-volume flow at outlet and the diffuser chokes, acting as a choked nozzle. As a result, it causes a very large reduction in the flow coefficient of the impeller. At speeds much lower than the design speed, the impeller is forced to operate at the left-hand, stall side of its characteristic: this will often produce separated flow at the casing. At high speeds, the impeller chokes and forces the diffuser to operate at high incidence with the consequent risk of instability; see Section 17.2.12.

It can be seen from Figure 11.19 that in the neighbourhood of stages A, B, C and D, an increase in either the work coefficient or the flow coefficient decreases the degree of reaction. A low reaction stage has a high proportion of its energy at the impeller outlet in the form of kinetic energy of the absolute velocity, and as a result the diffuser has a more difficult job to do. It is useful to plot (11.35) on to this figure to show the operating line of the stage as the work coefficient versus impeller outlet flow coefficient. An example of this is given in Figure 11.21, where the design point of the stage is shown as the white circle and the operating line of the work coefficient versus flow coefficient as the thick straight line through the design point, whereby a constant slip factor of $c_s/u_2 = 0.09$, a design work coefficient of $\lambda = c_{u2}/u_2 = 0.65$ and an outlet flow coefficient of $\phi_2 = c_{m2}/u_2 = 0.25$ have been assumed. Note that the change of scale of the axes causes the circles of constant reaction to become ellipses.

As the flow reduces along the operating line shown in Figure 11.21, both the work coefficient and the absolute flow angle increase. Conversely, as the flow increases along the operating line, both the work coefficient and the absolute flow angle decrease. This change in the work coefficient is the inherent feature of impeller backsweep, which provides stability to the operating characteristic of the stage. The design point and operating points near to design have similar degrees of reaction as the lines of constant degree of reaction run nearly parallel to the operating line at the design condition. Most impeller designs have a constant degree of reaction close to the design point, and the degree of reaction reduces both towards higher and lower flows further from the design point. This phenomenon of a nearly constant degree of

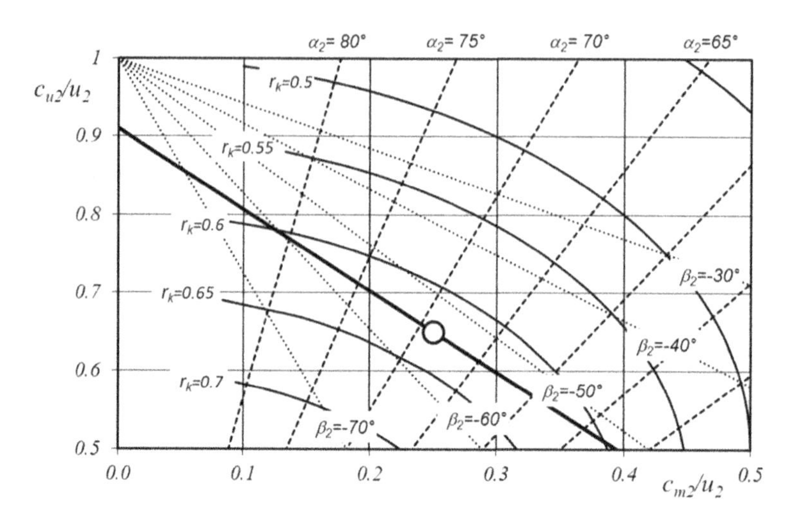

Figure 11.21 Design point and work input characteristic for a stage.

reaction at off-design is exploited in the map prediction method described by Casey and Robinson (2013).

As the flow coefficient decreases along the operating line, the degree of reaction decreases, indicating that at low flows the diffuser has an additional burden as the proportion of kinetic energy at diffuser inlet increases. However, it also indicates that at low flow a good diffuser has more potential to contribute to the pressure rise. For example, if the pressure rise coefficient in the diffuser remains constant at lower flows, then the diffuser has a higher pressure rise through the increase in its inlet kinetic energy. The nondimensional diffuser inlet kinetic energy is given from (11.48) as

$$\frac{c_2^2}{2u_2^2} = \lambda(1 - r_k). \tag{11.55}$$

The increase in the work coefficient and the decrease in reaction with lower flows allow the diffuser to contribute more effectively to the stability of the operating characteristic at these low flow conditions (Ribi, 1996). This applies to all stages with vaned and vaneless diffusers operating in the neighbourhood of stages A, B, C and D in Figure 11.19. In this way, the backsweep of the impeller also contributes to an increase in the pressure rise of the diffuser at low flows, provided that the diffuser can cope effectively with the incidence associated with the change in the inlet flow angle.

11.7 The Compressor Characteristic as Influenced by Losses and Work

11.7.1 The Influence of the Work Coefficient

The variation of the work coefficient with flow in a typical backswept impeller determines the basic nature of the shape of the compressor characteristic with an increasing pressure rise as the flow is reduced, as shown in Figure 11.12. The ideal pressure rise characteristic without any losses would be the same as the work coefficient characteristic, but the actual characteristic needs to take into account the various internal and parasitic losses of the stage. The work input taking into account the slip and the work input factor is given by (11.38). Taking into account the relationship between the global flow coefficient and the flow coefficient at impeller outlet in (11.45), this leads to

$$\lambda = (1 + C_{PIF}/\phi_{t1})\left(1 - \sigma + \frac{\phi_{t1}\tan\beta_2'}{\pi(b_2/D_2)\left[1+(\gamma-1)r_k\lambda M_{u2}^2\right]^{\frac{1}{n_{imp}-1}}}\right). \tag{11.56}$$

Equation (11.56) identifies five physical effects of the flow coefficient on the work coefficient for a backswept impeller, and these are shown schematically in the left-hand side of Figure 11.22, where they are labelled with the same letters as in the following list:

(a) An increase in flow causes a decrease in the work coefficient, and the slope of the work coefficient curve is related to the tangent of the impeller backsweep angle.
(b) The slip factor reduces the work below that based on the blade angle alone, which is best seen at zero flow, where the intersection on the axis is given by the shutoff value of the work coefficient as $1 - \sigma$.
(c) As the flow reduces, the disc friction causes a steady increase of the work input coefficient above the value predicted by the Euler equation, and this becomes noticeable at very low flows so that the work coefficient versus flow coefficient curve is no longer a straight line, as also shown in the measurements in Figure 11.12.
(d) For a fixed inlet flow coefficient, an increase in tip-speed Mach number increases the impeller outlet density, decreases the outlet flow volume and increases the work input coefficient.

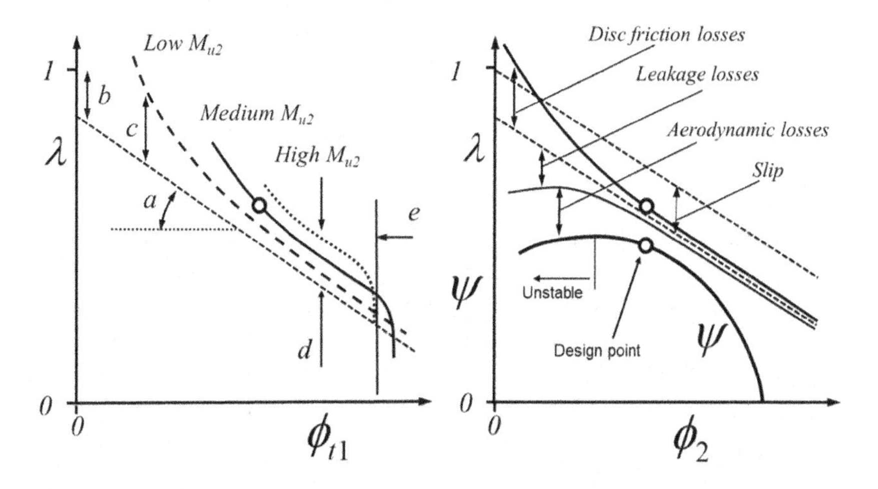

Figure 11.22 Left: effects a, b, c, d and e on the work coefficient versus inlet flow coefficient. Right: the anatomy of the pressure rise coefficient characteristic.

(e) On the high-speed characteristics, the work input falls off steeply at very high inlet flow due to the losses caused by choking of the flow in the impeller inlet, which increases the outlet flow volume, causing a sharp drop in the work coefficient.

These effects appear to be very well described by the simple equations given here, and this is also demonstrated in Section 18.3.

11.7.2 Influence of the Losses

In the determination of the pressure rise, the effect of the losses have also to be taken into account with the work characteristic. The profile losses increase with positive and negative incidence on both sides of the peak efficiency point near-zero incidence. The viscous losses due to friction in the flow channels increase as the flow rate increases, and there are additional losses due to the formation of shocks at the inlet at higher Mach numbers. In addition, there are parasitic and recirculation losses, which increase at low flows. The combination of these effects determines the shape of the pressure rise curve in the stage, as shown schematically in Figure 11.22 for a single tip-speed Mach number.

11.8 Guidelines to Detailed Impeller Design

11.8.1 Blade Loading Distribution in the Meridional Direction

There are a number of parameters at the designer's discretion which have an influence on the flow conditions within the impeller leading to a more uniform flow at the diffuser inlet and which have an effect on the stage efficiency. These include the meridional channel contour, the channel axial length, the blade lean or rake, the meridional blade angle distribution along the hub and the casing and the wrap of the blade in the circumferential direction. In some cases, the impeller may have different blade angles and even a slightly different impeller diameter on the hub and casing sides of the flow channel at impeller outlet.

The designer is free to choose the variation of the mean relative Mach number through the impeller between the values at the inlet and exit determined by the preliminary 1D design. The Mach number distribution can be controlled by the rate of change of blade angle and by the flow area and curvature of the meridional channel. Came and Robinson (1998) consider that there are broadly three strategies for this: (a) a small rate of change initially in the inducer, then increased diffusion later; (b) a linear reduction in the relative Mach number along the entire meridional length and (c) an initial high rate of change to quickly reduce the relative Mach number, then little diffusion of the mean flow around the bend from axial to radial.

The last option, (c), is roughly analogous to the use of front-loaded, controlled diffusion profiles in axial compressor design as outlined in Section 7.3.2; this approach can work well for designs at low pressure ratio with subsonic inducer inlet relative Mach number. The boundary layers are initially thin and therefore capable of sustaining a high level of diffusion. Then the flow is turned to the radial direction with little diffusion of the mean relative Mach number. This approach separates the diffusion and turning, which may also help to reduce losses. In high pressure ratio impellers, the blade needs careful profiling in view of the high levels of relative Mach number found at the casing inlet. It is necessary to limit the rate of change of the blade angle and the meridional curvature along the shroud, since either would lead to an increase in the peak suction surface Mach number ahead of the shock with a corresponding increase in loss and risk of shock-induced boundary layer separation.

The compromises are related to the three-dimensional nature of the flow. The casing streamline exhibits a deceleration of the relative flow, and the hub streamline experiences an acceleration of the relative flow. The problem is made more complex by the fact that any one aspect of the design, for instance the shape of the meridional channel, always interacts with other aspects, such as the blade shape or blade lean. The need to obtain an acceptable low stress level in the impeller and its blades and acceptable vibration characteristics also places constraints on the possible blade shapes. These constraints generally mean that the aerodynamic designer is free to choose the blade shape on the casing, which is the most critical in terms of inlet relative Mach number and de Haller number, but that the shape on the hub and on intermediate streamlines is then largely determined by mechanical considerations.

In Section 5.3.6, it is shown that the blade loading in terms of the difference in velocity level between the suction and pressure surface can be estimated using

$$\left(w_{ss} - w_{ps}\right) = \frac{2\pi}{Z}\frac{c_m}{w}\frac{\partial(rc_u)}{\partial m} = \cos\beta\frac{2\pi}{Z}\frac{\partial(rc_u)}{\partial m}. \tag{11.57}$$

This shows that the blade-to-blade loading reduces with the number of blades. In addition, the loading depends not only on the rate of change of the absolute tangential component of velocity in the impeller but also on the rate of change of the radius. Thus, in contrast to axial compressor stages where there is little change in radius, the radial movement of the flow through the impeller gives rise to a high blade-to-blade loading, from the Coriolis forces. Typically, the relative blade angle reduces from –60° to –30° through the impeller so that the effect of the changing blade angle also increases the loading in the rear of the impeller.

Equation (11.57) is the basis of an inverse design method by Zangeneh (1994), in which the designer specifies the swirl distribution through the impeller and the inverse method calculates the blade shape from this. Changing the distribution of the swirl, rc_u, changes the gradient in this equation; this can be used to control the blade-to-blade loading. The authors find, in their own designs, that it is more practical to control the loading by adjusting the rate at which the relative blade angle changes through the impeller on the hub and casing streamlines. Zangeneh et al. (1998) suggest that in order to reduce secondary flows, the casing should be front-loaded and the hub streamline should be aft-loaded.

For stages designed at a high tip-speed Mach number, an important issue is the control of the inlet Mach number to ensure that the flow does not accelerate into the impeller flow channels. If the flow is supersonic at the inducer inlet, this requires low blade-to-blade loading and thin blades at the front of the inducer, giving rise to relatively straight blades in the tip section. It also suggests that the meridional channel at the tip should also be fairly straight to avoid unwanted acceleration of the meridional flow due to curvature in this region. Most designs need to have a wide choke margin. Unfortunately, this requires more turning at the inlet to get the largest throat area between the blades and so a compromise is needed. For this reason, most transonic designs have fairly high incidence on the hub sections, which have low relative Mach numbers and accelerated flow. Some detail of the design of transonic inducers is given in Section 11.8.5. The use of splitter blades increases the throat area in the inducer where the loading is low, allows more turning in the middle of the impeller as the increased number of blades reduces the blade-to-blade loading and reduces the slip velocity, thus increasing the work.

A further important issue to be considered for all impellers is that of decelerating the flow on the blades without causing the flow to separate. The selection of the de Haller number in the design process is the most important aspect of the solution here, and it determines the overall amount of diffusion needed on the casing streamline. But how should this diffusion be distributed through the impeller? Unfortunately, there is no published work on radial machines which really clarifies this issue. This is mainly because changing this one aspect of the design automatically changes other features.

This problem of blade loading was examined by Dallenbach (1961). He considered the variation of the mean velocity through the blades, with the choices of a linear deceleration, a rapid initial deceleration and an initial slow or zero deceleration. This could be combined with possible variation of blade loading with the choices of a uniform loading, a front-loaded blade row with more turning in the inducer and a rear-loaded blade row with less turning in the inducer. He studied impellers with radial blade filaments where the shape of the blade at the casing completely determines the blade shape on the hub, so the loading distribution on the hub cannot be selected separately from that on the casing. Because of boundary layer considerations, he suggested that it is desirable to use a large deceleration in the front part of the impeller as a slower rate of deceleration leads to thicker boundary layers. Possible blade loading distributions as suggested by Dallenbach are shown in Figure 11.23. The blade-to-blade loading parameter, C_{btob}, is defined in Section 7.5.13. A high diffusion near the leading edge as shown in the loaded inducer case is not possible for stages with high inlet Mach number, as explained earlier in this section. Other examples are given by Morris and Kenny (1968), where they show how the splitter can be used to reduce the blade-to-blade loading in the rear part of the impeller.

Jansen and Kirschner (1974) considered different loading distributions for back-swept impellers using throughflow calculations. They made the point that practically no systematic experiments had been performed to determine the relative merits of different loading distributions. The throughflow calculations do not include the secondary flows or the clearance flows effects, and as such the real flow in the impeller may have regions of separation and be very different from the idealised flow

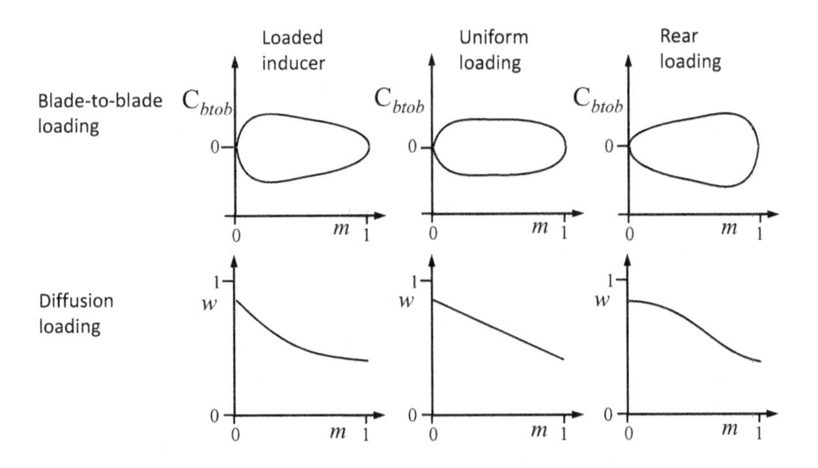

Figure 11.23 Overview of possible blade loading distributions.

distribution, especially to the rear of the impeller. Despite this weakness, the through-flow calculations provide loading diagrams that allow the designer to compare different designs. Jansen and Kirschner (1974) suggested that in most applications a constant linear loading distribution is quite satisfactory.

Several attempts to examine the effect of different loading distributions through experiments have subsequently been made. Mishima and Gyobu (1978) tested five different subsonic shrouded impellers with different diffusion levels (de Haller number) and a different meridional distribution of the blade loading. The three stages with the highest de Haller number of around 0.66 had the highest efficiency, and the best of these had the maximum diffusion near the leading edge. The worst stage had strong deceleration followed by constant velocity near the trailing edge. The linear distribution of loading shown in Figure 11.23 had an efficiency between the two extremes. Shibata et al. (2012) tested four different impellers and concluded that although aft and front loading contributed to the surge and choke margin, the different loadings had little influence on the efficiency. More recently, Diehl (2019) has suggested that in small compressor stages with an unavoidable high tip clearance a front-loaded design reduces the clearance losses, but taking into account the secondary flows he recommends a midloaded design for best overall efficiency.

Considerable research has been carried out on the most suitable diffusion rate in axial compressor blade sections. Controlled diffusion profiles have more turning and diffusion near the leading edge where the suction surface boundary layer is thin. Calvert and Ginder (1999) provide an overview of their design strategy for transonic fan blades at different spanwise positions and of the compromises that need to be made. At the tip with a high Mach number, the blades are thin and nearly straight in the inlet section to avoid acceleration in the supersonic flow. After the flow has reached the suction surface shock, more turning is then applied downstream of the shock. The subsonic blade sections on the hub are of the controlled diffusion type. The Calvert and Ginder strategy cannot be directly applied to radial wheels; the highly curved meridional channel and the large radius change along the hub contour in a radial stage makes this strategy less readily applicable.

For radial impellers, the flow channel is generally designed with a nearly linear change in the spanwise channel width from the inlet to the outlet. This leads to a more or less linear rate of deceleration of the mean meridional velocity. The curvature of the shroud naturally accelerates the flow on the casing to give a higher meridional velocity than on the hub. The curvature on the casing can be kept almost constant through the impeller so that there is no unwanted acceleration or deceleration of the meridional flow. The difference in velocities between the hub and the casing streamtubes can be reduced by increasing the axial length of the impeller to reduce the curvature. After setting up the flow channel, the blade angles are then selected. There is no turning in the inducer section for transonic impellers, and a more or less linear change in angle downstream of this to give a linear deceleration along the suction surface downstream of any shock or downstream of the throat. The velocity distribution should attempt to avoid unnecessary deceleration; the suction surface should not decelerate the flow near the casing only to reaccelerate it afterwards. The shock-related losses are a strong function of the aerodynamic loading at the tip section, and changes in the tip design can have strong effects on performance (Hazby and Xu, 2009b).

A similar strategy of controlling the meridional curvature is applied to try to avoid a strong deceleration in the inlet region on the hub. The flow on the hub streamline accelerates from low subsonic velocities at the inlet to the outlet. Accelerating relative flow on the hub contour suggests that this is less critical than the design of the tip section. Because of the accelerating flow, the hub can tolerate a larger incidence angle at the leading edge. In addition, a higher turning near the leading edge may be used to increase the throat area. The hub section is usually thickened to increase the natural frequency and reduce stresses on the blades. Splitters are used to reduce the blockage near the leading edge. The splitters on the hub do not need to be as thick as the main blades, and the use of thinner blades further reduces the blockage on the hub sections. In addition, the blade angle distribution along the hub section may be adjusted to give a more radial orientation of axial sections through the blade to reduce blade root bending stresses.

11.8.2 The Wrap, Rake and Lean Angles

An important geometrical parameter in the definition of the blade loading of the hub section is the tangential wrap angle of a blade, $\Delta\theta_{wrap}$, as shown in Figure 11.24. This is the difference between the circumferential coordinate of the blade at the leading edge and that at the trailing edge. It is related to the blade camber angle and the shape of meridional channel through the following equations:

$$\tan\beta = r\frac{d\theta}{dm}, \quad \theta_{wrap} = \int_{le}^{te} \frac{\tan\beta}{r} dm. \tag{11.58}$$

This integration is carried out in the meridional direction. The hub meridional section is naturally longer than the casing section and has a lower radius, so for the same blade angle distribution, the wrap of the hub section would be much larger than that of the

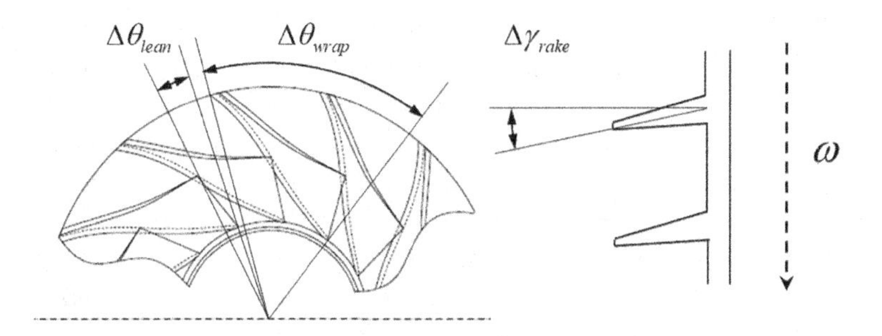

Figure 11.24 Tangential wrap angle, inlet lean angle and outlet rake angle.

Figure 11.25 Rake angle at the trailing edge (left: 45°, right: 25°).

casing. If the inlet of the inducer were purely radial, this would lead to a large lean of the blade at the trailing edge. In fact, the inlet angle on the hub is lower than that on the casing such that the hub blade angles are all lower than those at the casing. As a result, the lean effect is slightly less strong than this would suggest. Positive lean causes the tip of the blade at the trailing edge on the casing to precede that on the hub in its rotational direction as shown in Figure 11.25. In this way, the blade pressure surface leans towards the hub contour at the trailing edge.

The lean angle between the blade suction surface and the normal to the hub at trailing edge is called the rake angle, and two examples with positive rake are given in Figure 11.25. A positive rake angle has mechanical advantages in that the stresses in the blade roots of the impeller are lessened. It is for this reason that this is the usual form of rake to be found. Impellers with backsweep but no rake at the trailing edge do not have radial blade fibres, and the centrifugal loading causes a bending moment with increased stress in the blade roots. Applying a positive rake angle causes the blade fibres to be more nearly radial in the outlet region and reduces the bending stresses. An aerodynamic advantage of rake in either direction is that it reduces the excitation forces between the impeller and diffuser vanes, which usually have no lean.

The blade angle distribution of the casing section has priority over the hub and cannot normally be compromised as it is determined by the very strong aerodynamic considerations discussed in this section. The hub, which has accelerating flow, is considered to be less sensitive, and the designer may adjust the hub angle distribution in order to produce a rake that is mechanically sound and to ensure the blade fibres in the inducer are also radial. This is done by selecting lower values of the blade angle in the midchord region of the blade, leading to a typical blade angle distribution, as shown in Figure 11.7. This means that the blade on the hub generally has an *s*-shape in which it first turns towards the meridional direction and then turns back to create the required backsweep.

The high Mach number impeller in Figure 11.7 shows an impeller that has low turning in the inducer section due to the high relative Mach number and makes up for this with a higher rate of blade turning in the middle section. The process impeller in this figure is much shorter and has a high meridional curvature which reduces the flow velocity on the hub leading to the need for a marginally higher blade inlet angle (more negative) on the hub than on the casing. In more old-fashioned impellers, the blade angles on the casing reduce to values below the blade outlet angle and then the backsweep increases to the trailing edge.

11.8.3 Effect of Lean

The various causes of secondary flows are described in Section 5.6, and an illustration of their strength can be seen in the streaklines in Figure 11.17. A negative rake angle and negative lean of the blade reduces the secondary flows (Zangeneh et al., 1998). Negative lean causes the suction surface to be facing towards the hub and a component of the blade force to act outwards towards the casing. Since the flow must still follow the hub and casing, the blade force must be balanced by a pressure gradient, an outwards blade force then increases the pressure and reduces the velocity on the casing and vice versa on the hub. Negative lean can be used in low-speed fans and in pumps. However, negative lean is not generally suitable for high-speed centrifugal impellers as it increases the bending stresses in the blade roots near the impeller outlet. The effect of the hub to casing loading on the secondary flows are discussed by Zangeneh et al. (1998), where front loading on the casing and aft loading on the hub are recommended.

11.8.4 Control of Clearance Flows

The nature of the clearance flows in impellers is described in Section 5.7 and its influence on the losses in Section 10.4.4. It is not yet clear how best to control the losses produced by the tip clearance flows. In the first instance the key to weakening the effect of the clearance flows is to minimise the size of the clearance gaps. A high turning in the inducer tip section increases the clearance flow over the inducer, and the tip clearance vortex which develops may become quite strong. It may lead to possible vortex breakdown after the shock or farther downstream under the influence of the adverse pressure gradient in the flow channel (Hazby and Xu, 2009a).

As well as being a key factor contributing to the inefficiency of the stages, the clearance flow plays an important part in the onset of surge when this is being controlled by the impeller. The loading of the blades at the tip drives the strength of the overtip leakage vortex, and the blade count must have some influence on this. The trajectory of the vortex and how that interacts with the next blade it meets, either the next full blade or the splitter, are also relevant. Both effects magnify towards surge. Although this process seems amenable to optimisation using CFD, no systematic study seems so far to have been reported in the open literature. Howard and Ashrafizaadeh (1994) numerically studied the effects of blade lean for a centrifugal impeller and found that a compound lean could reduce leakage flow by unloading the blade tip and increasing total pressure ratio while keeping a constant efficiency.

11.8.5 Design of Transonic Inducers

In order to avoid high suction surface peak Mach numbers, the meridional flow channel of a transonic inducer should be designed with low curvature of the casing section and low curvature within the inducer. This tends to lead to a design with low choke margin on the casing section, but this can be compensated by higher incidence and more curvature of the hub section. As a typical impeller has to turn the flow from axial to radial, it is sometimes advisable to begin the radial-to-axial bend upstream of the impeller so that some of the curvature occurs in a region with no blades.

In transonic centrifugal impellers, the impact of aerodynamic shock waves on the flow in the inducer is significant and optimisation of the inducer, particularly the tip section, is critical in achieving high levels of efficiency. As discussed in Section 6.2.2, the flow structure in transonic passages is very sensitive to small changes in local blade profile, passage area and state of the surface boundary layers. This necessitates a good strategy for the choice of incidence, throat area and blade angle distribution.

The design of modern high-speed centrifugal impellers is heavily reliant on 3D viscous CFD. The initial throughflow analysis guides the designer through the comparison with previous experience on similar machines and establishes a good starting point for the shape of the meridional channel. The flow field and the state of the boundary layers are highly three dimensional such that large relative tip clearances result in severe blockage across the span, so working with two-dimensional streamtubes is far less plausible than in an axial machine.

To minimise the shock strength, the inducer tip section can be designed to operate close to the so-called unique incidence condition, which is where the shock is nearly attached to the thin leading edge of the main blade (see Figure 11.26). At low pressure ratios, there is an oblique shock attached to the leading edge of the main blade. The main passage shock, if it exists, is located deep inside the passage, as described in Section 6.4.3. The shock moves upstream with increasing back pressure; peak efficiency tends to occur just before the shock is expelled. The peak efficiency is strongly influenced by the shock-related losses but also by the energy transfer to the flow downstream of the inducer. As the mass flow is reduced, the inlet relative velocity falls. But the relative flow angle at the tip also reduces, which results in a higher preshock Mach number and a

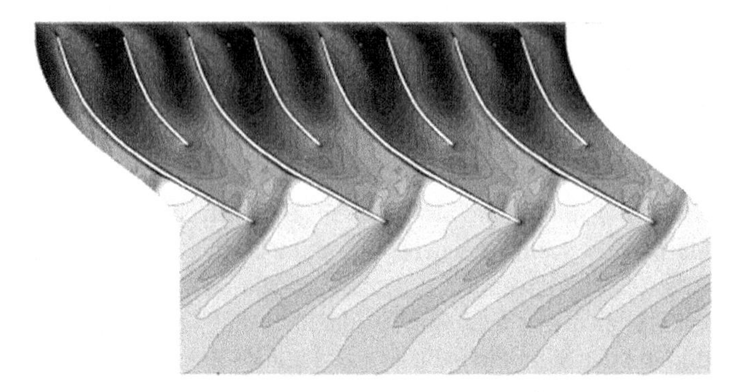

Figure 11.26 Mach number distribution at the tip of a transonic impeller at design.

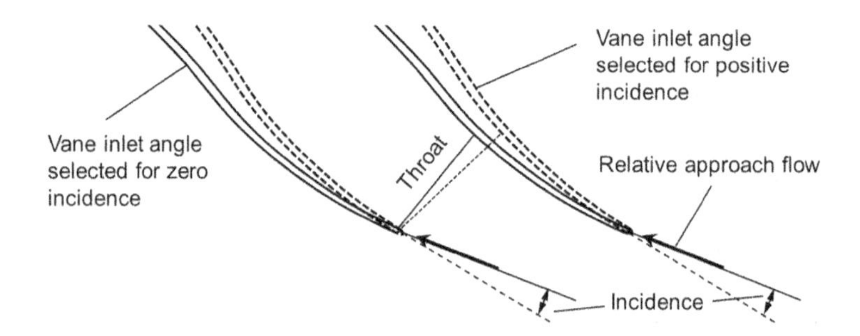

Figure 11.27 Increasing the throat on the hub with design at incidence.

stronger shock due to Prandtl–Mayer expansion of the supersonic flow at the leading edge of the blade. Near the stall point, the shock is normally expelled, increasing the loading at the leading edge of the blade. This in itself is a possible cause of stall.

If the tip section is designed to operate with an attached shock, it is effectively choked at the design point, resulting in a relatively steep constant speed pressure rise characteristic. Any increase in mass flow to achieve a higher choke margin at design speed can only be obtained by making the throat wider over the inner part of the blade. This is achieved by increasing the incidence or the curvature in the front part of the hub section (see Figure 11.27). From the 1D point of view, the relative flow accelerates along the hub of a typical centrifugal impeller given that the relative velocity at the impeller outlet is always higher than that at the impeller hub inlet. Also, the relative speed of the inlet flow at the hub is far lower than that at the shroud and therefore the hub is generally more tolerant of large incidences. It is not unusual to see turbocharger impellers operating with large incidence angles of 5°–10° at the hub at the design condition. This is because of the need to obtain a high flow capacity at high speed rather than for optimum aerodynamic performance. In ruled surface impellers, where the hub and tip sections are by definition connected using straight line elements, changes made to the hub directly impact the blade geometry in the middle and upper part of the passage. At the tip section, where the high-speed flow experiences deceleration from inlet to outlet, the impeller tip is normally arranged to operate with a low incidence (0.5°–1.5°) at the design or peak efficiency condition.

The flow capacity can also be increased by reducing the number of the impeller blades and therefore the metal blockage in the inducer. This also reduces the wetted area and the associated losses inside the passage. However, the length of the uncovered part of the passage increases as the number of the blades is reduced. As a result, some small turning needs to be maintained over a larger part of the blade in high-speed applications. This reduces the first flap frequency of the blades and may push in the direction of a thicker hub profile. Furthermore, the loading increases as the blade count is reduced, and this can result in flow separation and higher tip leakage losses.

11.8.6 Multipoint Design and Design Compromises

A key consideration when optimising a design is whether performance is to be maximised at a specific operating condition or whether the product depends on good performance over a range of flow and speed, as in a turbocharger application, as discussed in Section 17.3.1. To maximise performance at high speed and high Mach number, a small amount of turning is used in the front part of the blade to avoid excessive shock losses, followed by larger turning in the rear part of the blade. However, this results in a small throat area at the tip and reduces the choke flow. The blade also lacks sufficient camber to operate well under subsonic conditions at part speed. In other words, a gain in efficiency at high speed typically comes at the expense of a reduction in efficiency and perhaps surge margin at part-speed.

Compressors often need to cover a set of operating points whereby a compromise among the efficiency, flow capacity and operating range has to be achieved across the performance map. Hence single-point efficiency-based optimisation is not always possible or even desirable. Devices such as variable geometry and shroud recirculating bleed can assist (see Section 17.5), but these add cost and complexity and are not always appropriate technology depending on the application.

Although a single design point is generally used as an anchor for the aeromechanical development of the stage, the design optimisation is often heavily influenced by the off-design performance at low flow and low speed at one extreme and by high flow and high speed at the other. In a design for multiple-point operation, a compromise has to be found based on the expected performance map. It is for this reason that the method for map prediction described in Section 18.3, based only on the duty of the compressor, is so useful as it allows a suitable design point to be defined which gives some confidence that the final design will meet the conflicting off-design requirements.

11.9 Three-Dimensional Features

11.9.1 Ruled Surface Impellers

In general, impeller design is determined by the performance requirement, in terms of range, efficiency and manufacturing costs. The simplest form for a 3D blade for a centrifugal impeller is the use of radial blade filaments where the wrap angle or blade angle is specified at one section. The blade camber surface is completed by radial lines joining each point on the specified section with the machine axis, and the blade thickness is added normal to the camber surface. This style of design is only suitable for blades with no backsweep and is now only of historical interest; see Sections 5.8 and 11.6 and the example given in Figure 1.17.

The geometrical form that is most commonly used is the ruled surface approach, where separate camber lines are specified along the casing and the hub sections, so that the resultant surfaces after the application of thickness are joined by straight lines and the impeller can be machined by flank milling; see Section 14.5.1 and Figure 19.1. This leads to a three-dimensional blade shape, but the blade angles and thickness of the intermediate sections are interpolated by the ruled surface from the geometry of the hub and casing. This results in lower manufacturing costs in addition to reduced development effort as the blade is designed only at the hub and shroud sections. Nearly all the illustrations of impellers shown in this book are of this type. Some freedom is available to the designer to change the geometry of the intermediate sections by adjusting the orientation of the ruled surface. Blade lean can be introduced if the generating line section normal to the hub inclines to the circumferential direction, while blade sweep can be employed if the tip section is moved in the meridional direction relative to the hub section.

11.9.2 Free-Form versus Ruled-Surface Impeller

A completely free-form inducer removes the geometrical constraints imposed by the ruled surface and allows a nonuniform distribution of the thickness, blade angle and blade lean distribution in the intermediate sections of the blade. This type of free-form blade cannot be flank-milled but, with point-milling becoming more affordable, increasing attention is being paid to fully three-dimensional design of the impeller inducer especially in transonic compressor applications. An example of a fully free-form impeller is given in Figure 11.28.

A free-form blade has potential aerodynamic advantages but brings an overhead of a more difficult and time-consuming design process with challenges in both the manufacturing and the mechanical analysis. In high-speed impellers, where supersonic flow exists across most of the blade span at inlet, a more precise control of the geometry across the span is desirable and application of free-form blading can be considered. In this case, the distribution of blade angle, thickness, lean and sweep are specified independently at different spanwise locations. This gives full control of the geometry in 3D space but increases the manufacturing costs as the blades must be machined by point milling. Because of the greater freedom available to the designer, the design costs are likely to be higher and the ability to fine-tune an impeller for a specific application by trimming may be compromised. Recent publications describing the application of free-form impeller blade surfaces include Hazby et al. (2015), Hazby et al. (2017), Hazby and Robinson (2018), Hehn et al. (2018), Wittrock et al. (2018) and Wittrock et al. (2019).

Figure 11.28 A free-form transonic impeller. (image courtesy of David Wittrock and Institute of Propulsion Technology, DLR)

11.9.3 Leading-Edge Sweep

The concept of using leading-edge sweep to cut down shock losses in high-speed axial compressors has its roots in swept wing theory of aerofoils, where the shock strength is mitigated by reducing the component of Mach number normal to the wing; see Section 6.3.2. However, the effects of the leading-edge sweep in external wings differ from that in internal turbomachinery passages due to the presence of the end-walls. Denton and Xu (2002) show that the loading at the tip is affected by the loading of the blade section just under it. Moving the tip section forward reduces the loading at the tip as there is no blade section (zero loading) immediately under the tip section, as shown in Figure 11.29. Forward leading-edge sweep reduces the strength of the shock and leakage flows in transonic fans. Conversely, a backward sweep of the leading edge increases the tip loading in the front part of the passage.

Furthermore, the presence of the casing also affects the position of shock in swept blades as sketched in Figure 11.29 (Hah et al., 1998). At Mach number levels normally encountered in turbomachinery applications, the shock cannot be reflected from the casing and, in practice a shock always intersects the casing perpendicularly. Therefore, if the shock is assumed to sweep with the blade, the distance between the shock and the blade leading edge reduces when the blade is swept backward and increases when the blade is swept forward. Stall tends to occur when the shock is expelled with the result that backward sweep tends to reduce the operating range while forward sweep normally improves it.

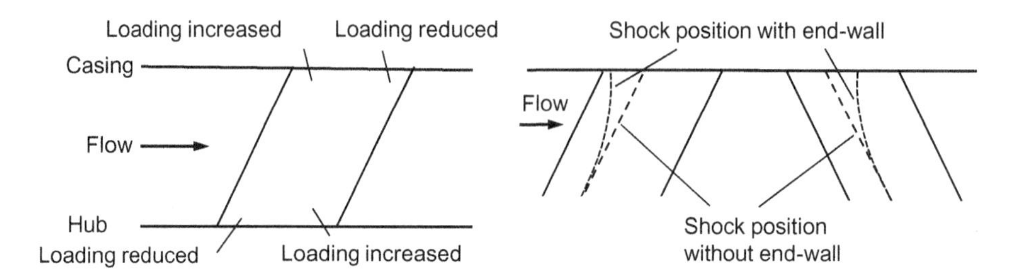

Figure 11.29 Effect of sweep on the loading near the end-walls and on the shock position.

Forward or positive sweep is generally useful in enhancing the efficiency and range of transonic impellers. Backward sweep often featured in early flank-milled impellers as a way of increasing the first flap frequency of the main blade. It may be more difficult to meet mechanical acceptance criteria with forward-swept blades.

11.10 Impeller Families

11.10.1 Industrial and Process Applications

In multistage industrial applications, it is common practice to design a series of stages in advance which can be adapted to meet the requirements of new applications. This reduces the engineering intensity of a new design for a new application. At defined fixed points within the typical flow coefficient range from 0.12 down to 0.005, certain base stages or master stages are designed and tested. At these fixed points, the performance and the geometry of the stages are completely engineered and validated, leading to a high reliability of the performance prediction accuracy for these stages. The geometry of the base impellers is then adapted to account for the diversity and complexity of different process compressor applications. A family of stages is generated for operation between these fixed points, and the performance of the members of the family can be obtained by interpolation, as described by Dalbert et al. (1988).

The need for this procedure can be seen in Figure 11.30, which shows the flow coefficient of each stage of a six-stage compressor in which all stages have the same impeller diameter. The first stage is designed with a flow coefficient of 0.12 in the knowledge that this should be above the optimum flow coefficient so that subsequent stages may be closer to the optimum. In this calculation, the performance of each stage is determined by the Aungier correlation for process stages with vaneless diffusers, as described in (10.56). Three cases are shown with different design tip-speed Mach numbers for the first stage of 0.3, 0.8 and 1.1. The compressor with a low tip-speed Mach number represents a light gas needs six high flow coefficient stages (with flow coefficients between 0.12 and 0.099) as there is only a small density and volume flow change across each stage. The compressor with a tip-speed Mach number of 0.8 in the first stage requires stages with high and intermediate flow coefficients from 0.12 to

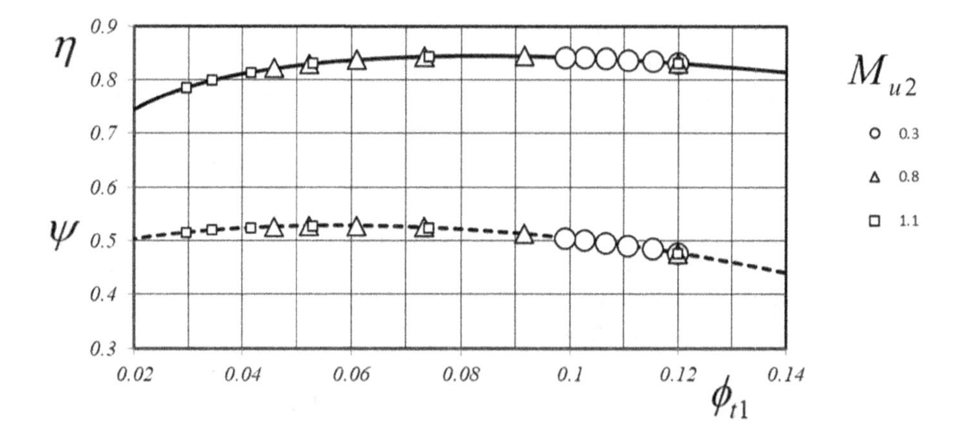

Figure 11.30 Effect of variation in tip-speed Mach number of the first stage on the flow coefficients of the individual stages in a six-stage compressor.

0.046. The high-speed machine represents a heavy gas and the last stage has a low flow coefficient of 0.03, so all types of stages are needed from high to moderately low flow coefficient.

In practice, the diameter of the stages may also be reduced though a compressor casing to increase the flow coefficients of the rear stage which would otherwise have a low efficiency. In integral gear compressors, the stages are individually on different shafts so that the rotational speed of the shafts and the diameter of the stages can be adjusted to ensure that more stages are closer to the optimum flow coefficient.

Depending on the tip-speed Mach number, the second and subsequent stages in Figure 11.30 are always different and require a different design flow coefficient. Without knowledge of the gas and the thermodynamic process of the compressor, it is not possible to fix in advance the schedule of the change in the flow coefficient. In fact, the inlet flow coefficient of intermediate stages can vary through an infinite number of values, depending on the tip-speed Mach number of the application. It is for this reason that the master stages need to be supplemented with a large number of additional derived stages for intermediate flow coefficients. Developing and validating a separate master stage for each possible flow coefficient would be totally uneconomic. In addition, the large possible variation in flow coefficient suggests that the adaptation of the flow for each stage needs to be done in small steps so that each can operate close to its best point with adequate surge margin.

There are several ways of dealing with this issue. In compressors with inlet guide vanes, the flow capacity can be adjusted by varying the angle of the inlet guide vanes. The stage pressure ratio can be adjusted by radial cropping of the impeller blades. In stages with vaned diffusers, an alternative to adjust the flow capacity is to adjust the inlet angle of the diffuser vanes. The most popular technique to adjust the flow capacity is by narrowing the width of the impeller flow passages by trimming the width of the channels, sometimes known as flow cuts, with corresponding reduction in passage height in the stationary components. One way in which this standardisation of the flow coefficients can be carried out is to select a constant ratio between the flow coefficients, or the outlet width ratios, of each derived stage such that a geometric sequence is produced. The selection of the geometric ratio determines the number of stages that are required to cover the whole region. Clearly many different ratios are possible. In the 1870s, the French army engineer Colonel Charles Renard proposed a set of preferred numbers for use with the metric system in order eliminate the number of ropes needed for dirigibles and balloons. His system was adopted in 1952 as international standard ISO 3. Renard's system of preferred numbers divides the interval from 1 to 10 into 5, 10, 20 or 40 steps. The factor between two consecutive numbers in a Renard series is constant (before rounding), namely the 5th, 10th, 20th or 40th root of 10 (approximately 1.58, 1.26, 1.12 and 1.06, respectively), which leads to a uniform geometric sequence. The numbers may be rounded to any arbitrary precision, as they are irrational. Clearly a variation in the flow coefficient of 12% between each member of the series would be unadvisable as each individual intermediate stage could then deviate by as much as 6% from its desired flow, with associated performance and operating range deficits. For this reason, most systems suggest using the R40 series with 6% steps for the selection of derived stages from the master stages. Table 11.1 gives the Renard R40 series.

The approach described in Dalbert et al. (1988) refers to an outlet width variation of impellers using the ISO standard R40 series, leading to a variation of around 6% in the outlet width for each member of the series. The 6% arises from the fact that $(1.05925)^{40} = 10$. This gives 40 members of the impeller family per decade, so a variation in flow coefficient from 0.125 to 0.0125 would include 40 impeller members, and the variation from 0.125 down to 0.005 would give rise to 56 members. In fact, Dalbert et al. (1988) describe this approach with regard to the selection of the outlet width ratios of the impellers rather than the flow coefficient, but given that the flow in trimmed stages varies nearly linearly with the outlet width ratio of the derived stages, the process is similar. The same system is often used for the change in shaft and impeller diameters of a series of machines of different size. Here the R20 series is often used with the result that the diameters of individual machines (frame size) and impellers vary by a ratio of 1.12. The frame size of a series of integral geared compressors described by Simon (1987) is also in a geometric ratio similar to a Renard R10 series, given by $(1.259)^{10} = 10$. In this case, the subdivision of the frame sizes and the grading of the impeller diameters was carried out to enable as many different identical impellers as possible to be used in different frame sizes.

In addition to the question of the size of the steps between the stages, the location of the master stages in the series needs to be considered. If a typical application is well

Table 11.1 A list of the Renard R40 series from 1.0 to 15.0.

1.00	1 06	1.12	1.18	1.25	1.32	1.40	1.50	1.60	1.70	1.80	1.90
2.00	2.12	2.24	2.36	2.50	2.65	2.80	3.00	3.15	3.35	3.55	3.75
4.00	4.25	4.50	4.75	5.00	5.30	5.60	6.00	6.30	6.70	7.10	7.50
8.00	8.50	9.00	9.50	10.0	10.6	11.2	11.8	12.5	13.2	14.0	15.0

known and reasonably constant for all machines, then a possibility would be to use thermodynamic calculations of the most common machine to choose the locations. For example, if in a given application the tip-speed Mach number of the first stage was always 0.8 and the process gas was always air, then the flow coefficients of the white triangles in Figure 11.30 could always be used for the master stages, with no intermediate derived stages needed. Unfortunately, the operating conditions and thermodynamic processes of most gas compressors are so variable that this is not possible. An alternative would be to, say, select every fifth or sixth member of the Renard series as the master stages. A modification of this approach is to make the steps between the master stages smaller for the high flow coefficient stages and larger for the low flow coefficient stages. This has two advantages for the high flow coefficient stages. First, as these tend to be relatively long compared with the low flow coefficient stages, there is a rotordynamical advantage in switching earlier to the low flow coefficient stages, which gives a shorter rotor. Second, the high flow coefficient stages tend to have lower performance than those near the optimum flow coefficient, which means that it is sensible to switch as soon as possible. A similar strategy can be found in the paper by Bygrave et al. (2010), showing that for high-pressure barrel compressors two families of 3D stages and four families of 2D impellers are used, with fewer members of the high flow coefficient stages. In the paper by Bygrave, there are seven 3D and twenty-six 2D impellers spread across six subfamilies for the whole range.

Even in other applications, such as gas turbines and turbochargers, it is useful to adapt an existing stage to be suitable for a range of flows in different applications by trimming the shroud contour and the diffuser width, see Rodgers (2001), Engeda (2007) and Dickmann (2013).

11.11 Impeller Trim

11.11.1 Shroud Cutting

In the Section 11.10, a distribution of the master stages is suggested. The adaptation strategies for the development of the derived stages from the master stages are now considered by means of trims or flow cuts. The derived stages (or adapted stages) are not engineered but are derived from each of the master stages by reducing the channel width. There are three broad approaches to this.

Shroud cutting is used typically for overhung stages with open impellers in which the blades are simply trimmed (or adapted) from the casing contour to produce a narrower flow channel. The blade shape and hub line remain constant, as shown in Figure 11.31. Typically, turbocharger compressor impellers are adapted in this way as it enables the same milling procedure for all of the stages in the series to be used. The notation for describing the trim in turbochargers is often as follows:

$$Trim = 100\frac{D_{1c}^2}{D_2^2} \tag{11.59}$$

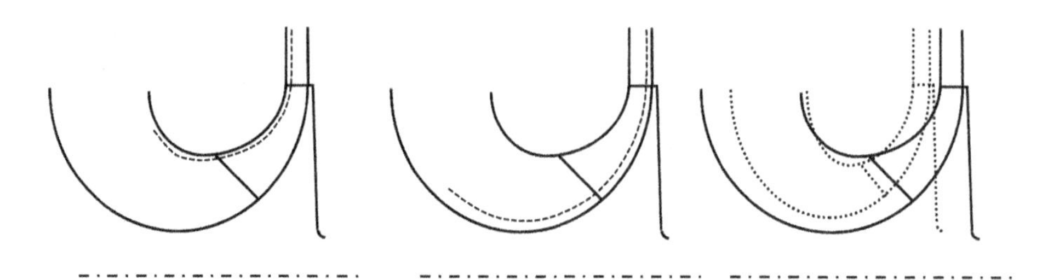

Figure 11.31 Schematic of type 1 adaptation with shroud cutting (left), type 2 adaptation with hub cutting (middle) and type 3 adaptation with parameterised geometry (right).

but can also be based on a simple ratio of the diameters. An impeller with a casing diameter of $0.7D_2$ has a trim of 49 according to (11.59).

The advantages of this procedure are that the hub remains constant for all stages so manufacturing of impeller and rotor can remain the same. The trim removes high Mach number tip sections and may, therefore, improve performance. The radius of curvature of the casing contour is increased at inlet, thus reducing high velocities near the shroud. At the same time, the centrifugal effect is raised (with higher radius ratio through the impeller) together with the degree of reaction. Given that the radius and flow area at the shroud are high, and the highest meridional velocities occur there, only small changes are needed to the shroud contour to produce a flow change. The disadvantages are a possible penalty due to a change in the aerodynamics, where a new design might have superior performance. There is no rotordynamic benefit from a thicker shaft waist diameter. The axial length of the shroud diaphragm or the casing changes with the trim, becoming longer.

11.11.2 Hub Cutting

This strategy is typically used for process stages with shrouded impellers. The shroud contour is retained but the meridional channel is adapted on the hub side. The blade shape and shroud line remain constant, as shown in Figure 11.31. Fabricated compressor impellers are normally adapted in this way. The blades may be welded first on to the shroud and are subsequently machined to size. A final weld is used to connect the blade to the hub.

The advantages of this procedure are that the shaft waist diameter becomes larger for narrower stages, which improves the rotordynamics. The critical tip sections are unchanged. The stage becomes shorter at lower flows while the axial length of the shroud diaphragm or casing remains the same. The disadvantages are a performance penalty due to the change in aerodynamics by removing hub sections which have low deceleration and decreasing the centrifugal effect.

11.11.3 Parameterised Geometry Definition of Intermediate Stages

This strategy is not generally used in practice. The blade shape and the channel shape both vary with the result that no geometrical features of the intermediate stages are the same as the master stages. This approach may be sensible with parameterised geometries and advantageous in situations where modern computer-aided design (CAD) processes coupled to milling are used to manufacture the stages, as shown in Figure 11.31. It is not known if this process has been used. The advantages are that all stages can be designed to be optimal so that there is no performance penalty due to the flow cut. The main disadvantage is that no geometrical features of intermediate stages are the same as the master stages, as all geometrical data for intermediate stages are interpolated from the master stages.

11.11.4 Determination of Contours for Hub or Shroud Cutting

For hub or shroud cutting, to what extent the contour needs to be trimmed has to be determined. This can be done following the process suggested by Dalbert et al. (1988). The hub or shroud contours of the intermediate impellers are determined using a contour derived from the streamlines of the throughflow calculation of the original master stage. This ensures that each stage has a similar aerodynamic flow in the throughflow calculation with the same inlet and outlet flow angles, incidence and blade loading. This similarity in the aerodynamics ensures that the characteristics of the derived stages are similar in shape and performance to those of the master stage.

An example is shown in Figure 11.32 for an impeller designed for a flow coefficient of $\phi_{t1} = 0.08$. The flow simulation here has been carried out with 17 streamlines with a variable distribution of the mass flow across the span. In this way, each streamline shown approximately represents the streamline needed for each of the derived stages, and this can either be carried out by a hub cut or a casing trim. The effect of the

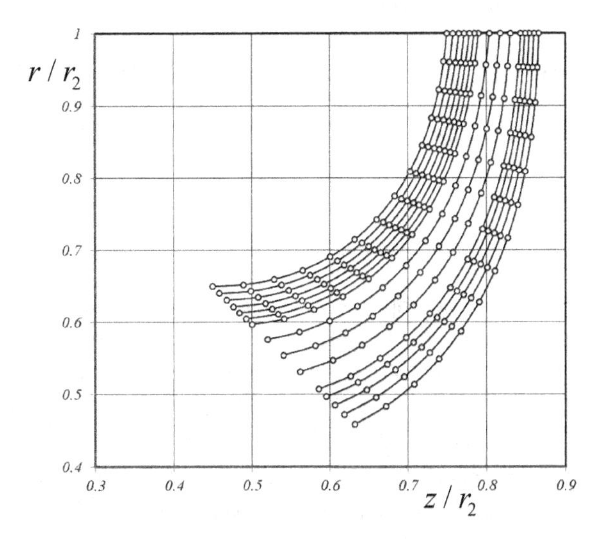

Figure 11.32 Meridional streamlines from a throughflow calculation of an impeller showing the basis for adaptation on the hub or casing contours.

meridional curvature on the velocity distribution across the span, and the different flow areas at different radii, give rise to different depths of cut for the different trims.

11.11.5 Impeller Outlet or Inlet Blade Cutback

In some situations, a standard impeller family with fixed impeller diameter is not always able to meet the pressure rise requirements of a particular application if this has side-streams or intermediate outlets. Here a further option is to take a member of the standard family and to cut back the diameter of the blades at impeller outlet. In some situations, it may be necessary to increase the flow of an existing compressor with the minimum possible amount of change. In this case, cutting back the leading edges of an impeller increases the blade angle and the throat area and makes a contribution to a higher flow rate.

11.11.6 Overfile and Underfile

A 2D impeller has the same blade inlet angle from hub to shroud along the leading edge of the blade. The inlet flow profile has a higher meridional velocity near the shroud due to the curvature of the flow channel and therefore zero incidence angle can only exist at one point along the leading edge. A simple way of correcting the mismatch of the blade with the flow is to file the suction or pressure of the leading edges to give different flow angles across the span. A similar procedure can be used to modify the blade outlet angle to match modified pressure rise requirements.

11.11.7 Other Issues Related to Stage Adaptation

A different way of adapting a base stage variation from the base stage is to change from a vaneless diffuser to a vaned diffuser to allow higher efficiency at the cost of operating range. Alternatively, the selection of different inlet throat areas for standardised vaned diffusers allows the stage to be adapted to higher Mach numbers and different gases, and provides a means of regulating the compressors. Generally, the impeller characteristics are wider than the diffuser characteristics, and this allows such changes to be made. Examples are given by Dalbert et al. (1988).

Another alternative is to include impellers in the series with a different level of area change across the impeller, as shown in Figure 11.18. Such impellers have a smaller or larger outlet width for the same inlet width to provide a means of adaptation to heavier and lighter gases with different tip-speed Mach numbers (Lüdtke, 2004). Where this change in the conicity of the channel is not implemented, the high pressure caused by the high tip-speed Mach numbers leads to a lower flow coefficient at impeller outlet and to a higher work coefficient at high tip-speed Mach numbers. Conversely, the work is reduced at lower tip-speed Mach numbers. This strategy changes the diffuser width, and with a vaned diffuser it changes the throat area and removes the differences in the matching of the diffuser and the impeller at different Mach numbers.

11.12 Comparison with Rotors of Other Machine Types

11.12.1 Comparison with Axial Compressor Stages

A comparison with axial compressor stages is of most interest in relation to the shape of the characteristics, efficiency, flow capacity and achievable pressure rise per stage. The axial compressor has a lower work coefficient, so that three stages are typically needed to achieve the pressure rise of a radial stage. The flow path in an axial compressor is less convoluted, with higher hydraulic diameters, allowing higher efficiencies per stage, which are often above 91% polytropic efficiency compared to 86% for typical large radial stages.

In axial compressors, which unless they have a rising hub line achieve their pressure rise purely by deceleration with no help from the centrifugal effect, a typical mean work coefficient is in the region of 0.2–0.3, using the tip speed of the rotor as reference velocity. Any attempt to go much above 0.45 leads to extreme difficulties with separated flow and high losses: the level of diffusion becomes high as the de Haller number falls below 0.7. Recent research examining this has been published by Dickens and Day (2011) and by Hall et al. (2012). Radial compressors can achieve much higher work input coefficients without excessive diffusion due to the centrifugal effect. In radial compressors, backsweep is not generally used to reduce the diffusion levels, although it does help in this respect. It is used to produce an inherently stable characteristic; see Section 11.6.3. In axial compressors, any reduction in work coefficient is generally made purely on the grounds of the diffusion levels and has no substantial effect on the steepness of the characteristic – this is steep enough already.

A further difference related to the work coefficient is that axial machines with steep work characteristics increase the blade loading with little change in the flow, so they are likely to meet a limiting loading, such as a diffusion factor limitation. Radial machines have less increase in loading with a large change in flow, and so are likely to meet a limit related to incidence or the large change in flow angle.

A further relevant difference to axial compressors is that in a radial impeller, about 75% of the pressure rise comes from the centrifugal effect. Because of this, the losses due to any excessive diffusion have less effect on the efficiency in radial impellers than in axial rotors. The improvement in efficiency of axial machines due to improved controlled diffusion profiles is known to have been around one percentage point. In radial impellers with only one quarter of the pressure rise due to diffusion of the relative flow, the benefit of improved profile design is less, and the performance is less sensitive to the detail of the blade shape.

In radial compressors with no inlet guide vanes, there is no inlet swirl. This means that the degree of reaction is not a parameter that the designer can use to influence performance, except in the sense that the outlet velocity triangle has an influence on reaction; see Figure 11.20. In multistage axial compressors, however, either the inlet guide vanes or the upstream stator from the previous stage are available to adjust the inlet swirl to a stage. This allows the axial compressor designer to vary the swirl angle from stage to stage and over the span, leading to designs with different vortex distributions (Horlock, 1958). Typically, a high swirl in the rotating direction will be used at the inlet to reduce the relative inlet Mach number at the rotor tip, and the swirl will be reduced steadily through the compressor. The degree of reaction of a typical multistage axial compressor thus increases stage by stage from a value of 0.5 at the inlet to 0.75–0.85 at the outlet, leading to less swirl in the outlet flow.

11.12.2 Comparison with Mixed Flow Compressors

Figure 11.6 indicates that a design for high flow coefficients with the minimum inlet relative Mach number requires a high casing diameter ratio, D_{c1}/D_2, above that of typical radial impellers which have a maximum value of $D_{c1}/D_2 = 0.73$. The effect on the impeller meridional channel of designing a stage with a lower outlet diameter but retaining the same inlet is sketched in Figure 11.33. As a consequence of the trends shown in this diagram, mixed flow compressors are broadly of three types depending on the direction of the discharge flow from the impeller outlet, which may be radial, diagonal or axial. In each case, the objective is to design a compressor for a very high flow coefficient with a low outlet diameter.

In trying to achieve a high flow coefficient, the designer tends to retain an axial impeller trailing edge if possible, giving the advantage of a radial outlet flow to the radial diffuser. A configuration with an impeller with a diagonal trailing edge is used when the inlet casing diameter ratio is higher than $D_{c1}/D_2 = 0.8$, as otherwise a very sharp bend in the impeller casing contour is needed if a purely radial impeller is retained (Hazby et al., 2014). This configuration usually retains the radial diffuser, or has a high meridional pitch angle in the diffuser of above 80°. The impeller is essentially a radial impeller with a high inlet diameter relative to the outlet diameter. The hub and casing sections are more similar than in a normal radial impeller, and this leads to diffusion being high on the hub section as well as on the casing section, unlike

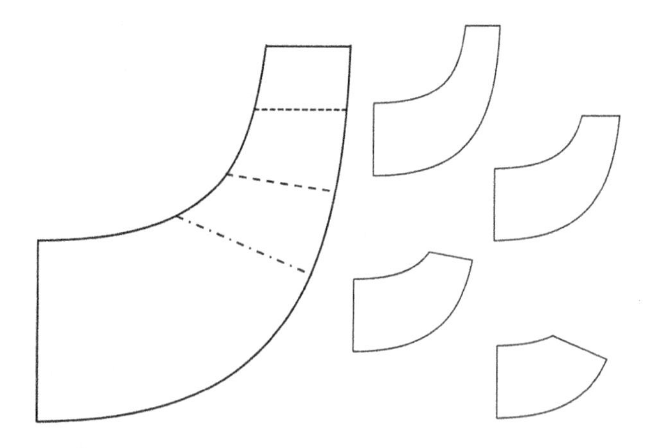

Figure 11.33 Reduction of impeller diameter for a fixed inlet leads to a mixed flow design.

in a normal radial design. The impeller has a diagonal outlet but is followed by a bend from the diagonal direction to the radial direction of the diffuser.

At higher values of the inlet casing diameter ratio, above $D_{c1}/D_2 = 0.8$, a mixed flow impeller is combined with a diagonal diffuser, and the impeller trailing edge may then have a high diagonal pitch angle from the axial direction, typically between 45° and 60°. In this case, the objective of such designs may also be to limit the radial extent of the compressor rather than to maximise the flow capacity.

In many applications with even higher inlet diameter ratios, the discharge flow is turned to the axial direction in order to reduce the radial extent of the compressor. The diffuser is then an axial deswirl vane. This type of mixed flow device is similar to a high aspect ratio axial compressor with a large radius change across the rotor. It is in many ways similar to the hub sections of a jet engine fan rotor, and similar design rules as outlined by Calvert and Ginder (1999) may be used. This design style is sometimes used as the last stage of an axial compressor for gas turbine applications where the axial outlet leads to the combustor. Such applications are not aimed primarily at maximising the flow coefficient as they often have high hub diameters. In some cases, only the hub section has a radius change and the tip section is at constant diameter. Where the deswirl vane is axial, this tends to have a very difficult duty, combining a high turning with high absolute inlet Mach number; see Section 13.8.

11.12.3 Comparison with Centrifugal Pump Impellers

At first sight, it might look as if the conditions under which the lowest risk of cavitation at the inlet to a pump is achieved are similar to the condition for minimum relative velocity in a compressor as described in Section 11.4.1. This would lead to an inlet blade angle of $\beta_{c1} = -60°$. However, the parameter used to identify the risk of cavitation at the impeller eye in a pump is not the absolute relative velocity but the net positive suction head (NPSH), defined as follows:

$$NPSH = (p_{inlet} - p_v)/\rho g, \tag{11.60}$$

where p_v is the vapour pressure of the liquid. The inlet static pressure leading to cavitation depends on both the relative velocity level at the inlet eye and on the static pressure lost in the inlet ducting up to this point. These pressure losses are determined by the absolute velocities in the stationary components rather than the relative velocity at the impeller inlet eye. To reduce these losses, low values of the meridional velocity are needed.

The simplest way of obtaining the optimum inlet angle of a pump uses the following equation for the static pressure at the inlet eye of the pump with no swirl:

$$p_{t,inlet} = p_1 + \rho c_{m1}^2/2 + \zeta \rho c_{m1}^2/2 = p_1 + \rho c_{m1}^2(1+\zeta)/2, \tag{11.61}$$

where the losses in the inlet are denoted by a loss coefficient, ζ, that is made nondimensional relative to the meridional velocity at the pump inlet eye. The pressure at the inlet at which cavitation inception occurs is given from (11.60) such that

$$p_{t,inlet} = p_v + \rho w_1^2 \sigma_i/2 + \rho c_{m1}^2 (1 + \zeta)/2. \quad (11.62)$$

The NPSH at cavitation inception is given by

$$gNPSH_{s,i} = \rho w_1^2 \sigma_i/2 + \rho c_{m1}^2 (1 + \zeta)/2. \quad (11.63)$$

The analysis itself is similar to that given in Section 11.4.1 – the NPSH is maximised by considering the area of the impeller inlet eye. After a lot of algebra, Pfleiderer and Petermann (1986) arrive at the following equation for a pump inlet, which is the equivalent of (11.9) for a compressor inlet:

$$\tan \beta_{c1} = -\sqrt{2\left(1 + \frac{1 + \zeta}{\sigma_i}\right)}. \quad (11.64)$$

Typical values of the ratio $(1 + \zeta)/\sigma_i$ lie between 2.5 and 6, and this leads to blade inlet angles of $\beta_{c1} = -70°$ to $-75°$. The flow coefficient at the inlet eye then becomes roughly

$$\frac{c_{m1}}{u_{c1}} = \frac{1}{\tan 72.5°} \approx 0.3. \quad (11.65)$$

The local inlet flow coefficient of a pump is thus roughly half that of a typical compressor inlet, and this leads to blade geometries in pumps that have very high stagger blades with inlet angles that are much larger and much more circumferential than those of compressors. The fact that the stagger is high means that the number of blades used is also reduced relative to compressors. In liquid fuel rocket pumps, where the fluid is even closer to the vapour pressure, the use of (5.39) leads to extremely high stagger blades in the inducer region to avoid cavitation.

One of the features of pumps at the impeller outlet which makes them highly efficient is the use of very high levels of backsweep. This is possible because the rotational speeds are normally not very high. The stresses in the metal produced by the centrifugal effects are proportional to the density of the metal and to the square of the tip speed: the pressure rise is also proportional to the mean fluid density and the square of the tip speed, as explained in Section 19.5. In other words, the stress and the pressure rise both scale in the same way. Because the density of liquids is typically three orders of magnitude larger than that of gases, pumps can produce a large pressure rise while experiencing modest centrifugal stresses in the metal. For pumps, large blade stresses are produced by the pressure of the fluid, while in compressors this component of the stress analysis is almost negligible.

The high density of the fluid implies that a pump can achieve a high pressure rise with a low speed and a relatively low work coefficient. In fact, the high backsweep combined with the low outlet flow coefficient in the velocity triangles, near to point C in Figure 11.20, means that the work input coefficient of pumps with a high backsweep is not significantly that much lower than that of typical compressors.

11.12.4 Comparison with Radial Turbine Impellers

In a typical turbocharger, the compressor and the turbine are at opposite ends of the shaft and at first sight appear very similar. The major difference, of course, is that the radial turbine has a radial inlet flow rather than a radial outlet flow, as dictated by the Euler equation. It also exhibits an accelerating flow which is inherently easier than the diffusing flow in a compressor. Because of the important issues related to the matching of turbines with compressors in turbochargers, which are discussed in Section 18.10, this section provides a short background to the design features of radial turbines.

A major difference is that the turbine of a turbocharger is usually designed with radial blades at its inlet, that is, with no backsweep. High temperatures giving stress limits, due to fatigue or creep, often compromise the design of radial turbines. In turbochargers and in micro gas turbines with no blade cooling, a nickel alloy with good strength at high temperatures is used. Despite this, the high inlet gas temperature causes significant weakness of the material properties so that the blades comprise of radial blade elements to avoid high bending stresses in the blades. In hydraulic radial turbines where this high-temperature stress problem is not present, the blades are not radial and are similar in appearance to backswept pump impellers. The typical turbocharger impeller closely resembles a radial compressor impeller for a very high flow coefficient with radial blades. It also has thicker blades in the blade roots to reduce stresses, so the number of blades is limited. In addition, it generally has no splitter blades and in order to reduce the moment of inertia it may also be fitted with scallops; see Figure 11.34.

Another compromise in the design of radial turbines for turbochargers is related to the demand for fast acceleration and response. The nickel alloy material of the impeller has a high density so that if the rotor requires a low inertia, this is achieved with the minimum possible impeller diameter. This is possible because the turbine

Figure 11.34 View of a typical turbine impeller for a turbocharger.

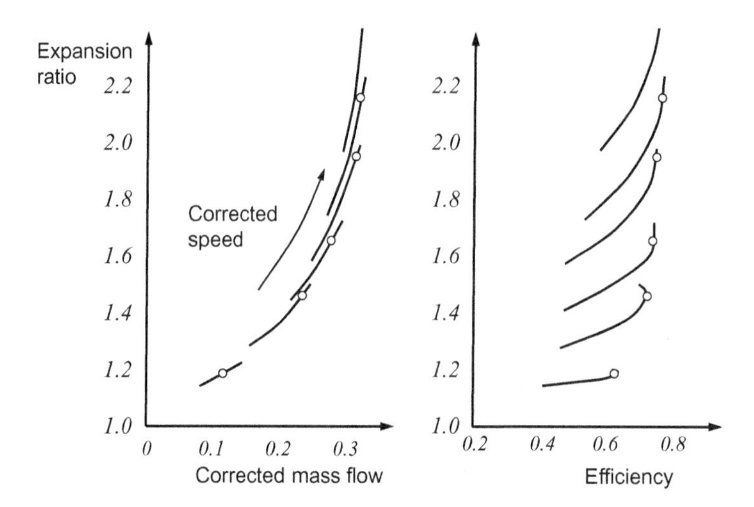

Figure 11.35 Typical performance map of a radial turbine at different speeds.

operates with a higher work coefficient than the compressor: the flow accelerates through the turbine, and there is no risk of flow separation so a work coefficient in the neighbourhood of $\lambda = 1$ is common. Although they operate at the same speed and are matched at a similar pressure ratio, the turbine diameter is invariably smaller than the compressor diameter, as explained in Section 18.10. This, in turn, tends to force the use of high flow coefficient designs with very high trims: the outlet casing diameter may typically be about 80–85% of the impeller diameter, whereas values above 73% are seldom for the inlet tip diameter in radial compressors.

A feature of radial turbines that is somewhat surprising for compressor engineers is that the speed lines in the performance map do not spread out as in a compressor map, but the different speed lines overlap, as shown in the sketch of a performance map given in Figure 11.35. The points shown on map are at the optimum efficiency of the stage for each speed. The best explanation of this phenomenon is given by Horlock (1966) and is the basis of the description given here. The approach is based on the linearity of the work coefficient with the flow coefficient based on the Euler equation, as given by (11.35) for compressors and which for a radial turbine can be written as

$$w_{turb} = u_1 c_{u1} - u_2 c_{u2}$$

$$\lambda_{turb} = \frac{w_{t12}}{u_1^2} = \frac{\Delta h_t}{u_1^2} = \frac{c_{u1}}{u_1} - \frac{u_2}{u_1}\frac{c_{u2}}{u_1}. \tag{11.66}$$

For convenience in this analysis, the turbine work output is considered a positive value, so the equation has reversed signs compared to the equation used previously for compressors. The location 1 is at the inlet to the turbine impeller, which is at the highest radius, and the tip speed at this position is used to define the work coefficient. Location 2 is the exducer of the turbine. Writing the velocity components at the turbine inlet in terms of the outlet flow angle of the inlet guide vane gives

$$c_{u1} = c_{m1} \tan \alpha_1, \quad c_{u1}/u_1 = (c_{m1}/u_1) \tan \alpha_1 = \phi_1 \tan \alpha_1. \tag{11.67}$$

A similar equation can be written at the turbine impeller outlet in terms of the relative flow angle from the exducer as

$$\begin{aligned} c_{u2} &= u_2 + c_{m2} \tan \beta_2 \\ c_{u2}/u_1 &= u_2/u_1 + (c_{m2}/u_1) \tan \beta_2 = u_2/u_1 + \phi_2 \tan \beta_2. \end{aligned} \tag{11.68}$$

Combining the three preceding equations gives the following equation for the turbine work coefficient:

$$\lambda_{turb} = -\left(\frac{u_2}{u_1}\right)^2 + \phi_2 \left[\frac{\phi_1}{\phi_2} \tan \alpha_1 - \frac{u_2}{u_1} \tan \beta_2\right]. \tag{11.69}$$

As in most turbomachinery analyses, the outlet angle of the flow from the stator, α_1, and that from the rotor, β_2, may be considered to be nearly constant. If in addition the density ratio across the turbine is taken to remain constant, then the ratio of inlet to outlet flow coefficients is constant. This is a reasonable approximation for small changes along each speed line at nearly constant pressure ratio. Under these circumstances, (11.69) reduces to

$$\lambda_{turb} = k_1 + k_2 \phi_2. \tag{11.70}$$

The value of the coefficients in (11.70) can be estimated from the turbine design. Parameter k_1 is the shutoff work coefficient at zero flow and given by

$$k_1 = \lambda_{shutoff} = -\left(\frac{u_2}{u_1}\right)^2 = -\left(\frac{D_2}{D_1}\right)^2 \approx -(0.5)^2 \approx 0.25. \tag{11.71}$$

Parameter k_2 can be determined from the condition that at peak efficiency the turbine is designed with a certain work coefficient, say $\lambda = 0.75$, and a certain outlet flow coefficient, say $\phi_2 = 0.25$, leading to a value for $k_2 \approx (0{,}75 + 0.25)/0.25 \approx 4$.

The nondimensional drop in total temperature across a turbine is given by a similar equation for the temperature rise in a compressor:

$$\frac{\Delta T_t}{T_{t1}} = (\gamma - 1)\lambda_{turb} \frac{u_1^2}{\gamma R T_{t1}} = (\gamma - 1)\lambda_{turb} M_{u1}^2, \tag{11.72}$$

where M_{u1} is the tip-speed Mach number of the turbine based on the tip speed at largest radius, which is at the inlet. Based on these equations, the temperature drop is given by

$$\frac{\Delta T_t}{T_{t1}} = k_1(\gamma - 1)M_{u1}^2 + k_2(\gamma - 1)M_{u1}(\phi_2 M_{u1}). \tag{11.73}$$

The parameter, $\phi_2 M_{u1}$, is proportional to the corrected mass flow (see Section 8.4.1) such that from this equation the temperature drop would be expected to be linear with the corrected mass flow for a given tip-speed Mach number, and the slope would vary

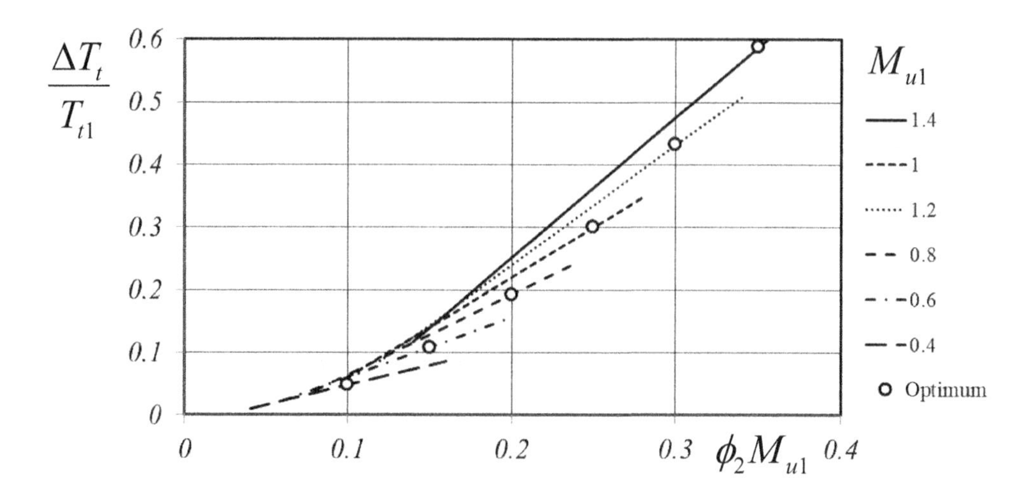

Figure 11.36 Nondimensional temperature drop of a turbine from (11.73).

with the tip-speed Mach number. Equation (11.73) is plotted in Figure 11.36 and demonstrates a remarkable similarity to the expansion ratio characteristics of a radial turbine, as shown in Figure 11.35.

The points shown on the characteristics in Figure 11.36 are for the optimum point of a turbine with a design point with an exducer outlet flow coefficient of $\phi_2 = 0.250$ and a work coefficient of $\lambda = 0.75$. At low flows, the characteristics for the temperature drop at each speed overlap because of the physics of the work input in the Euler equation. At high flows, the similarity in the expansion ratio characteristics in Figure 11.35 is also related to choking, which limits the corrected mass flow at high flows and speeds. In a radial turbine, it is often the nozzle guide vane that chokes first. This means that the mass-flow parameter is fixed with speed, and the speed lines at choke bunch together on the turbine map. If the nozzle guide vanes are variable and can be opened, then it is the impeller which causes choking, and the choke points on the speed lines vary with the speed, as in a compressor.

12 Diffuser Design

12.1 Overview

12.1.1 Introduction

This chapter considers the design parameters for the diffuser system immediately downstream of the impeller. The components of the stator system downstream of the diffuser are considered in Chapter 13, and Chapter 11 considers the design of the impeller. The function of the diffuser is to transform the kinetic energy at its inlet into a rise in the static pressure. In a typical centrifugal compressor, around 35–40% of the energy input to the impeller is available as kinetic energy at the diffuser inlet, so the effectiveness of the diffusion system is critically important to the performance of the whole stage.

Centrifugal compressors are usually fitted with either a vaned or a vaneless diffuser leading to a collector or volute. The diffuser meridional channel comprises an annular channel extending radially outwards from the impeller outlet, usually of the same axial width as the impeller outlet but sometimes with a decreasing width as the radius increases, known as a pinched diffuser. The axial width may occasionally increase with radius in a vaned diffuser. In some low-speed ventilator and pump applications, a collector system may be fitted directly around the impeller, thus obviating the need for a diffuser section.

The simplest diffuser system is a radial vaneless annular channel. In a vaneless diffuser, the flow is decelerated in two ways. The meridional component of the velocity is reduced by increasing the area of the channel with radius (conservation of mass). The circumferential component is reduced by the increasing radius in the diffuser (conservation of angular momentum). In a vaned diffuser, of which there are several types, there is a small vaneless region upstream of the diffuser vanes. The vanes themselves form flow channels designed to decelerate the flow more than is possible in a vaneless diffuser by turning the flow to a more radial direction.

The flow travels against the pressure gradient, which means that the avoidance of flow separation needs to be considered. The extensive data available in the form of Reneau charts on the performance of straight planar diffusers is of great value to provide guidelines on the suitable area ratio of vaned diffusers; see Section 7.4. In addition to decelerating the flow as efficiently as possible, the vaned diffuser needs to deal with the effect of the leading-edge incidence of the flow on to the vanes.

The impeller outlet flow is highly nonuniform and unsteady, such that the diffuser also has to cope with strong inlet disturbances. Furthermore, at high pressure ratios, typically above 4 in air, the flow at the diffuser inlet becomes supersonic.

12.1.2 Learning Objectives

- Be aware of the most important types of diffusing system, vaned and vaneless.
- Understand the use of the continuity equation and the conservation of angular momentum in the vaneless diffuser to achieve flow deceleration as the radius increases.
- Be aware of the different operating ranges of vaned and vaneless diffusers.
- Understand the different nature of choke in a vaned and a vaneless diffuser.
- Recognise the different types of vaned and channel diffusers: curved plate cascade diffusers, aerofoil cascade diffusers, low solidity diffusers, wedge diffusers, pipe diffusers and vane-island diffusers.
- Understand the effect of the different zones in the diffuser on the performance of vaned diffusers, especially the upstream vaneless space, the upstream semivaneless space and the diffuser flow channel.

12.2 Effect of the Flow at the Impeller Outlet

12.2.1 The Effect of the Impeller Velocity Triangle at the Design Point

The design of the impeller has a strong influence on the selection of the most suitable diffuser design. Figure 12.1 shows the values of c_{m2}/u_2 and c_{u2}/u_2 at

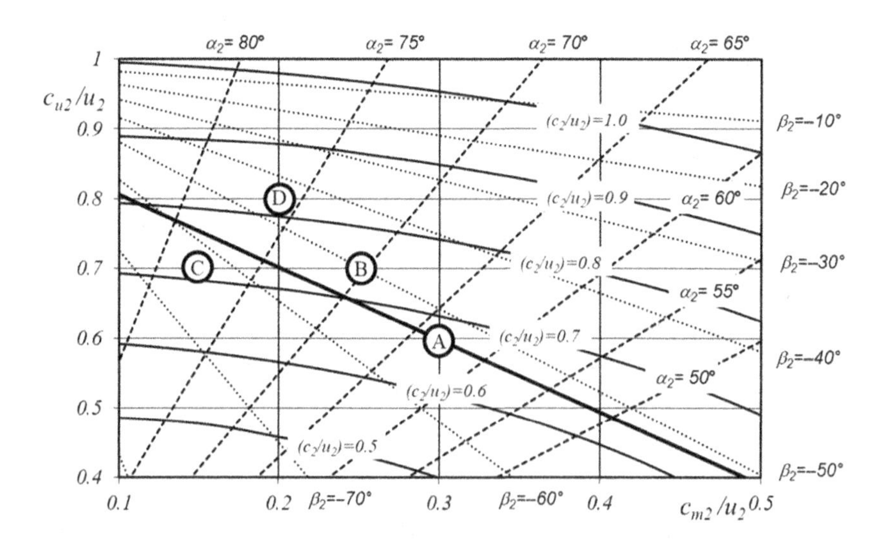

Figure 12.1 Absolute velocity level c_2/u_2 at diffuser inlet for different impeller designs.

impeller outlet for four characteristic impeller types, described in Chapter 11 and shown as A, B, C and D.

Figure 12.1 highlights the properties of the mean impeller outlet velocity triangle which are of most relevance to the diffuser. The ratio of the impeller outlet velocity to the impeller tip speed, c_2/u_2, is shown as a function of the work coefficient and the flow coefficient. If the complete region were depicted, the contours of constant c_2/u_2 would be circles with their centre at the origin (0,0). The figure also gives the lines of constant absolute and relative flow angles. Impeller D, with a high work coefficient and low backsweep, has a high absolute velocity at the impeller outlet and requires the most effective diffuser to achieve good performance – such that a vaned diffuser should be used. Impellers A and B have more backsweep and a lower absolute velocity at the outlet and can operate with a vaneless diffuser. Impeller A is a stage suitable for a vaneless diffuser with a low absolute flow angle into the diffuser, α_2, while impeller B is suitable for a vaned diffuser or a vaneless diffuser with pinch. The absolute outlet flow angle of impeller C with a high backsweep precludes its use with a vaneless diffuser, as explained later in this chapter. For the representative stages shown in Figure 12.1, the circumferential velocity component at the impeller outlet, c_{u2}/u_2, is roughly three times that of the meridional velocity component, c_{m2}/u_2. This implies that the circumferential component contains nine times more kinetic energy than the meridional component and indicates that the deceleration of the circumferential component of velocity plays a large role in the static pressure rise of the diffuser.

The operating line of impeller A for a variation of flow on the assumption of a constant slip factor of 0.11 is also included in Figure 12.1. This demonstrates that in a typical backswept impeller, as the flow rate decreases towards surge along the operating line, the absolute velocity at the diffuser inlet increases, adding to the kinetic energy available for the diffuser pressure recovery. This is important as it enables the diffuser to continue to secure a good pressure recovery even if its nondimensional pressure recovery coefficient decreases as it moves off design towards surge. An unsung feature of backsweep in the impeller is that this effect becomes stronger for designs with a low work coefficient. In stages with little backsweep and a high work input, such as stage D in this diagram, the diffuser inlet kinetic energy remains nearly constant along its operating line as flow is reduced, and in stages with no backsweep with an even higher work coefficient, the effect may reverse and the diffuser inlet kinetic energy falls towards low flow. This benefit of backsweep for the diffuser is strongest in stages with a low work coefficient, a low slip factor and a low flow coefficient at impeller outlet.

12.2.2 The Effect of the Flow Nonuniformity at the Impeller Outlet

The discussion of the jet-wake phenomenon in Section 5.8 highlights the complexity of the impeller outlet velocity profile as generated by the flow separations, secondary flows and the tip clearance flows in radial-bladed impellers. The velocity profiles of modern backswept designs are significantly more uniform at the impeller

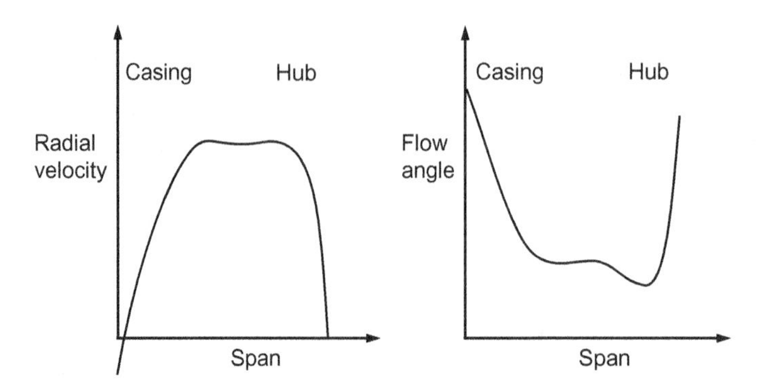

Figure 12.2 Spanwise flow angle and velocity distribution at the outlet of an impeller.

outlet than stages with no backsweep. Nevertheless, there is still a strong unsteady flow distortion in the diffuser inlet in both the tangential and the axial direction, and particularly in terms of accumulation of flow with low meridional velocity near the casing. Many publications suggest that the circumferential asymmetry perceived by the diffuser due to the distorted impeller exit flow does not strongly affect the pressure recovery of the downstream vaned diffuser (Cumpsty, 2004). This unsteadiness is, however, important as a source of vibration and noise (Raitor and Neise, 2008; Zemp et al., 2011).

The radial velocity profile at the diffuser inlet, with a large angle variation across the span, as shown in Figure 12.2, could reasonably be expected to have a determining effect on the diffuser performance. In fact, the effect of this velocity profile is much weaker than might be supposed and is usually ignored in most designs. Krain et al. (2007) describe an impeller which was designed to minimise the inhomogeneity of the impeller outlet flow, and measurements of this stage showed that this improved the impeller performance and the diffuser performance was not strongly affected. If the hub-strong meridional velocity profile was an important consideration, most vaned diffusers in real applications would have a skewed leading edge with different vane angles across the span to match the inlet flow angles. This distribution of inlet angle was put forward by Jansen and Rautenberg (1982) and others and both this and many other special geometries have often been tested in industrial applications; see Section 12.14. Few have led to any real improvement in aerodynamic performance, and none have become standard features of centrifugal compressors.

Steady CFD simulations with mixing planes connecting the rotating impeller and stationary diffuser domains (see Section 16.5) capture the mean effects of this variation in the radial flow angle at the impeller outlet well, and this variation is not substantially different with the time average of unsteady simulations (Robinson et al., 2012). The flow with a low radial velocity at the impeller outlet associated with secondary flows and the tip clearance vortex tends to be situated on the impeller suction surface at low flows, in the corner of the suction surface and casing at design conditions, and moves on to the casing at high flows (Johnson and Moore, 1983). The low momentum flow associated with the tip clearance leakage flow dominates the flow pattern and tends to concentrate near the casing and, in many cases, measurements and simulations show the existence of reverse flow near the casing, especially as the flow rate is reduced or with high clearance levels. A low swirl velocity in the relative frame of the impeller, for example in a wake, leads to a high swirl velocity in the absolute frame at the diffuser inlet, but the meridional component is the same in both frames of reference. This leads to a general trend for the outlet distortion of the flow in all modern designs: low meridional velocities and high flow angles near the casing (Yoshinaga et al., 1980; Robinson et al., 2012). Measurements in the vaneless diffuser show that the casing-side meridional velocity deficit persists well downstream of the diffuser inlet region (Schleer, 2006).

Many studies have focussed on how the diffuser performance is affected by this distorted flow at the impeller outlet. Inoue and Cumpsty (1984) tested vaned and vaneless diffuser configurations with a radial vaned impeller; these impellers could be expected to lead to the largest distortions due to the jet-wake effect. Rather surprisingly, they found that with a vaned diffuser the unsteady tangential flow distortion from the impeller was attenuated very rapidly in the inlet region of the diffuser vanes and had only a minor effect on the flow inside the vaned diffuser flow channels. After leaving the impeller, a strong tangential mixing takes place. This results in an intensive energy exchange and a fast homogenisation of the circumferential flow (Frigne and van den Braembussche, 1979). The measurements by Krain (1981) support this view and show a highly distorted, unsteady flow with local flow angle variations of up to 17° at diffuser inlet which decay to the throat and are not present in the diffuser channel. The measurements of Eisele et al. (1997) demonstrated that a similar decay occurs in a pump diffuser. Filipenco et al. (2000) argue from an experimental basis that the strong mixing from the circumferential unsteadiness in the diffuser inlet region plays an important role in the observed insensitivity of the pressure recovery in the diffuser to inlet flow field distortion. Cukurel et al. (2008) also concluded that for most operating conditions, the flow downstream of the throat in a pipe diffuser was relatively unaffected by the distortion of the flow at the impeller outlet.

12.2.3 The Effect of the Diffuser Inlet Flow Angle

The circumferential and axial distortion appears to have relatively little impact on the pressure recovery in the diffuser at the design point, although it does become important with regard to instabilities at lower flow. Many studies have identified that the key effect on the performance is the spanwise average flow angle at the diffuser inlet. Dalbert et al. (1988) and Casey (1994a) presented experimental data showing the performance of cascade diffusers as a function of the mean inlet flow angle. These data demonstrated that the performance curves for these diffusers correlate well with the mean diffuser incidence angle. Deniz et al. (2000) and Filipenco et al. (2000) also showed that the performance of a particular vaned diffuser is uniquely dependent on the average inlet flow angle. Filipenco et al. used a momentum-averaged inlet flow angle which under most flow conditions is very similar to the mass-flow average.

Everitt et al. (2017) have also argued that the mean incidence is the primary driver of diffuser performance in vaned diffusers. Similar results were found by Eisele et al. (1997) through laser measurements in a pump diffuser.

In many experimental studies on vaned diffusers, however, the vane inlet angle was adjusted by changing the stagger of the diffuser vanes to change the incidence. This also changed the throat area of the diffuser, and as a consequence the effect of the incidence is not always separated from the effect of the change in throat area. Swain (2005) and Rodgers (1961) show that the diffuser performance characteristics can also be unified in terms of the nearness to choke rather than the vane incidence. In most cases, it is the throat area which is the primary variable in matching the diffuser to the impeller; see Section 12.9. There is then little opportunity to make substantial changes in the geometry of the vane inlet angle while retaining the throat area.

12.2.4 The Effect of the Meridional Flow Channel at Diffuser Inlet

In some cases, the absolute flow angle at impeller outlet is too high to be compatible with a vaneless diffuser. In this case, the diffuser is equipped with a pinch at the diffuser inlet: the diffuser channel is reduced in width over the first few percent of the radial extent, as shown in Figure 12.3. This pinch increases the meridional velocity and reduces the radial flow angle. The pinch can be on the casing side, the hub side or on both sides of the channel. In high flow coefficient stages, it is general to place this on the casing side as it then forms a natural extension to the conicity of the impeller contour on the casing wall. A further reason for installing the pinch on this side of the channel is to accelerate fluid with low radial momentum from the impeller which concentrates in this location. In multistage compressors, it may be advisable to use a pinch on both sides of the channel to avoid a forward-facing step for the case that any axial movement of the impeller shaft relative to the casing occurs.

In situations where an impeller has to be matched to an existing diffuser vane, it is sometimes necessary to widen the flow channel to achieve the correct flow angle, leading to a backward-facing step or a shelf.

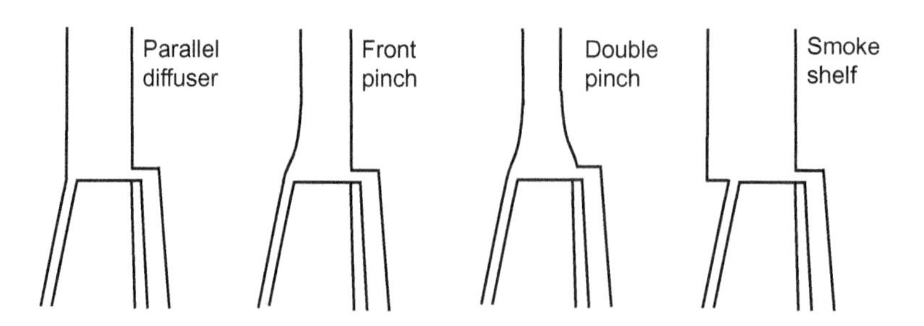

Figure 12.3 Types of channel transition from the impeller to the diffuser.

12.2.5 The Effect of the Impeller Outlet Mach Number

The importance of the inlet relative Mach number in terms of the ideal pressure recovery of planar diffusers is described in Section 7.4.3; an increase in the inlet Mach number causes an increase in the pressure recovery. This is a purely thermodynamic effect which also applies to centrifugal compressor diffusers. The increase in pressure recovery at higher inlet Mach numbers, however, leads to stronger adverse pressure gradients in the flow, making it more prone to separation. This impinges on the pressure recovery and adds to the losses. In addition, an increase in the Mach number narrows the loss bucket of a vaned diffuser leading to a smaller operating range, as discussed in Section 7.3.3.

The ratio of the absolute outlet velocity to the blade speed shown in Figure 12.1 also gives an insight into the absolute Mach number at diffuser inlet from the following:

$$M_{c2} = M_{u2}\frac{c_2}{u_2}\frac{a_{t1}}{a_2} = M_{u2}\frac{c_2}{u_2}\sqrt{\frac{T_{t1}}{T_2}} = \frac{c_2}{u_2}\frac{M_{u2}}{\left(1+(\gamma-1)\lambda r_k M_{u2}^2\right)^{1/2}}. \tag{12.1}$$

A design with a high diffuser inlet absolute velocity, c_2/u_2, such as impeller D in Figure 12.1, has a higher absolute Mach number at the diffuser inlet for a given tip-speed Mach number. Figure 12.4 shows the variation of the tip-speed Mach number and the diffuser inlet Mach number for impeller B for air with $c_{m2}/u_2 = 0.25$, $c_{u2}/u_2 = 0.7$ and $\eta_p = 0.81$. Here, the Mach number of the absolute inlet flow becomes supersonic at a pressure ratio of 4.3 and a tip-speed Mach number of 1.5. Thus, typical stages with an impeller of type B, operating with a pressure ratio above 4 in air, can be expected to have a supersonic diffuser inlet flow. The meridional Mach number is only 0.32 for this condition and is in the low subsonic region.

In many early turbocharger applications, centrifugal compressors for air were designed for a pressure ratio of close to 2 with vaneless diffusers that had a large radius ratio (Watson and Janota, 1982). These stages had an exceptionally wide

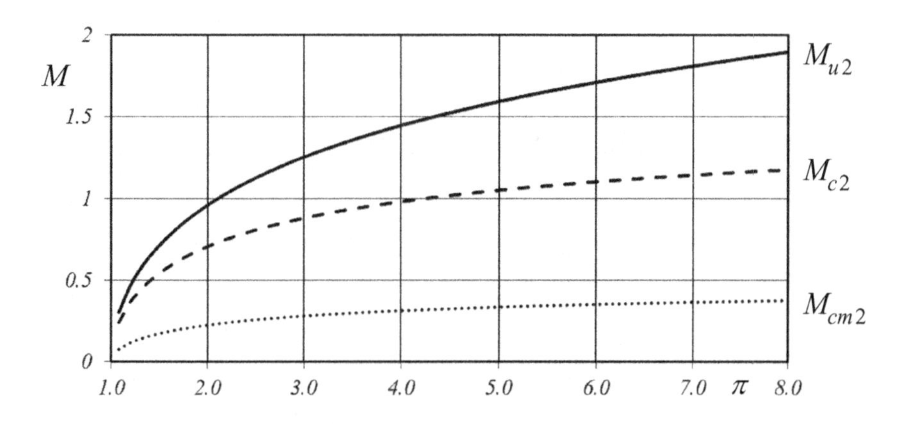

Figure 12.4 Mach numbers as a function of pressure ratio for stage B in air.

operating range from choke to stall with acceptable efficiency while still offering mechanical and aerodynamic simplicity. As pressure ratio requirements for turbochargers increased to around 4, vaned diffusers were introduced to reduce the stage overall diameter and frontal area and to improve performance. A large vaneless space was retained to obtain diffusion to subsonic Mach number at the cascade entry. When pressure ratios of 5–6 were attempted, the vaneless space required for subsonic leading-edge Mach number was prohibitive. Improved diffuser cascade types with thin leading edges were introduced to cope with supersonic inlet flow (Came, 1978). At high Mach numbers, a normal shock forms within the semivaneless space upstream of the throat section of the diffuser, similar to that described in Section 11.8.5 for transonic impeller inlets. The flow is then subsonic in the throat.

12.3 Vaneless Diffusers

A vaneless diffuser comprises a simple annular channel, without vanes, extending radially outwards from the impeller outlet. The channel is usually of the same width as the impeller but may have a decreasing width as the radius increases. In some applications, a pinched diffuser is used which narrows close to the impeller outlet and is followed by a parallel walled channel. The objective of the pinch is to increase the meridional velocity, and as the lowest meridional velocity is on the casing side, the pinch is usually applied on the casing; see Figure 12.3.

The vaneless diffuser produces a decrease in velocity through two mechanisms. First, the meridional channel area increases with radius and reduces the meridional velocity component of the fluid velocity. Second, in the absence of blade forces and also neglecting frictional forces, the angular momentum of the fluid remains constant and so the circumferential velocity component also decreases with radius. The radial extent of the vaneless diffuser then fixes the ideal pressure recovery that can be achieved. There are no simpler devices available in the whole of the turbomachinery world.

12.3.1 Ideal Pressure Recovery of an Incompressible Vaneless Diffuser

For incompressible flow in an ideal diffuser with no losses and with constant width, a very simple equation for the ideal pressure recovery coefficient of a vaneless diffuser can be attained. The equation of continuity becomes

$$c_m r = c_{m2} r_2, \quad c_m = c_{m2} r_2 / r. \tag{12.2}$$

The moment of momentum equation (the Euler equation applied across the diffuser with no work input from inlet to outlet) gives

$$c_u r = c_{u2} r_2, \quad c_u = c_{u2} r_2 / r. \tag{12.3}$$

Combining these equations shows that the flow angle in the diffuser remains constant as the radius increases, giving rise to a flow path known as a logarithmic spiral:

$$\tan\alpha = c_u / c_m = \tan\alpha_2. \tag{12.4}$$

If there are no losses, the total pressure remains constant and

$$p_t = p_2 + \tfrac{1}{2}\rho c_2^2 = p + \tfrac{1}{2}\rho c^2. \tag{12.5}$$

The ideal static pressure recovery coefficient is given by

$$Cp^{id} = \frac{p - p_2}{\tfrac{1}{2}\rho c_2^2} = \frac{\tfrac{1}{2}\rho c_2^2 - \tfrac{1}{2}\rho c^2}{\tfrac{1}{2}\rho c_2^2} = 1 - \frac{c^2}{c_2^2} = 1 - \frac{r_2^2}{r^2}. \tag{12.6}$$

Equation (12.6) highlights two important aspects of vaneless diffusers. First, the ideal static pressure recovery is a function of the diffuser radius ratio; see Figure 12.5. If the outlet radius is twice the inlet radius, the outlet velocity is halved, the kinetic energy drops to 25% and the ideal pressure recovery is 0.75 (shown as a square in Figure 12.5). This represents a very good diffuser given the ease of manufacture and simplicity of the design with no vanes. It is able to match the performance of an optimal vaned diffuser. The disadvantage is that the outer diameter of the stage becomes large, although there are many industrial applications where outer diameter is not limited and, in these cases, the increased outer diameter is not relevant. Vaneless diffusers with a radius ratio above 2 tend not be used because the friction losses in such long diffusers are high, and this increased length offers only a negligible increase in pressure recovery. Brown and Bradshaw (1947) showed that with a volute, no improvement is obtained with a radius ratio above 1.8, and in fact most vaneless diffusers accept a compromise at a ratio of 1.6 or below.

Where space is limited, a reasonable pressure rise can be attained with a more compact vaneless diffuser. For example, a radius ratio of 1.414, that is, $\sqrt{2}$ (shown as a circle in Figure 12.5) halves the kinetic energy and attains an ideal pressure recovery of 0.5. In most turbocharger compressors where both size and range are important, a relatively short vaneless diffuser is used upstream of the volute, and further diffusion is then achieved using a conical diffuser from the volute tongue to the outlet flange, as seen in Figures 1.14 and 1.16. Even in compressors with vaned diffusers, there is usually a short vaneless space upstream of the vanes with a typical radius ratio of around 1.1–1.15 (shown as the triangle in Figure 12.5). This gives a very useful ideal

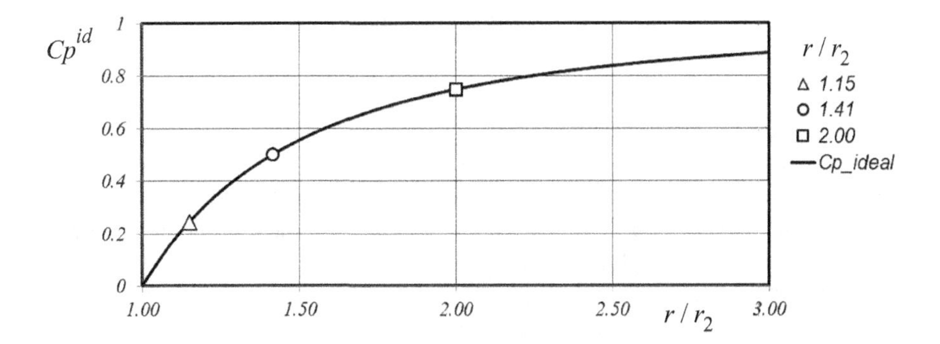

Figure 12.5 Ideal pressure recovery of a vaneless diffuser of different radius ratios.

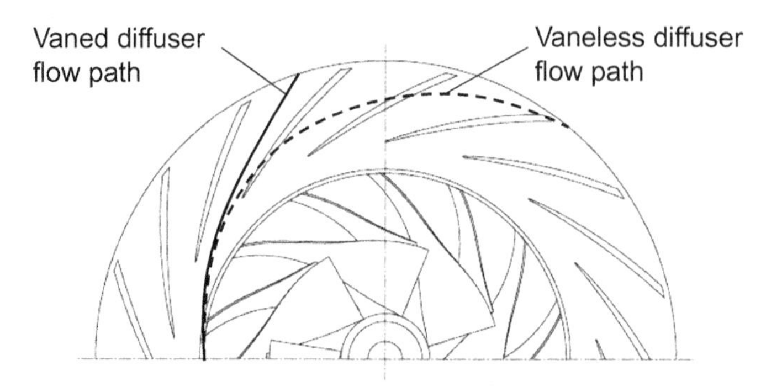

Figure 12.6 Flow path in a vaned and a vaneless diffuser.

static pressure recovery of 0.2–0.25 in the vaneless region where the highest velocity occurs and helps by reducing the absolute Mach number at entry to the vanes.

The second important feature of vaneless diffusers is that the ideal pressure recovery in (12.6) depends only on the radius ratio and not at all on the flow angle at inlet to the diffuser. Thus, when the operating point changes, causing a change in inlet flow angle to the diffuser, the vaneless diffuser still produces the same ideal pressure recovery. However, given that the viscous losses increase with the length of the flow path (see Section 10.3) and hence with an increase in flow angle, the flow angle does have an impact on the actual pressure recovery at large flow angles. Figure 12.6 shows the flow path through a vaneless diffuser compared to that of a vaned diffuser. This insensitivity to the inlet flow angle at angles which are close to 65° enables vaneless diffusers designed for this angle to operate effectively over a much wider range than vaned diffusers. Although a vaned diffuser may have a higher peak pressure rise coefficient than a vaneless diffuser, this only applies over a relatively narrow range of incidence. As a result, vaneless diffusers are invariably used in low-pressure ratio automotive turbocharger applications, where a wide operating range is needed to match the turbocharger with the engine over a large speed and flow range. Vaneless diffusers are also often used in industrial and process compressors with relatively small pressure ratio per stage and where overall package size is of less concern.

The flow path length, L, through a vaneless diffuser with a logarithmic spiral flow path is given by

$$L = (r - r_2)/\cos\alpha_2. \tag{12.7}$$

As the inlet flow angle increases, this is reflected in the length of the flow path. At a flow angle of 50°, the flow path length is 55% greater than the radial flow path. At a flow angle of 60°, the flow path length is twice the meridional path length, at 70° it becomes nearly three times as long and at 80° nearly six times as long. In vaneless diffusers at large flow angles above 75°, the increased path length causes a significant increase in the friction losses and the performance deteriorates to well below the ideal value given by the preceding analysis. The flow may even become unstable with a form of rotating stall so that angles above 65° at the design point should be avoided.

12.3.2 Surge or Rotating Stall in a Vaneless Diffuser

As discussed in the previous section, the ideal pressure rise of the diffuser is determined by the radius ratio and not by the flow angle. As the flow is reduced, the flow angle increases and the streamlines become more tangential. The radial pressure gradient remains the same but the diffuser wall boundary layers may then have insufficient radial momentum to overcome the radial pressure gradient. There is thus a strong tendency towards flow reversal in the radial direction within the boundary layers. At low flow, radial backflow may occur within the boundary layers and develop into rotating stall cells travelling backwards around the annulus. This is similar to rotating stall in a vaned cascade, as described in Section 17.2. Given the distortion of the flow at the impeller outlet with high flow angles on the casing side of the flow channel, the reverse flow will inevitably be on this side of the channel at the inlet to the diffuser. It may switch sides of the channel within the diffuser.

The overall pressure rise is not strongly affected by this radial flow reversal because at high flow angles the radial pressure rise is mainly generated by the swirl component of velocity (Greitzer et al., 2004). With a flow inlet angle of 71.6°, the circumferential velocity is three times the radial velocity ($\tan 71.6° = 3$) and, as mentioned earlier, the kinetic energy of the circumferential flow is $3^2 = 9$ times that of the radial flow. This also explains why vaneless diffusers of increasing width with radius are not often used – the circumferential component is by far the most relevant in the achievement of the pressure recovery, and this is only affected by the radius ratio and not the channel width.

The highly tangential flow in the diffuser inlet in many early industrial radial stages with vaneless diffusers led to significant problems with flow instability. There is an extensive literature on this subject, and many authors have attempted to identify an empirical value for the critical flow angle for reverse flow or carried out theoretical stability analyses of the flow. The critical flow angle lies between 75° and 85°. Some authors identify trends with the diffuser radius ratio, r_2/r_4, and others with the diffuser relative width ratio, b_2/r_2. Gao et al. (2007) and van den Braembussche (2019) provide an overview of the literature and find that there is a lot of scatter in the data with no reliable trends. Given the complex nature of the flow entering the diffuser, it is not surprising that none of the methods appears particularly reliable as a design tool. The most serious instability problems have occurred in industrial process compressors with high inlet pressure, where the instability of the rotating stall in the vaneless diffuser at high pressure levels causes subsynchronous vibrations of the shaft.

To hinder the onset of rotating stall in a vaneless diffuser, it is generally necessary to decrease the mean absolute flow angle in the diffuser above a critical value at the design point. This may be achieved as part of the impeller design process, or may be attained either by using a diffuser pinch at inlet or by decreasing the width of the diffuser channel through the diffuser. The diffuser inlet flow angle needs to be below or close to 65° at the design point – either by means of the impeller design, as in case A in Figure 12.1, or by the use of pinch at the diffuser inlet, as in case B in the same figure. This gives the vaneless diffuser a reasonable chance of avoiding flow instabilities down to 50% of the design flow. An even lower design flow angle is recommended in cases with a variable inlet guide vane which allows the flow to drop to very low values relative to that at the design point when the guide vanes are closed. Centrifugal pumps tend to have low meridional flow coefficients at impeller outlet and much higher inlet angles to the diffuser, rather like the impeller in case C in Figure 12.1, and for this reason rarely deploy vaneless diffusers.

12.3.3 Effect of Mach Number and Losses in a Vaneless Diffuser

For a compressible flow, the density increases through the diffuser, causing the flow angle to rise above the values associated with incompressible flow. In a parallel-walled diffuser, this may result in backflow and stall. A diffuser with converging walls, or a pinched diffuser, can be used to decrease the flow angle with increasing radius. This augments the radial component of velocity, thus creating a more radial flow path through the diffuser, which prevents backflow and separation at low flow rates. The losses in a vaneless diffuser tend to decrease the flow angle, through the action of friction on the swirl components and the increase in meridional velocity as a result of the losses.

Several authors have derived the governing equations for the one-dimensional description of the compressible nonisentropic flow in diffusers, including Traupel (2000) and Greitzer et al. (2004). These equations can be solved by a step-by-step integration process through the diffuser. By way of an example, the equations given by Traupel (2000) are reproduced here. The continuity equation in differential form for a radial diffuser becomes

$$c_m br\,d(\rho) + \rho br\,d(c_m) + c_m\rho\,d(br) = 0, \tag{12.8}$$

where b is the local width of the diffuser at radius r. The Euler equation becomes

$$\frac{d(rc_u)}{rc_u} = -\frac{rc_f\rho\sqrt{c_u^2 + c_m^2}}{\rho_2 c_{m2} b_2 r_2}d(r), \tag{12.9}$$

where c_f is the skin friction coefficient and suffix 2 refers to values at the impeller outlet. The second law in the Gibbs form becomes

$$\frac{d(p)}{\rho} = -\frac{d\left(c_u^2 + c_m^2\right)}{2} - \frac{c_d\rho\left(c_u^2 + c_m^2\right)^{3/2}}{\rho_2 c_{m2} b_2 r_2} rd(r), \tag{12.10}$$

where c_d is the local dissipation coefficient. Traupel recommends using a value for the skin friction coefficient for a fully developed pipe flow calculated at the same Reynolds number, relative roughness and hydraulic diameter as the vaneless diffuser, and that the dissipation coefficient should be taken to be $c_d = 0.0015 + c_f$,

Equation (12.8) shows that a change in the width of the diffuser only affects the meridional component and has no effect on the circumferential component. Equation (12.9) shows that the angular momentum decreases with the radius through the effect of friction. The friction factor is greater for narrow diffusers of small hydraulic diameter and causes a higher reduction in swirl with radius. This is also the case where there is a high swirl angle. Equation (12.10) shows that the main effect on the static pressure rise is the swirl of the flow and that in narrow diffusers the pressure rises less rapidly as the radius increases.

12.3.4 Choke in a Vaneless Diffuser

Although the absolute flow velocity at the impeller outlet may be supersonic, the meridional velocity remains subsonic. This is shown for the diffuser of stage B in Figure 12.4. The choking of the flow in an annulus without vanes is determined not by the absolute Mach number but by the meridional Mach number (Greitzer et al., 2004). As the meridional Mach number is normally subsonic, no choking can occur within a vaneless diffuser. A special case would be a vaneless diffuser with a very high pinch such that the meridional flow chokes. This configuration of diffuser is seldom used in practice, but examples of choked vaneless diffusers with very high pinch are given by Tamaki (2019).

The vaneless space upstream of vaned diffusers at high Mach numbers has a typical radius ratio of approximately 1.15, which allows a velocity deceleration of 15% before the vanes are reached. In this way, a vaneless diffuser can play a useful role in bringing the supersonic impeller outlet velocity down to a subsonic velocity without causing choking.

12.4 Vaned Diffusers

12.4.1 Comparison of Vaned and Vaneless Diffusers

A vaned diffuser allows for a greater pressure recovery in a given radius ratio than a vaneless diffuser; the vanes form separate flow channels to remove more swirl from the flow than is possible by the reduction of the circumferential velocity with radius purely through the Euler equation in a vaneless diffuser, as given in (12.3). The difference in the flow paths shown in Figure 12.6 demonstrates the difference in the swirl at outlet of a vaned and a vaneless diffuser. Where it is necessary to reduce the outer dimensions of a casing in order to build a less bulky machine, which requires less radial space and has less frontal area, a vaned diffuser with a smaller radius ratio is often preferred to a vaneless diffuser.

While a vaned diffuser achieves a larger pressure recovery than a vaneless diffuser in the same radial space, this is usually accompanied by a performance map with a narrower operating range. At high flows, a vaned diffuser may choke at the throat between the vanes, and at low flows it may stall due to large positive incidence or excessive pressure recovery in the vanes, leading to instabilities and surge. Another difference worth nothing between the performance maps of stages with vaned and vaneless diffusers is the trajectory in the map of the locus of peak efficiency at operating speeds below the design or optimum condition. For a vaned diffuser, the locus of the flow capacity at peak efficiency with speed is similar to a square law resistance line, typical of the working line in a power generation or gas turbine application. The trajectory of the optimum point across different speed lines in the compressor map for a vaneless diffuser falls far more steeply as flow is reduced, as discussed in Section 18.3.9.

12.4.2 Types of Vaned Diffusers

The design of vaned diffusers relies on similar principles to those discussed in Section 7.4 with regard to planar 1D channel diffusers. There are many types of vaned diffuser in use, and these can be roughly categorised as cascade diffusers, channel diffusers or pipe diffusers. Channel diffusers have wedge-shaped vanes with straight vane surfaces, and cascade diffusers use aerofoil shapes within the meridional channel defined by the end-wall geometry. Pipe diffusers have no real distinction between the blades and the end-walls but comprise of separate pipes or tubes for the flow.

The range of different cascade and channel diffusers is shown in Figure 12.7 (Hunziker, 1993). Different companies with different empirical databases and experience – or specific patents for diffuser design – tend to use different types of vaned diffusers, whereby there is a tendency for channel diffusers to be more common in the USA and cascade diffusers with aerofoil shapes in Europe. The construction of the machine may also play a part in the choice of vane type. In some configurations, it is

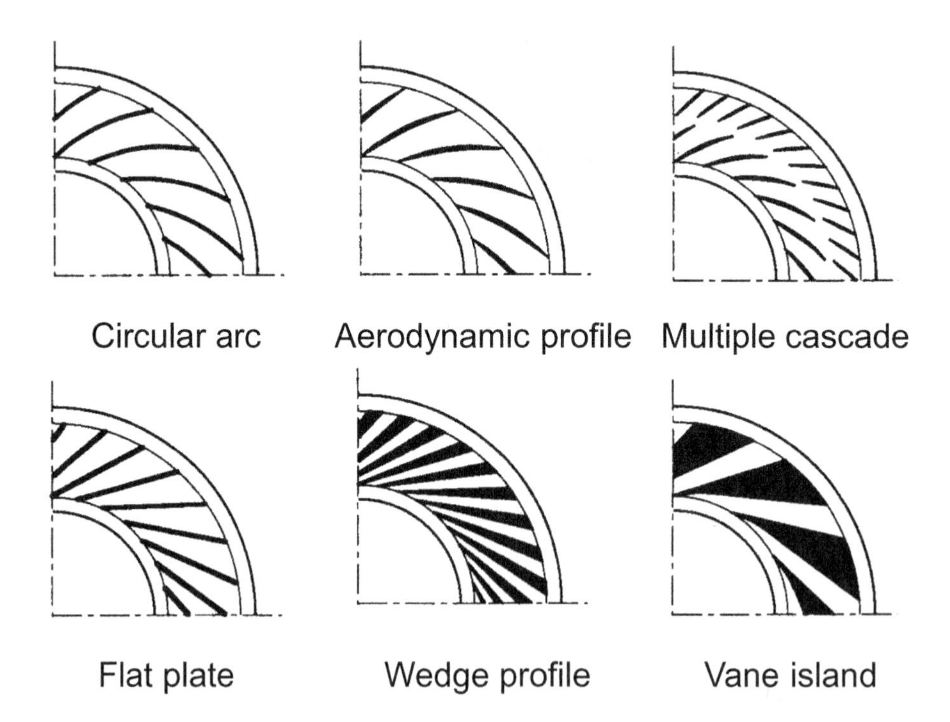

Figure 12.7 Cascade and channel diffusers. (image adapted with permission from Rene Hunziker)

necessary to pass bolts through the vanes, and this tends to steer the design towards wedge or vane-island styles. Many technical publications are available in the literature which compare the performance of the different types of vaned diffusers. Unfortunately, it is not always clear in these studies whether the diffuser throat area and the matching of the diffuser with the impeller is the same for each diffuser. This often makes a direct comparison of the different types very difficult.

12.4.3 Design of Vaned Diffusers

The design of vaned diffusers relies on similar principles to those discussed in Section 7.4 with regard to planar 1D channel diffusers. To recapitulate, the key parameters determining the pressure recovery coefficient of channel diffusers at high Reynolds number are

- Area ratio, which for a diffuser of constant width, b, is the exit width divided by inlet width, W_2/W_1 of the flow channel
- Diffusion length to inlet width ratio, L/W_1
- Divergence angle, 2θ
- Inlet blockage, δ^*/W_1
- Inlet Mach number, M_1

The aspect ratio and inlet flow distortion have a lesser impact. Moreover, the 1D data shows that the best geometry is mainly related to the area ratio and length-to-width ratio and is not largely a function of the blockage (Sovran and Klomp, 1967).

There are more geometrical parameters to be considered for the diffusers of centrifugal compressors. The main additional parameters, shown in Figures 12.8 and 12.9, are the following:

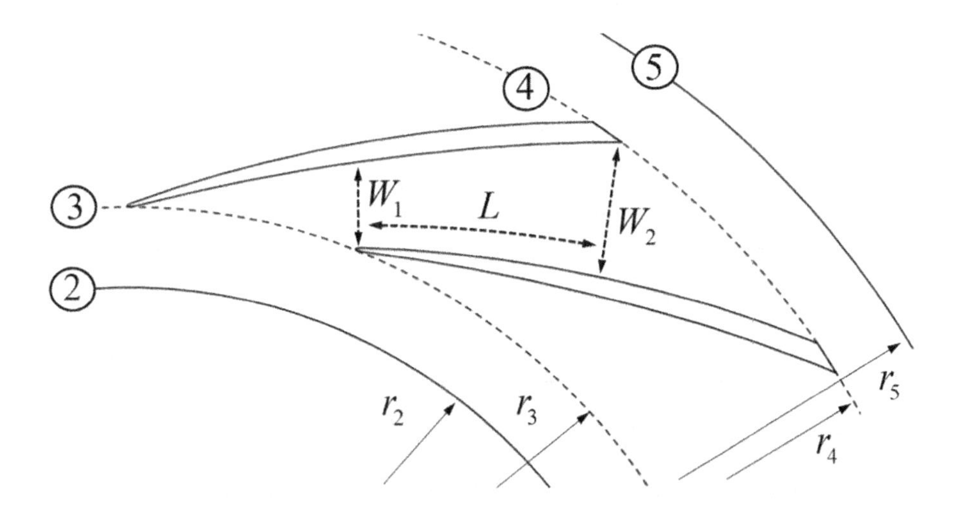

Figure 12.8 Geometry parameters of a cascade vaned diffuser.

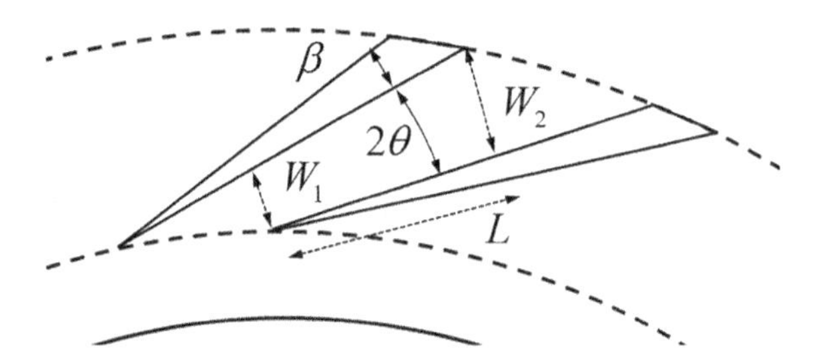

Figure 12.9 Geometry parameters of a wedge channel diffuser.

- Number of vanes, Z
- The vaneless space radius ratio (or the diffuser vane leading-edge radius ratio) – the radius to the leading edge of the diffuser vane divided by the impeller tip radius, r_3/r_2
- The diffuser vane trailing edge radius ratio (the radius to the trailing edge of the diffuser vane divided by the impeller tip radius), r_4/r_2
- The diffuser wall outlet radius ratio – the radius to the end of the diffuser channel divided by the impeller tip radius, r_5/r_2
- The shape of the vanes, including that of the vane leading and trailing edges
- Contouring of the vaneless space (pinch) and possible sidewall divergence
- The geometry of the downstream component (volute, plenum, etc.)

Fortunately, many of these parameters can be related directly to the planar diffuser geometrical parameters, allowing some very useful guidance to be obtained from typical Reneau charts, as is shown in the sections that follow.

The cascade design often consists of standard aerofoil shapes, such as a cascade of thin circular arc cambered aerofoils with rounded or elliptical leading edges (Ribi and Dalbert, 2000) or of classical aerofoil shapes from the NACA series with forward thickened vanes. Experiments with different profile thickness distributions have not found any significant advantage of NACA profiles with thicker leading edges at low Mach number (Kmecl et al., 1999). At transonic inlet Mach numbers, similar considerations apply to diffuser leading edges as with regard to impeller leading edges, as discussed in Section 6.4; so thick leading edges should be avoided as they lead to a strong bow shock. The lowest losses occur at an incidence where the shock is just attached to the thin leading edge. The blockage of the suction side of a thick leading edge causes a reduction in area for the flow and an associated acceleration to the throat and reduction in the choking mass-flow rate. Thin leading edges also reduce the strength of the shock, which mitigates the upstream influence of the vanes on the impeller and reduces possible high-cycle fatigue mechanical issues and noise. For these reasons, Herbert (1980) and Came (1978) developed aft-thickened cascade vanes with thin leading edges for high Mach numbers, as in Figure 12.8, a strategy also more recently applied by Gao et al. (2017) and Han et al. (2017) and remarkably similar to the diffuser of Elling in Figure 1.18.

As the objective of the vanes is to remove swirl, it may not be intuitively obvious that diffuser cascade vanes curve towards the impeller and not away from it. This is related to the fact that cascade vanes with no loading would lead to logarithmic spiral vane of constant vane angle, and these would be even more strongly curved towards the impeller than typical cascade vanes, as shown in Figure 12.6.

As an introduction to the many design parameters, it is useful to consider the most highly loaded cascade vanes that might routinely be used, which are simple straight vanes of thin flat plates. Thin straight vanes represent the limit of the wedge-type channel diffusers (with no vane wedge angle) and the limit of the aerofoil cascade type (with no vane curvature). The opening angle between such thin straight vanes is given by $2\theta = 2\pi/Z$, where Z is the number of diffuser vanes, so that 36 vanes would be required to achieve a respectable included angle of $2\theta = 10^\circ$ for good diffusion in a Reneau chart. In this case, however, the losses would be high due to the friction on the large number of vanes, and mechanical issues may arise with such thin vanes. If thicker vanes were used to alleviate the mechanical issues, then the losses would increase due to the higher velocity resulting from the blockage of the vanes. In some aeroengine applications requiring a high pressure ratio, however, diffusers with this high number of thin straight vanes have been studied, as shown in Figure 12.10 from Methel et al. (2016).

In most applications with fewer straight vanes, the adverse pressure gradient of straight vanes is too high due to the high opening angle, and the flow separates. To avoid the flow separation that would be associated with a smaller number of straight vanes, several different strategies are adopted. The first is to use wedge diffuser vanes which are made thicker towards the trailing edge to decrease the area ratio, as discussed in Section 12.10 and shown in Figure 12.9. The second is to use a cascade of curved diffuser vanes with a vane turning angle, ε, of 10°–16° (where ε is the difference between the inlet and outlet metal angles of the vanes), as discussed in

Figure 12.10 A flat plate diffuser with 35 thin vanes as used in an experimental research program by Methel et al. (2016). (image courtesy of the compressor research laboratory of Purdue University)

Figure 12.11 Low-solidity diffuser with vanes that do not overlap. (image courtesy of Siemens Energy)

Section 12.11 In some cases, a mixture of these two strategies is used with curved aft-thickened vanes, as shown in Figure 12.8. A further strategy is to reduce the loading by placing the vanes in a convergent meridional channel, and an example of a flat plate diffuser with a convergent meridional channel is given in Boyce (2002).

Yet a further design strategy is to use cascade diffusers equipped with very few short vanes that do not overlap, known as low-solidity diffusers, as shown in Figure 12.11 from Sorokes et al. (2018) and in Figure 12.23 from Tamaki (2019). The performance of short channel diffusers is poor, but the flow can withstand a higher opening angle without separation, as shown in Figure 7.15. These short diffusers achieve an operating range approaching that of vaneless diffusers but with a lower pressure recovery (Senoo et al., 1983; Hayami et al., 1990). The higher range is often attributed to the fact that such diffusers have no physical throat and are therefore unable to choke. However, due to the separation of the flow as choke is approached, they choke at a virtual throat bounded by the separation region, as discussed in Section 12.11.3.

In some highly loaded applications where nearly all of the swirl has to be removed, the cascade diffuser may be split into multiple rows of cascades, so that a new boundary layer can start at the leading edge of each row, as shown in the upper right of Figure 12.7. These are sometimes known as tandem diffusers. Pampreen (1972) pointed out that an advantage of a multiple cascade system is that it can accomplish the same diffusion as a single cambered vane but with a reduced diffusion factor per row. The problem with such diffusers is that although the individual blade profiles have lower loading, the end-wall boundary layers still have to cope with the complete adverse pressure rise and may be prone to separation.

Channel or wedge diffusers consist of flow channels which are usually straight and in which the flow is diffused as the area increases along the channels; see Figure 12.8. For a wedge diffuser, the effective included angle, 2θ, is simply a function of the number of vanes, Z, and the included angle, β, between the vanes:

$$2\theta = \frac{2\pi}{Z} - \beta. \tag{12.11}$$

This relationship emphasises why a flat plate design with $\beta = 0$ cannot be applied with a low vane count. For example, 17 vanes would imply a 2θ of over $21°$, well above the level normally associated with stable flow in diffusers. The wedge diffuser has a thicker vane towards the exit; where it has very thick vanes, it is often called a vane-island diffuser (see Figure 12.7). In some cases, the suction surface of the inlet part may be a logarithmic spiral up to the point where the throat is reached. The additional thickness downstream of the throat means that, in this case, the straight flow channel does not cause too much diffusion. The channel itself is similar to a planar diffuser channel. Extensive empirical data are available for this; see Section 7.4. The extra thickness is often used as a convenient route for through-bolting to attach the shroud casing to the backplate. The basic design issues relating to avoid separation in the wedge flow channels are similar to those for cascade diffusers. The large trailing-edge thickness leads to an abrupt increase in area at the end of the flow channel, but this also produces a static pressure rise and a loss somewhat similar to that of a step diffuser; see Section 7.4.6.

12.5 Ideal Pressure Recovery in a Vaned Diffuser

A simple 1D incompressible analysis of the ideal pressure recovery coefficient of a vaned diffuser shown in Figure 12.8, from location 2 at the impeller outlet to location 4 at the vane outlet, identifies many key features of this flow (Ribi and Dalbert, 2000). The ideal pressure recovery for the whole diffuser can be written as

$$Cp_{24}^{id} = \frac{p_4 - p_2}{\frac{1}{2}\rho c_2^2} = 1 - \left(\frac{A_2}{A_4}\right)^2. \tag{12.12}$$

The effective stream-tube flow area at the inlet depends on the absolute flow angle at the diffuser inlet and is given by

$$A_2 = b_2 h_2 = b_2 (2\pi r_2 / Z) \cos \alpha_2. \tag{12.13}$$

The inlet stream-tube area decreases as the compressor moves across its characteristic from choke to surge. Neglecting the effect of changes in flow deviation at the outlet, the effective area at the outlet is constant and is determined by the vanes

$$A_4 = b_4 h_4, \tag{12.14}$$

so that the ideal incompressible pressure recovery coefficient becomes

$$Cp_{24}^{id} = 1 - \left(\frac{2 b_2 \pi r_2 \cos \alpha_2}{Z b_4 h_4}\right)^2. \tag{12.15}$$

The ideal pressure recovery of the diffuser increases as the flow reduces in response to the larger absolute flow angle at diffuser inlet. This trend in the ideal pressure recovery can also be identified in the actual pressure recovery (see Figure 12.12, adapted from Ribi and Dalbert, 2000), whereby the losses cause a fall in performance from the ideal values at high positive and negative incidence. From this it can be seen that the dominant effect on the variation of the pressure recovery with flow is the change in the effective area ratio at the inlet to the upstream semivaneless space and the associated change in incidence with variation in flow.

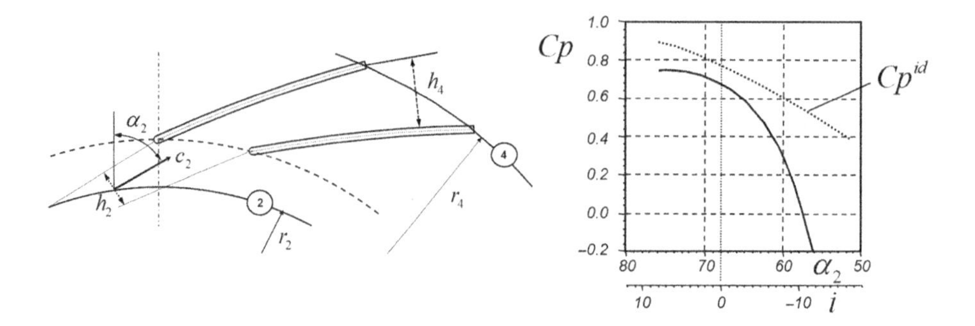

Figure 12.12 Ideal pressure recovery of a vaned diffuser with variation in inlet flow angle. (image adapted from Ribi and Dalbert (2000) using data from MAN Energy Solutions)

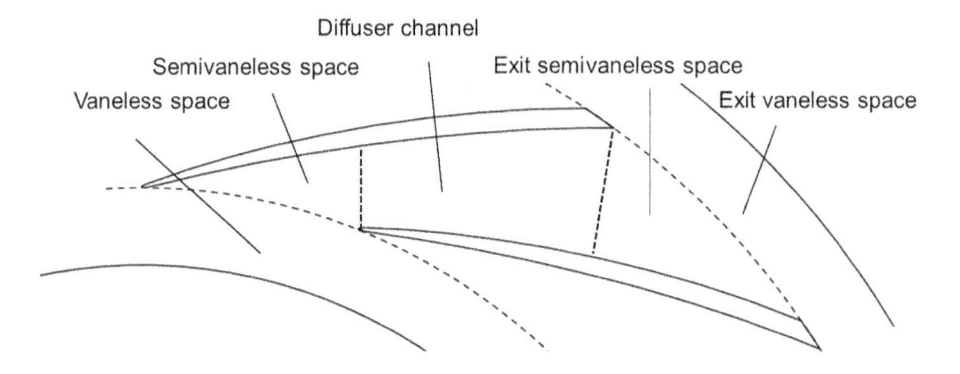

Figure 12.13 Vaned diffuser pressure recovery zones.

12.6 Zones of Pressure Recovery in a Vaned Diffuser

12.6.1 Pressure Recovery

The pressure recovery in a vaned diffuser can be split into distinct zones (Gao et al., 2017), shown in Figure 12.13 and described in this section. All of these regions have a part to play in the overall pressure recovery. In early studies, it was very common to neglect the effect of the downstream semivaneless and vaneless space because these are not important for conventional vaned diffusers which extend outwards to the diffuser channel outlet radius. These last two zones naturally have a lower pressure rise: the kinetic energy available to each subsequent zone of the diffuser falls as the kinetic energy drops; see Section 7.4.7. The downstream vaneless space can, however, become significant in low-solidity vaned diffusers with a low vane trailing-edge radius ratio followed by a large vaneless space to the diffuser channel outlet.

To highlight the potential for pressure recovery in the different zones, a useful starting point is the classical model of a planar channel diffuser and its pressure recovery coefficient depending on the area ratio and the diffusion length to width ratio, as in the Reneau chart; see Section 7.4.5. An attempt is made in the sections that follow to identify the pressure recovery capability of the different zones with the important geometrical properties of each zone separately.

12.6.2 The Vaneless Space

The upstream vaneless space from the impeller outlet to the diffuser inlet leading-edge radius is a classical vaneless diffuser with an unbounded flow path from the impeller exit to the diffuser inlet radius, where diffusion is mainly determined by the radius ratio of this region and to a lesser extent by the end-wall contour (pinch) and the mean flow angle.

12.6.3 The Semivaneless Space

The upstream semivaneless space from the diffuser leading-edge inlet radius to the throat is a semibounded flow path where diffusion is mainly governed by the vane suction-side geometry and the change in the stream-tube area to the throat. Gao et al. (2017) extend this zone to become a rectangular stream-tube partially overlapping with the vaneless space. In effect, a true demarcation between the vaneless and the semivaneless space is difficult.

12.6.4 The Diffuser Channel Downstream of the Throat

The diffusion in this flow channel is mainly set by the area ratio, the length-to-width ratio and the blockage at the throat, following classical planar diffuser guidelines as in the Reneau chart, discussed in Section 7.4.5. In some situations, this region is taken to include part of the upstream and downstream semivaneless space. This approach was taken by Yoshinaga et al. (1980) and is discussed further in Section 12.11. In low-solidity diffusers, there is sometimes no actual diffuser channel where the vanes overlap, or it becomes very short, so for these diffusers the approach of Yoshinaga et al. is extremely helpful.

12.6.5 The Exit SemiVaneless Space

The downstream semivaneless space is a semibounded flow path that starts at the suction-side trailing edge and ends at the pressure-side trailing edge. In this region, diffusion is mainly set by the vane exit angle, the vane pressure side and vane thickness at the trailing edge giving rise to a step diffuser. In conventional vaned diffusers, this region is very short and is usually considered to be part of the diffuser channel.

12.6.6 The Exit Vaneless Space

The downstream vaneless space is a conventional vaneless diffuser. Following the vaned diffuser channel, the flow angle is sufficiently small to give low losses. The angle remains nearly constant as the operating point changes as it is fixed by the blade outlet angle, such that the diffusion in this region is mainly set by the radius ratio. The matching of a vaned diffuser with a downstream component is made easier by the nearly constant flow outlet angle. This region is important in low-solidity diffusers.

12.7 Vaneless Space and Semivaneless Space

The upstream vaneless space is essentially a short vaneless diffuser where the physics of the 1D flow follows that already outlined in Section 12.3. The pressure recovery in this region is mainly determined by the radius ratio, as already given in (12.6), and not primarily by the flow angle. The exception is where the flow angle is larger than 75°. Here high losses may occur due to the long flow path and cause a lower pressure recovery as the angle increases.

The upstream semivaneless space is a stream-tube of fixed length and constant outlet width but with a variable inlet area due to the effect of incidence, as shown in Figure 12.14. Previous investigators suggested that the flow phenomena in this region are a critical factor regarding the overall stage stability (Hunziker and Gyarmathy, 1994) and in determining the pressure recovery (Inoue and Cumpsty, 1984; Elder and Gill, 1985). The area-ratio variation in Figure 12.14 is not fundamentally related to the geometry of the semivaneless space, but arises from the change in inlet stream-tube area with a change in the incidence as the inlet flow angle varies with a change in the flow rate. The effective stream-tube height at the diffuser leading-edge radius depends on the absolute flow angle at diffuser vane inlet and is given by

$$A_3 = b_3 h_3 = (2\pi r_3/Z) b_3 \cos\alpha_3. \tag{12.16}$$

The inlet stream-tube area decreases as the compressor flow is reduced. Neglecting the changes in blockage at the throat the effective throat area is constant. If the width of the diffuser remains constant, the effective area ratio for this semivaneless space becomes

$$\frac{A_{th}}{A_3} = \frac{Zo}{2\pi r_3 \cos\alpha_3}, \tag{12.17}$$

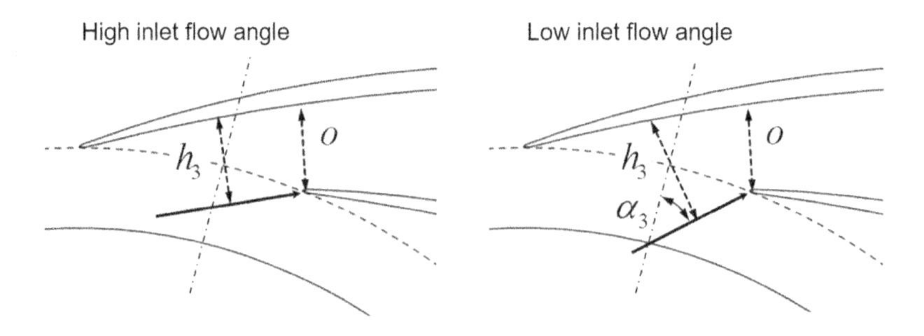

Figure 12.14 Change in area ratio of the semivaneless space with incidence.

where o is the throat width. For this stream-tube, the effective inlet area is determined by the operating condition, but the throat area is determined by the geometry; see Figure 12.14. The shape and length of the suction surface to the throat depends on the number of vanes, the distribution of vane angle in the inlet region and the thickness of the vane at the end of the semivaneless space. Herbert (1980) defines a wetted surface area of the semivaneless space and divides this by the throat area to derive a parameter based on the diffuser length to outlet width ratio for this stream-tube. A length to width parameter based on the inlet width is not appropriate when the inlet area varies (Koch, 1981). Following this suggestion, the diffuser length used here for comparison with the 1D Reneau diffuser chart data is the length of the suction surface from the leading edge to the throat. For straight vanes of zero thickness, this suction surface diffusion length depends on the vane inlet angle and the number of vanes as follows:

$$L = \frac{2r_3(1 - \cos(2\pi/Z))}{2 - 1/\cos^2(\alpha_3' + \pi/Z)}, \tag{12.18}$$

where Z is the number of vanes and α_3' the inlet vane angle. The throat width for the thin straight vanes is given by

$$o = L\tan(\alpha_3' + \pi/Z). \tag{12.19}$$

The length to outlet width parameter for the semivaneless space can then be calculated and a form of the Reneau chart for this region plotted. This is shown in Figure 12.15 for a vane of zero thickness with 19 straight diffuser vanes up to the throat.

In terms of a classical Reneau chart, the diffusion length to outlet width ratio of the diffuser in Figure 12.15 varies with the vane inlet angle and the area ratio varies with

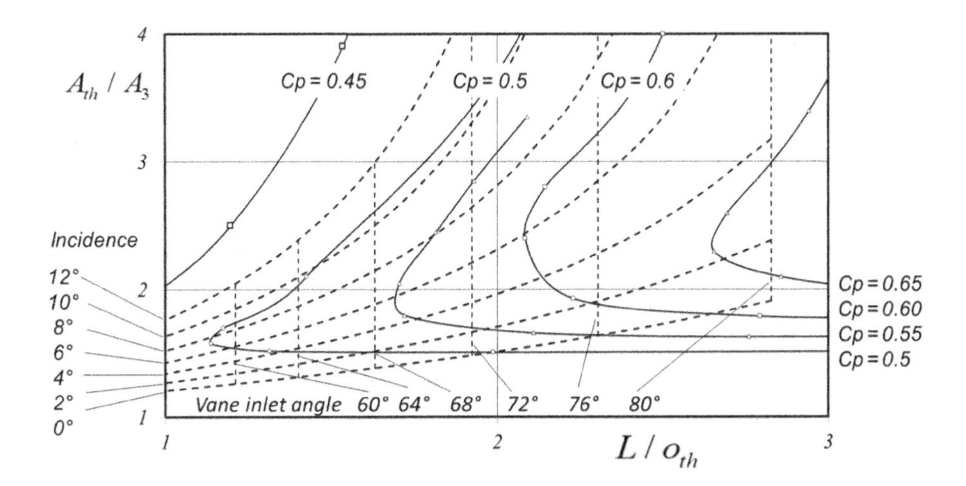

Figure 12.15 Reneau chart for the upstream semivaneless space (with pressure recovery test data from the Reneau chart for 3% blockage).

the incidence; this is equivalent to a vertical line in the chart as the incidence changes. Figure 12.15 includes lines of constant Cp for a blockage of 3% from the planar diffuser data of Reneau, accounting for the fact that the ordinate used is the length to outlet width ratio. This shows that the performance of the semivaneless space can be expected to be strongly dependent on the incidence.

Clearly, the pressure recovery of this semivaneless zone is low at 0° incidence as the effective area ratio is small. More importantly, as the incidence increases, the semivaneless space can be expected to have an increased pressure recovery and, in the present case, at around 5°–6° incidence the semivaneless space coincides with the ridge of high performance from the planar diffusers. Measurements in vaned diffusers show that the pressure recovery coefficient in this region may increase to 0.6. As the incidence increases above 6°, the 1D diffuser data from Reneau suggest that the pressure recovery coefficient of the semivaneless space starts to fall. This does not necessarily indicate an immediate decrease in the pressure rise, as the kinetic energy available at inlet to the diffuser continues to increase as the flow is reduced, as indicated in Figure 12.1. At an incidence of 12° in Figure 12.15, the pressure recovery coefficient is close to $Cp = 0.45$ and is approaching the line of fully developed stall in a planar diffuser so that the semivaneless space may become responsible for instability (Herbert, 1980). This discussion does not consider either the effect of Mach number or the effect of local incidence on the vanes; it may be expected that at high incidence with transonic Mach numbers that the pressure recovery is less than expected from the planar diffuser contours plotted in Figure 12.15, although the location of the peak pressure recovery is likely to be the same.

The designer can to a certain extent adjust the location of the semivaneless space in the Reneau chart by changing the width of the throat, the number of vanes, the vane inlet angle and the selection of the vane incidence at the design point. More vanes increase the length to outlet width ratio, and figures similar to that of Figure 12.15 could be produced for a different vane number. A smaller throat, which can be achieved by using a concave vane pressure surface in the inlet region, decreases the area ratio and increases the diffusion length. A larger vane inlet angle increases the diffusion length. Selecting a higher incidence angle at design positions the design point higher in the Reneau chart. The risk of stall at high incidence can be reduced by designing with a low incidence and with a smaller throat. In general, the matching of the impeller and the vaned diffuser is determined by the throat width (Casey and Rusch, 2014; Section 12.9). Thus, in a typical design the required throat width is known from the impeller design and is not a variable in the diffuser design process.

12.8 Blockage at the Throat in Diffuser Channels

The effective area of throat is limited by blockage due to boundary layer growth on vanes and end-walls, and this blockage is strongly dependant on the pressure rise in the upstream vaneless and semivaneless space. This pressure recovery, in turn, is mainly controlled by the incidence of the flow on to the vanes. Relatively simple 1D correlations for blockage, such as that of Kenny (1972), can describe this process reasonably well, even though this region is strongly influenced by the nonuniform, unsteady flow from the impeller. Recent unsteady computations and PIV measurements of this flow region confirm that the dominant factor is the incidence of the average flow onto the diffuser vanes (Shum et al., 2000; Cukurel et al., 2008; Robinson et al. 2012; Gao et al., 2017). Empirical correlations can be used to determine the blockage at the throat as a function of the pressure recovery in the vaneless and semivaneless space, based on a 1D analysis of test data in many diffusers, or on the basis of CFD simulations. A review of many correlations was carried out by Deniz (1997), and his suggestion for a curve fit for the throat blockage as a function of the pressure recovery in the semivaneless space is

$$B = 5.981C_p^4 - 0.642C_p^3 - 0.356C_p^2 + 0.12C_p + 0.087. \tag{12.20}$$

The trends from many publications are reproduced in Figure 12.16, showing that the blockage is small towards choke with accelerating flow in the inlet. At low incidence with no pressure rise in the semivaneless space, which might be considered to be the typical design point, a blockage of 3% is suggested by the trend in this figure.

CFD calculations can be used to define the diffuser pressure recovery coefficient, but these simulations generally have highly nonuniform diffuser inlet conditions as a result of the impeller flow patterns. Here, different averaging methods for the inlet total pressure distributions give slightly different results for the pressure recovery, and this may be the cause of some concern. The overall diffuser pressure recovery coefficient, based on a suitably averaged inlet total pressure, has been found to correlate well with the momentum-averaged flow angle into the diffuser (Deniz, 2000). Deniz showed that the generally accepted sensitivity of diffuser pressure recovery performance to inlet flow distortion and boundary layer blockage can be largely attributed to the inappropriate quantification of the average dynamic pressure at diffuser inlet. In fact, the experimental data which are the basis of the blockage correlation normally do not include detailed traverses in this region, and as a result the use of a simply defined blockage is commonplace in the diffuser literature.

Not all of the elements of the pressure rise which make up this pressure recovery vary with the flow angle. The vaneless space upstream of the leading edge (2–3) has a

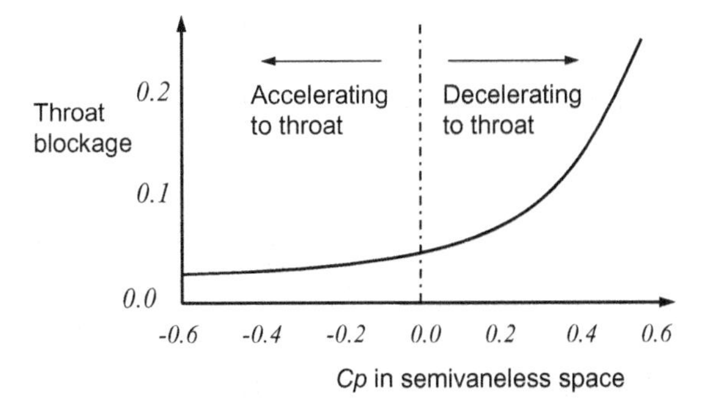

Figure 12.16 Correlation for the blockage at the diffuser throat.

constant pressure rise independent of the flow angle. The diffuser flow channel (th-4) has a fixed geometry and, apart from changes in the pressure recovery due to changes in the blockage at the throat, should produce a pressure recovery that is nearly constant with flow. It is clear from this that the variation of pressure recovery of a vaned diffuser with the flow angle, as shown in Figure 12.12, is essentially due to the performance of the upstream semivaneless space. The semivaneless space produces a variation in the pressure recovery with the flow angle, which becomes larger towards low flow due to the change in flow deceleration in this region. At very high flows, the pressure recovery in this zone is low – it acts a nozzle – while at low flows it is high because it acts as a diffuser, as shown in Figure 12.14.

12.9 Matching the Diffuser Throat with the Impeller

The choke mass flow of the impeller and that of the vaned diffuser depend directly on their respective throat areas and on the local total flow conditions, whereby the relative total conditions are needed for the impeller. At choke, the throat areas are not strongly affected by the blockage (see Figure 12.16), and both the impeller and the diffuser may be considered to have a similar low blockage at the throat. Due to the change of pressure and density between the inlet and outlet of the stage, either the throat area of the impeller or of the diffuser determines the maximum mass flow on each speed line. Tamaki et al. (1999) show that, in order to provide the best performance, the flow capacity of impeller and diffuser must match closely at the design Mach number. At higher Mach numbers, the impeller throat determines the maximum flow, and at lower Mach numbers the diffuser throat is determinant (Casey and Rusch, 2014).

Dixon and Hall (2010) provide some 1D equations that predict the maximum flow capacity of the impeller. Where there is no inlet swirl, these equations can be rearranged to give the maximum flow coefficient when the impeller chokes (see Casey and Schlegel, 2010), as

$$\phi_{t1}^{imp} = \frac{1}{M_{u2}} \frac{A_{imp}^{*}}{D_2^2} \frac{\left[1 + \{(\gamma - 1)/2\}\{D_1/D_2\}^2 M_{u2}^2\right]^{\frac{(\gamma+1)}{2(\gamma-1)}}}{[(\gamma + 1)/2]^{\frac{(\gamma+1)}{2(\gamma-1)}}} \tag{12.21}$$

based on the tip-speed Mach number, M_{u2}, the mean impeller inlet diameter, D_1, and the impeller throat area, A_i^*. Dixon and Hall (2010) also provide an equation for the maximum flow capacity of a vaned diffuser based on the total conditions at the impeller exit. The impeller exit total conditions can be calculated from the work coefficient, the tip-speed Mach number and efficiency of the impeller. The maximum flow coefficient if the diffuser causes choke can be derived as

$$\phi_{t1}^{diff} = \frac{1}{M_{u2}} \frac{A_{diff}^{*}}{D_2^2} \frac{\left[1 + (\gamma - 1)\lambda M_{u2}^2\right]^{\frac{(n+1)}{2(n-1)}}}{[(\gamma + 1)/2]^{\frac{(\gamma+1)}{2(\gamma-1)}}}, \tag{12.22}$$

where n is the polytropic exponent based on the impeller total–total efficiency.

The widest range achievable in a radial compressor generally occurs when the diffuser and the impeller choke at the same time (Tamaki et al., 1999; Rodgers, 2005). Operating range will be forfeited if the diffuser is designed for a larger choke flow, given that the impeller chokes at a similar flow and some surge margin will be lost due to incidence at the diffuser inlet. If the diffuser is designed for a lower choke flow, the diffuser rather than the impeller will limit the maximum flow and the range may be forfeited due to impeller stall.

From the preceding equations, an equation can be derived for the ratio of the impeller and the diffuser throat areas if both components reach choke at the same mass flow:

$$\frac{A^*_{diff}}{A^*_{imp}} = \frac{\left[1 + \{(\gamma - 1)/2\}\{D_1/D_2\}^2 M^2_{u2}\right]^{\frac{(\gamma+1)}{2(\gamma-1)}}}{\left[1 + (\gamma - 1)\lambda M^2_{u2}\right]^{\frac{(n+1)}{2(n-1)}}}. \tag{12.23}$$

Equation (12.23) is illustrated in Figure 12.17. Here the variation of the throat area ratio for a range of tip-speed Mach numbers and for an impeller with a total–static polytropic efficiency of 0.84 and a work coefficient of 0.6 in air with a variation of the impeller inlet diameter is given. Similar diagrams for the variation of the efficiency and the work coefficient show similar trends in that the major effect is the variation of the tip-speed Mach number, such that smaller diffuser throats are required for higher tip speeds.

For given values of M_{u2}, γ, D_1/D_2, n, and λ , the optimum area of the diffuser throat to match that of the impeller throat, A^*_{diff}/A^*_{imp}, can be calculated from (12.23). A stage with a matched diffuser at the design speed but which operates at higher speeds will become mismatched. As the diffuser is effectively too large at higher speeds, the stage will choke first in the impeller and may stall in the diffuser. Stages operating at lower than the design speed will also be mismatched and choke first in the diffuser, as the diffuser is too small and may stall in the impeller. In some designs, it may be a deliberate choice of the designer to match the stage at low speeds to favour the part-speed

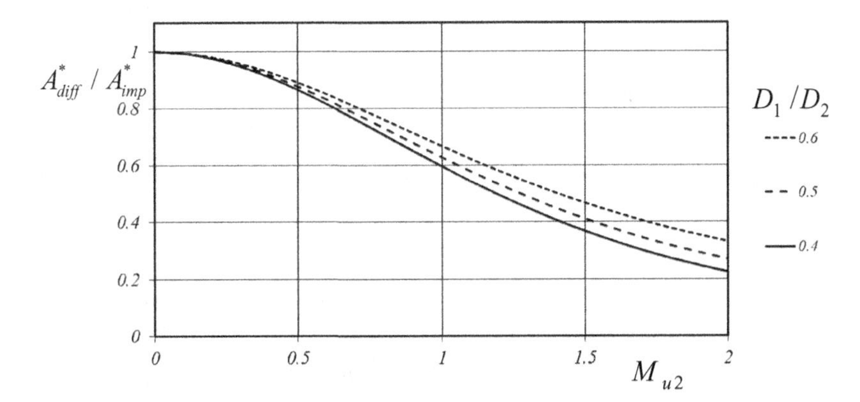

Figure 12.17 Ratio of diffuser and impeller throat area for a matched diffuser as a function of the tip-speed Mach number.

performance and operating range at the expense of the high-speed performance. Chapter 18 gives further discussion of impeller and diffuser matching and its effect on the stage characteristics.

Casey and Rusch (2014) collected data in the open literature and from proprietary sources for the ratio of diffuser to impeller throat area for a wide range of executed compressor stages. This database covers operational gas turbine, turbocharger and industrial process compressors with tip-speed Mach numbers ranging from 0.4 to 2.1, giving pressure ratios in air from 1.2 to 12. It includes a wide range of impeller types (variation in backsweep, splitters and no splitters, open and shrouded impellers) and diffuser types (wedge, cascade, low solidity and circular arc vanes). The predictions of (12.23) compared with the geometric design throat areas are shown in Figure 12.18. The tip-speed Mach number used for the comparison is for the design conditions, as provided by the engineers who supplied the data about the individual designs. As no information was available with regard to the impeller efficiency, the peak stage total–total efficiency at design was used to calculate the polytropic exponent for the impeller in (12.23).

Figure 12.18 shows that the area ratios predicted by this method are typically within ±5% of the geometrical area ratios of the completed designs and often more accurate than this. The scatter is due in part to the use of the stage efficiency rather than the impeller efficiency in the analysis and in part due to the simple 1D nature of the theory, which neglects the hub-to-casing and vane-to-vane variation of the flow velocities. In some cases, the area ratio used in the design may not be the true optimum: the design philosophy may have been flawed or biased to give a better high- or low-speed performance. The largest discrepancy in Figure 12.18, at an area ratio close to 0.5, is for a case where full design details were available and subsequent analysis of this case showed the tested diffuser to be poorly matched at the design speed as it was too small for the impeller.

The excellent agreement of the 1D theory with the design data of diffusers from many sources is surprising given the widely held view that diffuser matching is one

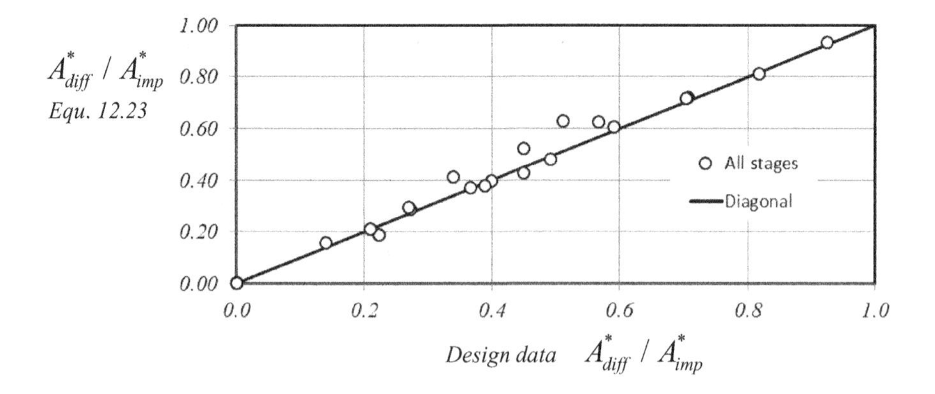

Figure 12.18 The theoretical diffuser to impeller throat area ratio from (12.23) compared to the design area ratio for a range of executed compressor stages.

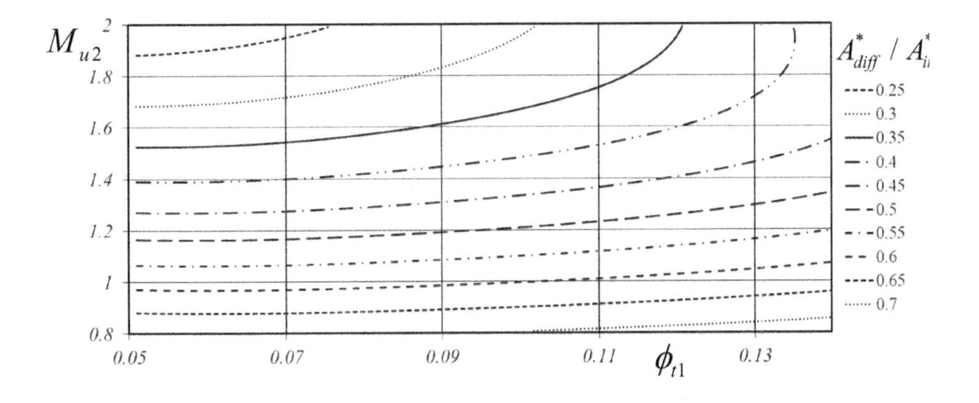

Figure 12.19 Design guideline for the ratio of diffuser to the impeller throat area.

of the most difficult aspects of radial compressor design. It would indeed appear that many features of the design may be crucial in determining the loss mechanisms and operating range of vaned diffusers, but nothing is as important for the designer than to match the throat areas of the impeller and the diffuser in order that they are able to pass the same mass flow at the choke conditions at design speed. Figure 12.19 represents a design guideline for the diffuser to impeller throat area ratio as a function of the tip-speed Mach number and the flow coefficient, following the approach of Casey and Rusch (2014). An increase in the tip-speed Mach number naturally causes a lower diffuser throat area to be required. The flow coefficient affects the efficiency of the impeller, and Figure 12.19 makes use of (10.55) for this. An increase in the flow coefficient also requires a higher inlet eye diameter, as discussed in Section 11.4.1, so that a larger diffuser throat area is also required. In this way, a trim of the impeller affects the matching of the impeller and the diffuser.

12.10 Wedge Diffuser Channels

The wedge diffuser channel downstream of the throat is very similar to a planar diffuser; see Figure 12.8. A Reneau chart for the channel geometry of wedge diffusers is shown in Figure 12.20. The data points shown in this figure represent the array of wedge diffusers tested by Rodgers (1982) and by Clements and Artt (1987, 1988). Unfortunately, the pressure recovery data for the channels of these diffusers was not published, but the diagram clearly shows the region where wedge diffusers are most often applied and demonstrates that it is broadly in line with the optimum geometry expected from planar diffusers. Clements and Artt (1987) have tested other wedge diffusers: these have a peak pressure recovery with opening angle between 10° and 16°, which is consistent with the Reneau chart and with similar charts produced by Runstadtler and Dean (1969) for higher Mach number flows.

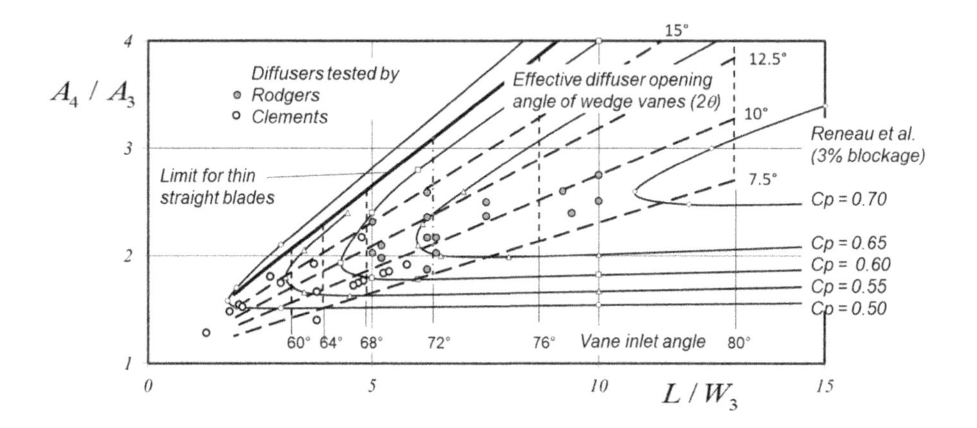

Figure 12.20 The Reneau chart for wedge diffusers.

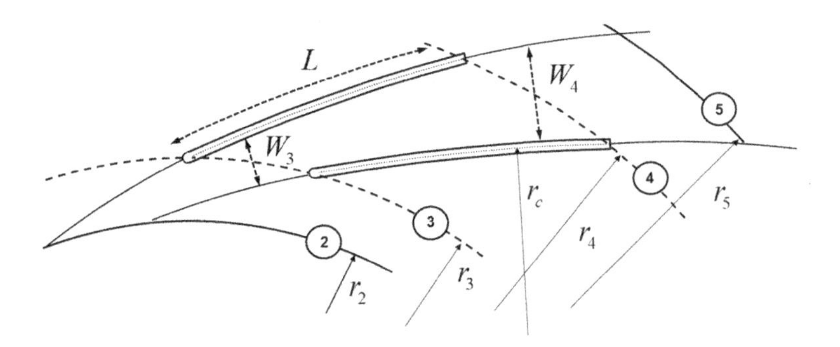

Figure 12.21 Example of a Reneau chart for a circular arc cascade diffuser with 19 vanes.

The optimum length to width ratio of the wedge flow channel is close to 4 and is smaller than expected for 1D planar diffusers (Clements and Artt, 1988). The lines in Figure 12.20 represent the area and nondimensional length of wedge diffusers for a range of included angles between the suction and pressure surface and inlet vane angles. In order that a direct comparison can be made with the cascade diffusers, the vane number and its radial extent are taken as the same as that in Figure 12.21. It is clear that an opening angle of 2θ of between 10° and 14° provides the best performance for wedge diffusers. A typical area ratio of the wedge diffuser channel is between 1.4 und 2.0. The diagram also shows the limit case of a wedge diffuser with thin straight blades, identifying that such a diffuser would be in a region of extremely low-pressure recovery, as with straight cascade diffusers.

Measurements and simulations generally show that the diffuser channel itself produces progressively less pressure rise towards the outlet, because further pressure rise is limited by the decrease in kinetic energy that has already taken place in the channel. The pressure recovery and the most suitable geometry of the diffuser channel can be estimated from planar diffuser data (Japikse and Baines, 1998; Section 7.4.2), whereby the blockage in the throat is critical for the performance of the diffuser (Section 12.8). The trailing edge of the wedge represents a step diffuser and may also contribute to a small additional pressure recovery (Section 7.4.6). A particular difficulty of the wedge diffuser is the abrupt cutoff trailing edge, which causes different separated flow structures in the downstream wake which are not generally amenable to accurate simulation (Smirnov et al., 2007; Gibson et al., 2017). Detailed stage performance measurements with a range of wedge diffusers are available in Ziegler et al. (2003a, 2003b).

12.11 Cascade Diffuser Channels

12.11.1 Cascade Diffuser Performance

The design of cascade diffuser flow channels can also be based on the Reneau chart for planar diffusers. The achievable pressure rise coefficient of the channel is again controlled by the area ratio and the nondimensional diffuser length. For cascade diffusers the nondimensional length and area ratio are determined by the choice of number of vanes, vane length, vane thickness and vane curvature. Preliminary designs of the channels use planar diffuser technology. The cascade diffuser geometry may also be defined by a coordinate transformation of an axial cascade (Scholz, 1977). This technique is in fact seldom used, as the physical spacing in the equivalent axial cascade is constant, while the spacing varies in the radial plane. Tamaki (2019) has, however, recently used a coordinate transformation to estimate the outlet flow angle and ideal pressure recovery of a cascade diffuser.

The extensive experimental work on cascade diffusers in a centrifugal compressor stage of Yoshinaga et al. (1980) showed that the peak pressure recovery of cascade diffusers in radial compressor applications occurs with a nondimensional diffuser length, L/W, that is less than that of typical planar diffusers. The peak recovery is found in a region of $4.8 < L/W < 6.2$ and an area ratio of nearly 2.0, with opening angle of 2θ between 9 to 10°, whereas the peak in a typical Reneau chart is at $L/W = 8$. Gao et al. (2017) have also found similar trends using parametric CFD simulations of low-solidity cascade diffusers.

The geometry used by Yoshinaga et al. to represent the cascade diffusers is shown Figure 12.22 and includes part of both the upstream and downstream semivaneless space. In this form, the area ratio and the diffusion length parameter are simple to calculate from the geometry of a circular arc vane with no thickness as

$$
\begin{aligned}
W_3 &= 2\pi r_3 \sin\left(\alpha_3'\right)/Z \\
W_4 &= 2\pi r_4 \sin\left(\alpha_4'\right)/Z \\
r_c &= \frac{r_4}{2}\frac{1-(r_3/r_4)^2}{\sin\alpha_4' - (r_3/r_4)\sin\alpha_3'} \\
L &= r_c\left[\tan^{-1}\left(\frac{\cos\alpha_4'}{r_c/r_3 - \sin\alpha_4'}\right) - \tan^{-1}\left(\frac{(r_3/r_4)\cos\alpha_3'}{r_c/r_3 - (r_3/r_4)\sin\alpha_3'}\right)\right].
\end{aligned}
\tag{12.24}
$$

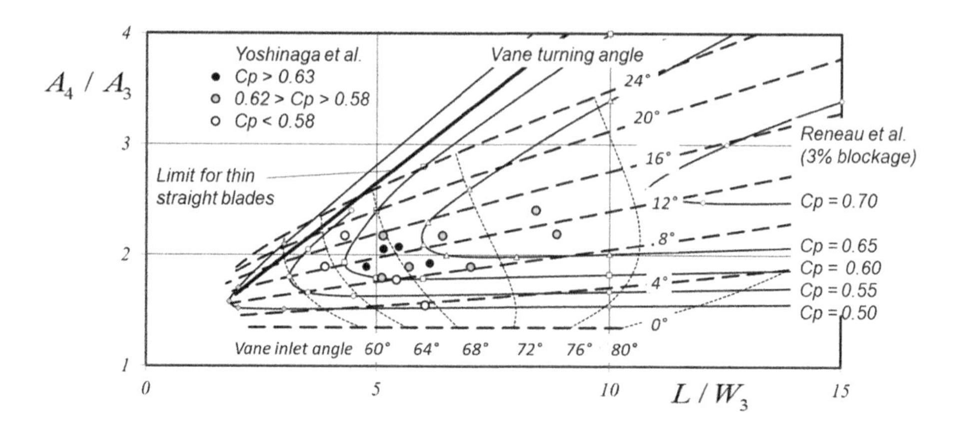

Figure 12.22 Geometrical parameters of a circular arc cascade diffuser channel as defined by Yoshinaga et al. (1980).

The advantage of this procedure is that the same geometry definition can be used for low-solidity diffusers with no covered passage as the diffusion length is taken to be the chord length rather than the length of the covered channel. The key parameters that determine the position of the cascade diffuser in the Reneau chart are the diffuser inlet and outlet radius ratios; the number of vanes, Z; the vane inlet angle, α_3'; and the vane turning angle, $\alpha_4' - \alpha_3'$, which is the difference between the inlet and outlet metal angles of the vanes.

Figure 12.21 shows a Reneau chart derived from the experimental pressure recovery data for 16 experimental test cases of Yoshinaga et al.; contours of pressure recovery cannot be drawn from so little data, but the darker shade of the experimental points indicates a higher level of pressure recovery. Also shown are the contours for the pressure recovery from the planar diffuser data of Reneau for 3% inlet blockage as an indication of the equivalent planar diffuser performance levels. The ridge of peak pressure recovery in the data agrees well with the ridge of data in the experiments of Yoshinaga et al., but the location of the peak values for the circular arc cascades in radial compressors is at a lower nondimensional diffuser length than the values giving peak pressure recovery in a planar diffuser. The cascades also exhibit a lower pressure recovery than the 2D diffusers, which may due to the complex inlet flow or may be partly attributed to the effect of curvature, as it is known that curved diffusers have less pressure recovery (Fox and Kline, 1962).

Figure 12.21 includes lines of constant turning angle for a typical vaned diffuser with 19 vanes, an inlet radius ratio of 1.15 and an outlet radius ratio of 1.55. A vane turning of 0° does not indicate a straight vane but implies a near-logarithmic spiral vane with no change in vane angle along its length and the same inlet and outlet angle; see Figure 12.6. An increase in the turning of the diffuser vanes makes these straighter and increases the area ratio. Figure 12.21 also shows the area ratio versus nondimensional length for straight vanes, which have a high deflection angle, and highlights the fact that such diffusers fall into an area with poor pressure recovery. This explains the need for wedge diffusers if the vanes are straight. An alternative is to use straight vanes in a converging diffuser channel with an outlet width lower than the inlet width such as that shown in Boyce (2002). Such a diffuser would have higher meridional velocity at the outlet with a low outlet flow angle. Figure 12.21 also shows lines of constant vane inlet angle, showing that a decrease in the vane inlet angle decreases the nondimensional length. The peak pressure recovery occurs with a turning angle close to 12°. For the typical impellers A, B, C and D in Figure 12.1, with their specific flow inlet angles, reasonable pressure recovery can be expected with a turning angle of the cascade diffuser of between 9° and 12°.

12.11.2 Cascade Diffuser Design Guidelines

The design of the vaneless space demands a value for the diffuser leading-edge radius ratio (r_3/r_2); typically this would be between 1.05 and 1.15. A smaller radial space may lead to higher efficiency (Robinson et al., 2012). However, while the performance improvement is appealing, smaller gaps may cause forced response and low cycle fatigue problems for the impeller. This is because the tip of the impeller disc runs through the static pressure potential field and expelled shock ahead of the diffuser vanes. The radial gap also provides a vaneless space for some diffusion of the flow. As a result, the diffuser vane incident Mach number is less than the absolute Mach number at impeller tip, and this can be beneficial for the operating range. The diffuser width can be matched to the impeller width by creating a pinch. This can best be included by continuing the meridional slope on the casing side from inside the impeller passage with a front pinch, as shown in Figure 12.3.

The diffuser outlet radius ratio is typically in the range $1.45 < r_4/r_2 < 1.55$ for conventional vaned diffuser stages, which is a compromise between performance benefits for long diffusers and space limitations. The smaller value, or even a lower value, is used where space is a primary consideration, but the value of 1.55 has been found to lead to slightly higher total–static stage efficiency and so is generally preferred. In some applications, a large overall radius ratio, r_5/r_2, is used, but the vane outlet radius, r_4/r_2, is less than this as in low-solidity diffusers (Gao et al., 2017). Clements and Artt (1988) showed experimentally that wedge diffusers in which the vanes were cut back from a large outlet radius so that they were followed by a long vaneless space performed just as well as diffuser vanes extending to the maximum radius. Zachau et al. (2009) also pointed out that a pipe diffuser may operate with the same performance even if the diffuser pipes are shortened.

Vane count is not a primary aerodynamic consideration in cascade diffuser design. It is mainly considered as a source of potential excitation for forced response, and so prime numbers are used which do not clash with the even number of impeller blades when a splitter is used. Typical vane counts are $Z = 15$, 17 and 19, though occasionally 13 and 23 are employed depending on the impeller eigenfrequencies. The prime number 17 is a good starting point, other vane counts being considered later in the design iterations once a structural analysis of the impeller has been carried out.

The key parameters considered in the design are the throat width, o_{th}, and the diffusion length to inlet width parameter, *L/W*, and this may sometimes be expressed as an equivalent opening angle for the covered passage, 2θ. The flow range of the diffuser normally dictates the flow range of the stage at moderate levels of impeller tip Mach number. There is an inlet flow angle which results in minimum loss for the diffuser just as for any other vane in a cascade. At lower angles, the vane will suffer higher loss due to negative incidence stall or choke, and at higher angles the vane will stall. It is important that the vane matches the impeller in order that its minimum loss occurs close to the impeller's peak efficiency. At low Mach number, simply matching the vane at close to zero incidence is adequate, but at higher Mach numbers it is important to assess the choke margin of the diffuser and match this with the choke margin of the impeller (Casey and Rusch, 2014; Section 12.9).

If the throat area is known, the remainder of the vane design is dictated by either consideration of the effective diffusion length of the covered passage or the opening angle of the diffuser, 2θ. This requires a geometric model of the diffuser with the appropriate channel dimensions (width, inlet and outlet radius ratio, vane count and thickness distribution) based on the geometry model described for impellers in Section 14.5. An initial approximation would be to set the vane angles for the expected absolute gas angle at inlet with approximately 10°–14° of deflection. The target throat area at the diffuser inlet can be modified by adjusting the inlet vane angle; the outlet angle can be adjusted to achieve the target opening angle ($9° < 2\theta < 12°$). The final selection of the vane outlet angle is also influenced by the type of downstream component, such as a volute or a return channel.

12.11.3 Low-Solidity Cascade Diffuser Channels

Low-solidity diffusers (LSD) are cascade diffusers with a low exit radius ratio and a small number of vanes leading to a large space-to-chord ratio (low solidity). Low-solidity diffusers represent an intermediate design between vaneless and vaned diffusers; see Figure 12.11. The performance levels of the whole diffusing system can be close to conventional high-solidity vaned diffusers but combined with an operating range close to that of vaneless diffusers (Senoo et al., 1983). Their great advantage is that in very low-solidity diffusers, there is no overlap of the vanes and no geometrical throat between the vanes; see Figure 12.11. This increases the operating range on the choke side of the characteristic. However, as shown by Tamaki (2019), choke does occur as a virtual throat is formed between the pressure-side trailing edge and the wake from the adjacent suction-side vane, as shown in Figure 12.23. In situations with only a short radial extent and no overlap, straight vanes are sometimes used, but short circular arc vanes are also common (Hayami et al., 1990; Kmecl et al., 1999).

Figure 12.24 shows an extension the Reneau chart of Figure 12.22 to include test data on low-solidity cascade diffusers. The performance data of the low-solidity diffusers from Reddy et al. (2004) and Engeda (2001) are shown as squares. The geometry has been taken from the guidelines of Grönman et al. (2013) and from Kim

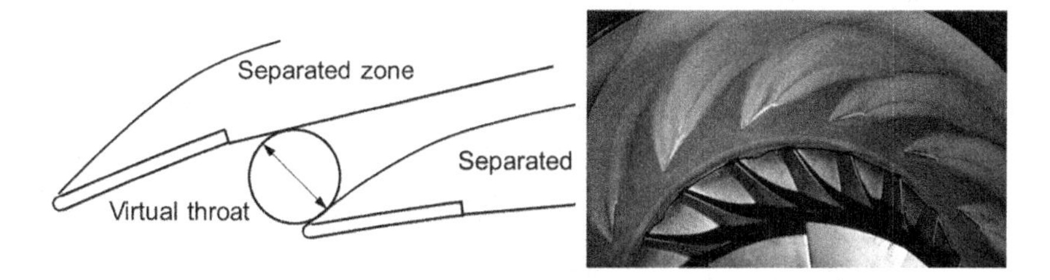

Figure 12.23 Flow visualisation and sketch of the virtual throat between the leading edge and the separation streamlines of a low-solidity diffuser at choke. (image adapted with permission from Hideaki Tamaki and IHI)

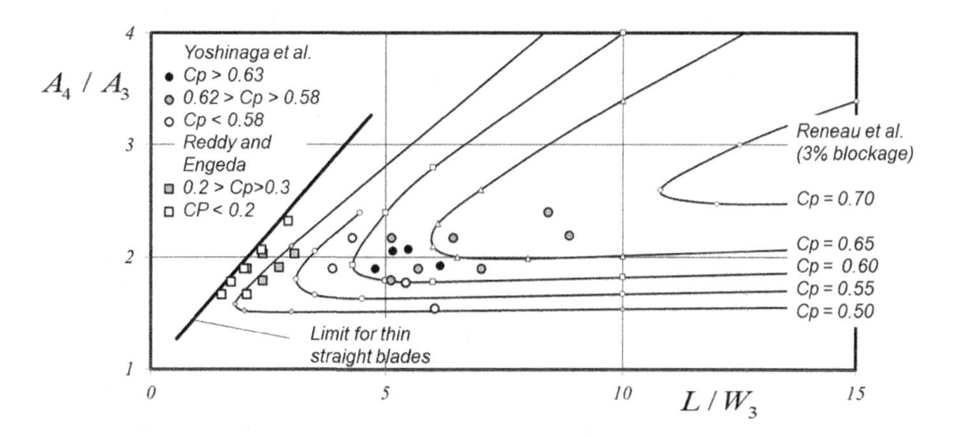

Figure 12.24 The Reneau chart for cascade and low-solidity diffusers.

et al. (2002), and the area ratio and diffusion length are based on 15 vanes, an inlet radius ratio of 1.1 and an outlet radius ratio of 1.25. The typical measured pressure recovery coefficient of these short cascades was between 0.15 and 0.3, which is consistent with their poor position in the Reneau chart. This indicates that low-solidity diffusers rely on an effective vaneless diffuser downstream of the vanes to produce a high overall pressure recovery.

In the experimental results given by Hayami et al. (1990), the outlet of the low-solidity vaned diffuser was at a radius ratio of approximately 1.25 and the downstream vaneless diffuser extended to a radius ratio of 2. The impeller flow outlet angle near to peak efficiency was about 70°. Under these conditions, a long, narrow vaneless diffuser would be expected to have high losses and to suffer from instability. In this case, a diffuser pinch of 50% was needed to stabilise the vaneless diffuser, and with this narrow channel the efficiency was poor. The low-solidity vaned diffuser with 11 vanes and no pinch achieved a higher overall pressure coefficient and improved the efficiency compared with the vaneless diffuser by 4–5 points, probably through lower losses from the higher flow angle in the vaneless diffuser downstream of the short vanes.

12.12 Pipe Diffusers

A further type of vaned diffuser is the pipe diffuser. This has circular holes arranged in a ring fitted in the diffuser annulus, each of which is followed by a straight conical flow channel. These openings are simple to make as a diffuser ring downstream of the impeller only needs to have sloped circular holes drilled in it to make the diffuser throat. The holes may be tapered or be followed downstream by discrete diffuser channels. Their initial use was probably related to their ease of manufacture (Reeves, 1977). The drilled holes are all tangent to the same tangency circle and lie in a radial plane. The drillings mutually intersect and a vaneless space is formed between the tangency circle and an outer circle called the leading-edge circle. This results in an array of symmetrically located elliptic ridges at the leading edges of the vanes which are strongly swept back.

Pipe diffusers are most often used for very large pressure ratios in aero gas turbines. The individual throats of the pipes can be arranged as separate discrete conical diffuser channels. These channels lead to channels or bends known as fish-tails. These provide a convenient way to distribute the flow from the individual diffuser channels to the combustion cans in a gas turbine engine. The ridges at the inlet generate vortices and the mixing caused by these are believed to improve the performance of the downstream channel. The conical diffusers have a larger hydraulic diameter than typical channel diffusers and have no internal corners, both of which features may contribute to reduced losses.

No data are given here but 1D Reneau charts can be produced for conical diffusers to support the design of pipe diffusers. However, the view that a discrete passage diffuser is a long, narrow channel with a 1D flow has limited utility owing to strong three-dimensional effects emanating from the inlet region. The pipe diffuser inlet region includes the vaneless space, the semivaneless space and a region known as the pseudovaneless space. The leading-edge region has a complex geometry and sharp cusplike edges formed by the intersection of the adjacent diffuser passages. The vane angle can be chosen to give a certain incidence; also, the throat area can be manipulated by making the cross section oval or rectangular to ensure optimum matching with a convenient number of pipes and avoiding a step change in meridional width between impeller and diffuser (Bennett et al., 2000). The pipe is a conical diffuser with its own length-to-width ratio and area ratio, and a section through this identifies that this is rather similar to a wedge diffuser.

Experiments show that two counterrotating vortices are generated in the diffuser inlet region, which continue into the diffuser channel and still exist downstream of the diffuser passage (Cukurel et al., 2008). Numerical analysis demonstrates that these vortices dominate the flow structure in the whole diffuser passages by shifting flows to certain locations. Near the front and back walls, the vortices shift high-momentum fluid from the suction side to the pressure side and suppress flow separation. As a result, however, the high-momentum fluid is not available in the central locations and may induce separation (Sun et al., 2017). Although pipe diffusers are used in the centrifugal compressor industry, they are normally found in stages of very high pressure ratio where the diffuser inlet Mach number is likely to be supersonic. At lower pressure ratios with lower diffuser inlet Mach numbers, pipes have not been so widely used, probably indicating that there is no significant performance advantage over the more conventional vanes discussed in this chapter. There are little data about design and performance in the literature, as most information is proprietary to specific companies. Some informative details can be found in the work of Grates et al. (2014) and other papers cited there.

12.13 Downstream Semivaneless Space and Vaneless Space

These zones are less important in conventional diffusers due to the diffusion already achieved by the time the flow reaches this point. In a conventional high-solidity diffuser, the downstream semivaneless space is simply a short extension of the vaned diffuser channel and does not have much influence on the pressure recovery. The main relevance of the downstream semivaneless space is in helping to set up the flow angle downstream of the diffuser, as the exit flow angle mainly depends on the geometry of this region. In stages with volutes, the volute losses are determined by the meridional velocity: the volute is designed to capture the swirl component and the radial velocities are dissipated, as discussed in Section 13.6. In these cases, it is less important to reduce the outlet flow angle. For a stage with a return channel or a downstream deswirl vane, however, this region can be used to further remove the swirl and thus reduce the turning required in the downstream component.

The downstream vaneless space is similar to the upstream vaneless space, and the physics of the vaneless diffuser applies, the pressure recovery mainly being determined by the radius ratio. At the low flow angle expected downstream of the vanes, the flow angle itself does not have a major influence on pressure recovery. In conventional vaned diffusers, this region is relatively short and makes no substantial contribution. In low-solidity diffusers, however, this region can be quite large, and the downstream vaneless diffuser pressure rise becomes relevant to the overall performance.

12.14 Special Cases

A number of unconventional vaned and vaneless diffuser designs can also be found in the literature. These special diffusers either attempt to account for the nonuniform angle distribution entering the diffuser from the impeller, or attempt to increase the mixing in the inlet region by modelling the complex inlet geometry of pipe diffusers. The problem with these ideas is that it is usually not clear from the publications whether some other problem in the design (inappropriate impeller design, mismatch of choke flows, false incidence, position of leading-edge shock, etc.) was not being corrected by these changes. As a result, it is unclear whether such approaches are universally applicable.

In the first category, there are diffusers with three-dimensional vane shapes that are modified to adapt the vane geometry to match the spanwise flow angle from the impeller. These include vanes with leading-edge recamber of the casing section (Jansen and Rautenberg, 1982) and with diffuser vanes with different stagger angles on the hub and casing section (Abdelhawab and Gerber, 2008). Some vaned diffusers have leading edges inclined in the meridional plane, which gives an angle variation with an otherwise axially constant blade profile. There are also diffusers with short part-span low-solidity vanes projecting from either the hub or shroud wall or on both walls (Issac et al., 2004; Anish and Sitaram, 2009). The objective of diffusers with leading-edge recamber is to align the blade leading edge to the flow direction so that generally the blade angle is higher on the casing side of the diffuser channel. The limitations of such modifications are that the whole pressure field in the diffuser inlet determines the flow pattern and the results of such geometrical modification are not always effective.

In the second category, there are vanes with scalloped leading edges, rectangular or v-shaped notches in the leading-edge generating vortices in the leading-edge region to enhance the mixing in this region, for example (Yoshinaga et al., 1980). These studies appear to show that leading-edge modifications can improve the efficiency of the centrifugal compression stages, but it remains unclear whether these modifications are always useful and universally applicable. The inclined leading edges do seem to be useful in reducing the interaction effects between impeller and diffuser and result in lower noise, but the overwhelming impression is that the most important aspect of vaned diffuser design for moderate to high pressure ratio stages is to achieve a good match on throat area between the impeller and the diffuser.

In vaneless diffusers, low-solidity vanes at the outlet can make the vaneless diffuser more stable at low flow (Abdel-Hamid, 1987). A vaneless diffuser with a rotating wall as an extension of the impeller hub or shroud, much like an impeller with a cutback trailing edge, has also been found to be helpful in some cases; see, for example, Sapiro (1983) and Seralathan and Chowdhury (2013).

13 Casing Component Design

13.1 Overview

13.1.1 Introduction

The objective of this chapter is to describe the essential design aspects of the stationary components within the casing – which guide the flow upstream of the impeller and downstream of the diffuser. Typically, the gas velocities in any pipes attached to a compressor are in the order of 20 m/s so that a high acceleration is present in the intake and a strong deceleration in the outlet. As the stationary components are integral parts of the compressor casing, a short description is provided of different casing configurations. The first of these components is the inlet nozzle, or suction pipe, upstream of the impeller. The function of the inlet nozzle is to accelerate the fluid from the low velocities at the compressor flange to the impeller inlet, keeping losses at a low level and distortion in the impeller velocity profile to a minimum. There is no energy transfer and the total enthalpy remains constant, but losses occur. The inlet duct to the impeller may be axial or radial, and it may be fitted with inlet guide vanes to change the swirl velocity at the inducer inlet to provide a way of controlling performance.

Downstream of the diffuser, there are different methods by which components can collect the flow and take it either to the outlet flange, a downstream component or an intermediate cooler. The scroll, or volute, is a component which wraps around the compressor and collects the flow leaving the diffuser of a single stage – or the last stage in a section of a multistage machine – bringing it to the outlet flange. In many designs, a further reduction of kinetic energy is carried out downstream of the scroll by means of a conical outlet diffuser. In multistage inline compressor applications, the scroll is replaced by a crossover bend and a vaned return channel to remove the swirl component of velocity and lead the flow to the impeller inlet of the next stage. In gas turbine applications, there are often axial exit guide vanes to remove the residual swirl at the diffuser outlet and to diffuse the flow to the low Mach number levels needed in the combustor. Special applications may also include stator components for the addition of a side-stream flow to the main core flow through the compressor, for gas offtakes or for a secondary inlet and outlet nozzle, allowing the flow to gain access to a cooler and be returned to the compressor.

The narrow side spaces between the rotating impeller disc and the stationary compressor casing includes sealing systems, typically labyrinth seals, to reduce the leakage flow. The leakage flow has a complex swirling nature, and this is important in determining the axial thrust of the compressor.

13.1.2 Learning Objectives

- Be aware of the applications of different casing configurations.
- Gain insight into the aerodynamic features of inlet nozzles, volutes, crossover bends, return channels and deswirl vanes.
- Understand the design and off-design performance of volutes.
- Have a knowledge of the flow in the disc cavities and their influence on the axial thrust.

13.2 Casing and Rotor Configurations

The configuration of the casing is important as it determines the form that the inlet and outlet nozzles can take. Single-stage centrifugal compressors with an overhung impeller, typical of turbochargers and refrigeration systems usually have axial inlets, and these may also be fitted with inlet guide vanes. In some applications, two-stage compressors are also overhung with the first stage having an axial inlet. In some multistage inline machines, the first stage is also overhung outside the bearing span and may then also have an axial inlet; see Figure 1.15. In integral geared compressors, individual stages are like single-stage compressors located at both ends of multiple pinions and driven from a central bull gear; see Figure 1.14

In multistage compressors, the number of impellers in a casing is usually limited by rotordynamic considerations, as discussed in Section 19.12. Therefore, the maximum aerodynamic work that can be generated in a single casing is limited. If more head is required, multiple casings that are driven by the same driver or by separate drivers have to be used. Another limitation for the pressure rise may be due to the temperature limitations of the compressor materials, whereby the discharge temperatures are often limited to around 175°C. If more pressure rise is required, the gas has to be cooled with intermediate coolers during the compression process, which has thermodynamic performance benefits as it approaches more closely the minimum work of an isothermal process; see Section 4.7.4.

The rotor of a multistage compressor consists of a shaft, a balancing drum and thrust collar and impellers. The rotor can be mounted from a single bearing for single- or two-stage machines or between bearings for a multistage machine, when it is known as a beam-style rotor. A compressor train may comprise up to three compressor casings powered by the same driver, possibly with a gearbox situated between the driver and the compressor train or between two of the compressors. Individual casings may contain multiple sections, each with its own suction and discharge nozzle. Figure 1.9 shows an inline compressor with nine impellers in a single section. Figure 13.1 shows a back-to-back configuration with nine impellers in two sections, each of which has its own radial inlet at either end of the compressor and discharge

Figure 13.1 Back-to-back compressor with nine stages in two sections. (image courtesy of Siemens Energy)

nozzles in the middle of the compressor. The impellers of the two sections are on the same shaft, with the impellers of the first section facing in the opposite direction to those of the second section; this assists in the task of thrust balancing; see Section 13.9. The eyes of the impellers of each section are generally oriented towards the shaft ends of the casings.

The impellers and thrust-balancing drum are shrunk-fit or pinned to the shaft. Sleeves may be placed between the impellers to prevent the shaft being directly exposed to certain gases. The balancing drum, which balances the axial thrust of each impeller, is designed to minimise the residual axial thrust in all working conditions. The residual thrust is usually borne by tilting pad thrust bearings. The shaft end seal is covered with a sleeve to protect the shaft from damage. A vibration sensor mounting section may be machined with the journal bearing.

The aerodynamic design of the suction and discharge nozzles and other internal stationary components is determined by the design of the casing. Diaphragms are placed in the casing of multistage machines to separate the flow between adjacent stages and to provide the inner and outer wall of the flow channel of the diffuser and return channel. The pressure difference across the diaphragm is generally small. In back-to-back designs, as shown in Figure 13.1, the pressure difference is larger across the intermediate diaphragm between the two sections and requires adequate sealing with the shaft.

For many applications, there is a choice between horizontally or vertically split casings. Stages with overhung impellers are generally vertically split, as in Figure 1.8. Horizontally split casings are used in multistage machines and comprise an upper and a lower part. These are bolted together with through-bolts at the split line, which is parallel to the axis. The impellers are mounted on the rotor between the bearings and, as a result, it is not possible to include an axial inlet to the first stage; see Figure 1.23. The main advantage of horizontally split configurations is their larger volume flow capacity and the ease of maintenance as the upper casing can be lifted vertically. The removable top casing is usually kept free of any pipe connections so that the pipework does not need to be disassembled to remove the casing. If the compressor is mounted on a table foundation, the suction nozzle is usually in the lower part of the casing. In some circumstances, the nozzle may be horizontal, in which case a tangential inlet nozzle is used.

A vertically split compressor, also known as a barrel compressor, is used for high-pressure applications and consists of an inner and an outer casing, the latter acting as a pressure vessel. The inner casing is horizontally split with diaphragms, stationary seals, rotating elements and shaft end seals. This module can be inserted or removed axially from the outer casing as a single piece, which simplifies compressor assembly and reduces turnaround times for maintenance and inspection, as shown in Figure 1.24. The inlet nozzle can be attached to the casing at any desired angle to the top, bottom or side of the casing in accordance with client specifications.

13.3 Inlet or Suction Nozzle

The inlet section of a centrifugal compressor is designed to lead the flow with low losses and with minimal distortion from the inlet flange to the annular channel at the eye of the first impeller. With an axial inlet flow, the only cause of disturbance to the inlet flow may occur in the pipework upstream of the inlet nozzle through the generation of secondary flows in the pipe bends, as documented by Shen et al. (2016), who computed the full piping system upstream of a single-stage overhung refrigeration impeller.

Multistage centrifugal compressors are normally equipped with symmetric radial suction nozzles or asymmetric tangential nozzles, which guide the flow from the inlet nozzle to the annular channel leading to the inlet eye of the impeller (Lüdtke, 2004). Multistage centrifugal pumps have similar inlet nozzle designs; examples of different configurations are given by van den Braembussche (2006a) and Gülich (2008). The objective of a radial inlet for a compressor is to minimise the total pressure loss across the inlet while distributing the flow as uniformly as possible compatible with minimum distortion to the eye of the first impeller. After passing through the circular area of the inlet nozzle, the flow is usually divided by an intake rib and then flows around the shaft before being diverted into the axial direction. The casing can either be annular with a similar flow area at different azimuthal positions or be shaped like a volute with a decrease in area away from the inlet nozzle. The inlet may also include baffle plates and vanes to reduce the risk of swirl being generated ahead of the impeller inlet eye. There may also be turning plates to turn the flow to the axial direction. The uniformity of the flow is improved when the eye is equipped with a bellmouth or a lip, both of which force the flow to enter the impeller eye more axially.

The flow patterns in such radial inlets are not axially symmetric, and the complex three-dimensional flow fields can cause large distortions in the inlet flow to the impeller. In earlier studies of such inlet casings, scale models were built to examine the flow field and adjust the geometry to guide the flow with minimum flow distortion. More recent studies rely on CFD rather than model tests, as documented by Flathers et al. (1996), Michelassi and Giachi (1997), Kim and Koch (2004) and Grimaldi and Michelassi (2018).

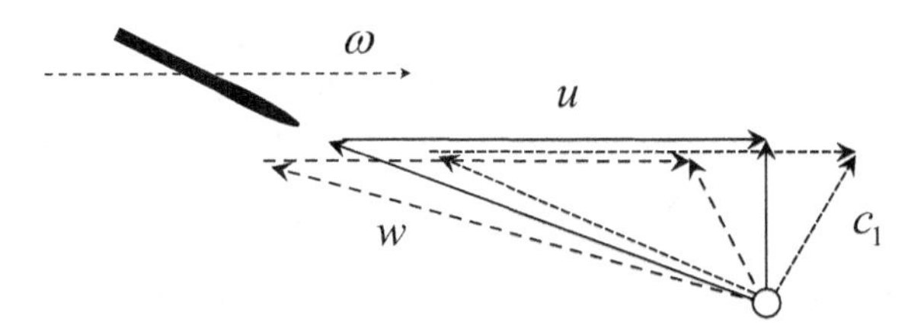

Figure 13.2 Effect of inlet distortion of the impeller inlet velocity triangle.

The studies show that even with flow straighteners, the flow delivered to alternative sides of the impeller inlet may have up to ±30° swirl in the absolute frame of reference. In the relative frame with a mean blade inlet angle of 60°, this value reduces to ±10°. The two sides of the impeller have to deal with intermittent disturbances with different incidence of up to ±10° and variation in the relative velocity magnitude; see Figure 13.2. This difference in the flow angle between the two sides reduces if there are inlet guide vanes and is not present at all in the stages following a well-designed return channel. Grimaldi and Michelassi (2018) show that this inlet distortion may account for both an increase in the impeller work coefficient of up to 3% and a similar drop in efficiency compared with that of a stage with no inlet distortion downstream of an optimum return channel.

13.4 Intermediate Inlet Nozzles

13.4.1 Inlet Volutes in Multistage Compressors

In inline multistage compressors with intercooling, the flow reenters the compressor after passing through a cooler. A typical configuration for this inlet flow nozzle is a reverse volute leading the flow with swirl to a series of inlet guide vanes which are designed to remove the swirl before it reaches the eye of the next impeller. The design procedure for such inlet volutes follows that given in Section 13.6.1 for outlet volutes.

13.4.2 Side-Stream Inlet

Radial turbocompressors in certain applications have one or more incoming side-streams (or sideloads) that mix with the core flow already in the compressor, as in Figure 1.17. Side-stream compressors are important in refrigeration applications and in the processing of heavy hydrocarbons and liquefied natural gas. The casing is designed to introduce an additional side-stream flow to the inlet of an internal stage of a multistage compressor. Depending on the compressor application, the side-stream flow may have up to three times as much mass flow as the core flow already in the compressor, but is usually much less than this.

In a vapour compression refrigeration process, using a two-stage centrifugal compressor with an economiser, the first- and the second-stage compressor impellers are on the same shaft but have different mass flow. The evaporated refrigerant gas at the low temperature of the cycle is compressed by the first stage to an intermediate pressure level. The liquid flow from the condenser passes through an expansion valve to the same intermediate pressure level; here some of the liquid evaporates and is known as flash gas. The liquid is further expanded to the evaporator pressure and, after evaporation, to the inlet of the first stage. The evaporated flash gas at an intermediate pressure enters the inlet of the second stage, where it mixes with the core flow and is compressed to the outlet pressure of the condenser. More complex processes in ethylene and polypropylene compressors may have up to three separate side-stream inlets in a single casing.

The design and operation of compressors with side-streams is challenging because flow entering the compressor needs to be mixed with the core flow in a manner that does not significantly affect the aerodynamic performance of the upstream and downstream stages. In such applications, the process typically dictates the side-stream flange pressure and the amount of side-stream flow, which may vary with operational conditions. The stages upstream of the side-stream inlet need to achieve the necessary pressure for the core flow. In addition, the downstream stage has to swallow the core flow with the side-stream flow, which is usually at a temperature different from the core flow. The strong secondary flows in impellers and the strong circumferential mixing at impeller outlet allow the impellers to cope with this temperature gradient.

There are several approaches to the design of the side-stream inlet depending on where the flow joins the mainstream. The side-stream may be configured as a tangential inlet volute delivering a swirling flow to the inlet of the return channel vanes from an upstream stage. The side-stream may enter in the side of the return channel within the return channel blading. Alternatively, the inlet may be an axisymmetric radial inlet removing the swirl completely and injecting the flow into the core flow downstream of the return channel vanes, either in the radial channel or, if space allows, axially upstream of the next impeller. The tangential inlet is more compact with a shorter axial length. It provides a longer flow path for mixing the side-stream and the core flow before the flow reaches the impeller leading edge. In addition, the flow enters at a point where the curvature is low so that issues related to the pressure gradients across the flow channel are reduced (Hardin, 2002; Koch et al., 2011).

13.5 Inlet Guide Vanes

Variable inlet guide vanes (VIGVs) may be placed in a radial inlet or in an axial inlet. These vanes modify the inlet velocity to the impeller by adding a circumferential swirl component to the throughflow velocity. With inlet swirl, the flow coefficient of the impeller at peak efficiency changes in order to retain similar levels of relative flow angle (incidence) on to the impeller leading edge, as discussed in Chapter 17. The efficiency of the stage varies in line with changes in the degree of reaction, the impeller inlet relative Mach number, the diffusion in the impeller and the matching between the impeller and the diffuser. At extreme closed settings, the guide vanes may also be used as a sort of inlet throttle valve. VIGVs enable a lower starting torque, which is particularly useful for a refrigeration compressor. Without this, a larger motor frame size may well be required. The VIGVs also bring the key benefit of a wider turndown range for nearly constant pressure service. Many applications require a turndown to below half of the nominal flow at ideally constant pressure rise (see Section 17.3.1), and VIGVs are a key component in this, supplemented where possible by variable speed.

The work input coefficient curves of an impeller with inlet guide vanes are not only moved to a different flow coefficient by the VIGVs, but also have a different slope. This is due to the effect of the inlet swirl on the gradient of the work input curve, which is similar to the effect of an increase in the backsweep angle. Owing to the steeper slope of the work coefficient, the stage is generally more stable at increased inlet swirl, which in turn improves the surge point on the characteristic. Section 17.5.4 describes the effect of inlet guide vanes on the velocity triangles and the performance map of stages with IGV control.

13.6 Outlet Volute

13.6.1 Design of the Scroll or Volute

The function of the volute or scroll, shown in Figure 13.3, is to collect the flow from the exit of the diffuser and lead it to the outlet pipe of the compressor with minimum impact on the flow in the compressor. Ideally there should be a uniform static pressure distribution at diffuser exit close to the nominal point and no strong once-per-revolution disturbance from any circumferential pressure gradients or from the tongue. The flange is typically offset from the axis of the machine and the flow is tangential to the rotational direction of the impeller.

In the design of the volute, the effect of friction is normally neglected, the volute area schedule is designed to have no pressure gradient in the circumferential direction and it is assumed that there is no flow disturbance from the tongue. Under these

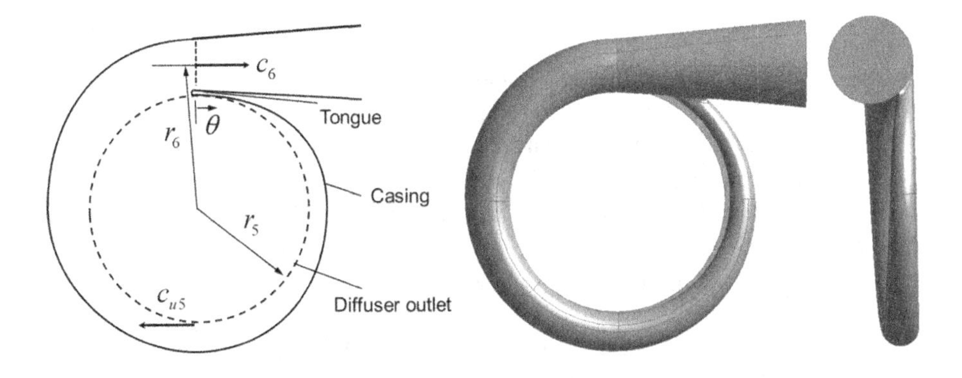

Figure 13.3 Left: a schematic view of the volute, Right: 3D views of the gas-swept surfaces of a volute.

approximations, the moment of momentum remains constant in the volute and the Euler equation gives the following relationship between the conditions at the diffuser outlet (denoted by state 5) and those at the volute throat (just downstream of the tongue at 360° and denoted by state 6):

$$c_{u5}r_5 = c_6 r_6, \quad c_6 = c_{u5}r_5/r_6. \tag{13.1}$$

The equation of continuity between these positions becomes

$$c_6 A_6 = 2\pi r_5 b_5 c_{m5}, \quad c_6 = 2\pi r_5 b_5 c_{m5}/A_6. \tag{13.2}$$

Combination of these equations shows that the area at the volute throat is given by the following:

$$A_6 = 2\pi r_6 b_5 c_{m5}/c_{u5} = 2\pi r_6 b_5/\tan\alpha_5. \tag{13.3}$$

This equation can also be applied piecewise around the volute to determine the local area for the given azimuth angle. However, given the simplified level of the analysis, the equation is generally used to determine the volute throat area at outlet, and then a linear variation in area around the volute is used. The area around the circumference may also be increased above this value to account for boundary layer growth.

Equation (13.3) can be rearranged as

$$A_6/r_6 = 2\pi b_5/\tan\alpha_5. \tag{13.4}$$

The parameter A/r relates the cross-sectional area at the exit of the volute divided by the radius from the turbomachine centre line to the centre of that area, so that this has units of length. The value of A/r is often used to denote the size of a given volute, as (13.4) makes it clear that the A/r ratio at the volute throat describes a specific geometric characteristic of the compressor housing. At the design point of the volute, A/r is defined by the diffuser outlet width and the flow angle at the diffuser outlet.

For a vaned diffuser with a given flow outlet angle, the design value of A/r is then related to the width of the diffuser. Thus, impellers with different outlet width, or different trims, can be matched to the appropriate volute if a standard series of components is used. When operating with a vaned diffuser, the inlet angle to the volute remains fairly constant for all operating points and the low flow angle at the outlet of a vaned diffuser leads to a low kinetic energy at the volute inlet. With a vaneless diffuser, the flow angle is larger and the volute has a higher inlet kinetic energy at inlet and is only matched to the exit flow angle of the vaneless diffuser at one operating condition. The difference in flow angle at diffuser exit between vaned and vaneless diffusers means the volutes are quite different in size for a given impeller.

13.6.2 Shape of the Volute Cross Sections

Different configurations for the shape of the volute cross sections are commonly used. In an external volute, the entire volute is positioned radially outside of the diffuser, and the increase in volute cross-sectional area with circumference leads to an increase in the outer diameter. To reduce the outer dimension, an internal volute – where a part of volute is at a lower radius than the exit of the diffuser – may be used. The increase in cross-sectional area involves a reduction in the inner radius, leading to an overhung design; in this case, a smaller throat area will be needed from (13.4). Both of these styles are commonly used, though the external type is preferred unless space is an issue. It is also possible that in some situations the volute may be designed with a constant outer radius, as this dimension may be part of a circular component of the casing. In addition, the volute geometry may be either symmetric to the midspan of the diffuser exit or to one side of the diffuser midplane.

The volute cross sections themselves vary in shape. A circular volute cross section provides the smallest wall area per unit cross-sectional area and so generates the least friction loss. This means that the volute will typically have a circular cross section. Rectangular, elliptical and triangular cross sections are also possible, as shown in Figures 1.9 and 13.1. A critical aspect of volute design is the proper generation of the shape of the tongue. Pressure disturbances from the tongue can have an impact on the flow field upstream of the impeller and influence both performance and range. A well-designed tongue with small relative thickness and low incidence will provide for the lowest losses and smallest disturbance.

The final part of the volute to be considered is the conical diffuser, which is downstream from the volute outlet to the pipe flange, as shown in Figures 1.8, 1.13 and 1.15. There can be a sizeable area ratio for this section, and performance can be lost if this conical diffuser is not long enough for good diffusion. Reneau charts suggesting expected pressure recovery of a conical diffuser for a given area ratio and length can be utilised to assess this component; see Section 7.4. The effective cone angle for this final diffusing component is often chosen to be about 7° based on Reneau charts, but the swirling flow means much higher angles such as 10°–12° can be used and have been successfully applied (Lüdtke, 2004).

13.6.3 Design and Off-Design Performance

At the design flow, the volute collects the tangential component of the velocity, which is decelerated with increasing radius as in a radial vaneless diffuser. The radial component of the velocity is not converted to pressure rise but is retained as a swirling flow in the volute and the outlet pipe, as shown in Figure 13.4. The vortex pattern associated with this swirl depends on the location of the volute with respect to the diffuser outlet. In a symmetrical volute, two counterrotating vortices are formed and in an asymmetric volute a single vortex is present. This swirling flow represents a source of leaving loss for this component, as the swirl is not converted into pressure rise. The loss coefficient at the design point related to this is lower with a high inlet flow angle than with a low inlet flow angle, as follows:

$$\zeta = c_{m5}^2/c_5^2 = \cos^2\alpha_5. \tag{13.5}$$

The typical absolute flow angle at the diffuser outlet for a vaneless diffuser is approximately 65°, which means that the leaving loss coefficient due to the dissipation of the radial component of velocity at the diffuser outlet is 18% of the available kinetic

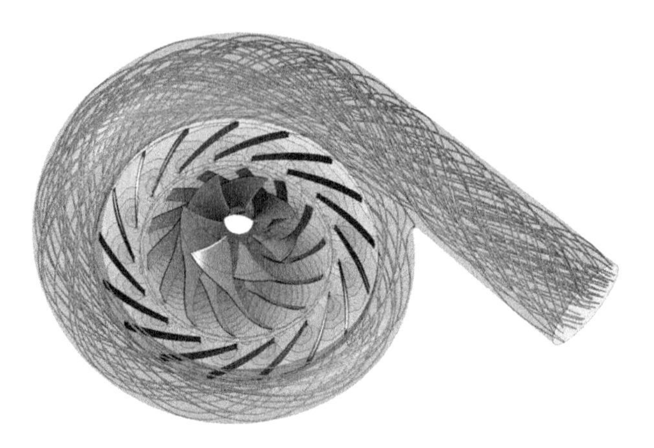

Figure 13.4 Streamlines in a volute at design from a CFD simulation.

energy (as $\cos^2 65 = 0.178$). With a vaned diffuser, a lower flow angle at diffuser outlet of 50° with a higher meridional velocity component can be expected such that the volute leaving loss coefficient is higher at 41% (as $\cos^2 50 = 0.413$), but the actual velocities and the kinetic energy are lower. This loss of kinetic energy is the most important source of loss at the design point of the volute and outweighs losses due to friction and 3D swirling flows and any deceleration in the volute, Hazby et al. (2020).

If the geometry of the volute throat and the diffuser exit is known, then the design flow angle for the volute is given by

$$\tan \alpha_d = 2\pi b_5 r_6 / A_6 \tag{13.6}$$

with the result that a volute with a known *A/r* value has a different design inlet flow angle depending on the width of the diffuser which precedes it. Experiments from many sources show that the minimum loss of the volute occurs close to the design inlet flow angle (Sherstyuk and Kosmin, 1969; Dalbert et al., 1988). The experiments show that the volute losses have the form of a loss bucket with the variation of the inlet flow angle. When operating with an inlet angle higher than the design angle, the meridional velocity is low and kinetic energy loss decreases, but the flow in the volute decelerates in the circumferential direction to the throat, which causes additional losses. When operating at low flow angles, the kinetic energy loss increases from the high meridional velocity, Hazby et al. (2020).

An example of the loss coefficients of a volute tested with a compressor stage having a vaneless diffuser of different outlet widths is given by Sherstyuk and Kosmin (1969). This volute has a circular cross section and a linear distribution of area circumferentially and was tested with six different diffuser widths. The measured loss coefficients of the volute are shown in Figure 13.5 as a function of the diffuser outlet flow angle for different diffuser outlet widths. This figure also shows the predictions of an unpublished loss model for volutes, based on that of Weber and Koronowski (1986). The volute itself has a fixed *A/r* ratio, and for each diffuser outlet width, (13.6) defines a different design flow angle for the volute, ranging from 45° to 59.5°.

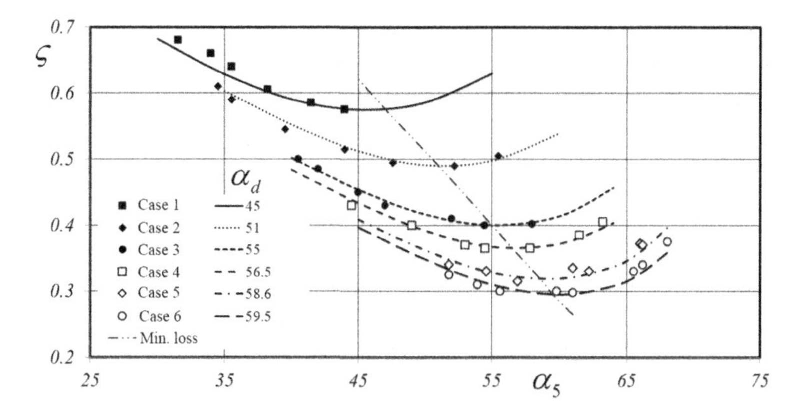

Figure 13.5 Loss coefficient variation with inlet flow angle of a single volute tested with a diffuser with six different outlet widths.

The experiments show that the variation of the volute losses with the inlet flow angle are like a loss bucket and that the minimum loss of each loss bucket is at or near the associated design flow angle, α_d. However, the minimum loss coefficient increases as the design flow angle decreases due to the larger loss of meridional kinetic energy at the design point, so that there is not a unique performance curve with a change in the inlet flow angle. From this, it can be seen that the loss coefficient of a volute can double when the volute is used with a diffuser leading to a low design flow angle. At the design flow angle, other sources of loss related to the nature of the swirling flow and to the frictional losses are much smaller than the kinetic energy loss in the swirling flow.

The absolute size of the volute scroll, in terms of *A/r* value, can have an effect on the performance map of the compressor. The losses that occur in the volute depend on the kinetic energy at the outlet of the diffuser. This is higher with a vaneless diffuser than with a vaned diffuser. The sensitivity of the stage total–total efficiency to a change in the volute loss coefficient can be derived as

$$\delta\eta_{stage}^{tt} = -\delta\zeta\left[\frac{(c_5/u_2)^2}{2\lambda}\right]. \tag{13.7}$$

The term in the square brackets is the ratio of kinetic energy at the volute inlet to the work input of the stage. This ratio tends to increase with the design flow coefficient and may reach 0.1 in high flow coefficient stages with a short vaneless diffuser.

If an undersized volute is used with a vaneless diffuser, the change in efficiency causes the performance curve to shift towards lower flows. If the volute is oversized, the performance curve is shifted towards higher flows. Sometimes this shift is made intentionally in order to fine-tune a machine to a desired flow range. In many applications, the volute casing has fixed dimensions even when operated with upstream stages of different outlet widths. In such cases, the volute is not perfectly matched to the upstream components. Historically, with vaned diffuser stages, particularly when the same volute may be used along with a series of trimmed stages, the volute was assessed using the concept of diffuser fractional recovery. The idea was that both the diffuser and the volute should carry a similar amount of pressure recovery:

$$\begin{aligned} \textit{Diffuser fractional recovery} &= \frac{\textit{diffuser area ratio}}{\textit{diffuser system area ratio}} \\ &= \frac{(\textit{diffuser exit area})/(\textit{diffuser inlet throat area})}{(\textit{volute passage area at the tongue})/(\textit{diffuser inlet throat area})} \\ &= \frac{\textit{diffuser exit area}}{\textit{volute passage area at the tongue}}. \end{aligned} \tag{13.8}$$

For the largest size in the family, the diffuser exit area should be significantly lower than the volute exit area. Conversely, when the volute is operating with the smallest diffuser for the smallest trimmed stage, the area ratio from diffuser exit to volute tongue should not be too high to avoid reaccelerating the flow and damaging the total–static pressure rise.

13.6.4 Volute Pressure Distortion

At low flows with high inlet angles, the volute is too large, with the result that the pressure rises with peripheral distance from the tongue. There is also risk of separation on the inner surface of the tongue. At high flows, the volute is effectively too small, and the flow accelerates around the periphery, leading to a fall in the pressure with distance from the tongue to the volute outlet. This pressure distortion can make itself felt in the inlet duct of the impeller, especially in turbocharger impellers with short vaneless diffusers.

At off-design conditions, the flow onto the volute tongue and the flow in the volute itself generates a circumferential pressure distortion around the periphery. Greitzer et al. (2004) show that if the circumferential pressure disturbance is approximated as a Fourier series, this leads to

$$p' = \sum_{k=-\infty}^{\infty} b_k e^{(2\pi|k|x/W)} e^{(2\pi iky/W)}, \tag{13.9}$$

where k is the harmonic number of an individual Fourier component, W is the circumferential spacing of the cause of the disturbance, y is the circumferential direction and x is the meridional direction. This equation can be applied to estimate the rate of decay due to a disturbance associated with a length scale equal to the blade spacing in a blade row but is also valid for disturbances with a larger length scale. This equation suggests that the largest circumferential disturbance at the location of the tongue, with $k = 1$ and with $W = 2\pi r_5$, reduces to 4% of its value at a distance of $W/2$ along the meridional direction and can therefore still be identified well upstream of the impeller inlet.

The disturbance in the symmetry of the flow may cause stall at the impeller inlet and contribute to a narrower operating range. With both vaneless and vaned diffusers, care is needed to avoid flow distortions near to the tongue, especially if for mechanical reasons the tongue is rather thick. Van den Braembussche (2019) provides more details of the types of disturbances that occur and their effect on the performance.

13.7 Return Channel System

13.7.1 Crossover Bend

In multistage inline machines, the scroll is replaced in the intermediate stages by a crossover bend to turn the flow inwards and a vaned return channel to remove the swirl and lead the flow to the inlet bend of the next stage, as shown in Figure 13.6. The crossover bend in a return channel system is a 180° bend in the meridional flow channel with a change in the mean flow angle from radially outwards to radially inwards. The flow in this bend has a typical flow angle of 65° from the meridional direction following a vaneless diffuser or 50° following a vaned diffuser, such that the swirl component of velocity is larger than the meridional component. The meridional section suggests that the flow passes a sharp bend, but thanks to the swirl the actual curvature of the streamlines seen by the flow is reduced.

CFD simulations of the flow in the bend show that there are strong secondary flows in the crossover bend often associated with flow separations. There is considerable agreement in the literature detailing flow calculations of crossover bends which demonstrate that the flow often separates on the casing wall at the inlet to the bend or at the inlet to the return channel on the hub side of the channel, where the flow also decelerates. Such simulations, however, seldom provide guidelines for the design. A simplified inviscid analysis, similar to that given in Section 5.3.5 for the effect of the impeller inlet curvature, can be used to estimate the spanwise velocity gradients expected in the crossover channel. Neglecting viscous forces and assuming there are no swirl gradients across the span, this simplified ductflow analysis leads to

$$c_m \frac{dc_m}{dq} = -\frac{c_m^2}{r_c} \quad \Rightarrow \quad \frac{dc_m}{dq} = -\frac{c_m}{r_c} \quad \Rightarrow \quad \frac{\Delta c_m}{c_m} \approx -\frac{b}{r_c}. \tag{13.10}$$

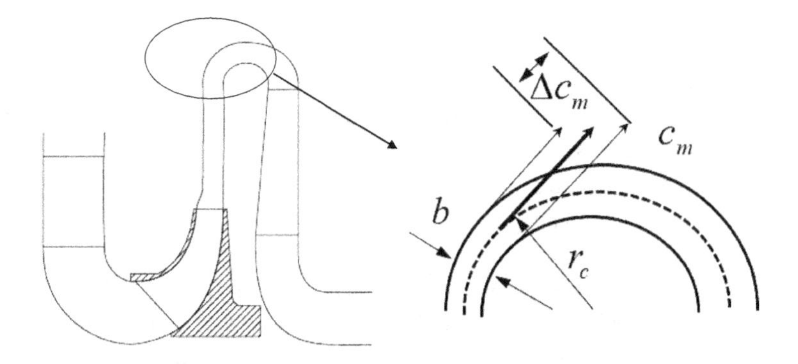

Figure 13.6 Inviscid velocity gradient in the crossover bend.

Equation (13.10) suggests that the ratio of the span to the radius of curvature is the relevant inviscid geometrical parameter for estimating the differences in inviscid hub and casing velocities in a crossover bend, as shown in Figure 13.6. The hub-to-casing meridional velocity difference is largest for wide channels with a low radius of curvature, so that for the same velocity difference across the channel a sharper bend can be considered for narrow stages.

It is clear, therefore, that higher velocities at the hub side are related to the lower pressure due to the curvature. On leaving the diffuser, the meridional velocity on the casing decelerates and is at risk of separation. Half of the velocity difference across the channel represents a decrease in the casing wall velocity, so that a de Haller number for the change in the meridional velocity at the entry region of the bend can be written as

$$c_{m2}/c_{m1} \approx 1 - b/2r_c. \tag{13.11}$$

The actual flow deceleration is less because of the strong swirl components of velocity. Assuming that the de Haller number for the meridional velocity needs to be more than 0.5 to avoid separation, the radius of curvature of a crossover bend should be larger than the width of the channel. This is possible to achieve in narrow stages. In the case of high flow coefficient impellers, this is more difficult because the axial length of the crossover bend would need to be greater. The use of a return channel in which the hub wall is conical towards the impeller, as in Figure 13.6, allows a longer crossover bend for a given length of stage and is recommended.

13.7.2 The Return Channel Outlet Width

Since a good flow distribution at impeller inlet is essential for good impeller performance, the design of the return channel in a multistage compressor requires consideration of the matching of the return channel with the downstream impeller. Any disturbance or separation in the inlet flow arising from the upstream return channel or in the 90° inlet bend to the downstream impeller may result in flow blockage at the impeller inlet. This will cause the impeller to operate below the performance measured in a test stand with an optimal inlet. To avoid such disturbances, the flow into the downstream impeller must have sufficient acceleration to avoid flow separation in the bend leading to the downstream impeller. As an approximate guideline, it is suggested that the mean velocity in the inlet bend should increase by 25–30% through the inlet bend into the downstream impeller; see Figure 13.7.

In this way, the outlet width of the return channel is determined by the inlet of the downstream stage. The problem here is that in multistage compressors, the flow coefficient of the downstream stage, and hence its inlet area, depends on the application. In the interest of a systematic standardisation of the stages, it is sensible for the return channel design to be associated with that of the upstream impeller rather than be changed for each application. Thus, the return channel outlet must be defined in such a way that it is suitable for all possible downstream stages. The alternative of modifying the return channel for each possible downstream stage

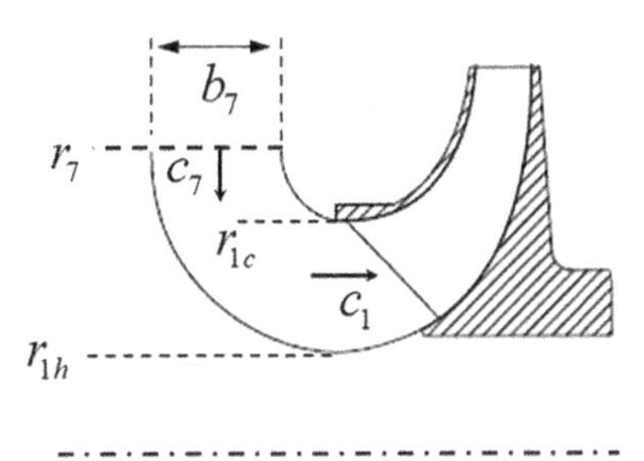

Figure 13.7 Acceleration of the meridional flow from the return channel outlet to the impeller inlet of the next stage.

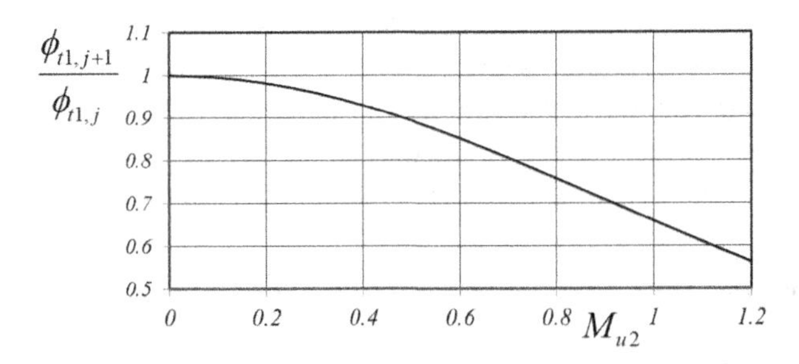

Figure 13.8 Ratio of the flow coefficient of successive stages with the tip-speed Mach number.

means that its performance may differ from that measured in a test stand with a particular return channel geometry.

If the compression is considered to take place as a polytropic process with exponent n, then the density ratio across a stage is given by

$$\Gamma = (\rho_{t2}/\rho_{t1}) = (T_{t2}/T_{t1})^{1/(n-1)} = \left(1 + (\gamma - 1)\lambda M_{u2}^2\right)^{1/(n-1)}. \tag{13.12}$$

The inlet flow coefficient of the upstream stage (suffix j) is defined as follows:

$$\phi_{t1,i} = \dot{V}/\left(D_2^2 u_2\right) = \dot{m}/\left(\rho_{t1} D_2^2 u_2\right). \tag{13.13}$$

Assuming a constant diameter for both stages, the inlet flow coefficient of the next stage (with a suffix $j + 1$) can be calculated from the density ratio as

$$\phi_{t1,i+1} = \dot{m}/\left(\rho_{t2} D_2^2 u_2\right) = (\rho_{t1}/\rho_{t2})\phi_{t1,j} = \phi_{t1,j}\left(1 + (\gamma - 1)\lambda M_{u2}^2\right)^{-1/(n-1)}. \tag{13.14}$$

This equation identifies that the nondimensional volume flow coefficient of the next stage depends on the density ratio and hence on the tip-speed Mach number, the efficiency, the work coefficient and the flow coefficient of the upstream stage.

As an example of the use of this equation, Figure 13.8 shows the ratio of flow coefficient of successive stages as a function of the tip-speed Mach number, with a work coefficient of 0.65 and a polytropic efficiency of 0.84. At a low tip-speed Mach number, there is barely any change in the density, with the result that there is no change in flow coefficient between stages. At higher tip-speed Mach numbers, the flow coefficient of the subsequent stage decreases.

The impeller inlet area varies strongly with the flow coefficient of the stage. The outlet radius of the return channel should be at a radius ratio $r_7/r_{ic} = 1.25$ compared with the casing inlet eye diameter of the next stage. This is to ensure that the inlet bend is smooth and that there are no sharp curvatures causing extreme velocities on the inner wall of the bend. An area ratio of $A_7/A_1 = 1.3$ from the return channel to the inlet eye gives a 30% mean acceleration ratio in the inlet bend. This acceleration is needed to avoid separation in the inlet bend; this could cause blockage at the inlet to the next impeller. If a series of stages are designed for a range of Mach numbers, the worst case is where the Mach number is low and there is a large downstream impeller leading to a large inlet area of the next stage. In a series of standard machines, it is necessary to design for this case to avoid too much deceleration into the inlet. Design for a high Mach number would lead to a smaller return channel outlet, which would then not be acceptable for the low-speed cases. Consequently, the return channel outlet width should usually be more than twice the outlet width of the upstream diffuser channel.

13.7.3 Schedule of Width Ratio

Section 13.7.2 shows the need for an increase in channel width across the crossover bend and the return channel. The alternatives available to the designer are either to accept a lower acceleration into the next impeller inlet or to modify the return channel for each application. The designer also needs to consider where to increase the channel width across the crossover bend and in the return channel blading. There are essentially three choices for the change in width schedule: an increase in width across the crossover bend, an increase in width across the return channel blading or a mixture of the two. Based on the discussion in Section 13.7.1, it is recommended that for high flow coefficient stages, the width of the crossover channel should be as small as possible and the radius of curvature as high as possible to reduce the velocity gradient across the channel. This can be achieved by having a constant width crossover channel and an increase in channel width across the return channel. The hub wall is then conical towards the impeller and allows a longer crossover bend for a given length of stage, as in Figure 1.23. As low flow coefficient stages with narrow channels do not need to reduce the velocity gradients, a sensible strategy is to increase the width of the channel across the crossover bend to achieve a parallel walled return channel, as in Figure 1.24. Intermediate flow coefficient stages may adopt one of the three options.

13.7.4 Return Channel Vanes

The return channel needs to remove the swirl from the flow ahead of the next stage. This requires a cascade of vanes with a high turning of typically 50°–60°; see Figure 13.9. However, as the radius decreases across the blade row, the meridional area decreases. As a consequence, the meridional velocity component accelerates, the

Figure 13.9 Radial impeller split diaphragm with return channel vanes. (image courtesy of MAN Energy Solutions)

circumferential component of velocity decelerates and the flow is characterised by low overall deceleration. A constraint on the number of vanes shown in this figure is such that no vane is cut by the split plane of the diaphragm.

The shape of the vanes and the channel determine the location of the throat in the vane. The thickness of the vanes is usually determined by the requirement to insert through-bolts to support the diaphragm between the diffuser and the return channel. A vane with initially high turning, as in a controlled diffusion aerofoil, may have a throat at the rear of the channel, and a rear-loaded vane may have the throat near the leading edge. The design of the return channel vanes should be such that the outlet angle is overturned from radial by a few degrees, typically 5°, to compensate for the deviation associated with the high gas deflection. If any swirl remains in the flow, this would act as unwanted preswirl which would lower the work input and the pressure rise of the downstream stage. Careful design of the return channel blading during the detailed design is therefore needed (Simon and Rothstein, 1983; Lenke and Simon, 1999). It may nevertheless be possible to use the same basic vane shape for a wide range of impellers so that the shape of the deswirl vane will be a common design feature for all stages across a range of flow coefficients and trims, with perhaps a separate vane shape for use with vaned and vaneless diffusers to account for the different inlet flow angles.

The leading edge of the return channel is close to the outlet of the crossover bend and may have a gradient of incidence across the vane if, as is usually the case, a 2D vane design is used. In some situations, especially in pumps, the return channel vanes are extended across the crossover bend with 3D vane shapes to avoid this problem (Gülich, 2008).

13.8 Deswirl Vanes

13.8.1 Axial Deswirl Vanes

In gas turbine applications and in some other situations where the radial extent of the diffusing system is limited, there is often a 90° bend at outlet from the radial

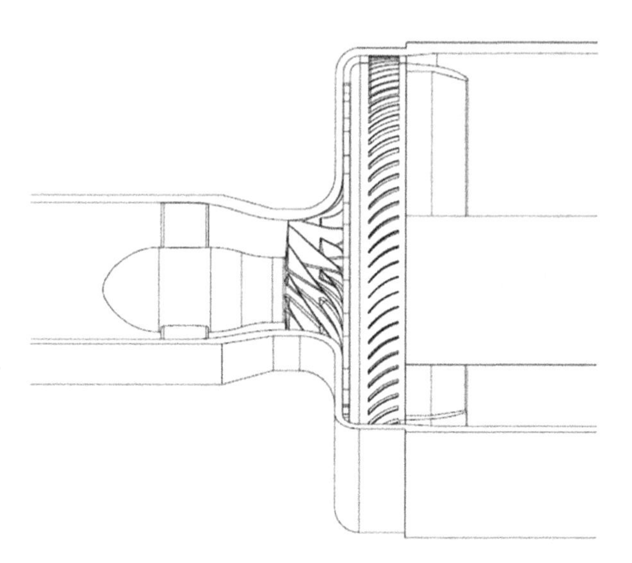

Figure 13.10 Radial stage with axial deswirl vanes.

diffuser channel returning the flow to the axial direction. Deswirl vanes are then used to remove the swirl in the axial exhaust duct and to diffuse the flow to the low velocity and swirl levels needed in the combustor; see Figure 13.10. The design of this blade row is similar to that of an axial compressor stator with large turning. If the flow leaves the vaned diffuser with a flow angle of 50°, then the de Haller number of the deswirl blade row is close to 0.65 (cos 50° = 0.643). This can be increased slightly by retaining a small amount of swirl in the flow of 15° or lowering the blade span across the blade row. In order to avoid separation, the deswirl blade row needs to be equipped with a considerable number of vanes to reduce the Lieblein diffusion factor to an acceptable level; see Section 7.5.10. As with all axial cascades, it is worth taking care over the detailed profile design. Controlled diffusion blades with forward curvature are ideal; the vanes are normally constant section and very high solidity. Because of the high deflection, a double row of vanes may sometimes be used.

13.8.2 Deswirl Vanes for Compressors with Integral Coolers

In compressors with integral coolers, the flow from the outlet of the rotationally symmetric diffuser needs to be guided to the inlet of the internal coolers and thus changed from an axisymmetric flow pattern to a linear flow pattern. Specially designed deswirl vanes are used for this purpose. Each individual flow channel has to be adapted to the local curvature and the requirement to distribute the air uniformly to the coolers. This makes severe demands on the aerodynamic design of the flow channels and the connecting manifolds. Some of the flow channels require high turning whereas others require no turning (Strub, 1974).

13.9 Axial Thrust

13.9.1 Balance of Forces

The axial thrust on the shaft results from the differential pressure across the impeller surfaces and the change in the axial momentum of the flow between the impeller inlet and outlet. The axial force due to the gas pressure acting on the impellers can be obtained by integrating the axial component of the pressure distribution on the rotor surfaces. The evaluation of the resultant force acting on each single stage is the starting point to determine the thrust. With reference to the control volume containing all the rotating walls (see Figure 13.11), this force balance can be expressed by

$$F = F_{mom} + F_{inlet} + F_{fs} + F_{bs}. \tag{13.15}$$

The term F_{fs} is the global force acting on the front shroud wall of the impeller while F_{bs} is relative to the impeller back space: both of these are evaluated considering the static pressure distribution along the rotating walls. F_{inlet} is generated by the pressure field at the impeller inlet section and is a function of the local static pressure value. F_{mom} is the momentum contribution in the axial direction and is calculated considering the mean value of inlet velocity and the mass flow. All these terms can be evaluated considering pressure levels, mass-flow conditions and the geometry of a particular stage. For each stage and at each operating point, the axial thrust can be calculated once the pressure field inside its components, including leakage cavities, is known. For a multistage compressor, the axial load can be obtained algebraically by adding the single-stage contributions. Such calculations can be done with the help of CFD simulations of the whole stage, including the secondary leakage flow paths. The challenge is to devise a procedure that will provide the resultant thrust value with moderate computational effort, dealing with the industrial requirements of accuracy and reduced time consumption.

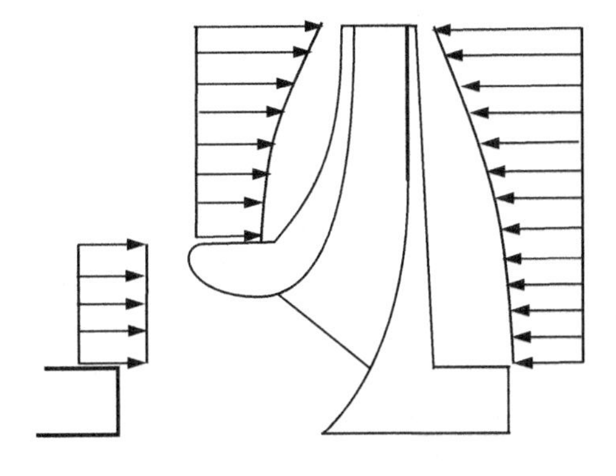

Figure 13.11 Pressure distribution acting externally to an impeller causing axial thrust.

13.9.2 Balance Piston

The combination of fluid dynamic forces acting on an impeller leads to an axial thrust which has to be taken up by a thrust bearing within the casing. In high-pressure, multistage, inline compressors, the cumulative thrust of all the impellers on the shaft is very large and has to be compensated by a so-called balance piston on the rotor. Discharge pressure is imposed on one end of the piston and suction pressure on the other, partially negating the unbalanced resultant axial force from to the impellers. The cross-sectional area of the balance piston is designed to reduce the axial thrust to a low residual level to lower the load on the axial thrust bearing. Overload of this bearing can lead to damage and failure of the compressor (Bidaut and Dessibourg, 2014; Baldassarre et al., 2014; Kurz et al., 2016).

13.9.3 Bulk Model for Swirling Flow in Cavities

For a precise determination of the axial thrust, it is necessary to know the radial pressure gradient in the space between the rotor and the casing. The axial cavities between the impeller and the casing, both on the impeller backface and on the outside of the cover in a shrouded wheel, experience a leakage flow in the direction of the pressure drop. Above the shroud, this leakage is directed radially inwards from the impeller tip to the impeller inlet. On the impeller backplate, this can be radially inwards for a single-stage compressor or the last stage of a multistage machine, but is radially outwards from the impeller outlet to the impeller tip in the case of a return channel stage. Without friction with the walls in the gap, the angular momentum of the flow entering the gap would be convected by the leakage flow so that the swirl would be constant, rc_u = constant. The friction on the stator walls tends to reduce the swirl in the leakage flow and, depending on the velocity of the impeller relative to the swirl velocity in the gap, the rotating impeller can either retard or increase the swirl.

The flow is assumed to be a rotating body which is effectively inviscid with boundary layers near the walls. In a typical application, the leakage flow in the axial gap is small. The rotation of the impeller disc leads to a radial flow in the end-wall boundary layer, and this is compensated by an inward radial flow on the stationary wall (see Figure 13.12), the balance being equal to the leakage mass flow (Hu et al., 2018).

An approximate model is to assume that the fluid between the rotating wheel and the casing rotates as an incompressible solid body with an angular velocity of ω_f, such that the swirl component of velocity is given by $c_u = r\omega_f$. The angular velocity of the rotor is ω such that the ratio of the angular velocity of the leakage flow and the rotor is given by $\beta = \omega_f/\omega$, which is known as the core rotation factor. From Chapter 4, the radial pressure gradient of the swirling fluid is given by

$$\frac{dp}{dr} = \rho\frac{c_u^2}{r} = \rho\frac{(r\omega_f)^2}{r} = \rho r\beta^2\omega^2. \tag{13.16}$$

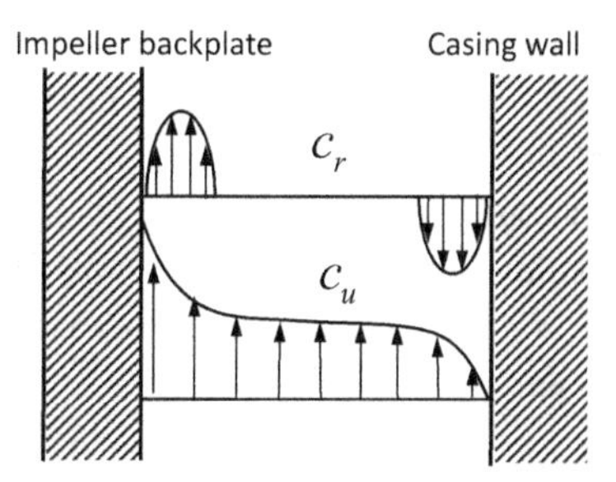

Figure 13.12 The flow distribution in the axial sidespace behind the impeller.

Assuming an incompressible flow, the pressure in the gap at a given radius can be obtained by integration and is given by

$$p_2 - p(r) = \frac{\rho r_2^2 \beta^2 \omega^2}{2}\left[1 - \left(\frac{r}{r_2}\right)^2\right]. \tag{13.17}$$

Baldassarre et al. (2014) provide examples of the good agreement between this approach and the measured pressure distribution in the backspace and above the shroud for a low flow coefficient and a high flow coefficient impeller. The axial thrust acting on the impeller can be obtained by integration of this pressure distribution on the backplate and the shroud. The core rotation factor can be determined from the radius ratio and the pressure difference from inlet to outlet of the rotor–stator gap as

$$\beta^2 = \frac{2(p_2 - p_1)}{\rho r_2^2 \omega^2\left[1 - \left(\frac{r_1}{r_2}\right)^2\right]}. \tag{13.18}$$

The torque on the rotor and on the stator can be calculated from the core rotation factor as

$$M_r = C_{Mr}\rho\left(1 - \beta^2\right)\omega^2 r_2^5, \quad M_s = C_{Ms}\rho\beta^2\omega^2 r_2^5. \tag{13.19}$$

This gives information of the disc friction on the impeller backplate and the shroud, and it can be seen that this depends on the core rotation factor of the leakage flow. The change in angular momentum of the leakage flow due to the combined effect of the stator and the rotor on the fluid in the gap is given by

$$\begin{aligned} M_r + M_s &= \dot{m}_l \Delta(r c_u) \\ C_{Mr}\rho\left(1 - \beta^2\right)\omega^2 r_2^5 + C_{Ms}\rho\beta^2\omega^2 r_2^5 &= \dot{m}_l \Delta(r c_u). \end{aligned} \tag{13.20}$$

And this can be used take account of the leakage flow in 1D performance calculations such as that given in Section 10.8.

13.9.4 Swirling Flow in the Rotor–Stator Cavities of a Shrouded Impeller

The model given in the preceding section assumes that the core rotation factor is constant at all radii, as if the fluid rotates as a solid body in the gap. This cannot be exactly correct, as the leakage flow from the impeller outlet has a swirl contribution that is different from this, typically with $\beta = 0.7$, and in a return channel stage with leakage from the outlet the leakage flow enters with no swirl with $\beta = 0$. A more exact model with a variation of the core rotation factor with radius can be derived by considering the angular momentum equation applied to a small cylindrical element of fluid of area A at radius r. The torque, M_r, exerted by the rotating disc on the fluid, and the torque exerted by the stator, M_s, are given by

$$M_r = \int_r r\tau_r dA, \quad M_s = \int_s r\tau s dA, \tag{13.21}$$

where the torque and the shear stress are taken to act in the rotational direction. The change in angular momentum of the fluid in the gap is determined by the torque exerted by the skin friction of the fluid on the two surfaces as follows:

$$M_r + M_s = \Delta(\dot{m}_l r c_u). \tag{13.22}$$

For the small cylindrical element of fluid, this can be written as

$$\tau_r r 2\pi r dr + \tau_s r 2\pi r dr M_s = \dot{m}_l d(rc_u), \tag{13.23}$$

where the leakage flow rate is outwards. The rate of change of the angular momentum of the leakage flow with radius is given by

$$\dot{m}_l \frac{d(rc_u)}{dr} = 2\pi r^2(\tau_r + \tau_s). \tag{13.24}$$

If the flow in the gap is considered to comprise of an inviscid core with no change in circumferential or radial velocity outside the boundary layers on the walls, the shear stress can be represented in terms of a skin friction coefficient as

$$\tau = ½c_f \rho c|c| \quad \text{where} \quad c = \omega r - c_u. \tag{13.25}$$

From this, (13.24) can be written as

$$\dot{m}\frac{d(rc_u)}{dr} = \pi r^2 \rho\left[-c_{fs}c_u^2 \pm c_{fr}(r\omega - c_u)^2\right], \tag{13.26}$$

where the positive sign is used if $r\omega > c_u$ and the negative sign if $r\omega < c_u$. Taking the case that $r\omega > c_u$ and assuming that the skin friction coefficient on both walls is the same, then

$$\frac{d(rc_u)}{dr} = \frac{\pi\omega^2 r^2 \rho c_f}{\dot{m}_l}\left[\left(1 - \frac{c_u}{r\omega}\right)^2 - \left(\frac{c_u}{r\omega}\right)^2\right]. \tag{13.27}$$

On substituting the local core rotation factor $\beta = c_u/r\omega$, which is the ratio of the local swirl velocity to the local impeller velocity, and $y = r/r_2$ into this equation, it can be simplified to

$$\frac{d\beta}{dy} = ay^2 - 2\beta\left[\frac{1}{y} + ay^2\right], \quad \text{where} \quad a = \frac{\pi\omega^2 r_2^2 \rho c_f}{\dot{m}}. \tag{13.28}$$

The parameter a is negative in a typical case with radial inward flow. This equation can be integrated from given initial conditions to determine how the core rotation parameter varies with the radius in the rotor and stator gaps, and combined with (5.12) this gives a more accurate estimate of the pressure distribution in the gap.

There are two interesting analytic solutions for (13.28). For the case that $ay^2 \gg 1/y$, then

$$\frac{d\beta}{dy} = -ay^2(1 - 2\beta), \quad \text{and} \quad \beta = \frac{k}{2}\left(1 - e^{2ay^3}\right), \tag{13.29}$$

where k is an integration constant. This solution represents the case where the leakage mass flow is small, and as a result the initial condition of the swirl entering the gap has only a small effect, and at small values of the radius ratio, the core rotation factor becomes 0.5. Another solution occurs for the case that the mass flow through the gap is large such that $ay^2 \ll 1/y$, then

$$\frac{d\beta}{dy} = \frac{2\beta}{y}, \quad \text{and} \quad \beta = \frac{k}{y}. \tag{13.30}$$

In this case, the friction of the walls has less effect and the swirl increases as the radius drops. In most real cases, the solution is between these two extremes and can be obtained by integration of (13.27).

13.9.5 Swirl Brakes

High-pressure centrifugal compressors sometimes experience vibrations due to rotor-dynamic instability. The main cause for the exciting forces that affects the stability is the tangential velocity component of the gas entering the many labyrinth seals throughout the machine (Baumann, 1999; Baldassarre et al., 2014). In order to control or limit these swirling flows, swirl brakes are generally implemented before the seals on the impeller shroud and at the inlet to the balance piston. Swirl brakes consist of a series of radial ribs placed upstream of the seal whose purpose is to reduce the tangential component (swirl) of the inlet gas flow to the seal. Simulations of these were first given by Mack et al. (1997) and comparison of simulations and measurements by Nielsen et al. (2001).

14 Geometry Definition

14.1 Overview

14.1.1 Introduction

The objective of this chapter is to describe the essential aspects of geometry definition of flow channels, blades and vanes in the annular parts of radial compressor flow channels. The impeller blades and flow channel make up a complex three-dimensional shape, and a general parametric method for defining such geometries using Bezier surfaces is described. The meridional channel is defined as a series of Bezier patches using a small number of parameters. This allows the channel to be defined quickly from the preliminary design information such as inlet and outlet radii, the axial length along the hub and the outlet width. The blade is defined as a series of blade elements associated with selected stream surfaces in the meridional plane. Each blade element is defined in terms of a distribution of blade camber angle and blade thickness distribution. Usually only two elements are used so that the impeller blade is typically defined as a ruled surface of straight lines joining points on the hub and the shroud contours.

A description of the geometry definition of asymmetric components such as the volute is given in Chapter 13.

14.1.2 Learning Objectives

Understand the transformation of the 3D geometry to its projection on to the 2D planes known as the (m, θ) plane and the (m', θ) plane.

- Have a knowledge of Bezier surfaces for impeller design.

14.2 Coordinate Systems for Turbomachinery

The geometry of impeller blades and the flow channel can be expressed as points in a Cartesian coordinate system, (x, y, z), representing the geometry as a grid of points on the surface of the meridional channel and on the surface of the blades, but this is not the most convenient method for its definition. Because of the rotational nature of turbomachinery, the hub and casing stream surfaces can best be described in the cylindrical coordinate system (r, z, θ), where z is in the direction of the machine axis and r is normal to this, that is, across the span of an axial flow channel. The meridional section of a machine is best defined as a series of points in the r, z plane, and the rotation of this around the z-axis defines the complete hub and casing surfaces, and a similar technique can be used to define any intermediate spanwise stream surfaces. The blading is also best generally defined as a series of blade elements associated with selected stream surfaces in the meridional plane. In axial machines, the blades are usually defined as blade elements along straight constant radius sections, but in radial machines blade elements are defined on curved sections at constant spanwise positions in the meridional channel.

Such meridional sections facilitate the visualisation of the geometry and data calculated in a 3D CFD flow field simulation as 2D planes. The meridional stream surfaces are often characterised by the arc length in the meridional plane, which is defined as

$$dm = \sqrt{(dr)^2 + (dz)^2}. \tag{14.1}$$

The meridional coordinate provides a coordinate transformation between a surface of revolution (r, z, θ) and a two-dimensional plane (m, θ) within the surface of revolution. The origin of the meridional coordinate is arbitrary. This transformation is, however, difficult to use in practice, as the two coordinates m and θ have different units. A better system would be to use an (m, $r\theta$) plane, but this also has its difficulties, as the transformation is not angle preserving. To overcome this, the m and $r\theta$ coordinates are often normalised with respect to radius. The normalised differential length is defined as m' and is given by

$$dm' = \frac{dm}{r}. \tag{14.2}$$

The transformation from the (r, z, θ) plane to the (m', θ) plane unrolls the geometry on to a 2D plane in a way that is angle preserving. A further length can be defined along the camber line of the blade in the (r, z, θ) plane as

$$ds = \sqrt{(dr)^2 + (dz)^2 + (rd\theta)^2}. \tag{14.3}$$

The integration of this from the leading edge to the trailing edge gives the blade camber length of a blade along the meridional surface.

These transformations allow the complex 3D geometry to be considered as a series of two-dimensional planes in which the meridional geometry is defined as axisymmetric stream surfaces at different spanwise positions across an annulus, and blade-to-blade stream surfaces are defined on these surfaces in the circumferential direction, as shown in Figure 14.1. The real flow in a radial compressor does not actually follow these stream surfaces particularly well, but defining the geometry in this way provides useful insights and understanding, and is very useful in the design process. This distinction between two types of stream surface dates back to the classical work of Wu (1952), where he denoted the blade-to-blade surface as an S1 surface and the bladelike surface in the meridional plane as the S2 surface.

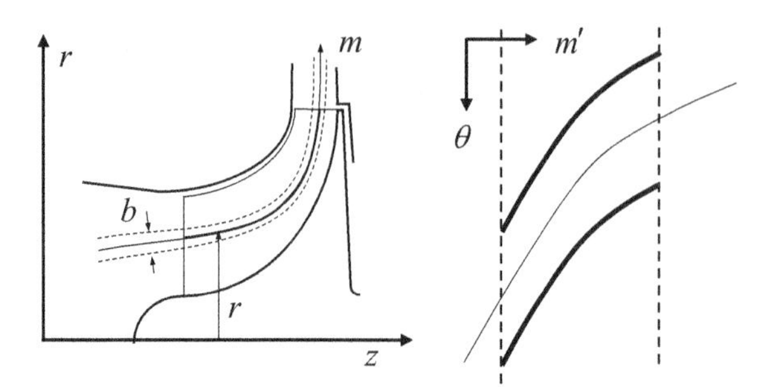

Figure 14.1 Meridional S2 stream surface (left) and axisymmetric S1 stream surface in a radial compressor (right).

14.3 Axisymmetric and Blade-to-Blade Stream Surfaces in Radial Compressors

The axisymmetric stream surfaces are shown for a radial compressor impeller in Figure 14.1. The annular nature of turbomachinery geometries with a cylindrical coordinate system (r, radial, z, axial and θ, circumferential), so that the coordinates of the axisymmetric streamline of the stream surface are denoted by (r, z), the local radius and the local rotor axis coordinate. In many situations, it is also convenient to make use of the stream surface thickness, b, in the spanwise direction. This varies through the stage, typically decreasing from the inlet to the outlet of an impeller.

In axial turbomachinery, with no change of radius, the blade-to-blade plane can usually be represented in terms of the meridional coordinate (m, where $m = z$) and the circumferential coordinate ($r\theta$) with no distortion of the blade shape, as the radius does not change with the axial position. This is not the case in a radial compressor, where the radius changes along the meridional surface. This radius change leads to a distortion of angles and lengths in a 2D representation, as in the Mercator projection of the world on to a flat plane. To demonstrate this, the real geometry of the cascade of blades in an axial (z = constant) section through the diffuser in Figure 14.2, and the same compressor stage is shown in Figure 14.3 in the (m', θ) plane.

Note that in this physical projection of the geometry, the area clearly increases through the blade row from inlet to outlet. The blades are separated by a constant value of the circumferential coordinate, $\theta = 2\pi/Z$, where Z is the blade number, but as the radius increases, the spacing between the blades also increases. This diagram shows that, as is typical for most radial diffusers, the blade angle decreases through the blade row to remove the swirl component of velocity. Also, the blades are not straight but curved towards the impeller, which is also typical. In a diffuser, the pressure surface of the blade is the convex surface away from the impeller and the suction surface of the blade is the concave surface facing the impeller.

In Figure 14.3 the blade-to-blade surface is shown in the (m', θ) plane, In the case of a purely radial cascade, the m' coordinate is given by $\ln(r)$. The transformation from physical space to the (m',θ) plane is angle preserving as the blade angle is given

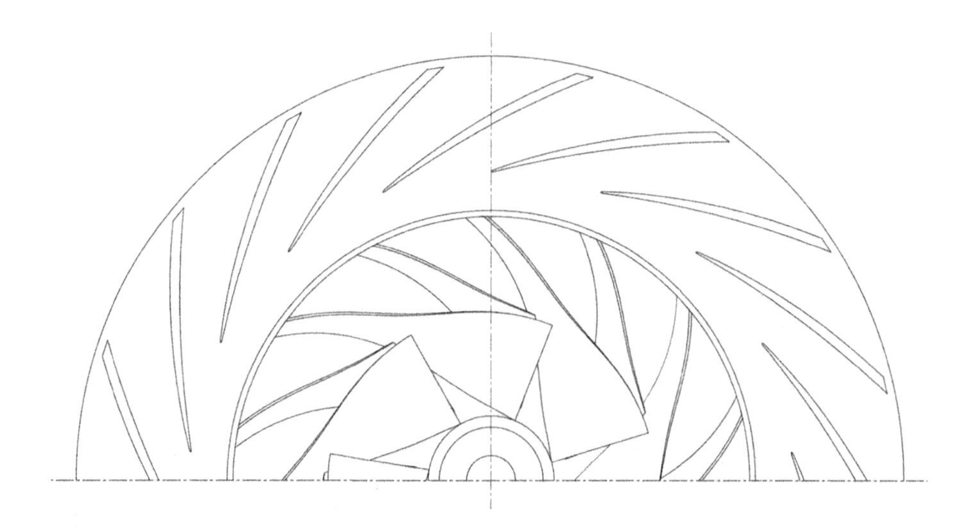

Figure 14.2 Radial stage with an axial section through a centrifugal compressor diffuser.

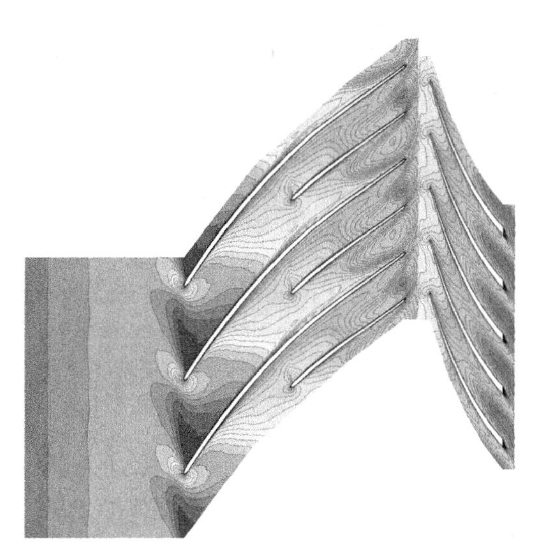

Figure 14.3 Flow calculation of an impeller and diffuser presented on a meridional section in the (m',θ) coordinates.

by $\tan\beta = rd\theta/dm$, but in this projection the blade appears to be curved in the direction away from the impeller. A blade with a constant flow angle called a logarithmic spiral would appear straight in the (m', θ) plane. Furthermore, the meridional m' coordinate becomes shortened as the radius increases in this projection. The example given in Figure 14.3 shows the foreshortening of the diffuser length relative to the impeller length in this type of presentation. Clearly, unlike with axial turbomachinery, great care is needed in the interpretation of the geometry shown in the blade-to-blade plane of radial compressor components. The (m',θ) plane is generally used to present flow contours from simulations of the flow at different spanwise positions across the flow channel as it allows a presentation of several blade passages and blade rows adjacent to each other.

14.4 Geometry Definition of Flow Channels

14.4.1 Geometry Definition

The preliminary design process provides an initial definition of the skeletal geometry of the blading and the annulus of the compressor. For the impeller, the bare bones of the design would be the inlet and outlet radii, the axial length along the hub and the outlet width, together with the inlet and outlet blade angles and the blade thicknesses on the hub and shroud. A geometry definition method puts flesh on these bones to make a detailed definition of the impeller so that it can be more easily be visualised and analysed.

Turbomachinery design systems then rely on design by analysis where the blading is assessed using CFD and finite element method (FEM) codes and iteratively refined (see Chapter 20) before the geometry is passed on to a computer-aided design (CAD) system for drawing and manufacture. The iterative process may be manual or may make use of optimisation methods. During this process, the shape of the blades and the flow channels is repeatedly adjusted until a suitable geometry is found. The criteria for what features are considered most suitable may involve aerodynamics aspects, including efficiency, pressure rise and operating range; mechanical aspects, such as stress levels in the blades and the hub and natural frequencies; and manufacturing aspects generally involving the practicability and cost. In some cases, noise generation may be an issue, and in others some possible shapes may have to be eliminated as they conflict with patents. This process of continual refinement of the shape can be time consuming, tedious and expensive, since at each stage the geometrical data for the necessary aerodynamic or mechanical analysis must be prepared. The task is greatly simplified when a system of geometry definition is available which includes appropriate interfaces for the transfer of data to the subsequent design processes and manufacture.

Traditional methods of geometry definition of blading for axial compressors involve the definition of the blade shape at several radial positions across the span. The individual sections are defined in terms of a camber line and thickness distribution along the chord, as discussed in Chapter 7. And the position of the sections relative to each other is defined by the stacking axis of the sections. In this process, the designer has a more direct relationship with the blade angles than with the actual blade Cartesian coordinates. Because of this, most design systems are based on a definition of blade camber angles along the chord, which are then integrated to find the coordinates of the camber line and then, with the thickness distribution, the coordinates of the suction and pressure surfaces. In radial machines, a similar system is used, but the camber line angles are defined along a meridional path through the impeller and there is no stacking axis other than the relative location of the sections in the circumferential direction which determines the lean of the blade.

The use of coordinate points along the surface to define the shape becomes awkward in regions of high curvature, such as the large change in curvature of the blades near the leading and trailing edges. Interpolation with polynomial curves between a series of individual points requires a special technique to avoid wiggles and overshoots in areas of large curvature. To avoid this, the blade surfaces are usually split into four parts. The suction and pressure surfaces are modelled by a string of coordinate points derived from the camber line and the normal thickness through which a suitable curve is fitted. The leading and trailing edges are modelled as separate circular arcs or ellipses, and in radial impellers the trailing edge is usually blunt and cut off in the circumferential direction. A very useful technique to obviate the need for a large number of data points is to use analytic functions or a smaller number of points which are joined by piecewise continuous segments; see, for example, the methods of Ginder and Calvert (1987) for transonic axial blade sections, Giesecke et al. (2017) for subsonic axial blade profiles and the methods of Smith and Merryweather (1973), Came (1978) and Casey (1983) for radial compressor blade sections.

Axial compressor design systems do not usually have any difficulty to deal with the detailed shape of the meridional flow path as this usually has little curvature and is simply a conical annulus defined as a series of coordinate points on the hub and casing along the axis. In radial turbomachinery, the emphasis shifts slightly from the definition of the blade profile shape to be more concerned with the shape of the annulus. Many different systems have been proposed to define the shape of the meridional channel from Lamé ovals, parabolic or elliptic shapes or simply a string of circular arcs. Since the advent of computational geometry for design of curves and shapes (of cars, ships, graphic design of fonts, and the curved lines within the drawing function of Microsoft PowerPoint) most curve definition systems use Bezier curves, B-splines or nonuniform rational B-splines (NURBS) systems.

The geometry definition method described here is that used in software known as VistaGEO by PCA Engineers Limited. The method is closely based on the computational geometry published by Casey (1983), where full details of the basic theory can be found. Newer implementations of this can also be found in several software systems developed by universities, individual companies and commercial software vendors, but the basic idea remains the same. In this method, the meridional channel is represented as a series of Bezier curves or Bezier patches, as explained later in this section. The blade sections are defined along the meridional surfaces and, in order to aid manufacture by flank milling, the blade surfaces are usually ruled surfaces of straight lines joining the hub and casing blade sections. For situations where a curved blade surface is required, Hazby et al. (2018), additional blade sections can be defined at intermediate spanwise positions, but then a more complex interpolation between the sections across the span is required. The surface is then no longer made up of straight lines and must be manufactured by point milling. The method as described here is sufficiently general that it can be applied to individual radial or mixed flow impellers;

to inlet guide vanes, diffuser vanes and return channel vanes; to multistage inline compressors; and even to axial compressor stages if needed.

14.4.2 Bernstein Polynomials

Bernstein polynomials are the foundation for building Bezier curves. The Bernstein polynomial of degree n associated with parameter u is defined as

$$\mathrm{B}_{n,i}(u)=\binom{n}{i}(u)^{i}(1-u)^{n-i}=\left[\frac{n!}{i!(n-i)!}\right](u)^{i}(1-u)^{n-1}. \tag{14.4}$$

The first term on the right is the binomial coefficient, which depends only on the degree n and the order i and is the well-known coefficient of the Pascal triangle. The first few Bernstein polynomials can be written as

$$\begin{aligned}
n=0:&\quad \mathrm{B}_{0,0}=1\\
n=1:&\quad \mathrm{B}_{1,0}=1-u,\qquad \mathrm{B}_{1,1}=u\\
n=2:&\quad \mathrm{B}_{2,0}=(1-u)^{2},\qquad \mathrm{B}_{2,1}=2u(1-u),\qquad \mathrm{B}_{2,2}=u^{2}\\
n=3:&\quad \mathrm{B}_{3,0}=(1-u)^{3},\qquad \mathrm{B}_{3,1}=3u(1-u)^{2},\qquad \mathrm{B}_{3,2}=3u^{2}(1-u),\qquad \mathrm{B}_{3,3}=u^{3}\\
n=4:&\quad \mathrm{B}_{4,0}=(1-u)^{4},\qquad \mathrm{B}_{4,1}=4u(1-u)^{3},\qquad \mathrm{B}_{4,2}=6u^{2}(1-u)^{2},\\
&\quad \mathrm{B}_{4,3}=4u^{3}(1-u),\quad \mathrm{B}_{4,4}=u^{4}.
\end{aligned} \tag{14.5}$$

It can be seen that the Bernstein polynomials are polynomials of degree n which vanish at the end points with $u=0$ and 1, and each polynomial is a single function of the parameter u over the entire interval from $u=0$ to 1, rather than a combination of piecewise functions. There are a total of $n+1$ nth degree Bernstein polynomials, and the coefficients of these successive polynomials are given by the elements of the Pascal triangle so that subsequent polynomials of higher order can easily be generated if it is necessary, as shown in Table 14.1.

14.4.3 Bezier Curves

A Bezier curve is a parametric representation of a space curve. The curve is specified by a series of points in space of which only the first and the last actually lie on the space curve

Table 14.1 Pascal triangle coefficients of Bezier polynomials up to degree 5.

Degree, n			Pascal triangle coefficients $\mathrm{B}_{n,i}$			
0	1	1	1	1	1	1
1	1	2	3	4	5	
2	1	3	6	10		
3	1	4	10			
4	1	5				
5	1					

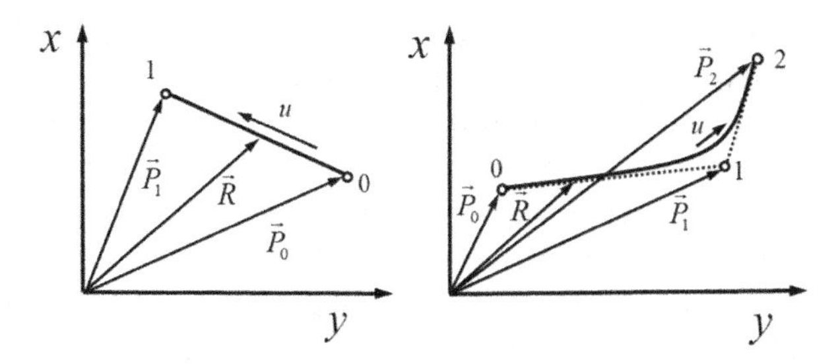

Figure 14.4 Bezier curve of degree $n = 1$ (left) and $n = 2$ (right).

they define. The points are known as the polygon points of the curve, and the figure constructed by connecting these points with straight lines is known as the Bezier polygon of the curve. The Bezier curve is the weighted average of the polygon points, whereby the weighting functions for the $n + 1$ points of the polygon are the $n + 1$ nth-degree Bezier polynomials. The simplest example is a straight line, shown in Figure 14.4, defined as a Bezier curve of degree $n = 1$, with two polygon points as follows:

$$\bar{R} = (1 - u)\bar{P}_0 + u\bar{P}_1, \tag{14.6}$$

where R is a vector of a point on the curve and P_0 and P_1 are the vectors of the polygon points (or knots) of the Bezier polygon (which in this case are also the end points of the line). This line could represent the trailing edge of a mixed flow impeller. As the parameter u varies from 0 to 1 along the curve, the vector R describes a straight line from point P_0 to P_1. The Bezier curve of degree $n = 2$ is defined by three polygon points:

$$\bar{R} = (1 - u)^2\bar{P}_0 + 2u(1 - u)\bar{P}_1 + u^2\bar{P}_2. \tag{14.7}$$

An example is shown in Figure 14.4, where it can be seen that the second polygon point does not lie on the curve, but that the tangents to the curve at the end points are in the direction of this point. This curve could represent the hub or shroud contour of the impeller, although usually a curve of higher degree is used for this. The Bezier curve of degree $n = 3$ requires four points to define it and can be written as

$$\bar{R} = (1 - u)^3\bar{P}_0 + 3u(1 - u)^2\bar{P}_1 + 3u^2(1 - u)\bar{P}_2 + u^3\bar{P}_4. \tag{14.8}$$

The form of these equations identifies that the Bezier curve is the weighted average of the polygon points defining the curve and that the weighting functions are the Bernstein polynomials.

The curve always goes through the first and last points of the polygon, and the direction of the start and end of the curve is always along the line joining the first two and the last two points. Similarly, the curvature at the end points is related to the position of the third point relative to a line joining the first two points, and the curvature is continuously smooth. It might, at first sight, appear to be a disadvantage that the only first and the last points actually lie on this curve. However, the curve has clear geometrical properties related to the position of the other points, and it is this aspect that makes Bezier curves ideal for constructing smooth curves that conform to a shape required by the definition of the meridional channel.

In particular for the design of radial compressor impeller, a simple Bezier curve can be generated that provides the required shape of the hub or casing with sufficient degree of freedom by using only four to five polygon points. In this application, the whole meridional channel can best be defined as a series of curves that are directly related to the skeletal geometry, such as single curves of degree three or four for the impeller casing and the hub walls, and linear elements of degree one for the diffuser walls. Using a string of curves in this way offers sufficient flexibility to define a complete stage of a multistage compressor or several stages if necessary, so that the necessary flow and mechanical analyses can be carried out. The coefficients of the Bezier polynomials are the same for the entire interval, so shifting a single polygon point changes the whole curve, and this is what makes it so useful in preliminary design. If one of the intermediate polygon points is shifted by a short distance, it attracts the curve towards it like a magnet. This may appear to be a restriction, but as will be shown later, this is not at all disadvantageous.

The method is not intended as a method to approximate an existing set of data points or to interpolate between a series of data points, for which other types of splines are generally more useful. One alternative would be to use B-splines or nonuniform rational B-splines (NURBS). The essential difference between a B-spline curve and a Bezier curve is that the degree of the curve and the number of polygon points are not related. The interpolating functions which form the curve from the weighted average of the polygon points in a B-spline act locally on only a few adjacent points and are dissimilar in a different region of a single curve. The interpolating functions are of relatively low degree, but are nonzero only over a part of the curve. This has advantages in defining complex curves using a single series of points, or knots as they are called, but it is not an important requirement in the definition of the typical geometries found in radial turbocompressor flow channels.

14.4.4 Bezier Surfaces and Bezier Patches

Two adjacent Bezier curves with a similar degree can be joined with a series of straight lines joining points with similar values of the parameter u. The straight lines can be defined in a similar way to (14.6), but with a parameter v varying from 0 to 1 to represent the fractional distance along this line between the curves. In this way, a parametric surface in space is defined, known as a Bezier surface, so that each arbitrary coordinate point (x, y, z) on this surface can be denoted by the parametric values u and v. The lines joining the two adjacent Bezier curves could also be curves in space, but this generality is a complication that is not needed for the definition of the meridional channel.

The meridional channel of an impeller is defined by joining the Bezier curves representing the hub ($v = 0$) and casing ($v = 1$) contours. Furthermore, the whole compressor stage can be defined by a progressive series of Bezier surfaces, or Bezier patches to define the whole meridional channel. Typically, the value of parameter u is

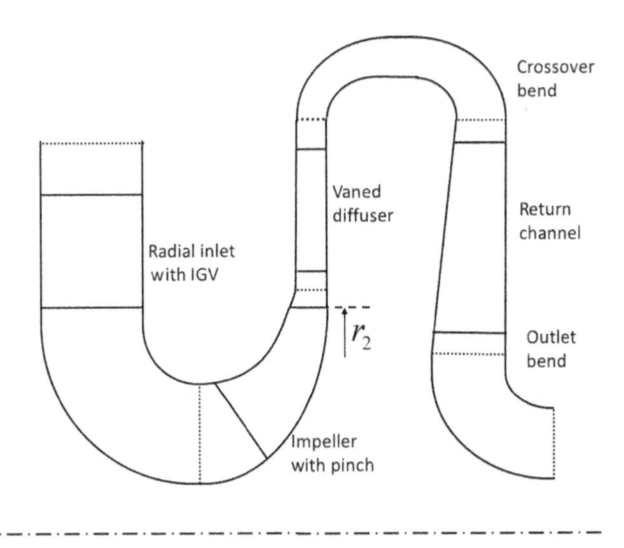

Figure 14.5 Bezier patches to define the meridional channel of a complete compressor stage.

taken to be 0 at the beginning of the first patch and to increase to n with n successive patches. The local value of u on each patch varies from 0 to 1, so for the nth patch the local value is taken as $u + 1 - n$ for the calculation of any local coordinates. Similarly, the value of v can be used to define an intermediate spanwise position, such as $v = 0.5$ for the midspan. Figure 14.5 shows an example of a series of patches defining the meridional channel of a compressor stage with a return channel, with separate patches for the radial inlet with IGV with the inlet bend, the impeller including the pinch at diffuser inlet, the diffuser channel, the crossover bend, the return channel with vanes and the outlet bend.

14.4.5 Meridional Channel as a Series of Bezier Patches

The meridional channel is defined by a string of Bezier patches in the r, z meridional plane of the impeller, thus defining a plane surface that is the meridional section through the compressor. The mathematical rotation of this plane surface around the z-axis defines the internal flow channel of the machine. The blades are defined separately. A series of polygon points define the geometry of the hub and the casing contour in the meridional plane as Bezier curves. A parameter u varies from 0 to 1 internally in each patch, but a global value is defined such that a value of 1.5 is on the second patch, 2.5 on the third patch and so on.

In most successful implementations of this method, the meridional geometry of the impeller is defined by templates based on separate subroutines for different types of impeller. These templates can be adapted to represent the different types of meridional channels that can be found, such as separate templates for a radial impeller with axial inlet, for a radial impeller with radial inlet, for a mixed flow compressor, a crossover bend, a return channel, a radial turbine and so on. The actual choice of parameters for

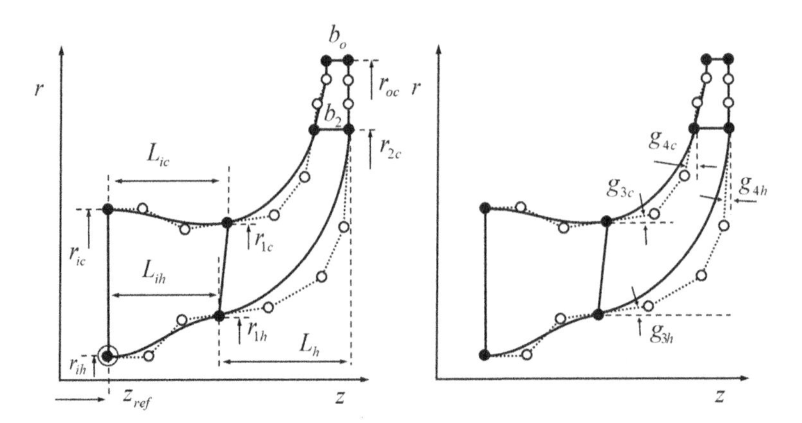

Figure 14.6 Template to define meridional geometry of a typical turbocharger stage.

such a system to define the precise shape of the impeller meridional channel is largely a matter of taste.

The template for the most typical centrifugal impeller with an axial inlet and a radial outlet comprises a series of three Bezier patches or segments as shown in Figure 14.6. The first patch is the inlet channel of the impeller, the second is the impeller itself, whereby in this case the leading and the trailing edges of the impeller are coincident with the patch boundaries on the hub and the casing, although this is not a requirement, and the third represents a vaneless diffuser channel with pinch.

Defining the meridional channel geometry in this way takes into account the fact that the impeller design is usually carried out in a stepwise process, beginning with a preliminary 1D design method. This means that first estimates of many key skeletal parameters are already fixed by the preliminary design optimisation process or by other constraints, and some of these no longer need to be changed during the subsequent design steps. For example, the choice of the impeller diameter, the inlet eye diameter and the outlet width generally result from the preliminary design process. In the implementation of this system as described here, there are four groups of parameters. In the first instance, global dimensions need to be defined, such as the axial location of the inlet plane on the hub, Z_{ref}, and the impeller radius, r_2, as these allow the geometry to be scaled in size and to be translated to a new position along the axis if required, but these are otherwise fixed by the application and the preliminary design.

All subsequent parameters are nondimensional and are defined either as angles, as a ratio to the impeller outlet radius or as a ratio to some other relevant dimension. The dimensionless geometry terms that follow can then be considered as parameters to define a family of impellers of a particular diameter and with a similar skeletal structure. This allows easy transfer of experience gained on a certain design with a certain diameter to be simply scaled into another design at a different size for a similar application.

The second group of parameters define the location of the patch corner points in the meridional channel and are the outcome of the preliminary design or the constraints of the specification. The following parameters are needed for the channel template shown in Figure 14.6:

(a) Inlet radius ratio on casing at duct inlet, r_{ic}/r_2
(b) Inlet radius ratio on hub at duct inlet, r_{ih}/r_2
(c) Axial length of the inlet duct, L_{ih}/r_2
(d) Impeller inlet eye radius ratio, r_{1c}/r_2
(e) Impeller inlet hub radius ratio, r_{1h}/r_2
(f) Axial length of impeller hub, L_h/r_2
(g) Impeller outlet width ratio, b_2/r_2
(h) Diffuser outlet radius ratio, r_o/r_2
(i) Diffuser outlet width ratio, b_o/r_2

The third group of parameters for the meridional geometry determine the slope of the meridional channel contours at the patch corner defining points at impeller inlet and outlet. Separate slope angles for each of the casing and hub walls at patch boundaries are defined. Note again that these angles may also be set by the preliminary design process or some other constraints, or one or more of them may be set to zero, so that in fact typically only three slope parameters need to be changed. If the Bezier patch includes internal polygon points, then these angles fix the direction in which the first internal point lies relative to the patch corner points. The most important of these angles, in the example shown in Figure 14.6, is the angle at the impeller outlet on the casing side, g_{4c}, which has a strong influence on the rate of change of area, and meridional flow velocity, in the flow channel at impeller outlet.

The fourth group of parameters for the meridional geometry determines the curvature of the meridional channel contours. Various shape factors along the hub and casing wall are used to determine the location of the internal Bezier patch polygon points. Each shape factor determines the location of the associated internal point as a fraction of an associated length in the channel. Roughly half of these parameters can be determined in advance and fixed for the design optimisation process so in fact typically only six free shape parameters are generally used for the shape of the meridional walls in the impeller region of the channel during the design process. These parameters are based on fractions of the length of the meridional channel, and experience shows that many of these fractions remain sensibly constant across a range of designs of impellers. Certain values of the shape parameters can be selected that cause the curvature of the channel to closely approximate that of an ellipse or a circle. This eases the task of a new design in that parameters optimised in an earlier design can be used as the starting values. In this way, one particular type of impeller can be cloned to make a new design. For example, if during the detailed design of a new impeller the axial length of the impeller needs to be decreased, then all of the intermediate points are also shifted by an appropriate amount to retain consistency.

Based on experience with the system described by Casey (1983), three internal polygon points are used to define the hub contour within the impeller, requiring four parameter values, whereas on the shroud only two free parameters are normally required. The shroud impeller contour is defined initially as a Bezier curve with two internal points, and this is converted to become one with three internal points to be consistent with the hub, using general rules related to Bezier curves which increase the order of a curve while maintaining its shape (Faux and Pratt, 1979). It would of course also be possible to define the shroud contour with three internal points, but this would require two additional free parameters, and the objective is to do a good design with the least number of free parameters. This also has advantages with automatic optimisation systems as the number of free parameters to be optimised is reduced (Casey et al., 2008).

14.5 Geometry Definition of Blades and Vanes

14.5.1 Ruled Surface

The impeller blade is defined as a ruled surface of straight lines joining points on the hub and the shroud contours. The orientation of these lines changes the shape of the intermediate blade even though the hub and casing geometry remains the same. In most cases, these lines are equidistant along the meridional channel of the impeller, between the leading edge and the trailing edge. Other orientations of the ruled surface can also be selected, such as equidistant along the actual blade length rather than the meridional length. The use of ruled surfaces is a standard technique for impeller design leading to simpler manufacture (through flank milling). This is not considered to be a severe limitation from the aerodynamics point of view, except for high pressure ratios with transonic inlet Mach numbers (Hazby et al., 2018).

In the case of an impeller with a splitter, the splitter leading-edge position is defined by the axial location of the leading edge on the hub and on the casing. The splitter leading edge is usually also a straight line, and the orientation of the ruled surface of the impeller is adapted to make this line one of the generating lines of the main blade. In this way, the splitter is a shortened version of the main blade with no leading-edge recamber (Came and Robinson, 1998). A separate definition of the splitter with its own camber and thickness distribution is also possible in most systems.

14.5.2 Camber Angle and Thickness Distribution

Another group of parameters defines the blade shape and the number of blades. These are defined as a camber surface and a superposed thickness distribution in a similar way to aerofoils; see Chapter 7. The hub and shroud blade sections are defined as distributions of blade camber angles (see Figure 14.7) and thickness normal to the camber line. These distributions can be specified as Bezier functions with a Bezier polygon or splines through points along the normalised meridional length, whereby the leading-edge and trailing-edge ellipses are defined as separate parameters. The blade outlet angle will sometimes not be a free parameter, as the 1D preliminary design will determine this, so there are about eight to 10 free parameters to determine the shape of the blade. These are also not completely free, as the mechanical constraints on the design may also require no radial blade lean at the leading edge

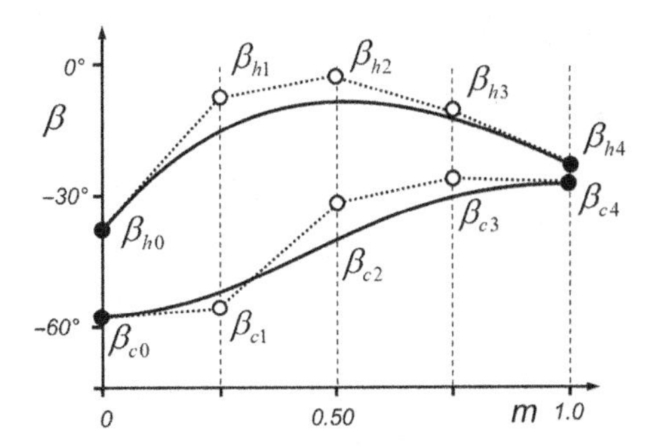

Figure 14.7 Blade angle distribution on the casing and hub streamlines.

and a certain rake at the trailing edge, as explained in Chapter 11. These constraints mean that actually there are fewer completely free parameters for the blade shape.

Usually the designer needs to achieve a specified lean angle (typically 0°) of the leading edge and a specified rake angle at the trailing edge (typically 25°). These values are usually known in advance from mechanical considerations. This description of the blade, with lean and rake parameters and the hub and shroud blade angle distribution, is overdetermined, so that some of the information specified by the designer has to be modified. The relative importance of the inlet and outlet angles and the angles along the shroud streamline in the diffusion process of the impeller suggests that these should be retained. A simple iteration can be incorporated which modifies the hub angle distribution to match the specified casing angle distribution, lean and rake angles. The modification makes use of the user-defined hub angle distribution as a guide but overrules this within the blade row to attain the required lean, keeping the inlet and the outlet blade angles the same so that incidence and work input are not affected.

The hub and casing thickness distribution is also defined as a Bezier polynomial or as a series of spline interpolated points, whereby the roundness and cutoff of the leading edge and trailing edge need to be defined by separate parameters. In the case of an impeller with a shroud, the inner shroud contour is the same as the casing contour of the blade. For an open impeller, the blades are generally defined as if they extended to the casing contour, and the tip clearance is applied as a trim to the blade, which may be variable from the leading to the trailing edge. In both cases, an additional modification is needed to apply the geometry of the fillet, which is clearly not part of the linear ruled surface definition.

14.5.3 Circular Arc Blades

In 2D impellers and diffusers with constant axial blade sections, it is useful to include a geometry definition based on circular arc blades, as already used in Chapter 12 for cascade diffusers. For given inlet and outlet radii and inlet and outlet vane angles, the radius of curvature of a circular arc vane, shown in Figure 12.21, is given by (12.24).

15 Throughflow Code for Radial Compressors

15.1 Overview

15.1.1 Introduction

The flow through a centrifugal impeller is turned in the circumferential direction by the action of the rotating blades and from the axial to the radial direction by the pressure gradients in the curved meridional channel. One of the simplest ways to analyse the velocity gradients of this complex flow is with an inviscid axisymmetric throughflow code. This allows the designer to control the curvature of both the blade and the meridional channel to achieve an optimum geometry with smooth relative velocity distributions. The method cannot examine detailed viscous effects and for this a CFD simulation is necessary.

The objective of this chapter is to describe a streamline curvature throughflow method which has been developed specially for radial machines, such as centrifugal compressor stages, pumps and radial turbines. The method can also be used for axial machines, both compressors and turbines, and for axial compressors with a centrifugal rear stage. The method is described in some detail as there are few sources in the literature with such complete descriptions of this method whereas there are many sources for description of 3D CFD methods.

The chapter begins with a general introduction to the throughflow method for readers who are not familiar with 2D throughflow methods. These 2D methods date back to the 1960s and were among the first digital computer codes for use in turbomachinery flow analysis. The chapter continues with a more detailed description of the streamline curvature throughflow equations. Here, the terms in the equations are discussed in detail in order to illustrate how the geometry controls the pressure and flow gradients both along and normal to the streamlines in the compressor.

15.1.2 Learning Objectives

- Understand the terms in the radial equilibrium throughflow equation and how they are determined by the geometry and flow gradients and so affect the meridional flow.
- Understand the basic numerical iterative scheme of streamline curvature methods for an iteration to mass flow or to a pressure ratio.
- Be aware of how the meridional grid spacing affects the speed and stability of computation in streamline curvature methods.

- Be aware both of the limitations of the throughflow method and of those areas where it is very useful.
- Understand how this method can most usefully be used in preliminary design.

15.2 A Preliminary Overview of the Throughflow Method

15.2.1 Quasi 3D Flow on S1 and S2 Stream Surfaces

Compressor designers striving to gain a better understanding of the flow use the best simulation tool available to them. For many years, the workhorse of the aerodynamic designer was a streamline curvature throughflow code. An overview of its use in the design of radial compressor stages for industrial compressors is given by Casey (1985). Analyses using such a code offer a good overview of the ideal velocity and pressure distributions through a component. They also highlight blade loading issues and identify risks of choking. Given the 3D viscous nature of flow, 2D inviscid codes cannot capture the finer details either of the viscous flow field or of the loss generation. Since the development of effective 3D CFD codes in the late 1980s (see Chapter 16), the throughflow codes have been relegated to act a preliminary design tool, and the final design is now always established with 3D CFD simulations. Nevertheless, throughflow codes can still play a vital role in the preliminary design. Although the results are not exact, especially with respect to 3D viscous effects, they nonetheless provide useful information that, with experience, can guide the improvement of a design with little computational effort. In effect, the use of a throughflow method weeds out the weakest design options very quickly and thus eliminates the need for more complex, time-consuming CFD simulations on poor designs.

The fluid dynamics equations that are solved in a throughflow method are the continuity equation, the energy equation (Euler equation of turbomachinery), a suitable equation of state and the inviscid momentum equation on a meridional stream surface. This stream surface is often known as the *S*2 meridional stream surface, in contrast to the intersecting *S*1 blade-to-blade surfaces at different spanwise positions. These surfaces were first defined in the throughflow theory of Wu (1952). In his classic paper, Wu envisaged an *S*1/*S*2 calculation procedure where several true *S*1 and *S*2 stream surfaces were iteratively computed, allowing a warped Q3D model of the flow with stream-surface twist. Wu's iterative approach is rarely used, as the errors associated with calculating an inviscid flow do not merit such a complex procedure. Instead, the meridional flow is considered as a circumferential average, and the equations are solved on a 2D plane which is the known as mean axisymmetric stream surface, as shown in Figure 14.1.

The flow in the meridional stream surface is obtained by averaging all flow properties in the circumferential direction. In essence, the flow is calculated on a mean bladelike surface between the blade rows. The velocity field is assumed uniform from blade to blade, but the blade forces derived from the tangential pressure gradients are considered. Note that the mean stream surface has an arbitrary location in the circumferential direction and that the flow direction on the mean stream surface is taken to be similar to that determined by the relative blade angles. These are provided as input data as no $S1$ calculations are available to calculate them. The flow blockage caused by the circumferential thickness of the blades increases the flow velocities within this surface, and this geometrical blockage is also part of the input data.

There are several numerical methods that can be used to solve the throughflow equations. These include streamline curvature methods, stream function methods, finite element methods, singularity methods and time-marching methods. A review of the particular methods in use during the main development period of throughflow codes for axial machines is provided by AGARD-CP-195 (1976), AGARD-AR-175 (1981) and Macchi (1985). Some information about the development of methods suitable for radial machines is provided in AGARD-CP-282 (1980).

Before the availability of digital computers, the throughflow equations were solved arduously using graphical methods. One of the earliest successful digital methods for throughflow calculations in radial impellers was the streamline curvature method of Katsanis (1966). Another technique was the matrix method (Marsh, 1968), which solved the stream function equations from the throughflow analysis of Wu (1952) using finite difference techniques on a fixed grid. Initially the matrix method as developed by Marsh could only calculate single radial impellers by means of a coordinate transformation as the stream function equation included terms which were divided by the axial velocity, which is zero in a purely radial flow at the impeller outlet. The method was improved by Bosman and Marsh (1974) to incorporate a more general derivation of the stream function, thus enabling stages to be calculated without a coordinate transformation. A singularity method for throughflow equations was developed by Ribaut (1968) and later improved (Ribaut, 1977, 1988), and this successfully calculated radial compressor flows at subsonic conditions. A finite element approach for calculating the throughflow of radial impellers was developed by Hirsch and Warzee (1976).

All of the preceding outlined methods suffered from Mach number limitations for flow fields where the relative Mach numbers were above one. The breakthrough, which has led to the widespread application of the streamline curvature method for situations with some supersonic flow, was the method developed by Denton (1978) for steam turbines. This streamline curvature method has become dominant through its relative simplicity and its superior ability to deal with supersonic flows. It is the basis of the method described here. Most attempts to remove the Mach number limitations have been pursued with time-marching CFD methods following the work of Spurr (1980) and Dawes (1990). This is still an active area of research for inviscid and viscous throughflow calculations (Damle et al., 1997; Sturmayr and Hirsch, 1999; Pacciani et al., 2017).

15.2.2 The Streamline Curvature Throughflow Method

The streamline curvature method uses fixed calculating stations which are roughly normal to the streamlines. The meridional grid is not fixed but is based on the streamlines of the mean circumferentially averaged flow in the meridional direction and changes continually during the computation to converge on the location of the final streamlines of the solution. This feature is what makes the method so appealing to the designer, as the meridional grid has a natural property of being oriented in the mean flow direction and generates a simplified 2D image of the complex 3D flow along the spanwise stream surfaces. The use of a grid involving meridional streamlines, all of which proceed smoothly from the inlet plane to the outlet plane of the calculation domain, assumes that there is no reverse flow. Operation points with reverse flow in the mean meridional flow (even at the design point) cannot normally be calculated by this method.

The fixed calculating stations are oriented with the blade row leading and trailing edges and are often known as quasiorthogonals, being nearly orthogonal to the streamlines. The calculating stations are often abbreviated to q-o's. The q-o's can be in duct regions, that is in the blade-free space upstream and downstream of blade rows, at the leading and trailing edges of the blade rows or internally within the blade rows; see Figure 15.1. The computational results are available only at the points where the meridional streamlines cross the q-o's.

In the first methods developed for axial throughflow calculations, internal q-o's within blade rows were not used, and such methods are now known as ductflow methods. The streamline curvature across a blade row is difficult to calculate with ductflow methods given that no details of the blockage of the blades can be included and there is only a single q-o at the leading and at the trailing edge. The more modern

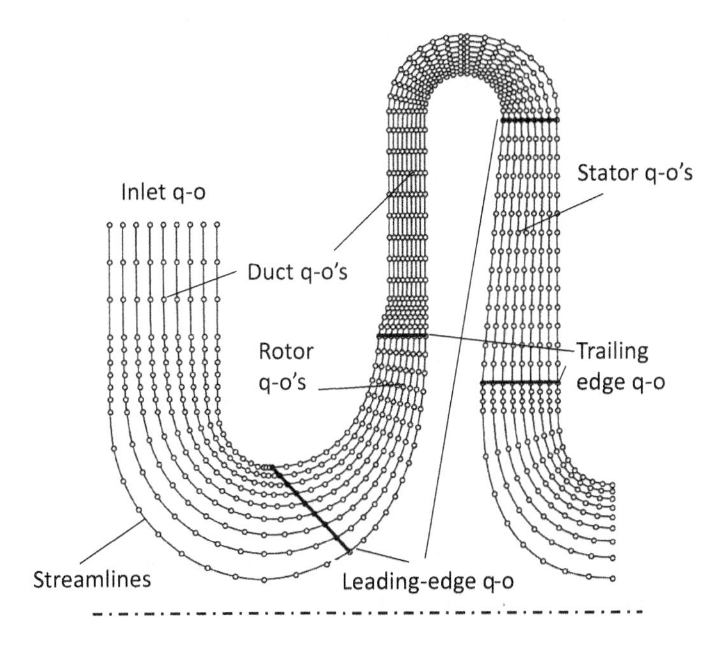

Figure 15.1 Meridional streamlines and quasiorthogonals in a throughflow calculation of a return channel compressor stage with a vaneless diffuser.

throughflow method described here uses a very crude grid for a radial impeller, typically with nine streamlines and 15 q-o's, giving roughly 150 grid nodes in a radial impeller, in comparison with modern CFD methods, which have around 500,000 nodes in a blade row. Often 5, 9 or 17 streamlines across the span are used giving a mass-flow distribution that is split into equal parts between each of the meridional streamlines. This conveniently leads to a streamline distribution which naturally gives details at the hub and casing and at locations with a fixed proportion of the spanwise mass flow distribution, including locations with 25%, 50% and 75% of the spanwise flow.

There are several key features that are important for throughflow calculations of radial machines. The calculation of highly curved annulus walls with 90° bends in the impeller must be possible (even 180° bends in the crossover channel). The code should provide a simple definition of axial and radial wall geometries and any combination of these, including radial inward flow in a return channel or a radial inlet. It should be possible to include any combination of blade row calculating stations, together with duct flow regions to allow multiple blade rows and multistage simulations. For some modern turbocompressor geometries, curved quasiorthogonal lines are needed to enable blades with sweep and curved leading and trailing edges to be modelled. In order to compute radial blade rows, or the radial channels of radial machines, axial calculating stations parallel to the machine axis are required. Compressible and incompressible fluids should be available, including real gas models and limited amounts of supersonic relative flow in blade rows. In blade rows with a sufficient number of internal planes, an approximation for the blade-to-blade flow field may also be calculated.

The streamline curvature solution method is iterative in terms of several variables (primarily the meridional velocity distribution, but also the density and other gas properties and the streamline locations). After each iteration, a new flow pattern is determined which allows the curvature of the streamlines to be recalculated; this is then used to repeat the cycle of calculations until a criterion for convergence is satisfied. All parameters progressively converge to a final solution within a fairly complex structure of nested iterations described in this chapter. In this convergence procedure, many variables lag behind the main iteration for changes in the meridional velocity. As a result, when updating the meridional velocity, many parameters are treated as constant. These parameters are then updated prior to the next iteration in line with the new estimate of the meridional velocity and curvature. This has no effect on the solution, provided that a converged solution for all variables is reached. An essential part of the streamline curvature method is that it is necessary to calculate the streamline locations on the q-o's in order to obtain the slope and curvature of the streamlines. The shape of the streamlines is often approximated by a curve fit through points of equal mass flow on neighbouring calculation planes.

The initial solution generally comprises a first guess for the streamline positions (typically by dividing the flow path into equal areas at each q-o) and for the flow variables at all the grid points (i.e., the junction points between the variable streamlines and the fixed q-o's). Once a first estimate of the streamline positions is in place, various terms in the radial equilibrium equation, such as the curvature of the streamlines or the rate of change of meridional velocity along the streamlines, can be determined. This first guess is then successively refined with each iteration and the cycle of calculations is repeated.

The meridional velocity distribution is determined using a velocity gradient equation along the fixed calculation stations in conjunction with the continuity equation. The equation for the gradient of the meridional velocity along a q-o is the most important equation of the method and is derived from the inviscid momentum equation in Section 15.4.4 as follows:

$$c_m \frac{dc_m}{dq} = \frac{dh_t}{dq} - T\frac{ds}{dq} - \frac{1}{2r^2}\frac{d\left(r^2 c_u^2\right)}{dq} + \sin\psi \; \frac{c_m^2}{r_c} + \cos\psi \; c_m \frac{\partial c_m}{\partial m} + \tan\gamma \frac{c_m}{r}\frac{\partial (r c_u)}{\partial m}. \tag{15.1}$$

The velocity gradient equation relates the local meridional velocity gradient at a point on the q-o to the shape and curvature of the streamlines, the orientation of the mean stream surface and the flow parameters and their gradients from the previous iteration. The geometrical terms are shown in Figure 15.2 and explained in more detail in Section 15.3. In its simplest form, this equation is the well-known simple radial equilibrium equation of turbomachinery flows describing the relationship between the radial gradient of the axial velocity and the gradient of the swirl in the radial direction, as discussed in Chapter 5. This equation goes back well over 70 years and can be found in different forms in many publications with more accessible derivations given by Cumpsty (2004) and Schobeiri (2005).

In its general form, the radial equilibrium equation considers flow gradients, streamline curvature and radial flows and allows a general arbitrary orientation of the q-o's. The key difference between the simple radial equilibrium equation and the more general equation is the inclusion of the streamline curvature terms, and it is these terms which give their name to the method. There are several forms of this velocity gradient equation, whereby in some versions the equation takes on different forms for the absolute and the relative reference frames. The form used here follows the method

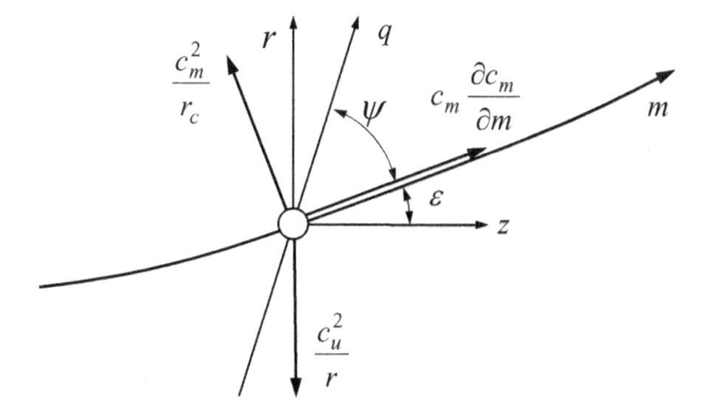

Figure 15.2 Meridional streamline showing the notation for the geometrical terms in the velocity gradient equation.

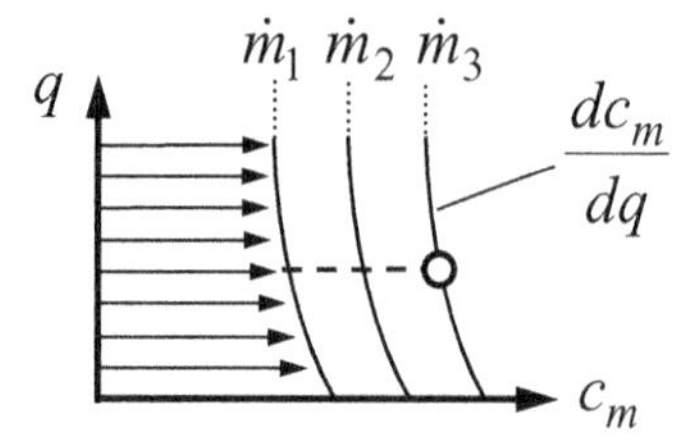

Figure 15.3 The continuity equation and velocity gradient equation determine the meridional velocity distribution.

of Denton (1978) but considers the blade force terms as described by Cumpsty (2004). It also includes special features for radial machines as treated by Casey and Roth (1984) and Casey and Robinson (2010).

The velocity gradient equation is solved in combination with a method for finding the correct velocity level on the mean streamline. This is to ensure that the velocity along the q-o satisfies the continuity equation and therefore gives the correct mass flow across the q-o, as shown in Figure 15.3. In the current method, the meridional velocity on the mean streamline is specified in the innermost iteration, integrated across the flow channel with the help of the meridional velocity gradient and then continually updated until the mass flow is correct. Here some care is needed to ensure that the method converges for all mass flows up to the choking mass flow. For this iteration, the density must be included in the continuity equation. In fact, there are two values of the density that can be used, depending on whether the flow is supersonic or subsonic (Marsh, 1968). The method described here follows that of Denton (1978) in that no attempt is made to distinguish between the two possible density values. During the innermost iteration for the mass flow, the density is simply taken as constant at the value from the previous iteration, and automatically takes on the supersonic or subsonic value as the iteration converges.

The meridional velocity distribution along the q-o determines the spacing of the streamlines of the meridional flow and hence the meridional streamline positions. These positions are continually updated as the program converges. It is not generally acceptable to use the new calculated positions of the streamlines directly from one iteration to the next; a relaxation factor (often much less than unity) is required to factor the streamline shifts to obtain convergence. A serious weakness of streamline curvature methods is that the necessary relaxation factor required becomes very small when the q-o's are closely spaced (Wilkinson, 1970). To obtain a more accurate result, more calculating planes have to be used, and this involves a significant increase in calculating time. For this reason, a coarse grid (compared with modern CFD computations) is used, leading to calculations that take just fractions of a second on a modern PC. The problem with closely spaced q-o's is related to the large errors in the estimated streamline curvature when even a small error in the streamline position is present. It is for this reason that many axial turbomachinery throughflow methods do not include internal calculation stations within blade rows as these automatically decrease the spacing between adjacent q-o's.

The streamline positions on each q-o can be used to interpolate new blade element data (such as blade angles and thickness) appropriate to the actual location of the streamlines and hence to determine the slopes, curvatures and derivatives of the flow parameters along the streamlines. These data are needed in the velocity gradient equation, (15.26). The accuracy and stability of streamline curvature methods are related to the prediction of the curvatures of the streamlines. There are a number of different numerical methods available to make this calculation; see Wilkinson (1970). In the present method, the curvature and flow gradients are calculated in a simple step-by-step parabolic curve using the coordinates of the streamline on the current, upstream and downstream q-o's, following Denton (1978).

In the region between blade rows, the total enthalpy and angular momentum may be considered to be conserved along the meridional streamlines from the previous q-o. The entropy may rise due to additional losses between the calculating stations. On the assumption that the flow follows the mean stream surface, the changes in momentum and enthalpy in a blade row are calculated from the Euler equation. The mean blade stream surface is roughly aligned with the camber surface of the blade. Nevertheless, the true fluid flow direction – considering the incidence and deviation of the flow from the mean blade surface – must be determined; this is generally done using empirical correlations for outlet angle, deviation or slip factor.

15.2.3 Design and Analysis Mode

In axial turbomachinery throughflow codes, it is common to make use of two different modes of operation known as the design and analysis modes. In the design mode, the spanwise work or pressure distribution for each rotor row or stage is specified together with the flow angles or swirl distribution at outlet to the stator blade rows. The code then determines the complete flow angles, velocity field and blade geometry making use of correlations for optimum incidence, deviation and losses. In the analysis mode, the geometry is specified and the code determines the flow field and losses from correlations. In radial compressors with many internal q-o's within the blade rows and a large turning of the flow in the radial direction, it is more appropriate to use the code in the design mode.

15.2.4 Losses and Deviation

Although these equations are inviscid and do not include frictional forces, the effect of the losses are included via the empirical changes to the entropy in the equation of state, such that the final solution has a density and pressure field consistent with the presence of losses in the flow. As explained in this chapter, the equations provide a clear insight into the physics of the inviscid flow and how the geometry controls the flow distribution. At the inlet plane (the first q-o), the user is generally required to specify the variation of total pressure, total temperature and angular momentum or flow angle, together with the gas data and information on the mass flow at the outlet plane. For calculations with choked flows or nearly choked flows, it is not possible to calculate with a specified mass flow, and it is necessary to calculate with specified outlet static pressure, such that the mass flow is a result of the simulation: this changes the nature of the iterations needed.

Additional empirical methods are used to provide data for the loss production and spanwise mixing, for the boundary layer blockage and for the deviation of the flow direction from the mean blade camber surface, so that the effect of losses and blade forces are considered. In fact, Denton (1978) suggests that 'in many applications, throughflow calculations are little more than vehicles for inclusion of empiricism in the form of loss, deviation and blockage correlations, and their accuracy is determined by the accuracy of the correlations rather than the numerics'. The three main effects of the empirical data are that in the equation of state, a change in the entropy leads to a pressure loss for a given value of the total enthalpy. In the continuity equation, the blockage due to the displacement effect of the wall boundary layers leads to a higher value of the meridional velocity. In the momentum equation, the deviation of the flow from the blade direction leads to a change in the swirl velocity calculated, and this is by far the most important empirical information needed in the method.

In this way, although the basic inviscid equation of motion used by the method is inherently incapable of predicting entropy rises through the machine, some effects of losses can be included. There are numerous possible combinations of data for this empirical information, based on various definitions of loss coefficients, dissipation coefficients, efficiencies and so on, as discussed in Chapter 10, and this leads to the largest source of confusion in the data preparation for throughflow codes and to a bewildering array of branching statements within a typical throughflow code to deal with the alternatives. The limitations of the empirical data were severe when most impellers were designed with a large jet-wake flow pattern within the impeller; see Chapter 5. As modern designs have overcome the problems associated with strong separations and jet-wake issues, these limitations are much less relevant today. The low meridional velocity on the casing at the impeller outlet, due to tip leakage which still exists in modern designs, is not captured at all by throughflow calculations, as in the inviscid calculation the curvature of the impeller dominates the flow distribution at impeller outlet and leads to the highest velocities near the casing (Casey and Roth, 1984).

15.2.5 Blade Surface Velocities

The mean stream surface provides the flow field in the meridional plane through the turbomachine. Since the engineer generally also needs information about the distributions of velocity, pressure and Mach number on the blade surfaces, the meridional solution needs to be combined with a blade-to-blade method. This may be a simple approximation or a more accurate solution of the blade-to-blade flow equations. The more accurate methods enable the exact location of the mean stream surface of the flow to be determined at different blade-to-blade planes, so that this can also lead to a further iteration in which the mean stream surface is no longer considered as fixed. Such iterative solutions are often known as S1/S2 solutions (the S1 surface being the blade-to-blade surface and the S2 surface being the meridional surface). Techniques of this nature were developed by Ribaut (1988) and Jennions and Stow (1985). This technique is mainly of historical interest only. As it was cumbersome and brought little additional improvement to the mean stream surface method, it has been replaced by fully 3D CFD solutions with mixing planes, as described in Chapter 16.

The simpler blade-to-blade methods generally use a linear approximation for the velocity variation from suction surface to pressure surface, following Stanitz and Prian (1951). These can only be expected to produce sensible results when a sufficient number of internal blade row calculating stations are included, five probably being the minimum for a good estimate of the blade-to-blade loading distribution.

15.2.6 Spanwise Mixing

A major shortcoming of the basic throughflow method is the neglect of the spanwise transport of angular momentum, energy and losses in the direction normal to the streamlines by turbulent mixing and other processes. By definition, a throughflow code assumes that the flow remains in concentric stream tubes as it passes through the turbomachine, and therefore that no mass transfer occurs across the meridional streamlines which are the stream-tube boundaries. In a duct region of a throughflow calculation enthalpy, angular momentum (swirl) and entropy are taken to be conserved along the meridional streamlines. Methods to take the spanwise mixing into account are described in Section 15.8.

15.3 Notation for the Blade Angles of the Velocity Gradient Equation

In the many different papers on streamline curvature theory, there are various notations used for the angles representing the special orientation of the streamlines, the calculating stations and the blade. The derivation offered here makes use of the notation for angles and their definitions as given in detail in Figure 15.4 and explained in the sections that follow.

- The distance along the q-o is denoted by q and along the meridional streamline by m.
- The inclination angle of the meridional flow to the axial direction, or the pitch angle, is given by ε.
- The slope angle of the q–o line to the meridional direction is denoted by ψ.
- The tangential coordinate at any point is given by the rotational coordinate system (r, θ, z) or by the intrinsic coordinates of the grid, (m, q, θ), where θ is known as the tangential wrap angle.
- The blade angle in the meridional plane at pitch angle, ε, is given as β, and the angle relative to the radial direction as γ_r (similar to the roll angle in a sailing ship) and relative to the axial direction as γ_z (similar to the yaw angle in a sailing ship).
- The intersection of the q-o with the blade gives a lean angle in the direction of the q-o of γ.
- The radius of curvature is denoted by r_c.

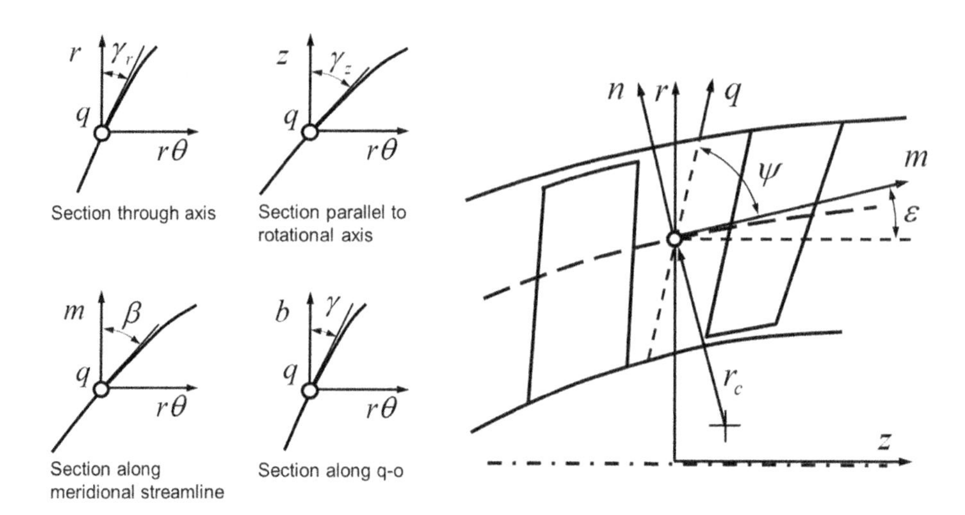

Figure 15.4 Notation for blade angles and for curvature in the meridional plane.

15.3.1 The Radial Lean Angle, γ_r

The radial lean angle is the blade lean to the radial direction and is a geometrical input parameter for the code along the q-o's within blade rows. If a point, q, on a radial impeller blade camber surface is considered the blade lean angle, γ_r, as shown in Figure 15.4, can be defined as the angle made by the blade fibres with a radial line at constant z, as follows:

$$\tan \gamma_r = r\left(\frac{\partial \theta}{\partial r}\right)_z. \tag{15.2}$$

In order to assist the understanding of this lean angle, the analogy to the angle known as roll of a sailing boat might help some readers. This blade lean to the radial direction is used as a geometrical input parameter for the method rather than the tangential wrap angle of the blade, θ, because any associated geometry definition method can calculate this more accurately than the throughflow method, which would need to use interpolation and differentiation along the sparse q-o's within blade rows.

The angle, γ_r, is zero for a blade comprising purely radial blade elements, and as such is generally close to zero for all axial blade rows – which tend to have only small lean angles, especially in rotors – and at centrifugal impeller leading edges or radial turbine trailing edges. It is nonzero for blades with lean. The lean angle can be constant along the span (straight lean) or can change along the blade height (compound lean). If the lean is against the direction of rotation, it is negative, and in the direction of rotation it is positive. Zero lean leads to a right angle between the hub and a radial line on the suction surface, positive lean leads to an obtuse angle and negative lean leads to an acute angle.

15.3.2 The Axial Lean Angle, γ_z

The axial blade angle is the blade angle measured from the axial direction in the direction of rotation and is also a geometrical input parameter for the code along the q-o's within blade rows for the same reason that the radial lean angle is specified. If a point, q, on a radial impeller blade camber surface is considered, as in Figure 15.4, the blade lean angle γ_z can be defined as the angle made by the blade fibres with an axial line at constant radius, as follows:

$$\tan \gamma_z = r\left(\frac{\partial \theta}{\partial z}\right)_r. \tag{15.3}$$

This lean angle is similar to the yaw of a boat.

In an axial blade element – or close to the leading edge of a typical radial impeller – this is the blade camber angle, and at the inlet to an axial blade row it is the blade inlet angle. In a typical rotor blade, the angle is negative, and in a typical stator it is positive. In an axial compressor rotor where the flow is turned by the blade rows towards the axial direction, the absolute values decrease from the leading edge to the trailing edge. By contrast in an axial turbine, where more swirl is added to the flow, the absolute values increase from the leading edge to the trailing edge. For a blade with a purely axial blade element – as in a typical 2D diffuser and at a radial impeller outlet with no rake angle – this angle is zero. In a radial impeller at the trailing edge, γ_z is the rake angle, as defined in Section 11.8.2.

15.3.3 The Meridional Streamline Inclination Angle or Pitch Angle, ε

This angle is the inclination of a meridional streamline to the axial direction, which on the hub and casing walls becomes the meridional slope angle of the walls.

$$\tan \varepsilon = \left(\frac{dr}{dz}\right) = \left(\frac{dr}{dm}\Big/\frac{dz}{dm}\right). \tag{15.4}$$

This angle corresponds to the pitch of a boat.

15.3.4 The Radius of Curvature, r_c

Note that the radius of curvature of the streamline is defined from the pitch angle as

$$\frac{1}{r_c} = \frac{\partial \varepsilon}{\partial m} \tag{15.5}$$

so that the curvature as shown in Figure 15.4 is negative as the meridional slope angle decreases along the streamline.

15.3.5 The Meridional Blade Angle, β

The lean angles γ_r and γ_z of the blade are properties of the blade geometry, but the effective blade angle in the meridional plane depends on the meridional pitch angle and can be determined as follows:

$$\tan\beta = r\left(\frac{d\theta}{dm}\right) = r\left[\left(\frac{\partial\theta}{\partial r}\right)_z \frac{dr}{dm} + \left(\frac{\partial\theta}{\partial z}\right)_r \frac{dz}{dm}\right]$$
$$\tan\beta = \left[\tan\gamma_r \sin\varepsilon + \tan\gamma_z \cos\varepsilon\right]. \tag{15.6}$$

This angle is zero for a blade in which the tangential wrap angle does not change in the meridional direction, so that $d\theta/dm = 0$, such as an axial strut in an axial channel or a radial strut in a radial channel. The tangential wrap angle is taken as positive in the direction of rotation so that in a typical impeller inlet, the value of the meridional blade angle is negative. At the impeller outlet with a backswept blade, the angle is also negative in the notation used here. The values are positive for a typical vaned diffuser.

Some features of the geometry of radial compressors can be examined with (15.6). An old-fashioned impeller with purely radial blade elements, $\gamma_r = 0°$, but with a rake angle at outlet of $\gamma_z = -40°$ and a meridional pitch angle of $\varepsilon = 70°$ on the casing, has an effective outlet angle of $\beta = -16°$ and an effective backsweep angle at this position of 16°. On the hub, which is likely to have a meridional pitch angle closer to $\varepsilon = 90°$, the backsweep would be 0° and the blade angle would be $\beta = 0°$, so that impellers with radial blade elements and negative rake in a sloped meridional channel have an effective backsweep, in this case a mean value of about 8°. Taking the case of an impeller outlet with a backsweep angle of $\beta = -30°$. If there is no axial lean or rake, then $\gamma_z = 0°$ at outlet, and for a radial pitch angle of $\varepsilon = 90°$ (15.6) indicates that the radial lean angle is also $\gamma_r = -30°$, so that the backswept blades do not have radial elements. In an impeller with the same backsweep, but with a rake angle at outlet of $\gamma_z = -30°$, the effect of the rake is to reduce the lean of the blade in the outlet region, which in turn reduces bending stresses in the blade root.

15.3.6 The Lean Angle of the Blade in Direction of the q-o, γ

The lean angles γ_r and γ_z of the blade combine with the meridional angle of the q-o, ψ, to give the lean angle of the blade in the direction of the q-o, as follows:

$$\tan\gamma = r\left(\frac{d\theta}{dq}\right) = r\left[\left(\frac{\partial\theta}{\partial r}\right)_z \frac{dr}{dq} + \left(\frac{\partial\theta}{\partial z}\right)_r \frac{dz}{dq}\right]$$
$$\tan\gamma = \left[\tan\gamma_r \sin(\psi+\varepsilon) + \tan\gamma_z \cos(\psi+\varepsilon)\right]. \tag{15.7}$$

15.4 The Throughflow Equation of Motion

15.4.1 Derivation from the Momentum Equation

There are many publications describing the derivation of the equations for the streamline curvature method, several of which are given in Section 15.2.2. Many of the first publications, however, are more than 60 years old, and so for completeness a fairly thorough derivation of the equations is included here. The benefit of discussing these equations in some detail is the insight they give as to how the geometry controls the pressure gradients and forces acting on the flow, leading to different meridional flow patterns in an impeller.

The basic equations solved by the throughflow method are the inviscid equations of motion with a body force per unit mass, F (see Greitzer et al., 2004 and Hirsch and Deconinck, 1985). If the flow field is assumed to be axisymmetric, so that all terms containing derivatives in the circumferential direction are omitted, these equations can be written in cylindrical coordinates (r, z) as

$$
\begin{aligned}
c_r \frac{\partial c_r}{\partial r} + c_z \frac{\partial c_r}{\partial z} - \frac{c_u^2}{r} &= F_r - \frac{1}{\rho}\frac{\partial p}{\partial r} \\
c_r \frac{\partial c_u}{\partial r} + c_z \frac{\partial c_u}{\partial z} + \frac{c_r c_u}{r} &= F_u \\
c_r \frac{\partial c_z}{\partial r} + c_z \frac{\partial c_z}{\partial z} &= F_z - \frac{1}{\rho}\frac{\partial p}{\partial z}.
\end{aligned}
\tag{15.8}
$$

The terms involving F are source terms representing the body forces which are needed to account for the force of the blades acting on the flow, as there are no blades in the axisymmetric flow field.

Streamline curvature methods work with a grid based on the meridional streamlines, where these are the streamlines of the circumferentially averaged flow in the meridional plane (r, z). Because of this, it is convenient to rewrite these equations for intrinsic coordinates in the meridional plane (m along the streamline and n normal to it) and in terms of the circumferential and meridional velocities (c_u and c_m), where

$$c_m^2 = c_z^2 + c_r^2, \quad \tan\varepsilon = \frac{c_r}{c_z}. \tag{15.9}$$

Note that by definition there is no component of velocity normal to the meridional streamline in an axisymmetric flow ($c_n = 0$).

This transformation to intrinsic coordinates yields the momentum equation in the stream surface normal direction

$$\frac{c_m^2}{r_c} - \frac{c_u^2}{r}\cos\varepsilon = F_n - \frac{1}{\rho}\frac{\partial p}{\partial n}\frac{1}{r_c}, \tag{15.10}$$

in the circumferential direction

$$\frac{c_m}{r}\frac{\partial (r c_u)}{\partial m} = F_u \tag{15.11}$$

and in the meridional direction

$$c_m \frac{\partial c_m}{\partial m} - \frac{c_u^2}{r}\sin\varepsilon = F_m - \frac{1}{\rho}\frac{\partial p}{\partial m}. \tag{15.12}$$

The blade force terms in these equations are mutually orthogonal: along the meridional direction, normal to the streamline in the meridional plane and in the circumferential direction. The second equation shows how the blade force in the circumferential direction is related directly to the rate of change of swirl along the meridional streamline. This allows the blade force term in the circumferential direction to be used directly in the estimate of the blade-to-blade loading (velocity and pressure distributions) from the gradient of the swirl distribution in the blade row; see Section 15.9.2. The other gradients can then be determined from the orientation of the mean stream surface.

Note that if the meridional slope angle of the streamline is zero, then these equations reduce to

$$\frac{c_m^2}{r_c} - \frac{c_u^2}{r} = F_r - \frac{1}{\rho}\frac{\partial p}{\partial r}, \tag{15.13}$$

in the radial direction

$$\frac{c_m}{r}\frac{\partial (rc_u)}{\partial z} = F_u, \tag{15.14}$$

in the circumferential direction and

$$c_m\frac{\partial c_m}{\partial z} = F_z - \frac{1}{\rho}\frac{\partial p}{\partial z} \tag{15.15}$$

in the axial direction.

These equations are the appropriate equations for an orthogonal grid. However, in a typical calculation grid, the streamlines are not exactly orthogonal (normal) to the computing stations – they are called quasiorthogonals to emphasise this. The location and orientation of the q-o calculating stations are specified as geometrical input and consider the orientation of the blades. The computing stations may be leading edges, trailing edges, internal stations or duct stations, and may be straight lines or any other arbitrary lines needed to match the blade geometry, such as a curved leading edge. The preceding equations also need to be formulated in terms of derivatives in the direction q along an arbitrary calculating station rather than in the direction n normal to the streamlines. To do this, the equations normal to and along the meridional streamline are combined to find an expression for the static pressure gradient along the direction of the quasiorthogonal calculating station. Here it is necessary to form the derivative in the q direction, which can be expressed using the chain rule as

$$\frac{d}{dq} = \frac{dn}{dq}\frac{\partial}{\partial n} + \frac{dm}{dq}\frac{\partial}{\partial m} = \sin\psi\frac{\partial}{\partial n} + \cos\psi\frac{\partial}{\partial m}, \tag{15.16}$$

where ψ is the angle between the meridional direction and the quasiorthogonal and is usually close to 90° as these lines are close to being orthogonal, as shown in Figure 15.4. Using this, we derive an equation for the pressure gradient along the q-o, as follows:

$$\begin{aligned}\frac{1}{\rho}\frac{dp}{dq} &= \sin\psi\frac{1}{\rho}\frac{\partial p}{\partial n}+\cos\psi\frac{1}{\rho}\frac{\partial p}{\partial m}\\ &= \sin\psi\left\{-\frac{c_m^2}{r_c}+\frac{c_u^2}{r}\cos\varepsilon+F_n\right\}+\cos(\psi)\left\{-c_m\frac{\partial c_m}{\partial m}+\frac{c_u^2}{r}\sin\varepsilon+F_m\right\}\\ &= \{\sin\psi\cos\varepsilon+\cos\psi\sin\varepsilon\}\frac{c_u^2}{r}-\sin\psi\frac{c_m^2}{r_c}-\cos\psi\; c_m\frac{\partial c_m}{\partial m}+\sin\psi F_n+\cos\psi F_m\\ &= \sin(\psi+\varepsilon)\frac{c_u^2}{r}-\sin\psi\frac{c_m^2}{r_c}-\cos\psi\; c_m\frac{\partial c_m}{\partial m}+\sin\psi F_n+\cos\psi F_m.\end{aligned} \tag{15.17}$$

This can be written as

$$-\sin(\psi+\varepsilon)\frac{c_u^2}{r}+\sin\psi\frac{c_m^2}{r_c}+\cos\psi\; c_m\frac{\partial c_m}{\partial m}=-\frac{1}{\rho}\frac{dp}{dq}+\sin\psi F_n+\cos\psi F_m. \tag{15.18}$$

This equation is where Denton (1978) starts his very short derivation of the radial equilibrium equation by observing, simply by inspection of Figure 15.2, that the left-hand side of this equation is the component of the total acceleration of a fluid particle along the q-o, and the right-hand side is the sum of the forces acting on the fluid particle when only the components along the q-o are considered. These must be in equilibrium.

15.4.2 Body Force Term

The component of the body force (representing the blade force) along the q-o direction is given by the last two terms in (15.18). If the blade orientation is such that the lean angle, γ, of the q-o is zero (which for an axial blade row implies radial blade elements), there is no component of the blade force along the q-o. For a blade with lean, the total blade force along the q-o can be written as

$$F_q = \sin\psi F_n + \cos\psi F_m = F_u \tan\gamma, \tag{15.19}$$

where γ is the lean angle of the q-o to the radial direction; this leads to

$$F_q = \tan\gamma\frac{c_m}{r}\frac{\partial(rc_u)}{\partial m}. \tag{15.20}$$

15.4.3 The Dissipation Force Term

The preceding momentum equations are based on an inviscid flow and are the usual equations used in the code. The use of a momentum equation which does not include any forces due to the friction and viscous flow is not entirely consistent with a flow calculation method that includes losses. To overcome this, a small dissipative force that opposes the velocity vector was included in the momentum equation (Horlock, 1971; Bosman and Marsh, 1974). This elegant move does not, however, improve the accuracy of the method. This is because the main sources of error are related to the estimates of the losses, blockage and deviation from correlations. In reality, the dissipative force is due to the boundary layers on the blade and channel surfaces and is, in any case, smeared out in the circumferential averaging process from the assumption of an axisymmetric flow. For an accurate estimate of the dissipative force, the spanwise distribution and meridional gradient of the entropy must be properly calculated so that its value is then tied with the distribution of the losses across the flow domain. This means that just including this extra term does not improve the accuracy.

15.4.4 Classical Radial Equilibrium Equation

Considering the blade force term, (15.18) can be used to obtain the radial equilibrium equation in the following form:

$$-\sin(\psi+\varepsilon)\frac{c_u^2}{r}+\sin\psi\frac{c_m^2}{r_c}+\cos\psi\, c_m\frac{\partial c_m}{\partial m}=-\frac{1}{\rho}\frac{dp}{dq}+\tan\gamma\frac{c_m}{r}\frac{\partial(rc_u)}{\partial m}. \tag{15.21}$$

In the usual derivation of the simple radial equilibrium equation, it is convenient to replace the static pressure gradients with total enthalpy and entropy gradients, as the change in total enthalpy can be derived from the Euler turbomachinery equation and the change in entropy from the specified losses. The preceding equations need to be adjusted considering the definition of total enthalpy and the entropy from the second law (the Gibbs equation). These can be combined to give the change in static pressure as a function of the changes in total enthalpy, entropy and velocity as follows:

$$\frac{1}{\rho}dp=dh_t-Tds-c_m dc_m-c_u dc_u. \tag{15.22}$$

With this, the gradient of the pressure along the q-o is given by

$$\frac{1}{\rho}\frac{dp}{dq}=\frac{dh_t}{dq}-T\frac{ds}{dq}-c_m\frac{dc_m}{dq}-c_u\frac{dc_u}{dq}. \tag{15.23}$$

With the help of this equation, the momentum equation along the quasiorthogonal can be written as

$$\begin{aligned}&-\sin(\psi+\varepsilon)\frac{c_u^2}{r}+\sin\psi\frac{c_m^2}{r_c}+\cos\psi\; c_m\frac{\partial c_m}{\partial m}.\\&=-\frac{dh_t}{dq}+T\frac{ds}{dq}+c_m\frac{dc_m}{dq}+c_u\frac{dc_u}{dq}+\tan\gamma\frac{c_m}{r}\frac{\partial(rc_u)}{\partial m}.\end{aligned} \tag{15.24}$$

Finally, noting that

$$\begin{aligned}\frac{1}{2r^2}\frac{d(r^2c_u^2)}{dq}&=\frac{1}{2r^2}\left\{r^2\frac{d(c_u^2)}{dq}+c_u^2\frac{d(r^2)}{dq}\right\}=\frac{1}{2r^2}\left\{r^2\frac{d(c_u^2)}{dq}+c_u^2 2r\frac{d(r)}{dq}\right\}\\&=\frac{1}{2}\frac{d(c_u^2)}{dq}+\frac{c_u^2}{r}\sin(\psi+\varepsilon)=c_u\frac{dc_u}{dq}+\frac{c_u^2}{r}\sin(\psi+\varepsilon),\end{aligned} \tag{15.25}$$

the classical radial equilibrium in its conventional form is derived as

$$c_m \frac{dc_m}{dq} = \frac{dh_t}{dq} - T\frac{ds}{dq} - \frac{1}{2r^2}\frac{d\left(r^2 c_u^2\right)}{dq} + \sin\psi \frac{c_m^2}{r_c} + \cos\psi \ c_m \frac{\partial c_m}{\partial m} + \tan\gamma \frac{c_m}{r}\frac{\partial(rc_u)}{\partial m}. \tag{15.26}$$

This is the velocity gradient equation that is used in the present method and is the basis of all streamline curvature throughflow methods. It is a nonlinear partial differential equation, and in the present method it is solved by finite difference techniques.

15.5 Streamline Curvature Velocity Gradient Equation

15.5.1 Different Forms of the Equation

An important difference between Denton´s throughflow method (Denton, 1978), on which the current method is based, and other methods is that the radial equilibrium equation is used directly in the form given by (15.26) and no further changes are made to distinguish between stationary and rotating blade rows or to distinguish between design and analysis situations. In a design situation, the swirl of the flow is specified, and in an analysis situation it is calculated from the blade angles. The terms in the equation are determined by various auxiliary equations, as described in this section, and are all determined to be consistent with the meridional velocity distribution from the previous iteration, whereby the swirl itself, rc_u, and not the circumferential velocity, c_u, is one of the parameters used in this process. This implies that for a calculation in a region with no hub, where the radius is zero, the value of c_u is indeterminate. To avoid this issue, such regions are considered to have a small radial hub of low blockage, rather like having a thin frictionless wire along the axis of the flow calculation domain. During the integration of the radial equilibrium equation, the computation always operates as if the swirl is specified and all the terms on the right-hand side are fixed from the previous iteration. In other methods, considerable complexity is added – alternative forms of this equation are generated by inserting the appropriate auxiliary equations directly into the radial equilibrium equation and deriving a new form of the radial equilibrium equation. This new equation is different for rotors and stators.

In these other forms of the radial equilibrium equation, the term representing the gradient of swirl along the q-o is often replaced by a term related to the gradient of a flow angle along the q-o. It is not clear that these forms of this equation have any particular advantage. They are certainly more complex, with some involving terms such as $1/(1 - M^2)$, where M is the local Mach number. Clearly, like the matrix method of Marsh (1968), these forms need to be avoided when the Mach number approaches sonic conditions. Denton (1978) states that these alternatives for the radial equilibrium equation are 'not obviously more accurate than the much simpler process of evaluating it from the previous iteration'.

15.5.2 Centrifugal Acceleration Due to Swirl

This component of the centrifugal acceleration along the q-o due to the angular (swirl) component of the velocity of the particle is given by the term

$$\cdots - \frac{1}{2r^2}\frac{d\left(r^2 c_u^2\right)}{dq}\cdots. \tag{15.27}$$

The value of this term is negative, that is, it is directed towards the machine axis, because the acceleration force is directed inwards to counteract the pressure gradient due to swirl. This term is the primary term in radial equilibrium calculations of axial machines leading to a radial pressure gradient from hub to shroud in a swirling annular flow; see Chapter 5. In the inlet to a radial impeller, the swirl is usually zero, so this term has no effect in the inlet, but as swirl is imparted to the flow within the impeller, the centrifugal acceleration begins to play a role in the distribution of the meridional velocity across the span. Near the outlet of a radial compressor, the swirl becomes more uniform across the span, and therefore this term then becomes small. The cross-channel pressure gradients due to the swirl in a radial flow channel are then negligible in their effect in the radial equilibrium equation.

In an impeller designed with no swirl at inlet, the centrifugal acceleration has no effect on the radial distribution of the flow at the inlet, but when the same impeller is used with an inlet guide vane this component will cause an acceleration of the meridional flow near the hub, which changes the incidence distribution of the impeller.

15.5.3 Centrifugal Acceleration Due to Curvature

This component of the centrifugal acceleration along the q-o due to the curvature of the meridional streamline is directed towards the centre of curvature of the streamline to counteract the pressure gradient due to the flow curvature and is given by

$$\cdots + \sin\psi\frac{c_m^2}{r_c} + \cdots. \tag{15.28}$$

The equation for the radius of curvature of a streamline in the meridional plane is given by (15.17). This radius of curvature is negative with streamline curvature as shown in Figure 15.4, where $d\varepsilon/dm < 0$. In such situations where the radius of curvature is negative, this term is negative and works inwards along the quasiorthogonal. For a typical radial impeller, the curvature is positive and the term works outwards along the quasiorthogonal, leading to a higher meridional velocity on the casing; see Section 5.3.5.

The angle ψ, which is the angle between the meridional direction and the q-o direction, is typically close to 90° as the q-o's are nearly orthogonal to the streamlines. As a result, the sine function in (15.28) is generally close to unity. The term includes the streamline curvature as the inverse of the radius of curvature, such that this becomes negligible on streamlines with no curvature. In multistage axial machines with little wall curvature, the curvature changes from positive to negative between the rotor and stator blade rows as the flow is redistributed due to the effect of radial equilibrium. In a radial impeller, its magnitude is positive and depends mainly on the curvature of the meridional channel walls of the annulus, leading to higher meridional velocities on the casing than on the hub.

The designer is able to lower the meridional velocity gradient due to this term by increasing the radius of curvature of the meridional contour. In cases with high wall curvature and low meridional velocity, this term can lead to very low velocity (and even reverse flow) on the hub and also to high incidence on the hub streamlines.

15.5.4 Normal Acceleration along the q-o

The component of the normal acceleration along the q-o due to the rate of change of velocity along the streamline is given by

$$\cdots + \cos\psi \; c_m \frac{\partial c_m}{\partial m} + \cdots. \tag{15.29}$$

The calculating stations or q-o's are more or less normal to the streamlines and the angle ψ is generally close to 90°. Because of this, this normal acceleration term will generally be small even if the flow is accelerating or decelerating in the direction of the flow so that this term has little effect on the meridional velocity distribution.

15.5.5 Pressure Force Term

The force due to the total pressure acting along the direction of the q-o is given by the terms

$$\cdots \frac{dh_t}{dq} - T\frac{ds}{dq} \cdots. \tag{15.30}$$

If the losses are small and evenly distributed across the span, the pressure force term is essentially related to the gradient of total enthalpy along the q-o. Thus, the rate at which work is done within the impeller on the different streamlines may affect the meridional velocity gradients of the flow. However, a higher rate of work input into a streamline leads to a higher swirl velocity such that this effect is counteracted by the swirl gradient term. Together the acceleration and the pressure gradient terms are known as the radial equilibrium terms and are usually smaller than the curvature terms in radial impellers.

15.5.6 Body Force Term

The component of the body force (representing the blade force) along the q-o direction is given by

$$\cdots \tan\gamma \frac{c_m}{r} \frac{\partial (r c_u)}{\partial m} \cdots. \tag{15.31}$$

Here γ is the lean angle of the q-o relative to the radial direction. If the blade orientation is such that the lean angle of the q-o is zero (which for an axial blade row or the inlet to a radial impeller implies radial blade elements), there is no component of the blade force along the q-o.

The main consequence of blade lean is a redistribution of the flow in the spanwise direction. This can influence the impeller outlet velocity distribution and the secondary flows. Since the flow must still follow the hub and casing, the blade force must be balanced by a pressure gradient; an outwards blade force then increases the pressure and reduces the velocity on the casing and vice versa on the hub. However, the use of lean is not generally suitable for high-speed centrifugal impellers as it increases the bending stresses in the blade roots near the impeller outlet.

15.5.7 Dissipation Force Term

The preceding momentum equations are based on an inviscid flow, and these are the usual equations used in the code. It has been mentioned previously that a small dissipative force that opposes the velocity vector can be included in (15.26), and these have been included in the code as an option that is not generally used. There are numerous derivations of these terms in the literature which show that the dissipation force term along the quasiorthogonal direction is of the form

$$F_d = \cos\psi \, T \frac{ds}{dm}. \tag{15.32}$$

In a flow with low losses and with a nearly orthogonal grid, $\psi = 90°$, the dissipation term is very small, and it is justified that it is neglected.

15.5.8 Modification of the Meridional Velocity Gradient

The spanwise meridional velocity gradient from the radial equilibrium equation determines the variation of the meridional velocity across the span. There are several situations in throughflow simulations where the gradient determined by the radial equilibrium equation must be ignored and replaced by another condition – especially for choked flows and for situations with a risk of flow reversal.

Under certain circumstances, the meridional velocity gradient equation becomes sufficiently large relative to the mean velocity level to cause a reverse flow. This is a physical problem related to the fact that a large radial pressure gradient in such a flow naturally causes low velocity on the hub or casing, and in a real machine, flow separation with reverse velocities occurs. This possibility is excluded in the streamline curvature method described here by setting the minimum meridional velocity to be equal to 5% of the mean streamline value. This means that the solution is no longer physically correct but it allows convergence with an appropriate warning so that the cause of the problem can be eliminated in the next design iteration.

In choked flows at the compressor inlet, the value of the meridional velocity on a particular streamline may be determined from the choked mass flow for this streamline, and the value from the radial equilibrium equation has to be ignored. In high-speed compressor flows, situations can occur where the radial equilibrium equation predicts a gradient in the meridional velocity that would lead to a mass flow at a particular section that is larger than that possible due to choking. In such cases, it is necessary to set a maximum value of the meridional velocity at the choked calculating station of the compressor near the leading edge, which is consistent with the maximum possible mass flow on the choked streamline. In radial turbines where choked flow occurs at the outlet, the variable supersonic deviation means that no such limit needs to be applied.

15.5.9 Integration of the Velocity Gradient Equation

It is necessary to integrate the radial equilibrium equation, (15.26), across the span in order to derive the variation of velocity along a calculating station. The procedure adopted is essentially that described by Denton (1978). The left-hand side of the radial equilibrium equation can be written as

$$c_m \frac{dc_m}{dq} = \frac{1}{2} \frac{dc_m^2}{dq} \tag{15.33}$$

so that the value of this at each grid point can be calculated. Note that in this process separate terms are calculated to represent the streamline curvature term (SCT), the radial equilibrium term (RET) and the blade force term (BFT), as follows:

$$\begin{aligned} SCT &= \sin\psi \frac{c_m^2}{r_c} \\ RET &= \frac{dh_t}{dq} - T\frac{ds}{dq} - \frac{1}{2r^2}\frac{d\left(r^2 c_u^2\right)}{dq} \\ BFT &= \tan\gamma \frac{c_m}{r}\frac{\partial(rc_u)}{\partial m}. \end{aligned} \tag{15.34}$$

It is sometimes useful to analyse the values of these individual terms at each point, as in some situations they can be helpful with the diagnosis of convergence problems. In this equation, the derivatives along the q-o are determined by finite difference equations. These take the mean value of the terms between the streamlines j and $j+1$. For example, the contribution of the entropy term is calculated from the gradient along the q-o using a forward difference as follows:

$$\left(T\frac{ds}{dq}\right)_{i,j} = \frac{1}{2}\left(T_{i,j+1} + T_{i,j}\right)\left(s_{i,j+1} - s_{i,j}\right)/\left(q_{j+1} - q_j\right), \tag{15.35}$$

and other gradient terms along the q-o are calculated in a similar way. The derivatives along the streamlines and the curvature terms and meridional flow angles are determined using a local curve fit of a parabola through three points along the streamline, following the procedure suggested by Denton (1978). In the internal region of the calculation domain, the three points are symmetric around the point under consideration and include the upstream and downstream points, as in a central difference method. At the inlet plane, two downstream points are taken, and at the outlet boundary two upstream points are taken. The parabolic interpolation routines are not given in detail here but are described by Denton (1978).

The velocity gradient equation is then integrated outwards and inwards from the midstreamline as follows:

$$\left(c_m^2\right)_{i,j+1} = \left(c_m^2\right)_{i,j} + \left(\frac{dc_m^2}{dq}\right)_{i,j}(q_{j+1} - q_j), \tag{15.36}$$

whereby the integration starts with the values at the mean streamline from the previous iteration (or from best guess initial condition or from an earlier solution). Traps are included to avoid negative velocities and to avoid values that would be larger than the choking condition. Here choking is calculated with the throat area and the corrected mass flow function.

Equation (15.36) provides the variation in c_m with the distance q along the q-o. But the level of c_m in this process may not satisfy the continuity equation

$$\int_{hub}^{casing} \rho c_m \sin(\psi)(2\pi r - Zt_u)(1 - B)dq = \dot{m}, \tag{15.37}$$

which considers the blockage of Z blades with circumferential thickness t_u and the boundary layers; here B is the fraction of the annulus blocked by the blade boundary layers. An iteration is then performed by adjusting the value of the meridional velocity on the mean streamline until the continuity equation is satisfied, as shown in Figure 15.3. During this process, the density is held constant at the value in the previous iteration. The continuity equation can then be thought of as providing the necessary constant of integration in the streamline curvature equation.

15.5.10 Update of Streamline Positions

The procedure outlined in the preceding section provides a new estimate of the velocity distribution on each quasiorthogonal. The mass-flow distribution related to this velocity distribution is calculated and then used to relocate the streamlines such that a certain mass flow is passed in each stream tube; see Figure 15.5. To prevent instability in the calculation, the changes in streamline position have to be relaxed. The streamlines, for example, may only be moved by 5% of the calculated distance to avoid instability. The aspect ratio of a cell (height, h, divided by streamwise extent, Δm) causes instability if it gets too high (above about $h/\Delta m = 15$) (Wilkinson, 1970).

The relationship between the relaxation factors needed and the aspect ratio of the computational grid in a through flow calculation was theoretically derived in the classic paper of Wilkinson (1970). His studies showed that close spacing of the quasiorthogonal lines required more damping. This is related to the fact that if the calculating stations are closer together, then a minor error in the displacement of a streamline can lead to a large curvature. In such a case, it is only possible to take a small part of the new solution forward to the next iteration, and so the number of iterations naturally increases.

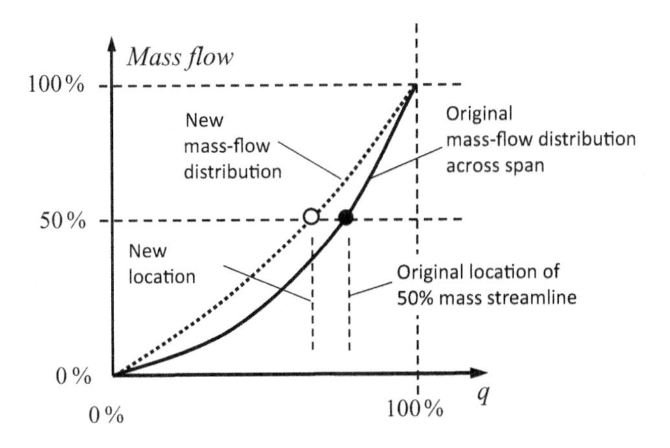

Figure 15.5 Recalculation of position of the mean stream surface enclosing 50% mass flow.

Wilkinson derived an equation of the following form to calculate the optimum relaxation factor for a certain grid aspect ratio:

$$RF = \frac{k_1}{1 + k_2 \left(\frac{h}{\Delta m} \right)^2}. \tag{15.38}$$

Wilkinson also showed that the value given here as k_2 is actually a function of both the Mach number and in blade rows also of the flow angle. The coefficient k_2 also depends of the method used to calculate the curvature of the streamlines. The aspect ratio of the grid, $h/\Delta m$, in this equation is the ratio of the calculating station length to the local meridional spacing of the grid lines.

A scheme to determine the damping factors has been developed which is robust and requires no data to be specified by the user. In the first instance, a local aspect ratio for each q-o is calculated based on the spacing on the midstreamline. The maximum value is also noted together with its location. The average aspect ratio of the whole domain is calculated as a general measure of the difficulty of convergence, and this is used to set an upper limit to the relaxation factor. The local damping factor for each q-o is calculated with the Wilkinson equation as given previously, but with $k_1 = 1.0$ and $k_2 = 0.125$. This damping factor is modified within the calculations if the absolute flow angle is large. Following this, the values of the damping factor near to a single cell with high aspect ratio are smoothed: experience shows that a single cell of high aspect ratio can be tolerated well.

A value of $RF = 0.01$ for the relaxation factor corresponds to an aspect ratio of around 15 and will lead to long calculation times since no more than 1% of the new solution can be taken into account in each iteration. In order to reduce an error in the initial solution to 0.01% of its initial value, the number of iterations required would be

$$n = \frac{\ln(0.0001)}{\ln(0.99)} = 916. \tag{15.39}$$

A value of 0.001 for this relaxation factor (aspect ratio near 50) would lead to 10 times the number of iterations. Values of the aspect ratio larger than 15 are thus not recommended. In cases where this is unavoidable, such as high aspect ratio blades with high span and small chord with internal calculating stations, then the method still converges but at a slower rate. If $RF = 0.1$ is selected as the limit, only a 10th of the number of iterations are needed. Between 100 and 200 iterations are usually sufficient to converge a radial compressor streamline curvature solution.

As in most numerical methods, a finer grid leading to a closer spacing of the quasiorthogonal lines and streamlines will lead to a higher numerical accuracy of the code. However, closer spacing of the quasiorthogonal lines for streamline curvature calculations causes instabilities in the convergence process. This can be overcome by increasing the numerical damping (lower relaxation factors), but this in turn leads to an increase in computational time so that a compromise is generally needed. With modern computers, this compromise is less relevant, as the time taken for convergence is usually only fractions of a second.

15.6 The Iterative Scheme

15.6.1 Iteration to Mass Flow

The iterative scheme of the streamline curvature method starts from a guessed initial condition, and to achieve convergence it involves a series of four nested iteration loops – the inner, middle and outer loops, together with an operating point loop if multiple operating points are calculated for map prediction, as illustrated in Figure 15.6. At the start, some initial guesses of flow velocities and streamline

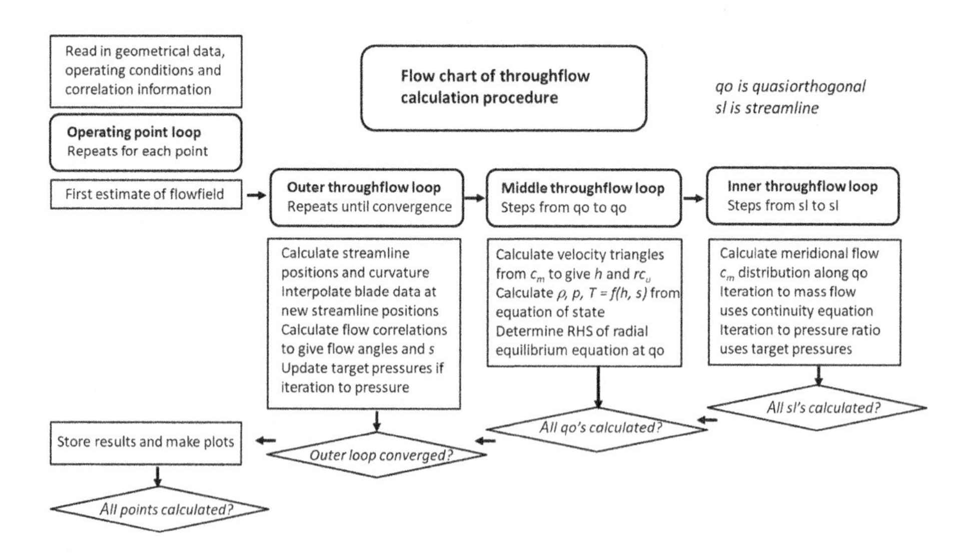

Figure 15.6 Flow chart of throughflow nested iterations.

locations are made, or these can be taken from an earlier calculation with a different flow condition or a similar geometry. The velocity distribution along each q-o and streamline is typically determined from the assumption of a uniform c_m or c_m/u and the streamline locations from uniform ρc_m. In the current method, a piecewise parabola is used along the streamlines to evaluate the meridional radius of curvature and the meridional pitch angle at each grid point. From specified values at the inlet plane, the method marches along the streamlines to determine the swirl, the total enthalpy and the entropy. Outside a blade row and up to a leading edge, the swirl is constant, and within the blade row up to the trailing edge, the flow angle is specified. The total enthalpy is determined from the changes in swirl in the rotor blades using the Euler equation. The entropy can be evaluated from loss or dissipation coefficients or specified polytropic efficiency.

These values allow the radial equilibrium equation to be evaluated along a q-o. The innermost iterative loop integrates this velocity gradient across each stream-tube to give a prediction of the current mass flow. The innermost loop carries out this integration using the current estimate of the gas density and the current estimate of the right-hand side of the radial equilibrium equation, both of which are taken to be constant in this inner loop at the values determined by the previous outer loop. The procedure starts with the value of the meridional velocity on the midstreamline from the previous outermost loop and integrates the radial equilibrium equation between the annulus walls to determine the mass flow consistent with this. This is compared to the specified mass flow for this q-o, and the value of the midstreamline meridional velocity is then adjusted accordingly and the iteration repeated until convergence.

The innermost iterative flow loop requires the meridional velocity and the density to be known at all streamline locations. The middle loop steps from one q-o to the next to determine the terms in the radial equilibrium equation for each streamline across the specific q-o and the local density for use by the inner loop. These values are based on the previous estimate of the meridional velocity distribution which was used in the outer loop to calculate the empirical information and the streamline curvature. Following the use of the inner loop to determine the new meridional velocity distribution along the q-o, the final step in the middle loop is to redistribute the streamlines on the q-o consistent with the new meridional velocities. The changes of streamline position are factored as explained earlier.

The outermost loop examines the whole flow field at the same time. It uses the newest estimate of the meridional velocities and the streamline positions to determine the streamline curvature, streamline slopes and streamline acceleration terms in the radial equilibrium equation. The appropriate blade angles and geometrical information are then interpolated based on the new streamline positions. The final step is to calculate the empirical information – losses, blockage and deviation – to be consistent with the new streamline positions and the values of Mach number, loading and incidence from the last middle iteration. The control then passes to the innermost mass-flow loop to determine the new meridional velocity distribution.

15.6.2 Iteration to Pressure Ratio

The preceding procedure is used in the basic streamline curvature method, in which the mass flow is prescribed. The method will not converge if a mass flow greater than the choking mass flow is prescribed. In such cases, it is essential to specify the overall pressure ratio, allowing the calculation itself to take choking into account and thus to correctly determine the choked mass flow. The iteration procedure for prescribed pressure ratio follows the innovative 'target pressure' suggestion of Denton (1978), with details as described in Came (1995) and Casey and Robinson (2010). This is very relevant for the design of radial turbines but is not as relevant for the design of radial compressors, so it is not described in full detail here.

In an iteration with a prescribed pressure ratio, trailing edge 'target pressures' are defined from the specified overall pressure ratio together with a guess of the inlet mass flow. At all trailing edges, an iteration on the midspan meridional velocity is used to equalise the midspan local static pressure and the target pressure rather than to achieve a specified mass flow. At other planes, the velocity is determined with the mass-flow iteration from the continuity equation. Once the target pressures have been achieved within a specified limit, the target pressures for all trailing edges are updated to try to achieve the correct mass flow imbalance on the next downstream trailing edge. The inlet mass flow is also updated to be consistent with the mass flow through the first blade row; the process is then repeated.

15.7 Empirical Modifications

15.7.1 Deviation and Incidence

The angles in the general streamline curvature equation are the angles of the mean stream surface of the flow, which means it is these angles that actually need to be specified. They are not identical to the angles of the mean blade camber surface owing to the effects of incidence at a leading edge and deviation or slip close to a trailing edge.

The current throughflow method computes the flow based on the geometry of the mean camber surface of the blade, which is specified in the geometry input file. The true mean stream surface is not congruent with the mean camber surface, and the differences have to be taken into account by empirical modifications. This is done in the present code by the use of blending functions for the change of blade angle in the blade row. The basic idea of these procedures is a combination of the use of blending factors on the swirl as outlined by Traupel (2000) and so-called 'departure angles' as described by Smith (2002). These effectively allow the blades to be partly transparent to the flow, such that the flow angle differs from the blade angle. The most obvious example here is at a blade row inlet with incidence, where the flow just upstream of the blade has a different flow angle to the blade inlet angle. In the forward part of the blade, the mean flow angle will differ from the blade angle due to the incidence effects. If these effects are ignored and it is assumed that the flow is immediately congruent with the blade at the leading edge, then the effects of incidence do not appear in the blade-to-blade loading estimates, but on entry to the blade the leading edge actually has an infinite loading level.

In order to account for the changes in angular momentum of the flow within a blade row, the swirl relative to the blade row is calculated by the procedure of Traupel as follows:

$$(w_u)_{i,j} = (c_m)_{i,j} \tan\left(\beta'_{i,j} + i_{le}(\gamma_{in})_{i,j} + \delta_{te}(\gamma_{out})_{i,j}\right). \tag{15.40}$$

This equation takes the blade angle and modifies it to account for the incidence at the leading edge and the deviation at the trailing edge. The incidence and deviation modification terms are steadily reduced within the blade row by blending factors. The inlet blending factor, γ_{in}, decreases from unity at the leading edge to zero within the blade row so that the history of the upstream incidence progressively disappears. The outlet blending factor, γ_{out}, increases from zero within the blade row to unity at the trailing edge with the result that the flow deviation builds up steadily in the blade row. A similar equation can be derived when using the departure angle method of Smith, and both methods are available in the code.

Suggestions for the values of the blending factors have been given by Wilkinson (1969) and by Liu et al. (2000). In the inlet region, it is assumed that the incidence effects disappear at a meridional distance from the leading edge corresponding to about one blade pitch, or at a fractional meridional distance of

$$m'_{i,\,in} = \frac{2\pi r_{i\_le,j}}{Z m_{j,\,bl}} \tag{15.41}$$

from the inlet. If the blades are very widely spaced and this value is larger than 0.5, then a value of 0.5 is used. The blending factor at inlet is given by

$$(\gamma_{in})_{i,j} = 1 + a\left(\frac{m'_{i,j}}{m'_{in}}\right) + b\left(\frac{m'_{i,j}}{m'_{in}}\right)^2 + c\left(\frac{m'_{i,j}}{m'_{in}}\right)^3, \tag{15.42}$$

where $a = -3$, $b = 3$ and $c = -1$ following Wilkinson (1969). A similar approach is used to determine the growth of the deviation towards the trailing edge; again, the earliest point at which the deviation can have an effect is taken as 50% of the blade length. An advantage of this blending technique in high-speed compressors is that the blade is effectively slightly transparent to the flow at the inlet and lower Mach numbers result, increasing the stability of the method.

15.7.2 Losses

Many methods have been included in the throughflow calculation procedures to account for the effects of irreversibility. The simplest method is that suggested by Marsh (1968), in which a local small-scale static–static polytropic efficiency is specified. This is done using the formulation based on entropy as already given in Chapter 4:

$$\eta_p = \frac{s_{1h} - s_2}{s_{1h} - s_1}. \tag{15.43}$$

The state $1h$ is denoted by conditions (h_2, p_1), with the final enthalpy of the process but the pressure of the initial state. This would be the state that would be achieved if all the work input in the compression process was dissipated by the internal losses such that no pressure rise occurs. On this basis, the change in entropy from one grid point $(i - 1, j)$ to the next (i, j) along the meridional streamline is given by

$$\begin{aligned} s_2 &= s_{i,j} = s_1 + \left(1 - \eta_p\right)\left(s_{1h} - s_1\right) \\ \eta_p &= \eta_{p,(i,j)} \\ s_1 &= s_{i-1,j} \\ s_{1h} &= f\left(p_{i-1,j}, h_{i,j}\right). \end{aligned} \tag{15.44}$$

Another very effective technique for estimating the viscous losses at the design point in radial compressors is to apply a dissipation loss coefficient, as described in Chapter 10, which leads to a change in entropy of

$$ds = \frac{\rho c_d P w^3}{T\dot{m}} dl. \tag{15.45}$$

for a region of length dl in the flow channel. This term can be determined for each surface of the flow path from one q-o to the next in a throughflow calculation to determine the entropy production. Correlations for the losses with variation of the flow rate in radial compressors following the method of Casey and Rusch (2014) are also included in the code.

15.7.3 Blockage

Empirical information about the end-wall and blade boundary layer blockage can be included in the calculations by including values for the blockage terms in the continuity equation, (15.37). The general technique adopted is to calculate without blockage for large radial machines at high Reynolds numbers and to increase the blockage for small machines at lower Reynolds numbers. The value of the blockage needed can be determined from calculations of experimental data on similar compressors.

15.8 Spanwise Mixing

A major shortcoming of the basic streamline curvature method is the neglect of spanwise transport of angular momentum, energy and losses in the direction normal to the streamlines, as briefly discussed in Section 15.2.6. In reality, there are several mechanisms that lead to an apparent spanwise transport of fluid relative to flow on the mean streamlines:

- Nonaxisymmetric blade-to-blade stream surfaces as a result of streamwise vorticity shed by the blades (leading to stream-surface twist), which is averaged out in the mean stream surface
- Secondary flows in the end-wall boundary layers and in the blade boundary layers, which can only be accounted for by a 3D viscous method
- Wake momentum transport downstream of blade rows
- Tip-clearance flows with tip-clearance vortices
- Turbulent diffusion

Where these effects are not modelled, throughflow codes in which realistic loss levels are specified for the end-wall regions often predict unrealistic profiles of the loss distribution after several stages. This is because there is no mechanism for the high losses generated near the end-walls to be mixed out. The simplest approach to this problem is to specify unrealistic loss distributions across the span, thus avoiding high levels in the end-walls. In axial compressor and turbine correlations, there is usually a separate profile loss and an additional boundary layer and clearance loss near the end-walls. Most useful methods find an approximation to distribute the end-wall losses across the span in a realistic way, so that they act as if they are already mixed. In radial compressors, a useful approximation involves specifying a mean-line value of loss coefficient or efficiency and to assume that the entropy generated by the losses is the same on each stream-tube. This approximates to a complete mixing of the entropy distribution across the span. In the current method, a modern spanwise mixing method is also included, as outlined in the following subsections, together with an additional strong mixing at impeller trailing edges.

The experience of the many authors who have worked on this problem (AGARD-AR-175, 1981; Adkins and Smith, 1982; Gallimore, 1986; Lewis, 1994) is that spanwise mixing is needed to mix out the high losses close to the end-walls but that the method for including the mixing is not particularly important, so that a relatively crude physical model will suffice for this second-order effect.

15.8.1 Denton Mixing Model

The spanwise mixing model of Denton, as published in AGARD-AR-175 (1981), assumes that because of spanwise mixing, there is no perfect convection of conserved properties along streamlines. The values of entropy, angular momentum (swirl) and total enthalpy on a particular q-o in a duct flow are determined mainly by the values on the same streamline at the upstream q-o and partly by the values transferred from the adjacent streamlines through turbulent mixing. A fraction, $1 - f$, is convected along the streamlines, and a fraction, $\frac{1}{2}f$, is transferred from each of the two adjacent streamlines, where f is a mixing factor with a value less than unity; see Figure 15.7. In this procedure, the value of the parameter, P, at a particular station is first calculated

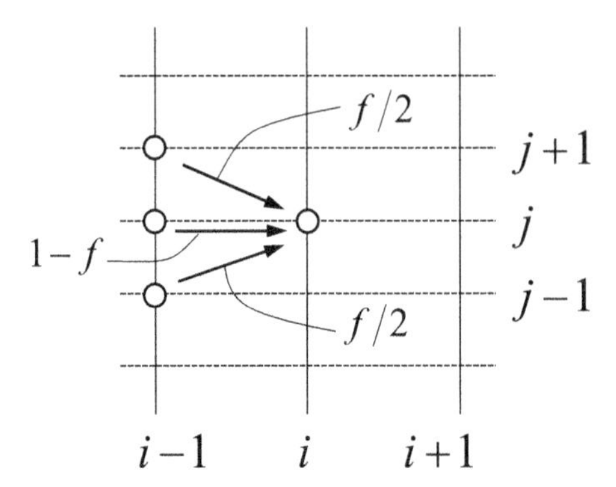

Figure 15.7 Spanwise mixing model of Denton.

on the assumption of no mixing, and this is then modified by a small amount to take account of mixing, where the modification is given by

$$\Delta P_{i,j} = (-f)P_{i-1,j} + (f/2)P_{i-1,j+1} + (f/2)P_{i-1,j-1}. \tag{15.46}$$

In this way, a fraction, f, of the conserved parameter, P, diffuses away from the streamline under consideration, j, between the previous q-o, $i-1$, and the current q-o, i, ($f/2$ to the upper streamline and $f/2$ to the lower streamline) while a fraction, $f/2$, of the values on the upper, streamline, $j+1$, and lower streamline, $j-1$, diffuses to the streamline, j. This can also be written in the following form:

$$\begin{aligned}\Delta P_{i,j} &= (f/2)\left(P_{i-1,j+1} - P_{i-1,j}\right) - (f/2)\left(P_{i-1,j} - P_{i-1,j-1}\right)\\ &= (f/2)\left[\left(P_{i-1,j+1} - P_{i-1,j}\right) - \left(P_{i-1,j} - P_{i-1,j-1}\right)\right].\end{aligned} \tag{15.47}$$

Writing the difference between the parameter, P, on adjacent streamlines, one obtains that

$$\begin{aligned}\delta P_{i-1,j+1} &= P_{i-1,j+1} - P_{i-1,j}\\ \delta P_{i-1,j} &= P_{i-1,j} - P_{i-1,j-1}\\ \Delta P_{i,j} &= (f/2)\left(\delta P_{i-1,j+1} - \delta P_{i-1,j}\right).\end{aligned} \tag{15.48}$$

If the parameter, P, has a positive gradient along the quasiorthogonal, there is a positive contribution to the value from the upper streamline and a negative contribution from the lower streamline due to mixing. If the gradient is constant, then this leads to no change in the parameter, P.

This simple idea is very effective in replicating mixing as the flow proceeds downstream. Denton included this model in an attempt to force mixing to take place where this happens, but was fully aware that the model may be physically unrealistic. This model is unsatisfactory as soon as the meridional grid spacing or the number of streamlines is changed as then it requires a different value of the mixing factor f to produce the same level of spanwise mixing. Denton suggested simply that a value of $f = 0.5$ should be used to overcome entropy buildup in multi-blade-row calculations. In fact, this buildup can be neutralised by using a turbulent diffusion equation to determine the strength of the mixing factor f, as explained in the following section.

15.8.2 Diffusion Equation Mixing

Following the approaches given by Lewis and Gallimore, it was discovered that the spanwise mixing of a parameter P, where P may be angular momentum, total enthalpy or entropy, is best determined by a diffusion equation of the type

$$c_m \frac{\partial P}{\partial m} = \varepsilon \frac{\partial^2 P}{\partial q^2}, \tag{15.49}$$

where m is the meridional direction, q is the spanwise direction, P is the parameter undergoing spanwise mixing and ε is a diffusion coefficient. For simplicity in this model, q is taken as the distance along the quasiorthogonal rather than the exact spanwise direction. The change in the parameter P due to spanwise mixing by diffusion for a small step, δm, along the meridional streamline is given by

$$\Delta P = \frac{\delta m}{c_m} \varepsilon \frac{\partial^2 P}{\partial q^2}. \tag{15.50}$$

Note that this equation correctly predicts that no spanwise mixing occurs with a diffusion coefficient of zero. The second derivative along the q-o can be written as

$$\frac{\partial^2 P}{\partial q^2} = \frac{\partial}{\partial q}\left(\frac{\partial P}{\partial q}\right). \tag{15.51}$$

This is important as it shows that if there is no change in the gradient of parameter, P, along the q-o, the second derivative is zero and again no spanwise mixing takes place. Thus, in a flow with a constant value across the span there is no effect on parameter P, even if spanwise mixing occurs. In addition, if there is a constant gradient across the span there is no effect on the parameter, P, as the spanwise transfer due to diffusion of P from the streamline with the higher value is compensated for by the spanwise transfer from the adjacent streamline with the lower value.

An estimate for the value of this second derivative can be found from

$$\frac{\partial^2 P}{\partial q^2_{i,j}} = \frac{\partial}{\partial q}\left(\frac{\partial P}{\partial q}\right)_{i,j} = \frac{2}{q_{i,j+1} - q_{i,j-i}}\left(\frac{P_{i-1,j+1} - P_{i-1,j}}{q_{i,j+1} - q_{i,j}} - \frac{P_{i-1,j} - P_{i-1,j-1}}{q_{i,j} - q_{i,j-1}}\right), \tag{15.52}$$

and on the assumption that the streamlines are evenly spaced, we obtain

$$q_{i,j+1} - q_{i,j-i} = 2\left(q_{i,j+1} - q_{i,j}\right) = 2\left(q_{i,j} - q_{i,j-1}\right) = 2\delta q_{i,j}. \tag{15.53}$$

The second derivative of P along the q-o can then be approximated by

$$\frac{\partial^2 P}{\partial q_{i,j}^2} \approx \frac{1}{\delta q_{i,j}^2}\left[\left(P_{i-1,j+1} - P_{i-1,j}\right) - \left(P_{i-1,j} - P_{i-1,j-1}\right)\right]. \tag{15.54}$$

The difference between the parameter P on adjacent streamlines, as given in (15.48), can be written as

$$\frac{\partial^2 P}{\partial q_{i,j}^2} = \frac{1}{\delta q_{i,j}^2}\left(\delta P_{i-1,j+1} - \delta P_{i-1,j}\right). \tag{15.55}$$

Thus, the value of the change in parameter, P, due to spanwise mixing becomes

$$\Delta P_{i,j} = \frac{\delta m_{i,j}\,\varepsilon_{i,j}}{c_{m,i,j}\,\delta q_{i,j}^2}\left(\delta P_{i-1,j+1} - \delta P_{i-1,j}\right). \tag{15.56}$$

This includes a positive contribution transferred from the upper and the lower stream-tubes and a loss due to diffusion to these stream-tubes. Equation (15.48), given for the Denton mixing algorithm, is formally identical. As a consequence, the simple mixing algorithm suggested by Denton is equivalent to the solution of the diffusion equation if Denton's mixing factor is directly related to the physical diffusion coefficient as follows:

$$f/2 = \frac{\delta m_{i,j}\,\varepsilon_{i,j}}{c_{m,i,j}\,\delta q_{i,j}^2} = \left(\frac{\varepsilon_{i,j}/c_{m,i,j}}{\delta q_{i,j}}\right) \Big/ \left(\frac{\delta q_{i,j}}{\delta m_{i,j}}\right). \tag{15.57}$$

It is clear that the actual value of the Denton mixing factor needs to be changed throughout the flow field to consider the spacing of the quasiorthogonals, the spacing of the streamlines and the value of the local meridional velocity. Any modelling of the turbulent diffusion which fails to do this will not offer a realistic model of turbulent diffusion. Stronger turbulent diffusion, wider spacing of the quasiorthogonals, narrower spacing of the streamlines or a lower value of the meridional velocity all require a higher value of the mixing factor, f.

The first term of (15.57) is clearly related to the spanwise transport of parameter P under the influence of diffusivity across the streamlines and represents the rate of spread of parameter P across the streamlines. The second term is the tangent of the angle between adjacent neighbouring points in the grid and represents the spread of the grid. In this way, it can be seen that the mixing factor, f, needs to be adjusted to account for the spread of parameter P relative to the spread of the grid. Further details can be found in Casey and Robinson (2010).

The data given by Gallimore and Lewis identify a physically realistic value for the diffusion coefficient ε scaled with the mean meridional velocity and the stage meridional length, L_m, for axial turbines and compressors as

$$0.001 > \frac{\varepsilon}{c_m L_m} > 0.004 \tag{15.58}$$

whereby there is considerable scatter between different machines and different operating points. The larger value is needed where spanwise mixing is high due to deterministic effects (secondary flows, stream-surface twist, etc.) and the lower value is required to account for pure turbulent mixing, which is higher in compressors than in turbines.

15.8.3 Trailing-Edge Mixing at the Outlet of a Compressor Radial Impeller

Many early publications on radial impellers discussed the high level of mixing that occurs just downstream of the impeller trailing edge, and these discussions are well summarised by Cumpsty (2004). The real outlet flow is not uniform and may even have a jet-wake structure at the impeller outlet; see Chapter 5. Any circumferential distortion of the impeller outlet flow mixes out rapidly in the radial direction, but a distortion across the passage in the axial direction is retained. It is not possible to model this in a throughflow code without a complex specification of losses and work distribution at the outlet, and no data for an effective model of this can be found in the technical literature. The approach taken here is to assume that strong mixing takes place at the trailing-edge plane of an impeller, which is in any case quite narrow, and that the specification of a slip factor on the mean streamline defines the outlet swirl velocity on all the streamlines at the impeller outlet. Admittedly, this rather drastic approximation amounts to little more than a 1D model of the impeller outlet flow, but it is an effective and robust means of analysing the global matching of the diffuser with the impeller on a mean-streamline basis.

15.9 Pressure Gradient from Blade Force

15.9.1 Circumferential Pressure Gradient

The solution on the mean stream surface provides the flow field in the meridional plane through the turbomachine, so that this needs to be combined with a blade-to-blade method to the find blade surface velocities. In traditional S1/S2 methods, this is done with the help of an additional S1 blade-to-blade method. The method described here makes use of blade internal calculating stations with a simple linear approximation of the blade-to-blade flow derived from the circumferential blade force.

First, the mean stream surface analysis previously described determines the pitch-wise average properties (static pressure p, swirl and meridional velocities c_m and c_u, and density ρ) on each streamline and on each quasiorthogonal at the midposition between the blades. The meridional flow equations are solved as if there is no variation in circumferential flow. In addition, the pressure forces provided by the blades are replaced by a distributed body force within the flow. This circumferential blade force may be calculated from

$$rF_u = c_m \frac{\partial(rc_u)}{\partial m}. \tag{15.59}$$

If an element of fluid between two blades is considered the circumferential force per unit, mass can be determined from the gradient of the pressure force acting on the fluid by the blades, as follows:

$$-\frac{\partial p}{\partial \theta} = \rho r F_u. \tag{15.60}$$

These two equations in combination provide a solution for the circumferential pressure gradient that is consistent with the distributed blade force

$$-\frac{\partial p}{\partial \theta} = (\rho c_m)\frac{\partial (r c_u)}{\partial m}. \tag{15.61}$$

The term for the meridional swirl gradient in this equation has already appeared in the radial equilibrium equation, so no additional effort is needed to determine it. The swirl gradient term depends on both the blade angles and the deviation of the flow from the blade direction. The algorithm for calculating the meridional gradient in the swirl relies on a parabola through three points. At the trailing edge, a value of zero is imposed to be consistent with the Kutta condition, and in the blade row the parabola is based on central differences. At the leading edge, a forward difference algorithm is used.

15.9.2 Linear Blade-to-Blade Approximation

If the gradient in pressure between averaged across the pitch, then the difference between the static pressure on the suction and the pressure surface of the blade is obtained as

$$(p_{ps} - p_{ss}) \approx \rho c_m \Delta\theta \, \frac{\partial (r c_u)}{\partial m}. \tag{15.62}$$

If the blade-to-blade pressure difference is assumed to be small relative to the relative stagnation pressure, then for this region the variations in properties in the relative system can be considered to be incompressible and loss-free. From the Bernoulli equation, the following expression for the velocity difference between the suction and pressure surfaces is obtained:

$$\Delta p_{ps-ss} = p_{ps} - p_{ss} \approx \frac{\rho}{2}\left(w_{ss}^2 - w_{ps}^2\right), \tag{15.63}$$

where the relative velocities are relative to the respective blade row (absolute velocities in a stator). The velocity term can be expanded and on replacing the sum of suction surface and pressure surface velocities by the midpassage velocity, the pressure difference is obtained as

$$\Delta p_{ps-ss} \approx \frac{\rho}{2}\left(w_{ss}^2 - w_{ps}^2\right) = \frac{\rho}{2}\left(w_{ss} + w_{ps}\right)\left(w_{ss} - w_{ps}\right) = \rho w\left(w_{ss} - w_{ps}\right). \tag{15.64}$$

Thus, the difference between the suction and pressure surface velocities can be estimated as

$$\left(w_{ss} - w_{ps}\right) \approx \Delta p_{ps-ss}/(\rho w) = \Delta\theta(\rho c_m)/(\rho w)\frac{\partial (r c_u)}{\partial m} = \Delta\theta\frac{c_m}{w}\,\frac{\partial (r c_u)}{\partial m}, \tag{15.65}$$

and the final algorithm for the suction and pressure surface velocities is given by

$$w_{ss} = w + (\Delta\theta/2)\frac{c_m}{w}\frac{\partial(rc_u)}{\partial m} w_{ps} = w - (\Delta\theta/2)\frac{c_m}{w}\frac{\partial(rc_u)}{\partial m}, \tag{15.66}$$

where $\Delta\theta$ is the angular spacing between the suction and pressure surface of the blades, as given by

$$\Delta\theta = (2\pi r - Zt_u)/Z. \tag{15.67}$$

for Z blades with circumferential or tangential thickness t_u, where this is related to the normal thickness as follows:

$$t_u = t/\cos\alpha. \tag{15.68}$$

It should be noted that simply doubling the number of blades, as occurs at a splitter leading edge, simply halves the annular distance between the blades and halves the difference in pressure between suction and pressure surfaces. This crude approximation cannot hope to capture any details of the different loadings between the splitter and main vane loadings but provides a simple rough estimate of the effect of the splitter on the loading distribution and velocity distribution on the blade surfaces.

15.10 Choking

15.10.1 Choking in a Throughflow Code

To be useful in preliminary design, the throughflow code needs to identify how close a particular operating point is to choke. This is useful in a code intended for design purposes as it facilitates the identification of possible choking problems relatively early in the design process and allows the designer to adjust the geometry to ensure that sufficient margin to choke is available before carrying out more complex CFD simulations. It is also useful if the method can calculate an operating point that is fully choked. This enables the component responsible for choke to be identified, and if this is done at different rotational speeds it identifies the change in axial and radial matching of the blade rows with speed.

A calculation of fully choked conditions is only possible using a specified pressure ratio rather than a specified mass flow. Simulations in which the specified mass flow exceeds the choking mass flow would lead to a physically impossible solution and generally do not converge with the method described here. There are many solutions with different pressure ratios available at the choking mass flow, as at choke the performance characteristic of the stage is a vertical speed line. As choking is intimately related to the throat areas between two blade rows, any proper estimate of the choking flow relies on accurate estimates of the throat areas.

It must be said that the throughflow method is not ideal for choked blade rows: the mean stream surface equations average out the flow in the circumferential direction and thus do not adequately account for any high Mach numbers on the suction surface of blades. In addition, the shocks that occur in turbomachinery flows are generally not oriented in the circumferential direction, so they are smeared out in the circumferential averaging of the flow. Furthermore, the throat itself is not oriented to the circumferential direction and may cross several q-o's. Nevertheless, despite these serious limitations, an attempt has been made to model choking in the blade rows in this method by the use of a virtual throat, as outlined in the following subsection. The maximum flow and the additional losses related to shocks are considered in the overall predicted performance by using 1D correlations. In this way, the method includes aspects of choking that are compatible with the level of empiricism of typical 1D calculation methods. The simulations are generally more successful than 1D methods in that they take account of the variation of Mach number over the span.

15.10.2 Choking in Compressor Blade Rows

Choking in a compressor blade row can occur through three main mechanisms. First, if the inlet flow is subsonic, choking will occur if the Mach number reaches unity at the throat between the blades. This can occur at subsonic inlet Mach numbers with high negative incidence or at low incidence with very thick blades, leading to high blade blockage and a small throat area. Both cases give rise to an acceleration from a subsonic flow to a Mach number of one at the throat. As the flow accelerates to the throat, there are no large losses caused by this process other than incidence effects.

Second, if the inlet flow is supersonic at high inlet Mach numbers (say $M > 1.2$), choke may occur due to unique incidence. At unique incidence, the flow in the inlet remains supersonic up to and including the throat and until it reaches a normal shock or oblique shock in the compressor flow channel. The flow is choked upstream of the throat by the supersonic expansion wave between the leading edge of the upper blade and the suction side of the lower blade in the flow channel between two blades. Incidences lower than the unique incidence condition are not possible, as they would imply a higher mass flow than this choking mass flow. If the back pressure is reduced under these conditions, the shock in the flow passage moves farther backwards, but there is no change in the mass flow and no change in the flow upstream of the shock.

Third, if the inlet flow is supersonic, then the blade can also choke at the throat. If the flow chokes at the throat under these circumstances, this implies that first there is a detached shock from the suction surface of the blade to upstream of the adjacent blade. The supersonic inlet flow becomes subsonic at this shock and then reaccelerates to be supersonic at the throat. The relative total pressure at the throat is thus lower than that at the inlet because of the losses across the shock, and these losses need to be modelled as if they occur in a normal shock. This operation condition is usually avoided in the design process.

15.10.3 Choking of an Individual Stream-Tube in an Annulus

This section follows the analysis of isentropic compressible flow given in Chapter 6. Consider first one-dimensional isentropic flow of a perfect gas through a stream-tube of cross-sectional area A with constant total pressure and temperature. From classical one-dimensional gas dynamics, the corrected mass flow per unit area, Φ, can be written as a function of the local Mach number M as

$$\Phi = \frac{\dot{m}}{A\rho_t a_t} = \frac{M}{\left(1 + \frac{1}{2}(\gamma - 1)M^2\right)^{\frac{(\gamma+1)}{2(\gamma-1)}}}. \tag{15.69}$$

As the stream-tube area changes along the channel, the corrected mass flow plotted against Mach number reaches a maximum value with a Mach number of one at the throat with area A^*, and is given by

$$\Phi_{\max} = \frac{\dot{m}_{\max}}{A^*\rho_t a_t} = \left(\frac{2}{\gamma + 1}\right)^{\frac{\gamma+1}{2(\gamma-1)}}. \tag{15.70}$$

For a given value of the corrected mass flow below the maximum value, there are two possible values of the Mach number that could be possible. Subsonic and supersonic flows both have a value of the mass flow that is less than the maximum value.

Division of the two preceding equations gives the ratio of the local mass flux (mass flow per unit area) to the mass flux required for choke in the 1D stream-tube as

$$\frac{\dot{m}/A}{\dot{m}_{\max}/A^*} = M\left(\frac{2 + (\gamma - 1)M^2}{\gamma + 1}\right)^{-\frac{\gamma+1}{2(\gamma-1)}}. \tag{15.71}$$

For a constant mass flow along a given stream-tube, this is also the ratio of the area at the throat to the local area A^*/A. The local area would need to become smaller by this ratio to achieve a Mach number of one at this location. In a one-dimensional stream-tube of varying cross-sectional area, this area ratio identifies the nearness to choke of the local stream-tube of area A. A value of 0.95, for example, would imply that a further reduction in area of 5% at this location would choke the flow. Values above unity are not physically possible. Values below unity are associated with both subsonic and supersonic flows and values very close to unity with transonic flows with a Mach number near to one. This parameter can be used as an indication of the closeness of the local flow to choked conditions at this location with this flow area. The information about the nearness to choke is, of course, also present in the value of the relative Mach number. However, given the flatness of the corrected mass-flow curve around $M = 1$, the area ratio is often a more useful parameter to the designer as it directly indicates the sensitivity of the flow to changes in area in this region.

These values relate the local Mach numbers to the local area of the flow and define the mass flow that is possible if the local area were to achieve a Mach number of one, that is, if the local area becomes the throat. In other words, if we assume that the local area A is a throat, then the equations give the maximum mass flow that could pass through this throat. However, choking occurs when the local Mach number at the real throat, which has a different area and a different location, reaches unity, and it is quite possible that no calculating plane is exactly at the throat position. Clearly some care is needed in using the preceding equations to identify the choking mass flow in a throughflow calculation, and at choke the value of A/A^* may not be unity anywhere in the flow field.

A useful example here is that of a one-dimensional flow through a turbomachine annulus shaped as a Laval nozzle with a reduction in flow area to a throat followed by an increase in area downstream of the throat. The annulus has a number of calculating planes both upstream and downstream of the throat, but there is no plane exactly at the location of the throat. If the flow is entirely subsonic, the values of A/A^* are all less than unity, thus indicating that the flow is not choked. If the mass flow is now increased such that the throat is just on the verge of choking, the values of A/A^* are still less than unity at all calculating planes in the flow, but no more mass flow is possible as the throat is now choked. By lowering the back pressure in the real nozzle, the mass flow will cease to increase. The calculating stations upstream of the throat do not change their conditions and would still have a value of A/A^* less than one, but it is now possible to expand the flow supersonically downstream of the throat. Even under these conditions, the value of A/A* at the local calculating plane downstream of the throat does not show that the flow is choked and on the supersonic branch of the mass-flow curve. Clearly the throat must be included in the calculation even though there is no calculating plane at this location.

15.10.4 Choking at a Virtual Throat

In an adiabatic flow, the passage from the subsonic to the supersonic branch of a channel flow always requires an intermediate throat at which the Mach number is unity. As already discussed, an accurate consideration of the flow at the throat is not possible in a streamline curvature throughflow model of turbomachinery blade rows. The throat represents the minimum distance between blade rows and is generally not oriented in the circumferential direction, whereas the calculating planes are by definition in the circumferential direction. Because of this, it is not possible to identify the throat position exactly with a particular calculating plane, and it may in fact lie between or even across two calculating planes. It may even be on different q-o planes across the span such that it is not really part of the calculating grid at all. Denton (1978) suggested that it is sensible in turbine calculations to avoid having any internal blade row calculating planes in the neighbourhood of the throat and to assume that the throat is coincident with the trailing edge. This assumption is also used in the current method, and the throat is taken as just downstream of the leading-edge plane in compressors and just upstream of the trailing-edge plane in turbines. This throat is called here a virtual throat as it is not part of the calculating grid.

In compressor throughflow calculations, it is typical that choking occurs at the throat between two blades, and it is then the throat width or the size of the smallest gap between adjacent blades which limits the mass flow. The throat width is defined as the shortest distance between the blades along a meridional stream-tube and is given the symbol o, where o signifies opening. A diagram showing the throat at different positions across the span at the inlet to an impeller is given in Figure 11.9. The throat is not a calculating plane of the throughflow calculation and is in some way smeared across the calculating planes, as it is located at a different meridional position on the suction and pressure surfaces and from the hub to the casing. It can nevertheless be used to assess whether the stream-tube concerned is choked and to limit the mass flow accordingly. For this reason, accurate geometrical information about the throat width in the flow channel needs to be available to the code.

If the calculating plane just upstream of the throat is considered and isentropic flow between this plane and the throat is assumed, then the maximum mass flow at the throat is given by

$$\dot{m}_{\max, th} = Zob\rho_t a_t \left(\frac{\gamma+1}{2}\right)^{-\frac{\gamma+1}{2(\gamma-1)}}, \tag{15.72}$$

where b is the spanwise width of the stream-tube, that is, normal to the meridional flow direction and not the length along the quasiorthogonal, and Z is the number of blades. The throat area is less than the flow area available on the leading-edge calculating plane. The choking mass flow at the throat is then less than the maximum mass flow at the leading edge in the ratio of the areas concerned, as follows:

$$\frac{\dot{m}_{\max, th}}{\dot{m}_{\max, le}} = \frac{Zob}{A_{le}}. \tag{15.73}$$

The relevant area of the stream-tube is the flow area normal to the flow direction at the leading edge. If s is the circumferential distance between the blades, that is, the circumferential spacing from suction to pressure side, the area available for the relative flow is given by

$$A_{le} = sbZ\cos\beta = (2\pi r - Zt_u)b\cos\beta, \tag{15.74}$$

where β is the local relative flow angle, Z is the number of blades and t_u is the local circumferential blade thickness. It is clear that the value of the mass-flow ratio in (15.73) can be used to identify the nearness to choke of the real flow. A value above one implies that the flow is not choked and more mass flow can be passed through this stream-tube. Values of one indicate that the throat is choked. In theory, a value below one is not possible. Such values can, however, occur for several reasons. First, the throat is not properly included in the calculation due to its orientation and the real effective throat area may not be known. Second, any losses near to the throat are not included in the calculation. Third, the mass flow may be specified to be above the choking mass flow or the flow distribution in an unchoked flow leads to some streamlines being choked.

15.10.5 The Choke Limit of the Meridional Velocity, c_m

Clearly the mass flow at the leading edge cannot be greater than the choking mass flow of the adjacent throat so that the value of the meridional velocity at the throat calculating plane is limited to have a maximum value of

$$(c_m)_{\max, le} = \frac{\dot{m}_{\max, th}}{\rho A_{le}}. \tag{15.75}$$

At a compressor blade row throat, the value of c_m as given in this equation is checked during the inner iteration for mass flow in the iterative procedure. If the local value is found to be above the maximum value of a choked flow, then a limit on the meridional velocity on this stream-tube is applied. Note that applying this limit implies that the velocity gradient equation no longer controls the gradient of the meridional velocity in this choked region. Given that this gradient equation is the theoretical basis of the streamline curvature method, it should be clear that this feature can only be an approximate way of including such effects, and cannot be expected to provide very accurate solutions. The limit on the mass flow at the inlet leads to a change in the curvature of the streamlines and thus affects the upstream flow to the impeller. An example is provided by Casey and Robinson (2010).

15.10.6 Choking in Q3D Isentropic Flow

A throughflow calculation has a Q3D flow across the span, which can be considered as a series of individual 1D stream-tubes across the flow channel. The maximum mass flow of each individual stream-tube can be analysed on a one-dimensional basis using the preceding equations. The maximum possible mass flow is related to the maximum mass flow when choke occurs at a nearby throat. The integration across the span leads to

$$\dot{m}_{\max} = \int (1 - B) b \dot{m}_{\max,th} \sin \psi \cdot dq. \tag{15.76}$$

The equations for choking given earlier are adapted for the duct flows, compressor flows and turbine flows. The estimates of choking are calculated, along with the losses and the deviation angles, in the outermost iteration loop, and the results are then passed to the innermost mass-flow iteration loop where they are used to limit the maximum values of the meridional velocity based on the choking conditions. The following cases are of particular interest.

15.10.7 Choking of an Individual Stream-Tube in an Unbladed Annulus

The preceding equations can be used to identify whether a particular stream-tube in the flow in a duct region is choked. If a stream-tube is choked, that is, operating at the maximum mass flow, the area of this stream-tube is given by $A*$ such that there is a maximum value of the meridional velocity at this location given by

$$(c_m)_{\max} = \frac{\dot{m}}{A*} = \frac{a_t \rho_t}{\rho} \left(\frac{\gamma + 1}{2} \right)^{-\frac{\gamma+1}{2(\gamma-1)}}. \tag{15.77}$$

In an unbladed annulus, there is no problem with choking until the mass-averaged Mach number exceeds unity as described in the following subsection, individual stream-tubes can be supersonic.

15.10.8 Choking of the Annulus as a Whole

In the flow in a turbomachinery annulus, it is possible that local streamlines may be choked, but that the flow across the annulus as a whole is not choked. This occurs, for example, when the casing stream-tubes are choked through high Mach numbers at the blade rotor tip, giving rise to choke in the tip sections, but more flow can still pass through the hub stream-tubes which are still subsonic. It can also happen that the subsonic flow at the root is choked owing to the thick blades used here, but the tip flow is not choked even if it is supersonic. This can be analysed from the continuity equation in the form $\dot{m} = A\rho c_m = A\rho c \cos\beta$, which can also be written in terms of the local stagnation pressure and temperature as

$$\dot{m} = Aa_t\rho_t M\left(1 + \frac{\gamma - 1}{2}M^2\right)^{\frac{1}{2}-\frac{\gamma}{\gamma-1}} \cos\beta, \tag{15.78}$$

where $M = c/a$, the local Mach number; β is the flow angle; and the meridional component of velocity is

$$c_m = \frac{\dot{m}}{\rho A} = a_t M\left(1 + \frac{\gamma - 1}{2}M^2\right)^{-\frac{1}{2}} \cos\beta. \tag{15.79}$$

Differentiating these equations logarithmically, and dividing them, if β is fixed, as it is in a blade row, leads to

$$\frac{d\dot{m}}{dc_m} = \frac{\dot{m}}{c_m}\left(1 - M^2\right), \tag{15.80}$$

which gives the familiar result that the annulus chokes at a relative Mach number of $M = 1$.

If, on the other hand, the circumferential component of velocity is fixed, as it is at a duct plane or leading edge, we obtain

$$\frac{d\dot{m}}{dc_m} = \frac{\dot{m}}{c_m}\left(1 - M^2\cos^2\beta\right) = \frac{\dot{m}}{c_m}\left(1 - M_m^2\right) \tag{15.81}$$

such that the annulus chokes when the *meridional* Mach number is 1. This is also the case within blade rows when operating in the design mode when c_u is fixed.

In the two-dimensional case, where the pressure, temperature and velocity vary across the span, similar equations can be derived if the stagnation pressure is constant, β is fixed and the flow is in radial equilibrium:

$$\frac{d\dot{m}}{dc_{m,mid}} = \frac{1}{c_{m,mid}}\int \left(1 - M^2\right)d\dot{m}. \tag{15.82}$$

This shows that the flow effectively becomes choked, not when the velocity becomes locally supersonic anywhere, but when the mass-weighted M^2 or M_m^2 becomes supersonic. This equation is used to identify the Mach number at which choking in an annulus occurs. during the outer streamline curvature iteration. The parameters used as a measure of the nearness to choking of the annulus constitute a nondimensional parameter given by

$$\frac{d\dot{m}/\dot{m}}{dc_{m,mid}/c_{m,mid}} = \frac{d\dot{m}}{dc_{m,mid}}\frac{c_{m,mid}}{\dot{m}} = \frac{1}{\dot{m}}\int\left(1 - M^2\right)d\dot{m} \tag{15.83}$$

and the effective mean Mach number for choking of the annulus, which can be calculated from this as

$$M_{eff} = \sqrt{1 - \frac{d\dot{m}}{dc_{m,mid}}\frac{c_{m,mid}}{\dot{m}}}. \tag{15.84}$$

Values of $M_{eff} < 1$ indicate that the annulus as a whole is in a 1D sense still on the subsonic branch of choke and can pass more flow, even if individual stream-tubes in the annulus are already supersonic. Values of $M_{eff} > 1$ show that the flow is beyond choke and, in the sense of a 1D stream-tube, is on the supersonic branch despite the fact that there may still be stream-tubes in the annulus with subsonic flow. As the value of M_{eff} approaches, one a change in the meridional velocity on the mean streamline produces no change in the mass flow. The annulus choking calculation simply identifies by the value of M_{eff} the branch of the mass flow versus Mach number curve on which the q-o is situated. The actual limiting mass flow is determined by the effects of the throats on each individual stream-tube, which are at a different location, and so the calculation of M_{eff} provides no information on the actual limiting mass flow.

15.10.9 Choking by Unique Incidence

Choking by unique incidence is also dealt with by applying a limit on the meridional velocity. Assuming a correlation is available for the unique incidence, i_u, or that this is known from other data, then the maximum flow angle at the leading-edge q-o on a particular streamline can be calculated from the blade angle. The swirl velocity upstream of the leading edge is known, and this allows the maximum value of the meridional velocity due to unique incidence to be estimated from the blade inlet angle, as follows:

$$c_{m,\max,ui} = c_u/\tan\left(\beta_1' + i_u\right) \tag{15.85}$$

for a stator and a similar equation with the relative swirl velocity for a rotor. If this value is less than that which would occur due to choking at the throat, then this is used to limit the meridional velocity at the leading-edge plane on this stream-tube.

A crude correlation for the unique incidence in degrees on thin, straight blades has been derived from graphical data given in Bölcs (2005), as follows:

$$M_{w1} > 1.2: \quad i_u = 3(M_{w1} - 1) + |\beta_1 - 60°|/10. \tag{15.86}$$

Note that these limits on the meridional velocity are only applied at the inlet plane upstream of the throat. At downstream planes, there is no limit applied to the meridional velocity as this would hinder the development of a supersonic flow downstream of the throat, so at the downstream stations the Mach number may automatically become supersonic again. In some blade rows calculated with many internal blade calculating planes, the second and even the third calculating plane may actually be upstream of the effective throat, so an error arises from this. In some difficult calculations of supersonic inlets which are difficult to converge, it has been discovered that it might be worthwhile to omit stations close to leading edge, and this is probably related to the uncertainty in the actual location of the throat.

15.10.10 Losses in a Choked Compressor Blade Row

In a choked blade row, the losses within the bladed region are no longer a function of the mass flow, which is constant, but become a function of the back pressure. This is rather like the situation in a choked inviscid 1D converging-diverging Laval nozzle, where the back pressure determines the location and strength of the shock and the level of losses that occur are a function of the shock strength. At lower back pressures in a 1D Laval nozzle, the shock moves backwards in the diverging channel and becomes stronger with more losses.

This process in an impeller is modelled by the method. The previous equations are first used to identify the maximum mass flow at the leading-edge plane. If the flow is choked and the calculation is at a specified pressure ratio, then additional losses need to be generated within the blade row to so that the mass flow at outlet (where the target pressure is specified) matches the choked mass flow calculated at the inlet plane. Without additional losses, the outlet density would be too high and the mass flow on the blade trailing edge would be too large. The additional losses are distributed evenly across the span and uniformly downstream of the first calculating plane whose annulus is unchoked with $M_{eff} < 0$. In each iteration, the additional losses are determined from the condition needed to correct the mass flow at the trailing edge. This algorithm assumes that when the target pressure is achieved ($dp = 0$), then from the Gibbs function we have

$$Tds = dh - vdp, \quad ds = dh/T = c_p(dT/T). \tag{15.87}$$

The equation for an ideal gas can be differentiated to give

$$pv = RT, \quad \frac{d\rho}{\rho} = \frac{dp}{p} - \frac{dT}{T}, \tag{15.88}$$

and if $dp = 0$, these equations can be combined to give

$$ds = -c_p(d\rho/\rho). \tag{15.89}$$

The error in the density is assumed to be related to the error in the mass flow at the trailing edge, so we obtain that the additional losses that are needed to match the mass flow at the trailing edge with the choked mass flow at the inlet can be estimated from the trailing-edge mass-flow error as

$$\Delta s \approx -c_p(\Delta \dot{m}/\dot{m}). \tag{15.90}$$

The error in the mass flow at the trailing edge is thus used to update the losses within the blade row until both the mass flow and the target pressure at the trailing edge are correct. In this process, the change in the additional losses is damped in each iteration. The additional losses in this process are determined by the program and are in addition to any losses that may be specified by the user or determined by the specified correlations.

The blade sections with $M_{eff} > 0$ do not have extra losses added, as this would mean that the flow pattern changes with back pressure, which is not the case in a supersonic flow downstream of the throat and upstream of the shock location. This procedure automatically increases the velocity levels towards the impeller outlet with decreasing back pressure until the trailing edge reaches supersonic flow at low back pressure. At this point, the algorithm for a supersonic exit flow from a turbine determines a supersonic deviation.

15.10.11 Choking of a Stream-Tube at a Turbine Blade Row Outlet

Choking at a turbine or inlet guide vane outlet is dealt with in a rather similar way to the compressor inlet and leads to the phenomenon of supersonic deviation. It is assumed that the choking of a turbine always occurs at the throat, which is taken to be close to the turbine blade row trailing-edge plane. In their choking calculations, which were generally for blade rows without internal planes, Denton (1978) and Came (1995) assumed that there are no relative total pressure losses along a stream-tube between the turbine leading-edge inlet plane and the turbine throat. In the current code with internal blade row calculating stations, it is assumed that there are no losses from the next upstream quasiorthogonal to the throat. Based on these flow conditions, the maximum mass flow per unit stream-tube width in the stream-tube can be estimated from

$$\dot{m}'_{\max,th} = \frac{o p_{t,rel}}{\sqrt{RT_{t,rel}/\gamma}} \left(\frac{\gamma+1}{2}\right)^{-\frac{\gamma+1}{2(\gamma-1)}}, \tag{15.91}$$

as with the compressor case, where o is the effective throat width of the stream-tube and $p_{t,rel}$ and $T_{t,rel}$ are the relative total pressure and temperature at the throat (taken from the upstream q-o).

Note that in radial turbines and in turbines with a high flare, there may be a radius change between the stations, and this implies that the radius change is needed to determine the local relative total pressure and temperature at the throat. These can be determined from the values at the upstream calculating plane on the assumption of adiabatic isentropic flow from the condition that the rothalpy is conserved in the impeller, where the rothalpy is defined as in Section 2.6.4 as

$$I = h_t - uc_u = h + c^2/2 - uc_u = h_{t,rel} - u^2/2. \tag{15.92}$$

At the upstream plane, the rothalpy, I, is determined from the local conditions. The throat is assumed to be close to the trailing-edge plane, so that the swirl velocity and the blade speed is the same as at the trailing-edge plane of the turbine. This determines the value of the relative total enthalpy at the throat consistent with the work input from upstream, as follows:

$$\begin{aligned} I_{th} &= I_{up} = h_{t,rel} - u_{te}^2/2 \\ h_{t,rel} &= I_{up} + u_{te}^2/2, \end{aligned} \tag{15.93}$$

and the total relative temperature at the throat can then be determined from the enthalpy. On the assumption of no losses from the upstream plane, the relative total pressure at the throat from the pressure at the inlet plane is

$$p_{t,rel} = p_{up}\left(\frac{T_{t,rel}}{T_{up}}\right)^{\frac{\gamma}{\gamma-1}}. \tag{15.94}$$

For a stator blade row such as an inlet guide vane, these equations reduce to the usual equations for the stagnation pressure and stagnation temperature.

The choking mass flow is calculated by integrating the maximum mass flow per unit stream-tube width across the flow channel at the location of the outlet plane. This integration assumes that the pitch angle of the meridional velocity and the stream-tube width does not change between the throat and the trailing edge, giving

$$\dot{m}_{\max,th} = \int \left[o(1-B)\dot{m}'_{\max,th}\sin\psi\right] dq. \tag{15.95}$$

At the outlet to the blade row, the effective stream-tube area is given by

$$A' = Zs\cos\beta = 2\pi r\cos\beta \tag{15.96}$$

as with the compressor case, so the mass flow per stream-tube width at the trailing-edge plane can also be written in terms of the local Mach number here as

$$\dot{m}' = 2\pi r\cos\beta\frac{p_{t,rel,te}}{\sqrt{RT_{t,rel,te}/\gamma}}M_{w,te}\left(1+\frac{\gamma-1}{2}M_{w,te}^2\right)^{-\frac{\gamma+1}{2(\gamma-1)}}, \tag{15.97}$$

where $M_{w,te}$ is the relative Mach number at the turbine trailing-edge plane.

15.10.12 Supersonic Deviation at Turbine Outlet

If the stream-tube in a turbine blade row is choked at the throat and the outlet flow is supersonic, then the maximum possible mass flow in this stream-tube provides an upper limit on the flow angle that is possible of

$$\cos\beta_{\max} = \frac{\dot{m}'_{\max,th}}{2\pi r\frac{p_{t,rel,te}}{\sqrt{RT_{t,rel,te}/\gamma}}M_{w,te}\left(1+\frac{\gamma-1}{2}M_{w,te}^2\right)^{-\frac{\gamma+1}{2(\gamma-1)}}}. \tag{15.98}$$

Higher outlet flow angles are not possible, as they would imply a higher mass flow than the maximum possible choking mass flow. Thus, independent of the correlations used for the turbine outlet angle, as soon as the outlet flow is supersonic, the exit angle is determined by the outlet Mach number via the continuity equation and is not predetermined from deviation angle correlations. This mechanism for choking produces an upper limit on the flow angle similar to the unique incidence condition at a compressor inlet, and a minimum value of the meridional velocity similar to the maximum value at the compressor inlet.

15.11 Use of Throughflow in Design

To be useful in design, a throughflow code should be robust enough to converge even if the geometry being calculated is not finalised or if the user makes errors in the input data. With a converged solution, a bad design or bad data can be quickly identified. Even a simulation that is not fully converged may be useful to identify issues arising from bad input data. Because of its speed of computation, a robust throughflow code can also be successfully linked to a geometry definition method for the stage, similar to that described in Chapter 14. This allows the designer to quickly iterate on the 3D geometry of the complete stage to produce a preliminary design that meets the required incidence and loading requirements. The throughflow code can also be coupled to an optimiser, and the code described here has been combined with an evolutionary algorithm for the design of both radial compressor impellers and radial turbine impellers (Casey et al., 2008; Cox et al., 2010).

The throughflow code should also be able to produce a series of diagrams in a standard format so that different stages can be quickly compared with each other. Throughflow codes with loss correlations can also be used to calculated performance maps, and standard correlations for axial turbines and axial compressors are available in the code. The authors tend not to use the throughflow code for performance prediction in radial compressor design, as the map prediction method described in Chapter 18 is generally used and requires far less data input. The main use of the throughflow code in the design process is to set up a compressor geometry with acceptable incidence, a good blade loading distribution and no unnecessary acceleration due to curvature of the blade walls; also, it meets the geometry constraints and performance requirements of the specification in terms of the pressure ratio, flow rate and choke margin. This preliminary design from the throughflow calculations becomes the geometry that is then analysed in detail and further improved with CFD; see Chapter 16.

The importance of the throughflow in the design of the inlet of a radial wheel is shown in Figure 5.5. The curvature of the meridional channel affects the spanwise distribution of the meridional velocity, and without an accurate calculation of this, it is not easy for the designer to set up the variation of the inlet blade angles to attain the appropriate incidence levels.

An example of the velocity distribution within an impeller calculated with a throughflow code is given in Figure 15.8, adapted from Casey (1985), where it is

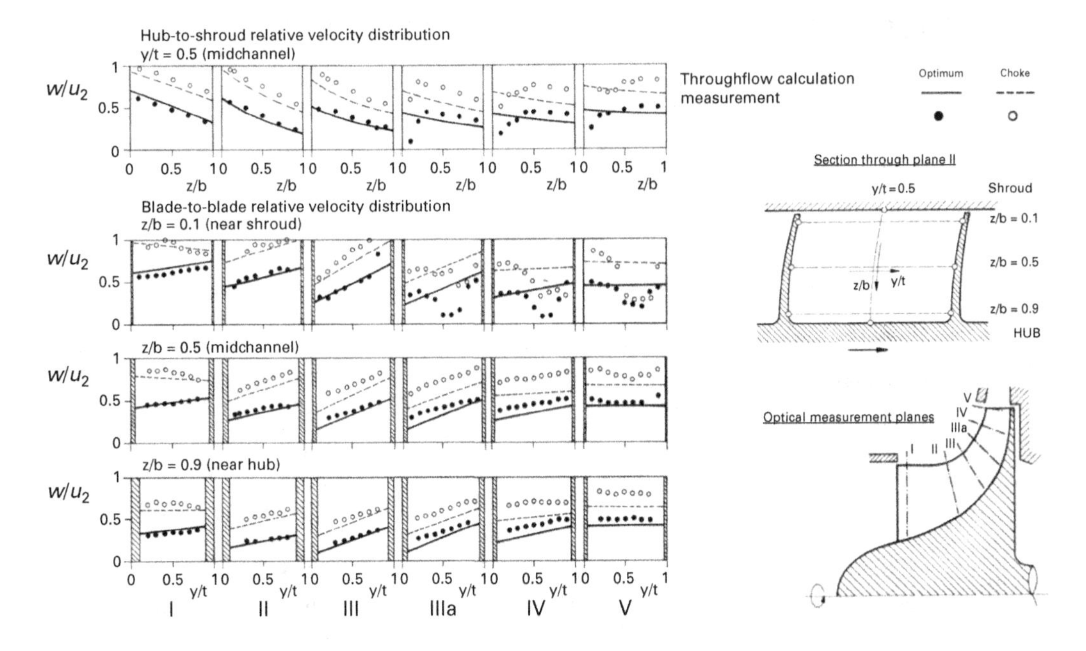

Figure 15.8 Throughflow calculation of the backswept Eckardt impeller A compared with measurements at two operating points. (image adapted with permission of VDI)

compared to the measurements of Eckardt on a backswept impeller at operating points near peak efficiency and near choke. This impeller is the impeller A tested by Eckardt, as described in Chapter 5. This impeller suffers from flow separation, as it turns the flow to the radial direction before backsweep is applied closer to the impeller outlet. Despite the flow separation, the agreement between the measurements and the throughflow calculations is very good, especially in the inlet region of the impeller, where the pronounced jet-wake flow pattern of this impeller has not developed. The blade loading distribution at inlet due to operation near choke reverses as would be expected at high negative incidence. These results show that the inviscid throughflow method with a simple linear blade-to-blade approximation is adequate to estimate the blade-to-blade and hub-to-shroud velocity distributions during the preliminary design. In essence, it predicts the dominant gradients in velocity shown in Figure 5.6 without aspects such as the boundary layers and leakage flows. In this case, larger errors occur where the jet wake pattern causes additional blockage in the flow channels but the trends without this viscous effect are very reasonable.

16 Computational Fluid Dynamics

16.1 Overview

16.1.1 Introduction

The objective of this chapter is to provide an up-to-date introduction to modern computational fluid dynamics (CFD) methods as used in an industrial design context for the development of components in radial turbocompressors. Today, CFD simulations of whole compressor stages using the Reynolds-averaged Navier–Stokes equations (RANS), with a steady-state rotor–stator interaction model between the blade rows and a suitable turbulence model, are the basis of aerodynamic design calculations. In terms of the computational effort needed for a rapid turnaround consistent within the constraints of typical design processes, this is currently the only practical design technique. It provides the ability to analyse and interpret the large ensemble of related simulations in a time-critical period (typically 24 hours). More sophisticated high-performance research calculations are also described.

CFD technology has gone from infancy to maturity over the last 40 years, and some of this historical process is described in the first part of this chapter. This is followed by sections describing the basis of most modern CFD methods. Compared to more general applications of CFD, turbomachinery simulations require additional, specialised technologies including special boundary conditions and methods to handle the relative motion between the compressor impeller and stationary parts, as well as specialised postprocessing. The engineer requires an understanding of the source of errors and the importance of validating the CFD methods before using CFD as a design tool. The prediction of the turbulent flow is one of the main approximations in CFD simulations, and an overview of turbulence models is given. Another important piece of pragmatism and potential source of error is the use of averaging planes between individual blade rows, and these are described. Some guidelines for the use of CFD in radial compressor design are provided.

The CFD solver requires a number of important supporting technologies to be useful, such as geometry generation, mesh generation and postprocessing. In the evolution of an effective design system, an efficient and user-friendly route through these processes is almost as important as the solver itself. Examples of results from radial compressor simulations are given in other chapters to highlight aspects of the designs. These results include transonic flow, tip clearance flows, stage calculations, impeller and diffuser flow patterns in radial compressors and the flow in the nonbladed parts of the radial machines, such as outlet casings and impeller-casing gaps.

16.1.2 Learning Objectives

- Know the meaning and significance of the important terms describing the numerical methods.
- Understand the difference between RANS, URANS, LES and DNS calculations.
- Appreciate the difference between zero-, one- and two-equation turbulence models and the most influential models of each type in practical use.
- Know the source of different types of error in CFD simulations and how to actively manage these errors through quality control.
- See the practical aspects of the application of CFD in the design process, where several related simulations are required in a time-critical period.

16.2 Historical Background to Turbomachinery CFD

16.2.1 The Revolution in Design Tools

For most of the last 50 years, the growth in the performance of computers has closely followed Moore's law, which was related initially to the number of transistors that can be mounted on a single integrated circuit; Moore, 1965, but is now widely applied to performance levels of computers. Moore's law predicts a doubling of computer performance every 18 months. This spectacular growth in performance has driven the development and use of numerical methods in all technical and scientific disciplines. Through this rise in computing power, the designer now has available mathematical models of increasing complexity and with a much finer numerical resolution of the dynamics of the systems modelled. Because of these developments, the design of turbocompressors has undergone an enormous transformation over the last 40 years, and especially during the careers of the authors of this book, who were both involved with CFD for turbocompressor design in an industrial environment during its formative period (Robinson, 1991; Casey, 1992). Both authors are users and interpreters of CFD with RANS equations rather than developers of such tools. We have watched the progress of CFD from two-dimensional inviscid calculations around single aerofoils through 3D Euler calculations to 3D unsteady Navier–Stokes solutions in three dimensions for multiple blade rows. Cumpsty (2010) states that each stage of this development was a wonder and, like him, the authors strove to make the most of the results with the best methods available to them for their designs.

With the success of CFD, the emphasis in compressor development has shifted from being mainly an experimental activity with extensive prototype and model testing to being an exercise in numerical flow simulation regarding the prediction of the flow, the performance and also the mechanical integrity. The growth of CFD correlates with the many economic benefits of applying CFD simulations in the design process. CFD leads to shorter design cycles at a lower cost because there is less dependence on the manufacture of physical hardware for testing. CFD also offers a more accurate prediction of turbomachinery flow behaviour and loss prediction than is possible with any other form of analysis. Higher performance becomes possible because CFD provides a better understanding of the global features of the flows such as incidence, blade loading, velocity levels, mean flow direction, secondary flows and unsteady effects. It also provides insight into the weak features of the flow such as flow separation and other unwanted loss mechanisms.

If applied with care and with proper consideration of the sources of numerical error, CFD has the ability to quickly assess the impact of changes to the geometry and thus remove undesirable flow features. Although absolute quantitative accuracy for efficiency prediction is usually not possible, CFD allows competing designs to be analysed, assessed and evaluated on a comparative back-to-back basis. This feature allows for more bold designs with less risk; these designs may even be outside the range of previous experience. Engineers can develop new designs which more closely approach the natural physical limits controlling fluid dynamic behaviour (in compressors these are frictional losses, flow separation and choking). Such designs enable improved compressors with higher aerodynamic loading and flow capacity. The use of these tools reduces development costs by reducing rig and machine tests, lowers production cost and weight by reducing part count, reduces manufacturing costs through simpler geometry of a smaller size, and lowers operating cost by increasing performance and extending life.

Today, design optimisation of all turbocompressor components is carried out with the help of CFD before final prototype testing. Tests are still generally carried out to confirm that the design has attained the guaranteed performance levels. In areas where the guarantees or risks related to nonachievement of specified performance levels are low, CFD has already eliminated experimental verification, and some changes in designs may be accepted purely based on CFD simulations. In research and development on compressors, experimental and numerical investigations are highly complementary. Experiments are less detailed in the information that they provide but are still essential to determine the location of any flow instabilities and to validate the simulations. Despite this, computations provide much more detailed information about the flow not usually available in the measurements. CFD also provides guidance in the design of the experiments themselves (which operating point to focus on, where to take measurements and so forth). In this way, CFD simulations not only reduce the number of experiments required but also improve the effectiveness of any experimental investigations.

16.2.2 Emergence of CFD for Compressor Design

Development of CFD was first driven by aerospace applications in the 1960s, following the start of commercial jet transport and the subsequent need to understand transonic aerodynamics. Excellent turbomachinery performance and mechanical integrity combined with low weight, low pollution and low noise are crucial for both military and civil transport aircraft, and so it was natural that aeroengine applications were at the forefront of CFD developments (Smith, 2002). In the 1960s, the simple radial equilibrium equations described in Chapter 4 were often solved by hand. During the late 1970s and 1980s, the bulk of the routine aerodynamic design work for turbomachinery designs was carried out with throughflow and blade-to-blade methods as outlined in Chapter 15. Throughflow calculations are based on an axisymmetric model of the flow in a compressor and rely heavily on empirical input but are still widely used for the rapid preliminary design of turbocompressors.

Most industrial turbomachinery organisations made their first steps towards the use of CFD in the 1970s and 1980s. The first applications made use of potential flow methods, stream function methods and finite element methods for solving the inviscid Euler equations for blade-to-blade flows; see the review by Gostelow (1984). These inviscid calculations were combined with boundary layer methods in viscous-inviscid zonal codes. These methods are still useful in the design process for rapidly assessing details of blade loading and blade boundary layers in axial machines. However, as they are limited to a 2D view of the flow, they will often fail to predict the flow in a radial compressor accurately. In recent years, MISES, a Q3D inviscid-viscous interaction method, replaced many of the earlier codes in profile design for axial turbomachinery (Drela et al., 1986; Drela and Youngren, 1991).

16.2.3 Academic Codes

In the 1980s, there was a move towards the use of 3D codes developed in universities, many of which became available commercially. The initial steps towards 3D simulations were encouraging enough for further development. However, the available hardware at the time usually limited the size of the grid and the accuracy of the predictions that could be achieved, with around 10,000 elements for a 3D viscous blade row calculation. The number of elements for a blade row increased progressively with the capacity of the available computers to around 50,000 in the 1990s. These days, a mesh with 200,000–400,000 elements is typical of a 3D calculation using wall functions and 1,000,000 or more nodes is needed for a mesh which resolves the boundary layers and transonic flow features. Only a limited number of numerical schemes have been applied to turbomachinery flows; Lakshminarayana (1996), Denton and Dawes (1998) and Chen (2010) provide overviews.

The breakthrough publications demonstrating the use of CFD in radial compressors were certainly those of Moore and Moore (1981) and Moore et al. (1984) using a pressure-based code. Although the computational grid used was less than 2000 elements, some details of the complex flow field were made clear for the first time. These papers are essential reading for radial compressor CFD practitioners particularly with regard to the detailed analysis of the CFD results. Together with other publications of the same research group mentioned in Chapter 5, these authors showed the real opportunity there was to explain the complex flow patterns and to understand the losses in centrifugal compressors with CFD methods.

Denton (1975, 1983, 1986) and Dawes (1987 and 1988) developed important density-based turbomachinery time-marching codes in the Whittle Laboratory at Cambridge University. These codes used a control volume approach with a structured skewed grid, with spanwise planes oriented between the axisymmetric casing and hub geometry, streamwise planes oriented with the blades and pitchwise planes oriented in the circumferential direction. The upstream and downstream faces of a skewed element had no projected area in the circumferential direction. This arrangement made it unnecessary to calculate fluxes and forces in this direction on these faces, thus reducing computational effort. The streamwise grid lines were aligned closely with the flow direction: this is known to reduce discretisation errors in the calculations. Both the Denton and the Dawes codes were used widely in the radial compressor industry. Dawes (1987) published his evaluation of a backswept radial compressor that had been tested by Eckardt (1979). An early example comparing the Denton and Dawes codes and validating them against measurements on a radial impeller was given by Casey et al. (1992) on a different compressor that had also been tested by Eckardt (1979).

Denton continued his pioneering work and was the primary source of many developments, in particular the development of stage calculations with rotor–stator interfaces (Denton, 1992). The Fortran source of one of the later versions of his codes with multiblock capability, MULTALL, has since been made freely available as open source code (Denton, 2017). He also developed the code, TBLOCK, which was often used in radial compressor research, for example by Hazby and Xu (2009). The Cambridge Whittle Laboratory remains active in this area with a powerful tool called Turbostream, based on TBLOCK but running in parallel on many parallel core processors as found in graphics cards (Brandvik and Pullan, 2011).

The Dawes BTOB3D code became a standard design and analysis tool for radial compressor impellers in the 1990s in many industries, national research institutes and universities and enabled the improvement of many radial turbomachinery designs (Krain and Hoffman, 1989; Goto, 1992; Sorokes, 1993; Dalbert et al., 1999; Came and Robinson, 1998). Dawes also developed a multistage calculation capability (Dawes, 1992) and a code called NEWT (Dawes, 1992), using an unstructured tetrahedral mesh which allows application in the analysis of the casing components with complex geometry.

Other universities and institutes were also active in the development of CFD codes during this formative time but mainly for more general fluid dynamics problems. Imperial College, London developed a general-purpose code, PHOENICS (Parabolic Hyperbolic or Elliptic Numerical Integration Code Series), based on the SIMPLE (Semi-Implicit Method) algorithm of Patankar (1980), using a pressure-correction scheme. Other codes from the Imperial College stable were STAR-CD (Gosman and Johns, 1978) and its successor STAR-CCM+. Sheffield University developed the general-purpose code FLUENT with an industrial partner. Illinois Institute of Technology worked on FIDAP (Fluid Dynamics Analysis Package). The UK Atomic Energy Authority developed FLOW3D and CFX. Turbomachinery specific codes were developed by the Brussels Free University, which produced FINE/Turbo, now marketed by NUMECA International. Raithby in the University of Waterloo in Ontario worked with his students on a finite volume method (Galpin et al., 1985). The latter uses an element-based finite volume approach, where the control volumes are formed from the mesh and functional variation within each element is based on element shape functions. Together they produced TASCflow, marketed at the time by Advanced Scientific Computing which later evolved into the basis for ANSYS CFX, which is a widely used CFD code for turbomachinery.

16.2.4 Codes from Government Organisations and Large Turbomachinery Companies

Many research organisations around the world developed their own turbomachinery CFD codes, some public and some for internal research only. NASA developed several CFD codes for analysis of flows in turbomachinery. Among these, the SWIFT code from NASA solves the Navier–Stokes and energy equations on body-fitted grids using an explicit finite-difference scheme. In the early 1990s, the DLR in Germany developed a code called TRACE (Turbomachinery Research Aerodynamic Computational Environment) in order to calculate and investigate the complex flows in turbomachinery. The solver is based on a cell-centred finite volume approach developed by DLR and MTU Aero Engines AG. ONERA started in 1997 to develop the *elsA* software for complex external and internal flow aerodynamics and multi-disciplinary applications. The flow equations are also solved by a cell-centred finite-volume method and an efficient multigrid technique to accelerate convergence.

Similarly, the largest gas turbine engine manufacturers developed their own proprietary codes. The General Electric Company developed its inhouse CFD solver named TACOMA, Rolls-Royce Ltd developed Hydra and Pratt and Whitney the code Y237.

16.2.5 Difficulties in the Early Years

Newcomers to this field are likely to be unaware of the many hurdles that were overcome to get this far with CFD. But the uncertainties that still prevailed back in the 1980s and 1990s are highlighted well by the introductory remarks made in a paper by Moore et al. (1984):

> There are perhaps three categories of flow calculations; a) Those that we do not believe, b) Those that give results which we do believe, but we knew already, and c) Those that give results which we do believe, which we could not obtain by measurement, we could not previously quantitively predict and that we did not foresee. In the development of three-dimensional viscous flow calculation methods for turbomachinery design, we have been attempting to reach category (b), in the hope of progressing to category (c).

This was almost 40 years ago. CFD and CFD practices have come a long way since then. CFD practitioners in industry have essentially arrived at category (b) in this list on a daily basis, and there are many examples of category (c) successes. Working with CFD in the early years was like trying to climb a previously unconquered mountain, with many uncertain routes that might have led to category (b), but no one being sure which route would be most successful. The individual developers of each method were, however, always very confident that their own method was the most appropriate. Professor Brian Spalding of Imperial College captured the essence of this when he quipped that in some quarters, CFD stood for *Cheats, Frauds and Deceivers* (Spalding, 1998).

The early difficulties with CFD simulations in turbomachinery were clearly demonstrated by the American Society of Mechanical Engineers (ASME) test case held in 1995 for the NASA Rotor 37, where 11 participants submitted blind simulations of a transonic compressor rotor – blind in the sense that the measurement data were published only after the simulations had been submitted (Strazisar and Denton, 1995). The simulations submitted illustrated that very large differences were obtained by different users of the same code, as well as by different codes. This test case is important for radial compressors since many features of the flow and shock structure at the inlet to a transonic axial compressor rotor are very similar to those in a transonic impeller inducer. The test case used a single blade row as multiple blade row calculations were not widely available at that time. The overall accuracy of the prediction of the efficiency for all methods deployed (different codes, different turbulence models, different grids, different users of the same code, etc.) was $\pm$ 2%; some codes performed better than this (Denton, 1997).

There was also considerable variability in the Rotor 37 results with regard to prediction of a flow separation on the hub; none of the codes predicted this well. Subsequent CFD sensitivity studies with the same test data have identified that this separation was most probably associated with a lack of knowledge related to a small leakage flow through the rotor root on the hub (Dunham, 1998). It was not possible to repeat the tests without the hub leakage flow as the rotor had disintegrated in a subsequent test program. Denton (2010) and Casey (2004) describe some other studies from this period where blind comparisons of different CFD codes were made with experimental test data.

In Europe, the European Research Community on Flow, Turbulence and Combustion (ERCOFTAC) took the lead in attempting to analyse the uncertainty in CFD by setting up a special interest group (SIG) on problems of flow simulation in turbomachinery. The turbomachinery SIG established a series of test cases, and participants brought their solutions to these cases in a series of annual meetings. This CFD problem was not just related to turbomachinery, so ERCOFTAC set up a more general SIG on quality and trust in industrial CFD calculations. This body considered the types of errors in the simulations and published best-practice guidelines (Casey and Wintergerste, 2000), giving practical advice on achieving high-quality industrial CFD simulations using the RANS equations. As a follow-on, ERCOFTAC organised a thematic network for quality and trust in the industrial application of CFD, known as QNET-CFD, which included a network related to turbomachinery simulations (Casey, 2004).

CFD simulation methods have improved by orders of magnitude since their commercial inception 40 years ago. At most of the important turbomachinery conferences during the early period, there were many sessions devoted solely to the subject of different numerical methods. Almost all of the CFD sessions at turbomachinery conferences these days focus on investigating a physical flow phenomenon identified in measurements or a topic related to design improvements achieved with the help of CFD.

16.2.6 Commercial CFD Software

Individual students and professors involved in the development of CFD codes in universities set up commercial companies to market their codes, and this market has now grown to over one billion dollars per year in license fees. Many companies have merged, and several have been taken over by others. ANSYS Inc., which was previously best known for finite element mechanical calculations, acquired the FLUENT, TASCFLOW and CFX codes into its product range. These days, ANSYS-CFX is a major commercial supplier of CFD for turbomachinery computations, together with NUMECA International.

In the first applications of CFD, its use was technology driven – it provided new insights into the engineering science of turbomachinery, leading to better designs. In recent years, it has matured to become an established and trusted design tool, and issues related to its integration into the overall design process with links with the preliminary design, CAD and the mechanical calculations have become essential. The drive to reduce costs and to increase performance with new design features ensures that further improvements in simulation methods and their integration have high priority. The crucial competitive advantage of CFD in the radial compressor industry is only possible through the acceleration of the design process. The need for acceleration means that considerable development effort takes place on methods integration (in computer-aided engineering [CAE] systems for concurrent engineering) within aerospace and turbomachinery companies.

The pressure on costs has meant that there has been a steady move towards the use of commercial software packages for turbomachinery design in the industry as few companies can now afford to develop (and maintain) CFD software tools. Commercial CFD packages distribute the total costs of development among many users, and so become less expensive than the inhouse development of codes. The disadvantage for commercial companies producing turbomachinery designs is that competitors now often use the same software packages for their designs. In such cases, the company that has invested most resources in the validation of the code and its models for their machines and establishment of best-practice procedures will still have a significant competitive advantage. The skill and training of the engineers is also an important factor, as a CFD simulation can give the flow field, to a certain accuracy, but says nothing about how to improve the performance, range, efficiency and other key success factors for a good design. A skilful designer has the ability to determine the first pass of a suitable geometry without CFD and then to improve the design on the basis of the knowledge gained from the simulation.

Individual companies often have private agreements with the code suppliers allowing them to develop and validate special features and models within the framework of the commercial code for their applications. Universities and research institutes, which have less commercial pressure, still carry out development work on new models of relevance to turbomachinery. However, these models generally have to be incorporated into the commercial packages to achieve wide acceptability. The numerical aspects of commercial packages may lag those of academic codes, but the convenience of use (preprocessing and postprocessing) are generally much higher for the commercial packages. This aspect may be less attractive for those focussing on the design of turbomachinery stages where robustness, convenience and throughput of iterations on geometry are highly valued. The high cost of commercial CFD software licenses has pushed some recent academic developments towards open source software for turbomachinery applications, such as OpenFOAM (Hanimann et al., 2014).

16.3 The Governing Equations

16.3.1 Navier–Stokes Equations

Computational fluid dynamics solves the Navier–Stokes equations, as discussed in Chapter 5. These are partial differential equations including the conservation of mass and momentum, plus the total energy conservation equation, in order to predict the fluid flow within a defined volume.

Conservation of mass

$$\frac{\partial \rho}{\partial t}+\frac{\partial(\rho u_i)}{\partial x_i}=\dot{S}_i^{mass} \tag{16.1}$$

Conservation of momentum

$$\frac{\partial(\rho u_i)}{\partial t}+\frac{\partial(\rho u_i u_j)}{\partial x_i}=-\frac{\partial p}{\partial x_i}+\frac{\partial \tau_{ij}}{\partial x_i}+\dot{S}_i^{mom} \tag{16.2}$$

Conservation of total energy

$$\frac{\partial(\rho h_t)}{\partial t}-\frac{\partial p}{\partial t}+\frac{\partial(\rho h_t u_i)}{\partial x_j}=-\frac{\partial q_j}{\partial x_j}+\frac{\partial}{\partial x_j}\left(\tau_{tj} u_i\right)+\dot{S}_i^{energy} \tag{16.3}$$

In the preceding five equations (three component momentum equations), there are five principle unknowns: the components of the fluid velocities u in three directions, the fluid static pressure p, and the total enthalpy h_t. Source terms, denoted by S, are included due to additional physical effects, such as centrifugal and Coriolis forces due to system rotation, in each of the conservation equations. The density ρ is determined from the thermal equation of state, for example the ideal gas law.

The stress tensor τ_{ij} is expressed in terms of the fluid velocity gradients. For a Newtonian fluid, commonly assumed for gas compressors, the stress is proportional to the strain through the effective fluid dynamic viscosity, μ_{eff}, which is the sum of the eddy and molecular viscosity:

$$\mu_{eff} = \mu_{turb} + \mu, \quad \tau_{ij} = \mu_{eff}\left(\frac{\partial u_i}{\partial x_j} + \frac{\partial u_j}{\partial x_i}\right) - \frac{2}{3}\mu_{eff}\delta_{ij}\frac{\partial u_j}{\partial x_j}. \tag{16.4}$$

The eddy viscosity, μ_{turb}, is determined by a suitable turbulence model, as described in Section 16.6. Likewise, the heat flux, q_j is proportional to the temperature gradient through an effective fluid conductivity, λ_{eff},

$$q_j = -\lambda_{eff}\frac{\partial T}{\partial x_j}, \tag{16.5}$$

but this is not so important in turbocompressor applications. The static temperature, T, is computed from the static enthalpy, h, using the calorific equation of state, Chapter 3. In rotating frames of reference, the relative velocity is solved. The momentum equations include the additional Coriolis and centripetal acceleration as source terms, as discussed in Chapter 5. In rotors, the rothalpy, I, is advected in the energy equation, rather than the total enthalpy, and the viscous work term in a rotor with tip clearance is included as a source term; see Chapter 2.

16.4 The Modern Numerical Method

16.4.1 The Finite Volume Method

Many CFD techniques exist and the literature on this subject is vast, just as is the number of researchers who have made lasting contributions. This section is intended to provide a guide to the terminology for an engineer and for a more scholarly description of CFD the books by Patankar (1980), Hirsch (2007) and Ferziger and Peric (2002) are suggested.

There are three main families of numerical methods for solving the Navier–Stokes equation: finite difference methods, finite element methods and finite volume methods. Other methods have been developed but are not so well represented in the literature for turbomachinery applications, such as spectral methods, boundary element methods, vorticity-based methods and lattice gas methods. The finite difference method is the oldest of the methods with the publication of a calculation of the viscous flow around a cylinder by Thom in 1933. It discretises the governing equations whereby the first and second derivatives of a parameter ϕ are approximated by truncated Taylor series expansions

$$\phi(x) = \phi(x_i) + (x - x_i)\left(\frac{\partial \phi}{\partial x}\right)_i + \frac{(x - x_i)^2}{2!}\left(\frac{\partial^2 \phi}{\partial x^2}\right)_i + \frac{(x - x_i)^3}{3!}\left(\frac{\partial^3 \phi}{\partial x^3}\right)_i + \cdots. \tag{16.6}$$

Using these expansions, approximate expressions for the derivatives at point x_i can be obtained. There are several choices for the first derivative as follows:

$$\begin{aligned}
&\text{Forward difference}: &&\left(\frac{\partial \phi}{\partial x}\right)_i \approx \left[\frac{\phi_{i+1} - \phi_i}{x_{i+1} - x_i}\right] \\
&\text{Backward difference}: &&\left(\frac{\partial \phi}{\partial x}\right)_i \approx \left[\frac{\phi_i - \phi_{i-1}}{x_i - x_{i-1}}\right] \\
&\text{Central difference}: &&\left(\frac{\partial \phi}{\partial x}\right)_i \approx \left[\frac{\phi_{i+1} - \phi_{i-1}}{x_{i+1} - x_{i-1}}\right].
\end{aligned} \tag{16.7}$$

and even more for the second derivative. The backward difference formula is also known as an upwind difference as it only includes information from an upwind cell. The finite difference approximation leads to an algebraic equation at each grid node depending on the value at that node and at adjacent nodes. The number of unknowns and equations has to be equal, so there is a large set of algebraic equations for a computational grid which must be solved numerically. A disadvantage of the finite difference method is that it requires the mesh to be structured with nodes located on individual grid lines which do not join or intersect with others in the same direction.

The finite element method originates from the field of structural analysis, where the computational domain is divided into many types of small subdomains, called elements. The equations are solved for each individual element type using different forms of parameter variation within the elements. After each element has been analysed the equations are reassembled to analyse the whole domain. In fact, the finite element technique is seldom used in fluid dynamic analysis, but it is gaining in popularity as a choice for high-order schemes.

The most popular approach used by CFD to solve the governing partial differential equations is the finite volume method. In this method, the partial differential equations are integrated over control volumes defined in a mesh which defines the control volume boundaries. The divergence theorem is used to convert the volume integral to a surface integral for all terms in divergence form. These terms are the advection terms related to transport by the fluid, and the diffusion terms due to molecular agitation and turbulence. As an example, consider the continuity equation, (16.1), which for a control volume can be integrated over the volume and expressed in this form as

$$\frac{\partial(\rho)}{\partial t} + \nabla \bullet \left(\rho \vec{u}\right) = \dot{S}_i^{mass}, \quad \int_V \frac{\partial \rho}{\partial t} dV + \int_V \nabla \bullet \left(\rho \vec{u}\right) dV = \int_V \dot{S}_i^{mass} dV. \tag{16.8}$$

Applying the divergence theorem, the volume integrals can be converted to surface integrals, giving

$$\int_V \frac{\partial \rho}{\partial t} dV + \int_S \rho \vec{u} \bullet d\vec{S} = \int_V \dot{S}_i^{mass} dV, \tag{16.9}$$

where the subscript V and S stand for the volume and surface of the cell. If the values of the unknowns within this cell are taken as constant, this leads to

$$\frac{\partial \rho}{\partial t} V + \sum_{faces} \rho \vec{u} \bullet \Delta \vec{S} = \dot{S}_i^{mass} V. \tag{16.10}$$

For a steady flow with no source terms, this reduces to the continuity equation as discussed in Chapter 5. Similar equations for each of the momentum equations and for the energy equation can be determined. The major advantage of the finite volume approach is that conservation is strongly and discretely imposed independent of the mesh resolution (e.g., the mass flowing in equals mass flowing out of the volume). In addition, the method does not limit the cell shape or the grid structure.

16.4.2 Pressure-Based versus Density-Based

The five governing equations have five unknowns. All modern CFD methods solve for the three cartesian velocity components, most solve for total enthalpy (or rothalpy in rotating blade rows) and a few choose static enthalpy or static temperature. There is a choice to be made for the fifth unknown, either pressure or density can be selected, and the CFD community divides into two groups on this choice. A cursory look at the mass conservation equation suggests density should be the unknown. But if the density is constant, as in a liquid, then density no longer appears in the continuity equation, yet pressure remains an unknown in the momentum equations. One group declares pressure as the fifth unknown, the so-called pressure-based methods, while a second group declares density as the fifth unknown, hence density-based methods. Pressure-based methods use the thermal equation of state to compute density from pressure, $\rho = f(p,T)$, whereas density-based methods invert the equation of state to compute pressure knowing the density $p = f(\rho,T)$. The density-based approach fails if the equation of state does not, or only weakly, depends on pressure for example for flows where $\rho = f(T)$ or $\rho = constant$.

In both families of methods, the velocity field is computed from the momentum equations. In density-based solvers, the continuity equation is used to compute the density field while the pressure field is determined from the equation of state. In pressure-based solvers, the pressure field is computed by solving a pressure or pressure correction equation formulated from a combination of the continuity and momentum equations. The density field is then computed from the equation of state.

Pressure-based methods were originally developed for incompressible flows but were extended to address a wide range of flow conditions, including compressible flow at all Mach numbers (Patankar, 1980). The original family of pressure-based algorithms, known as SIMPLE (semi-implicit method for pressure linked equations), solves the conservation equations in a sequential fashion, that is, only solving for a single scalar quantity while keeping all other properties constant. Coupled solution of the momentum and mass equations improves numerical robustness, which was pioneered for pressure-based solvers in the late 1980s and is now commonly used in many commercial and proprietary CFD codes. Density-based time-marching methods originated in the aeronautics industry for the simulation of compressible flows (MacCormack, 1969), and for a long time were the dominant method used in the simulation of transonic and supersonic flows in aerodynamics applications. Methods involving artificial compressibility have been developed to adapt the density-based methods for operation at low speeds and in incompressible flows.

Both density-based methods and pressure-based methods solve the identical equations using a similar control volume approach, with the key difference being the choice of the dependent variable in (16.1)–(16.3). Fortunately, there is no evidence to suggest that the choice of algorithm has any effect on a possible converged result; nonetheless, it may influence the computational effort involved and of course the numerical accuracy as a function of mesh resolution. Both approaches can give excellent CFD predictions. For turbomachinery applications, both pressure-based and density-based solvers are in use in different research groups. All the major commercial CFD codes, however, use pressure-based methods as this approach robustly handles regimes of incompressible, low- and high-speed flow without having to introduce artificial compressibility into the fluid. In modern radial compressor CFD analysis, more and more details of the geometry are included in the analysis, such as front and rear leakage paths, seals and so forth, where the local gas velocities can be very low, thereby favouring the pressure-based methods.

16.4.3 Computational Grid

Any CFD analysis starts with a spatial discretisation of the geometry of the region of interest into a computational mesh of hexahedral, tetrahedral and arbitrary polyhedral grids. The computational domain needs to be discretised using grid cells that should provide an adequate resolution of the geometrical and expected flow features. Several kinds of mesh topology are available. The most frequently used approach is to use body-fitted grids where the cell surfaces follow a curved domain surface as a sequence of flat facets and satisfy the geometrical constraints imposed by the domain boundary.

In a structured grid, the mesh cells in a block are addressed by a triplet of indices (ijk). The mesh topology is straightforward because cells adjacent to a given face are identified by the indices. Cell edges form continuous mesh lines which start and end on opposite block faces. In this case, the cells have a hexahedral shape, and this type of grid is needed for finite difference methods. In an unstructured grid, the meshes are allowed to be assembled cell by cell freely without considering continuity of mesh lines. Hence, the mesh topology for each cell face needs to be stored in a table. The most typical cell shape is the tetrahedron, but any other form including hexahedral cells is possible. These grids may have matching cell faces, but for reasons of mesh modularity they may have nonconformal interfaces. This occurs often in turbomachinery stage calculations when different components have a different blade number and the circumferential spacing of the adjacent blade passages differs.

A special case of an unstructured grid is the block-structured grid where for the sake of flexibility the mesh is assembled from a number of structured blocks attached to each other. Attachments may be regular, that is, cell faces of adjacent blocks match, or arbitrary (a nonconformal attachment without matching cell faces). Similarly, a hybrid grid may combine different element types, such as tetrahedra, hexahedra, prisms and pyramids. In a chimera grid, structured mesh blocks are placed freely in the domain to fit the geometrical boundaries and to satisfy resolution requirements, and the blocks may overlap in three dimensions.

The grids must be fine enough to provide an adequate resolution of the important flow features, as well as geometrical features. This may be achieved by local grid refinement. The accuracy of the simulation increases more than linearly with increasing number of cells, that is, with decreasing cell size. However, due to limitations imposed by the increased computer storage and run-time, some compromise in mesh size is nearly always inevitable in a design calculation. In addition to grid density, the quality of a mesh depends on various criteria such as the shape of the cells (aspect ratio, skewness, warp angle or included angle of adjacent faces), distance of cell faces from boundaries or spatial distribution of cell sizes. Typical grids for a centrifugal compressor stage simulation are shown in Figure 16.1. The left-hand picture shows the domain of the impeller and the diffuser with a body-fitted hexahedral grid on the surface, where only one passage (including splitter blade) of the impeller and the diffuser is meshed. The right-hand side shows the impeller, diffuser and volute together with the tetrahedral grid for the volute. The volute includes an addition with a convergence at the outlet to help avoid reverse flow on the outlet plane during the simulations. Figure 16.2 shows the grid for a labyrinth seal.

It is hard to understate the importance of the mesh generation to the robustness, efficiency and accuracy of a radial compressor CFD analysis. Many decisions affect the mesh generation process: availability of mesh generation software, mesh style (hexahedral boundary fitted style generally still critical for efficient and accurate turbomachinery blade row simulations), mesh size, the ability to mix different mesh styles within a stage, turbulence model selection, engineering requirements for the analysis, geometric complexity, cost and wall clock time factors, available computational resources (modest or large cluster) and mesh quality metrics, to name only the major factors. The unstructured methods in CFD are popular in industry due to the attraction of semiautomated meshing coupled with the capability to resolve boundary layers efficiently through use of high aspect ratio prism cells near the walls.

Figure 16.1 Computational grids for a stage calculation with volute.

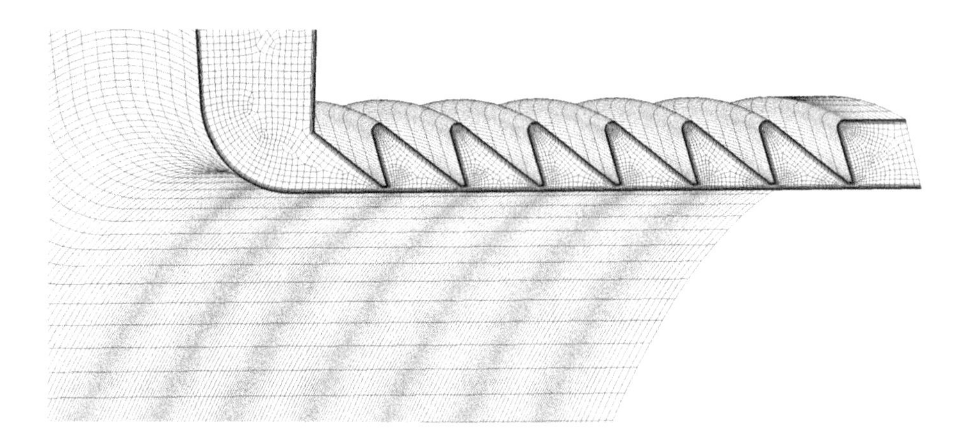

Figure 16.2 Computational grid for a labyrinth seal.

Hexahedral grids are unique, as they are the only element that supports *two* aspect ratios while maintaining orthogonality (which governs both quality and robustness). The aspect ratio at the walls can be made large to allow grids to capture the boundary layer. The aspect ratio and the grid stretching in the spanwise and/or chordwise direction can be made large as required to resolve end-wall boundary layers, as well as economising on cell count parallel to the blade chord. Radial compressors have particularly long blades relative to span, hence dual aspect ratio mesh elements are mandatory for computational efficiency. Meshing of a tip gap region is even more stringent, requiring a very high geometric aspect ratio between tip gap and meridional length to capture the large change in the velocity profile over the tip gap.

Tetrahedral grids with triangular surface elements are seldom used in radial compressor blade row simulations owing to their lower computational efficiency and a requirement for many more grid cells to achieve a certain level of accuracy. The natural alignment of mesh lines with the principle flow direction from a body-fitted hexahedral mesh also gives a large computational advantage in terms of cost for a given accuracy, relative to a tetrahedral mesh. Numerical errors are minimised when the main flow direction aligns well with the mesh, and there tends to be some cancellation of discretisation error across a hexahedral mesh. For a tetrahedral mesh, there is no preferential alignment direction and no beneficial cancellation of error. However, in a complex geometry, such as a volute, a tetrahedral mesh is often used as it can be automatically generated by grid generation tools, though it is normally supplemented by hexahedra adjacent to the walls.

An efficient mesh for computation is critical during a radial compressor design process as, for each design being considered, hundreds of CFD simulations may be needed. Typically the designer considers multiple operating points (from five to 10) from choke to stall, on multiple speed lines (from three to six), to explore the performance map of the compressor, giving 15–60 or more simulations in total for each design. This level of coverage will normally be a final step in the design process, to generate a map to supersede or complement the preliminary map produced by 1D methods. During design iterations, typically two or three critical speed lines are calculated on each iteration. It is important to realise that although the designer normally has a nominal or design point in mind, it forms only part of the assessment which must consider the other important factors such as proximity to surge and choke and normally part-speed, off-design performance.

As CFD has become more mature and more easily deployed, it has become practicable to include more features of the details of a stage, such as secondary leakage paths, and casing treatments. Perhaps not surprisingly, the more detail the model includes, the better the agreement with test data; see, for example, Hazby et al. (2019). However, for the majority of design iterations, a simplified model is applied such as that shown in Figure 16.1. This provides a good compromise between throughput and accuracy and is generally reliable in the assessment of iterative changes in geometry, accepting normally a small offset in, for example, the predicted absolute level of efficiency.

16.4.4 Some Aspects of the Internal Methods of the Code

The numerical methods used internally within the codes rely on details that are often not needed by the turbocompressor designer but are fundamental to the robustness and accuracy of the solutions. Some of these are briefly touched on here, but more details can usually be found in the code manuals, from CFD code developers and in the textbooks already mentioned. The convergence properties and the robustness of the code is the result of all of the choices made by the code developer. Different aspects of this are robustness to using a bad mesh, robustness to a bad initial guess, robustness to an inexperienced user and robustness to solver input settings.

In the finite volume method, a control volume is defined over which the differential equations are integrated. There are two kinds of control volume approaches: the cell-centred approach or the cell-vertex approach. In the cell-centred approach, the unknowns are stored at the centre of cells, and the control volumes are the mesh cells. In the cell vertex approach, the mesh lines define elements and nodes whereby the unknowns are stored at the vertices of the elements.

One of the key components of the finite volume method is the approximation of the fluxes across the faces of the control volume, which is required to compute the surface integrals and determines the numerical accuracy of the method. These schemes can broadly be categorised into two groups: upwind biased and central biased. Upwind biased methods start with some interpolated value for the scalar ϕ upstream of the cell face, and then correct this value using estimates of the nearby gradients of ϕ. The correction to the upwind biased value is made with care, assessing the size of the local gradients, as well as the current values for the surrounding variables. The correction is limited by algorithms to balance the trade-off between numerical robustness, accuracy and strict boundedness. In areas of low gradients, the full correction is used and the scheme is second-order accurate and robust. In local areas of high gradients, the correction will be reduced, even to zero, so that locally the scheme reduces to a first-order upwind scheme in space, thereby remaining robust and accurate. The idea is that the volume of the domain where the upwind biased method has reduced to pure upwind is small, and reduces quickly with mesh refinement. It is important to never use or trust a CFD simulation that has been obtained with a first-order upwind biased advection scheme. Almost without fail, first-order advection simulations have high levels of numerical diffusion, and the predictions simply cannot be trusted, even under aggressive mesh refinement. A first-order advection scheme is the quickest and most robust way to obtain a CFD prediction of no value whatsoever.

Central biased schemes begin with a centred interpolation for ϕ, and then explicitly add a nonlinear diffusionlike term to the equations to control the negative impact of central difference numerics. The main idea is to control the amount of numerical diffusion directly, adding only the amount needed locally to overcome numerical challenges of central differencing. Details regarding the specifics of advection discretisation schemes are available for the interested reader in a plethora of publications, as given at the beginning of this chapter. The CFD practitioner should be aware of the significant influence the choice of the advection discretisation scheme has on the robustness and accuracy of predictions. The most accurate advection numerics that the CFD solver offers should be selected, but also the one that remains stable and convergent.

The equations are solved in an iterative manner where at each iteration a new value of the variable at the centre of a cell is calculated from the conservation equation. It is common to apply relaxation such that the new value for variable ϕ in cell P is taken as a fraction of the predicted value and the old value

$$\phi_p^{new} = \phi_p^{old} + RF\left(\phi_p^{predicted} - \phi_p^{old}\right) = \phi_p^{old} + RF\delta\phi_p, \tag{16.11}$$

where RF is the relaxation factor and $\delta\phi_p$ is the change in ϕ for a given iteration. With a value of $RF < 1$, the equations are underrelaxed, which slows the calculation but increases the stability of the calculation with less chance of divergence or oscillations. The iterative process is repeated until the change in the variable from one iteration to the next, $\delta\phi_p$, becomes so small that the solution can be considered converged. It is clearly seen that the predicted value for ϕ is independent of the value chosen for RF, assuming convergence is attained.

Iterations are used to advance a solution through a sequence of steps from a starting state to a final, converged state. This is true for both when the solution is a final steady-state result and when the solution is one step in an unsteady-state calculation. In both cases, the iteration steps resemble a timelike process and can be either explicit or implicit. Explicit methods calculate the state of a system at a later time from the state of the system at the current time, while implicit methods find a solution by solving an equation involving both the current state of the system and the later one.

The base iterative solver in a modern CFD code is accelerated by some technique, most commonly based on multigrid techniques. Multigrid solver acceleration methods solve the equations on a sequence of grids going from fine to coarse. The influence of boundaries and faraway points is more easily transmitted to the interior with coarse meshes than on fine meshes. In coarse meshes, grid points are closer together in the computational space and have fewer computational cells between any two spatial locations. Fine meshes give more accurate representations of the solution, ultimately solving the original (finest) mesh system of equations.

A fluid boundary is an external surface of a fluid domain where the solver recognises user-defined boundary conditions. In this way, the boundary conditions separate the finite computational space from the unbounded environment around the turbomachine. A viscous wall is a no-slip boundary impenetrable to the fluid flow and where the fluid velocity is zero relative to the wall. At an inlet, the flow is expected to enter the domain, and associated conditions of inlet total pressure and total temperature are usually defined. At an outlet boundary, the flow is expected to leave the computational domain and the static pressure, the outlet mass flow or the outlet corrected mass flow is defined. A symmetry plane in the domain identifies that the geometry and the flow is expected to be symmetrical about this plane. A periodic boundary condition is similar to a symmetry plane except that the symmetry is periodic around the annulus. This allows the calculation of a single blade row to represent the full annulus.

16.4.5 Boundary Conditions

Radial turbocompressor simulations require special strategies to compute across the full speed line, to determine choked mass flow, as well as to estimate the numerical stall point. Inlet total pressure and temperature and inlet flow angle are almost universally chosen as inlet boundary conditions, but the outlet boundary condition can be challenging for a few reasons. To predict the choked flow condition, one can set the outlet static pressure, typically at the inlet total pressure level if no other information is provided. This combination of conditions is satisfactory on the steep part of the radial compressor characteristic towards choke, but becomes less stable towards stall and surge as the operating characteristic of pressure rise versus flow becomes less steep, even ‘flat’, with little change pressure as flow is reduced; see Chapter 11. Away from choke, it is desirable to specify the outlet mass flow, in order to uniquely characterise operating points across the flatter part of the compressor speed line. Attempts to simulate across the speed line by changing the outlet static pressure is generally an exercise in frustration, as the pressure ratio versus mass flow curve is insensitive to the outlet pressure over a range of mass flow, and may even be multiply valued; two compressor mass-flow rates can generate the same pressure ratio. The risk of stall is best identified as the peak total–static pressure ratio achieved, as the mass flow is lowered. It is often possible to predict numerically the compressor flow at mass flows below the peak pressure ratio when the mass flow is specified, which is a much more controlled process than trying to define the stall point as the lowest outlet pressure at which the CFD simulation does not catastrophically diverge.

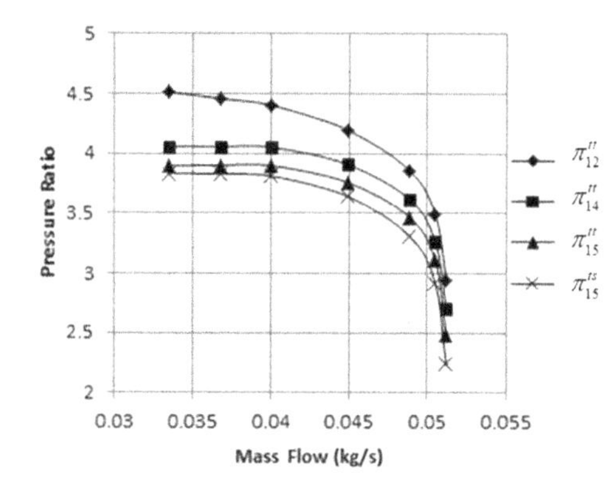

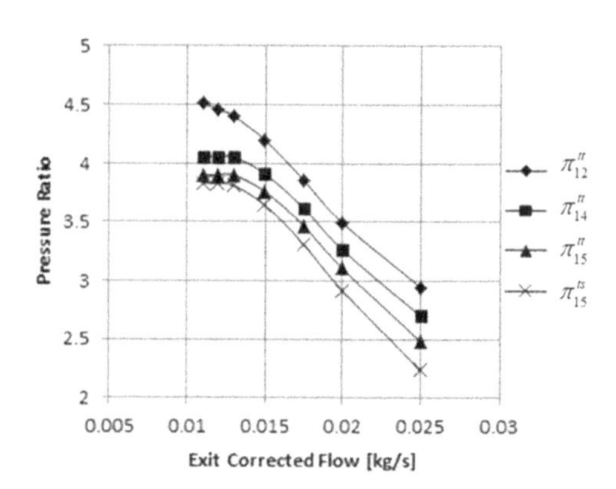

Figure 16.3 Pressure ratio along a speed line at different positions in the stage plotted against inlet corrected mass flow and outlet corrected mass flow (2 is impeller outlet, 4 is diffuser outlet and 5 is the volute outlet).

Probably the most useful outlet boundary condition for centrifugal compressor analysis is to specify the exit corrected mass flow at the stage outlet, analogous to a constant throttle setting:

$$\dot{m}_{exit,corr} = \dot{m}\frac{\sqrt{T_{t2}/T_{ref}}}{p_{t2}/p_{ref}}, \tag{16.12}$$

assuming that this boundary condition is available in the code. Normally the reference conditions are taken as the stage inlet total conditions. Constraining the exit corrected mass flow at the outlet is optimal as it behaves like a pressure constraint at choked conditions, but like a mass-flow constraint across the rest of the speed line down to stall. It also offers a soft startup from simple initial conditions prior to the compressor generating some pressure rise. This avoids the common issue of a mass-flow outlet boundary condition that enforces that the mass is evacuated at the outlet before sufficient fluid has arrived at the outlet, pulling the pressure down to near vacuum conditions within the domain and often causing the CFD solver to crash. An example is given in Figure 16.3, showing that it is easier to calculate the performance map with the outlet corrected mass-flow boundary condition. It has the virtue that the range of the exit-corrected mass flow across each constant speed characteristic is similar, so as speed is varied, the same outlet flow conditions can be used, which simplifies the setup for automatic full-map prediction.

16.5 Stage Calculations with Interface Planes

16.5.1 Steady-State Interface Planes

A typical blade row calculation examines a single passage of the machine between adjacent blades and applies periodicity on pitchwise boundaries of the blade row within the domain. It does not allow for the exact influence of upstream and downstream blade rows, other than through the specification of radial pressure profiles on the upstream and downstream boundaries, perhaps obtained from a throughflow calculation. Before the turn of the millennium, this limitation was removed by sophisticated multi-blade-row 3D Navier–Stokes calculations with averaging of variables on the planes between the blade rows. These were developed from the original idea of Denton to the level where they were also being applied routinely in design; see the papers of Denton (1992), Dawes (1992) and Galpin (1995). Averaging-plane methods solve all blade rows simultaneously, exchanging spanwise distributions of averaged flow quantities at a common grid interface between the blade rows. Since averaging-plane methods often use mixed-out averages, they are commonly referred to as mixing-plane methods or stage interface methods.

There are several different techniques to enable the computational domains of individual blade rows to be used to make a multiple blade row calculation. In multiple frame of reference codes, each component is simulated in its local frame of reference. Depending on the implementation, either interface or boundary conditions between the different frames of reference connect the control volume faces in the two different frames of reference. A common approach is to use virtual ghost cells between the interfaces, that is, a set of fictitious cells adjacent to the boundary faces between the two frames of reference. At each iteration, the values of the interface flux variables are updated in such a way that the interpolated value on the boundary face corresponds to the boundary condition. There are two special features of this implementation. The first is that it includes a virtual 2D control surface between the upstream and downstream regions of the interface, so that the circumferential static pressure variations in both regions are determined by the local domain and only the mean spanwise value is determined by the circumferential averaging of fluxes that takes place there. Secondly, it allows flow to traverse the interface in both directions. There are a variety of mixing-plane interface models, each with their own tricks and modelling assumptions, depending on the CFD code.

The most robust interface method is a frozen rotor calculation. This is a steady-state calculation method in the rotating and the stationary frames which simply converts the circumferential component of velocity from relative to absolute at the interface between the domains. It needs to take into account any change in grid topology and pitch between the two frames. It is most accurate where the circumferential variation is small. A meridional velocity deficit such as a wake downstream of an impeller in the rotational frame remains present in the absolute frame. The transfer across the interface is, however, not realistic since the wake remains in a fixed circumferential position as it passes through a downstream component; see the top part of Figure 16.4. There are then several possible frozen rotor solutions, and the computational results are sensitive to the relative circumferential position of the components. One of these solutions is shown in Figure 16.4, where the wake passes between the downstream blades, as highlighted by the entropy contours. In another solution with a different relative position of the upstream and downstream blade rows, the wake would move directly on to the leading edge of the downstream blade.

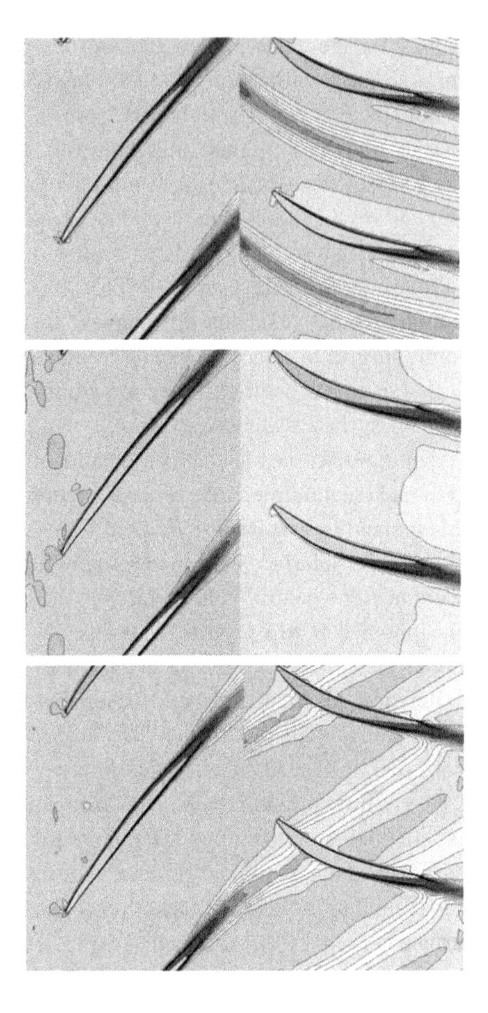

Figure 16.4 Entropy field of an axial compressor frozen rotor simulation (top), a mixing-plane simulation (middle) and a transient blade row calculation (bottom). (images courtesy of Paul Galpin of ISIMQ)

A frozen rotor interface between impellers and diffusers is not a good choice, even in the case of a vaned diffuser. The trajectory of the wakes in the diffuser is entirely incorrect when that interface is applied. The wakes lag the rotor in the measured data, and with the stage interface; with a frozen rotor, they leave in the direction of rotation downstream of the interface.

It is general good practice to use a stage interface, or mixing plane, between the rotating and stationary domains, which in some manner mixes the flow circumferentially as it passes from the rotating to the stationary domains. This scheme is a really useful approximation for many design situations and is computationally very efficient. It leads to a solution as shown in the middle of Figure 16.4; this approach now dominates CFD turbomachinery design procedures. Sophisticated mixing-plane treatments have been developed and are the mainstay of current multistage designs in industry. The methods mix out the enthalpy and entropy between the blade rows but allow the flow direction and the flow velocities upstream and downstream of the mixing plane to be circumferentially nonuniform. The unsteady effects are completely neglected by these methods. Fritsch and Giles (1995) identified that the mixing loss generated by a mixing plane can be larger or smaller than that occurring when the flow mixes out gradually as it moves downstream, so it may give a contribution to any uncertainty in the absolute prediction of the losses. The mixing-plane approach has contributed to the development of many successful well-matched designs of the individual blade rows in radial turbocompressors (Robinson et al., 2012), but provides no information on the unsteady interactions.

16.5.2 Unsteady Calculations

In reality, there is always a transient interaction between components and there may be unsteady vortex shedding at trailing edges. Denton (2010) points out that the flow at a trailing edge with a finite thickness is always unsteady with vortex shedding from both surfaces and that this imposes the flow direction. There is no tendency for the stagnation point to shift on to the pressure or suction surface at a trailing edge, but this may happen with steady simulations with rounded trailing edges.

The wakes from the upstream blade trailing edge pass downstream and interact unsteadily with the leading edges of downstream blade rows, and the pressure field around leading edges or shocks of any downstream blade rows affect the upstream blade row. Halstead et al. (1997) have shown that the unsteady wakes traversing downstream blade rows have a significant effect of the location and type of transition; see Section 5.5.6. Unsteady information requires a transient rotor-stator (TRS) calculation, but since blade numbers are usually deliberately chosen to avoid common multiples, to set up an accurate transient calculation with fewer than full 360° coverage is not normally practicable and is computationally very intense. In the last 15 years, new transient blade row methods have been developed which are capable of reproducing the transient behaviour of a full wheel from the simulation of a few passages. These methods promise greater simulation fidelity for unsteady flows at a much lower computational cost. Furthermore, transient methods, such as time transformation, also return information on pressure fluctuations which may be used to assess the mechanical impact of radial spacing on the compressor design. A snapshot of the entropy field for such an unsteady simulation for two blade rows is shown in Figure 16.4.

The large difference between the unsteady flow and the flow with mixing planes suggests that it may be unwise to ever wholly rely on a steady calculation and that the use of steady RANS simulations with mixing planes still fails to account for a lot of the real loss-producing processes. Bulot and Trebinjac (2009) provide an example which shows the difference in losses predicted by a steady and an unsteady CFD simulation for a radial compressor stage with a pressure ratio of 6.5 with a close proximity of the impeller and the vaned diffuser ($r_3/r_2 = 1.05$). The difference in entropy between the steady and unsteady calculations is situated in a region near the diffuser leading edge, where the shock structure on the diffuser vane leading edge is strongly affected by the passage of the impeller blades. The work of Medic et al (2014) shows in a similar high pressure ratio stage, also with close proximity of the impeller and the vaned diffuser ($r_3/r_2 = 1.05$), that it is necessary to carry out an unsteady calculation to determine the mean characteristic of the stage. Robinson et al. (2012) show that the unsteady effects are far less serious if the diffuser vanes are less closely spaced to the impeller (at $r_3/r_2 = 1.15$). The computational effort, however, for a full 360° transient simulation increases by a factor of 50 so that other forms of unsteady simulations such as time transformation are needed. The important aspect of unsteady CFD simulations is their use to predict pressure fluctuations leading to mechanical vibrations rather than their use to assess designs at their best point. At the time of writing, the use of unsteady methods in design iterations is not attractive due to the computational effort. As with other computationally limiting constraints, this may well change as more computer power becomes available.

16.6 Turbulence Models

16.6.1 The Chief Outstanding Difficulty

Most compressor flows of practical engineering interest are turbulent such that turbulent mixing usually dominates the behaviour of the flow. Turbulence plays a crucial part in the determination of many relevant engineering parameters, such as frictional drag, dissipation losses, heat transfer, flow separation, the transition from laminar to turbulent flow, the thickness and blockage of boundary layers, the extent of secondary flows and the spreading of leakage jets and the growth of wakes. A wide variety of models for turbulence have appeared over the years, as different developers have tried to introduce improvements to the models available. It is hardly surprising that most early technical papers on turbulence in turbomachinery applications were unable to provide clear recommendations on the most suitable turbulence model (Gregory-Smith and Crossland, 2001). It is still worth bearing in mind that some of the most successful exponents of CFD in the design environment have worked with the simplest of turbulence models; see Section 16.2.3.

The choice of the turbulence model and the interpretation of its performance by establishing bounds on key predicted parameters are far from trivial. A fuller introduction to the subject of turbulence and turbulence modelling can be obtained by consulting standard reference texts on the subject. The objective here is to describe the background to the most popular turbulence models used in turbomachinery calculations.

16.6.2 Turbulent or Eddy Viscosity

The most common turbulence modelling approach relies on the concept of a turbulent or eddy viscosity, μ_t. This viscosity relates the turbulent stresses appearing in the

RANS equations to the averaged velocity gradients in analogy to the classical interpretation of viscous stresses in laminar flow through the fluid viscosity, μ, as outlined in Section 5.2.6. For example, in a two-dimensional shear layer, the turbulent shear stress is given as

$$\tau_{turb} = \mu_{turb} \frac{\partial c}{\partial y}, \tag{16.13}$$

where c is the averaged streamwise velocity and y is the cross-stream coordinate. Unlike its molecular counterpart, the turbulent viscosity is not a fluid property but a function of the state of turbulence. From dimensional considerations,

$$\mu_{turb}/\rho \sim VL, \tag{16.14}$$

where V is a velocity scale and L is a length scale of the energy-containing turbulent motions (related to the mixing length in so-called mixing length models). The state of turbulence determines both the velocity scale V and the length scale L, and, over the years, many prescriptions for V and L have been proposed.

16.6.3 Algebraic (or Zero-Equation) Models

The most straightforward definition of V and L is with the so-called algebraic (or zero-equation) class of models. These assume that V and L can be related by algebraic equations to the local properties of the flow, which is a straightforward process for two-dimensional boundary layer flows but can often be difficult in geometrically complex configurations. A version sometimes used in turbomachinery calculations is that from Baldwin and Lomax (1978). For example, in a simple wake or free shear layer, V is often taken as proportional to the velocity difference across the flow and L is taken as constant and proportional to the width of the layer. In a boundary layer close to the wall, V is given as $L \cdot \partial c/\partial y$ (or $L \cdot \Omega$ where Ω is the magnitude of the vorticity) and L is related to the wall-normal distance from the wall (y-direction). The outer part of the boundary layer is treated similarly to a wake.

Algebraic models of turbulence have the virtue of simplicity and are applicable to two-dimensional shear flows, such as attached boundary layers, jets and wakes. For more complex flows where the state of turbulence is not locally determined but is related to the upstream history of the flow, a more sophisticated prescription is required. Algebraic turbulence models are based on boundary layer concepts, including shear layer thickness, distance from a wall and velocity differences across the layer. These quantities cannot easily be computed in a Navier–Stokes code and introduce significant additional uncertainty into the computation, as the solutions are dependent on implementation details. This is the main reason why these models are generally not recommended for turbomachinery applications of RANS methods.

The model of Baldwin and Lomax (1978) is an algebraic two-layer turbulence model with the turbulent eddy viscosity defined by one expression in the inner layers close to the wetted surfaces and a different expression in the outer layers. The distinguishing feature of their model is the use of a function constructed from the product of vorticity and distance from the nearest surface. This function has a sharp peak at a certain distance from the wall, and its location determines the changes in the equations for the mixing length in the inner and outer layer. The BTOB3D code for turbomachinery of Dawes used the method of the Baldwin–Lomax model to calculate the turbulent viscosity.

16.6.4 One-Equation Models

The one-equation models attempt to improve on the zero-equation models by using a turbulent eddy viscosity that no longer depends purely on the local flow conditions but considers the flow history, that is, where the flow has come from. The majority of approaches seek to determine V and L separately and then to construct μ_t/ρ as the product of V and L. Almost without exception, V is identified with $k^{1/2}$, where k is the kinetic energy per unit mass of fluid arising from the turbulent fluctuations in velocity around the averaged velocity. A transport equation for k can be derived from the Navier-Stokes equations and this is the single transport equation in the one-equation model. This equation is closed, i.e. reduced to a form involving only calculated variables, by introducing modelling assumptions. This model thereby furnishes a prescription for the velocity scale, V, which accounts, at least partially, for nonlocal effects.

Spalart and Allmaras (1992) devised a formulation of a one-equation model specifically for aerodynamic flows. It determines the turbulent viscosity directly from a single transport equation for μ_t. This model is quite successful for practical turbulent flows in external aerofoil applications. Although the model is not well-suited for more general flows it is quite widely used in industrial turbomachinery applications.

16.6.5 Two-Equation Models

For general applications, it is usual to solve two separate transport equations to determine V and L, hence the name of the model. In combination with the transport equation for the turbulent kinetic energy, k, an additional transport equation is solved for a quantity which determines the length scale, L. This class of models is the best-known and the most widely used in industrial applications since it is the simplest level of closure which does not require geometry or flow regime–dependent input. Prior to the turn of the century, the most prevalent model was the k-ε model, where ε is the rate at which turbulent energy is dissipated to smaller eddies (Launder and Spalding, 1974). A modelled transport equation for ε is solved and L is then determined from the term $C_\mu\, k^{3/2}/\varepsilon$, where C_μ is a constant. This method is included in almost every commercial and inhouse code and is still heavily used in industrial applications. It has the merits of robustness and is useful for centrifugal compressor cases where the viscous effects are relevant, but the flow tends to be heavily influenced by three-dimensional effects and rotational pressure gradients.

The second widely used type of two-equation model is the k-ω model, where ω is a frequency of the large eddies (Wilcox, 1998). A modelled transport equation for ω is solved and L is then determined from $k^{1/2}/\omega$. The k-ω model performs very well close to walls in boundary layer flows, particularly under strong adverse pressure gradients (hence its popularity in aerospace applications). However, it is susceptible to the freestream value of ω, and unless great care is taken in setting this value, spurious results may be obtained in both boundary layer and free shear flows. The k-ε model is less sensitive to freestream values but is often inadequate in adverse pressure gradients.

Many different variants for circumventing this problem are found in the literature. An attractive option, the BSL (baseline) model has been proposed by Menter (1994) using a turbulence model, which retains the properties of k-ω close to the wall and gradually blends into the k-ε model away from the wall. This model has been shown to eliminate the freestream sensitivity problem without sacrificing the k-ω near-wall performance. The BSL model performs similarly to the k-ε model on a mesh that does not resolve the near-wall region, but gives significant advantages over the k-ε model when the near-wall region is fully resolved, removing the need for additional empirical models for unresolved steep flow gradients inside the boundary layer.

Various attempts have been made to modify two-equation turbulence models to account for strong nonequilibrium effects, such as when the flow approaches separation. The so-called SST (shear stress transport) variation of Menter's BSL model (Menter, 1996) leads to marked improvements in performance for nonequilibrium boundary layer regions such as those found behind shocks or close to separation. Menter et al. (2003) provide an overview of the success of this model in a wide range of applications. It is popular and very effective for relatively well-ordered flows, such as in axial flow machinery. However, such modifications should not be viewed as a universal cure. The SST model is less able to deal with flow recovery following reattachment. A further improvement of this is the addition of a rotation-curvature correction. This version of the model (SST-CC) has been extensively tested on a wide range of both wall-bounded and free shear turbulent flows with system rotation and/or streamline curvature, including centrifugal compressor impellers (Smirnov et al., 2007). Gibson et al. (2017) have also found that the SST-CC improves the accuracy of the original SST model for modelling the flow in an impeller.

Since there is such a variety of options and, in the opinion of the authors, no clear 'winner' for universal design applications, it is important for users to build up their own calibration material, based on stages that they know well, have designed and ideally tested so that they feel comfortable with the simulations being carried out.

16.6.6 Other Models

Radial compressor simulations generally do not include laminar to turbulent transition in the calculations, it being assumed that transition occurs at the leading edge. However, a correlation-based model for transition which has been used in axial compressors is available, Langtry and Menter (2005). Turbulence and transition modelling are active areas of research and many other models have been developed; these are not referred to here as they are not generally used for turbocompressor design calculations.

The preceding turbulence models are based on the eddy viscosity concept and assume that the turbulent stresses are linearly related to the rate of strain by a scalar turbulent viscosity, and that the principal strain directions are aligned to the principal stress directions. This is reasonable for fairly simple states of strain in aerodynamic flows, especially when the model constants have been carefully calibrated from similar classes of flows, but may prove inadequate for modelling complex strain fields arising from the action of swirl, body forces such as buoyancy or extreme geometrical complexity. Under such circumstances, a more subtle relationship between stress and strain must be invoked. The so-called Reynolds stress transport models (RSTM) dispense with notion of turbulent viscosity, and determine the turbulent stresses directly by solving a transport equation for each stress component. This requires the solution of six additional coupled equations, together with an equation for ε to provide a length scale. These methods are also sometimes referred to as second moment closure (SMC) models. Such methods are seldom used in radial compressor calculations as they are expensive to compute with prevailing design resources.

16.7 Quality and Trust

16.7.1 Errors and Uncertainties

Distinctions between types of errors and uncertainties occurring in a CFD computation formed the basis for the main chapters of the best-practice guidelines given by Casey and Wintergerste (2000). The following sections use their categories for describing the errors.

16.7.2 Round-Off Errors

Round-off errors occur when the difference between two values of a parameter is below the machine accuracy of the computer. This arises from the limited number of computer digits available for the storage of a given number. The impact of round-off errors on the quality of the simulation results depends on a number of factors, including the mesh size, the mesh aspect ratio and the internal details of how the CFD solver assembles the equations. Round-off error generally goes as the square of the mesh aspect ratio – due to the terms arising from the integration of second derivatives in the partial differential equations. Computer memory and disk space scales by a factor of two comparing single to double precision, which is not generally an issue on modern workstations or clusters. Of greater significance is that, in practice, the computational time increases significantly with double precision, which can be up to two times slower (code and computer hardware dependent). The main cause is the RAM memory look-up time as twice as much data have to be stored and retrieved for all operations, whereas the low-level time for the actual mathematics operations may be independent of the precision in modern CPUs.

16.7.3 Iteration or Convergence Error

Before considering other errors, it is necessary to ensure that the simulation has converged and that iteration errors are negligibly small. The equations solved by CFD methods consist of large systems of nonlinear equations, hence all CFD solvers use iterative methods, starting with an initial approximation to the flow solution and iterating to a final result. The iterative solution process repeats until the change in the variables from one iteration to the next becomes sufficiently small that the solution can be considered converged. At convergence, all the conservation equations (continuity, momentum, energy, etc.) are obeyed in all control volumes to a specified tolerance. Residuals are used to measure any imbalance in the conservation equations, obtained by substituting the current solution estimate into the control volume equations. For steady-state simulations, when the solution no longer changes with additional iterations, the mass, momentum and energy equations are solved. If the iterative process is incomplete, however, then the control volume equations are not solved and numerical errors arise due to this lack of convergence. It is common to monitor the scaled residuals, that is, the error relative to the local value of a property, and to require these to be on the order of 10^{-3} to 10^{-4} or less for convergence.

In order to properly judge convergence, it is important for the user to understand exactly what residual is reported, as this is CFD code dependent. Some codes report the raw residuals (having dimensions of the control volume equations), whereas others normalise the residuals. If the residuals are not normalised, then the absolute level of the residuals required for convergence will change based on the geometric scale of the simulation domain, whereas normalised residuals attempt to remove the scale of the problem, hence providing dimensionless and thus more universal residual definitions. Additionally, a single number is usually presented for the residual for each equation as well, usually the root-mean-square (RMS) of the residual of all control volumes in the simulation. Some codes also provide the location and value of the maximum residual.

Convergence errors arise either as a result of an insufficient number of iterations or because the numerical methods used are inadequate and do not allow the solution algorithm to complete its progress to the final converged solution. For example, a steady-state solution may not exist for a particular set of flow conditions, such as a compressor operating at low flow in the stalled region, and attempts to iterate to convergence may also stall with constant or oscillating nonzero residual values. In an ideal simulation, residuals converge monotonically. In practice, this is often not the case. Residuals may decrease with iteration and plateau at a finite value (or even increase for some iterations) before finally decreasing to the minimum values possible. It is also not unusual for the residuals to reach a terminal minimum value, but the solution itself still slowly evolves with iteration: the small residual imbalances can still drive a slow change to the simulation prediction.

As a first indicator of convergence, it is necessary to achieve sufficiently small residuals. But how small must the residuals actually be? This is a difficult question which is CFD code specific, as well as problem specific. For these reasons, it is *critical* to monitor and judge convergence not only by monitoring residuals but by monitoring additional important quantities such as flow balances, mass-flow balance from inlet to outlet, as well as overall momentum and energy flow balances together with all key performance indicators for the flow. In a compressor simulation, typical performance indicators include the inlet to outlet mass flow, mass averaged inlet to outlet pressure and temperature ratio, the total torque and power, the efficiency, and perhaps additional performance indicators.

Ideally all performance indicators should be monitored with each iteration, so their convergence can be observed. If this is not possible, the simulation can be run to a specified residual level and each of the performance indicators can be evaluated. Then the simulation can continue running to a lower residual level and the performance indicators can be evaluated again and compared to the previously recorded values. The process is repeated until the performance indicators have converged to the required levels. In addition to overall performance indicators, two-dimensional and/or three-dimensional plots of solved variables such as pressures, velocities and temperatures should be made as a function of residual convergence level. Some CFD codes permit animations of field variables as a function of iteration, which are highly valuable to graphically visualise the impact of convergence level on the predicted solution. A good compressor design office should know what is possible regarding monitoring of convergence for the simulation software in use and take advantage of the capabilities offered. Simply monitoring residual level is necessary but often insufficient for properly judging iterative convergence.

16.7.4 Discretisation or Numerical Error

Numerical errors arise from the discrete approximations that are inevitable in any numerical method. This type of error arises in all numerical methods and is related to the approximation of a continually varying parameter in space by polynomial or other functions for the variation across a grid cell. In first-order schemes, for example, the parameter is taken as constant across a specific region. This type of error also arises from the artificial viscosity or smoothing that is necessary to stabilise some codes. The most significant errors often occur in regions where both the second derivatives and the grid spacing are large (Denton, 2010), thus it is essential that the grids are fine enough in these regions of the flow and second-order schemes are essential.

These errors arise due to the difference between the exact solution of the modelled equations and a numerical solution with a limited resolution in time and space. For consistent discretisation schemes, the higher the number of grid cells, the closer the results are to the exact solution of the modelled equations, but, in fact, both the fineness of the grid and the distribution of the grid points affect the result. In short, discretisation errors arise where numerical approximations do not supply an exact solution to the equations we are trying to solve.

16.7.5 Model Uncertainty

These are the uncertainties due to the difference between the real flow and the exact solution of the model equations. They cause errors because the exact governing flow equations are not solved but are replaced with a simplified model of reality. The most well-publicised errors in this category are those arising from turbulence modelling, as described earlier in this section. However, there are other model errors. Examples of these which are relevant to turbocompressors are the simplification of an equation of state of a real gas to that of an ideal gas, the errors associated with the mixing out of the flow at a stage interface, the neglect of unsteady-state effects in a steady simulation and the neglect of laminar to turbulent transition. The modelling errors have been reduced over the years as CFD has developed, and no doubt they will continue to do so.

The numerical and model errors are not always isolated in publications. Some turbulence models were developed and calibrated on coarse grids with relatively large numerical errors. Their predictive capability then apparently got worse on fine grids because the numerical errors were reduced. Often turbulence modelling has been blamed for all problems but, in cases with relatively coarse grids from the past, the real culprit was the numerical errors (Gregory-Smith and Crossland, 2001).

The designer can actually influence the model uncertainty through design decisions. A well-designed impeller should not have a high deceleration as denoted by a low de Haller number, as described in Section 7.5.4, so that the limitations of some turbulence models with regard to the prediction of separation becomes less relevant. Impellers can be designed with thin cutoff trailing edges to avoid any risk of the simulation allowing the trailing-edge stagnation point to move on to the suction or the pressure surface. A vane-island or wedge diffuser with a thick trailing edge can be avoided to eliminate the difficulties of simulation of the unsteady flow downstream of the thick trailing edge. The leading edge of a vaned diffuser can be placed farther downstream in the diffuser channel to reduce the strong unsteady interactions with the flow at the impeller outlet.

16.7.6 Application Uncertainties

Inaccuracies also occur when the precise data needed for the simulation are not available. This type of systematic error is usually related to lack of knowledge either of the boundary conditions or of the geometry, or by ignoring surrounding components or features that turn out to have a noticeable effect. As with physical models, the accuracy of the CFD solution is only as good as the initial/boundary conditions provided to the numerical model; for example, if the flow is supplied to a compressor by a very long pipe, the simulation should use a fully developed profile for velocity rather than assume uniform conditions, which are suitable for an inlet from the ambient environment.

Uncertain geometry may apply to blade profiles, blade fillets and corners, tip gaps, thermal or mechanical deformation of the blades in operation or even unknown leakage under the rotor root, as in the case of the ASME Rotor 37 test case. Uncertain geometry does not represent an error in the CFD method itself but does have a considerable influence on the accuracy of CFD predictions. In many turbocompressor problems, the geometries are extremely complex and require considerable effort to specify them correctly for a computer simulation. There are many sources of error which can arise in this process, including the following:

- Neglect of changes in the geometry that occurred during the design or testing process. The geometry of a tested component, such as a change in the pinch of a diffuser, may get modified during the testing procedure. If these modifications are not added to the original drawings, the validation against test data will be with the wrong geometry.
- The CAD geometry definition is insufficiently complete for flow simulation. Some surfaces and curves may not meet at the intended end point locations due to different levels of accuracy in different parts of the CAD model. Other curves may be duplicated.
- The geometry may not be manufactured within the tolerances given on the drawing, particularly regarding to small geometry features, such as the rounded shape of turbomachinery leading edges and clearance gaps.
- The change of shape due to flow or mechanical loading may not have been considered such that the running geometry is not the same as was intended. For example, the twist of an inducer due to centrifugal forces and the change in tip clearance as the result of thermal or mechanical loads may have been neglected.
- The geometry of the surface or its roughness, especially the shape of the leading edges, may have changed during use due to wear, erosion or dirt fouling. As a result, the actual performance and geometry are different from design intent.
- Small details of the turbomachinery geometry may have been omitted or are unknown, such as tip clearance gaps, roughness on the walls, welding fillet radii and leakage paths beneath blade roots.
- The blade shape of an impeller with straight line generators may be correctly defined on the hub and casing lines. The exact shape of the blade, however, depends on the orientation of the blade generating lines on an impeller blade surface, and these may not have been specified correctly.
- The blade shape may be defined as a series of points along the surface. If these are too far apart in a region of high curvature, an interpolation between these points is required to avoid simulating a nonsmooth surface.
- The blade profile may be defined as a combination of a camber surface and a thickness distribution, which may be either normal to the camber surface or in the circumferential direction.
- Seal and tip clearances are seldom known accurately and may be affected by erosion in service. Small variations of the tip clearance gap can have a dramatic impact on the overall machine performance, both surge line and efficiency, and this is exacerbated for physically small stages.

Unknown boundary conditions also play a role in the simulations. Small errors in the initial and boundary conditions can influence the solutions. In radial compressors, the user needs to be aware of the possibility of a recirculation on the walls of the diffuser with flow inadvertently entering at the diffuser outflow boundary. This inflow at an outlet boundary location may lead to difficulties in obtaining a stable solution or even to an incorrect solution. If it is not possible to avoid this by relocating the position of the outlet boundary in the domain, one possibility is to restrict the flow area at the outlet, provided that the outflow boundary is not near the region of interest. If the outflow boundary condition allows flow to reenter the domain, appropriate boundary conditions should be imposed for all transported variables on this boundary. Hysteresis effects may occur either numerically, physically or both, causing a simulation at a specific operating point to differ depending on whether the initial conditions were from a higher or lower mass flow.

The specification of turbulence properties at the inlet is another aspect of application uncertainty for turbomachinery. The use of a turbulence model (other than an algebraic model) requires that turbulence properties at a domain inlet region need to be specified. Validated quantities should be used because their magnitude can significantly influence the results. If no data are available, sensitivity tests with different simulations are required to determine the influence of the choice of boundary conditions.

16.7.7 User Errors

Many errors also arise from user mistakes and carelessness of the user. Such errors generally decrease with increasing experience of the user or the availability of standard computational procedures, but they cannot be eliminated. User errors are reduced by adequate training, enough resources assigned to the project and good quality management procedures. Two important aspects are the overoptimistic expectation of project managers that the CFD results will be available on time and right the first time, and a novice user's blind faith in computer methods despite lack of experience or lack of understanding of the limitations of CFD. The errors can be reduced by proper quality management procedures with checklists, user training, interaction and review with more experienced expert users. Users should scrutinise the results to check that they are realistic and always study the details of the CFD solution to try to understand the underlying physics. It is always valuable to take a step back, having done some calculations on a particular case, and to consider whether the extent of domains modelled, the approach to meshing and selection of boundary conditions and interfaces still looks sound, and the code is given a fair chance to capture the factors likely to influence performance. Especially in modelling quite complex situations, decisions will be made by the user to make the analysis more tractable, and it is useful to revisit these and to be prepared to take another approach once results are available for interpretation.

16.7.8 Code Errors

Errors also occur due to bugs in the software, unintended programming errors in the implementation of models or very rare compiler errors on the computer hardware. Such errors are often difficult to find, as CFD software is highly sophisticated, typically involving millions of lines of code for a commercial product. Computers are very unforgiving. Even a relatively simple typing error that might easily be overlooked on this page, such as a j for an i, can have disastrous consequences if incorporated into a line of code. The main cause of software errors are bugs in the program, whereby the code then solves equations which are not the actual equations specified by the user. These errors can be significantly reduced by introducing verification procedures for code development. Also, the more users of a code, the more chance of identifying errors caused by bugs.

Coding errors are also inadvertently introduced from version to version of the code, despite efforts by developers, commercial and academic, to avoid such pitfalls. This adds an additional onus on users of CFD software to verify the correct functioning of a new software version. This is a nontrivial task, as the software may intentionally give different answers with a new release. A different answer might be the result of a bug that was resolved or a newly introduced bug in the new release. It is difficult to separate the two effects besides hard work, experience, and a judgemental eye towards a new release of the software. A company-specific automated regression deck is an extremely good approach to managing software change and identifying changes in simulation results, intended or otherwise, in a proactive manner. Finally, it is good practice (actually mandatory) to stay with the same software version throughout a given design project. It is good practice for users to maintain their own validation cases to check that new releases of any code give consistent results to the previous version.

16.8 Checklists for the Design Process with CFD

16.8.1 Design Rules

In a design iteration, a balance must be struck between what can be done when time and resource constraints are relevant and what could theoretically be achieved with the highest levels of analysis. In radial compressor design, this balance leads to the use of CFD in two different ways. A candidate stage design is first analysed with a simple methodology; that is, a single passage calculation for each blade row using a mixing plane between the blade row and with a simplified housing geometry. This simplified geometry may neglect leakage flow paths and parasitic losses in the end spaces of the compressor and omit the downstream volute or return channel. Such calculations can be carried out rapidly on modern workstations and can be used to optimise blade and channel shapes. This simplified CFD methodology allows a simple grid to be used, and an operating point can be calculated very rapidly consistent with the timescale of the design process. Optimisation of the design may involve recamber of leading and trailing edges (Came and Robinson, 1998), optimisation of the location of the splitter leading edge (Lohmberg et al., 2003), and complex stacking of the blades to adjust the location of the shocks with sweep, dihedral and lean (Hazby, 2018). It would be inconceivable to attempt these changes without the availability of 3D viscous CFD and accurate mechanical calculations to assess the associated structural integrity (Hazby and Robinson, 2018). In the design phase, it may also be possible to make use of a simplified gas model by approximating the gas to an ideal gas.

This simplified methodology, with no parasitic losses and without including the downstream components to the outlet flange, does not capture the real losses accurately enough to allow efficiency predictions of sufficient fidelity (Hazby et al., 2019). Following the optimisation of the blade geometry, a more detailed CFD analysis using the appropriate gas equations, with a more thorough geometrical model comprising the entire geometry with few simplifications, is usually carried out. This gives more accurate predictions of efficiency but requires more effort for the grid generation and the solution.

In this process, predictions of the performance and the CFD predictions of the flow are of crucial importance to the designer. The flow field prediction allows an assessment of the weak points in the flow, through an examination of aerodynamic criteria relating to incidence levels, loading parameters, velocity and pressure distributions and boundary layer development. It is often possible to identify desirable or undesirable features of the flow even when their direct effects on performance cannot be quantified. CFD provides the tool to achieve intended consequences in the flow patterns. The key aerodynamic aspects used by the designer to assess and improve the quality of a turbomachinery blading design are provided in the following checklist with reference to more detailed discussion in earlier chapters:

- Avoid poor incidence onto blading.
- Reduce dissipation losses on wetted surfaces.
- Avoid kinetic energy loss.
- Avoid flow separation.
- Avoid unnecessary acceleration and deceleration in the flow by choosing a smooth geometrical path that avoids area reductions and discontinuities.
- Check established blade loading criteria (the de Haller number for impellers).
- Provide a uniform distribution of flow onto downstream blade rows.
- Minimise the preshock Mach numbers in transonic and supersonic flow.

Some of these requirements may often conflict with one another so that compromises based on experience are generally needed; further details are provided by Casey (1994b).

16.8.2 Assessing a Design with CFD

There are three separate questions related to the use of CFD for turbomachinery design that are important for the turbocompressor designer: How can the designer achieve the best possible quality of CFD simulations? What is the useful trustworthy range of application of the chosen models, especially a turbulence model? How can the designer use the CFD simulations to confidently discriminate between different designs, despite numerical, physical model and possible systematic errors present in the CFD simulation?

The first issue is the quality of the CFD simulations themselves. The priority in this case is ensuring that the CFD simulation is as accurate as possible given practical constraints such as time, computing resources and the goals of the simulation results for the design phase. This requires careful verification of the solution methods and defined numerical procedures with respect to selection of grids, and is related to iterative and grid convergence. These procedures ensure that the code solves the model equations correctly and consistently from simulation to simulation. Such verification is always needed for a new application. In most situations, the experience already gained in a compressor design office on previous designs, or advice from a code supplier, allows code settings and grid types to be determined from standards developed on earlier simulations of successful compressor designs.

The second issue is the extent to which the results can be trusted. As designers are more and more exposed to CFD results and less and less to experimental results, it is vital that they understand when CFD results can be trusted and when they cannot. This knowledge is particularly important when CFD is combined with optimisation software to produce an 'optimum' design within certain constraints. The optimisation algorithms may exploit weaknesses in the CFD and search for a nonoptimum geometry so that care is needed in interpretation. Has the process searched the full design space or found a local optimum? Does the method acknowledge the multiple acceptance criteria that are carefully balanced by the designer, such as surge margin, choke margin, part-load performance and mechanical difficulty?

The critical aspect is knowledge about how well the various models represent the real world by validating the models and procedures and testing them for turbomachinery applications against reliable test data. In most design situations, the CFD calculation being made is similar to many made in the past, and experience gathered on other similar compressors, or from technical publications, can be used to provide guidance. In this process, it is often the case that experience outweighs innovation. A model that converges rapidly and where there is wide experience of its use for design, such as the standard k-ε turbulence model, may be used in preference to a newer model available in an updated version of the code. The main weakness of the standard k-ε turbulence model is that the flow tends to remain attached in a decelerating flow, when in reality the flow may be separated. But the use of design rules involving classical loading parameters, as discussed in Chapter 7, allows the designer to avoid this issue. Careful calibration of the newer tools with earlier experience may allow these to become the standard calculation model at some point.

The third issue is how to make best use of CFD to ensure reliable engineering design decisions are made. The central aspect is the definition of processes that enable the engineer to live with the known errors and possible weaknesses in the models and still make the right engineering decisions. The most robust approach is to use CFD in a comparative sense to assess changes in design from a trusted and well-developed basic design which is known to achieve the required performance levels, or working from a relevant baseline of known geometry and performance.

The most difficult aspect of performance prediction for compressors of all types has historically been prediction of the surge line. Even today, 1D methods cover this perhaps most reliably via empiricism, as discussed in Chapter 18. However, as CFD becomes more widely used and is seen to be reliable around the design condition, the next step is to wonder to what extent it can predict the absolute surge line for a specific geometry. The fact that surge is a complex phenomenon (see Chapter 17), involving factors other than the basic design of the impeller and diffuser of the compressor stage, immediately adds complexity to this process. Factors influencing the boundary in practice include, inter alia, the state of the inlet flow; radial and circumferential distortion; the enclosed volume of the downstream components; any circumferential nonuniformity in geometry or clearance and transient effects. Since in most cases design and development are carried out on simpler models, the best that can be expected is a comparison between one design iteration and the next. Numerical surge in CFD often occurs as the low momentum flow from a separation at the casing is expelled from the inducer. The codes typically will continue to run stably to lower flows, but there take a step in the predicted total–static pressure rise. A pragmatic guide is to monitor the peak total–static pressure rise and arbitrarily to take that as the 'surge' point.

Because of the uncertainty in performance prediction, the good designer examines only the differences in the simulations for the various competing designs and uses these differences to rank the designs. One of the simulations would ideally involve a previous similar design for which the performance is known from quality test data available to assess the possible risks with the code. Good designers use CFD to copy features into a new design from an earlier good design; for example, a long assessment of the blade loading distribution is not needed for each individual new design, as previous successful experience can act as a guide. Here a qualitative understanding of the flow may be more useful than a quantitative prediction (Denton, 1997).

16.8.3 Interpretation of the CFD Data

The challenge for engineers working with CFD is how to analyse the computation efficiently and effectively given the ever-increasing quantity of simulation data (Pullan, 2017). The postprocessing of the CFD results for turbomachinery simulations has undergone some standardisation in the design process within recent years (Garrison and Cooper, 2009). Designers now make full use of the wide variety of visual tools available in modern CFD packages to interpret the flow field that has been calculated, whereby a good physical understanding of the application and experience with similar simulations are the best guide as to which plots to make. The following are typical of the data and methods which are used to interrogate the CFD results:

- Convergence of target parameters (flow turning, work input, efficiency, loss), residuals and global balance of conserved variables
- Two-dimensional contour plots of the distribution of Mach number, pressure, velocity magnitude, velocity vectors and entropy in representative 2D cutting planes (blade-to-blade and throughflow) or on the wetted surfaces
- Line diagrams of parameters (whereby circumferentially mass-averaged mean values of relevant parameters are often plotted against meridional flow direction or the span)
- Identification of the mean flow process in a Mollier diagram

- Pressure distribution and isentropic Mach number distribution along a blade section
- Three-dimensional visualisations of streamlines and 3D views of isosurfaces, such as regions of low pressure or high Mach number
- Surface streak lines
- Identification methods to visualise vortex structures, such as line integral convolution (LIC)
- Video sequences of unsteady-state simulations

Some of these techniques are shown in the examples of CFD simulations given in other chapters of the book.

17 Compressor Instability and Control

17.1 Overview

17.1.1 Introduction

Centrifugal compressors are most effective when running at their design capacity and speed, with the velocity triangles well matched to the blade inlet angles. In many situations, the application requires operation off-design, either with more flow or speed in an overload or overspeed condition, or with a lower flow and lower speed at part-load or part-speed. Unfortunately, operating at part-load can cause serious instabilities in the compressor aerodynamics, even leading to damage to the compressor. The operating range with stable flow decreases with increasing pressure ratio at high speeds, as discussed in Chapter 18. Experimental testing of compressor applications is still needed to determine whether stable compressor operation is possible in the expected application regime. The objective of this chapter is to describe some aspects of the physics of unstable flows in compressors, to outline their consequences on the operating characteristics and to describe some control methods to extend the region of stable operation.

An inherent feature of centrifugal compressors is that on a speed line the flow capacity decreases as the delivery pressure increases. The characteristic pressure–volume curve has a negative slope at the design point, and its steepness depends on the value of the design work coefficient (see Chapter 11). When the delivery pressure increases safe operation continues until a point of instability is reached. At this point, small pressure fluctuations grow into rotating stall, which is a local unsteady stalled condition rotating within the blading, or into much larger global disturbances known as surge involving the whole compression system. This may be mild surge in which the flow repeatedly recovers without reverse flow in the compressor, or it may be deep surge which involves backflow through the compressor before the flow recovers to start another surge cycle.

Surge and rotating stall are to be avoided as they result in noise and excessive vibration and can damage the compressor. Flow control systems that allow the compressor to function at part-load without becoming unstable are needed. Methods to extend the stable operating range of compressors by control with variable speed, variable geometry, passive recirculation systems and other regulation devices are described.

17.1.2 Learning Objectives

- Understand the distinction between surge and rotating stall, and between spikes and modes.
- Understand the effect of system characteristics on the type of instability, as included in the Greitzer *B* parameter.
- Be aware of different types of system pressure versus volume flow characteristics.
- Understand the different systems available for generating a wider performance map.
- Be able to select the appropriate type of control to meet different system requirements.
- Be able to describe the matching of individual stages in a multistage compressor such as which stages may be responsible for stall or choke at different conditions.

17.2 Instabilities in Compressors

17.2.1 Instability

In axial compressors, intensive research for over 75 years, and in particular the systematic experiments of Day, Greitzer and others for over 45 years, has provided clear distinctions of the different types of compressor instability and its inception. Much of the description given here is adapted from the many publications of Day and Greitzer, from Day (1976) to Day (2016), and Greitzer (1976) to Greitzer et al. (2004).

The performance of a compressor is determined by its performance map describing how pressure rise varies with volume flow at different speeds. This performance map shows how the pressure rise of the device changes with flow at different speeds and is described in more detail in Chapter 18. The operating range of a radial compressor stage on a speed line is limited by choking of the flow channels at high flow with low delivery pressure and by the onset of instabilities (surge or rotating stall) at low flow with high delivery pressure.

At high flow, no further increase in mass flow at a particular speed is possible when the Mach number in one of the components reaches unity and the compressor chokes (see Chapter 6). In radial compressors with a vaneless diffuser, the impeller throat inlet area will invariably determine the choking mass flow, but with a vaned diffuser it may be the diffuser that chokes first, especially at lower Mach numbers. When operating at part-load, the reduced volume flow along a speed line leads to a change in the velocity triangles, with increased incidence on to blades, a higher work coefficient and a higher outlet pressure, as discussed in Chapters 7 and 11. If the flow through the compressor is decreased further, the work input continues to increase, but the pressure–volume characteristic becomes horizontal due to higher losses at part-load. The higher loading and increased incidence may cause small flow separations in the flow channels. The first small fluctuations in the flow are precursors to more radical instability. The precursors have been found to be of two different types, which are categorised as modes and spikes, and if they grow unhindered, they nudge the compressor into a fully unstable mode of operation.

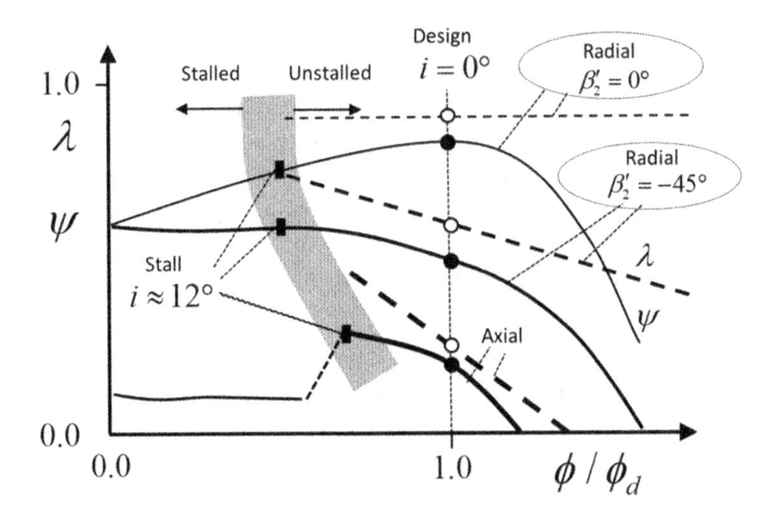

Figure 17.1 Sketch of idealised compressor work coefficient and pressure rise characteristics down to shutoff for an axial stage and radial stages with and without backsweep

At low flow, the fluctuations grow and rotate around the annulus in a form known as rotating stall. Large separations with the onset of stall cause increased losses and flow blockage with an abrupt loss of work input and pressure rise. The performance of radial and axial stages in the stalled region can often be measured in low-speed ventilators and pumps, and in high-speed compressors where a throttle or valve is placed immediately downstream of the compressor.

The step in the pressure rise characteristic of an axial stage at stall, as shown in Figure 17.1, represents what happens when an axial compressor operates with and without stall. It is sometimes associated with a hysteresis, in that it occurs at a different flow rate when the flow is decreasing than when it disappears as the flow increases. In most radial compressors, there is no abrupt loss in pressure as the flow reduces and separations occur. This feature is also shown in Figure 17.1 for a backswept and a radial vaned impeller. The characteristic curve remains continuous but loses its negative slope at low flows when the outlet pressure reaches a peak and then begins to fall steadily at lower flows. With no backsweep, the peak of the pressure rise is reached at the design flow rate, as discussed in Chapter 11.

Stalled operation is identified by pressure or volume flow fluctuations, higher noise, increased blade vibrations and higher rotor and casing vibrations. If such fluctuations are allowed to continue, they even lead to possible destruction of the compressor through fatigue; see Chapter 19. With rotating stall, the disturbances are generally of a higher frequency than the rotational speed and are confined to the compressor. With surge, they involve the whole compression system and its pipework and have a lower frequency and much larger amplitude.

Rotating stall and surge are unacceptable operating conditions in most compressor applications, such as gas turbines, turbochargers and industrial compressors. They are generally not completely independent phenomena, as the change in the slope of the pressure rise characteristic caused by the onset of rotating stall is often sufficient to initiate surge. In many real situations, each surge cycle begins with rotating stall, as this is what causes the pressure rise through the machine to drop. This loss of pumping capability allows the flow through the machine to decelerate rapidly, sometimes reversing. When the pressure behind the compressor lowers, the stall clears and the compressor outlet begins to pump up to full pressure again. The process then repeats itself in a surge cycle. Understanding of the boundary between the stable and the unstable operating region of a compressor is crucial to the safe operation of compressors and the definition of a control strategy.

17.2.2 Differences between Axial and Radial Compressor Characteristics

Although rotating stall was first reported in a centrifugal compressor (Cheshire, 1945), much of the basic research work on compressor instability has been carried out in axial compressors. For this reason, it is worthwhile recalling the important differences between axial and radial machines with respect to their performance characteristics already identified in other chapters of the book. First, in a radial compressor, the centrifugal effect provides a large proportion of the design pressure rise, typically 50% of that in the stage and 75% of that in the impeller; see Chapter 2. As the flow reduces, the pressure rise due to the centrifugal effect remains the same and is still present after the onset of stall. Because of this, a stalled centrifugal machine experiences a pressure rise not much different from that at the design point. From data given by Gülich (2008) on radial pumps, the shutoff pressure rise coefficient at zero flow is given by $\psi = \Delta p/\rho u_2^2 \approx 0.65$ in low flow coefficient designs, the value decreasing in designs with a higher flow coefficient and larger inlet radius ratio. This decrease is related to the fact that high flow coefficient designs experience a smaller centrifugal effect as the inlet casing diameter relative to the impeller tip diameter increases with the flow coefficient. By using a throttle valve immediately downstream of the compressor outlet, Fink et al. (1992) measured the whole characteristic of a stage with an impeller with no backsweep down to zero flow on several speed lines and found a shutoff value of $\psi = \Delta p/\rho u_2^2 \approx 0.5$ at all speeds. The centrifugal effect is not present in an axial compressor where the onset of stall causes a larger deficit in the pressure rise characteristic, as shown in Figure 17.1. In fact, in an axial compressor the stalled pressure rise is largely independent of the compressor design and a stalled axial compressor stage at zero flow achieves a pressure rise coefficient per stage of $\psi = \Delta p/\rho u^2 \approx 0.11$ (Day et al., 1978).

Second, another relevant difference is that radial compressors have a much larger work coefficient compared to axial compressors, and as a consequence they have much flatter pressure rise versus volume flow characteristics; see Chapter 11. In a radial compressor with no backsweep, the work remains nearly constant with flow. In a more typical radial compressor with a backswept impeller, a decrease in flow of 10% produces a 3.5% increase in work coefficient, whereas in a typical axial compressor a 10% decrease in flow increases the work coefficient by about 30%, which is 10 times larger. Third, it is an inherent property of the typical velocity triangles of the rotor inlet that when the flow is reduced the incidence on the blades increases more rapidly in axial than in radial compressors (see Chapter 7). A 10% reduction in flow causes an increase in incidence of 2.5° in a typical radial stage but an additional incidence of 4° in a typical axial stage. Figure 17.1 shows the location of an incidence of 12° on the axial and the centrifugal characteristics, which in this sketch is nominally taken as the stall point.

Each of these effects result in wider and flatter characteristics in radial than in axial compressors, and it would not be surprising if the onset of surge and instability in the two types of machine are different. In a recent review, Day (2016) mentions that the severity of a surge event will be different in the two types of compressor. This difference in stalling behaviour between continuous and discontinuous characteristics, that is, between small and large pressure drops at the point of stall, has not always been properly appreciated and has not really been studied in the literature on this subject.

17.2.3 Surge

Surge is an unsteady operating condition involving oscillation of both the pressure rise and the mass flow through the entire compression system. Surge can be sufficiently severe that it involves reverse flow in the compressor. When the surge point is reached, the flow through the compressor reduces and the pressure ratio falls, until a momentary equilibrium at a lower flow is established. Under this condition, compressor operation moves from the surge point to some mass-flow operating point below it on the performance curve. When the compressor continues to operate against sustained system pressure, the operating point moves back up the characteristic until surge occurs again, as shown in Figure 17.2.

The frequency and intensity of the surge cycle depends strongly on the volume of enclosed gas in the compression system, but typically surge requires many rotor revolutions to go through one cycle, giving frequencies of just a small percent of the rotor frequency, typically of the order of 3–10 Hz. Fully developed surge gives rise

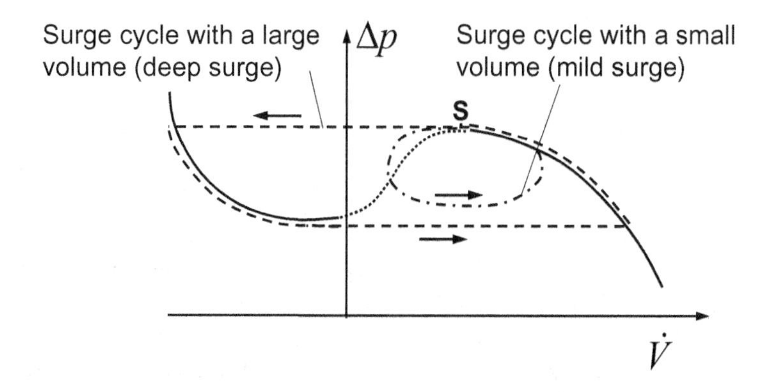

Figure 17.2 Compression system surge cycles – mild and deep surge.

to disturbances in the mass flow of around 50% of the design flow, with mild surge and over 100% with deep surge. The fluctuations in pressure are of the order of the dynamic head defined by the blade speed ($\rho u_2^2/2$). Continued operation under surge conditions is not possible without risk to the integrity of the machine, as the mass flow is not constant and the pressure fluctuations are similar in magnitude to the pressure rise produced by the compressor in normal operating points. The blades need to be designed to withstand the forces of the fluid in a sudden backflow in order to retain integrity during surge. This is one advantage of the more robust shrouded impeller designs used in industrial compressors.

The nature of the characteristics also gives rise to differences in the nature of surge in axial and radial compressors. In a centrifugal compressor having a characteristic like that in Figure 17.1, the surging process is likely to produce a flow oscillation which is limited in amplitude and will be of simple harmonic form, that is, the mass-flow/pressure rise trajectory on the performance map will be circular. The axial compressor, with its dramatic drop in pressure rise at the point of stall, is likely to produce a performance trajectory which is more rectangular in shape and may be extensive enough to include reversed flow. There are many uncertainties in the prediction of surge associated with operating conditions. These include inlet distortion, transient throttle changes, transient geometry changes (if the tip clearance changes following changes in speed) and compressor mechanical damage (including blade erosion and effects of large foreign object ingestion) (Ribi, 1996).

17.2.4 Rotating Stall

Rotating stall, on the other hand, is a severely asymmetric distribution of throughflow velocity around the compressor annulus at some position in the compressor, but with a steady annulus-averaged throughflow through the machine. It takes the form of single or multiple waves, known as stall cells, that propagate in the circumferential direction at a fraction of the rotor speed. Rotating stall cells typically rotate at between 20% and 70% of the rotor frequency. When a rotating stall cell occurs, the increased losses often lead to a large drop in the mean pressure rise of an axial compressor stage but not such a noticeable drop in a radial compressor; see Figure 17.1. This is a risk to mechanical integrity as the high-frequency fluctuations may cause blade vibrations and lead to high cycle fatigue if resonance with a blade natural frequency occurs, as described in Chapter 19.

The onset of rotating stall can be understood from the description given by Emmons et al. (1955), as shown in Figure 17.3. This involves a stall in a particular flow channel which is analogous to wing stall in external flow. As the flow rate is reduced, the compressor delivers a higher and higher pressure with increased incidence on the blades until, suddenly, the flow field is partly disrupted by a stall in one flow channel. When the flow separates, then the associated stall causes blockage in this channel and diverts the flow to the adjacent flow passages, leading to an increase of incidence on one of the adjacent passages, which leads to a new separation and a stall in this passage There is a decrease in incidence on the other adjacent passage with

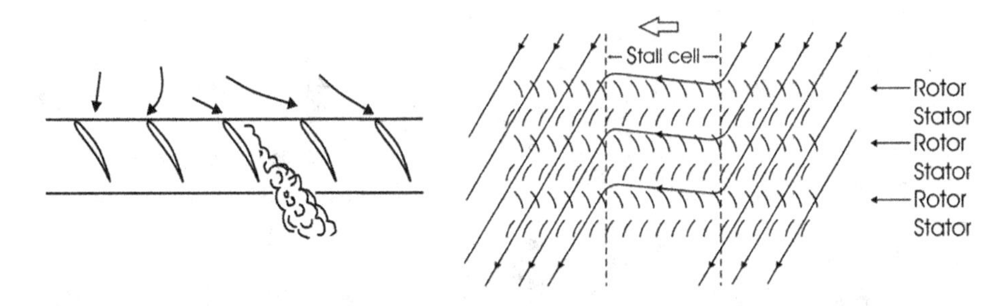

Figure 17.3 Sketch of a rotating stall cell at inception and a mature stall cell. (images courtesy of Ivor Day)

reattachment of the flow and a tendency for the flow to unstall. In this way, the rotating stall cell propagates around the annulus.

The blockage of the rotating stall may also comprise several stall cells around the circumference. It can occur both in an impeller or in a vaned diffuser where it is triggered by flow separation from the leading edge, but it can also occur in regions with no blades. At low flow in radial compressors with a vaneless diffuser, the rotating stall occurs in the diffuser, even though there are no blades. This is a well-known and much-studied phenomenon; see Chapter 12. Generally, the onset of vaneless diffuser rotating stall is considered to be related to a critical absolute flow angle at inlet to the diffuser greater than 80°, which leads to a long spiral flow path for the flow through the diffuser. Below the critical flow angle, the flow at some point around the circumference reverses, and the flow blockage from the reverse flow allows the flow in other parts of the annulus to pass through the diffuser with a lower flow angle and shorter flow path. At low flow rates which lead to high flow angles, the compressor seems to try to avoid extremely long flow paths in that it closes off part of the annulus flow area in the diffuser by forming a blockage. The volume flow is no longer uniformly distributed around the annulus but splits into regions with relatively healthy flow conditions and blocked stall-cell regions with strongly disturbed turbulent flow.

In axial compressors, part span stall cells occur over part of the span near the tip and also as multiple cells which are equally spaced around the annulus. These tend to occur in the long blade rows at the front of an axial compressor at low flow and rotate with 50–80% of rotor speed with only a small effect on performance and may even go unnoticed. Full span stall cells occur over the whole span usually as single stall cells. These tend to rotate with 20–50% of rotor speed. The pressure loss may become so severe that operation may not be possible and the compressor needs to be shut down. A fully developed full-span rotating stall involves flow blockage over several blade rows with active momentum exchange between the blocked and unblocked regions, involving transfer of healthy flow into the cell at its trailing edge and out of the cell at its leading edge. Day et al. (1978) showed that in multistage axial compressors, a fully developed stall cell extends axially through the machine and is of constant width through the machine. As the flow rate is reduced, the width of the stall cell increases and the region of healthy unstalled flow becomes smaller.

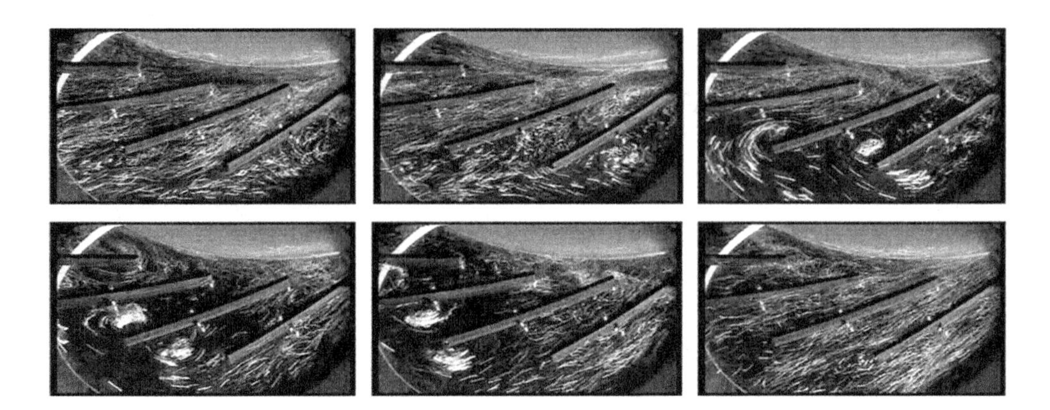

Figure 17.4 Stills from a video of a rotating stall in a vaned diffuser as the stall cell passes by (Staubli et al., 2001). (Video courtesy of Georg Gyarmathy and Thomas Staubli)

In radial stages with a vaned diffuser at low speeds, rotating stall will first occur in the inducer and the similarity of the inducer to long axial blade rows suggests that this will be part-span stall. At high speeds, it will first occur in the diffuser and may initiate surge. As in an axial compressor, a fully developed stall cell of a radial compressor extends through the blade rows in the meridional direction. Based on flow visualisation of the stall cell in a transparent model compressor operating with water, Staubli et al. (2001) were able to follow the path of finely dispersed air bubbles introduced into the flow, as shown in Figure 17.4. The first frame shows steady forward flow through the diffuser vanes. Streamlines appear as a series of points due to the illumination of the bubbles with a high-speed stroboscope. In frame 2, backflow has started in the lower diffuser channel. In frame 3, this generates a starting vortex that begins to block the channel flow in the diffuser channel, and a new vortex begins to appear in the adjacent channel. Frames 4 and 5 show how the diffuser channel flow is blocked by the vortices, and shows how these vortices are washed out of the diffuser trailing edge. Frame 6 shows the steady healthy flow after the passage of the stall cell, and the process then repeats from frame 1 with the passage of the next stall cell.

A fuller picture of the activities in the diffuser is given by Staubli et al. (2001) and a sketch of the flow patterns in the frame of reference of a mature rotating stall cell from these measurements is given by Gyarmathy (1996) and is given heuristically in Figure 17.5. There is no net forward flow in the stall cell, and the circumferential extent of the stall cell grows as the volume flow reduces.

Gyarmathy (1996) points out that many builds of test compressors with different impellers and diffusers show similar stall cell propagation speeds and strong pressure fluctuations in the vaneless space between the impeller and the diffuser. Gyarmathy carried out an idealised analysis of the momentum exchange between the fluid mass involved in the stall cell as it exchanges momentum across the vaneless space between the impeller and the diffuser. He developed the following approximate equation for the propagation speed ratio of the cell:

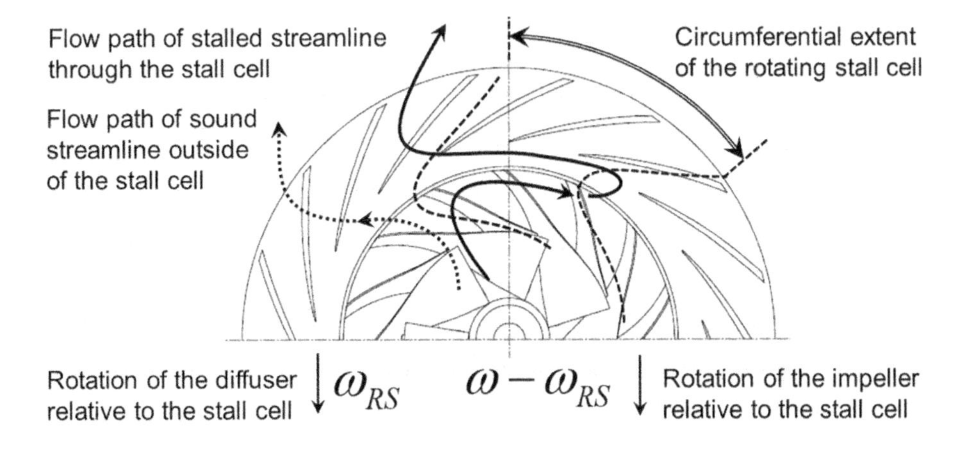

Figure 17.5 Sketch of the streamline path through a rotating stall cell in the frame of reference of the stall cell rotating at a speed of ω_{RS}.

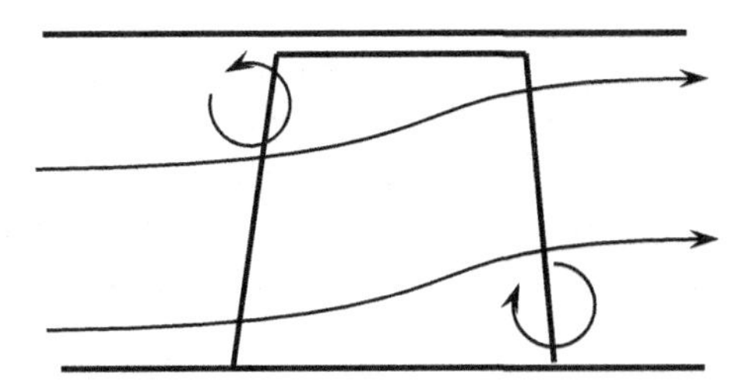

Figure 17.6 Recirculation regions in axial flow fans at reduced flow.

$$\Omega_{RS} = \frac{\omega_{RS}}{\omega} = \frac{\sin^2 \alpha}{\sin^2 \alpha + \mu \sin^2 \beta}. \quad (17.1)$$

In this equation, α is the stagger of the stator, β the stagger angle of the rotor in the simplified analysis which used straight blades and μ is the ratio of the volume of fluid contained in the diffuser channels and in the rotor channels. The analysis agreed satisfactorily with the propagation speeds of those compressors that were analysed.

17.2.5 Inlet Recirculation

In centrifugal pumps and axial fans, which have low Mach numbers and do not surge but can operate down to zero flow, recirculation regions are known to occur at the impeller inlet and exit between the onset of stall and zero flow (Eck, 1972; Gülich, 2008), as shown in Figure 17.6. Ribaud (1987) identified that at low speeds with low flow a recirculation at the inlet and the outlet can occur in a radial compressor. In many cases, a radial compressor stage operating stably without surge at low flows

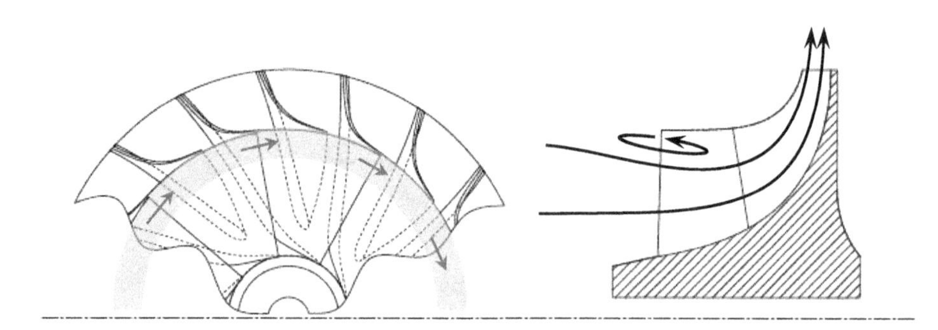

Figure 17.7 Sketch of inlet flow recirculation in the inducer of a radial compressor

can exhibit an inlet recirculation. Under these conditions, the flow is generally steady but involves reverse flow, leaving the compressor inlet eye at the tip of the impeller and moving backwards into the inlet duct before recirculating forwards into the impeller, as shown in Figure 17.7. The flow in the upper part of the recirculation bubble features a strong absolute swirl component in the direction of rotation as the relative flow direction of the recirculating flow is in the direction of rotation. This flow pattern is sometimes known as a ring stall. The main forward flow through the impeller enters close to the hub and passes under the recirculation region at the impeller inlet eye before leaving the impeller normally at the impeller exit. Ribaud (1987) provides data showing that the depth of the recirculating zone at impeller inlet increases as the flow reduces.

A study of inlet recirculation has recently been published by Schreiber (2017). He shows that as the flow rate is reduced, the inlet recirculation remains stable and extends both upstream and downstream of the impeller eye. A large region at flow rates lower than design conditions is affected by inlet recirculation, and the phenomenon occurs well before the compressor goes into surge. In terms of flow range, inlet recirculation is present over about 40% of the compressor map width at low rotational speeds in this configuration. This value decreases as the rotor speed is increased. It is present beyond when the inlet flow becomes transonic, but as soon as shocks are present, the recirculation region diminishes significantly. At higher speeds, the recirculation cannot be identified as the compressor surges before it can become established. Schreiber (2017), Harley et al. (2014) and Qiu et al. (2008) have developed mean-line models for prediction of the onset of recirculation and the losses associated with it.

When inlet recirculation occurs, the mean pressure rise of the centrifugal compressor does not necessarily drop, and it is possible for a compressor stage to operate stably under these conditions as the mean mass flow is constant through the compressor. In fact, it may even be beneficial to the pressure rise that the inlet recirculation forms. Firstly, if a recirculation forms, then rotating stall in the impeller appears to be inhibited, at least at first. At the inducer inlet, the tip sections operate with higher losses due to the recirculation, but the incidence levels close to the hub (and in the whole region below the recirculation zone) are reduced. In addition, the forward flow enters the impeller at a lower mean radius, giving a higher effective radius ratio of the impeller (outlet radius/mean inlet radius). This gives an increased contribution of the centrifugal effect to compensate for the extra losses and may tend to increase the pressure rise.

Several methods of enhancing the stability of radial compressors at low flows operate by enhancing the natural tendency of an impeller to generate inlet flow recirculation and are described in Section 17.5.

17.2.6 Other Instability Mechanisms in Compressors

A wide range of additional flow-induced vibration and noise mechanisms can be found in the turbocompressor literature. The most important of these are vortex shedding and acoustic resonance. Vortex shedding from bluff bodies generally occurs at a fixed Strouhal number so that the frequency varies linearly with the flow velocity. The periodic detachment of the flow from the two surfaces into the wake at a cutoff trailing edge causes an unsteady pressure and velocity field. The frequency is related to the distance between the shear layers from each side of the body, so this distance together with the fluid velocity at the trailing edge is generally used to define the appropriate Strouhal number. The instability may lock into the mechanical natural frequency of a component and cause higher vibrations over a small range of velocities. Such unsteady situations are known to cause vibration issues in centrifugal pumps and hydraulic turbines, where trailing-edge shapes and thicknesses may need to be modified to change the Strouhal number to avoid resonance (Gülich, 2008).

Acoustic resonance occurs when the acoustic mode of inlet or outlet ducts or the natural acoustic frequency of a cavity in the compressor becomes excited by the gas flow, either by vortex shedding or by rotor–stator interaction. An overview of this is given by König et al. (2009). Such instabilities may occur due to the acoustic patterns of the inlet pipes of compressors (Ziada et al., 2002), in the outlet volute (Tsujimoto et al., 2013) and may be related to the acoustic modes of the side cavities (Petry et al., 2010).

17.2.7 Conditions for Compression System Instability

Consider an operating point on the negatively sloped part of a speed line of a compressor operating with a closely coupled throttle and no storage volume, as in Figure 17.8. A small excursion of the flow rate from this operating point to a lower flow rate moves the operating point of the compressor up its characteristic and the throttle resistance to a lower pressure drop. The compressor delivery pressure is now slightly higher than that required by the exit throttle, so there is a pressure difference available to drive extra flow through the exit throttle. The initial excursion is thus unsustainable, and the mass flow will return to its original equilibrium position.

A slight excursion of the operating point to a higher mass flow yields the opposite result: more flow gives less driving pressure, but the throttle resistance increases,

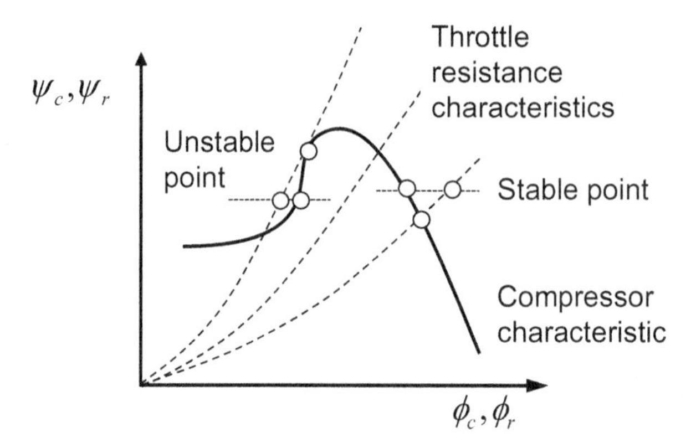

Figure 17.8 Stable and unstable condition.

which would then induce less flow. The initial flow excursion is self-correcting and the operating point is stable. Similar considerations for different operating points at different throttle settings lead to the following condition for static stability. For static stability, the slope of the system resistance characteristic has to be steeper than the compressor characteristic, as shown in Figure 17.8.

$$\left.\frac{d\psi}{d\phi}\right|_r > \left.\frac{d\psi}{d\phi}\right|_c . \tag{17.2}$$

For a system with a large system volume between the compressor and the throttle, the mass flow of the compressor and the throttle may be different as mass is stored in the system. A small shift to lower volume flow in the compressor does not reduce the pressure in the plenum, so the system becomes unstable at the peak of the compressor pressure rise characteristic and the condition for dynamic stability becomes

$$\left.\frac{d\psi}{d\phi}\right|_c < 0. \tag{17.3}$$

In most industrial applications with a large system volume, the second of these criteria is the relevant one to avoid system instability. Greitzer et al. (2004) provide further information on the concepts of static and dynamic stability. The slope of the compressor characteristic not only determines whether instabilities will grow but also determines the amount of the energy available to reinforce their growth.

17.2.8 The Greitzer B Parameter

Whether the performance drop caused by the onset of stall is sufficient to cause surge is generally determined by compression system dynamics, in which the whole system can be likened to a resonator, and in which the slope of the compressor pressure-volume characteristic is a primary factor. The key parameter in the system compression dynamics that is used to measure the likelihood of rotating stall or surge is a stability parameter known as the Greitzer B parameter, as defined by Greitzer (1981):

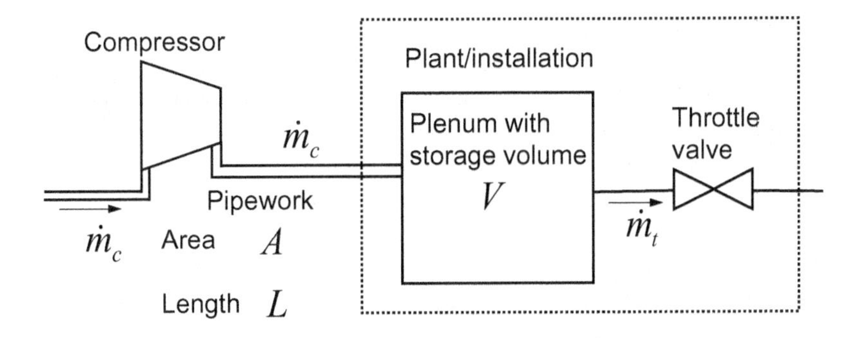

Figure 17.9 Compression system model for consideration of surge.

$$B = \frac{u}{2a}\sqrt{\frac{V}{AL}}. \tag{17.4}$$

This is based on a lumped-parameter model of a compression system as shown in Figure 17.9. In this theory, the whole compression system piping is modelled as a single volume (volume V) connected to the compressor by a duct (with duct area and length A and L), and with ambient speed of sound a and blade speed u. A large value of the B parameter (that is, high rotor speeds and large plenum volumes with ducts of short length and low area) tends to give rise to surge, and a low value of the B parameter (low rotor speeds with low connected volume and ducts of long length with high area) tends to give rise to rotating stall. Unfortunately, there is no single critical value of this parameter, because the critical values depend on the compressor pressure rise and flow coefficient.

The surge cycle depends strongly on the value of the Greitzer B parameter. For a compressor attached to a small downstream volume, the filling and emptying of the volume occurs at a high rate. The small volume is then more quickly emptied, as the mass flow into it is less than the mass flow out of it, and the compressor recovers before it gets to reverse flow, as shown in Figure 17.2. This is known as mild surge. With a large value of the B parameter, the compressor may go into deep surge with backflow through the compressor until the volume has emptied. In industrial compressors and in gas turbines, there are generally large connected volumes, such as the combustion chamber or the process to which the gas is being delivered. This gives a large Greitzer B parameter for the compression system, leading to deep surge in situations with a positively sloped pressure-volume characteristic. As the flow rate is reduced to the point where the pressure of the stage first drops below the peak value, the flow in the large connected volume at a higher pressure empties itself through the compressor in a deep surge cycle. Because of this, in industrial compressor applications, the international standards for conducting acceptance tests (ASME performance test code PTC10, 1997; API standard 617, 2016) defines the instability point on the characteristic as the point with the highest total to static pressure rise. Operation to the left of this point is forbidden, even if test data during an acceptance test with low connected volumes can be obtained in this region.

The Greitzer B parameter includes the blade speed and the speed of sound in a ratio similar to a blade-speed Mach number (u/a). This indicates that for high speeds (or fluids with a low speed of sound), the B parameter increases, so that surge is more likely. At low speeds, the B parameter is smaller, and stable operation with rotating stall is possible without surge. In fact, pumps and simple ventilator fans which operate at low speeds and low Mach numbers often do not experience surge at all and can be operated down to zero flow through a series of unstable operating points. Because of the effect of speed on the B parameter, it is possible that different speed lines in a compressor map show different sensitivities to the effect of geometrical changes on the surge onset. This has been demonstrated by Ribi and Gyarmathy (1993) and Zheng et al. (2017) with tests at different speeds of a compressor with a vaned diffuser. These experiments identified that stall occurs at low speed, mild surge occurs in the midspeed range and the system enters deep surge at high speed. Many turbocharger compressor maps exhibit this pattern of behaviour, with wide characteristics containing regions of positive gradient at low speeds and much narrower ones with only negative gradients at high speeds. In turbocharger compressors with a relatively small volume connected to the compressor (low Greitzer B parameter), stable operation may be possible in the positive sloped part of the characteristic at low speed, but a high level of sensitivity of the surge line position to small changes in geometry in this operating region can be expected.

It is important not only to understand the compression system dynamics of stall and surge, but also the effect of operating conditions on these dynamics. A key issue is inlet distortion, or nonuniformity of the flow entering the compressor. Distortion can be introduced by bends and turns in the ducts leading to the compressor, by inlet recirculation or by nonuniform circumferential pressure distortions caused by asymmetric components, such as a radial inlet or the outlet volute. The nonuniformities can be either circumferential or radial, and typically stall instabilities are exacerbated more severely by the former than the latter.

17.2.9 Stability of Components

A further effect relevant for the compression system dynamics is the pressure rise characteristics of the individual components. In a compressor, several components act together to produce the pressure rise, mainly the impeller and the diffuser system. To assess the influence of single components, Dean (1971) defines a stability parameter, by splitting the whole pressure ratio into the product of those of the stage components:

$$\pi_{\text{stage}} = \pi_{\text{inlet}}\pi_{\text{impeller}}\pi_{\text{diffuser inlet}}\pi_{\text{diffuser channel}}\pi_{\text{outlet}}. \tag{17.5}$$

With the aid of logarithmic differentiation, it can be seen that

$$\frac{1}{\pi_{stage}}\frac{\partial \pi_{stage}}{\partial \dot{m}} = \sum_i \frac{1}{\pi_i}\frac{\partial \pi_i}{\partial \dot{m}}. \tag{17.6}$$

Dean defines these terms to be the stability parameter for the stage and for each component, i, and this results in the following expression:

$$\frac{1}{\pi_{\text{stage}}}\frac{\partial \pi_{\text{stage}}}{\partial \dot{m}} = SP_{\text{stage}} = SP_{\text{inlet}} + SP_{\text{impeller}} + SP_{\text{diffuser inlet}} + SP_{\text{diffuser channel}} + SP_{\text{outlet}}. \tag{17.7}$$

Components with a positive stability parameter are destabilising, and those with a negative value are stabilising. Similar arguments are given by Greitzer (1981).

In a modern radial compressor, the impeller has a negatively sloped characteristic (due to the backsweep) and helps the stability of the system. The inlet to the diffuser also has a negative slope which increases the stability, as discussed in Chapter 12. The vaned diffuser channel is generally destabilising. Ribi and Gyarmathy (1993) carried out measurements at $M_{u2} = 0.9$ on a vaned diffuser stage operating close to its design condition, with $\alpha'_{2d} = 65°$, and with a backsweep of $\beta'_2 = -60°$. They conclude that the impeller has a stabilising effect over the whole of its characteristic, and the diffuser and volute are stabilising at high flow but slightly destabilising at low flow. Tests at higher speeds show that the diffuser becomes mismatched as it is too large for higher pressure ratio, and is forced to operate at high incidence, which is destabilising at higher speeds. Aspects of the matching of the impeller and diffuser at different speeds are discussed in Chapter 18. Tests with different diffuser vane setting angles identified that, at high diffuser vane angles, leading to low or negative incidence in the diffuser, the diffuser becomes strongly stabilising for the stage and that this allows the impeller to operate with inducer tip recirculation and even rotating stall at low flows without surge.

It is possible in some situations that a positively sloped characteristic of one component, which might by itself cause instability, can be overruled by the introduction of a closely coupled component with a negative slope (Longley and Hynes, 1990). In compressors with vaned diffusers, this leads to different responsibility for surge at different speeds. When operating at low speed, the diffuser is too small for the impeller, acts as a throttle and forces the impeller inducer into stall. The compressor then goes successively from stable flow to stall and deep surge with decreasing mass-flow rate. The instability initiates at the impeller inlet, and any flow control method needs to focus on this region when attempting to enhance the compressor stability at low speed. At high speed, the compressor goes from stable flow to mild surge and then to deep surge with decreasing mass-flow rate. The instability process initiates at the diffuser inlet region as this is forced to operate at low flow and high incidence. To improve the stability of the compressor at high speed, the diffuser inlet region deserves most attention.

In multistage compressors, similar effects can occur at different speeds, as discussed in Chapter 18, so that at low speeds a rear stage may operate on the normal part of its characteristic, but towards choke it may provide sufficient stability to the system dynamics that the front stages continue to operate in stall on the positive sloped part of their characteristic. In this way, it is possible that a restriction to the flow or a throttle placed close to the impeller inlet or outlet, or a component operating in choke, can lead to operation of another stage in rotating stall instead of surge.

Vaneless diffusers can also give rise to rotating stall, in which blocked regions of flow, or multiple stall cells, rotate around the circumference. As there are no blades, this cannot be related to incidence or blade stall effects. Vaneless diffuser rotating stall generally occurs as a result of a flow separation of the boundary layer on the diffuser wall and consequent reverse flow. This kind of stall has been studied theoretically and experimentally by Jansen (1964), Senoo et al. (1977), Abdel-Hamid (1980) and others. Van den Braembussche (2019) provides a good review of current knowledge.

Another consideration relevant to the onset of stall is the tip clearance of the impeller. A key parameter on the stability of axial compressors is the tip- clearance gap and this can have an enormous impact on stability in axial compressors (Koch, 1976). Structural and thermal loads can cause deformation of the casing, leading to temporary rubbing of the rotor blade tips so that the tip gap increases. Or to avoid this, the design may make use of a high clearance in the first place. This can lead to inlet distortion in the flow into the diffuser and a premature onset of rotating stall. Work on radial machines shows that the tip clearance effect on surge flow is less strong than in axial machines, but the pressure ratio at surge drops with increasing clearance due to the losses. The flow rate at which instability occurs in radial stages is, however, not strongly affected by the tip clearance (Eisenlohr and Chladek, 1994; Schleer et al., 2008; Jaatinen-Varri et al., 2013). The fact that most of the pressure rise comes from the centrifugal effect may make radial stages less sensitive to clearance effects than axial machines where the clearance has a strong effect on the onset of instability (Koch and Smith, 1976; Freeman, 1985).

17.2.10 Rotating Stall Inception in Axial Compressors

Stall inception is the transient process by which the flow through a compressor changes from steady to unsteady flow. In axial compressors, there are two ways in which this can happen, and these different types of stall inception can be identified by transient measurements of the pressure fluctuations on the casing. The first type is called a mode. Modes are related to a modal perturbation of the pressure field, as shown theoretically by Moore and Greitzer (1986) and first demonstrated by McDougall et al. (1990). A small amplitude, wavelike perturbation is set up circumferentially in the flow, which then grows into a stall cell in a particular blade channel and develops into the pattern of a fully developed rotating stall. The small circumferential perturbations to the flow (on a wavelength of the order of the machine circumference) grow exponentially in amplitude over tens of revolutions. The second type of pressure fluctuations are called spikes and are short-length scale disturbances that develop with a wavelength similar to the blade pitch (Day, 1993). A small spike appears without apparent warning rotating around the annulus at a relatively high speed, say at 70% of rotor speed, which slows down to about 40% when the stall cell is fully formed. In the last few years, the onset of rotating stall in axial compressors has become amenable to analysis with modern CFD methods. The simulations have identified that in axial compressors spikes are, in fact, the low pressure in the core of vortex tubes formed when a starting vortex is shed from the leading edge of the

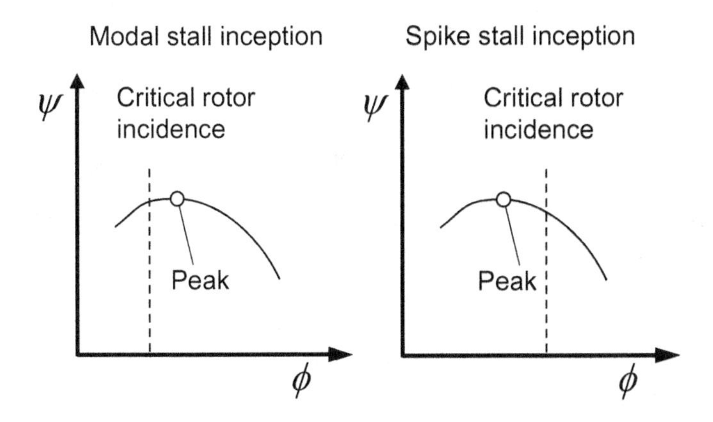

Figure 17.10 Modal stall and spike stall inception.

compressor blade when it stalls at high incidence and starts travelling around the circumference (Pullan et al., 2015).

Camp and Day (1998) have identified in axial compressors that modes tend to occur if the peak of the pressure rise characteristic is reached before the onset of flow separation and so are related to the growth of instabilities on a region of positive slope of the pressure rise characteristic ($d\psi/d\phi > 0$). Spikes tend to occur if flow separation occurs before the peak of the characteristic is reached on the negative sloped part of the characteristic ($d\psi/d\phi < 0$) as illustrated in Figure 17.10.

17.2.11 Stall Inception in Radial Compressors

An impeller designed with no backsweep reaches the peak of its pressure rise characteristic at peak efficiency, as described in Chapter 11. At this condition, the incidence is close to zero. The stage can be expected to show modal instability at lower flows on the positive sloped part of the characteristic. As the backsweep of an impeller is increased, the peak pressure rise moves to a lower flow compared to the peak efficiency point so that a critical incidence might be reached before the peak pressure rise. This would give rise to spike stall inception on the negative sloped characteristic.

Spikes and modes have been identified in measurements by Spakovsky and Roduner (2009), whereby spikes were found at the design configuration and modes were found when the design configuration was modified with bleed at the diffuser inlet. Modal stall inception has been identified in experiments with a backswept impeller by Lawless and Fleeter (1993) and Oakes et al. (1999). Godard et al. (2017) were unable to identify any precursors prior to stall that could be interpreted either as spikes or as modes; the rotating stall started as a small disturbance and simply got stronger until it was fully developed. It would appear that the stall inception in radial machines is more complex than that of axial machines and requires further research.

17.2.12 The Effect of Impeller and Diffuser Matching on Stability

An interesting diagram showing the different regimes of stall and surge for the impeller and the vaned diffuser in a centrifugal compressor stage was developed by Yoshinaka (1977) based on a 1D analysis of his experimental data. An adaptation of the Yoshinaka diagram is given in Figure 17.11, based on equations given in Chapter 18. The plot presented here shows the trajectory of diffuser incidence versus impeller casing incidence, calculated at different flow rates and on different speed lines. The original Yoshinaka diagram is based on the diffuser inlet angle versus the flow coefficient at the impeller eye, but using the incidence for both components makes this more general. To what extent the diagram is representative of all compressors is not clear, but it provides some useful insights into the mismatch of the impeller and the diffuser at different speeds.

In the example given in Figure 17.11, the design point is for a tip-speed Mach number of 1.4, a flow coefficient of 0.08 and a work coefficient of 0.6 with a slip factor of 0.1. The outlet flow conditions on different speed lines are calculated on the assumption that the Euler work coefficient is given by (18.15). The value of the work coefficient, degree of reaction and the polytropic exponent in the large bracket of (18.15) are taken as the values at the design point without considering the effects of a change in the flow on these values, and the power input factor is taken as zero to give the Euler work.

In Figure 17.11 the design point is assumed to have zero incidence in the impeller and in the diffuser and is marked as a white circle on the speed-line for a tip-speed Mach number of $M_{u2} = 1.4$. The different speed lines are a ray of lines around this point whereby increasing flow with smaller incidence is to the left. The line at the design speed passes through the design point and the incidence of the impeller and the diffuser both increase as the flow decreases. The effect of the density change at

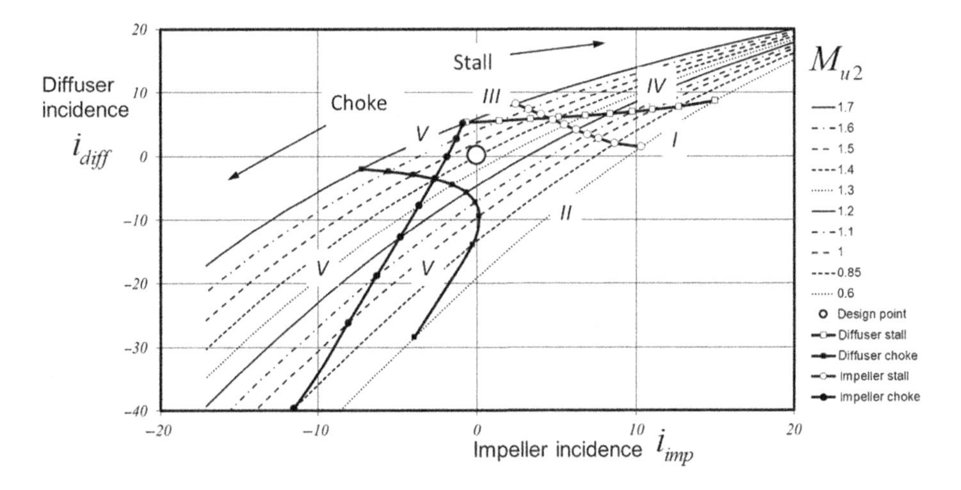

Figure 17.11 Locus of stall and choke incidence on different speed lines.

different speeds causes the higher speed lines to have higher diffuser inlet flow angles than the lower speed-lines leading to a higher diffuser incidence. The effects of choking are not included in these calculated speed lines which extend beyond the choking mass flow and only the locus of the expected choking condition is shown on them, as in the original Yoshinaka diagram.

Moving to the left along the design speed line is equivalent to increasing the flow and moving towards choke. In the example shown in Figure 17.11, the design strategy is such that the diffuser and the impeller both reach choke together at the design speed. Under such conditions, the locus of the lines for diffuser choke and for impeller choke cross this speed line at the same position and both choke at the same flow. At higher tip-speed Mach numbers, the impeller causes choke, and at lower tip-speed Mach numbers the diffuser causes choke. The choke lines given in this figure are calculated from the 1D equations for choke given in Chapter 18. No operation is actually possible in zone *V* (because either the impeller or the diffuser chokes).

Moving to the right along the design speed line in Figure 17.11 is equivalent to moving towards surge. Lines showing the approximate location of the diffuser and the impeller stall are also included in the diagram. In this diagram, the stalling conditions are calculated on the basis of a nominal stalling incidence for the diffuser and for the impeller casing streamline. The stalling incidence is assumed to decrease with an increase in the Mach number as follows:

$$\begin{aligned} i_{imp}^{stall} &= 14 - 10M_{w1c} \\ i_{diff}^{stall} &= 14 - 12M_{c2}. \end{aligned} \tag{17.8}$$

where the diffuser and impeller incidences are given in degrees.

With these data, the diffuser and the impeller stall together on the design speed line. The impeller tends to stall before the diffuser at low speeds, and the diffuser tends to stall before the impeller at high speeds. Yoshinaka (1977) pointed out from his experiments that, despite the impeller exceeding its stall incidence at low speeds, the stage tended to become unstable only when the diffuser stalls as well. Similarly, despite the diffuser exceeding its stall incidence at high speeds, the stage remained stable until the impeller stalls. Yoshinaka presented experimental data to show that operation is possible in zone *I* (because only the impeller stalls), zone *II* (because neither component stalls) and in zone *III* (because only the diffuser stalls). Surge occurs in zone *IV* (where both the impeller and the diffuser stall).

The background to this complex behaviour is that surge is a system phenomenon. Impeller or diffuser stall in themselves do not necessarily cause a massive drop in the pressure rise capability of a radial stage; the pressure continues to rise with a fall in flow beyond the instability point, and this may give sufficient system stability that the stage can operate without surge even if the flow in one of the components is unstable. At low speeds, the instability in the impeller is often a recirculating ring stall, but the stage can continue operation as the diffuser is at low or negative incidence, and this tends to stabilise the compressor. Other researchers have identified that the transition from zone *I* with impeller stall at low speeds to zone *III* with diffuser stall at higher speeds may cause a kink in the surge line (Jansen and Rautenberg, 1982; Hunziker, 1993). The diffuser instability is usually a rotating stall. The situation changes if both components become unstable, and then the system stability is completely lost in zone *IV*.

17.3 Off-Design Operation of Radial Compressors

17.3.1 Types of System Requirements

The control system of a compressor must ensure that, for a certain delivery pressure ratio and flow of the compression system, the operating point remains on the stable part of the compressor characteristic even during any unexpected excursion of the pressure required by the system. When possible, the user tries to operate the compressor near to its best efficiency point, with a suitable safety margin to the stability limit such that the compressor does not become unstable and operate with rotating stall or surge. In some situations, the need for adequate safety margin between the rated point and the stall point, may make it necessary that the rated point is at a higher flow than the peak efficiency point, and this is especially the case in impellers with low backsweep as a higher work coefficient naturally decreases the range; see Chapter 11.

Lüdtke (2004) categorises three main types of system pressure versus volume requirements for purposes of process and industrial compressor control. These are shown schematically as the trajectory of the normalised pressure rise against the normalised volume flow in Figure 17.12 and labelled as Category A, B and C. For completeness, category D and E are added, which are associated with turbocharger applications.

Category A is a compressor delivering to a more or less constant process pressure together with any system resistance in the associated pipework. The pressure needed is nearly constant for all flows but drops slightly at lower flows due to the lower losses in

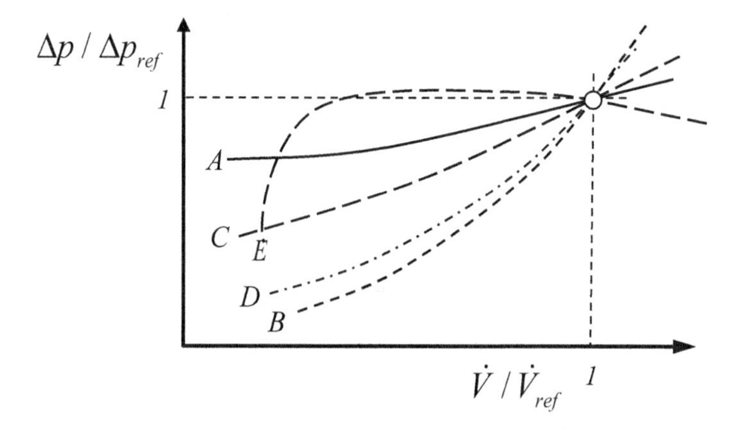

Figure 17.12 Different system requirements in terms of normalised pressure versus flow characteristics relative to reference condition close to the design point.

the ducts. An example would be a compressor maintaining a constant pressure for a compressed air system in a factory. Another example would be a compressor for a wastewater treatment plant where the compressed air is forced to enter the water treatment tank against a constant head of water for all flow rates. Category A is probably the most difficult pressure-volume requirement for a process compressor to match, as a high pressure is required at both high and low flows, and this cannot be achieved with variable speed control alone. Such system requirements can be met with a vaned diffuser control, but at low flows the efficiency of such systems will drop. A better option with a higher efficiency is obtained using a combination of variable speed and variable inlet guide vanes, using the guide vanes to lower the flow rate and the speed to increase the pressure ratio.

Category B is a compressor delivering the compressed fluid against a parabolic system resistance curve, which would be the case required to overcome system resistance in a process, in a throttle in process plant, or even to supply a turbine in a gas turbine process. The parabolic nature of the system resistance curve naturally blends to control with variable speed as the compressor also has an inherent characteristic that is close to the parabolic fan law. In some cases, variable inlet guide vanes may be sufficient. In high-pressure turbocharging of diesel engines for power generation, where the generator runs at constant speed (typically 50 or 60 Hz) independent of the load, the turbocharger has to match a parabolic resistance curve. In turbochargers for diesel engines used to drive ship's propellers, the low flow requirement is at a slightly higher pressure as the speed changes with the load (Baines, 2005), shown as category D here.

Category C is a compressor delivering the fluid to a system with a system characteristic between those of categories A and B. An example would be a compressor in a refrigeration process compressing the refrigerant fluid from the evaporator to the condenser. At low flows, the refrigeration system has a lower heat transfer, and both the evaporator and condenser become more effective with a lower approach temperature, leading to a smaller temperature rise in the process and hence smaller pressure rise between the two. This applies to many different refrigeration and heat pump processes such as liquefaction of air and liquified natural gas.

Category E is the expected operating line of a turbocharger operating at medium speed to meet the boost requirements of the engine at design, but with an increased pressure ratio at lower flow to meet the peak torque requirements of the engine during acceleration and a higher flow at high speeds, see Section 18.10. Turbochargers tend to use no variable geometry in the compressor and attempt to meet this requirement with passive control devices.

17.3.2 Safety Margin Definitions

There are several definitions related to the operating range between the rated point of a compressor compared to the surge point in use in the compressor industry. These find use in different applications as a means of defining the safety margin of the rated operating point. The different definitions are given in (17.9) and shown

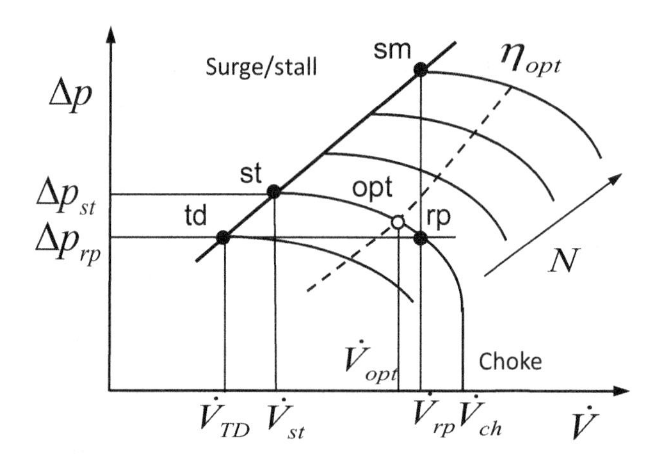

Figure 17.13 Operating points in the compressor map relating to safety margin definitions.

schematically in Figure 17.13. The first three of these relate to the volume flow along a constant speed characteristic and define a nondimensional ratio between the flow at the onset of instability (*st*), the rated point flow (*rp*) and the choke flow (*ch*) at constant speed. The fourth parameter, the turndown ratio (*td*), refers to the volume flow at the turndown point on a different speed characteristic (or one with different settings of variable geometry blades) relative to the volume flow at the rated point. The turndown point is at the same pressure ratio as the rated point. The fifth parameter is the surge margin (*sm*) and refers to the pressure ratio on the surge line at the volume flow of the rated point, and so is at a higher speed. This is often used in gas turbine compressors. Finally, the head rise to surge refers to the increase in the aerodynamic work (polytropic head) between surge and the rated point, and a positive value indicates that the characteristic is negatively sloped at the rated point.

$$\text{range} = \left(\frac{\dot{V}_{ch} - \dot{V}_{st}}{\dot{V}_{ch}}\right), \qquad \text{stability} = \left(\frac{\dot{V}_{rp} - \dot{V}_{st}}{\dot{V}_{rp}}\right),$$

$$\text{choke margin} = \left(\frac{\dot{V}_{ch}}{\dot{V}_{rp}}\right), \qquad \text{turn down} = \left(\frac{\dot{V}_{rp} - \dot{V}_{td}}{\dot{V}_{rp}}\right), \tag{17.9}$$

$$\text{surge margin} = \left(\frac{\pi_{sm} - \pi_{rp}}{\pi_{rp}}\right), \qquad \text{head rise to surge} = \left(\frac{y_{12,st} - y_{12.rp}}{y_{12,rp}}\right).$$

The preceding definitions suffer from being either related to a change in the volume flow or a change in the pressure ratio or head rise. A more modern definition of the surge margin is given as

$$\text{surge margin} = \left(\dot{m}_{rp}\pi_s/\dot{m}_s\pi_{rp}\right) - 1, \tag{17.10}$$

where rp is the rated point condition, s is the surge condition, π is the pressure ratio and m is the mass-flow rate, and this combines both a change in flow and a change in pressure ratio.

17.4 Typical Operating Range of Single-Stage Radial Turbocompressors

17.4.1 Field Experience

During the design of a radial turbocompressor, the compressor engineer has to balance the requirements of achieving the maximum flow at choke and the minimum possible flow before instabilities begin on a given speed line. This inevitably requires some compromise. The physics of the flow limits what is possible, and the design choices that lead to stable characteristics at low flow (such as low incidence on the blades) tend to reduce the maximum possible flow (as the throat area becomes smaller). As a measure of the operating range that can be achieved between surge and choke in a radial turbocompressor, the parameter called range is often used:

$$\text{range} = \frac{\dot{m}_{ch} - \dot{m}_{st}}{\dot{m}_{ch}} = 1 - \frac{\phi_{st}}{\phi_{ch}}. \tag{17.11}$$

A large value indicates a wide operating range, and a low value suggests a small range between stability and choke. For radial compressor stages operating between a pressure ratio of 2–3, excellent state-of-the-art values of the operating range between choke and surge, as defined by (17.11), are between 0.4 and 0.6, the higher value being associated with the lower pressure ratios and lower tip-speed Mach numbers. Stability control methods are able to increase this range, as discussed in Section 17.5.

The trend of field experience available to the authors with regard to operating range is shown in Figure 17.14. This diagram includes measurements on more than

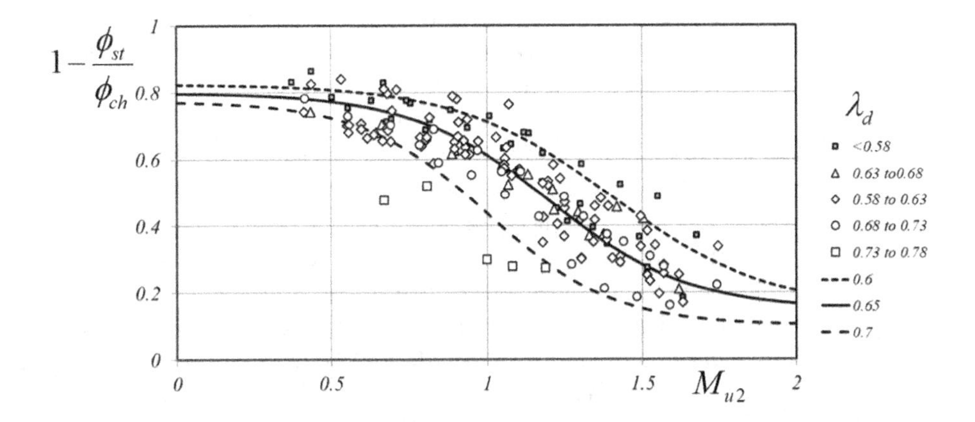

Figure 17.14 Operating range of turbocharger stages as a function of Mach number for different design work coefficients.

50 turbocharger stages with vaned and vaneless diffusers, each tested at several speeds. The data for this figure have been taken from the most reliable data sets of those stages analysed in the work of Casey and Robinson (2013) and Casey and Rusch (2014), together with additional stages that have subsequently been analysed. The stability point in this diagram has been taken from the peak total–static pressure rise of the stages and not from the measured surge condition. The choke point is not always reached at maximum flow in a test measurement at low speeds and has been calculated with the assistance of the correlations of the two aforementioned publications and the equations in Chapters 11 and 12.

There is quite a lot of scatter in these data as it is a survey of data taken from many separate sources using many different stages with different experimental rigs, measurement techniques and test procedures. In some cases, the stages may have been optimised for maximum range and others for maximum efficiency, which also leads to a variation in the operating range (Ibaraki et al., 2012). The dominant effect in the data is that the operating range decreases strongly as the rotational speed and pressure ratio increase. Part of this reduction in range is due to the high gas velocities and Mach numbers in the flow at higher speeds causing the compressor to choke earlier. In addition, the higher Mach numbers in the flow also cause higher losses when operating away from the design point so that the flow range from the design point to the peak of the characteristic becomes smaller. This is consistent with knowledge of the operating range of axial cascades, as given in the alpha-Mach diagrams of Chapter 7. Stages at low speed with vaneless diffusers tend to have a wider range to surge than is shown in this diagram, which is based only on the volume flow at the peak of the pressure rise characteristic.

A less strong effect than the tip-speed Mach number in this diagram is the design work coefficient of the stages, λ_d, showing that the peak pressure rise of their characteristics arises at a higher flow for stages with a higher work coefficient, leading to a narrower range. The stages with vaned diffusers included in this analysis all have a design work coefficient in the region $0.65 < \lambda_d < 0.78$ and have the lowest range, but interestingly this is not substantially lower than the few vaneless diffuser cases analysed with a high work coefficient. In many cases, the stages with vaneless diffusers operated in a stable manner below the peak pressure rise on the characteristic. A rough trend of this effect is shown in the tentative lines for the effect of different design work coefficient in the figure. At low Mach numbers, the upper flow limit is not caused by choking, but by increased losses and blockage at high negative incidence, and at high Mach numbers the range cannot be less than zero. Together these constraints lead to a form of an s-curve for the range variation with Mach number as outlined in the aforementioned publications.

An interesting issue of great relevance to the designer is whether the effect of the Mach number is stronger than the effect of the work coefficient. For a given design pressure ratio, the trade-off between work coefficient and Mach number is given in Figure 8.5. Designing for a higher work coefficient involves reducing the backsweep or reducing the slip factor by the use of many blades, but the associated reduction in Mach number may be more beneficial on the range than the increase in the work

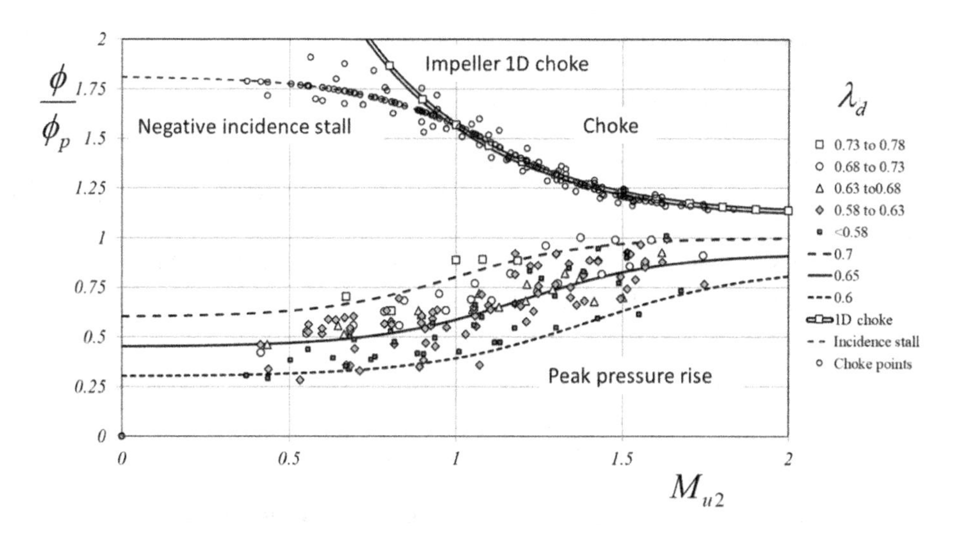

Figure 17.15 A flow coefficient versus Mach number diagram for turbocharger stages with vaneless diffusers.

coefficient. The impellers which are most affected by this are those designed for aeroengine core compressors, and the images of such impellers given in Chapter 1 indicate that designers tend to use many blades with a high work coefficient for such impellers.

The data for the vaneless turbocharger stages are shown again in Figure 17.15 in a different form. This shows separately the choke and surge flow limits relative to the peak efficiency flow, ϕ_c/ϕ_p and ϕ_s/ϕ_p in a flow coefficient versus Mach number diagram, which is similar in concept to an alpha-Mach diagram, as discussed in Chapter 7. For vaneless stages, it is the impeller that causes choke so that the relationship for isentropic choke given in Chapter 11 can be compared to the trend in the measurements. The impeller 1D choke relationship agrees reasonably well with the data on the choke of the stages and identifies that the range to choke tends towards a plateau at high Mach numbers. At low Mach numbers, choke does not occur and the upper flow range is then controlled by blockage caused by negative incidence stall at the impeller inlet. The surge margin shows the same trend as in Figure 17.14: the stages with a lower work coefficient achieve the lowest flow at the peak of their pressure rise characteristics. This is to be expected from the effect of the work coefficient on the steepness of the characteristic, as discussed in Chapter 11. The effect of recirculating bleed flows (see Section 17.5.7) is to increase the range by lowering the flow at the peak of the characteristic. Typically, this increases the operating range of high work stages to be similar to those of the low work stages.

Similar trends to those shown in Figure 17.15 can be observed with industrial process compressor stages, but these tend to have thicker blades giving a smaller range by about 10%, mainly due to the lower choke margin (Casey and Robinson, 2013).

17.5 Stability Control and Enhancement

17.5.1 The Power Relative to That at Design

To improve the surge line of compressors, various forms of surge control have been suggested. The most popular make use of variable geometry or variable speed. A general overview can be obtained by considering the equation for the power of the machine in terms of the nondimensional variables describing the duty, speed and inlet density. A form relative to that at the design point (denoted here by the suffix *d*) can be expressed as

$$\frac{P}{P_d} = \frac{\rho\phi\lambda n^3}{\rho_d\phi_d\lambda_d n_d^3}. \tag{17.12}$$

From this equation, the background to the different control methods can be described as follows. Speed control operates directly on the pressure rise and power through the change in rotational speed, $P/P_d = n^3/n_d^3$. An inlet throttle valve operates through a change in the inlet density of the gas, $P/P_d = \rho/\rho_d$. Control through bleed flows operates on the flow capacity somewhere within the stage, $P/P_d = \phi/\phi_d$. This may involve blow-off of unwanted fluid, recirculating bleed flows and intermediate bleed at impeller outlet to control the volume flow in the diffuser at part-speed. Control with variable inlet guide vanes changes the work coefficient and the flow capacity through the effect of adjustable blade stagger on the inlet velocity triangles, $P/P_d = (\lambda\phi)/(\lambda_d\phi_d)$. Control with variable diffuser vanes changes the flow capacity through the effect of adjustable diffuser blade stagger on the outlet velocity triangles, $P/P_d = (\phi)/(\phi_d)$. These individual control strategies may be combined.

An overview of the performance of the main types of compressor control in terms of their effect on the head rise characteristics has been given by Casey and Marty (1985) and in more detail by Brun and Kurz (2019). This is shown in Figure 17.16, where it can be seen that a variable diffuser enables a higher head rise at low flows, and variable inlet guide vanes provide a head rise between that of variable diffuser and

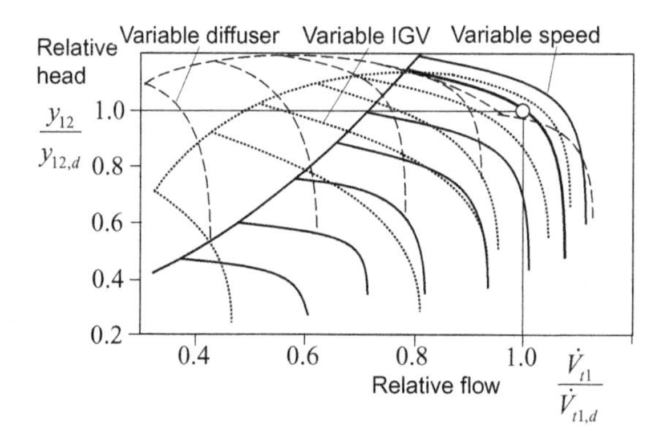

Figure 17.16 The effect of the control strategy on the head rise characteristic. (image adapted from Casey and Marty, 1985, courtesy of the Institute of Refrigeration, www.ior.org.uk)

variable speed at low flows. These different types of control may also be combined, and the combination of diffuser vane control and inlet guide vane control can be very effective for a constant speed machine (Simon et al., 1987). All the different forms of control lead to a reduction in efficiency at part-load and sometimes it is more economical in process compressor applications to split the duty between several compressors so that a part-load operating point is then met by shutting down one of the available units (Kurz, 2019).

17.5.2 Variable Speed Control

Speed control allows the whole stable operating range of the compressor map to be used. Broadly, the volume flow is proportional to speed, and the head is proportional to the square of speed. The stable operating points are at the cross section of the speed lines with the system characteristics, so if the system has a parabolic resistance characteristic, the operating points may be close to the peak efficiency point at all speeds. This control method has relatively high part-load efficiencies as the compressor produces only the required head, and the speed acts on all stages of the machine. Because of this, it is the first choice of all control modes. However, a driver with a variable speed is needed. Performance maps with variable speed are described in Chapter 18.

17.5.3 Suction and Discharge Throttling

In a constant speed machine, only a single speed line is available. A throttle in the discharge line allows the apparent system resistance to be changed. This reduces the discharge pressure, but the mass flow and power of the compressor are unchanged. There are low investment costs but high operating costs in the part-load region due to low part-load efficiency. The compressor continues to generate the whole pressure rise, which is dissipated in the throttle.

Suction throttling with a butterfly valve is in principle similar to discharge throttling. The compressor is operated with a low inlet pressure and a low inlet density compared to the rated point. Suction throttling has the advantage over discharge throttling that the density of the gas at compressor inlet reduces so that the compressor mass flow is reduced and the outlet pressure and the power consumption are reduced.

17.5.4 Adjustable Inlet Guide Vanes

Variable inlet guide vanes (often denoted as VIGVs) change the impeller inlet velocity by adding a circumferential swirl component to the throughflow velocity. With inlet swirl, the flow coefficient of the impeller at peak efficiency point changes in order to retain similar levels of relative flow angle (incidence) on to the impeller leading edge. This lowers the flow capacity and reduces the work input so the characteristics shift to lower flow and lower pressure ratio. The efficiency of the stage changes owing to the changes in the degree of reaction, the impeller inlet relative Mach number, the diffusion in the impeller and the matching between the impeller and the diffuser. At extreme closed settings, the guide vanes may also be used as an inlet throttle valve.

The work input coefficient curves of the impeller are not only shifted to a different flow coefficient, but also have a different slope due to the effect of the inlet swirl on the gradient of the work input curve, which is similar to the effect of an increase in the backsweep angle (see Chapter 11). Owing to the steeper slope of the work coefficient the stage is generally more stable at increased swirl, and the surge point on the characteristic generally improves at increased swirl. Some of the features of the effect of the inlet guide vanes on the performance can be identified by an analysis of the velocity triangles with the help of the continuity equation and the Euler equation.

Different inlet guide vane types (radial, axial with or without centre-body, as shown in Figure 17.17) can lead to different changes in the inlet flow angles at the impeller inlet eye for a specific change in the inlet guide vane setting angle. A 1D analysis using the continuity equation, $V = Ac_m$, and the conservation of angular momentum in a duct flow, $rc_u = constant$, provides useful insight. The analysis from the outlet of the inlet guide vane (suffix *igv*) to the impeller inlet plane (suffix 1) leads to following equation for the absolute inlet flow angle, α_1, as a function of the blade angle, α_{igv}, at the guide vane outlet:

$$\begin{aligned} \tan\alpha_1 &= K\tan\alpha_{igv} \\ K &= (A_1/r_1)/(A_{igv}/r_{igv}). \end{aligned} \tag{17.13}$$

In this equation, A_1 is the cross-sectional area at impeller inlet, r_1 is the rms radius at impeller inlet, A_{igv} is the cross-sectional area at the outlet of inlet guide vane and r_{igv} is rms radius at the outlet of the inlet guide vane. With an axial inlet with the same area and radius at the guide vane outlet as at the impeller inlet, the parameter $K = 1$. A decrease in radius from the outlet of the inlet guide vane to the impeller inlet produces an increase in the swirl component of the velocity at impeller inlet and magnifies the effect of the guide vane setting angle. A decrease in area from the guide vane to the impeller inlet counteracts this. For a radial inlet in a typical industrial impeller with the guide vane placed at a high radius, the parameter K may increase to $K = 1.5$, and thus the radial inlet guide vane is 50% more effective at changing the flow angle at impeller inlet. A radial inlet guide vane does not need to be closed so far as an axial inlet guide vane to achieve the same effect at the impeller inlet.

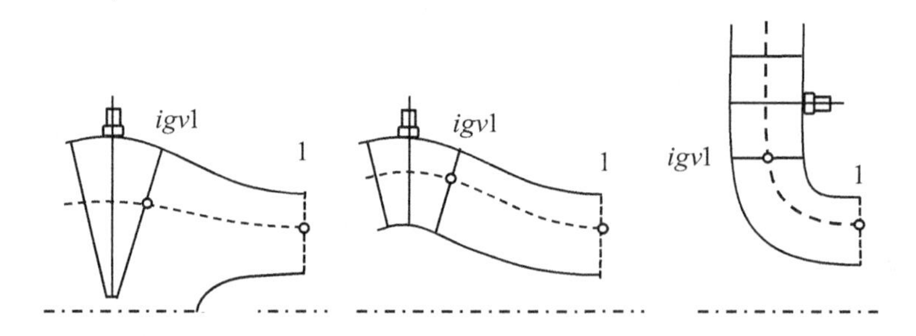

Figure 17.17 Inlet region of a compressor stage with three types of IGV control.

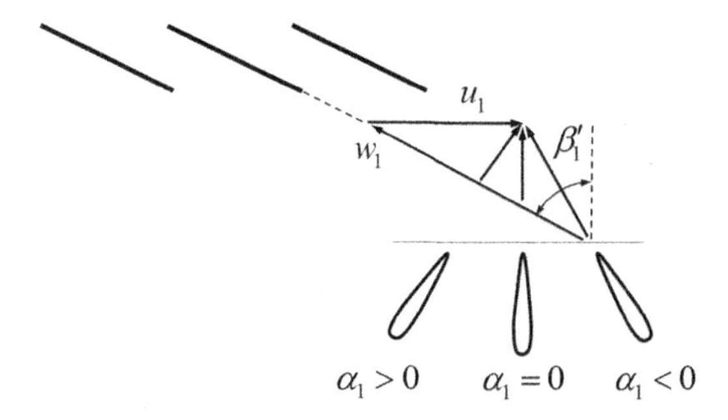

Figure 17.18 Inlet velocity triangles with 0° incidence and positive and negative swirl from an inlet guide vane.

The peak efficiency point of the stage is related to low incidence at the impeller leading edge. To retain near-zero impeller incidence, the meridional velocity at inlet adjusts as the inlet swirl changes; see Figure 17.18. The velocity triangle at the impeller inlet leads to the following expression:

$$c_{u1} = u_1 + w_{u1}, \quad \tan\alpha_1 = u_1/c_{m1} + \tan\beta_1' \\ c_{m1}/u_1 = 1/(\tan\alpha_1 - \tan\beta_1'). \tag{17.14}$$

It can be expected that the inlet flow coefficient of the compressor varies together with the meridional velocity at impeller inlet, so that the change in flow rate of the stage at peak efficiency (superscript p) with swirl from an inlet guide vane can be obtained as

$$\frac{\phi_{t1}^{p,\alpha}}{\phi_{t1}^{p,0}} = \frac{\tan(-\beta_1')}{\tan\alpha_1 + \tan(-\beta_1')}. \tag{17.15}$$

Figure 17.19 shows the shift in flow at peak efficiency for a range of measurements on different stages with axial and radial inlet guide vanes and compares this with the estimate from (17.15). Although the method takes no account of choking or effects due to the matching between the diffuser and the impeller, it can be seen that the 1D equation predicts the trends reasonably well. The shift in flow due to inlet guide vanes is broadly controlled by the properties of the impeller inlet velocity triangle. The actual shift in the characteristic is less than that predicted by (17.15) as the model does not take into account that the diffuser inlet also has an effect on the flow capacity. In general, it is possible to drop the flow to zero with closed guide vanes and to increase the inlet flow by a small amount with negative inlet guide vane settings.

The enthalpy rise per unit mass across an impeller is determined by the Euler equation (see Chapter 2), and including the effect of the swirl at inlet, the nondimensional work coefficient is given by

$$\lambda = \frac{\Delta h_t}{u_2^2} = \frac{c_{u2}}{u_2} - \frac{u_1}{u_2}\frac{c_{u1}}{u_2}. \tag{17.16}$$

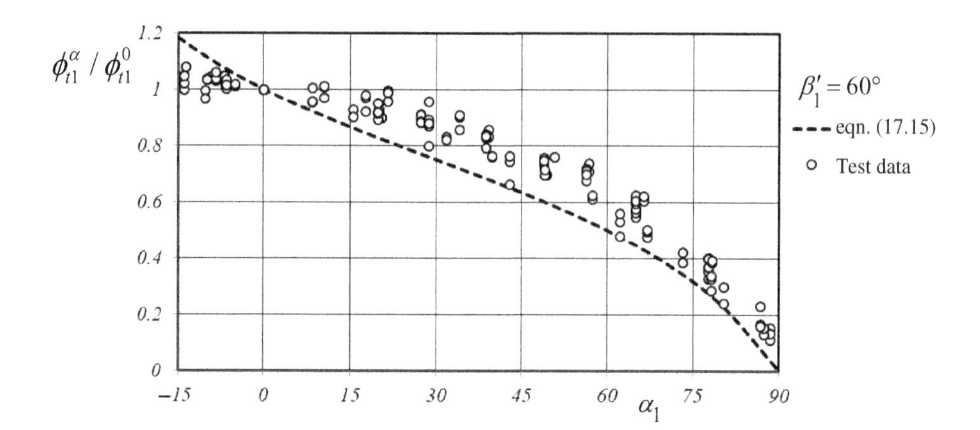

Figure 17.19 Shift in flow at peak efficiency with the swirl angle at impeller inlet.

From the inlet velocity triangle and the outlet velocity triangle taking into account the slip velocity, the following equation is obtained:

$$\lambda = 1 - \frac{c_s}{u_2} - \frac{c_{m2}}{u_2}\tan\beta_2' - \frac{u_1}{u_2}\frac{c_{m1}}{u_2}\tan\alpha_1. \tag{17.17}$$

This equation can be written as a correction to the work input with zero inlet swirl for the effect of the inlet swirl as follows:

$$\lambda^{\alpha} = 1 - \frac{c_s}{u_2} - \frac{c_{m2}}{u_2}\tan\beta_2' - \frac{u_1}{u_2}\frac{c_{m1}}{u_2}\tan\alpha_1 = \lambda^0 - \frac{u_1}{u_2}\frac{c_{m1}}{u_2}\tan\alpha_1. \tag{17.18}$$

This provides the basis of an additive correction equation for the effect of inlet swirl on the work input:

$$\lambda^0 - \lambda^{\alpha} = \frac{u_1}{u_2}\frac{\rho_{t1}}{\rho_1}\frac{D_2^2}{A_1}\phi_{t1}\tan\alpha_1 \approx \frac{D_1}{D_2}\frac{4\phi_{t1}}{\pi}\frac{D_2^2}{A_1}\tan\alpha_1. \tag{17.19}$$

The correction to the work coefficient is for a change in the guide vane setting is linear with the tangent of the inlet swirl angle and proportional to the inlet flow coefficient. Figure 17.20 shows that this agrees well with test data on an industrial compressor stage.

The important features of control with inlet guide vanes are that medium part-load efficiency is obtained, except for the incidence losses in the VIGV as it closes. This may be reduced by having a two-part guide vane or an asymmetric guide vane profile, as discussed by Coppinger and Swain (2000) and others.

17.5.5 Adjustable Diffuser Vanes

Diffuser vane control is well suited to flat system requirements, category A from Figure 17.12. The impeller produces a higher work input at low flows and counteracts any additional losses in the impeller, which is forced to operate at incidence and may

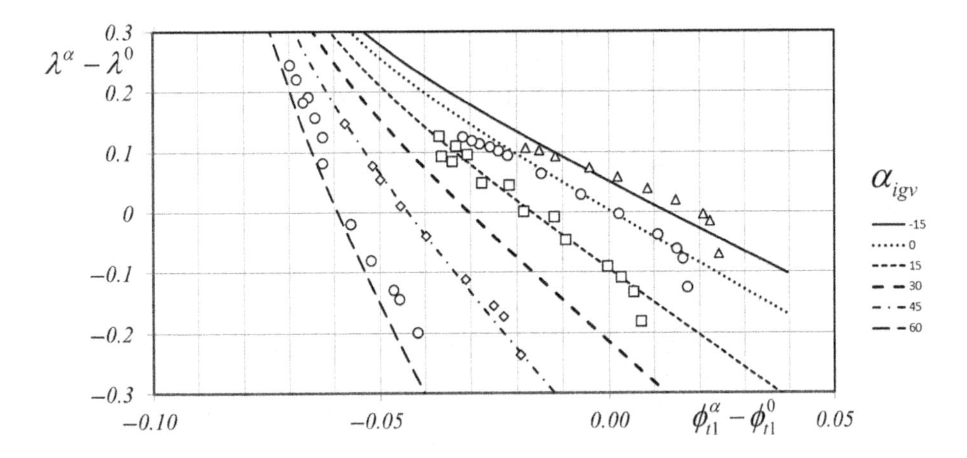

Figure 17.20 Prediction of the change in work coefficient characteristic with a change in the IGV angle.

experience inducer stall at low flows (Simon et al., 1987). The increased impeller losses cause the stage to have lower efficiency at part-load. If it is the diffuser which causes choke, variable diffuser vanes can also be used to increase the maximum flow but at a lower pressure ratio.

Useful insight can be gained with a 1D analysis of the velocity triangle. From the outlet velocity triangle in Chapter 11, it can be shown that

$$\phi_2 = \frac{\lambda}{\tan\alpha_2} = \frac{1 - \sigma + \phi_2 \tan\beta_2'}{\tan\alpha_2}, \tag{17.20}$$

which can be rearranged as

$$\phi_2 = \frac{1 - \sigma}{\tan\alpha_2 - \tan\beta_2'}. \tag{17.21}$$

This equation can be rewritten at the design point, with suffix d, and with the assumption of a constant slip velocity, σ, and the further assumption that the outlet flow coefficient is proportional to the inlet flow coefficient along a speed line, one obtains

$$\frac{\phi_{t1}^{p,\alpha}}{\phi_{t1}^{p,d}} \approx \frac{\phi_2}{\phi_{2d}} = \frac{\tan\alpha_{2d} - \tan\beta_2'}{\tan\alpha_2 - \tan\beta_2'}. \tag{17.22}$$

This equation provides an approximate relation for the shift in flow coefficient at peak efficiency for a change in the setting angle of the diffuser at inlet. A comparison between this 1D equation and the measurements from Simon et al. (1987), Dalbert et al. (1988) and Jiang and Whitfield (1992) shows that the trends are correctly predicted by this simple analysis, but due to the limited data available this comparison is not given here. The shift to higher and lower flows with a variable diffuser is controlled by the properties of the impeller outlet velocity triangle, and the matching of the impeller with the diffuser is discussed further in Sections 12.9 and 18.7. This flow shift is accompanied by a lower or a higher impeller work input as determined by the slope of the impeller work input characteristic so that with a large backsweep, the pressure ratio tends to increase as the flow is reduced, and with a small backsweep the pressure ratio remains constant as the flow is reduced.

17.5.6 Bypass and Bleed Regulation

In this control strategy part of the mass flow is removed from the compressor to retain stability. In blow-off control, excess fluid is simply exhausted to the environment with high losses, and this is not really possible with gases other than air. The same effect can be obtained using recirculating bleed in which the compressed gas is throttled and diverted back to the compressor inlet, but as the outlet gas is hot it needs to be cooled in the return flow. Intermediate bleed is often used in compressors to assist startup of multistage machines and to aid recovery from locked-in stall. Bleed can also be taken from the impeller outlet in single-stage applications to aid operation at part-speed conditions. This is a simple control strategy, but with extreme high losses, so it is only suitable for temporary operation (typically for startup) (Raw, 1986). In some applications, bleed air near the impeller exit is often used for secondary airflow systems (Spakovsky and Roduner, 2009).

17.5.7 Impeller Recirculating Bleed

Recirculating bleed is a system to increase the range of compressors by allowing a recirculating flow to occur near the impeller eye. This is a particularly common and effective type of casing treatment for radial compressor impellers, and there are many different custom-tailored treatments available. It is not a method of control but a device to widen the compressor map to lower flows on each speed line. Leakage slots are installed in the casing wall just downstream of the inducer, and these allow the leakage flow to recirculate back to the compressor inlet at part flow operating points. Such systems are also known as map width enhancement (MWE), ported shroud, shroud bleed slot, internal bleed system, stabilisers and self-recirculating casing treatment.

The typical recirculating bleed system involves recirculation of the flow from the impeller from a slot in the annulus wall downstream of the impeller throat, through a chamber above the impeller inlet eye, back to the impeller inlet. This may improve stability at low flows by enhancing inlet recirculation at the impeller eye. It can also increase the flow rate at high flows by allowing some extra flow to enter the impeller through the slots from the chamber downstream of the impeller throat. Such systems have found application in many compressors, but the exact mechanism of surge improvement through self-recirculating casing treatments is still an active area of research.

The first system was reported by Fisher (1988). The system patented by Fisher achieved more flow at choke and less flow at surge than the stage without treatment. The slot was positioned downstream of the inducer throat. The slot area was about

25% of the inducer annular area and inclined at 25° from the radial direction to encourage the inflow at choke. Leakage flow from the slot into the impeller near to choke increases the choke flow, and the recirculation flow out of the slot near to surge lowers the total mass flow at stage inlet. The measured maps showed that the slope of the pressure characteristics near surge became more negative and were thus more stable. Many different systems are available, partly related to their increased effectiveness, and partly due to the innovations required to overcome patents on other earlier systems.

Assistance with the design of a recirculating bleed system can be obtained from the work of Jung and Pelton (2016). Their numerical studies suggest that for maximum benefit, the bleed slot should be placed such that there is near-zero recirculation at the design point, and this means the slot needs to be near the impeller throat. The slot area should be close to 23% of the inducer eye area. Vanes placed within the recirculating cavity should attempt to remove as much swirl as possible.

There are several possible mechanisms which account for the effectiveness of the recirculating bleed systems. First, there is the global mechanism related to a change in the actual inlet mass flow as some of the internal flow recirculates. Second, the bleed flow includes swirl either from the impeller or from vanes in the bypass duct and changes the velocity triangles at the inlet eye of the impeller. Third, there is a local fluid dynamic effect as the recirculating flow through the bleed slot also removes vortical structures from the boundary layer and clearance jet near the casing sections of the impeller flow.

The global principle of recirculating casing treatment is shown in Figure 17.21, as described by Hunziker et al. (2001). The direction of the flow in the bleed channel depends on the difference in casing static pressure between the bleed slot over the impeller and the inlet. Near choke, the inducer acts as an accelerating nozzle, with a low pressure in the throat area of the inducer tip so the flow enters the bleed channel in the direction of the main flow. If the bleed slot is downstream of the throat, the inlet mass flow will be higher than the flow passing through the choked throat and therefore the flow capacity will be increased.

In stages with vaned diffusers, the flow at low speeds is limited by the choking of the diffuser, the impeller always operates at the left of its characteristics, away from

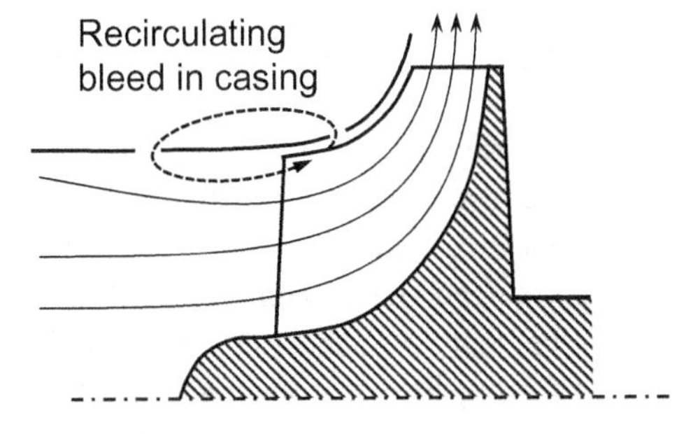

Figure 17.21 Sketch of the basic operating mechanism of the recirculating bleed slot.

choke, and the bleed system does not influence the stage maximum flow capacity. The static pressure at the inducer tip (near the throat) increases with lower volume flow. At low flows, the local pressure at the inducer tip becomes higher than the upstream pressure, forcing part of the flow to recirculate upstream. As a result, more flow passes through the inducer, which reduces the incidence and therefore stabilises the flow in the inducer section. At high speeds, the tip section of the impeller may operate close to choke condition over the whole pressure ratio range. This can eliminate the recirculating bleed flow as the static pressure at the downstream slot never becomes higher than the upstream pressure.

The flow entering the bleed slot from the impeller has a considerable component of circumferential velocity in the direction of rotation due to the work done by the impeller up to this location. This swirl component of velocity is largely conserved in the channel and results in positive preswirl (in the direction of rotation), which reduces the incidence and the relative Mach number at the blade tip which can reduce the inducer losses. However, aerodynamic losses are generated by frictional and mixing losses in the bleed system. Hunziker et al. (2001) observed that the application of a bleed system reduced the relative Mach number at the impeller tip, although more flow passes through the inducer as it has a higher axial velocity. They also observed a small increase in the relative Mach number near the hub which decreases the hub incidence and helps suppress any inducer separation at the hub.

The effect of swirl in the recirculating flow has been studied by Tamaki (2012), who investigated two types of recirculating casing treatment in a 5.7 pressure ratio centrifugal compressor designed for marine applications. A conventional bleed system and a system with vanes to create a negative preswirl were examined. In contrast to the conventional recirculating bleed system, which creates positive preswirl and reduces the relative Mach number at the impeller leading edge, the vaned system creates negative swirl (opposite to the direction of rotation), which results in higher incidence and relative Mach number compared to the conventional design. Adding vanes to the bypass duct clearly extended the operating range, especially at high speeds.

Many authors have investigated the local fluid dynamic effects of bleed systems, and research in this area is very active, as the removal of vortical structures near the casing of the impeller may be advantageous in reducing separation or flow blockage (Sivagnanasundaram, 2010; Chen and Lei, 2013; He and Zheng, 2019).

17.5.8 Variable Trim

Variable trim has been described by Grigoriadis et al. (2012) and is sketched in Figure 17.22. This method of control shifts the operating range of a compressor stage to lower flow by the insertion of a variable conical insert immediately upstream of the impeller eye. The impeller remains unchanged, but the conical insert acts as a pinch in the casing flow channel upstream of the impeller so that the effective casing diameter upstream of the impeller is less than the impeller inlet tip diameter. The conical trim reduces the inlet cross section and shifts the map to lower flows as no flow enters at the tip. The mechanism for the conical channel is designed so that the trim can be retracted

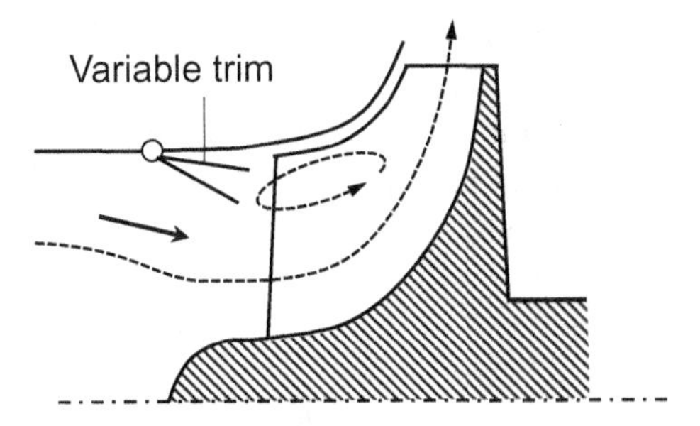

Figure 17.22 Sketch of the operating principle of variable trim.

at operating points requiring a higher flow (Karstadt et al., 2018). This device forces the impeller to operate at low flow with a fixed inlet recirculation upstream of the impeller eye and can be controlled by the depth of insertion of the conical element causing the recirculation.

17.5.9 Passive Control Systems with Slots and Chambers

The most effective of the passive systems in axial compressors are called casing treatments in which slots, grooves or chambers attached to the annulus wall influence the flow in the compressor components to decrease the surge flow, and a review is given in Hathaway (2007). These have found limited use in radial compressors because their effectiveness at increasing range mostly correlates with the fall in efficiency that they produce. In some of these casing treatments, grooves in the casing interact with the flow in the impeller or diffuser to reduce the blockage caused by flow separation so that the fall in pressure rise caused by the flow separation is reduced. In this way, casing treatments not only affect the onset of the instability but also cause a change in the slope of the pressure rise characteristic.

Two recent studies show that axial grooves at the impeller inlet are an effective method of improving the surge margin with a limited effect on efficiency, while working within a restricted design space (Harley et al., 2017; Leichtfuss et al., 2019). The grooves extend from over the impeller inlet region to upstream of the inlet eye. They appear to work by introducing a recirculating flow in the grooves, and this shifts the surge limit at high speeds to a lower flow and causes the characteristic to have a negative slope. At low speeds, the surge line may shift to higher flow rates.

17.5.10 Bleed or Injection within the Diffuser

Casing treatments in vaned and vaneless diffusers have been developed which consist of circumferential chambers near the inlet of the diffuser or along the whole radial extent of the diffuser. These are connected to the diffuser channel by small holes or slots and hinder the onset of rotating stall presumably by reducing pressure disturbances in this region (Bolleter, 1980; Raw, 1986; Galloway, 2018). In some cases, the porous throat is also part of a recirculation or injection system for the flow (Skoch et al., 2003; Galloway et al., 2018; Pakle and Jiang, 2018).

In other research, CFD has identified the existence of separation in the end-wall flow and that suction slots in the diffuser wall can be used to remove the end-wall boundary layer in this region (Marsan et al., 2014)

17.5.11 Active Control Systems

Active control systems to avoid instability are also a subject of active research. These require sophisticated instrumentation to identify the initial growth of instabilities and some form of actuation mechanism to remove them before the surge occurs. Systems have been developed using jets of air and using mechanical actuation of vane angles, but it is not known if any practical application of this technology in radial turbocompressors has been found.

18 Maps and Matching

18.1 Overview

18.1.1 Introduction

This chapter discusses how the compressor performance varies with rotational speed in terms of the change in the key nondimensional parameters. A relatively simple method of map prediction for a single stage is presented; this determines the map from the design duty of the compressor stage and offers good estimates of the performance map of a well-designed stage using only limited information about the geometry. This is particularly useful during the preliminary design process when the geometry or the detailed design is not available. As in earlier chapters, one-dimensional (1D) steady flow analysis is used to generate a basic idea of the effects of the important parameters on the off-design performance.

The prediction of the performance of multistage compressors is then discussed; this uses the principle of stage-stacking calculations. There are a number of interesting aspects of the matching of individual stages in a multistage compressor which can be identified by this method, for example, which stages may be responsible for stall or choke at different conditions.

Other matching issues related to compressor applications are studied. First, the matching of an impeller with a vaned diffuser at different tip-speed Mach numbers is considered. Second, the matching of a compressor with a turbine in a micro gas turbine is described. Finally, the matching of a centrifugal compressor with a centripetal turbine in a turbocharger for a combustion engine is examined. This shows the limitations of a conventional turbocharger with no variable geometry and demonstrates the advantages of the use of a turbine bypass valve or a turbine with variable inlet guide vanes to improve the performance.

18.1.2 Learning Objectives

Understand the use of normalised characteristics to describe the efficiency variation of a compressor stage with flow and tip-speed Mach number and their use with the Euler equation to predict performance maps.

- Have a good knowledge of the stage-stacking method for prediction of the performance of multistage compressors.

- Understand the matching of an impeller with a vaned diffuser at different tip-speed Mach numbers.
- Understand the matching of a centrifugal compressor with a centripetal turbine in a micro gas turbine and with the turbine and engine in a turbocharger.

18.2 Methods of Map Prediction

18.2.1 Different Map Generation Methods

The performance of a compressor is described by the pressure rise versus volume flow and efficiency versus volume flow characteristics at different speeds. This is expressed in a compressor map, discussed in Chapter 8. Despite the high quality of modern CFD simulations (see Chapter 16), a test program is still used to determine the quality of a new design in terms of demonstrating the guaranteed map performance for the customer, as discussed in Chapter 20. Because of the need for suitable hardware and test stands, this testing is expensive and time consuming. Although CFD simulations of the flow path can be carried out very quickly, such simulations are seldom accurate enough to provide exact characteristics, as they often exclude details of the flow such as the volute, leakage paths, disc friction and blade root fillets. In addition, uncertainties always remain as to the predicted location of the surge line. It very challenging to predict the operating range of a compressor either using empirical models or even with high-fidelity simulation tools. Empirical approaches are still widely used for estimating compressor surge margin during the design cycle (Lou and Key, 2019).

An important requirement during the preliminary design of a centrifugal compressor stage is the calculation of a reliable performance map as a guide to both the expected operating flow range and the sensitivity of the final design to speed variations. With this information, the designer can assess the suitability of the design for the application. To do this, the required operating points are plotted into the expected compressor map in order to identify the efficiency and the margin to surge and choke of the different operating points. This overview may then lead to the selection of the most suitable design point. For example, it is possible to check if a new design will provide adequate efficiency at the design pressure ratio and at other operating points, enough surge margin on the off-design low-speed characteristics and sufficient choke margin and pressure ratio at high speeds.

Map prediction during the preliminary design is complex, as the final detailed geometry covering throat areas, blade angles, blade number, blade thicknesses is not available. In addition, many aspects of the aerodynamic design, for example diffusion levels and flow angles, have not generally been finalised at this point in the design process. An additional difficulty is that the performance map has to be generated early on, typically during engineering discussions of a proposed new development. Sometimes several expected maps for various design points are required to aid the decisions to be made. The most accurate methods of determining the map for competing design solutions, by first completing the design and then making CFD simulations or even measurements, are not possible in this time frame.

In some cases, a new stage may be similar to a stage that has previously been tested. The map of the existing stage can be quickly scaled using the conventional similarity parameters from Chapter 8 to provide a good estimate of the expected performance of a new stage. The scaling approach is, however, limited in that it is only possible if such data are available. It will not work well if the design point of the new stage is very different from previous designs, especially if the backsweep and work coefficient have changed, as this changes the slope of the work versus flow characteristic. In addition, the scaling of maps does not provide a parametric description of the speed lines, no speed lines at higher or lower speeds than those tested nor a useful reference based on the best experience derived from many different stages.

In some situations, it is common to use empirical correlation-based methods for predicting individual loss sources and to combine these to calculate performance maps. Such methods require fairly detailed information of the stage geometry, at least on a one-dimensional (1D) basis. Examples are given by Herbert (1980), Oh et al. (1997), Aungier (2000) and Swain (2005) and are briefly covered in Chapter 10. However, the authors' experience with such methods is that they often require fairly tedious tweaking of coefficients in the empirical models to generate a satisfactory performance map and that the characteristic curves that they generate tend to have unreliable shapes.

This chapter describes an engineering approach to estimating the performance map for such preliminary design applications. This approach is based on the global correlation procedure developed for this purpose by Casey and Robinson (2013) for vaneless diffuser stages and subsequently refined by Casey and Rusch (2014) for vaned diffuser stages. The method uses the experience gained from the analysis of many measured test characteristics from numerous sources. The test data available to the authors for this analysis rarely included details of the geometry other than the impeller tip diameter and the type of diffuser. It is often possible to obtain measured test characteristics from manufacturers but very difficult to get detailed information about the geometry. For this reason, the method developed is not based primarily on the geometry.

The approach is based on the fact that from experience it is known that well-designed state-of-the-art compressor stages for a particular aerodynamic duty tend to have fundamentally similar shapes of their performance maps. This indicates that the duty itself is an excellent guide to the form of the performance map. Several engineers given the task of designing a stage for a specified duty are likely to produce fairly similar designs with fairly similar performance maps because, to produce a good design, they rely on similar design rules. Provided that a technique based on the design duty data is reliable, it can provide very useful information in the preliminary design process and can indicate the expected target performance levels that can be achieved in different parts of the map without first designing the stage. The method described here has proved to be more reliable than might be expected given that it has no information about blade angles and the meridional channel.

The method deploys four key nondimensional parameters describing the duty at the design point to determine the stage characteristics and from these the performance map. These are the key nondimensional parameters as described in Chapter 8, as follows:

- The global volume flow coefficient, ϕ_{t1}
- The stage work coefficient, λ
- The stage tip-speed Mach number, M_{u2}
- The expected design point polytropic efficiency, η

The first three values may be determined quite easily from the design specification, and the efficiency may be based on experience or correlations, for example those given in Chapter 10.

Using this method, it has been found that a single set of correlations is not able to predict the maps for all types of stages, despite the fact that the same analytic equations can be used. In order to predict the performance of different types of stages, it is necessary to categorise them into four broad types: turbocharger impellers and process-style impellers, vaned diffusers and vaneless diffusers. Turbocharger-style impellers, which are similar to gas turbine impellers, have an axial inlet designed with an inducer and are relatively long, $L/D_2 > 0.3$. They have unshrouded open impellers and are equipped with thin blades and splitter vanes in the inducer to make them suitable for transonic flow. Process-style impellers in multistage compressors are shrouded and shorter. They rarely have an inducer, and the leading edge is often in the bend from the axial to the radial direction. They also have thick blades, generally with no splitter vanes, and are usually designed for lower tip-speed Mach numbers. Both of these styles of impeller design can be combined with vaned or vaneless diffusers. Drawings of the different type of impellers are provided in Chapter 1.

The basis of the method is that stage characteristics, that is, nondimensional curves for the efficiency, the pressure rise coefficient and the work coefficient versus the flow coefficient, can be generated using analytic equations in the functional form

$$\eta, \lambda, \psi = f(\phi_{t1}, M_{u2}) = f\left(\dot{V}_{t1}/D_2^2 u_2, \quad u_2/a_{t1}\right). \tag{18.1}$$

A series of such flow characteristics at different tip-speed Mach numbers can then be used to calculate the efficiency, power and pressure rise at different speeds in the form of a performance map for a single stage or combined by a stage-stacking technique for a multistage compressor. This procedure is based on the inherent nondimensional stage characteristics (work coefficient and efficiency as a function of flow coefficient) and can be used for different gases, different inlet conditions and different sizes. Appropriate correction terms may be needed for the effect of the Reynolds number or of the isentropic exponent on the characteristic, as discussed in Chapter 8. This allows the performance map of a single-stage or a multistage compressor to be generated covering the efficiency, pressure ratio and power as a function of the corrected volume flow and the corrected speed, as follows:

$$\eta, \pi, P = f\left(\dot{V}_{t1,corr}, n_{corr}\right). \tag{18.2}$$

18.3 Map Prediction for Single Stages

18.3.1 Normalised Efficiency Characteristics

The determination of the characteristics is based on the fact that efficiency characteristics – other than variations that depend on the Mach number levels in the flow – have similar shapes and forms for similar types of radial stages when plotted in suitable nondimensional coordinates. Thus, if it is possible to determine the performance of a stage in terms of nondimensional parameters at the optimum point, then this, together with the Euler equation for the work, can be used for scaling typical nondimensional efficiency characteristics to produce a performance map. As a result, a performance calculation at the optimum point may be made using high-quality correlations for the design point. The off-design performance can be determined on the basis of typical normalised off-design characteristics. This means that detailed loss correlations for all aspects of the aerodynamics of the stage become unnecessary, especially at off-design, where such correlations are increasingly weak and uncertain.

Several very effective axial compressor stage-stacking techniques also use this type of system, for example Howell and Calvert (1978). In these techniques, the variation of efficiency away from the best point is assumed to follow a fairly similar shape for all characteristics. The concept presented here was first demonstrated for centrifugal compressors by Rodgers (1961). Rodgers showed that there are three slightly different forms of such normalised characteristics. These can be used to define the efficiency characteristics of the stage, of the impeller and of the diffuser. In the curves produced by Rodgers, the choke point is used as the basis for making the flow characteristics nondimensional such that the horizontal axis on the map represents the flow divided by the flow at choke. Other systems are possible, such as that of Casey (1994a), where the peak efficiency point is used as a basis for making the flow nondimensional for both impeller and diffuser characteristics. The former is certainly preferable at high speed, and the latter may be preferable at low Mach number levels where choking is less relevant.

For a good model of the stage, impeller and diffuser characteristics, different definitions of the flow and geometry parameters may be required. With appropriate parameters, suitable plots of the nondimensional performance can be obtained from experimental data. The performance variation with flow can then be modelled by an algebraic equation, and the experimental data may be used to guide the selection of suitable equations and coefficients. The method of Swain (2005) is based on the individual impeller and diffuser characteristics. Unfortunately, most experimental data covering radial compressor performance are obtained from flange-to-flange measurements, with the result that separate impeller and diffuser characteristics are difficult to obtain. For this reason, the method described here is based on the flange-to-flange stage characteristics.

The basis of the method is to calculate the values of two dependent variables describing the stage performance (the polytropic efficiency and the work coefficient)

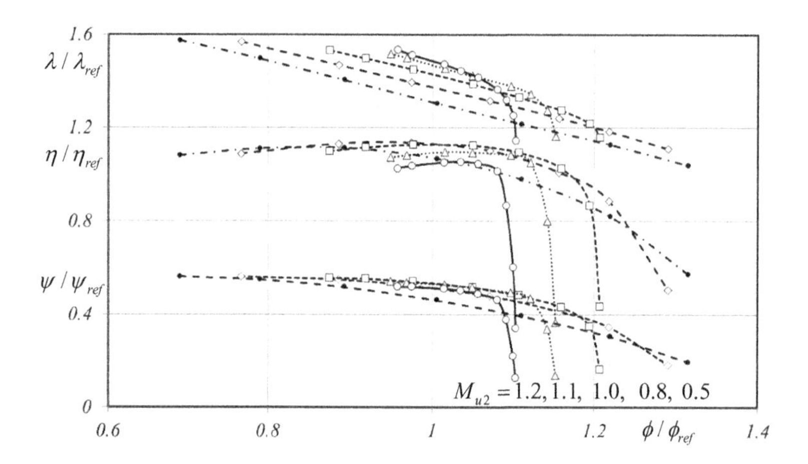

Figure 18.1 The work coefficient, stage polytropic efficiency and pressure rise characteristic of a vaned process compressor stage at different tip-speed Mach numbers (data courtesy of MAN Energy Solutions).

for specific values of the independent variables (the flow coefficient and tip-speed Mach number) and to combine these to give the pressure rise coefficient

$$\eta, \lambda = f(\phi, M_{u2}), \quad \psi = \eta\lambda. \tag{18.3}$$

An example of the measured variation of the stage efficiency, the work coefficient and the pressure rise coefficient versus the flow coefficient at different tip-speed Mach numbers for an industrial compressor stage with a vaned diffuser is given in Figure 18.1. This is adapted from Dalbert et al. (1988) whereby the reference values for the different characteristics are selected to displace the curves to avoid overlap. In this section of the book, some of the subscripts have been dropped for clarity; η denotes the polytropic efficiency, and ϕ denotes the inlet global flow coefficient. The subscript p gives the value at the peak efficiency point of the characteristic, while subscript d denotes the design point. All other thermodynamic performance information – such as isentropic efficiency, pressure ratio, volume flow and mass flow – can then be calculated from the variables in (18.3) using the thermodynamic relationships defining these nondimensional parameters. Speed lines are generated for specific values of the tip-speed Mach number by varying the flow coefficient and the full map comprises an array of speed lines with different tip-speed Mach numbers. Examples of predicted stage characteristics and the associated performance map for a vaneless turbocharger compressor stage are given in Figure 8.8.

18.3.2 Variation of Efficiency with Flow Coefficient and Tip-Speed Mach Number

The variation of the pressure rise coefficient with tip-speed Mach number and flow is shown in Figure 18.1 and can be seen to be quite complex. The variation of the efficiency with flow is also complex but is more amenable to generating a suitable model. The basis of the efficiency calculation described here is to split measured variation of efficiency with the flow coefficient and the tip-speed Mach number into separate parts, as described by Casey and Robinson (2006) and as sketched in Figure 18.2.

This complex variation is similar to the variation of losses with incidence and Mach number in compressor cascades (see Section 7.3.3) and can be converted into three separate effects which are easier to model individually. First, there is the change in the shape and width of the normalised efficiency characteristic with the normalised flow coefficient for different tip-speed Mach numbers: the characteristic becomes narrower as the tip-speed Mach number increases, as shown in Figure 18.2. Second, the peak efficiency tends to increase with speed towards the design tip-speed Mach number – with its optimum matching between the components and higher Reynolds number – before it drops at higher tip-speed Mach numbers due to component mismatch and higher losses, as shown in Figure 18.3. In parallel to this, the change in the matching of the impeller and the diffuser with speed (see Section 18.7) causes an increase in the flow coefficient at peak efficiency with tip-speed Mach number, as shown in Figure 18.3.

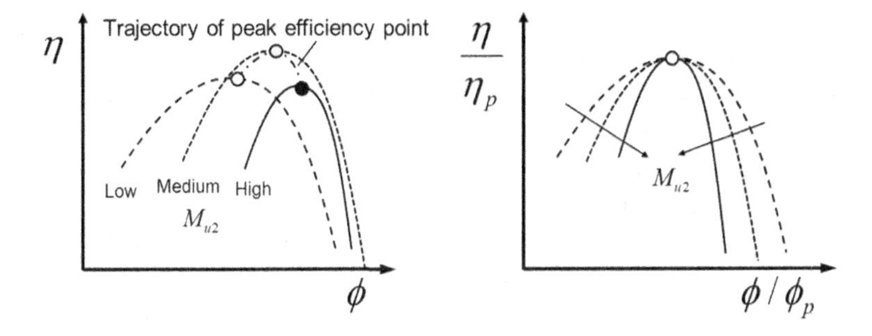

Figure 18.2 Sketch of the efficiency characteristics of a stage for low, medium and high tip-speed Mach numbers (left) and the variation of normalised efficiency with normalised flow coefficient with tip-speed Mach number (right).

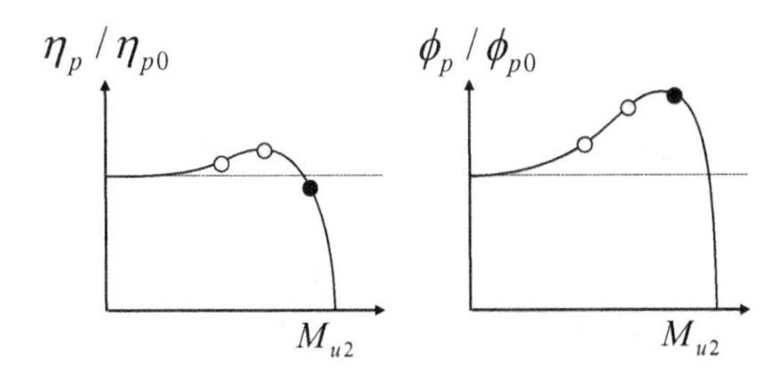

Figure 18.3 Variation of the peak efficiency and the flow coefficient at peak efficiency with the tip-speed Mach number.

If the peak efficiency, η_p, and flow coefficient at which it occurs, ϕ_p, are used to normalise the efficiency variation with flow, this leads to a functional equation of the form

$$\eta/\eta_p = f(\phi/\phi_p, M_{u2}). \tag{18.4}$$

In Figure 18.3, the efficiency, η_{p0}, and the flow coefficient at peak efficiency, ϕ_{p0}, for a fictitious zero Mach number speed line are used for the normalisation, giving functional equations of the form

$$\eta_p = \eta_{p0} f(M_{u2}), \quad \phi_p = \phi_{p0} f(M_{u2}). \tag{18.5}$$

In compressor maps, measurements are not usually made much below a tip-speed Mach number of 0.3; this provides a good guide to the curves at a fictitious Mach number of zero as the flow is effectively incompressible at this speed. Tests on low-speed centrifugal fans identify almost no change in the shape of the characteristics below a tip-speed Mach number of 0.3 (Wolfe et al., 2015).

The empirical equations describing the functional dependency of efficiency with flow include several nondimensional parameters, a considerable number of variable coefficients and some fixed constants described later in this chapter. The same analytical equations are used for turbocharger and process compressor stages with vaned and vaned diffusers but with different values of the coefficients for the different machine types. With different values of the coefficients and parameters in the functions, these equations have been used by Wolfe et al. (2015) to model the maps of low-speed centrifugal fans. However, at low tip-speeds, there is no need for a variation in the coefficients with Mach number.

A key feature of the shape of the characteristics at different speeds is that the impeller speed has a direct influence on both the pressure ratio and the density ratio at a point within the stage. For this reason, the local volume flow at the impeller outlet depends not only on the flow at inlet but also on the density rise that has occurred up to this point in the stage. As a result, the performance of a downstream component, for example a diffuser, which is dependent on the local volume flow can only be estimated if compression up to that point is known. This implies that the matching and performance of the components changes with the tip-speed Mach number. For incompressible flow, the density remains constant through the whole stage and no mismatch occurs as the speed changes.

18.3.3 Equations for the Variation in Efficiency from Surge to Choke

The variation of the efficiency with flow and with tip-speed Mach number is modelled by equations first developed by Swain (2005) and by similar equations given by Casey (1994a). The variation of stage efficiency with flow along a speed line is a modified form of an elliptic curve:

$$\frac{x^z}{a^z} + \frac{y^z}{b^z} = 1, \quad \frac{x}{a} = \left[1 - \left(\frac{y}{b}\right)^z\right]^{1/z}. \tag{18.6}$$

If $z = 2$, this is the equation for an ellipse and if in addition $a = b$ it is that for a circle. Where $a = b$ and $z > 2$ the same equation, which has recently been dubbed 'squircle', defines a mathematical shape between a circle and a square. This is used by Apple in IOS7 and onwards to define the outer shape of icons on the iPhone; using a value of $z = 5$ gives the appearance of a square in which the corners have been cropped by a circle. An increase in the parameter z causes the circular shape to become progressively squarer. In the application here, where $a \neq b$, the equation is called a 'super-ellipse' (Guasti, 1992). A further development of this equation with different exponents for the terms of x and y may also be suitable for modelling the variation of efficiency with flow with more accuracy, but this has not been pursued here as the experimental data available to the authors do not justify further complexity.

Following Swain (2005), for flows below the peak efficiency point, this equation has a variable exponent, D, to give the characteristic curves for the efficiency ratio as a function of flow ratio:

$$\phi < \phi_p, \quad \eta = \eta_p\left[1 - \left(\phi/\phi_p\right)^D\right]^{1/D}. \tag{18.7}$$

This equation has the natural physical property that the efficiency is zero at zero flow coefficient. The efficiency has a maximum at the flow coefficient for peak efficiency, η_p and the curves are horizontal at this point. To match the test data, the coefficient D takes different values for different tip-speed Mach numbers. As a result, the shape of the characteristics in the low-flow region are to a small extent dependent on the Mach number. For example, for turbocharger stages with vaneless diffusers the typical value of D varies from 2.1 on the low-speed characteristics to 1.7 on the high-speed characteristics: this is clearly related to the increase in the strength of the incidence losses at high Mach numbers. The equations do not include any specific details of flow separation or instability at low flows, and they retain a simple ellipselike shape down to zero flow. This is a weakness in the prediction of the onset of instability: in reality, relatively small changes in the slope of the pressure rise variation with flow can lead to the onset of compression system instability.

For flows above the peak efficiency, a different exponent, H, is used. The equations are further modified to take account of the fact that when the maximum flow is reached, the efficiency may not be zero:

$$\phi \geq \phi_p, \quad \frac{\eta}{\eta_p} = (1 - G) + G\left[1 - \left(\frac{\phi - \phi_p}{\phi_c - \phi_p}\right)^H\right]^{1/H}. \tag{18.8}$$

This equation has a similar shape to the equations below the peak efficiency point, and this leads to a smooth transition between the two arcs around the peak efficiency. The curves automatically have the maximum efficiency at the flow coefficient for peak efficiency and have the maximum flow with a vertical characteristic at the flow coefficient for choke. The coefficients H and G and the ratio of the flow coefficient at peak efficiency to that at choke, ϕ_p/ϕ_c, are not constant but also vary with the tip-speed Mach number to give narrower characteristics of a different shape as the Mach number increases.

Exponent H has a similar function to exponent D in (18.7). It takes into account the fundamentally different shapes of efficiency characteristics for low Mach number and high Mach number stages. Low Mach number stages tend to have a smooth drop in efficiency ratio – related to incidence losses – as the flow increases above peak efficiency. A value of H near to 2, giving elliptical curves, is needed to match low-speed experimental data. High Mach number stages tend to have a small plateau of high efficiency close to the peak efficiency point; they then drop much more sharply into choke due to the transonic nature of the impeller inlet flow. A value of $H = 3.5$ is needed to match experimental data on the shape of high Mach number characteristics.

Equation (18.8) includes an additional change to (18.7) to account for the fact that the location of the maximum flow at choke is not necessarily at zero efficiency but rather at an efficiency ratio of $(1 - G)$. High Mach number stages reach choke at relatively high efficiency (with $0 > G > 1$) or even at peak efficiency if all sections operate at unique incidence conditions. The vertical choked characteristics start immediately below this point rather than when zero efficiency is reached. By contrast, low Mach number impellers fall to an efficiency of zero well before the maximum choke flow (with $G > 1$) is reached, as high losses due to negative incidence stall occur before choke; see Chapter 17. The flow then continues to increase below the point of zero efficiency. Good agreement between these equations and the test data is shown in the papers of Casey and Robinson (2013) and Casey and Rusch (2014). An example of this agreement is given in Figure 18.4, where this example was not used to determine the coefficients in the equations.

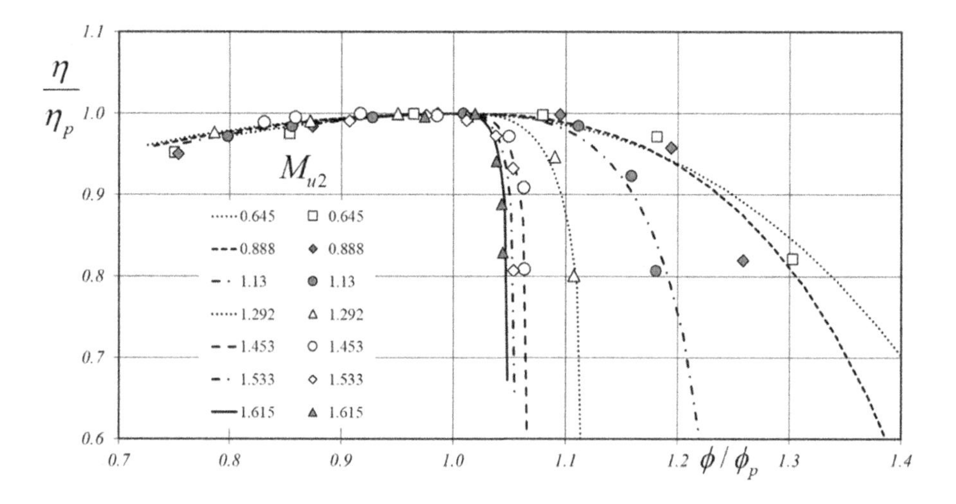

Figure 18.4 Variation in the normalised efficiency with the normalised flow coefficient at different tip-speed Mach numbers in a turbocharger compressor with a vaned diffuser.

The variation in the values of D, G and H are blended between the low-speed values and the high-speed values using an s-shaped function as follows:

$$\begin{aligned} D &= (1-p)D_l + pD_u \\ G &= (1-p)G_l + pG_u \\ H &= (1-p)H_l + pH_u \\ p &= \frac{1}{1+e^{-t}}, \quad t = (M_{u2} - B)(AM_{u2} + C). \end{aligned} \tag{18.9}$$

The values of the coefficients in the s-shaped function are constants. Parameter p in this equation is the s-shaped logistic function, and it varies from 0 to 1 with increasing tip-speed Mach numbers. P blends the values of the coefficients between the lower asymptote at low tip-speed Mach numbers (subscript l) and the upper asymptote at high tip-speed Mach numbers (subscript u). At low Mach numbers, the variation with Mach number is small as p remains close to 0; this is physically realistic as the flow is effectively incompressible up to $M = 0.3$. At high Mach number, the equations also attain a constant asymptotic value with p close to 1. This matches experience and is supported by the constant value of the nondimensional impeller choking parameter at high Mach numbers, as discussed in Chapter 11. With a value of $p = 0.5$ with $t = 0$ at a Mach number of $M_{u2} = B$, the transition between these extreme values reaches the halfway value between the upper and lower asymptotic values. The rate of transition around this value is determined by the constants A and C. Coefficient B in (18.9) can be envisaged as a measure of the transonic capability of the stage or as a sort of critical tip-speed Mach number of the design. The analysis of test data shows that typical turbocharger stages tend to require a value of B of around 1.1 or 1.2 in order for their characteristics to be modelled. Process style stages in multistage compressors require a value of B of around 0.8–0.9. In all the stages that have been examined, a good model for the transition over the range of Mach numbers is obtained with values of A between 0 and 1 and of C between 4 and 5.

18.3.4 Variation of the Peak Efficiency with Mach Number

The Casey and Robinson (2013) method does not include any feature that allows the coefficients in the equations to be related to the design Mach number of the stage. The assumption of the original method is that the tip-speed Mach number with the very best peak efficiency and optimal matching is the one of most interest, and so the coefficients are adjusted to match the experimentally determined characteristics of this tip-speed Mach number. Stages with poor matching between the impeller and the diffuser are ignored in this procedure. During further development of the method on the basis of additional data from a considerable number of sources, modification of the coefficients allows for the effect of the matching of the impeller with the diffuser (Casey and Rusch, 2014).

The matching effect is strongest for vaned diffuser stages. If a design is carried out for a low tip-speed Mach number, the impeller and diffuser design is then adapted to achieve peak efficiency near the design speed. Here the diffuser throat area represents a significant fraction of the throat area of the impeller, as discussed in Chapter 12. An increase in the tip-speed Mach number causes the diffuser to become too large because the specific outlet volume from the impeller drops with higher speed and therefore the diffuser becomes less well matched to the impeller. This leads to a fall in efficiency, as the Mach number is increased above the design speed. If, however, the design is carried out for a higher tip-speed Mach number, a much smaller diffuser throat area matched at the higher speed is obtained. In this second case, the peak efficiency should be closer to the higher design speed, and the efficiency will fall as the speed decreases. Thus, the variation in the efficiency with tip-speed Mach number will be different in each case and is not just a function of the tip-speed Mach number but also of the design speed Mach number due to the different diffuser throat areas that are needed. These throat areas control the matching between the components and the performance at different speeds.

The principle of the correlation for this effect, developed for vaned and vaneless diffusers, is shown in Figure 18.5 for vaned diffusers in turbocharger style stages. The lines represent the correlation for the variation in peak efficiency with tip-speed Mach number relative to that for an incompressible flow with low Mach numbers. Depending on the matching with the diffuser and its effect on the design Mach number of the stage, the peak efficiency occurs at a different Mach number. The highest value is found close to the design tip-speed Mach number with optimal matching between the impeller and the diffuser. As such, a machine designed for a high tip-speed Mach number has a different trend to that designed for a low-tip-speed Mach number. Based on this, a number of the coefficients in the equations are made to be a function of the design Mach number (Casey and Rusch, 2014).

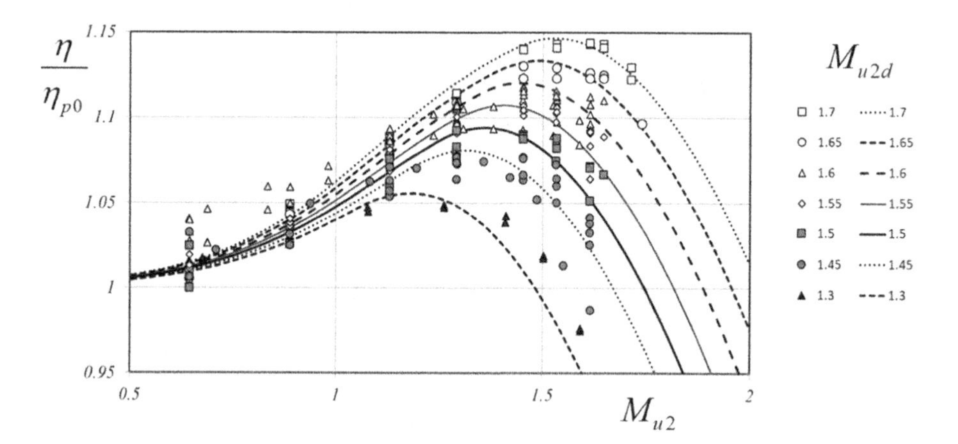

Figure 18.5 Variation of normalised efficiency with the tip-speed Mach number for turbocharger vaned diffuser stages with different design Mach numbers.

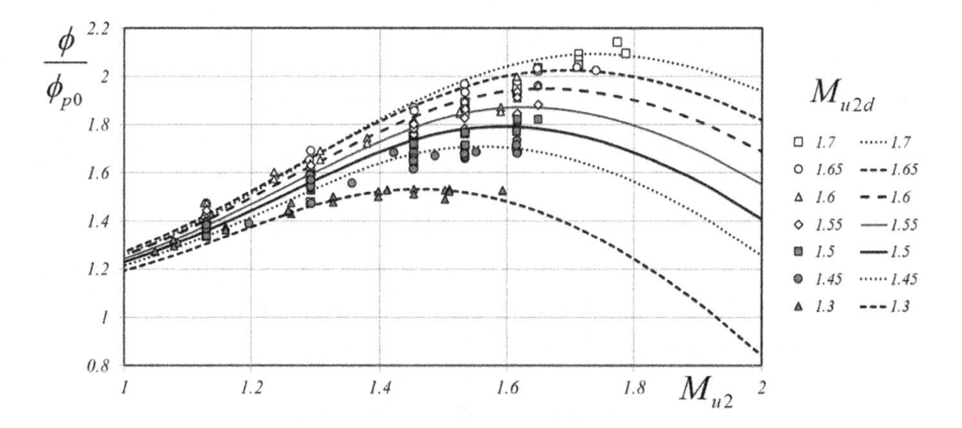

Figure 18.6 Variation of the normalised flow coefficient with the tip-speed Mach number for different design Mach numbers of a vaned diffuser stage.

18.3.5 Variation of the Flow Coefficient at Peak Efficiency with Mach Number

A similar effect occurs with the flow shift – the location of the flow coefficient at peak efficiency at different speeds relative to that at low speeds with nearly incompressible flow. If the diffuser is too small, it tends to act as a choked nozzle at low speeds and causes a large shift in the flow coefficient to lower values. To take this effect into account, some of the coefficients for the flow shift are a function not only of the actual tip-speed operating Mach number but also of the design Mach number, as shown in Figure 18.6.

18.3.6 Location of the Surge Line and the Choke Line

One of the most difficult aspects of compressor flows is the prediction of the stability limit due to the onset of rotating stall or surge. The difficulty is exacerbated by the fact that different components (diffuser or impeller) may be responsible for the instabilities, different phenomena may occur (rotating stall, ring stall and surge), test procedures to identify instabilities may differ and the compression system itself may play a role, as outlined in Chapter 17. With such a large number of effects it is not surprising that any systematic comparison of different stages shows a wide variability in the operating range of compressor stages. It is clearly not possible in a preliminary design method with no detailed information on the internal aerodynamics, geometry and loading of the stage to predict the exact stability point. Even with more detailed knowledge of the geometry and the internal flow, and even with the help of computational fluid dynamics, it is still exceptionally difficult to predict the onset of instability with any certainty.

The most pragmatic approach in the framework of this method is to derive a logically consistent set of equations for the variation of the flow at the stability point and then to suggest a range of typical coefficients that can be used in these equations to express an optimistic, realistic and pessimistic stability line. The equations used for the variation of the flow coefficient on the surge line as a function of Mach number are a development of the equations for the normalised flow ratio for the flow at peak efficiency, given in Figure 18.6. The variation of the choke line is modelled in a similar way and adjusts the width of the characteristic above the peak efficiency flow, whereas the change in the surge line simply determines the instability point and does not change the shape of the curve. Different values for these coefficients are used for different surge and choke lines to represent the optimistic, realistic and pessimistic estimates. Examples of the operating range compared to experimental data using these equations are given in Chapter 17. The addition of optimistic and pessimistic curves provides an approximate way to incorporate poor designs into the system, such that stages with a small impeller throat area relative to the inducer inlet area can be considered to choke early. In the same way, the improvement in the surge line due to a recirculation device can be taken into account.

18.3.7 Variation of the Work Coefficient with Flow and Mach Number

The second part of the functional relationship of (18.3) is the variation of the work input coefficient with the flow and the tip-speed Mach number. This variation is more complex than described in Chapter 11, as the work coefficient is influenced by the matching with the diffuser such that if the diffuser chokes first the impeller cannot choke. Under the assumption that the flow has no swirl at inlet to the impeller, the Euler work can be estimated from the velocity triangles. This work is directly related to the flow coefficient at impeller outlet, the assumed slip factor and the impeller outlet blade angle.

The basic equations needed are described in Chapter 11 and repeated here for convenience, as follows:

$$\lambda_{Euler} = 1 - \sigma + \phi_2 \tan\beta_2' \tag{18.10}$$

$$\lambda = (1 + C_{PIF}/\phi_{t1})\lambda_{Euler} \tag{18.11}$$

$$\phi_2 = \phi_{t1}\frac{1}{\pi}\frac{D_2}{b_2}\frac{\rho_{t1}}{\rho_2} \tag{18.12}$$

$$r_k = 1 - \left(\lambda_{Euler}^2 + \phi_2^2\right)/2\lambda \tag{18.13}$$

$$\rho_2/\rho_{t1} = \left[1 + (\gamma - 1)r_k\lambda M_{u2}^2\right]^{\frac{1}{n-1}}. \tag{18.14}$$

Taken together, these equations lead to the following equation for the work coefficient as

$$\lambda = (1 + C_{PIF}/\phi_{t1})\left(1 - \sigma + \frac{\phi_{t1}\tan\beta_2'}{\pi(b_2/D_2)\left[1 + (\gamma - 1)r_k\lambda M_{u2}^2\right]^{\frac{1}{n_{imp}-1}}}\right). \tag{18.15}$$

In the method of Casey and Robinson (2013), the value of the work coefficient and the polytropic exponent in the term in the large bracket in (18.15) is taken as that at the design point. The design point degree of reaction is specified by the user at around 0.6; see Chapter 11. The Euler work coefficient then varies linearly with inlet flow and, in this earlier model, none of the physical effects of choking in the impeller are taken into account, with the result that the sharp fall in work coefficient towards choke on the high-speed characteristics is neglected, as indictated in Figure 18.1.

Casey and Rusch (2014) use a different procedure to take choking in the impeller into account. In this, it is necessary for one additional piece of information to be specified at the design point. This can be the degree of reaction; the impeller back-sweep; the impeller outlet width ratio, b_2/D_2; the design value of the impeller outlet flow coefficient; the impeller absolute flow angle at outlet; or the de Haller number on the shroud. In the equations as written here, it is assumed that the additional information is the impeller blade outlet angle.

Using the equations from Section 11.7.1, the impeller outlet flow coefficient at the design point is given by

$$\phi_{2d} = (\lambda_{Euler\_d} - 1 + \sigma)/\tan\left(\beta_2'\right). \tag{18.16}$$

This allows the design degree of reaction to be determined:

$$r_{kd} = 1 - \left(\lambda_{Euler\_d}^2 + \phi_{2d}^2\right)/2\lambda_d. \tag{18.17}$$

The design point total-to-static density ratio across the impeller can be calculated using

$$(\rho_2/\rho_{t1})_d = \left[1 + (\gamma - 1) r_{kd} \lambda_d M_d^2\right]^{\frac{1}{n_d - 1}}. \tag{18.18}$$

The value of the impeller polytropic exponent in this equation is taken to be similar to the stage design point polytropic exponent, as in most stages the stage polytropic efficiency and the impeller total-static efficiencies are similar in value, as explained in Chapter 10. From this, the approximate design value of the impeller outlet width ratio can be determined:

$$(b_2/D_2)_d = (\phi_d/\pi\phi_{2d})(\rho_{t1}/\rho_2)_d. \tag{18.19}$$

This value should be a close approximation to the actual outlet width ratio, any difference being due to the fact that the impeller total-static efficiency is assumed to be the same as the stage total-to-total efficiency. In the calculations described here, this difference is not relevant, as the impeller outlet width ratio is only used directly in (18.15) as a fixed constant value that is consistent with the other design values. The knowledge of the outlet flow coefficient and the work coefficient also allows the exit velocity triangle to be defined. The flow angles consistent with the design data are given by

$$\tan\alpha_{2d} = \lambda_{Euler\_d}/\phi_{2d}, \qquad \tan\beta_{2d} = (\lambda_{Euler\_d} - 1)/\phi_{2d}. \tag{18.20}$$

In the determination of the work coefficient at other points on the characteristic, (18.10) to (18.14) are used in turn with a small iteration. In this iteration, the design point value of the work coefficient and the degree of reaction are used as starting points. The density ratio (18.14) and the outlet flow coefficient (18.12) are calculated in order to derive new values of the work coefficient (18.10 and 18.11) and the degree of reaction (18.13). The iteration is repeated until convergence for each point on the characteristic is achieved.

The impeller total-static polytropic exponent is also needed in (18.14). As a simple expedient, Casey and Robinson (2013) assume that this exponent has the value of the peak total–total efficiency point on each characteristic. This approach is adequate where the impeller determines the choke flow, because in such cases the change in the assumed impeller polytropic exponent correctly falls at high flow and causes the work coefficient to reduce at high flows. This is suitable for vaneless diffusers. But in stages with small vaned diffusers where the diffuser causes choke, the impeller efficiency actually remains high while the stage efficiency falls, as shown in Figure 4.10. The assumption that the impeller efficiency is the same as the stage efficiency is not at all correct as a means of calculating the polytropic exponent for the impeller density ratio when the diffuser chokes. A simple way to deal with this has been adopted by assuming that on the low-speed speed lines below the design tip-speed Mach number, where the diffuser chokes and the impeller efficiency remains high, the impeller efficiency is the same as the stage efficiency for all points. The algorithm to calculate the polytropic exponent in this case is described by Casey and Rusch (2014).

18.3.8 Validation of Maps Predicted with This Method

The objective of this map prediction method is to predict a performance map over the whole speed range from the of values a number of key parameters related to the compressor design, obviating the need to first carry out a detailed design of the stage. The method deploys simple models for the stage characteristics that give the variation of efficiency and work input as a function of flow coefficient for varying tip-speed Mach numbers of each individual stage. It makes use of empirical coefficients that have been adjusted to match the measured performance of a wide range of different types of single-stage radial compressors. The predictive capability of the method is illustrated with the two compressor maps shown in Figures 18.7 and 18.8. Figure 18.7 shows the performance map of a compressor stage with a vaned diffuser for use in a marine turbocharger with a design pressure ratio near to 4.5. Figure 18.8 shows the performance map of a compressor stage with a vaneless diffuser for an automotive turbocharger with a design pressure ratio of 2.55. It is clear from the good agreement that the performance map of a well-designed compressor stage is largely determined by the nondimensional parameters describing the duty of the compressor. Similar good agreement is found for process stages with vaned and vaneless diffusers.

18.3.9 The Difference in the Characteristics of Vaned and Vaneless Stages

The efficiency and work coefficient versus flow coefficient characteristics at different speeds for a vaned stage and a vaneless stage predicted with this method are given in

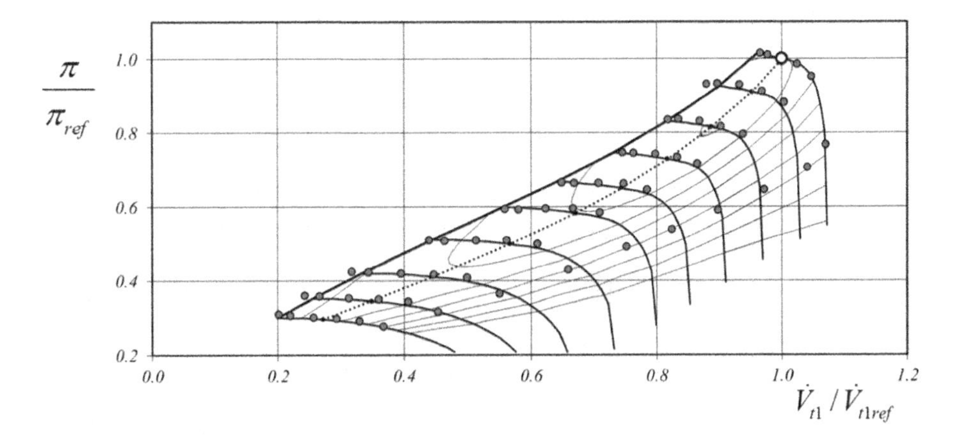

Figure 18.7 Predicted and measured performance map of a compressor stage with a vaned diffuser for a marine turbocharger stage.

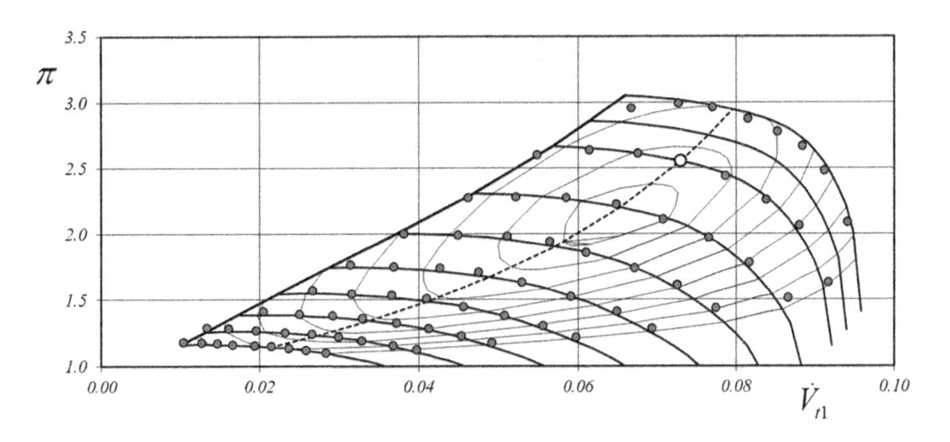

Figure 18.8 Predicted and measured performance map of a compressor stage with a vaneless diffuser for an automotive turbocharger.

Figures 18.9 and 18.10. Both stages are designed with a flow coefficient of $\phi_{t1} = 0.09$ and represent typical designs for turbocharger compressor applications with a vaned and a vaneless diffuser. The vaned stage in Figure 18.9 is a stage similar to stage B from Chapter 11 with a design pressure ratio close to 4.6 at a tip-speed Mach number of 1.58 and a work coefficient of $\lambda = 0.7$ and a backsweep of 25°. The vaneless stage in Figure 18.10 is similar to stage A from Chapter 11 with a design pressure ratio close to 2.4 at a tip-speed Mach number of 1.23 and a work coefficient of $\lambda = 0.6$ and a backsweep of 40°.

The efficiency speed lines are narrower for the vaned stage with a high design Mach number and a high work coefficient than for the vaneless stage at all speeds. At all speeds below the design speed, the diffuser in the vaned stage chokes, so there is no evidence of a sharp drop in the work coefficient at high flow coefficient. As the speed

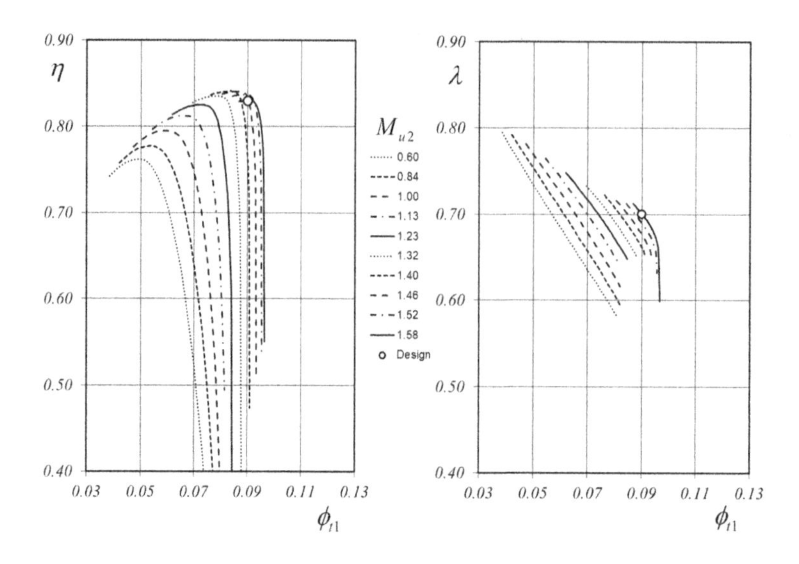

Figure 18.9 Predicted variation of work coefficient and efficiency versus flow coefficient at different tip-speed Mach numbers for a turbocharger stage with a vaned diffuser.

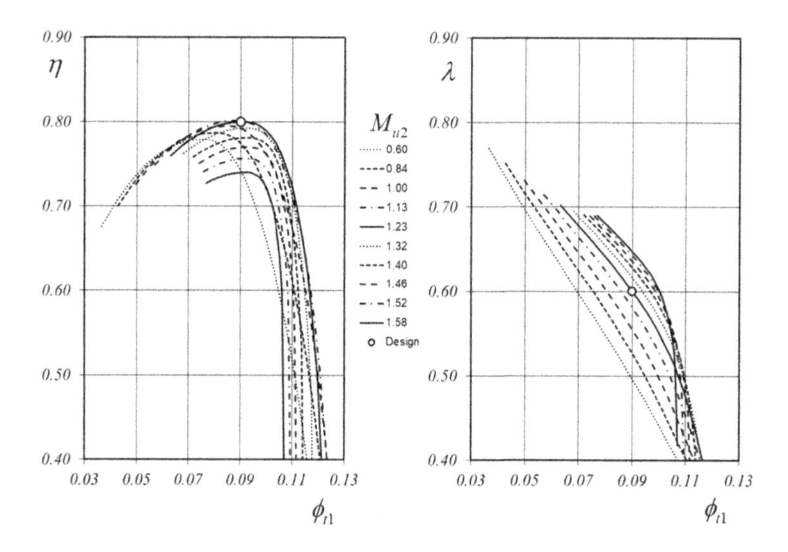

Figure 18.10 Predicted variation of work coefficient and efficiency versus flow coefficient at different tip-speed Mach numbers for a turbocharger stage with a vaneless diffuser.

decreases from the design speed, the work coefficient and the efficiency speed lines of the vaned stage shift strongly to lower flow coefficients due to the choked diffuser, limiting the maximum flow. The combined trend is that the work coefficient remains roughly constant at the peak efficiency operating point on all speed lines. This causes the pressure ratio speed lines at different speeds in a map with a vaned stage to shift to lower flows as the speed is decreased.

The vaneless case shows that at high flow coefficient on all but the lowest speed lines, the impeller chokes, as evidenced by the sharp drop in work coefficient at high flow coefficients. The work coefficient curves bunch closer together than in the vaned stage as operation with a vaneless diffuser, which does not choke, does not force the impeller to operate at such an extreme low flow coefficient at low speeds. The pressure ratio speed lines in a compressor map with a vaneless diffuser do not shift as strongly to lower flows as in a map with a vaned diffuser but remain much closer to the design flow coefficient as the speed changes.

18.4 Apparent Efficiency Due to Heat Transfer Effects

In turbocharger stages, heat is transferred from the hot turbine parts to the compressor, which causes an additional temperature rise of the gas above that produced by the work input from the impeller (Baines et al., 2010). As the efficiency and work input in typical turbocharger test rigs are determined from the temperature rise, this leads to an apparent increase in the work input and an apparent fall in the efficiency for the compressor. This effect is particularly strong at low speeds where the work input is low and the heat transfer becomes a larger proportion of the temperature rise. There is no effect on the pressure rise, as the effects on the work and the efficiency counteract each other and the real thermodynamic effects of the heat transfer are small. The calculation method described earlier first derives the performance without heat transfer and then applies a correction for the effect of heat transfer to give the apparent values with heat transfer. The equations for the corrections have been derived by Casey and Fesich (2010) and tested against extensive experimental data by Sirakov and Casey (2012). The correction for the apparent work coefficient is

$$\lambda_a = \lambda + k_c \frac{1}{\phi_{t1} M_{u2}^3}. \tag{18.21}$$

The empirical heat transfer coefficient, k_c, can be adjusted to match the experimental data and has a typical value of 0.002–0.004 in turbochargers but in other applications with no heat transfer is zero. The related correction for the apparent efficiency is given by Sirakov and Casey (2012) as

$$\eta_{pa} = \eta_p \frac{\lambda}{\lambda_a}. \tag{18.22}$$

This method allows data gathered under diabatic test conditions to be converted to performance maps for adiabatic conditions, without the need for direct measurement of heat transfer. Examples of the prediction of this effect compared with test data using these equations are provided by Sirakov and Casey (2012). Figure 18.11 shows the apparent work coefficient and efficiency characteristics of the stage in Figure 18.10 with a heat transfer coefficient of $k_c = 0.003$. On the low-speed characteristics, this effect can easily lead to an apparent efficiency 20 percentage points below the actual efficiency level.

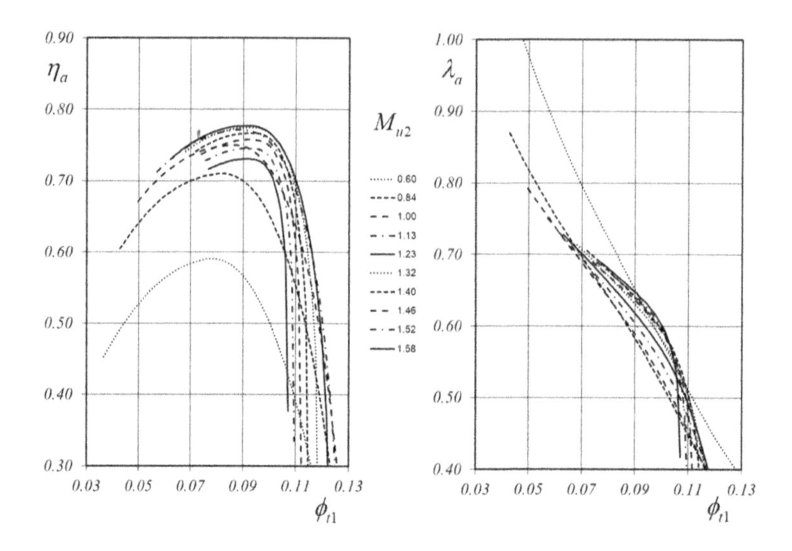

Figure 18.11 The apparent work coefficient and apparent efficiency versus flow coefficient for a turbocharger stage with a vaneless diffuser and a heat transfer coefficient of $k_c = 0.003$.

18.5 Extrapolation of a Measured Map

18.5.1 Extrapolation to Low Pressure Ratio

Constant speed lines on compressor performance maps from typical test rigs usually have no test data at low pressure ratio and high flow. This is because the system resistance of the test rig precludes operation in this region – the high flow leads to a large pressure drop in the system, and the compressor cannot deliver this pressure at high flows. The speed lines on the maps are generally measured down to a pressure ratio corresponding to about 50% efficiency. No data are available at lower pressure ratios than this. In the operation of multistage compressors there are situations where the matching with other stages forces one of the stages deep into choke causing it to operate in this region. In turbochargers too, the engine sometimes forces the turbocharger compressor to operate in this region. A typical example is when stationary thrust is needed with a cold engine. The turbine power is low as the engine exhaust gas is cold and the engine then sucks the flow through the compressor as if it were a throttle. This may also occur when the engine is used for braking for longer periods when coasting downhill with no fuel being burnt or during mismatched operations in two-stage compression systems. Whole-engine simulation systems need to extend the test data in some way to include the low-pressure region of the map. The model described in Section 18.3 is able to predict this region of the map in the process of estimating the performance of the whole map. In some turbocharger modelling systems, however, a physically based method of extending a measured performance map by estimating the performance at low pressure ratios from measurements made in the rest of the map is needed.

A procedure for this extrapolation for stages with vaneless diffusers has been described by Casey and Schlegel (2010). The test data over the measured range of speed and flow are used to estimate the work transfer and loss characteristics of the impeller. The effective impeller throat area can be determined from the choking mass flow on the high-speed characteristics with the help of the impeller choke equation in Chapter 11. The Euler equation justifies a linear extrapolation of the work coefficient to higher values of the outlet flow coefficient, much in the same way as Section 18.3.7. The extrapolation of the losses as the impeller moves into choked operation is more difficult. Casey and Schlegel (2010) examine several ways of doing this and recommended a technique based on the density ratio to extrapolate the losses rather than estimating the losses to determine the density ratio. This method does not include any correlations for the losses but accounts for choking losses within the impeller in a similar way to determining the choking of a one-dimensional Laval nozzle with different back pressure, as discussed in Chapter 6. In a choked 1D Laval nozzle, the back pressure determines the location of the shock, and the shock strength determines the losses. In fact, in a choked Laval nozzle the shock moves to adjust its strength, and hence the losses, to match the imposed back pressure. A similar situation arises in a radial impeller with a very low back pressure.

At low speeds and at higher speeds away from choke, the ratio of the inlet and outlet flow coefficients in a radial impeller near the peak efficiency point is nearly constant as the density ratio remains more or less constant, from (18.12). In this nearly linear region, the relationship between the inlet and outlet flow coefficients can be described by the following equation:

$$\phi_{t1} = l\phi_2. \tag{18.23}$$

where the slope, l, is determined from the measurements near the blend point. At higher speeds within the choked region, the ratio of the inlet and outlet flow coefficients is determined by the density ratio. This is because the inlet flow coefficient is limited by its maximum choking value. This behaviour is shown in Figure 18.12 for a typical medium-speed characteristic. The relationship between the inlet and outlet flow coefficients in this diagram is determined by the test data and by two asymptotes: the high-flow asymptote with a constant value of inlet flow coefficient, and the low-flow asymptote with a constant ratio of inlet and outlet flow coefficients or a constant density ratio. A high-speed characteristic, which would typically reach the choke point, would be biased to the choke asymptote when plotted in this way. The original method of Casey and Schlegel (2010) has subsequently been improved by a suggestion of Spasov (2009), giving a simpler equation for the required behaviour to blend these two asymptotes as follows:

$$\phi_{t1} = ke^{-m\phi_2} + n. \tag{18.24}$$

The three unknown coefficients k, m and n are found by satisfying boundary conditions for the curve above the blend point between the test data and the extrapolated curve – the slope and value of the test data at lower outlet flow coefficients should match the curve – and the curve should match the choke asymptote at higher outlet flow coefficients.

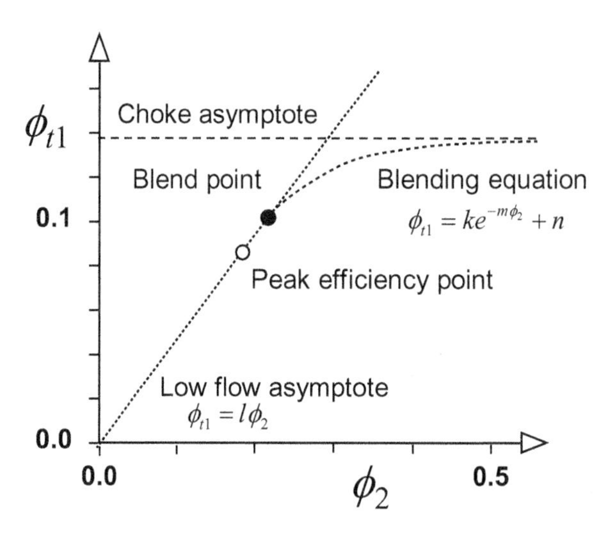

Figure 18.12 Relationship between inlet and outlet flow coefficient as the impeller chokes.

The asymptote at high values of the outlet flow coefficient is used to determine n:

$$\phi_2 = \infty, \qquad \phi_{t1} = \phi_{ch} \quad \Rightarrow \quad n = \phi_{ch}. \tag{18.25}$$

The value of the flow coefficient at the blend point leads to

$$\phi_{t1} = \phi_b, \;\; \phi_{2b} = \phi_b/l \;\; \Rightarrow \;\; \phi_b = ke^{-m\phi_b/l} + n \;\; \Rightarrow \;\; \phi_b - \phi_{ch} = ke^{-m\phi_b/l} \tag{18.26}$$

The slope of the equations at the blend point are the same and determines l:

$$\begin{aligned} &\frac{d\phi_{t1}}{d\phi_2} = -mke^{-m\phi_2} \\ &\phi_{t1} = \phi_b, \quad \phi_{2b} = \phi_b/l \quad \Rightarrow \quad \left(\frac{d\phi_{t1}}{d\phi_2}\right)_b = l \quad \Rightarrow \quad l = -mke^{-m\phi_b/l}. \end{aligned} \tag{18.27}$$

Division of the last two equations gives an equation for m:

$$(\phi_b - \phi_{ch})/l = -1/m \quad \Rightarrow \quad m = -l/(\phi_b - \phi_{ch}) \quad \Rightarrow \quad m = l/(\phi_{ch} - \phi_b). \tag{18.28}$$

This leads to an equation for k:

$$(\phi_b - \phi_{ch}) = ke^{-\phi_b/(\phi_{ch} - \phi_b)} \Rightarrow \quad k = \frac{(\phi_b - \phi_{ch})}{e^{-\phi_b/(\phi_{ch} - \phi_b)}}. \tag{18.29}$$

The outlet flow coefficient for a given inlet flow coefficient in the extrapolated region is given by

$$\phi_2 = -\frac{1}{m}\text{In}\frac{\phi - \phi_{ch}}{k}. \tag{18.30}$$

The stage performance at low pressure ratios in the extrapolated region can then be obtained by recombining the extrapolated loss and the work characteristics. The

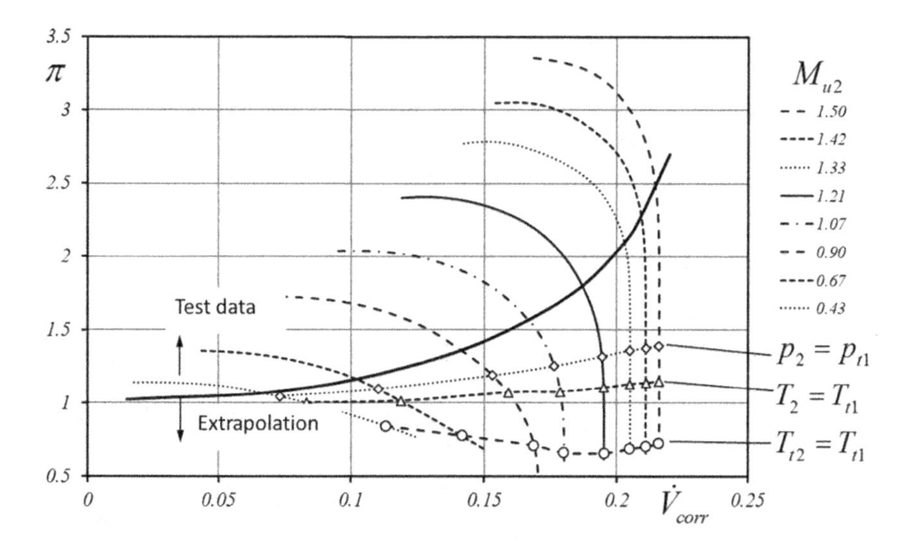

Figure 18.13 A performance map of a stage with a vaneless diffuser extrapolated from test data to low pressure ratios.

efficiency in this region is calculated from the density ratio, and the work is calculated from the equations in Section 18.3.7.

An example of a performance map for a vaneless stage extrapolated using this procedure beyond the last measurement points on each speed line is shown in Figure 18.13. On each speed line, the full line that crosses the speed lines represents the last point on each characteristic where measurements are available. The map above this line shows the region where measurements are available, and at all pressure ratios below this line the characteristics are extrapolated. Figure 18.13 also shows various limits of operation in the extrapolated low pressure ratio region described in more detail by Casey and Schlegel (2010). These are of interest with regard to operation at extreme off-design points.

The first of these limits is the locus of operating points on each characteristic at which the outlet static pressure of the impeller is equal to the inlet total pressure such that the degree of reaction (in terms of pressure) is zero. At this point, there are high losses in the impeller and together with the high kinetic energy at the impeller outlet no static pressure rise occurs in the impeller. Operation below this pressure in a turbocharger can cause oil-blow-by in the compressor as the pressure at impeller outlet is then lower than that of the oil in the engine sump, with the result that oil may be sucked into the impeller. The next limit shown is the locus of points where the impeller outlet static temperature is the same as the inlet total temperature. This indicates that there is no static enthalpy rise in the impeller and that the kinematic degree of reaction is then zero. Along this line, all of the work input into the impeller appears as kinetic energy at the impeller outlet. Just below this locus is the horizontal line where the total-to-total pressure ratio is unity, at which point the work input is completely dissipated by the losses leading to no pressure rise, a total pressure ratio of unity and a total–total stage efficiency of zero. In this case, the extrapolations have been continued below a pressure ratio of unity so that the point on the characteristic is reached where the impeller outlet total temperature becomes the same as the inlet total temperature. Below this line, the work input is negative and the compressor then operates as a rather poor turbine – with radial outward flow and excessive kinetic energy in the outlet.

18.5.2 Extrapolation to Negative Flow

Considering the magnitude of the flow and the pressure rise, the complete compressor map includes four quadrants with positive and negative values of these two variables. Normal operation is in the first quadrant with positive flow direction and a positive pressure rise, and the extrapolation considered in Section 18.5.1 extends the map into the fourth quadrant with forward flow, positive rotational speed and negative pressure rise. Gülich (2008) points out that for centrifugal pumps considering positive and negative values of the rotational speed, the flow rate, the pressure rise and the torque, there are in fact 16 different combinations, whereby eight of them are significant in practice as they can occur in transient or abnormal operating conditions. Little information is available in the technical literature with respect to compressor maps in these different quadrants, but measured characteristics in all four quadrants on a low-speed axial fan are provided by Gill et al. (2007).

An important issue is the extrapolation of a measured performance map into the second quadrant with negative flow and positive pressure rise. Operation in the second quadrant can occur during surge excursions as a consequence of an extreme discharge pressure. Rudimentary knowledge of the map in this region is needed for the sizing of surge protection devices. For modelling such surge events, it is necessary to extrapolate the map into this region. A method that is often used in this region is that of Gravdahl and Egeland (1998), which assumes that each speed line has the form of a cubic equation. Further methods based on 1D velocity triangles with loss modelling and on CFD simulations are discussed by Belardini et al. (2016).

18.6 Map Prediction for Multiple Stages

18.6.1 Basic Equations for Stage Stacking

Prediction of the performance map of a multistage centrifugal compressor is typically carried out using a method known as stage stacking. This relies on knowledge of the individual stage characteristics of the compressor stages in a standard nondimensional form, which are stored as an array of data for interpolation purposes. These characteristics may be obtained from test data, CFD data or correlations, or they can be generated from the system described in Section 18.3. For each speed line (actually tip-speed Mach number line) of the characteristic curve of each stage, there are a series of points representing different values of the inlet flow coefficient, and at each of these points the polytropic efficiency and the work coefficient are defined. The pressure coefficient (or polytropic head coefficient) is defined as the product of the efficiency and work coefficient so that the data can be represented as a matrix of information:

$$\eta_{ijk}, \lambda_{ijk}, \psi_{ijk}, \phi_{t1,ijk}, M_{u2,jk} = f\left(\text{stage}_k, \text{point}_j, \text{speed}_i\right). \tag{18.31}$$

In order to calculate the performance map, additional information with regard to the type of gas equations to be used, the inlet gas conditions and the mass flow of the compressor is needed, together with the impeller diameter and the rotational speed of the shaft for each stage of the compressor.

The fundamental basis of the stage-stacking procedure is that for each stage, the mass flow at inlet is given. From the inlet density, which can be calculated from the inlet pressure and temperature, the volume flow at inlet to the stage can be calculated from the mass flow. This can be used to determine the nondimensional flow coefficient at inlet to the stage, and from this the operating point of the stage at the given speed can be interpolated from the associated nondimensional stage characteristics. The pressure ratio and temperature ratio of the stage can then be determined by means of a thermodynamic calculation based on the nondimensional work coefficient and the efficiency. This thermodynamic calculation allows the outlet pressure and temperature of the stage to be determined so that the density of the gas at the inlet of the next stage is then known. This, in turn, determines the inlet volume flow of the next stage so the procedure can be repeated from stage to stage (or in the usual terminology, the characteristics can be stacked from stage to stage) to determine the overall performance of the machine. The description given here assumes that the calculation is carried out with a real gas having a constant value of the real gas factor Z (which is 1 for an ideal gas) or for a real gas defined as equations, as described in Chapter 3. The basic equations describing the performance of an individual stage are given in this section, whereby they can be applied to any stage of the machine, so that each equation is used separately to define values for each stage of the machine in turn. Some parameters have a different value at both the inlet and outlet location of the stage, whereby the value at the outlet is then usually identical to the value at the inlet of the next stage.

In the calculation of a specific operating point, it is assumed that the values given in Table 18.1 are known at the inlet to a particular stage. The calculating procedure for each stage is then as defined in Table 18.2. Following this procedure, the outlet temperature and outlet pressure of a particular stage are then known, and the procedure can then be repeated for the next stage by equating these to the inlet pressure and temperature of the next stage. This is repeated for all stages to determine the overall performance of the machine. Slight modifications of this procedure are made to allow for bleed flows, side-stream flows, any change in the rotational speed of stages on different shafts and the presence of an outlet cooler with pressure losses.

Table 18.1 Inlet data for the stage-stacking calculation.

Inlet total pressure	$[N/m^2]$	p_1
Inlet total temperature	$[K]$	T_1
Inlet mass flow	$[kg/s]$	$\dot{m}_1$
Rotational speed	$[RPM]$	N
Gas constant	$[J/kgK]$	R
Real gas factor	$[-]$	Z
Isentropic exponent	$[-]$	γ
Impeller outlet diameter	$[m]$	D_2

Table 18.2 Calculation procedure for stage stacking of an individual stage.

Parameter	Unit	Gas with constant Z	Real gas functions
Angular velocity of shaft	$[rad/s]$	$\omega = 2\pi N_k/60$	
Impeller tip-speed	$[m/s]$	$u_2 = \omega r_2 = \omega D_2/2$	
Inlet density	$[kg/m^3]$	$\rho_{t1} = p_{t1}/ZRT_{t1}$	$\rho_{t1} = \rho(p_{t1}, T_{t1})$
Inlet enthalpy	$[J/kg]$		$h_{t1} = h(p_{t1}, T_{t1})$
Inlet entropy	$[J/kgK]$		$s_{t1} = s(p_{t1}, T_{t1})$
Specific heat at constant pressure	$[J/kgK]$	$c_p = ZR\gamma/(\gamma - 1)$	$c_p = c_p(p_{t1}, T_{t1})$
Inlet volume flow	$[m^3/s]$	$\dot{V}_{t1} = \dot{m}/\rho_{t1}$	
Rotor inlet flow coefficient	$[-]$	$\phi_{t1} = \dot{V}_{t1}/u_2 D_2^2$	
Stage inlet speed of sound	$[m/s]$	$a_{t1} = \sqrt{\gamma ZRT_{t1}}$	$a_{t1} = a(p_{t1}, T_{t1})$
Rotor tip-speed Mach number	$[-]$	$M_{u2} = u_2/a_{t1}$	
Stage polytropic efficiency	$[-]$	$\eta_p = \eta_p(\phi_{t1}, M_{u2})$, by interpolation	
Rotor work coefficient	$[-]$	$\lambda = \lambda(\phi_{t1}, M_{u2})$, by interpolation	
Stage total pressure rise coefficient	$[-]$	$\psi = \eta_p \lambda$	
Polytropic head rise	$[J/kg]$	$y_{12} = \psi u_2^2 = \eta_p \lambda u_2^2$	
Total enthalpy rise	$[J/kg]$	$h_{t12} = \lambda u_2^2$	
Outlet enthalpy	$[J/kg]$	$h_{t2} = h_{t1} + h_{t12}$	
Outlet entropy	$[J/kgK]$	$s_2 = \eta_p s_1 + \left(1 - \eta_p\right) s(h_{t2}, p_{t1})$	
Outlet total temperature of stage	$[K]$	$T_{t2} = T_{t1} + h_{t12}/c_p$	$T_{t2} = T(h_{t2}, s_{t2})$
Total temperature ratio	$[-]$	$\tau_t = T_{t2}/T_{t1}$	
Total outlet pressure	$[-]$	$p_{t2} = p_{t1}\tau_t^{\eta_p\gamma/(\gamma-1)}$	$p_{t2} = p(h_{t2}, s_{t2})$
Total pressure ratio	$[N/m^2]$	$\pi_t = p_{t2}/p_{t1}$	
Power required	$[W]$	$P = \dot{m} h_{t12}$	

18.6.2 Interpolation in the Stage Characteristics

During the stage-stacking calculations, a task that often repeats itself is the calculation of the work coefficient λ and polytropic efficiency η_p for a given value of the inlet flow coefficient ϕ_{t1} tip-speed Mach number M_{u2} as described in functional form in (18.3). This calculation is solved by means of an interpolation in the stage characteristics. Some methods of carrying out this interpolation may cause difficulties because the curves become nearly vertical on the choke line.

To aid this interpolation, a system known as β lines is often used in turbocharger maps, as described by Baines (2005). The β lines are a fan of straight lines with a value from 0 to 1 which radiate from the position (0,1) on a pressure ratio versus corrected flow map. The values of the parameters at the point where the speed lines are crossed by the β lines allow sensible interpolation, and even extrapolation, against the value of β in the map. Other systems are possible. A simple system developed by the authors is described as follows.

Although the interpolation is really a two-dimensional interpolation, it can be carried out as two successive one-dimensional interpolations. First with the help of a parameter that varies along the length of each characteristic from the choke point to the surge point, a series of new points are defined equidistant along each speed line. A suitable smooth curve fitting spline function for this is the Akima spline, following Akima (1970). The Akima spline is built from piecewise third-order polynomials in which only the neighbouring points are used to determine the local coefficients of the interpolation polynomial.

The first interpolation defines a new virtual characteristic for the required tip-speed Mach number from the defining points at other Mach numbers and then the second interpolation is completed by interpolating for the efficiency and work coefficient at a certain flow coefficient in the defining points of this virtual characteristic. This procedure has the advantage that each separate interpolation is easier to control, and wild wiggles in the characteristics are less likely to occur. Provided that the defining points of the characteristic are reasonable uniformly spaced along the characteristic curve, very good experience has been made with this procedure. It has the slight disadvantage that each speed characteristic first needs to be defined with the same number of points.

18.6.3 Determination of Surge, Choke and Peak Efficiency Points

The procedure outlined in the previous section can be used to define the performance for a specific value of inlet mass flow associated with given speed, inlet gas conditions and gas properties. During this process, the local value of the flow coefficient of any stage is used to calculate a surge margin (*SM*) and a choke margin (*CM*) for each stage, defined as follows:

$$\begin{aligned} SM &= \phi_1/\phi_{1,j=1} \\ CM &= \phi_1/\phi_{1,j=n_p}. \end{aligned} \tag{18.32}$$

These values are the relative values of the flow capacity compared to that of the first point on the characteristic and that on the last point of the characteristic. If at the end of this calculation any particular stage is found to have an *SM* which is less than 1, then the operating point is on the unstable side of the surge line.

If any stage has a value of CM that is greater than 1, then this point lies to the choke side of the last point on the characteristic. The actual surge and choke points are then determined by searching for the mass flow in which $SM = 1$ and $CM = 1$ for any stage within the machine. A similar iteration can be used to identify the location of the machine peak efficiency point relative to the peak efficiency of the individual characteristics.

18.7 Matching of the Diffuser with the Impeller

18.7.1 The Effect of Impeller and Diffuser Matching on Choke

The equation for the matching of a vaned diffuser with an impeller in terms of matching the diffuser and impeller throat areas is given in Chapter 12. The final equation describing the relative areas of the throats if both attain choking at the same time is derived to be

$$\frac{A^*_{diff}}{A^*_{imp}} = \frac{\left[1 + \{(\gamma - 1)/2\}\{D_1/D_2\}^2 M^2_{u2}\right]^{\frac{(\gamma+1)}{2(\gamma-1)}}}{\left[1 + (\gamma - 1)\lambda M^2_{u2}\right]^{\frac{(n+1)}{2(n-1)}}}. \tag{18.33}$$

For a situation where the design is completed and the throat areas are known, this equation can be used to determine the tip-speed Mach number at which both components choke simultaneously at the same mass flow. In what follows, this value is denoted as the nominal design tip-speed Mach number, M_{u2d}, and this characterises how the matching controls the shape of the characteristics, which is sketched in Figure 18.14. On the speed lines below the nominal design tip-speed Mach number the diffuser will choke before the impeller so that the impeller cannot go into choke. On the speed line at the nominal design tip-speed Mach number, and on those above this, the impeller will determine choke.

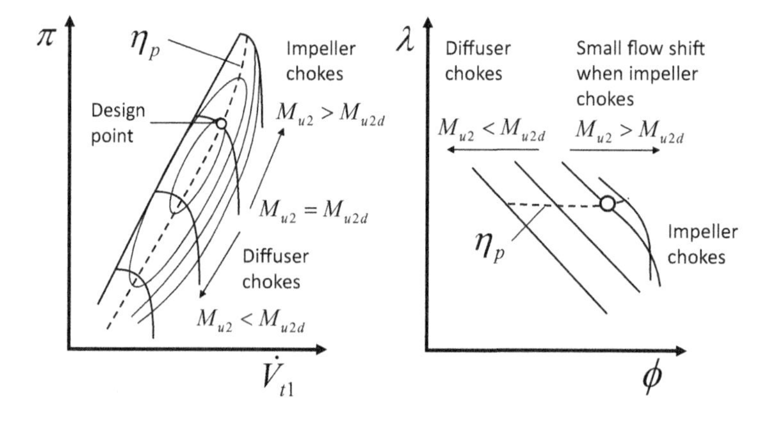

Figure 18.14 The nominal design Mach number affects the map and the work coefficient characteristic of a vaned compressor stage.

At low speeds, the diffuser chokes before the impeller, and the diffuser then acts as a choked nozzle limiting the mass flow and causing the mass flow to reduce strongly as the tip-speed Mach number reduces, as shown in Figure 18.9. The impeller cannot reach its choking mass flow at low speeds so that the work coefficient versus flow coefficient characteristic remains a nearly straight line. At high flow rates on the speed line at the nominal design tip-speed Mach number. the impeller begins to choke as well as the diffuser. The work coefficient characteristic begins to fall sharply at high flows, and in test data this can be used to identify the speed at which the nominal design tip-speed Mach number is attained. A further increase in the tip-speed Mach number does not cause a large increase in the mass-flow function because, as shown in Chapter 11, the choking inlet flow coefficient of the impeller remains nearly constant as the tip-speed Mach number increases. Nevertheless, when the impeller chokes, the increase in impeller losses causes an increase in volume flow at the impeller outlet such that the diffuser also chokes, leading to a very steep map at high flows. These effects combine to cause the peak efficiency point of the work coefficient characteristic to be approximately at a similar work coefficient for all flows.

18.7.2 The Effect of Impeller and Diffuser Matching on Surge

An interesting diagram showing the different regimes of stall and surge for the impeller and the vaned diffuser in a centrifugal compressor stage was developed by Yoshinaka (1977), and an extension to this is described in Section 17.2.12.

18.8 Matching in Multistage Compressors

18.8.1 Idealised Example of Multistage Compressor Matching

The performance of multistage centrifugal compressors is determined by the cumulative performance of the individual stages. Figure 18.15 provides an illustration of some of the interesting interactions that can occur between the performance of the whole machine and the characteristics of its stages. In this somewhat idealised example, taken from Casey and Marty (1985), the performance of a five-stage process compressor using impellers of the same diameter and of the same family has been calculated. The effect of the Mach number on the stage characteristic has, for convenience, been neglected.

Figure 18.15 shows the performance of the machine and the operating points of the individual stages for different inlet volume flows and rotational speeds. Each column of this figure represents a certain rotational speed and each row contains the following diagrams:

- Row A: pressure–volume (π - V) map of the machine
- Row B: operating points of the individual stages on their efficiency versus flow coefficient characteristic curves
- Row C: operating points of the stages on their pressure coefficient versus flow coefficient characteristics.

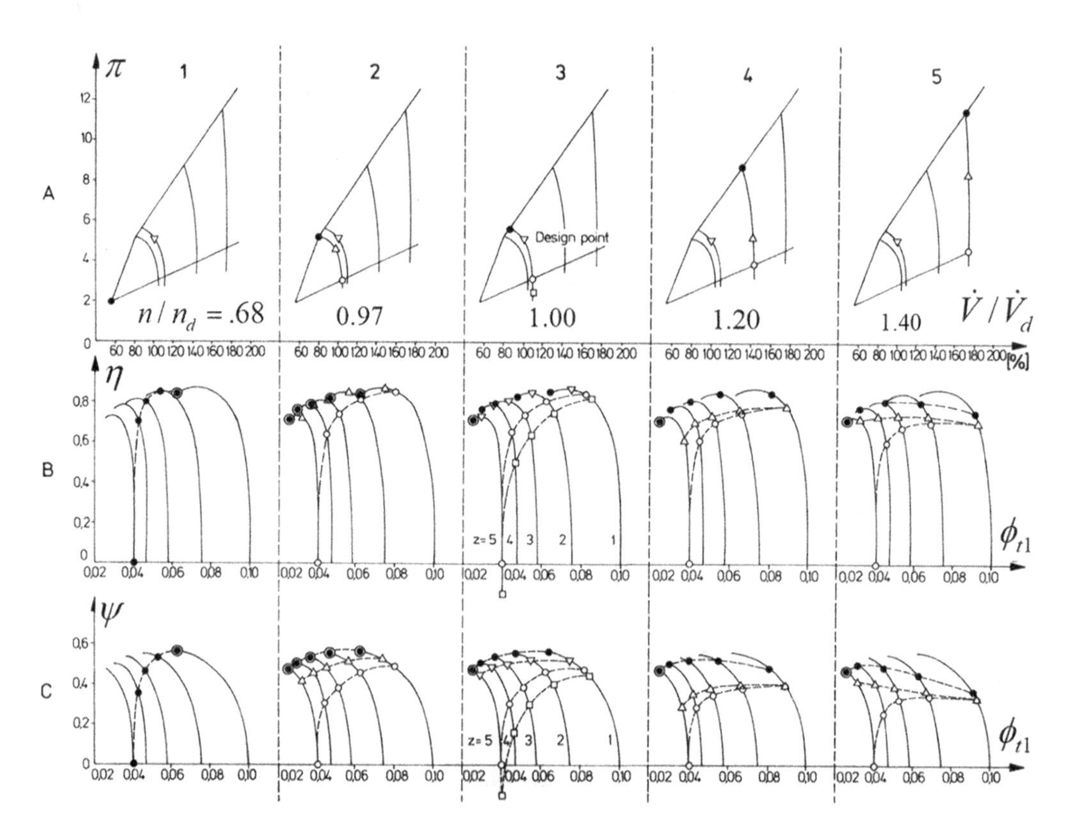

Figure 18.15 Matching of the stages in a five-stage compressor at different speeds. (image courtesy of the Institute of Refrigeration, www.ior.org.u)

The diagrams in column 3 show the operating characteristics of the compressor stages at its design speed, $n/n_d = 1.00$. At the design point (▽), all of the stages operate at their best efficiency point – the stage efficiency and the pressure coefficient reduce towards the end stages because here the flow coefficients are low and the impellers are narrow. If the inlet volume flow is reduced below the design point, then the first stage moves to the left along its characteristic and produces more pressure rise than at design. As a consequence, the second stage receives proportionately less volume than at its design point and lies farther to the left on its own stage characteristic. This behaviour continues through all the stages. When the first stage has a certain minimum inlet volume, the last stage lies so far to the left of its characteristic that it reaches its limit of stability, and the machine can go into surge (•). With an increase in inlet volume, the first stage moves to the right along its characteristic. Each successive stage receives progressively more volume flow than at its design point as the pressure ratio of the upstream stage decreases. At a certain maximum inlet volume flow, the last stage produces no pressure rise at all and operates at its choke point (○). Any further increase in inlet volume flow is impossible, although the machine can still operate at lower pressure ratios since the last stage can move into a region of expansion with an efficiency of less than zero (□).

Columns 4 and 5 show the operation of the machine at speeds above its design speed or with a gas which has a lower speed of sound. With a moderate increase in tip-speed Mach number to $n/n_d = 1.20$, column 4 shows the stages each produce more pressure ratio than at the design speed. A relatively small change in volume flow in the first stage is amplified by succeeding stages. The first stage can now only move a short way along its characteristic while the last stage moves all the way from surge to choke. The machine characteristic becomes steeper than at the design speed. Column 5 shows that the machine characteristic would become even steeper at a speed of $n/n_d = 1.40$ of design. This is really only of academic interest, however, as the influence of the Mach number on the characteristic curves of the stages has not been taken into account, and the mechanical limit on the tip speed is likely to have been exceeded!

Columns 1 and 2 show the performance at speeds less than design or with a lighter gas. Column 2 illustrates that at $n/n_d = 0.97$ of design speed, all of the stages of this machine operate near their stability point when the machine surges. Below this speed, the first stage becomes responsible for the surge of the machine, giving rise to the kink in the surge line on the machine characteristic. At much lower speeds, $n/n_d = 0.68$, column 1 shows that a configuration occurs where the first stage operates at its stability point while the last stage is in choke so that there is no operating range at all.

This very simplified example highlights some of the important features of the design of multistage compressors. It shows clearly that exact stage characteristics for the individual stages are needed in order to avoid an accumulation of errors in the design. Some of the complications of real machines that have not been included are the following:

- Each stage is usually of a different type with different characteristic curves and perhaps different work coefficients.
- The effect of the Mach number is not negligible. and the rear stages would have wider characteristics than the first stages because of the lower Mach numbers.
- The best efficiency points of each stage and the design point of the machine do not necessarily coincide, for example to increase the low-speed surge margin the peak efficiency of the rear stage may be placed closer to surge at the design point.

In practice, stage-stacking methods are usually combined both with interpolation procedures for the characteristics of the individual stages, with correction techniques for the change in performance with variable inlet guide vanes and Reynolds number and with the appropriate real gas equations for the gas properties. This means that sophisticated software becomes essential. An example of the different matching of the stages in a multistage compressor based on experimental analysis is given by Falomi et al. (2016).

18.9 Matching of the Compressor to a Turbine in a Gas Turbine

18.9.1 Efficiency of a Gas Turbine

An important application of centrifugal compressors is in small single-shaft gas turbines and microturbines. After specifying the design conditions and the gas turbine

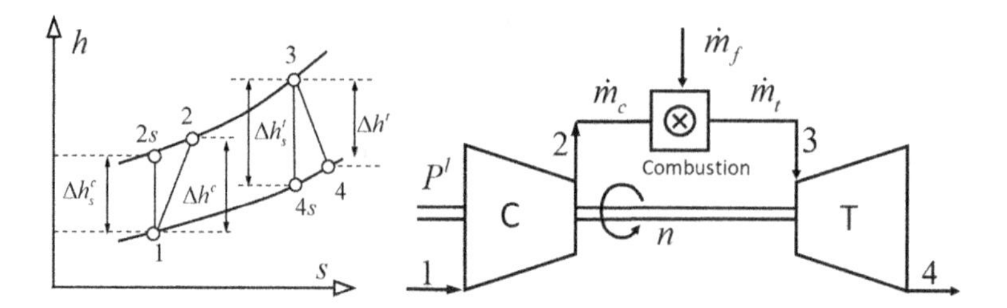

Figure 18.16 Simple cycle gas turbine cyle without recuperation.

cycle, the components can then be designed to deliver the required power at the normal running speed. The importance of the compressor efficiency in the overall efficiency of a gas turbine is well indicated from a thermodynamic analysis of the simple open cycle gas turbine process as shown in Figure 18.16.

The isentropic efficiencies of the components and the pressure ratio of the compressor and the expansion ratio of the turbine are given by

$$\begin{aligned} \pi^c &= p_2/p_1 > 1, \qquad \eta^c = \Delta h_s^c/\Delta h^c \\ \varepsilon^t &= p_3/p_4 > 1, \qquad \eta^t = \Delta h^t/\Delta h_s^t, \end{aligned} \tag{18.34}$$

and the definitions of the enthalpy changes are shown in the h-s diagrams of Figure 18.16. The compressor pressure ratio is the outlet pressure divided by the inlet pressure and the turbine expansion ratio is the inlet pressure divided by the outlet pressure. In this simple example, these are considered to be the same with no pressure losses in the combustor.

The specific work required by the compressor is given by

$$w_{t12} = h_2 - h_1 = c_p(T_2 - T_1) = c_p(T_{2s} - T_1)/\eta^c, \tag{18.35}$$

where η^c is the isentropic efficiency of the compressor. The combustion gases drive the turbine, and expansion through the turbine produces the work needed to drive the compressor as

$$|w_{t34}| = h_3 - h_4 = c_p(T_3 - T_4) = c_p(T_3 - T_{4s})\eta^t, \tag{18.36}$$

where η^t is the isentropic efficiency of the turbine. Assuming that the compressor outlet pressure and the turbine inlet pressure are the same, that there are no losses in the inlet and outlet ducts and that the mass flow of fuel is negligible, then the overall gas turbine efficiency is given by

$$\eta_{GT} = \frac{|w_{t34}| - w_{t12}}{q_{23}} = \frac{[\eta^c\eta^t(T_3/T_1)(1/\pi^m) - 1](\pi^m - 1)}{\eta^c(T_3/T_1 - 1) - (\pi^m - 1)}, \tag{18.37}$$

where $m = (\gamma - 1)/\gamma$ is the isentropic pressure exponent. The useful power produced by the gas turbine at the coupling is given by

$$\frac{w_{GT}}{c_p T_1} = \frac{1}{\eta_V}[\eta^c\eta^t(T_3/T_1)(1/\pi^m) - 1](\pi^m - 1). \tag{18.38}$$

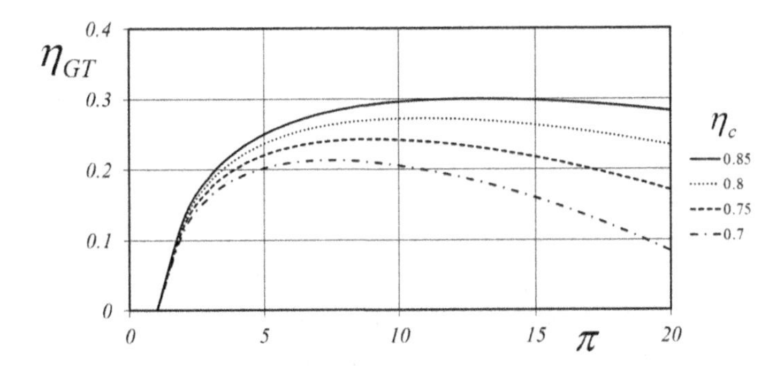

Figure 18.17 Efficiency of a simple cycle gas turbine as a function of pressure ratio for different compressor efficiencies, a turbine efficiency of 0.85 and a turbine entry temperature of 1200 K.

The variation of the gas turbine efficiency with the compressor efficiency for the simple cycle is shown in Figure 18.17, whereby the data are adjusted to be typical of the combustion temperature (1200 K) and the efficiencies of microturbines. With the conditions of a microturbine, the optimum gas turbine efficiency is given at relatively low pressure ratios, whereby differentiation of (18.37) shows that the optimum pressure ratio in a simple cycle is given by

$$\pi_{\eta,opt} = \frac{\eta^c\eta^t(T_3/T_1) - \sqrt{(T_3/T_1)\eta^c\eta^t B(1+\eta^c B - T_3/T_1\eta^c\eta^t)}}{\eta^c\eta^t(T_3/T_1) - \eta^c B},$$

$$B = (T_3/T_1 - 1). \tag{18.39}$$

The work required for compression is a large proportion of the turbine output power, such that a small improvement in compressor efficiency leads to a much larger change in the gas turbine efficiency and a higher power output. Similar sensitivity arises for a helicopter gas turbine engine. In order to improve the efficiency of microturbines above this level, a recuperated heat exchange cycle is generally used as this has a higher efficiency and has its peak efficiency at lower pressure ratios suitable for single-stage radial compressors.

18.9.2 Off-Design Performance

An important issue is the matching to the turbine when running at different speeds with different combustion temperatures. The basic approach to such calculations is described in some detail by Saravanamuttoo et al. (2009) and by Cumpsty and Heyes (2018). One example is described briefly here as an introduction to Section 18.10, which uses a similar approach to describe the matching of a turbine and a compressor to an internal combustion engine.

The simple gas turbine cycle is shown in Figure 18.16. As the turbine and the compressor are on a single shaft, there are several conditions which must be met. First, it is clear that the turbine and compressor operate at the same speeds on the same shaft:

$$n^c = n^t = n. \tag{18.40}$$

Second. the turbine has to deliver the power to drive the compressor and the attached load, which may be a generator, so that

$$P^c + P^l = P^t. \tag{18.41}$$

Third, the mass flow through the different components must satisfy the continuity equation, so that

$$\dot{m}^c + \dot{m}^f = \dot{m}^t. \tag{18.42}$$

In simple calculations, it is assumed that the mass flow of the fuel, $\dot{m}^f$, is negligible and the turbine and compressor mass flows are the same. The final matching condition is that the gas turbine cycle must be physically possible with the same inlet and outlet pressures:

$$p_1 = p_4. \tag{18.43}$$

Following these conditions, a further approximation may be made: that the turbine operates above the critical pressure ratio with a choked inlet nozzle such that the nondimensional mass-flow function is known; see Chapters 6 and 11. Assuming that the gas is an ideal gas with the properties of air, the mass flow then satisfies the following condition:

$$\frac{\dot{m}^t}{A_3 \rho_{t3} a_{t3}} = \left(\frac{2}{\gamma+1}\right)^{\frac{(\gamma+1)}{2(\gamma-1)}} = 0.5787. \tag{18.44}$$

On rearrangement, and assuming there is no pressure loss in the combustor, this equation becomes

$$\frac{\dot{m}\sqrt{RT_{t3}}}{p_{t3}A_3} = const. \approx 0.685, \quad \frac{\dot{m}\sqrt{RT_{t1}}}{p_{t1}A_1}\frac{p_{t1}}{p_{t2}}\frac{p_{t2}}{p_{t3}}\frac{A_1}{A_3}\sqrt{\frac{T_{t3}}{T_{t1}}} = K$$
$$\frac{p_{t2}}{p_{t1}} = K\sqrt{\frac{T_{t3}}{T_{t1}}}\frac{\dot{m}\sqrt{RT_{t1}}}{p_{t1}A_1}. \tag{18.45}$$

In this way, the condition of mass continuity allows the equilibrium operating point of the turbine to be plotted into the compressor characteristic, as shown in Figure 18.18.

Equation (18.45) shows that with the approximations discussed earlier, the running line of the turbine is a straight line in the compressor map when this is plotted as the pressure ratio versus the compressor nondimensional mass-flow function. At different firing temperatures for a given speed, more power becomes available for the output load. For a generator application, the gas turbine is allowed to run up in speed until the required generator speed is reached, and then the load is increased through increased firing temperature. It is important that the running line does not cross the surge line of the compressor.

18.10 Matching Issues of a Compressor in a Turbocharger

18.10.1 Turbocharger Efficiency

The most common application of centrifugal compressors is in a turbocharger coupled with a radial turbine. Downsizing of combustion engines by turbocharging

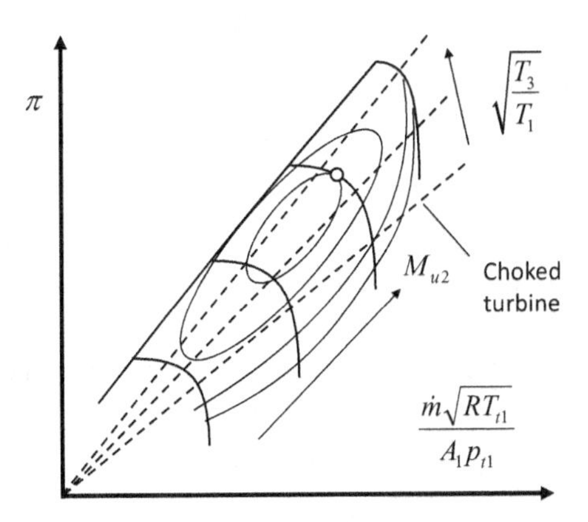

Figure 18.18 Equilibrium running lines of a choked turbine in the compressor map.

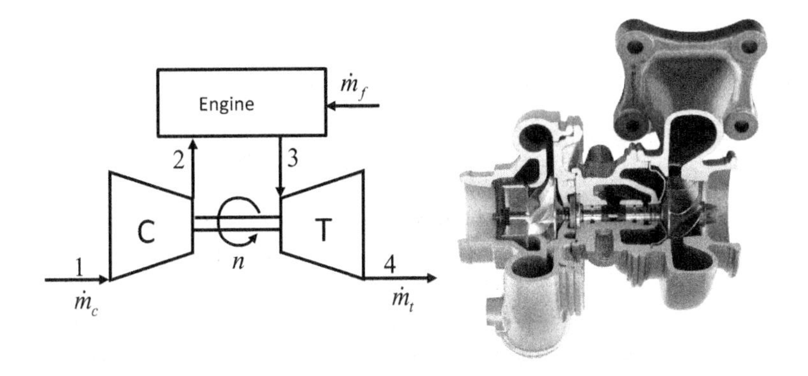

Figure 18.19 Typical turbocharger arrangement and a sectioned view of a turbocharger for a commercial vehicle. (image courtesy of BorgWarner Turbo Systems Engineering GmbH)

is an effective way to reduce CO_2 emissions in automotive engines, and the matching of the turbocharger components with the engine is a key aspect to the optimisation of the process. A similar approach to that described for a gas turbine in Section 18.9, involving a mass balance and a power balance, is used here to highlight the matching issues of a compressor to a turbine in a turbocharger. The notation used is that the components are denoted by superscripts with compressor, turbine and engine denoted by superscript c, t and e. The inlet and outlet ducts are denoted by i and o, mechanical losses by m and the whole turbocharger by TC. Gas conditions (only total quantities) at different locations are denoted by subscripts with the inlet and outlet of compressor denoted by subscripts 1 and 2 and the inlet and outlet of turbine denoted by 3 and 4. The basic turbocharger cycle with the engine is shown in Figure 18.19.

The isentropic efficiencies of the components and the pressure ratio of the compressor and the expansion ratio of the turbine are given by

$$\begin{aligned} \pi^c &= p_2/p_1 > 1, \qquad \eta^c = \Delta h_s^c/\Delta h^c \\ \varepsilon^t &= p_3/p_4 > 1, \qquad \eta^t = \Delta h^t/\Delta h_s^t \end{aligned} \tag{18.46}$$

and the definitions of the enthalpy changes are as shown in the *h-s* diagrams of Figure 18.16.

From this information, it is possible to define the turbocharger efficiency, as given by Watson and Janota (1982). The minimum possible specific energy for isentropic compression over the compressor from total states 1 to 2*s* is

$$\Delta h_s^c = h_{2s} - h_1 \tag{18.47}$$

and the maximum possible specific energy that can be extracted from isentropic expansion over turbine from total states 3 to 4*s* is

$$\Delta h_s^t = -(h_{4s} - h_3). \tag{18.48}$$

If it is assumed that the mass flow in compressor is same as that in turbine, then the efficiency of the turbocharger can be written as

$$\eta_{TC} = \frac{\Delta h_s^c}{\left|\Delta h_s^t\right|} = \frac{\Delta h_s^c}{\Delta h^c}\frac{\Delta h^c}{\left|\Delta h^t\right|}\frac{\left|\Delta h^t\right|}{\left|\Delta h_s^t\right|} = \eta^c\eta^m\eta^t. \tag{18.49}$$

The turbocharger efficiency is the product of the compressor isentropic efficiency, the turbine isentropic efficiency and the mechanical efficiency (representing the bearing losses). For convenience, the turbine efficiency definition often includes the mechanical losses. The turbocharger efficiency is directly proportional to the compressor efficiency.

The power produced by the turbine drives the compressor and overcomes the mechanical bearing losses at constant speed, so that the power balance gives rise to the following equation:

$$\begin{aligned} P^c &= \eta^m\left|P^t\right|, \quad \dot{m}^c\Delta h_t^c = \eta^m\dot{m}^t\left|\Delta h_t^t\right| \\ \dot{m}^c\lambda^c(\omega r^c)^2 &= \eta^m\dot{m}^t\left|\lambda^t\right|(\omega r^t)^2 \end{aligned} \tag{18.50}$$

and from this a simple equation for the relative size of the turbocharger components without bypass is obtained as

$$(r^c/r^t)^2 = \eta^m\dot{m}^t\left|\lambda^t\right|/\dot{m}^c\lambda^c. \tag{18.51}$$

With typical design values for a radial compressor and for a radial turbine,

$$\dot{m}^t \approx 1.04\cdot\dot{m}^c, \quad \left|\lambda^t\right| \approx 0.85, \quad \lambda^c \approx 0.65, \quad \eta^m \approx 0.97, \quad r^c/r^t \approx 1.15. \tag{18.52}$$

This shows that because the work coefficient of a radial turbine is typically larger than that of the compressor, then the compressor outer diameter needs to be larger than the radial turbine to match the power of the two components at the design point, as can be seen in Figure 18.19.

18.10.2 The Compressor Boost Pressure Ratio

The complete matching conditions for the compressor and the turbine in a turbocharger are that the rotational speed of compressor and turbine are the same and different from that of the engine, with the superscript e:

$$n^c = n^t = n = \omega/2\pi, \quad n^e \neq n. \tag{18.53}$$

The turbine power drives the compressor and overcomes any mechanical bearing losses. Any imbalance causes a change in speed until equilibrium is reached:

$$\eta^m |P^t| = P^c + (I^c + I^t)\frac{d\omega}{dt}. \tag{18.54}$$

The continuity equation has to be satisfied, taking into account the mass of fuel (f) and any bypass flows (b):

$$\dot{m}^c + \dot{m}^f = \dot{m}^t + \dot{m}^b. \tag{18.55}$$

The whole process has to be physically possible such that the pressure at the exhaust of the turbine is the same as at inlet (ambient), taking into account inlet (i) and outlet (o) pressure losses and pressure ratio across the engine:

$$p_{ambient} = p_1^c + \Delta p^i = p_4^t - \Delta p^o. \tag{18.56}$$

Furthermore, each component operates at its own nondimensional point in its performance map (pressure or expansion ratio versus corrected mass flow), and at the same time the engine determines the required engine load and speed.

The power balance can be expressed in terms of the component efficiencies at steady speed and leads to (18.40) as given previously. The isentropic relations can be used to derive the temperature ratios across the turbine and the compressor as

$$\begin{aligned} T_{2s}/T_1 &= (p_{2s}/p_1)^{(\gamma^c-1)/\gamma^c} = (p_2/p_1)^{(\gamma^c-1)/\gamma^c} = (\pi^c)^{(\gamma^c-1)/\gamma^c} \\ T_{4s}/T_3 &= (p_{4s}/p_3)^{(\gamma^t-1)/\gamma^t} = (p_4/p_3)^{(\gamma^t-1)/\gamma^t} = (1/\varepsilon^t)^{(\gamma^t-1)/\gamma^t}. \end{aligned} \tag{18.57}$$

This can be used to define the isentropic enthalpy change across the compressor and the turbine as follows:

$$\begin{aligned} \Delta h_s^c &= c_p^c(T_{2s} - T_1) = c_p^c T_1 (T_{2s}/T_1 - 1) = c_p^c T_1 \left[(\pi^c)^{(\gamma_c-1)/\gamma_c} - 1\right] \\ \Delta h_s^t &= c_p^t(T_3 - T_{4s}) = c_p^t T_3 (1 - T_{4s}/T_3) = c_p^t T_3 \left[1 - (1/\varepsilon^t)^{(\gamma_t-1)/\gamma_t}\right]. \end{aligned} \tag{18.58}$$

This allows the power balance to be written in terms of pressure ratios

$$\dot{m}^c c_p^c T_1 \left[(\pi^c)^{(\gamma^c-1)/\gamma^c} - 1\right] = \eta_{TC} \dot{m}^t c_p^t T_3 \left[1 - (1/\varepsilon^t)^{(\gamma^t-1)/\gamma^t}\right], \tag{18.59}$$

and the turbocharger efficiency can then be defined in terms of pressure and expansion ratios of the components as

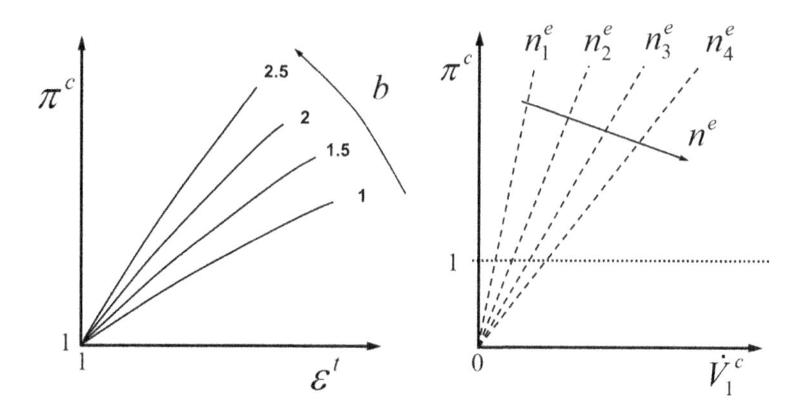

Figure 18.20 Relationship between compressor pressure ratio and turbine expansion ratio (left) and the engine volume flow characteristic in compressor map coordinates (right).

$$\eta_{TC} = \frac{\dot{m}^c}{\dot{m}^t}\frac{c_p^c}{c_p^t}\frac{T_1}{T_3}\frac{\left[(\pi^c)^{(\gamma^c-1)/\gamma^c} - 1\right]}{\left[1-(1/\varepsilon^t)^{(\gamma^t-1)/\gamma^t}\right]}. \tag{18.60}$$

This approach is equivalent to a control volume analysis and can be used on any turbocharger system. It allows the turbocharger efficiency to be determined from measured data of the temperatures and pressures obtained on test rigs at the inlet and the outlet of the turbocharger.

If the turbocharger efficiency is known at a particular operating point, it allows the boost pressure ratio to be linked to the turbine pressure ratio as follows:

$$\pi^c = \frac{p_2}{p_1} = \left(1+b\left[1-(1/\varepsilon^t)^{(\gamma^t-1)/\gamma^t}\right]\right)^{\gamma^c/(\gamma^c-1)}, \quad b = \eta_{TC}\frac{\dot{m}^t}{\dot{m}^c}\frac{c_p^t}{c_p^c}\frac{T_3}{T_1}. \tag{18.61}$$

This equation is sometimes known as the main turbocharger equation, and b as the dimensionless boost parameter. The important consequence of this power balance is that it shows how the compressor boost pressure ratio is determined, as shown in Figure 18.20. A higher boost pressure ratio is attained if the overall turbocharger efficiency is high, requiring high component efficiencies and a high mechanical efficiency of the shaft with low bearing losses. A higher turbine expansion ratio increases the boost, as does the ratio of the turbine inlet temperature to the compressor inlet air temperature.

18.10.3 The Engine Operation in the Compressor Map

The engine is a positive displacement device. The mass-flow rate of the air entering the inlet manifold is proportional to the product of the engine speed, swept volume, inlet manifold density and the volumetric efficiency, giving

$$\dot{m}_2 = V_{sw}\eta_{vol}n^e\rho_2/a, \tag{18.62}$$

where $a = 1$ for a two-stroke engine and $a = 2$ for a four-stroke engine. The volume flow rate of the air at the compressor intake is given by

$$\dot{V}_1 = \frac{\dot{m}}{\rho_1} = \frac{\rho_2}{\rho_1}\frac{V_{sw}\eta_{vol}n^e}{a} = \frac{p_2 T_1}{p_1 T_2}\frac{V_{sw}\eta_{vol}n^e}{a}. \tag{18.63}$$

This allows the swallowing capacity of the engine to be plotted in the compressor map, showing that there is a linear relationship between the compressor boost pressure ratio and the engine volume flow rate at the compressor inlet:

$$\pi^c = \frac{p_2}{p_1} \approx \dot{V}_1 \frac{T_2}{T_1}\frac{a}{V_{sw}\eta_{vol}n^e}. \tag{18.64}$$

The engine swallowing characteristic can then be plotted into the compressor map as a series of straight lines originating at the origin (0,0), as shown in Figure 18.20.

The consequences of these relationships are that there is a good match of the engine characteristic with the compressor map as there is a nearly linear relationship with the boost pressure ratio. The operating range of the compressor limits the engine speed range with a risk of surge at low engine speeds and a risk of choke at high engine speeds. A compressor map with a wide range is extremely important for gasoline engines with wide speed range.

For high overall efficiency of the turbocharger, matching between the compressor wheel and the turbine wheel is very important. The specification of an operating point in the compressor map results in a specific turbocharger speed. The turbine must be adjusted in such a way that it works within this operating range with the greatest possible efficiency. There are three typical ways of doing this: a conventional turbocharger system, a system with a turbine bypass valve and a system with a turbine with variable inlet guide vanes. An interesting pictorial presentation of the matching between the compressor and turbine based on the main turbocharger equation, (18.61), was presented by Engels (1990) and has subsequently been extended by Casey (2006). In this presentation, the matching of the compressor with the turbine is described in terms of diagrams of the individual performance maps of the compressor and the turbine and the physical relationship between the compressor pressure ratio and the turbine pressure ratio depending on the value of the boost parameter b in (18.61).

18.10.4 A Conventional Turbocharger with Fixed Turbine Geometry

Consider first a conventional system with fixed turbine geometry (with no bypass valve or inlet guide vanes) in which the engine and turbine are well matched at a medium speed condition indicated by the symbol ⊙, as shown in Figure 18.21. The engine operates at a certain speed and mass flow, requires a certain boost pressure and delivers gas at a certain exhaust temperature to the turbine. The compressor is selected in the centre of its map with a high efficiency at the required boost pressure and design mass flow and a map with operating range giving margin for surge, choke and overspeed. The main turbocharger equation determines the turbine expansion ratio from the compressor pressure ratio, given the engine exhaust temperature, which is

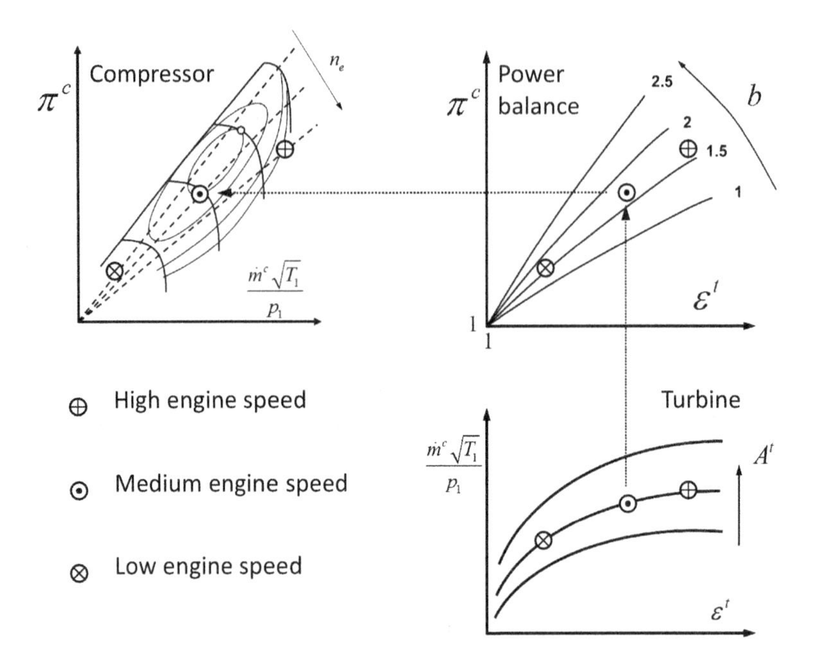

Figure 18.21 Matching of the compressor and turbine in a conventional turbocharger.

included in the boost parameter b. The value of b also depends on the turbocharger efficiency, so iteration may be needed. The turbine is then selected. Its inlet pressure is determined from outlet pressure of the engine and the expansion ratio from the ambient pressure. The corrected mass-flow parameter of the turbine is determined from engine data and inlet pressure. This allows a turbine to be selected that gives the required expansion ratio and corrected mass flow, as follows:

$$\dot{m}^t_{corr} = \frac{\dot{m}^t\sqrt{T_3}}{p_3} = \frac{p_4}{p_3}\frac{\dot{m}^t\sqrt{T_3}}{p_4} = \frac{1}{\varepsilon^t}\frac{\dot{m}^t\sqrt{T_3}}{p_4}. \tag{18.65}$$

At the midspeed operating point where the components have been selected, they are well matched. The corrected mass-flow parameter of the turbine varies on a unique characteristic with its expansion ratio according to (18.65), and is shown in the bottom-right of Figure 18.21. The compressor pressure ratio is related to the turbine expansion ratio via the power balance curve, shown in the top-right. The boost parameter b is proportional to the turbocharger efficiency and to the engine exhaust temperature. Different values are relevant for different engine conditions. This will be high at high engine load.

At operating conditions away from the design condition, the mass flow of the engine is proportional to the engine speed and the inlet air density. At low engine speeds denoted with the symbol (⊗), there is a low corrected flow in the turbine, the turbine power is low and the boost pressure ratio of the compressor is very low. At high engine speeds denoted with the symbol (⊕), there is a high flow in the turbine. The corrected mass flow of turbine is limited by choke so that the turbine inlet pressure must rise to pass the required corrected mass flow, and the expansion ratio increases. There is a high turbine power available giving a very high boost pressure ratio in the compressor.

In a conventional turbocharger with a fixed geometry, the turbine flow capacity is too small at the high speed. The higher flow necessitates a higher turbine inlet pressure, and this results in a high compressor boost pressure. During operation at low engine speed with fixed turbine geometry, the flow capacity of the turbine is too high and the turbine inlet pressure and power will be low, resulting in a low compressor boost pressure. This may lead to engine detonation problems in a gasoline engine or to high structural loads in a diesel engine.

18.10.5 A Turbocharger with a Wastegate (Turbine Bypass Valve)

The solutions to these issues are to limit the boost by control of turbine either with a bypass valve, known as a wastegate, or with a variable geometry turbine. The wastegate simply matches the turbine with the compressor at low speed and dumps the excess mass flow at high speed. This is shown schematically in Figure 18.22. The turbocharger is matched to the engine to provide good boost at low engine speed, leading to a small turbine sized for low-speed requirements. The power of the turbine at high engine speed is controlled by reducing the mass flow through the turbine in that the wastegate opens to bypass a fraction of the exhaust gas around the turbine. This gives a constant boost pressure and roughly constant torque characteristic at high speed. Much of the engine exhaust flow expands without doing work in the turbine, so there is a reduction in system efficiency.

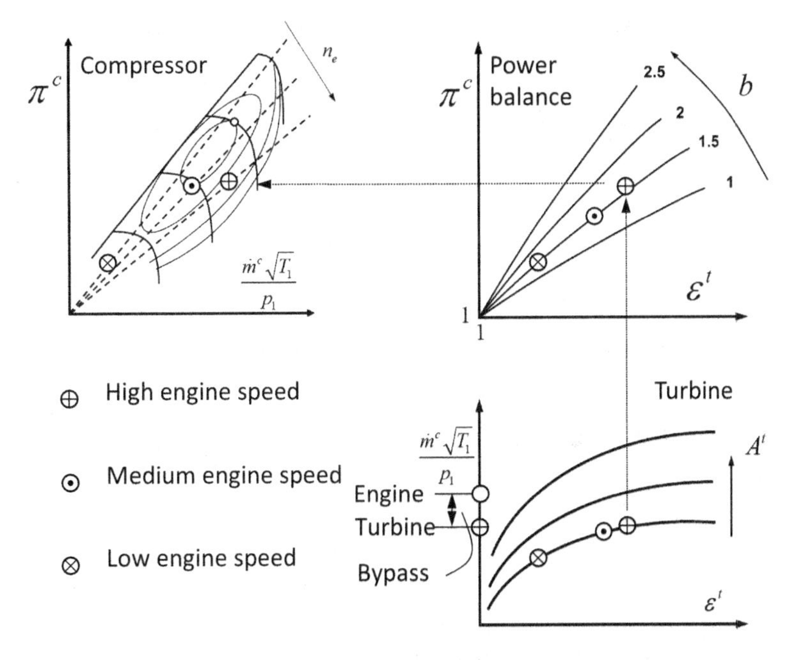

Figure 18.22 Matching of the compressor and turbine in a turbocharger with a wastegate.

18.10.6 A Turbocharger with a Variable Geometry Turbine

The use of a variable geometry turbine with variable inlet guide vanes allows the matching to be adjusted with speed, as is shown schematically in Figure 18.23. The turbine is sized with a maximum nozzle area, which allows it to pass full engine flow at maximum speed with correct boost. At low engine speed, turbine capacity is reduced by decreasing turbine throat area so that a high boost pressure can be obtained at low speeds. The turbine nozzle area can be changed by using variable guide vanes or by the movement of a sidewall. This limits boost at high engine speed and provides high boost at low engine speed, where the compressor surge margin needs to be high to avoid surge.

Matching with a turbine with variable geometry allows variable flow capacity without the wasteful expansion of bypass gas, which can be eliminated, although a gasoline engine may still need a wastegate at high engine speed. The turbine pressure ratio varies little over the engine speed range, and this results in a relatively flat compressor boost curve. If the compressor allows this without surge, the low-speed boost can be high, as there is no need to compromise on the minimum turbine flow. The turbine rotor experiences a wide range of inlet flow angles, and there may be a loss in turbine efficiency at extreme conditions, although this is expected to be less wasteful than a wastegate. Variable geometry tends to be used on diesel engines, where the lower engine exhaust temperature allows a robust variable geometry system (e.g., variable stagger vanes) upstream of the turbine rotor. The engine speed range is lower, so a wastegate can be eliminated. A cheap, reliable gasoline engine system is still to be developed.

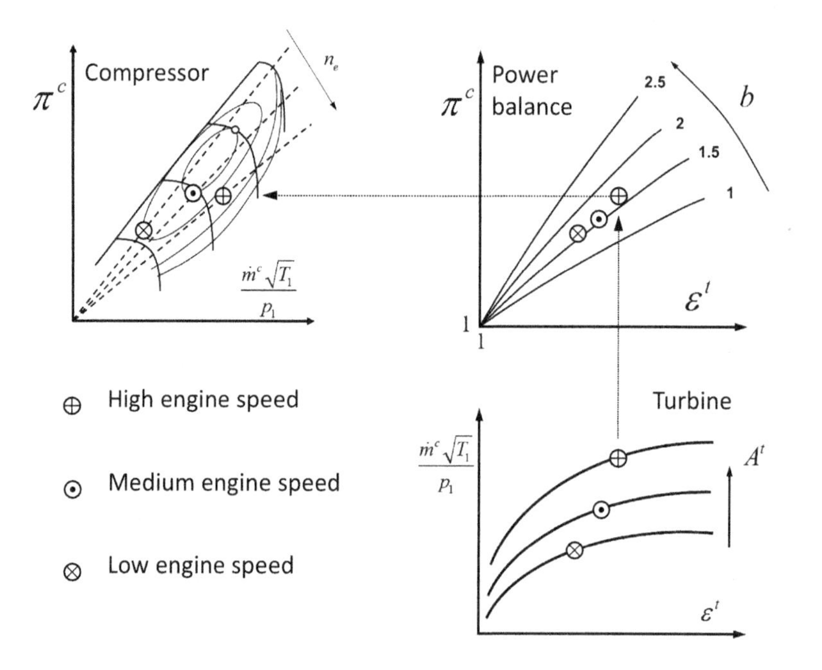

Figure 18.23 Matching of a compressor and turbine in a turbocharger with a variable guide vanes.

19 Structural Integrity

19.1 Overview

19.1.1 Introduction

The objective of this chapter is to describe the essential aspects of the structural mechanics, vibration dynamics and rotordynamics of compressors, including impeller manufacture, impeller mechanical integrity and impeller blade and disc vibrations.

The essential background knowledge required to understand the key structural issues is given with some simple analytic solutions. These identify the effects of blade thickness and blade taper on the stresses and the vibration characteristics. Such analytical solutions are only available for the simplest geometrical cases but give an insight into the fundamentals; complex computer-based methods are used for the more detailed mechanical analysis of blades, discs and rotors. An overview of impeller manufacture is also given as this strongly influences the mechanical integrity. Subsequent sections discuss mechanical design guidelines for impellers and discs.

A brief introduction to rotordynamics is also presented with the intention of familiarising the reader to the most important concepts and the engineering of robust vibration-free compressor shafts. The fundamental knowledge required to understand the key issues in rotordynamics are briefly introduced with an analytic solution of a simplified rotordynamic system. The characteristics of this simple rotor assembly, known as the Jeffcott single mass rotor, are examined using the equations of motion. Some limits on multistage compressor design introduced by rotordynamics can be derived even on the basis of the simple Jeffcott rotor; for example, the shaft should be thick and short to achieve a high critical speed. More complex rotor assemblies again need to be analysed with computer-based methods which are readily available.

19.1.2 Learning Objectives

- Appreciate the implications of different methods of impeller manufacture such as flank milling, point-milling, casting and fabrication by welding.
- Understand why the relevant material parameter in determining its strength is the material yield stress divided by its density.
- Be aware of the different modes of vibration involving bending modes, torsional modes and disc modes.

- Understand the use of the Goodman diagram for fatigue and the Campbell diagram for vibrations
- Understand why the change in material properties does not usually cause a change in the vibration characteristics.
- Recognise the importance of back-face design for stresses in open impellers.
- Be aware of the effects of rotor length on the rotordynamics.

19.2 Open or Closed Impellers

The most critical design choice with regard to the structural integrity for the compressor designer is whether to use an open or a closed (shrouded) impeller. The trend in most compressor development is to strive for more compact products with fewer stages and hence, where possible, to use single stages at high rotational speed with a high specific swallowing capacity. The highest operating speeds can be attained with open impellers, so in order to develop a compact single-stage product the impeller is normally shroudless.

For industrial applications with high overall pressure ratios, or for applications with interstage cooling, multistage machines are needed. The key determinant on whether to use shrouds is the number of stages on a shaft. In a single-stage, two-stage or an integral gear compressor, open or closed impellers may be used. In a multistage inline machine with more than two stages, shrouded impellers are usually used as it is not easily possible to control the clearance between open impellers and the casing during operation. An exception is shown in the three-stage machine with open impellers in Figure 1.4. The design must take into account sufficient movement to allow for manufacturing tolerances in the components, tolerances in the build and mechanical and thermal deflections. With open impellers the blades must withstand the frictional torque loads if the casing and the impeller were to rub. If all of the stages are open, then either large clearances are needed which would lead to high losses or the casing components need to be designed to compensate for any possible axial movement of the shaft.

The next most important factors are the stresses induced in the impeller, and the effect of the operating temperature on the strength of the material. The mechanical stresses in a compressor are primarily the result of centrifugal forces developed from the rotational speed and, to a lesser extent, stresses developed by the fluid acting on the blades. The centrifugal forces apply both to the rotating blades and the discs of the impeller, and the fluid forces also apply to the stator components. A high rotational speed increases the centrifugal loading of the impeller causing a rise in the mean stress level of the material and typical areas of concern are the blade fillet stresses, the bore stress and the backplate stress. A high specific flow rate requires long inducer blades which are vulnerable to additional cyclic stresses from forced vibration, and damage to impellers is sometimes linked to high-cycle-fatigue (HCF) blade damage as a direct result of vibration loading.

Dynamic stresses occur because of blade vibrations from unsteadiness in the flow, temperature gradients and changes in the operating conditions of the compressor.

A high degree of operational safety is expected of the impellers, as mechanical damage leads to failure and outage of expensive plant and to high costs. In most compressor applications, the centrifugal load dominates over the steady aerodynamic pressure loading. In general, a shrouded impeller will have much higher stresses due to the increased mass and the high mean centre of gravity imposed by the shroud. A shroud increases the static loading and usually requires thicker blades to support it, so it has an effect on the performance. But a shrouded configuration may be more robust against vibrations as the shroud increases the stiffness of the impeller, and its natural frequencies may be increased outside a range where they could become excited. An open impeller tends to have relatively low natural frequencies, and stiffness and the assessment of the dynamic stress of the blades then becomes more important. A shrouded impeller also tends to generate a lower axial thrust than an open impeller.

19.3 Impeller Manufacturing and Materials

19.3.1 Manufacturing of Open Impellers

From the aerodynamic point of view, a manufacturing process is needed that is accurate with respect to those parameters which are most sensitive in determining impeller performance. These are the inlet blade angle, blade thickness, the backsweep angle, the surface finish, the tip clearance and the throat area, as discussed in Chapters 10 and 11. From the point of view of structural integrity, the manufacturing process should ensure good mechanical strength of the material and dimensional accuracy so that the calculated static and dynamic stresses of the design are not exceeded. In addition, the manufacturing process should avoid microscopic cracks, which might grow under cyclic stresses, leading to sudden failure in fatigue. Microscopic cutter marks may tend to attract deposits leading to fouling and in some gases cause stress corrosion (Childs and Noronha, 1999).

Open impellers are conventionally manufactured by multiaxis milling from a solid forged disc in which the finishing of the blade surfaces are milled using the side knives of the milling tool, as shown in Figure 19.1. The flank milling process allows a fast production of impellers, and the surface of the blades is of high quality, providing both dimensional accuracy with regard to the blade shape and thickness and good surface finish. In a milled impeller, the final machined hub surface may contain small ridges, as shown in Figure 11.24, whose width and depth depends on the number of cuts and the size of the milling cutter. There appears to be no large detriment to the performance from these ridges provided the cusps are small and they are orientated in the direction of flow, Childs and Noronha (1999). CFD simulations show that small vortices are generated by the ridges, and these can affect the secondary flow structure in the impeller. Lei and Lixin (2015) have argued that if the riblets are 0.1 mm in height in an impeller with a diameter of 240 mm, these actually reduce the impeller losses and confirmed this through CFD

Figure 19.1 An open impeller during flank milling. (image courtesy of Werner Jahnen of SWSTech AG)

simulations and experiments. In any case, milled impellers appear to be more efficient than typical cast impellers, Cousins et al. (2014), as they tend to be smoother in the direction of the flow.

For flank milling, the impeller has to be designed with a straight-line ruled surface on the suction and pressure surfaces. To avoid large rotation of the milled object during milling it is normal that the straight-line generators of the blade are not parallel to the trailing edge but inclined to this by 15°–40°. The minimum passage width between the blades on the impeller hub near the leading edge determines the largest size of the slender milling tools, and this is naturally larger in impellers with splitter blades. A three-dimensional simulation of the impeller geometry, and the cutter motion is usually made by the milling software to determine if the tools can access all points in the flow path without interference or collision with other blades.

High flow coefficient impellers with long blades at the inducer leading edge increase the surface area that must be milled, and the volume of material that must be removed so that these require slender tools which cannot cut as quickly since the spindle and tool are prone to deflection. The strength during milling of the very thin blades, and the thin outer radial backplate of the impeller needs to be examined to avoid chatter vibrations. The milling cutter is typically conical in shape, as this makes it more robust and reduces deflections of the tool, as shown in Figure 19.1. Additional rough cuts with thicker tools may be used to reduce blade deflection during the finishing cuts. Point milling may be used with more complex curved free-form shapes needed for the blades of impellers in transonic flow (see Chapter 11) and for leading edges and fillet radii. The alternatives to milling, often used with very large steel impellers, is for impeller blades to be forged into shape and then welded with a continuous weld to the hub disc, which is more sustainable and less energy intensive (Peng et al., 2017). Increasing interest is being shown in additive manufacturing for large impellers and in 3D printing for components of test rigs (Meier et al., 2019).

19.3.2 Manufacturing of Shrouded Impellers

The manufacturing methods available for the efficient production of complete impellers with a shroud are milling, electrodischarge erosion from the forged material and casting. One possibility is to mill the passages in the impeller from the leading edge and the trailing edge using a technique known as side entry milling. Not all shrouded centrifugal impellers can be machined from a single forging of solid material followed by milling of the flow channels as it is difficult for the milling tool to reach blade and end-wall surfaces in the enclosed passages near the midlength of the blade channel. In some impellers, this is not possible without collision with the hub, shroud disc or adjacent blade surfaces, so such a technique requires special blade shapes optimised for manufacture and point milling has to be used. The material that remains after milling, or the full blade channel, can be machined by electical discharge machining or electrochemical machining using form electrodes. These methods do not show geometrical restrictions. However, machining times are long compared to milling times. The shroud and impeller can be manufactured in two parts by milling and riveting, milling and brazing, milling and welding and welding and welding. Typically, the blades are integrally machined to either the hub or the shroud and are then welded to the shroud or the hub. Precision casting is often used when a large number of similar impellers are needed, Cousins et al. (2014). In some cases, slots are milled in the shroud to match the location of the blades, and the shroud is welded to the blades through these slots.

19.3.3 Materials

The link between the structural analysis and the actual strength of an impeller is the mechanical properties of the impeller material. It is important to have good knowledge of the material properties so the mechanical properties of all forgings for impellers should be checked with a tensile test. Fatigue may occur when the material is subject to fluctuating stresses such that the material fails at a stress level lower than its ultimate yield strength. As fatigue life is usually one of the design criteria, fatigue data also need to be obtained by means of fatigue tests. An overview of material properties for industrial compressor components in including suitable coatings is provided by Dowson et al. (2008).

Tests are carried out on material specimens prior to and after manufacture to prove the required material properties have been obtained. On completion of manufacture, the dimensions of the impeller are checked, after which the impeller is then tested for cracks, using dye penetration or magnetic particle methods. The impeller is then initially balanced and run at overspeed. This speed is at least 15% higher than the maximum expected speed. After overspeeding, the impeller is once again checked for any permanent deformation, for dimensional accuracy and tested for cracks. On successful completion of these tests, the impeller may be rebalanced.

The rotational speed of the spin test is determined by international standards for compressor testing; see Chapter 20. This overspeed test confirms that the impellers

Table 19.1 Typical compressor material properties.

Material	Density, ρ [kg/m^3]	Proof stress, $\sigma_{0.2}$ [MPa]	Ultimate tensile stress (UTS), σ_u, [MPa]	$\sigma_{0.2}/\rho$
Aluminium alloy	2700	250	290	0.0926
Titanium alloy	4430	830	900	0.1874
Steel alloy	7700	880	930	0.1143

have the specified mechanical properties. It also opens possible hidden cracks at the surface and thus makes them visible during nondestructive testing. A spin test at a speed above design introduces local prestresses in peak stress zones where the material yields. Due to the compressive prestresses generated in the overspeed test, the maximum tensile stresses in operation will actually be lower than had an overspeed test not taken place.

The allowable tip speed can be established for all impeller types with a finite element analysis making use of the static strength of the material and the fatigue data. The manufacturing methods need to be taken into account and lead to lower tip speeds for brazed impellers. The materials in common use for compressors are aluminium alloys, titanium alloys and a variety of steel alloys. The typical room temperature properties of the most common materials are presented in Table 19.1. In principle and in terms of specific strength, any of these could be used for many impeller designs. Cost, ease of machining and availability of detailed properties in the public domain often guide the designer to choose one material over another.

19.3.4 Heat Treatment

Steel impellers usually make use of a chromium nickel molybdenum alloy steel, typically with 12–13% chromium such as X3CrNiMo13-4, to give high tensile strength, good hardenability, ductility, weldability, toughness, wear resistance and good fatigue strength. The final material properties are tuned by means of heat treatment with a quenching and tempering process. The quenching process heats the material to a temperature of 950 °C with a soaking time of two hours, and then the material is cooled rapidly in an oil bath. The tempering processes involves annealing by reheating to a lower temperature for several hours.

Titanium impellers usually make use of an aluminium alloy titanium, such as Ti6Al-4V. Castings usually meet the required properties, but forged material leads to higher tensile stress and an improved high-cycle-fatigue capability. Fatigue strength and tensile stresses are improved further by heat treatment with a suitable quenching and annealing process. Aluminium impellers are produced from Al-Si-Mg-Cu alloys. With most aluminium alloys, the best strength is also obtained by a tempering heat treatment.

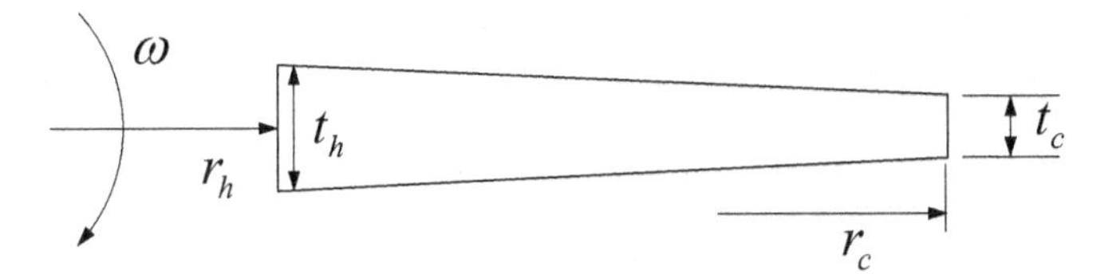

Figure 19.2 A tapered radial blade section.

19.4 Introduction to Static Blade Loading of Impellers

19.4.1 Static Radial Force and Stresses Due to Rotation

Figure 19.2 shows a sketch of a tapered radial beam representing a typical blade section of the inducer blade of a compressor impeller, which is not attached to the shaft. The thickness distribution decreases linearly with radius and is given by

$$t = t_h + A(r - r_h), \quad A = \frac{t_c - t_h}{r_c - r_h}, \tag{19.1}$$

where c represents the tip of the blade at the casing and h the root of the blade at the hub. The elemental centrifugal force on a small radial section of the blade is given by

$$\delta F = \rho\omega^2 rth\delta r, \tag{19.2}$$

where ρ is the material density, ω is the angular velocity, h is the depth of the blade section and r is the local radius. Hence the tangential force at the root of the blade is given by

$$\begin{aligned} F_t &= \int dF = \int \rho\omega^2 rthdr = \rho\omega^2 h\left[t_h\frac{r^2}{2} + A\frac{r^3}{3} - Ar_h\frac{r^2}{2}\right]_{r_h}^{r_c} \\ F_t &= \rho\omega^2 h\left[t_h\left(r_c^2 - r_h^2\right)/2 + \frac{A}{6}\left(2r_c^2 + r_h^3 - 3r_h r_c^2\right)\right]. \end{aligned} \tag{19.3}$$

For a blade of uniform thickness, $A = 0$, the radial force at the root of the uniform beam is given by

$$F_u = \rho\omega^2 ht_h\left(r_c^2 - r_h^2\right)/2 = \rho u_2^2 ht_h\left[\left(\frac{r_c}{r_2}\right)^2 - \left(\frac{r_h}{r_2}\right)^2\right], \tag{19.4}$$

where the impeller tip radius, r_2, and impeller tip speed, u_2, are used to nondimensionalise the terms. In the tapered beam, $A < 0$, and the tapered beam has a lower root force than that for the uniform beam. For example, if the blade tapers in width from a thickness of $t_h/r_2 = 0.03$ at the hub to one third of this, $t_c/r_2 = 0.01$ at the tip for an impeller with a hub radius of $r_h/r_2 = 0.25$ and a casing radius of $r_c/r_2 = 0.65$, then $A = -0.05$. The force at the blade root for the tapered and the uniform blade with a constant thickness is then given by

$$F_u = 0.285 \cdot \rho u_2^2 h t_h, \quad F_t = 0.12094 \cdot \rho u_2^2 h t_h, \tag{19.5}$$

such that the tapered blade has a radial force that is 42% of the uniform thickness blade.

The radial stress at the hub is given by

$$\sigma_h = F_t / h t_h. \tag{19.6}$$

For a blade of constant thickness, this leads to a stress of

$$\sigma_{\theta h} = \rho u_2^2 \left[\left(\frac{r_c}{r_2} \right)^2 - \left(\frac{r_h}{r_2} \right)^2 \right]. \tag{19.7}$$

This can be rearranged to give the tip speed for a given stress level as

$$u_2 = \sqrt{\frac{(\sigma/\rho)}{\left[\left(\frac{r_c}{r_2} \right)^2 - \left(\frac{r_h}{r_2} \right)^2 \right]}}. \tag{19.8}$$

This simple blade representation is probably a reasonable approximation for an axial compressor blade but cannot be expected to accurately represent a real impeller with a curved meridional channel and curved blades which often have no purely radial elements. The lean relative to a radial line introduces bending stresses in the blade when centrifugal forces are applied. Two important effects , however, can be seen from (19.8); the thickness of the blade does not affect the root stress of a radial element blade, and the relevant material parameter is the material yield stress divided by its density, as listed in Table 2.1. Using the values of material properties in Table 2.1, then on the basis of (19.8), this leads to the fact that a steel blade should be able attain a tip speed 11% higher than an aluminium one, and a titanium blade 42% higher than an aluminium one. With taper, as discussed earlier, all three cases could achieve a further 19% higher blade speed. The blade thickness at the root also has a large influence on the resulting stress in the blade roots: the thicker the blade, the lower the root stress. From (19.7), the stresses in the blade are a function of the square of the tip speed, so that in rig testing, the condition of thermodynamic similarity based on similarity of the tip-speed Mach number leads to similar stresses in a rig test as in the full-size machine when the test is carried out with the same gas.

19.4.2 Stresses in a Blade Due to Lean and Bending

In order to reduce the bending stresses in the blade root due to the amount of radial lean in the blades of open impellers, the hub blade angle distribution may be modified and combined with rake at the trailing edge. This is done by selecting lower values of the blade angle in the middle of the blade along the hub, as discussed in Chapter 11. This technique can be used to ensure nearly radial blade elements near to the leading edge and in the inducer. The backward leaning blades near the trailing edge are then more affected by the lean, and this causes bending stresses in the fillets at high speed. The key geometrical factors affecting the fillet stress in the blade near the trailing edge are the hub pitch angle, the thickness of the hub backplate and the fillet radius.

19.4.3 Stresses in a Rotating Disc

For a solid rotating disc of radius, b, with a uniform thickness. the radial and tangential stress is given by

$$\sigma_{r\omega} = \frac{3+\nu}{8}\rho\omega^2\left(b^2 - r^2\right)$$
$$\sigma_{\theta\omega} = \frac{3+\nu}{8}\rho\omega^2 b^2 - \frac{1+3\nu}{8}\rho\omega^2 r^2. \quad (19.9)$$

At the centre of the disc, with $r = 0$, the stresses are given by

$$\sigma_{r\omega} = \sigma_{\theta\omega} = \frac{3+\nu}{8}\rho\omega^2 b^2, \quad (19.10)$$

and at the outer edge of the disc, with $r = b$, the stresses are

$$\sigma_{r\omega} = 0, \quad \sigma_{\theta\omega} = \frac{1-\nu}{4}\rho\omega^2 b^2. \quad (19.11)$$

The radial displacement at the outer edge of the disc, with $r = b$, is given by

$$u_{\omega} = \frac{1-\nu}{4E}\rho\omega^2 b^3. \quad (19.12)$$

For a disc with a central hole having a bore radius, a, the radial and tangential stress is given by

$$\sigma_{r\omega} = \frac{3+\nu}{8}\rho\omega^2\left(b^2 + a^2 - \frac{a^2 b^2}{r^2} - r^2\right)$$
$$\sigma_{\theta\omega} = \frac{3+\nu}{8}\rho\omega^2\left(b^2 + a^2 + \frac{a^2 b^2}{r^2} - \frac{1+3\nu}{3+\nu}r^2\right). \quad (19.13)$$

The maximum tangential stress is at the bore where

$$\sigma_{r\omega a} = \frac{3+\nu}{4}\rho\omega^2\left(b^2 + \frac{1-\nu}{3+\nu}a^2\right) \quad (19.14)$$

and the tangential stress at the rim is given by

$$\sigma_{r\omega b} = \frac{3+\nu}{4}\rho\omega^2\left(a^2 - \frac{1-\nu}{3+\nu}b^2\right). \quad (19.15)$$

The additional radial and tangential stress in the disc due to an external tensile pressure, σ_b, resulting from the centrifugal force of the blades is given by

$$\sigma_{rp} = \frac{\sigma_b b^2}{(b^2 - a^2)}\left(1 - \frac{a^2}{r^2}\right)$$

$$\sigma_{\theta p} = \frac{\sigma_b b^2}{(b^2 - a^2)}\left(1 + \frac{a^2}{r^2}\right) \tag{19.16}$$

$$\sigma_b \cong \frac{Z_{bl} F_t}{2\pi b},$$

where Z_{bl} is the number of blades.

In the case of a disc of variable thickness, the determination of the stresses can be approximated by considering the disc to comprise a number of uniform discs of different uniform thickness and considering the conditions at the boundaries between these discs. Finite element techniques can be employed to calculate an optimum disc profile for a given design stress in a fairly automated manner, as discussed in Section 19.10. The disc stresses lead to high stresses in the bore of the impeller and to locally high stresses in the hub backplate.

19.4.4 Effect of Mechanical Forces on the Blade Shape

The stacking of the blade profiles in the inducer of centrifugal impellers is different from their axial counterparts in the sense that impellers are often designed with zero or a small lean at the leading edge to minimise the stress in the hub/leading edge where the blade height is largest. The higher inlet blade angle at the tip section (typically –60°) compared to that at the hub (typically –40°) results in a negative lean (lean that is opposite to the direction of rotation) in the front part of the inducer section. The centrifugal load causes the blade tip to move in the direction of rotation under mechanical loading and results in blade twist. as opposed to the untwist experienced by transonic axial rotors. The net effect is an increase in the inlet blade angle and reduction of the throat area in the tip section of the inducer. This can reduce the impeller choke flow and also move the shock farther into the passage.

The effect of the pressure load is, however, in the opposite direction to the centrifugal load, whereby the blade tip section is pushed against the direction of rotation. The radial tip clearance at the inlet of the impeller is reduced under running conditions due to centrifugal forces. However, the size of the tip gap at the trailing edge can reduce or increase depending on the geometry of the blade and the disc. A study of these effects has been made by Hazby et al. (2015).

19.5 Introduction to Dynamic Blade Loading of Impellers

19.5.1 Source of Vibrations

In operation, compressors experience transient aerodynamic forces excited by the flow and pressure distribution from stator components, such as inlet guide vanes, diffuser vanes, return channel vanes and asymmetric inlet and outlet geometries. In some situations, such forces can also be generated by seals and by acoustic resonance in the space between the impeller backplate or shroud and the casing components (König et al., 2009). The blade rows induce spatial nonuniformities in the flow field such as those due to potential effects, wakes and shocks. The motion of an object through a nonuniform pressure field leads to a continuous temporal unsteadiness of the pressure on the blade surface which translates into an unsteady force. Resonances of the transient forces with the natural frequencies of an impeller disc and its blades can occur. The frequency of the pressure fluctuations related to stationary spatial disturbances changes with the rotational speed so that resonances occur at particular speeds.

The most important vibrations that can occur in compressors are blade vibrations, disc vibrations and rotordynamic vibrations, the latter of which will be considered in Section 19.12. These cause an increased mechanical loading of the components and may be a source of noise or even of catastrophic failure. The major influences on the dynamic behaviour of the components are the natural frequencies of the components, the source of excitement and the damping of the vibrations. The allowable mechanical loading is determined from the amplitude of the vibrations leading to a dynamic loading, from the material properties, and from the static loading of the components. To examine the dynamic loading, it is possible to consider the blade alone, the blade joined to the hub disc and the complete shrouded impeller with a cover disc if this is present.

Vibrations can be of different types depending on the type of interaction of the flow field with the system. The two main types of failure mechanism with respect to vibrations are failure due to overload in low cycle fatigue (LCF) and failure due to high cycle fatigue (HCF); see Section 19.8. The different vibration phenomena are referred to as forced response, flutter or nonsynchronous vibrations. Forced response vibrations in compressors are caused by aerodynamic interaction of the flow with the components. The damping of such vibrations is due to the material properties (which is usually low) and to aerodynamic damping forces which depend on the operating point. Flutter is a self-excited form of vibration of blades.

A nonsynchronous vibration (NSV) denotes a fluid dynamic instability with a frequency that that locks onto a structural resonant mode. A common example would be when the vortex shedding frequency of a fluid dynamic instability becomes close to a natural frequency of vibration of the structure. The fluid instability exists without the motion of the structure but the locking on occurs abruptly when the frequency of the inherent flow unsteadiness gets close enough to a natural frequency of the structure and continues even if no resonance occurs.

19.5.2 A Brief Introduction to Mechanical Vibrations

The basic background to the understanding of mechanical vibrations is the linear mass-spring-damper system having a single degree of freedom with a forced exciting force, as given in standard texts on mechanical vibrations, such as Thomson (1996).

This system is described by the differential equation giving the equation for the displacement of the mass:

$$m\ddot{x} + c\dot{x} + kx = F(t). \tag{19.17}$$

where x is the displacement, m represents the mass of the system, k is the coefficient of the linear elastic restoring force acting against and proportional to the displacement, c is the damping coefficient of the viscous damping force which is proportional to the velocity of the mass and F is the exciting force.

Free vibrations of the undamped system (that is, with $F = 0$ and $c = 0$) leads to a sinusoidal solution of this equation of the form

$$x = c_1 \sin \omega_n t + c_2 \cos \omega_n t, \quad \omega_n = \sqrt{\frac{k}{m}}, \tag{19.18}$$

where the natural frequency of the system, f_n, as the number of cycles per second, is given by $f_n = 2\pi\omega_n$. In real situations, the mechanical system is more complex than this, with many degrees of freedom, which can be thought of separate masses joined by springs leading to many different natural frequencies of the coupled system representing the natural oscillation of the system in different vibration modes.

In a typical mechanical system, there will always be some damping present to dissipate the energy of the vibration, and then the solution of (19.17) for free vibrations of the damped system (with $F = 0$) may be expressed in the form

$$x = e^{st}. \tag{19.19}$$

Substituting this into (19.17) shows that this equation represents a valid solution for all values of t when

$$s^2 + \frac{c}{m}s + \frac{k}{m} = 0, \quad s_{1,2} = -\frac{c}{2m} \pm \sqrt{\left(\frac{c}{2m}\right)^2 - \frac{k}{m}} = -\frac{c}{2m} \pm \sqrt{\left(\frac{c}{2m}\right)^2 - \omega_n^2}. \tag{19.20}$$

The behaviour of the damped system depends on the two values of s determined by the radical, the term in the square root. Following Thomson (1996), a critical reference quantity for the damping coefficient, c_c,, can be determined which reduces the radical to zero such that

$$\left(\frac{c_c}{2m}\right)^2 - \omega_n^2 = 0, \quad c_c = 2m\omega_n \tag{19.21}$$

and the actual damping of the system can be expressed in terms of a nondimensional damping ratio, $\zeta = c/c_c$, such that (19.20) becomes

$$s_{1,2} = \omega_n\left(-\varsigma \pm \sqrt{\zeta^2 - 1}\right) \tag{19.22}$$

and the differential equation for free vibrations of the damped system becomes

$$\ddot{x} + 2\zeta\omega_n\dot{x} + \omega_n^2 x = 0. \tag{19.23}$$

The solution of (19.23) is of two types; for $\zeta > 1$, the solution is an overdamped vibration in which an initial displacement of the mass subsides exponentially, or for $\zeta < 1$ the solution represents an underdamped vibration in which an initial displacement leads to a sinusoidal motion in which the amplitude of the successive peaks reduces exponentially with time. The damped oscillation is often referred to in terms of the natural logarithm of two successive amplitudes known as the logarithmic decrement, which is given by

$$\delta = \frac{2\pi\zeta}{\sqrt{1-\zeta^2}}. \tag{19.24}$$

The preceding discussion refers to free vibrations of a damped mass-spring system governed by (19.17), but in most cases of interest the vibrations are excited by a harmonic force in which $F(t) = F_0 \sin \omega t$, where the frequency ω is related to the rotational speed of the machine. The frequencies which may excite the blade vibrations may be high-order disturbances, which are a multiple of the blade rotational frequency, or low-order disturbances occurring at a fraction of the blade rotational frequency. If the frequency of the exciting force corresponds to the natural frequency of the system, ω_n, then the system is said to be in resonance and the amplitude of the displacement can grow exponentially until limited by the system damping, failure occurs, or until some limit cycle oscillation is attained. In a more complex system, the displacement grows if the frequency of the exciting force equals one of the many natural frequencies of the system and the exciting force profile has the same shape as the associated mode shape of the vibration.

The solution for the vibration amplitude may be assumed to be $x = X\sin(\omega t - \phi)$, and Thomson (1996) shows that this may be expressed in a nondimensional form as

$$\frac{Xk}{F_0} = \frac{1}{\sqrt{\left[1-(\omega/\omega_n)^2\right]^2 + [2\zeta(\omega/\omega_n)]^2}}, \qquad \tan\phi = \frac{2\zeta(\omega/\omega_n)}{1-(\omega/\omega_n)^2}, \tag{19.25}$$

where is ϕ the phase angle between the imposed force and the resulting vibration, Xk/F_0, is the transmissibility function or magnification factor as shown in Figure 19.3. These curves show that the damping factor has a large influence on the amplitude near to resonance, especially if the damping ratio is small. In centrifugal compressors, these vibration modes apply to the blades themselves and the circular disc to which these are attached, and there may also be more complex coupled modes in which both the disc and the blades are involved. At high frequencies, the amplitude of the forced vibration approaches 0.

The importance of forced vibrations in a compressor is that the frequency of the vibrations may be very high, leading to a very high number of vibration cycles in a short time. For example, with an impeller at a rotational speed of 60,000 rpm, the frequency of the most important fourth excitation order disturbance is 4(60,000/60) = 4000 cycles per second, which implies 14.4 million cycles per hour. In this way, if the blades operate in resonance at high amplitude, they may quickly succumb to high cycle fatigue and disintegrate; see Section 19.8.

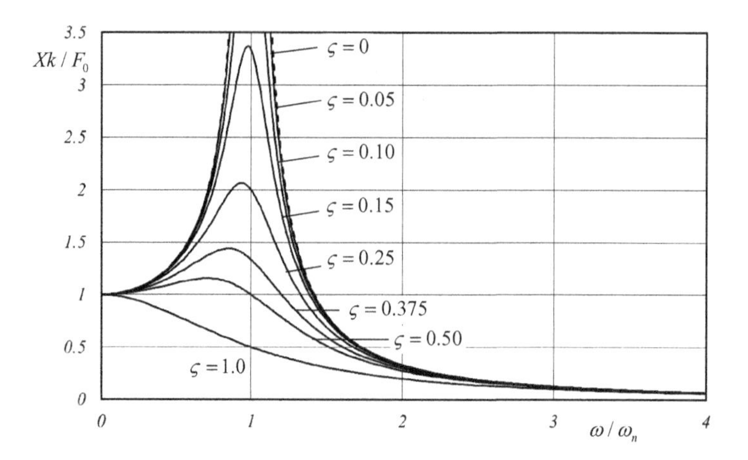

Figure 19.3 Dimensionless amplitude of the forced response of a mass-spring-damper system versus frequency ratio, (19.25).

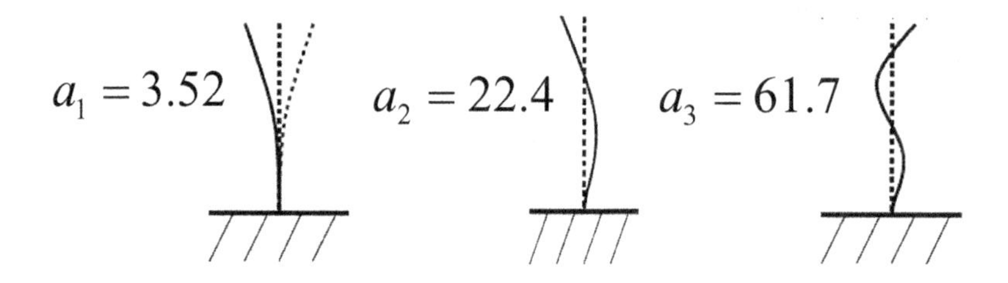

Figure 19.4 Bending modes of vibration of a cantilevered beam.

19.5.3 Vibrations of a Blade

As an introduction to the vibration of a blade, it is useful to consider the natural vibration of a uniform cantilevered beam which is free at one and clamped at the other, for which the natural frequency is given by

$$f_n = \frac{a_n}{2\pi l^2}\sqrt{\frac{EI}{\rho A}} = \frac{a_n}{2\pi l^2}\sqrt{\frac{Ect^3}{12\rho ct}} = \frac{a_n t}{2\pi l^2}\sqrt{\frac{E}{12\rho}}, \tag{19.26}$$

where E is the Young's modulus, $I = ct^3/12$ is the second moment of area of the section, c is the width of the beam (equivalent to chord of the blade in this analogy), t is the thickness of the beam, EI is the bending stiffness of the section, l is the beam length, $A = ct$ is the cross-sectional area of the beam, ρ is the density of the material and the value of a_n takes on different values for different modes of vibration of the beam, as given in Figure 19.4 for the first three bending vibration modes. Note that the ratio E/ρ describes the role of the material in the frequency. The ratio of E/ρ is similar

Table 19.2 Further compressor material properties.

Material [MPa]	Density, ρ [kg/m^3]	Young's modulus E	$\sqrt{E/\rho}$ [m/s]
Aluminium alloy	2700	71,700	5153
Titanium alloy	4430	114,000	5073
Steel alloy	7700	200,000	5096

for the three typical materials in Table 19.2, the square root of the ratio differs by only a percent or so, hence a change in material results in almost no change to the natural frequencies for a given geometry.

Equation (19.26) shows that the frequency of beam vibrations is inversely proportional to the square of the blade length and linearly proportional to the thickness. The natural frequency decreases with the square of the length of the blade so that a blade that is 10% shorter has a 20% higher natural frequency. The effect of increasing the thickness of the blade by 10% is to increase all the natural frequencies by 10%. It can be shown that the effect of introducing a taper in the blade thickness is to increase the frequency of mode 1 but reduce the frequency of mode 2 (Woods, 2018). In this way, the thickness and the taper of a blade can be used by the designer to adjust its natural frequencies in order to avoid resonance.

In a model test where a geometric scale factor is used to reduce the size of the test object, both the length and the thickness of the blade are decreased by the scale factor, and the frequencies then increase linearly with the size of the model. The rotational speed needed to retain thermodynamic similarity of the tip-speed Mach number in a smaller model also increases linearly with the scale factor of the model so that the blade resonance frequencies retain a fixed ratio to the operational speeds. The linear relationship between size and frequencies is obviously convenient in simple scaling of geometry to adapt an impeller from one application to another.

In addition to the vibrations due to bending of the beam torsional vibrations exist for which analytic solutions are also available for simple blade shapes. The complexity of the blade vibration modes involving bending and torsion in a radial impeller can be seen in Figure 19.5 as determined by numerical simulation known as a modal analysis. In this presentation, the leading edge of the blade is at the bottom and the casing contour on the left of each blade. The darker shading shows the amplitude of the first five mode shapes of an impeller blade designed near the optimum flow coefficient and where the thickness distribution has been designed to satisfy the static stress requirements and ensures that the frequency of the first bending mode, often known as first flap, is above 10% above the first excitation order of the machine (1EO). These mode shapes have been calculated with a simplified approach assuming a cantilever attachment of the blade to a rigid hub disc so there are no displacements on the hub side of the blade, as in Figure 19.4. Assuming a rigid hub results in an overestimate of the blade frequencies, typically by 5%, but the computational speed and ease of setup of such a calculation means this gives rapid feedback to the designer during the aerodynamic optimisation of the

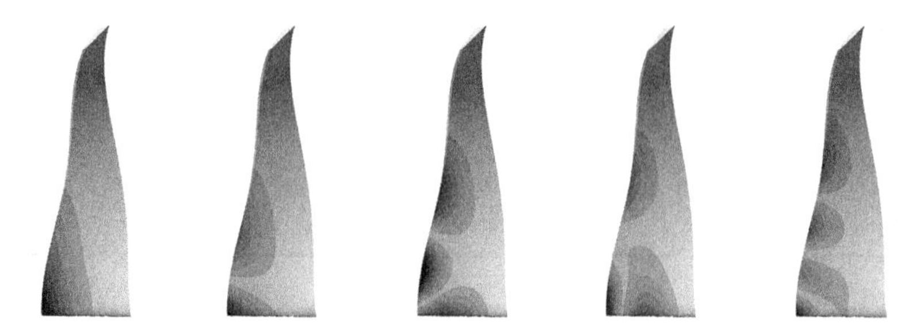

Figure 19.5 Vibration mode shapes of an impeller blade viewed from the side with the fixed hub on the right and leading edge at the bottom.

vane. In this case, the blade is relatively long from the leading edge to the trailing edge compared to the blade span. Most modes involve a bending of the leading edge at the tip where the blade has the largest span.

19.5.4 Vibrations of Circular Discs

There are many vibration mode shapes of circular plates, and for a simple disc these can be represented using nodal lines and nodal circles. Nodal diameters and nodal circles consist of stationary points for a specific mode shape. The shape of these patterns depends only on the associated frequency, and the amplitude depends on the exciting force as with the vibrations described in Section 19.5.2. The higher modes are generally more difficult to excite. In an impeller, the disc modes and the blade vibration modes combine to cause more complex vibration patterns.

19.5.5 Vibrations of Bladed-Disc Structures

Complex vibration modes involving blades and discs may be found in complete impellers. In shrouded impellers, the relevant natural frequencies are high due to the stiffening effect of the combined blade with its shroud and hub disc. Examples of such coupled vibration modes and the damage that this can cause to the impeller back plate are given in Kushner et al. (2000).

19.5.6 Mistuning

During the design, the impeller is considered to be a periodic structure consisting of a finite number of identical substructures (sectors) making up a closed circular structure. Each sector can be identified as one blade, together with a splitter blade if present, and the slice of disc and shroud (if present) where the blade is connected. The structural analysis examines only one sector by applying a constraint of cyclic symmetry at the sector interfaces, thus saving computational time. The consequence is that every sector is assumed to vibrate with the same oscillation amplitude and frequencies. In practice, however, there are always small differences among the structural properties of individual blades, which destroy the cyclic symmetry of the bladed disc assembly. These structural irregularities may derive from manufacturing tolerances; deviations in material properties, or nonuniform operational wear asymmetry; different contact conditions at the blade roots; and modifications introduced during maintenance and repair processes. This causes an unpredictable response of the system where some blades can vibrate with larger amplitudes than in the tuned symmetric case. The random disorder affecting the rotationally periodic structure is called mistuning (Ewins, 1969).

Mistuning is known to have a potentially dramatic effect on the vibratory behaviour of the rotor, since it can lead to spatial localisation of the vibration energy as the vibration energy in a bladed disc becomes confined to a single blade or to a few blades rather than being uniformly distributed throughout the system. This phenomenon can be explained by viewing the vibration energy of the system as a circumferentially travelling wave. In a perfectly tuned system, the wave propagates through each identical disc–blade sector, yielding uniform vibration amplitudes that differ only in phase. In the mistuned case, however, the structural irregularities may cause the travelling wave to be partially reflected at each sector. This can lead to confinement of vibration energy to a small region of the assembly. As a result, certain blades may experience forced response amplitudes and stresses that are substantially larger than those predicted by an analysis of the nominally symmetric design.

19.5.7 Achieving Safe Designs Due to Vibrations

The important guidelines to avoiding damage due to vibrations are

- Avoid excitation of resonant frequencies in the speed range of operation by the design of the impeller and vanes.
- Reduce the amplitude of any excitation force to a minimum.
- Increase damping.
- Alter system dynamics to make system more robust (for example with shrouds).
- Avoid occurrences of negative aerodynamic damping and self-excited vibrations.
- Reduce the static stresses to lower the relevance of the contribution of any additional vibratory stresses.
- Build with good concentricity between rotating and static parts that remain centred throughout the operating range.
- Use more HCF-tolerant material.

Resonances usually occur when the rotating speed, or a multiple of the rotating speed, coincides with one of the natural frequencies of the rotor. Resonances rarely pose serious problems, unless the steady operating point lies very close to a critical natural frequency. Efforts are made in the design phase to create an impeller geometry with a relatively wide range of resonance-free speeds around the normal operating point. The designer can influence the natural frequencies through the thickness of the blade, the variation of the thickness across the span and the curvature of the blade. In variable speed machines, the passage through critical speeds, if made rapidly enough, is not a severe condition, but not all operating speeds may be allowed.

Self-excited vibrations are oscillations in which a component vibrates at one of its natural frequencies and in which, due to some positive feedback mechanism, energy from an external source (usually the fluid) is absorbed into the vibrational mode. Exact balancing does not remove this type of vibration. Once initiated, if damping is insufficient, the vibration will increase exponentially in amplitude until some non-linear mechanism intervenes, or until rubbing occurs. Self-excited vibrations are also called 'rotordynamic instabilities' or 'subsynchronous vibrations'. Chapter 17 describes a common cause of subsynchronous vibration in centrifugal compressor stages with vaneless diffusers as these can give rise to a rotating stall instability with the frequency of the excitation force being a fraction of the rotor speed.

It is essential to carefully evaluate the resonance situation for an unshrouded compressor impeller; the higher the flow coefficient, the more important this becomes. Because of the freestanding blades, vibration modes are more easily excited, and potentially harmful resonances must be avoided. It is not yet possible to accurately predict the transient aerodynamic forces in a compressor, and the decision with regard to prohibited and allowable resonances is still often based on the manufacturer's empirical experience.

In contrast, shrouded impellers are not designed to avoid these resonances. Due to their inherently stiff design, the natural frequencies are much higher and the stresses remain below the fatigue limit, even when impellers are operated at a resonance point. Designing shrouded impellers of industrial compressors to avoid resonances is not only unnecessary, but it would be hardly feasible as the different trims to adapt them to other flow coefficients give rise to a range of exit widths for a single impeller type.

In open radial compressor impellers, experience shows that the most damaging blade vibrations may be avoided if the lowest natural frequency of the impeller is greater than the fourth excitation order (4EO) at the maximum speed condition. The international API standard 617 (2014) specifies a requirement for axial compressors that the blade natural frequencies shall not coincide with any source of excitation from 10% below minimum allowable speed up to 10% above maximum continuous speed (MCS). Some manufacturers arbitrarily aim for a higher margin, for example of order four to five times the rotor frequency, where the impeller is subject to significant cycling in speed, for example in automotive turbocharging. If such a margin is not feasible, blading should be designed with stress levels low enough to allow unrestricted operation, at any specified operating speed for the minimum service life established for the machine. This can be verified by plotting Goodman diagrams or their equivalent, as discussed in Section 19.8. If the lowest natural frequency of an impeller blade is excited by pressure disturbances, it can cause high cycle fatigue, leading to cracks in the blade or disintegration of the tip of the blade near the leading edge. Examples are given by Zemp (2012).

19.6 Computational Methods

The mechanical design of impellers makes use of powerful general-purpose finite element programs such as ANSYS-mechanical and Abaqus. Impellers are rotationally periodic, and it is sufficient to analyse a single periodic sector of an impeller. This provides efficient

Figure 19.6 Calculated stress levels in a medium flow coefficient shrouded impeller.

computation and simple postprocessing and reduces the required storage capacity. In addition to the important centrifugal loads, gas loads also act upon the impellers but in most compressor applications these are less important and may usually be neglected.

Based on static finite element analyses, impellers are analysed to determine the allowable tip speed of an impeller or designed to achieve a specified maximum tip speed. Generally, the investigations start with a simple linear elastic analysis of the stresses induced by the rotation. An important result of this analysis is the impeller's maximum stress amplitude for the start/stop cycle of the compressor. Figure 19.6 shows the stress field of a medium flow shrouded impeller under centrifugal load. High stresses typically occur in the transition zones between hub and blades and between shroud and blades. A more complex nonlinear elastoplastic analysis is carried out to find the impeller bursting speed. As the rotational speed increases, the stresses start to exceed the yield strength of the material, first locally and then in large regions of the impeller. The impeller stiffness and ability to carry additional centrifugal load then drops considerably. Any further speed increase would result in the explosion of the impeller. With the elastoplastic stress–strain curve of the material, the true plastic deformation can be considered.

A modal analysis is used to predict the natural frequencies and mode shapes of the impeller. A range of approaches to the prediction are possible, starting with a simplified analysis of the blade alone assuming rigid attachment at the hub. Ignoring the stiffness of the hub will result in an overestimation of the blade frequencies, typically by 5%, but the computational speed and ease of setup means this can give useful almost real-time feedback to the designer as the aerodynamic optimisation of the vane is carried out. The next step is then to consider the full wheel, and including the effects of stress stiffening (Section 19.9.2). This more complex and more time-consuming analysis permits examination of the coupled vibrations of blade and disc. Figure 19.7 shows the calculated displacements in the fifth bending mode (4 nodal diameters, or ND) for an unshrouded impeller.

19.7 Design Data for Mechanical Analysis

19.7.1 Design Cycles

The number of design cycles refers to the number of major cycles the impeller is required to undergo during its design life. For a constant speed machine, such as a

Figure 19.7 Calculated displacements in the fifth mode (4ND) for a high flow coefficient open impeller.

microturbine or an industrial compressor application, this is usually fairly straightforward to specify and could be based on the number of start/stop cycles over a given design life, for example, two cycles per day for a design life of 10 years. For turbocharger applications, the specification of the design cyclic duty is more complex since, in general, it operates as a variable speed machine. For an automotive application, at least two extreme operating scenarios may be identified, which could be a motorway duty and a city or town duty. For the former case, the turbocharger would be operating between two fairly narrow speed and temperature ranges, whereas for the latter the turbocharger would be required to respond to a wide range of engine speeds and temperature consistent with frequent vehicle stops and starts. Under such varying operating conditions, turbocharger manufacturers have recorded actual turbocharger operation, such as speed and temperature, during different types of operating scenarios. These data may be processed to determine the number of cycles at a number of specific speed bands to enable the application of accumulative fatigue damage models.

19.7.2 Design Life

The design life specifies the time over which the impeller is to operate, which is usually, but not always, identical to the specified machine design life. This is an important parameter for high-temperature applications where the impeller material could undergo creep damage but is generally less of a concern for compressors.

19.7.3 Design Speed

The design speed is usually defined as the maximum or 100% operating speed of the machine. However, it is usual practice to allow for the following overspeed conditions at which structural failure must not occur:

- A malfunction of the speed control system
- A prescribed overspeed 'proof test' after manufacture and after assembly

A design safety factor on speed, SF_N, is applied to account for these overspeed conditions, and for constant speed machines it is usual practice for $SF_N = 1.05$, which equates to a 10% increase in stress.

19.7.4 Design Temperature

The design temperature used for initial design is based on the static gas temperature at impeller outlet, which is often simply determined based on 1D equations given in Chapter 10 or can be postprocessed from a CFD solution. The static impeller outlet temperature is a good estimate of the metal temperature to be used to establish the design stress of the impeller. As an example, consider a turbocharger compressor with a pressure ratio of 4.7, an inlet temperature of 20°C and an efficiency of 80%; then the total temperature at impeller outlet is 224°C and with a degree of reaction of 60% the static outlet temperature of the impeller is 143°C. Zheng et al (2013) show how the ultimate tensile strength of a particular aluminium alloy reduces from 412 MPa to 345 MPa at 160°C. The reduction in ultimate tensile stress of aluminium with temperature in this case decreases the maximum pressure ratio of this particular turbocharger impeller from 4.6 to 4.2. Steel and titanium impellers are less sensitive to the temperature of operation.

19.7.5 Design Stresses

The design stresses are derived from properties that define the strength of a material. The impeller needs an adequate speed margin to the bursting speed to prevent sudden failures of hub and shroud, as well as plastic deformation of the impeller during the overspeed test. LCF stresses caused by start/stop cycles must be low enough to ensure a sufficiently long life of the impellers. In some applications with a risk of stress corrosion, the peak stress in the impeller must be limited (including the prestresses from the overspeed test).

19.7.6 Yield Stress

The maximum allowable stress in an impeller is based on a design factor, F, times the minimum yield or more commonly the minimum 0.2% proof stress, $\sigma_{0.2}$, of the material at the design temperature. The design factor, F, is a safety factor where $F < 1$ to account for the variations in material properties and uncertainties implicit in the design process. The value of F is typically taken to be 0.9 and ensures that the maximum calculated stress at the design speed is below the yield or 0.2% proof stress.

19.7.7 Disc Burst

A considerable body of experimental data has shown that for parallel-sided discs, the speed at which it will burst tangentially depends almost entirely on the area-weighted mean hoop stress, As the wheel overspeeds beyond the point where the maximum stressed region begins to yield, the increased centrifugal load will primarily be absorbed by the remaining elastic regions of the disc. As the speed is increased further, the proportion of the wheel remaining elastic reduces until the ultimate tensile strength of the material has been reached at all points of the cross section. A further increase in speed theoretically results in complete fracture. This is what happens with an ideally ductile and non-notch-sensitive material. However, in practice, actual discs burst when the stress attains a value of between 75% and 100% of the ultimate tensile strength. Some evaluation procedures define the burst margin as the ratio to the design speed at which the area-weighted mean hoop stress $\bar{\sigma}_\theta$ becomes equal to some safety factor times the ultimate tensile strength of the material. The safety factor, F, is usually taken to be between 0.7 and 0.75 and an acceptable burst margin should be >1.4, that is, the burst speed should be at least 1.4 times greater than the design speed.

$$\frac{N_b}{N_d} = \sqrt{\frac{F \cdot \sigma_u}{\bar{\sigma}_\theta}}, \tag{19.27}$$

where N_b is the burst speed, N_d is the mechanical design speed and σ_u is the ultimate tensile stress. The mechanical design speed is often taken as 105% of the aerodynamic design speed.

The area-weighted mean hoop stress may be computed from the results of a finite element axisymmetric stress analysis in the following manner:

$$\bar{\sigma}_\theta = \frac{1}{A_d} \sum_{i=1}^{i=n} \sigma_\theta \Delta A, \tag{19.28}$$

where A_d is the cross-sectional area of the disc and n is the number of elements constituting the disc.

In practice, most discs are subjected to additional centrifugal radial forces at the rim or gas swept surface from the blading, and to account for the more general stress state that ensues, the volume-weighted mean equivalent stress is sometimes used in place of the area-weighted mean hoop stress. It is computed in the following manner from the results of a 3D finite element stress analysis:

$$\bar{\sigma}_{eq} = \frac{1}{V_d} \sum_{i=1}^{i=n} \sigma_{eq} \Delta V, \tag{19.29}$$

where V_d is the volume of the disc and n is the number of elements constituting the disc. Most finite element (FE) calculation programs have postprocessing tools to enable the volume-weighted mean equivalent stress for a set of elements to be calculated.

19.7.8 Creep

For loads acting over a long period of time at high temperature, continuous deformation takes place until creep rupture failure occurs. As previously mentioned, this tends not to be relevant for compressors. For most radial turbine applications, where the inlet gas temperature can be between 650°C and 900°C, the creep strength of the material must be considered in determining a design stress.

19.7.9 Fouling, Corrosion and Erosion

Erosion, corrosion and fouling cause degradation of compressor performance and can result in a shorter operating life through the effect on load-carrying components and clearance gaps (Brun and Kurz, 2019). The change of the shape of the aerodynamic components of the rotor and the stator reduces performance, particularly at leading edges, which are most susceptible to erosion due to high relative velocities, and the erosion of blade tips and labyrinth seals. Fouling causes blockage of the flow paths and can change the flow capacity of a component and the matching of components and stages.

Erosion is the wear of components due to abrasive particle or liquid drops in the gas path. Dust and other airborne particles cause solid-particle erosion, and liquid drops cause liquid droplet erosion. Several studies show that the erosive potential of particles or drops in the flow causes erosion at a rate that is proportional to the fourth to fifth power of the fluid velocity for ductile materials (Ahmad et al., 2009). Solid-particle erosion can be controlled through the use of filters or design of the intake ducts to include particle separators to remove the airborne particles. Erosion rates can be combatted by means of material selection and the use of suitable surface coatings in sensitive areas. Fouling can be combatted by the use of Teflon and other polytetra-fluoroethylene (PTFE) coatings (Dowson et al., 2008; Brun and Kurz, 2019).

19.8 Assessment of Fatigue

19.8.1 Definitions

Impellers are subjected to fluctuating stresses and, under these conditions, material failure can occur at magnitudes that are generally much lower than would occur for static loading. Fatigue is often characterised by low cycle fatigue (LCF) and high cycle fatigue (HCF). LCF occurs when the impeller undergoes high-amplitude stress changes at low frequency. These are usually associated with the start/stop cycle of the compressor or infrequent changes in the operating speed. Fatigue damage is evidenced by crack initiation in regions of high stress concentration, where the yield stress can be exceeded and inelastic material behaviour occurs. Regions of high stress concentration are introduced by material defects or geometric features such as sharp corners or grooves. Characteristic damage caused by LCF is described by Christmann et al. (2010).

HCF is characterised by low-amplitude, high-frequency elastic strains and occurs when a vibrating impeller undergoes small but continuous fluctuating stresses. The failure mechanism is that microscopic cracks begin in regions of high stress, and these grow during each cycle to reach a critical size which then propagates, suddenly causing a fracture. The different methods available to assess the risk of failure through HCF and LCF are described by Woods (2018) and summarised in the following subsections. These subsections describe the different methods available to assess the risk of failure through HCF and LCF.

19.8.2 The Goodman Diagram

In the majority of cases, an alternating stress is accompanied by a mean stress, and it is well known that this will influence the fatigue life. The work of Goodman (1919) has been used extensively for many years in this respect because of its simplicity of use, but is well known to be excessively conservative for tensile mean stresses.

The problem of HCF is illustrated in the Goodman diagram, which defines an endurance limit when static and vibratory stresses act together. Static stresses arise from centrifugal forces or gas bending forces acting on the structure. Vibratory stresses are introduced due to forced response from aerodynamic excitation. Depending on the amplitude of the two stress contributors, the stress loading may be within the endurance limit, that is, below the endurance line as indicated by point *A*. With an increase in static stress due to higher centrifugal loading, the static stress may move the overall stresses into a regime where the blade would fail as indicated by *B*. Unsteady blade excitation cause an increase in vibratory stress as indicated by point *C*. In any event, operation outside the endurance limit represents a failure mechanism. With reference to Figure 19.8, the allowable alternating stress, σ_a, is given by

$$\sigma_a = \sigma_{ar}\left(1 - \frac{\sigma_m}{\sigma_u}\right), \tag{19.30}$$

where σ_u is the ultimate tensile strength of the material and σ_m is the mean static stress.

19.9 Vibrational Considerations

19.9.1 Mode Shapes of Rotationally Periodic Structures

Consider an impeller disc with a number of blades equally distributed on its periphery with the centre clamped to a shaft which is allowed to vibrate. In such a system, there

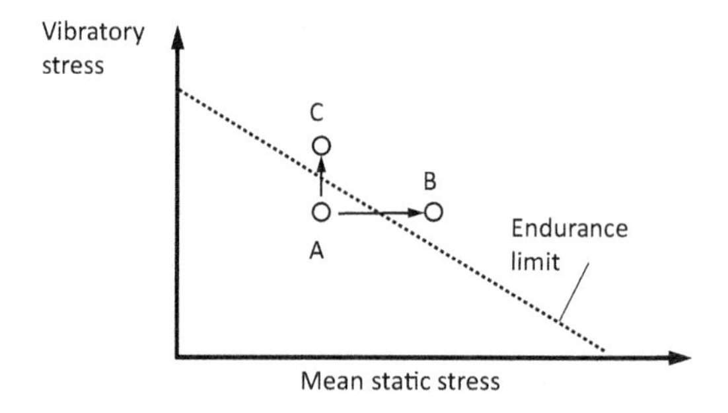

Figure 19.8 Goodman diagram.

are infinitely many natural modes of vibration; examples of a number of these modes corresponding to the zeroth nodal up to the sixth nodal diameter (ND) disc modes are illustrated in Figure 19.9. With the exception of the zeroth nodal diameter mode, the circumference of the disc divides into a number of equal parts with the deflection normal to the disc appearing as a sinusoidal wave around the circumference.

A radial impeller is a structure having rotational periodicity: once the geometry has been defined for a sector, the remainder of the structure can be obtained by repeated rotations through the sector angle, ψ, where $\psi = 2\pi/N$, N being the number of sectors, which is typically the number of main blades. This is illustrated in Figure 19.10, where for an impeller with a total of eight main and eight splitter vanes, $N = 8$.

The property of rotational periodicity can be exploited using complex constraints in the FE simulation to enable a series of analyses to be conducted on one sector only.

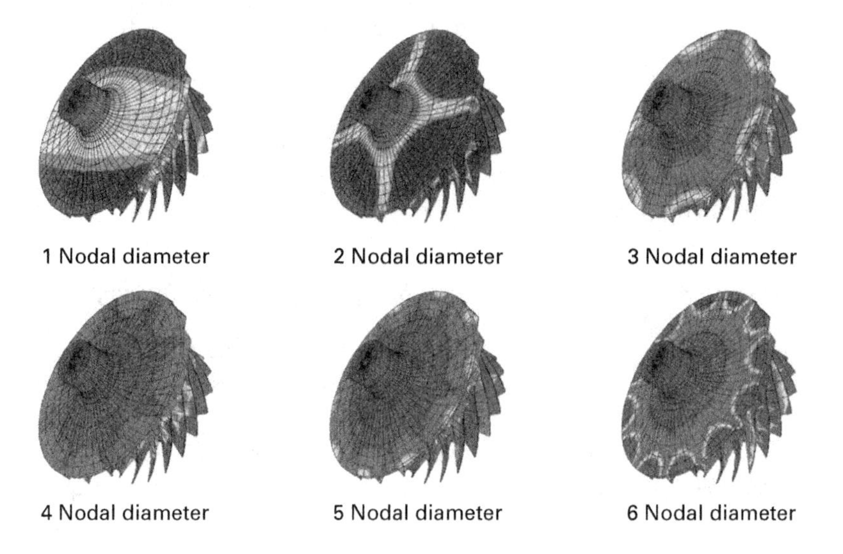

Figure 19.9 Nodal vibration patterns of an impeller disc from a principle component analysis (PCA) stress analysis code.

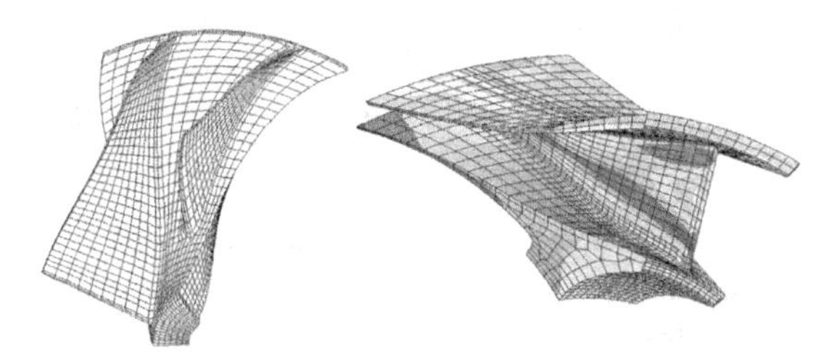

Figure 19.10 Impeller periodic sectors for mechanical analysis.

These sector analyses enable the characteristics of the complete structure to be described. Calculations which include the hub and the whole disc show that blade frequencies are affected by attachment to the hub; a large number of vibration modes exist which cannot be predicted using a freestanding blade element alone.

19.9.2 Centrifugal Stress Stiffening

The steady-state tensile stresses developed by centrifugal force act to increase the stiffness of the body and as a consequence increase the resonant frequencies. It was first described in turbomachinery as centrifugal stiffening (Campbell, 1924). Since the stress is predominantly a function of the square of the blade speed, the natural frequency due to stress stiffening will also change in a quadratic fashion with speed.

It is usual practice to determine the natural frequencies at both the zero and the 100% speed condition taking into account the stress stiffening. From these results, the frequencies at intermediate speeds may be approximated in the following manner:

$$\omega_n^2 = \omega_0^2 + (n/N)^2\left(\omega_{100}^2 - \omega_0^2\right), \tag{19.31}$$

where ω_n is the frequency at the speed n compared to the frequency at 0% speed condition, ω_0, at the 100% speed condition, ω_{100}, and N is the rotational speed at the 100% rotational speed.

19.9.3 Avoidance of Dangerous Resonances in Centrifugal Stages

Excitation due to general unsteadiness is primarily the result of unsteady aerodynamic forcing. A typical vibration analysis identifies the natural frequencies in resonance with the machine running speeds multiplied by discrete excitation sources such as vane and blade passing frequencies, producing so-called excitation orders (EO), sometimes called engine orders. An EO of five is a frequency equal to five times that of the running speed. Low-order modes which are composed of low excitation order (LEO) harmonics are known to cause low-order mode vibrations. Low-order vibrations may also be developed through the shaft from equipment attached to the machine, such as gearboxes, alternators and pumps, and from resonance in any attached spaces between the impeller and the casing. LEO forced response excites low nodal diameter fundamental blade modes which exhibit increased vibration levels.

Industrial experience suggests that any loss of symmetry can give rise to LEO forced response. With respect to centrifugal compressors the following are the most significant factors:

- Casing support struts, of which there are often three to four
- Inlet flow distortion from bends or components in the inlet duct
- In gas turbine applications, the number of individual combustor cans downstream of the compressor

- Inherent nonuniform pressure distribution of the flow field in an outlet volute and the once-per-rev distortion arising from the volute tongue
- Excitation from close-coupled static blade rows such as inlet guide vanes and vaned diffusers

Experience has indicated that damaging blade vibration in compressors may usually be avoided if the lowest vane natural frequency is greater than the fourth excitation order (4EO) at the maximum speed condition. However, although to clear 4/rev is good design practice, it may be impossible to achieve in combination with acceptable aerodynamic performance for very high flow coefficient impellers. Designers may decide to relax this based on experience of the likely excitation sources in a specific machine – for example, to clear the third excitation order (3EO) with the first flap frequency sitting between 3EO and 4EO at the design speed. There are precedents for this in, for example, low hub-tip ratio axial fan and compressor blades which typically have much lower natural frequencies than centrifugal impellers.

19.9.4 Campbell Diagram

The Campbell diagram is used to examine the location of the natural frequencies relative to the frequencies associated with low excitation order vibrations. The diagram maps rotational speed on the horizontal axis and frequency on the vertical axis. The relevant natural frequencies and the frequencies of exciting forces are plotted together in the diagram. The natural frequencies are plotted as a function of the rotational speed together with the frequencies associated with specific excitation orders. The coincidence of the natural frequencies with the exciting frequencies is often taken as the definition of resonance. Figure 19.11 provides an example of a

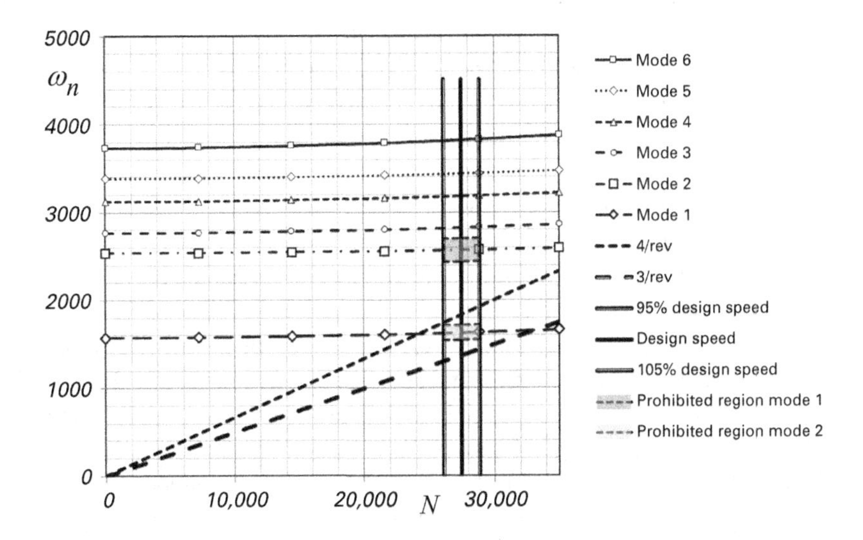

Figure 19.11 Example of a Campbell diagram calculated for 4ND modes.

Campbell diagram used to evaluate LEO excitation. The diagram plots the results of a modal analysis for an impeller conducted for the fourth nodal diameter (4ND) group of frequencies. The vibratory modes are the near-horizontal lines (not completely horizontal but parabolic due to stress stiffening), and the synchronous excitation orders are the sloping lines.

It is usual practice to conduct the evaluation over a speed range of ±5% of the design speed. However, for some applications, such as a motor-driven compressor connected to a national grid, the speed range may be reduced. For variable speed applications, the +5% maximum design speed is generally applied. To account for blade frequency variation due to manufacturing tolerances and material property variations, it is usual design practice to consider blade frequency ranges of ±5% of the calculated values. The combination of the preceding factors defines prohibited regions, which an excitation order line must not encroach, as indicated in Figure 19.11.

As the Campbell diagram shows, it is not possible to avoid all potential resonances at all times during the operation of a compressor. It must pass through resonant conditions during starting, stopping or changes in operating speed. Intersection at speeds well below the maximum are generally less important due to the low level of static stress and there is less energy available to excite them.

19.9.5 Interference Diagrams

The Campbell diagram presents the variation of natural frequency with rotational speed for a specific engine order or nodal diameter and a separate diagram is needed to assess the criticality for each individual nodal diameter. In the modal assessment of bladed disc assemblies, the interference or SAFE diagram compares not only the frequencies of the exciting harmonics and the natural frequencies but also the shape of these harmonics and the normal mode shapes at a specific speed. Separate diagrams are needed for each speed of operation, whereby only the position of the excitation orders change with rotational speed in the diagram.

An interference diagram is a plot for a certain operating speed which provides a visual analysis of possible resonant frequencies taking into account that each resonant frequency depends on the nodal diameter. The interference diagram is described in detail by Singh et al. (1988) and many subsequent publications. It is often referred to as the SAFE interference diagram (Singh's Advanced Frequency Evaluation). A good example of its use in the assessment of vibrations in radial impellers is given by Bidaut and Baumann (2012).

The resonant frequencies include distortion of the blades as well as the disc and hub and of the shroud (if present). Some impeller vibration modes are dominated by blade motion and some by hub or shroud motion, but all modes exhibit the characteristics of the nodal diameter and nodal circle. An interference diagram is usually presented as an underlying plot of natural frequencies plotted against number of nodal diameters leading to lines for each mode which are very approximately horizontal and are distributed vertically for different higher-order modes. The different excitation orders are plotted separately for a given speed, and this leads to a zig-zag line with increasing

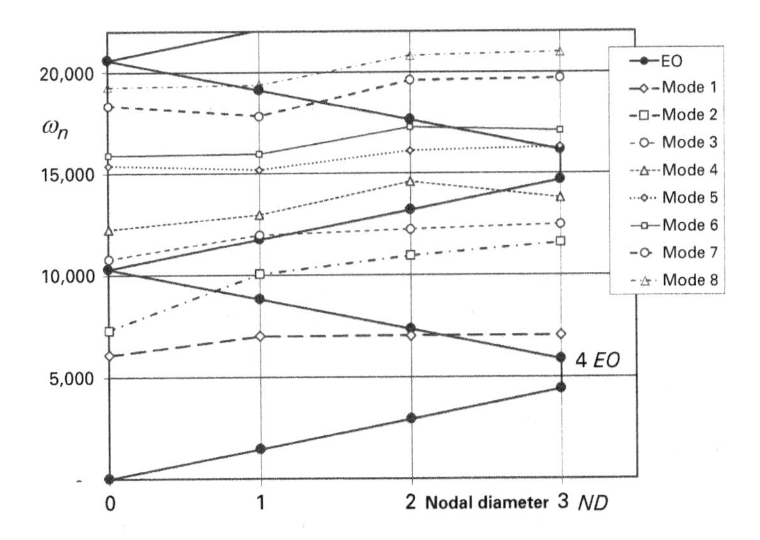

Figure 19.12 Example of an interference diagram for an open impeller with seven periodic sectors.

frequency. An example of an interference diagram for an open impeller with seven periodic sections in the impeller is given in Figure 19.12.

The zig-zag line represents the excitation order at the design speed, and in this case the impeller has seven sectors; these form a zig-zag line reversing direction at 3.5 and seven nodes and so on. The frequency of vibration of each mode for given nodal diameters are given and the risk of the design can be assessed by the nearness of the modes to the line representing the excitation orders. In radial impallers, it is usual to ensure that the fourth excitation order is below the first excitation mode.

19.9.6 High-Order Resonances

The forced response due to excitation generated as the impeller blades pass by the diffuser vanes or the inlet guide vanes may lead to lead to higher-order resonances. Flow variations of this type are primarily caused by inlet guide vane blade wakes or by the upstream pressure field from the diffuser vanes, both of which are experienced by the rotor blades as time-varying synchronous forces with a frequency or periodicity related to the rotational speed and the blade number.

As a general rule, two conditions are required to excite an assembly mode:

- The excitation frequency which is a multiple of the engine speed must coincide with a blade natural frequency.
- The excitation pattern must match the associated nodal diameter and modal pattern.

The primary flow wakes emanating from an upstream blade comprise the effects both of the potential flow field and of wakes. The magnitude of the pressure fluctuations due to the potential flow field decays exponentially with the length of the vaneless space between the blades; those due to wakes do not decay as rapidly.

For axial machines and for the axial vaneless space between the inlet guide vanes and the impeller, it is common practice to apply the following design guideline:

$$\Delta z \geq 0.25s, \tag{19.32}$$

where Δz is the axial space between adjacent blade rows and s is the circumferential spacing between the blades. For radial machines, the radial proximity of the diffuser vane leading edge and the impeller tip are subject to the following:

$$\Delta r \geq 0.15r_2, \tag{19.33}$$

where Δr is the radial gap between the impeller and the diffuser vanes and r_2 is the impeller tip radius.

For a constant speed machine, the number of stator vanes may be chosen to avoid a potential passing frequency resonance. If, however, it is assumed that – due to variations in material properties and the geometric manufacturing tolerance – the actual natural frequency can lie within $\pm 5\%$ of the computed value, the close spacing of the higher-order natural frequencies makes it extremely difficult to avoid a potential resonance. For a variable speed machine, a potential resonance is impossible to avoid. For this reason, a limit on the length of the vaneless space is specified in accordance with (19.32) and (19.33) in order to minimise the effects of blade passing resonances.

A practical difficulty is in the accurate prediction of high-order, more complex modes by FE analysis. These are felt to be less accurate than predictions of the simpler low-order modes, which have been well validated by experiment.

19.9.7 Experimentally Determined Campbell Diagrams

The Campbell diagram can also be determined from measurement of blade and disc vibrations during a sweep of speed at a constant throttle setting. This is also known as a waterfall plot. An example is given in the analysis by Kammerer (2009) of the vibration signal of a single strain gauge on a full blade during a test run at a single operating condition. The plot includes the lines for the vibratory modes and the engine order together with the local vibration amplitude at different frequencies. This illustrates the frequency-dependent loading of the full blade at the position of the applied strain gauge for the overall speed range relevant to the operation of the compressor. The Campbell diagram presented is only valid for one throttle position. For a different throttle position, that is, in the part-load region, other excitation forces occur in part which give rise to a different amplitude distribution. Since the strain gauge records the strain amplitude at a specific location on the blade, the response at other regions on the blade cannot be deduced from these results. However, with knowledge of the location and orientation of the strain gauge and the results of a finite element modal analysis, the response at other, perhaps more critical, regions may be deduced.

19.10 Disc Design

19.10.1 Introduction

Centrifugal loading due to high operating speeds is the principal load experienced by a compressor impeller. The objective of the disc design in an open impeller is to determine a hub disc profile that is able to support the blades (and shroud if present) and to satisfy the design stress and life requirements. A compressor hub disc will sometimes have a bore which accommodates a tie bar to attach the disc to the main shaft. In this case, the peak stresses of the system occur in the region of the bore; see Section 19.4.3. In a shrouded impeller, the design of the disc includes the determination of both the shroud and hub profile.

19.10.2 Impeller Hub Disc

For an impeller with a bore, the maximum stresses due to centrifugal loading generally occur in the bore, the back-face of the disc and the blend radius between the blade and gas swept surface of the disc, as indicated from the stress distribution of a typical high flow coefficient open impeller. The magnitude of these stresses is a function of ρu_2^2, where ρ is the material density and u_2 is the blade tip speed. In this case, there are high stresses on the disc back-face and also in the fillet radius of the blade near the trailing edge. Adjusting the shape of the backplate leads to a different deflection of the tip, an effect known as the umbrella or the flowering effect. In this way, the design of the backplate also affects the maintenance of the clearance gap at the impeller tip or the risk of a mechanical rub at different speeds.

A typical generic meridional disc profile is illustrated in Figure 19.13. The vane geometry, which is determined from aerodynamic considerations, defines profiles of the hub and shroud lines; for an impeller without a shroud, the hub and tip radii of the disc, r_2 and r_1, are defined. Hence, for a given vane, one of the principal geometric parameters which can be varied is the bore radius, r_3, whereby from Section 19.4.3, the tangential hoop stress at the bore is a function of $r_2^2 + Cr_3^2$, where C is a constant. Thus, for a given rim radius, r_2, the bore stress increases with an increase in bore radius. Additional parameters of consequence are the back-face geometry, which is defined by a combination of the variables r_4, r_5, z_2, r_b and the thickness of the rim, z_1.

To demonstrate the importance of the design of the backplate, a series of stress analyses have been conducted for a typical aluminium compressor impeller with a variety of meridional disc hub profiles, as shown in Figure 19.13. In this study, the effect of variations in the profile of the disc back-face is investigated while keeping all the other disc geometric parameters, such as the rim radius, r_1, constant.

A summary of the results for different back-face geometry is presented in Table 19.3 for an impeller with a rim radius, r_2 of 26.25 mm, an operating speed of 140,000 rpm and a tip speed of 385 m/s. The results of this study indicate that a minimum inertia disc is obtained by adopting a disc back-face design similar to profile (a). However, the back-face and fillet radius stresses are then the highest of all the

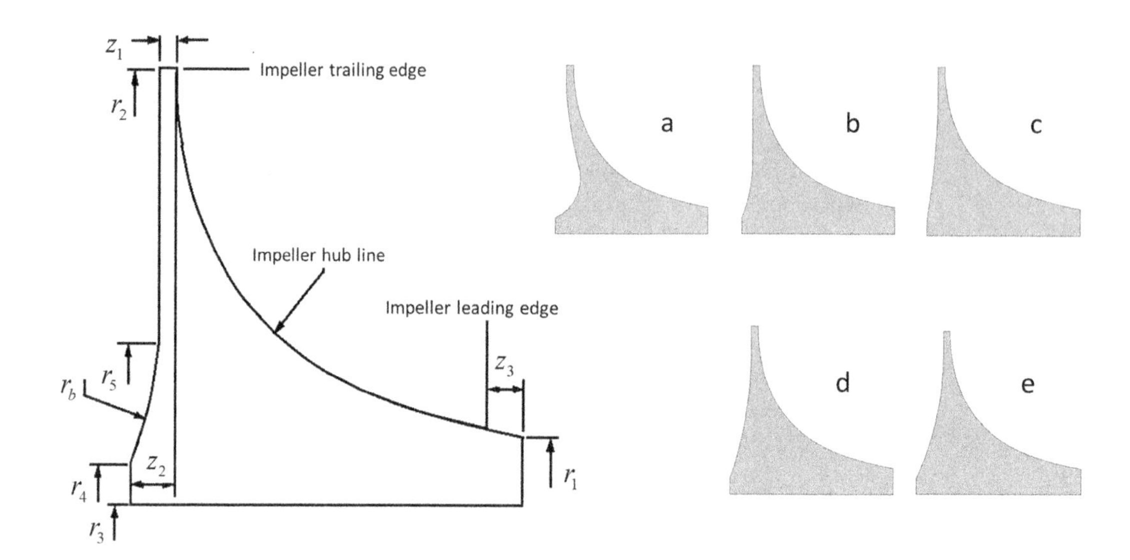

Figure 19.13 Parameterisation of impeller disc profile and disc profiles with varying back-face geometry.

Table 19.3 Stress and displacement results with the variation of the back-face.

Back – face type	Max bore stress (MPa)	Max back-face stress (MPa)	Max fillet radius stress (MPa)	VWA stress (MPa)	Vane tip axial displacement (mm)
a	178	204	188	75.912	–0.0360
b	183	125	160	68.996	–0.0420
c	193	106	156	68.028	0.0016
d	169	95	151	66.356	0.0098
e	164	89	148	65.493	0.0140

profiles studied; also, profile (a) has the largest axial displacement. To satisfy maximum stress requirements, there may be a need to modify the geometry of the back-face towards profile design (b). The axial vane tip displacement is negative in all cases. This leads to an increase in the tip clearance with speed and is influenced by the design of the back-face. Gradual stress reductions in all the primary maximum stressed regions are achieved by increasing the axial length, z_2; however, this also increases the vane tip clearance.

The results of changes in bore radius, while maintaining a disc profile conforming to profile design (b), are given in Table 19.4. The results demonstrate the rapid increase in bore stress with bore radius. With the exception of the weighted-volume average (WVA) von Mises stress, the remaining stress and displacement measures remain essentially unchanged with the bore radius.

Table 19.4 Effect of a variation in bore radius.

R3 (mm)	Max bore stress (MPa)	Max back-face stress (MPa)	Max fillet radius stress (MPa)	WVA stress (MPa)	Vane tip axial displacement (mm)
2.00	177	125	159	66.535	–0.0041
2.55	183	125	160	68.996	–0.0042
3.00	195	125	163	74.641	–0.0045

For many applications a good pragmatic approach to the disc is to choose a backface of type (b) or (c) with a backface extension z_2 of about 5–6% of the tip diameter. This reasonably balanced design leads to moderate bore stress and tip deflection and keeps the zone of maximum bore stress away from the downstream contact area.

19.10.3 Impeller Shroud

Movement outwards under rotational forces is supported by the shroud itself through an increase in the tangential hoop stress. When attached to leaned blades, the outward movement leads to bending stresses in the root zones where the blades are attached to the hub and shroud. There is no general rule for the design of the shroud with respect to stress, with the exception of the shroud thickness, which is kept as low as possible in order to reduce the mass and hence the centrifugal force acting on the blades and the whole structure. An example is shown in Figure 1.11.

19.10.4 Thermal Stresses

Thermal stresses are also introduced into the disc with the temperature rise from the gas being compressed. This temperature gradient induces compressive stresses at the rim of the disc; at the bore or centreline, tensile stresses are introduced. The temperature distribution on the gas swept surfaces of the vane and hub can be determined for an FE analysis by mapping the results of a CFD analysis to the FE model of the impeller.

19.11 Assembly Designs

19.11.1 Axial Tie Bolt

The majority of compressor impellers in turbochargers are attached to the main shaft via an axial tie bolt. If a tie bolt arrangement is used, the assembly load provided by the bolt needs to be determined to ensure a positive drive at the running speed. A typical example of such an assembly is illustrated in Figure 19.14. The compressor impeller is assembled to engage with mechanical teeth or the equivalent at B. An alloy

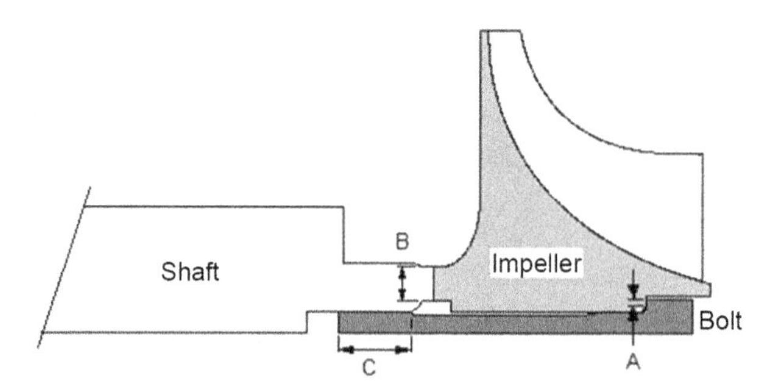

Figure 19.14 Typical tie bolt assembly.

steel or titanium alloy centre bolt, which screws into an axial threaded hole machined in the impeller shaft, C, is pretensioned by applying a specified torque at the bolt head. To simulate this assembly, a two-dimensional axisymmetric finite element model comprising the centre bolt, the impeller disc and the relevant section of shaft can be developed and a series of linear elastic stress analyses conducted to simulate a typical operational cycle. The cycle may comprise the assembly condition with a prescribed axial centre bolt load at uniform ambient temperature, and a cold running condition taking into account the additional effects of centrifugal load. The steady-state running condition may also be simulated in order to determine the effects of centrifugal load and thermal expansion. The assembly steady-state temperature distribution can be calculated from the temperatures along the gas swept surface at the hub line of the impeller and estimates and/or measured temperatures in other areas, particularly in the vicinity of the main shaft. In some applications, a simulation of the hot zero-speed condition may be needed.

19.11.2 Shrink Fit

In multistage process compressors and in some single-stage applications, a shrink fit of the disc onto a shaft may be used. The shrink fit needs to be determined to maintain a positive drive at the running speed and should be designed with sufficient margin in order to withstand also any additional thermal load which appears in transient conditions (such as startup, shutdown, surge, etc.). The shrink fit offers a uniform stress distribution over the whole circumference and a constant self-centring effect. It is designed so that sufficient shrinkage still exists to transmit the torque and the axial thrust once the bore has expanded due to centrifugal force at maximum speed. The impellers can be removed without damage by heating whenever necessary.

A design involving either shrink fitting the impeller onto a shaft or fitting a tapered bore on to a tapered shaft is illustrated in Figure 19.15. This is a particular case, and in some applications the shrink is along the whole length of the impeller. Radial contact of sufficient magnitude to transmit the torque to the shaft is usually designed to occur over two distinct axial regions, such as c_1 and c_2. However, since c_2 is close to the region where the maximum impeller bore stress occurs, there is a

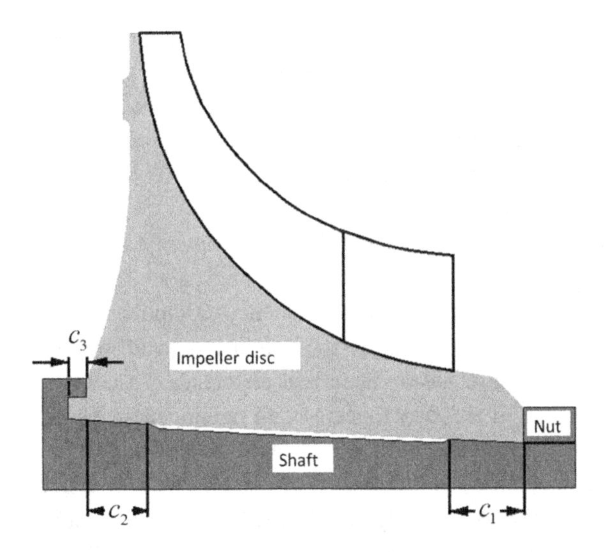

Figure 19.15 Shrink fit assembly.

possibility of loss of radial contact, so careful design is necessary to prevent this from occurring. An alternative is to provide a positive torsional engagement, for example in region c_3.

19.12 Rotordynamics

19.12.1 An Introduction to Compressor Rotordynamics

The branch of engineering science that examines the vibrations and whirling motion of rotating shafts is known as rotordynamics. The first paper on the whirling motion of a rotor was published by Rankine (1869), where he examined the relationship between centrifugal and restoring forces in a whirling rotor and concluded that operation above the first critical speed is impossible. This conclusion turned out to be wrong. Further information on the historical development of rotordynamics and on the first rotor shafts that ran above the first critical speed in the Laval and Parsons steam turbines is given by Gasch et al. (2006) and Vance et al. (2010). Nearly all modern compressors operate above the first critical speed, which is still considered to be the most important vibratory mode in the system. Continuous operation at or near the critical speeds is avoided by ensuring a certain safety margin between the operating speed and the nearest critical speed of the system.

The intention of this section is to raise the awareness of aerodynamic designers to the issues involved. For those who would like to pursue this field of study, many specialised books are available. This brief introduction to rotordynamics is presented with the intention of familiarising the reader to the most important concepts and the engineering of robust vibration-free compressor shafts. The fundamental knowledge required to understand the key issues in rotordynamics is briefly introduced with an analytic solution of a simplified rotordynamic system. The characteristics of this simple rotor assembly, known as the Jeffcott single mass rotor, are examined using the equations of motion. Some limits on multistage compressor design from rotordynamics can be derived even on the basis of the simple Jeffcott rotor; for example, the shaft should be thick and short.

Analytical solutions are only available for the simplest cases so that complex computer-based simulations are used for more detailed vibration analysis of rotors. Synchronous vibrations have a dominant frequency component the same as the rotating speed of the shaft. Subsynchronous vibration, or whirling, has a dominant frequency below the operating speed. The main elements of a compressor rotordynamic system are the shaft of the rotor together with the impellers, bearings and seals. The shaft supports the impellers, and the bearings support the rotating shaft. The bearings provide damping for stabilisation of the system and provide a limit to the rotor vibration. Sealing systems not only prevent leakage flows but may also influence the damping of vibrations and, in some cases, may even be the cause of additional fluid-induced rotordynamic effects. The objective of the designer is to reduce vibrational energy as much as possible, to ensure smooth running without noise, to extend machine life and reduce costs by preventing failure due to whirling at critical speeds. Whirling at the critical speed is not the only possible cause of damage. Lateral vibration should, in general, also be kept low to avoid contacts between the rotor and the casing, known as rubs, in locations with small clearance and to keep bearing forces low. Shaft stresses are also an issue for lateral vibrations, but torsional vibrations in turbocompressor trains normally can cause higher stresses.

19.12.2 Rotor Instability

The rotordynamics of centrifugal turbocompressors is affected by the impellers transferring energy from the rotor to the process fluid and by the complete shaft which drives the rotation of the impeller via gears and a coupling. Additional components involved are the bearings and the seals used to reduce internal leakage by separating regions of high and low pressure while allowing the rotor to rotate freely. A rotordynamic analysis includes not only the shaft and the impellers but also the bearings and the seals. The bearings not only support the rotating shaft but also provide damping and stiffness to stabilise the system and to limit the rotor vibration. Seals not only prevent undesired leakage flows but may also influence the damping of the system and, in some cases, may even be the cause of additional fluid-induced rotordynamic effects.

Many factors contribute to energy being transferred from rotation of the shaft to cause lateral vibrations. The most well known of these is rotor unbalance, when the centre of mass of the rotor does not coincide with the rotational axis. The rotor unbalance acts like an external centrifugal force and excites vibrations synchronous with the rotational speed, similar to 1EO vibrations of the blades. In some cases, the unbalance may also excite vibrations at higher excitation orders. Free vibration takes place when the system vibrates due to forces inherent in the system itself and occur at the characteristic natural frequencies of the system and subside at a rate dependent on the damping. Free vibrations may also occur during a change in the system operation point or speed of rotation.

Forced vibration occurs due to external forces, which are usually related to transient or periodic aerodynamic activity in the compressor, such as rotating stall or surge as described in Chapter 18. API standard 617 (2016) identifies some additional sources for these periodic disturbances to be blade and vane passing frequencies, gear tooth meshing, coupling misalignment, loose rotor components, acoustic resonance and aerodynamic cross-coupling forces. If the external stimulus is periodic, the system will vibrate at some combination of the frequency of excitation and its multiples, with amplitudes dependent on the position of the excitation frequencies to the system natural frequencies. For nonperiodic forces, the response of the system will again be at some combination of its natural frequencies.

A further type of rotor instability is a self-excited vibration of the rotor components which also occurs close to one of the natural frequencies. In centrifugal compressor rotors, these vibrations can be driven by fluid dynamic instabilities in the seals and bearings by a mechanical rub causing a hot-spot on the shaft or by cross-coupling stiffnesses in bearings and seals.

Various modes of vibration may be present in the rotor shaft, such as torsional vibration, longitudinal vibration along the axis and lateral (bending) vibration of the shaft. Torsional vibration is twisting of the rotor statically and dynamically about its rotational axis. Torsional modes are typically lowly damped, which causes high response amplitudes while crossing critical speeds. When electrical motors are used to drive compressor trains, torsional excitations are introduced to the shaft system during startups or during electrical fault conditions such as short circuits. The response amplitudes of the shaft system to such resonant conditions might exceed multiples of the shaft nominal torque. This makes the torsional vibration analysis a fundamental step in the layout of a compression train to ensure the structural integrity of the shaft components and to guarantee safe operation of the equipment. The API standard 617 gives requirements and recommendations for good practice of torsional analyses.

Lateral vibrations manifest themselves as shaft vibrations, usually measured by means of noncontacting eddy current probes, and as casing or bearing pedestal vibrations, usually measured by seismic velocity or acceleration probes. The objective of a lateral rotordynamic analysis is to predict accurately the expected vibration amplitudes, the mode shapes, the resonance frequencies and the stability behaviour of the system in its intended operation. This knowledge enables an engineer to understand the cause of any vibration and take action to reduce it to an acceptable limit.

Detailed vibration analyses are carried out to ensure the operational availability of all modern compressors. The most widely accepted requirements for rotordynamic analyses and acceptance criteria are given in the API standard 617. Lateral rotordynamic analyses normally consist of an unbalance response analysis to determine the critical speeds and a complex eigenvalue analysis to assess the stability behaviour of the rotor in the intended service.

19.12.3 Unbalance Response Analysis

These analyses compute the lowest lateral critical natural speeds based on the bearing stiffness and damping to ensure that the critical speeds are sufficiently separated from the desired running speed range. The required separation margin is normally a function of the amplification factor of the respective critical speed. The speed range of modern turbo-compressors is normally situated between the first and the second lateral critical speed.

The unbalance response analysis allows assessment of the sensitivity of the rotor system to possible unbalance distributions of the rotor, such as residual unbalances of the mounted impellers, fouling (deposits) on the rotor or an additional unbalance caused by the removal and remounting of a coupling hub. These and many more are examples of synchronous harmonic calculations where the exciting force is at a fixed location on the rotor. But there are also examples of nonsynchronous harmonic excitations where the force is rotating at a speed different from that of the rotor, such as the resulting forces from a rotating stall, which normally exhibits frequencies in the subsynchronous region.

The bearing properties have a distinct influence on the frequencies and the amplification factors of the critical speeds. It is often necessary to assess the sensitivity of the rotor system with respect to bearing property variations, such as the following:

- Bearing clearance – nominal, minimum/maximum manufacturing tolerance
- Bearing pad curvature (preload) – nominal, minimum/maximum manufacturing tolerance
- Lube oil supply temperature – nominal, minimum/maximum allowable

Detailed vibration analyses are carried out to ensure the operational availability of all modern compressors. Both torsional and lateral analyses are standard analyses in a project. Stability normally is a lateral issue. These analyses compute the first and second lateral critical natural frequencies based on the bearing stiffness and damping together with the stability and torsional analysis. A rotordynamic analysis allows the precise determination of the critical natural frequencies, which are principally lateral vibrations of the rotor caused by unbalance forces. The first lateral critical speed is generally considered to be the most important resonance mode. Many multistage compressors have long, thin shafts to accommodate the impellers, seals and bearings, leading to fairly flexible rotors with a first lateral critical speed that is usually below the operational speed of the shaft. Because of this, multistage compressors are usually operated at a speed between the first and second lateral critical speeds. In fixed speed machines, continuous operation near any of the critical speeds is avoided by maintaining a speed margin of 15% between these and the operating speed. In variable speed machines, where the operational speed may be close to the critical speed or may need to pass through this speed to a different operating condition, the damping levels need to be sufficiently high to avoid severe levels of vibration near the critical speeds.

Forced vibration occurs due to external forces, which are usually related to transient or periodic aerodynamic activity in the compressor, such as rotating stall or surge as described in Chapter 18. API 617 identifies some possible additional sources for these periodic disturbances to be coupling misalignment, loose rotor components, acoustic resonance and aerodynamic cross-coupling forces. Frequencies due to blade and vane passing and to gear tooth meshing are usually too high to be of relevance to the rotordynamics. If the external stimulus is periodic, the system will vibrate at some combination of the frequency of excitation and its multiples, with amplitudes dependent on the position of the excitation frequencies to the system natural frequencies. For nonperiodic forces, the response of the system will again be at some combination of its natural frequencies.

A further type of rotor instability is a self-excited vibration of the rotor components. Self-excitation is a stability issue which does not occur close to one of the natural frequencies but actually excite a natural mode with a vibrational amplitude. In centrifugal compressor rotors, these vibrations can be driven by fluid dynamic instabilities in the seals and bearings. A mechanical rub causing a hot-spot can lead to a thermal instability and is a synchronous vibration. They cause the self-excitation of natural modes of the shaft or by cross-coupling stiffnesses in bearings and seals.

19.12.4 The Jeffcott Single Mass Rotor

Rotordynamic systems have complex dynamics for which analytical solutions are only possible in the simplest cases. With modern computational power, numerical solutions for 2D rotordynamic analyses are readily available. However, these numerical analyses do not provide the insight that can be obtained from analytical solutions of simple systems, such as that given in this section, which is adapted from the descriptions of Yoon et al. (2013).

The vibration theory for the Jeffcott single mass rotor was first developed by Föppl but named after the publication of Jeffcott (1919). Both publications employed a simplified rotor/bearing system, commonly known as the Jeffcott rotor or the Laval shaft, which is analogous to the mass-spring-damper system used to analyse mechanical vibrations in blades; see Section 19.9. The Jeffcott rotor is a simplified lumped parameter model used to solve the equations of motion analytically. It is a mathematical idealisation with several simplifications that may not reflect actual rotordynamics.

The Jeffcott single mass rotor model comprises a rigid rotor disc rotating at angular frequency ω, as shown in Figure 19.16. The shaft is considered to have no mass but

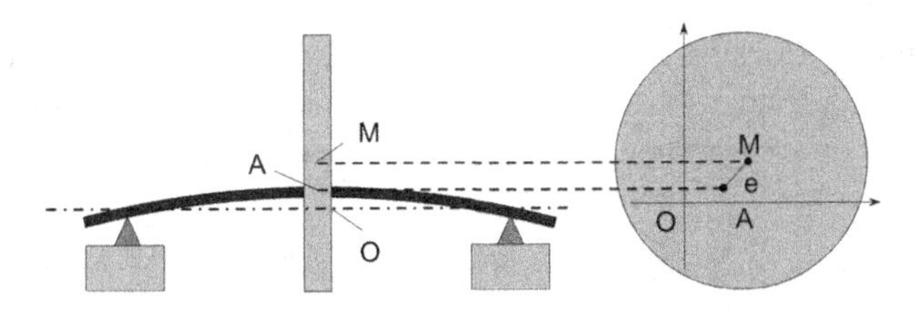

Figure 19.16 The Jeffcott rotor model with rigid bearings.

supports a disc of mass m at its centre and acts as a spring restoring any displacements of the disc. The disc can only move in the plane defined by axes 1 and 2 within the disc and the mass centre M and centre of rotation of the shaft A are different points separated by the eccentricity of the mass, e. The bearings are considered to be rigid. The coordinates of the shaft centre A, (y_1, y_2), relative to the axis defined by the bearing supports, O, are not the same as the coordinates of the centre of mass, (z_1, z_2), of the disc, which is eccentric to the rotational axis through O. The rotor position is defined by

$$z_1 = y_1 + e\cos\phi, \quad z_2 = y_2 + e\sin\phi. \tag{19.34}$$

When rotating at constant speed, the forces acting on the disc are its inertia, the spring forces due to the flexible shaft with a spring coefficient of k and damping forces with a damping coefficient of c, leading to an equation of motion as follows:

$$m\ddot{z}_1 + c\dot{y}_1 + ky_1 = 0, \quad \text{and} \quad m\ddot{z}_2 + c\dot{y}_2 + ky_2 = -mg. \tag{19.35}$$

The damping coefficient, c, is a feature of the model to study its affect, but the model does not have any defined damping related to any shaft structural damping, fluid damping acting on the rotor from the flow and the damping added by the bearings. The lateral bending stiffness, k, at the centre of a simply supported uniform beam is given by $k = 48EI/l^3$, where E is the elastic modulus of the beam, l is *the* length between the bearings and I is the moment of inertia, which for a uniform cylindrical shaft of diameter D is given by $I = \pi D^4/64$.

Substituting (19.34) into (19.35) and assuming a constant speed of rotation, the following equations of motion for the displacement of the disc are obtained:

$$\begin{aligned} \ddot{y}_1 + (c/m)\dot{y}_1 + (k/m)y_1 &= e\omega^2\cos\omega t \\ \ddot{y}_2 + (c/m)\dot{y}_2 + (k/m)y_2 &= e\omega^2\sin\omega t - g. \end{aligned} \tag{19.36}$$

Following the approach given in Section 20.4.3 for a vibrating beam, the equations for the motion with no eccentricity of the mass become

$$\begin{aligned} \ddot{y}_1 + (c/m)\dot{y}_1 + (k/m)y_1 &= 0 \\ \ddot{y}_2 + (c/m)\dot{y}_2 + (k/m)y_2 &= 0. \end{aligned} \tag{19.37}$$

The solution to this second-order homogeneous system takes the form of

$$y_1 = A_1e^{st} \quad \text{and} \quad y_2 = A_2e^{st} \tag{19.38}$$

for a complex value of s and where the values of the constants A_1 and A_2 depend on the initial conditions. Considering the undamped equations with $c/m = 0$ gives rise to the characteristic equation $s^2 + k/m = 0$ with the solution $s = \pm\,\omega_n$ and lead to an undamped natural frequency known as the critical speed of the rotor of

$$\omega_n = \sqrt{\frac{k}{m}} = \sqrt{\frac{48EI}{l^3m}} = \sqrt{\frac{48E\pi D^4}{64l^3m}} = \frac{D^2}{l^2}\sqrt{\frac{3El\pi}{4m}}. \tag{19.39}$$

For a shaft with no damping and no eccentricity, these equations are similar to the equations of motion of a vibrating beam; see Section 19.5.3.

This analysis shows clearly that the critical frequency increases with an increase in the diameter of the shaft, with a decrease in the length of the shaft and with a decrease in the eccentric mass of the disc. For a high critical frequency, the shaft should be thick and short, and any unbalance of the impellers should be small. The unbalance does not influence the critical speed, but it influences the vibration level. It also identifies that in multistage process compressor applications, there is a strong pressure from rotordynamic considerations for the stages to be as short and as light as possible with a thicker shaft. For this reason, in comparison with typical single-stage impellers with a ratio of hub diameter to impeller diameter of $0.25 > D_{h1}/D_2 > 0.3$ and a typical axial length ratio of $L_{imp}/D_2 > 0.35$, multistage impellers have a higher hub diameter ratio of $0.35 > D_{h1}/D_2 > 0.4$ with much shorter impellers, $L_{imp}/D_2 < 0.25$.

Considering the Jeffcott rotor with a nonzero damping, the solution of (19.37) has a characteristic equation and a solution of

$$s^2 + \frac{c}{m}s + \frac{k}{m} = 0, \quad s_{1,2} = -\frac{c}{2m} \pm i\sqrt{\left(\frac{c}{2m}\right)^2 - \frac{k}{m}} = -\frac{c}{2m} \pm i\sqrt{\left(\frac{c}{2m}\right)^2 - \omega_n^2}, \tag{19.40}$$

which is similar to the simplified blade vibration equation (19.20). The behaviour of the damped system depends on the two values of s determined by the radical, the term in the square root. As with the blade vibrations, a critical reference quantity for the damping coefficient, c_c, can be determined which reduces the radical to zero such that

$$\left(\frac{c_c}{2m}\right)^2 - \omega_n^2 = 0, \quad c_c = 2m\omega_n, \tag{19.41}$$

and the actual damping of the system can be expressed in terms of a nondimensional damping ratio, $\zeta = c/c_c$, such that

$$s_{1,2} = -\varsigma\omega_n \pm j\omega_n\sqrt{1-\zeta^2}. \tag{19.42}$$

This is written in this way as normally the rotorbearing system is underdamped, leading to damped oscillations with a rate of decay of $e^{-\zeta\omega t}$. The typical values of the damping coefficient are between $0.015 < \zeta < 0.3$. API standard 617 requires a damping ratio of 0.016 (corresponding to log. dec. of 0.1) for safe operation. The dimensionless amplitude of any forced vibrations and the phase angle are then similar to those already given in Figure 19.3 for the vibration of a mass-spring-damper system and as shown in Figure 19.17. When the damping ratio is small, the amplitude ratio increases rapidly near the critical speed as the unbalance forces excite the rotor resonance mode. For high frequencies, the amplitude ratio approaches 1.

The full equation (19.36) has a steady-state solution given by

$$y_1(t) = r\cos(\omega t - \phi), \quad \text{and} \quad y_2(t) = r\sin(\omega t - \phi) - y_g, \tag{19.43}$$

where values of the amplitude of the disc displacement, r, the phase angle, ϕ, and the defection of the shaft due to gravity, y_g, are given by

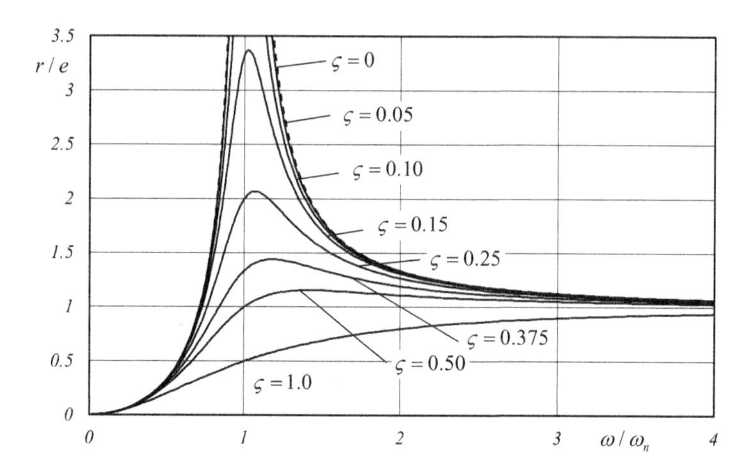

Figure 19.17 Dimensionless amplitude of the forced response of the Jeffcott rotor versus frequency ratio, (19.44).

$$\frac{r}{e} = \frac{(\omega/\omega_n)^2}{\sqrt{\left[1-(\omega/\omega_n)^2\right]^2 + (2\zeta\omega_n)^2}}, \quad \tan\phi = \frac{2\zeta\omega_n}{1-(\omega/\omega_n)^2}, \quad \text{and} \quad y_g = \frac{mg}{k}. \tag{19.44}$$

An important consequence of this analysis is that compressor rotors must be well balanced, such that any mass unbalance acting on the rotor to cause vibrations is small. To achieve this, high-speed balancing of complex rotors is usually used. The mechanical damping of the rotor is usually small and the amplitude of the lateral oscillation increases significantly while operating close to the critical speed. For high-speed compressors, it may happen that the critical speed of a rotor occurs within the normal operating speed range of the machine, and the damping needs to be sufficient to avoid large vibrations, or the critical speed should be passed quickly. One of the important effects, which can be shown by this simple example, is the self-centring effect when running above a critical speed. This was detected by Laval and is what Rankine failed to notice.

The analytic model presented in this section is an oversimplification of real cases, as it does not take into account complex aspects. The rotor has several different masses on the shaft, the shaft has its own mass, the bearings have different arrangements such that the mass is not necessarily central, the bearings are flexible and not rigid and any internal damping of the rotor or gyroscopic effects are not present. This leads to several critical speeds and requires more complete numerical models for the estimation of vibration characteristics.

19.13 Rotor Modelling

19.13.1 Theoretical Model Development

Since the work of Jeffcott (1919), many different approaches to the simulation and modelling of the stability of rotors have been developed; see Gasch et al. (2006). The most prevalent method used today for rotordynamics analysis is the finite element method. Although the structural properties of the rotors and their supports could well be modelled by any general finite element program, special engineering tools with computational details of bearings, seals and supports are used. Important examples of specialist rotordynamic software are MADYN 2000, XLRotor and Dyrobes. The advantage of specialised rotordynamic tools is the handling of bearings and seals and their characteristics. They are speed dependent and load dependent, and in some cases are even frequency dependent. In the latter case, they can no longer be handled with damping and stiffness coefficients. Magnetic bearings, for example, can also not be correctly treated just by coefficients.

An important issue is the interaction of fluid forces in the fluid film bearings and seals with the rotors (Dietzen and Nordmann, 1987; Childs, 1993). Fluid forces were originally modelled as stiffness and damping coefficients, but thanks to more sophisticated analysis tools and measurements, it recently has been recognised that this approach is too restrictive (Schmied, 2019).

19.13.2 Rotor Stability

As mentioned previously, there are three types of lateral rotordynamic vibration: forced vibration due to unbalance, forced vibration due to external forces (such as rotating stall) and self-excited vibration (usually referred to as stability). Rotordynamic stability in centrifugal compressors depends on forces in the bearings, seals and leakage flow passages in the impeller side spaces that can be stabilising or destabilising. The stabilising forces include the bearing and seal damping, which dissipate energy. The destabilising forces arise from cross-coupling mechanisms, which cause tangential forces that increase rotor vibration amplitudes.

Destabilising forces in compressors extract energy from the process gas and rotation to excite a rotor vibration mode (usually the first forward whirling mode). Brun and Kurz (2019) point out that if these forces grow large enough and overcome the damping forces in the rotor system, the vibration amplitude of that mode will grow unbounded and can cause operational issues, unplanned shutdowns or even a lack of operability. The fluid film bearings support the rotor through hydrodynamic lubrication. As a shaft rotates, the lubricant creates a hydrodynamic fluid pressure between the bearing surface and rotor that supports the rotor. Lubrication models are used to predict this fluid film pressure profile for the different bearing types and provide the necessary information on the stiffness and the damping of these elements. Seals are employed to prevent gas leakage between the compressor stages, and the compressible flow in these seals generate lateral forces that act on the rotor in the form of stiffness and damping,

The destabilising forces are commonly called cross-coupling forces. These forces cause an orthogonal forward displacement of the rotor for each normal rotor displacement. If the force is sufficient in magnitude to overcome the damping in the system, an instability occurs. Stability is a function of rotor geometry, bearing-to-shaft stiffness ratio and the hydrodynamics of the bearings and seals.

19.13.3 Rotordynamic Modelling

In modern rotordynamic analysis, the compressor rotor shaft is modelled as a series of smaller finite elements that appropriately capture its shape with steps in the shaft diameter and the attachment positions of the impellers, thrust bearings and other components which add mass or stiffness to the shaft. Examples showing the cross section of a shaft, and different possible finite element models of the shaft with different levels of detail are given by Vance et al. (2010) and shown in Figure 19.18.

The impellers are modelled as rigid discs at the appropriate location. The different models show the inclusion of the contribution of the impeller hubs to the bending stiffness of the shaft, the inclusion of foundation stiffness and the inclusion of a flexible coupling. The elements of the impeller hubs are treated as additional layers overlaid on the elements of the underlying shaft. According to the API standard 617 (2016), a foundation model should be included if the flexibility of the foundation is less than 3.5 times the bearing stiffness.

More complex analyses are possible. An analysis by Biliotti et al. (2014) describe a rotordynamic model of a beam compressor, including the following features. Each

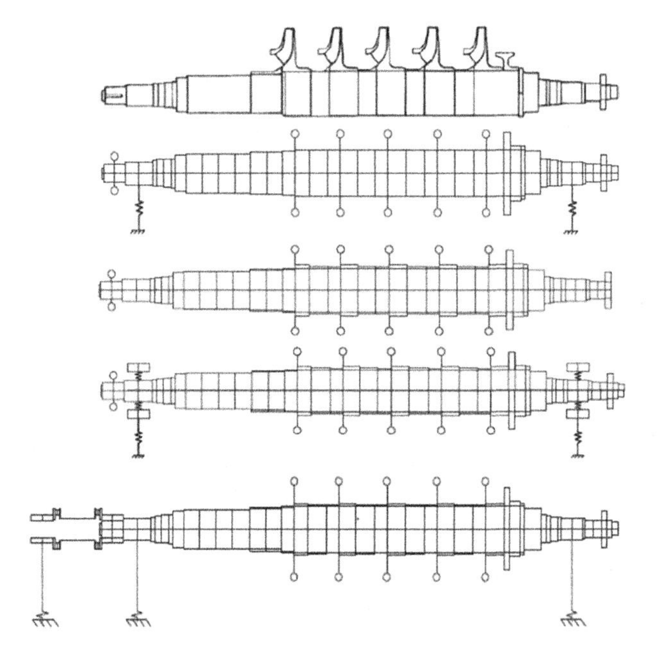

Figure 19.18 Rotor cross section with impellers and different levels of finite element modelling.

impeller is modelled as an inertial element with the relevant lumped mass placed in the centre of gravity. The flexible coupling, the thrust collar, balance piston and dry gas seals are also modelled as inertial elements with a lumped mass placed in the relevant centre of gravity. Hydrodynamic tilting pad type bearings are modelled through a pair of stiffness and damping matrices. Internal seals are modelled at their respective locations through the relevant dynamic coefficients. This model has been used to assess the effects of rotating stall on the vibration characteristic of the rotor.

The simplest model of the fluid film forces acting between a bearing and rotor represents these as a system of stiffness and damping coefficient for springs (representing stiffness) and dampers (representing damping). These rotordynamic coefficients are a function of several parameters, including bearing type, bearing geometry, oil properties, loading and rotational speed. The stiffness and damping forces from the bearings are calculated and applied to the rotor model in the lateral rotordynamic analysis.

Rotordynamic stability is assessed through the calculation of damped natural frequencies and the associated logarithmic decrements. For centrifugal compressors, it is necessary to include potential destabilising forces from all relevant components, including journal bearings, seals and impellers. Internal damping is due primarily to friction at rotor component interfaces.

19.13.4 Rotordynamic Specifications for Compressors

Both the International Organization for Standardization (ISO) and the American Petroleum Institute (API) published sets of specifications developed for different types of turbomachine used in industrial applications, although the API standards are largely preferred in the chemical and petrochemical industries.

20 Development and Testing

20.1 Overview

20.1.1 Introduction

The objective of this chapter is to cover the essential aspects of the modern design and development process for radial flow turbocompressor stages. A general overview of the design process is given. This is followed by a description of the different phases of the design process, including the conceptual design of the compressor type, the preliminary design of the components, the geometry specification of the ducts and blade rows, the blade-to-blade design, the throughflow design, 3D CFD performance analysis and FEM mechanical analysis. The description of the design process draws on the detailed information already given in the earlier chapters. For completeness, the use of inverse methods and optimisation methods in design is also discussed.

The final decision about the quality of a design is made through compressor testing. The chapter concludes with a section on the testing of compressors; this includes a discussion of the different types of tests, testing methods, standards, guidelines and procedures. Information about suitable instrumentation is also provided. Finally, there is a short review of recent experimental studies from some of the most active research groups with experimental rigs.

20.1.2 Learning Objectives

- Have a knowledge of the different phases of centrifugal compressor design.
- Understand the different types of compressor tests.
- Have an awareness of the international standards for compressor testing.
- Be able to explain the different types of instrumentation used in compressor testing.

20.2 Design and Development for Centrifugal Compressors

20.2.1 Design Process

The design process for radial compressors in high-technology products such as gas turbines, turbochargers and industrial compressors must ensure that new designs reach very good performance with acceptable mechanical integrity. Over the last two decades, optimisation of the fluid and structural aspects above the high level already achieved has become ever more challenging. On the other hand, new issues have also come to the fore. Emphasis is now placed on the ability to design quickly, close to 'right-first-time' and at low cost. These aspects shorten the time for product entry into service and also improve competitiveness. One approach in terms of achieving these objectives is the use an integrated design system which steps logically through the 1D, 2D and 3D design tools during the process of design optimisation. Each of these tools is part of an effective screening process which successively limits the number of designs to be considered at each stage (Robinson et al., 2011). In this way, unsuitable designs can be eliminated early in the design process using simple methods prior to using the more complex analysis tools. This provides more time for the engineer to examine real issues with the latest analysis techniques rather than wasting time analysing in unnecessary detail what might prove to be completely unsuitable designs. The integrated design process is rewarded by a quicker, cheaper and more reliable design, which often also performs better. The integrated design system should ideally be so quick that it is possible to examine most design issues with a first pass geometry using a run-through of all the steps before committing to making a final design.

The aerodynamic and mechanical design of radial turbomachinery components for most applications generally follows a particular pattern. The design process can be traced through several distinct phases, as shown in Figure 20.1 and as described in the sections that follow. The integrated design process described here is based on that described by Robinson et al. (2011) and used by the authors. Similar systems are available as inhouse software systems in most turbocompressor companies and also from consultants and software vendors.

In some applications, the design process is used to define a final product that can be delivered to many different customers without adaptation, such as in jet engines, refrigeration plants and turbochargers. The design of industrial and process compressors requires more flexibility, as often the compressor needs to be custom-designed to order or adapted for a particular customer and application.

20.2.2 Conceptual Design

The first step in the design process is the conceptual design of the machine that is needed to meet the demands and specification of the process in terms of pressure ratio, volume flow, operating range, efficiency and rotational speed. At this stage of the design process, a decision is made between the different types of compressors that are available – turbocompressors (axial, radial or mixed flow) or positive displacement machines. In most situations met by the authors, this decision is based on previous experience. On the basis of the parameters specific speed and specific diameter, Balje (1962) attempted to generalise the conceptual design for single stages in a series of selection diagrams, as discussed in Chapter 9.

In industrial applications, positive displacement machines (reciprocating, or rotary screw) are usually used in situations where the low inlet flow would require very small

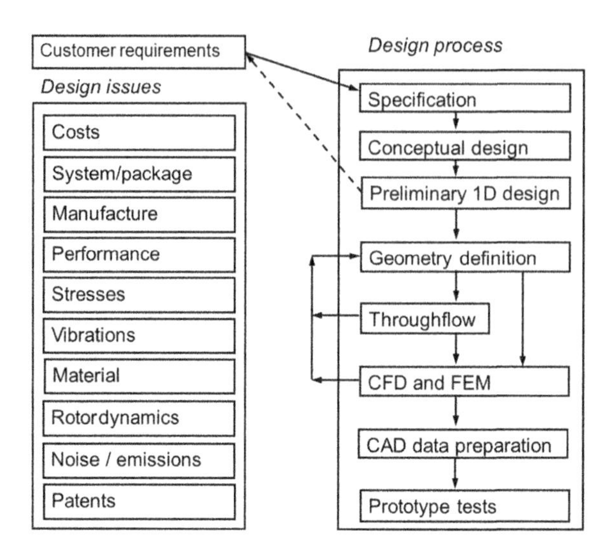

Figure 20.1 Typical radial turbomachinery design system.

impellers with narrow flow channels or when the discharge pressure is extremely high. Turbocompressors have limitations due to surge at low flows, and so in some applications with a wide range of flow several centrifugal compressors operating in series may be used. Alternatively, positive displacement machines which do not suffer from surge can also be deployed to provide a wider operating range. If the compression process has a wide fluctuation in gas composition, it may be impractical to use turbocompressors. This is, however, possible where a change in speed or a changeout of compressor internals within the casing is allowed for.

Axial compressors are suitable for industrial applications where the inlet volume flow is large (from 50,000 m^3/hr to 2,000,000 m^3/hr) and the discharge pressures are low (below 20 bar). Axial flow compressors have higher compression efficiencies (about 4–5% above those of the best centrifugal configuration) but have less variation in flow capacity, as discussed in Chapter 11. Axial machines have a straight axial flow path and – with adjustable stator guide vanes – may achieve high efficiency operation over a wide capacity range, even with a constant speed driver. Centrifugal compressors tend to be used for the intermediate flows (from 500 m^3/hr to 500,000 m^3/hr) and where equipment reliability and robustness are of key importance. Many aspects of the conceptual design process for the choice of the type of compressor are empirical and some may even be arbitrary. They are based on engineering experience, system or installation limitations, costs and other factors of which the designer must be aware. However, to be able to proceed with a detailed design, a fairly complete conceptual form of the machine must be generated.

In this phase, it is necessary to confirm that the type of machine to be employed can successfully attain the specification required of it. This process is generally iterative, as the selection of a suitable machine to meet the process requirements has to be brought into line with engineering experience as to what is practical and achievable. Most turbocompressor companies know where their compressors are not competitive in particular applications and derive the requirements for the development of these machines and for new stages from this knowledge. These days there are only a few situations where completely new machine types are developed.

In most turbocompressor applications, aerodynamic aspects are crucially important, in terms of both a stable operating range and the efficient performance of the compressor over a suitably defined range of operating conditions. In industrial applications, the choice of machine type is also based on very practical issues such as the physical limitations or the weight and size of the compressor, rotordynamic issues related to the number of stages and issues related to the availability of a suitable driver at the speed and power required. In addition, other important issues are noise and other emissions, manufacturability, maintainability, reliability, repairability, mechanical integrity, durability, integration with the process application (such as side-streams) and ease of packaging and delivery.

For large industrial compressors, economic considerations related to the investment costs and the operating costs in a cost–benefit analysis are the driver for the customers. The investment costs, I, (in \$ on a 2009 basis) for large industrial compressors increase with the power consumption, P (in kW) and can be very roughly estimated from a cost function as

$$I = 9000 * (P)^{0.6} + 20,000 \tag{20.1}$$

as given by Morandin et al. (2013). In energy intensive machines, the running costs over the lifetime of the machine may well be many times larger than the initial investment. This is taken into account in investment decisions by using a nominal cost per kW of power over the lifetime of the machine, for example in applications for air-liquefaction equipment the cost per kW saved through improved efficiency is currently valued as \$5000/kW. A new machine with a development that leads to a lower power consumption may be selected even if it is more expensive than an alternative machine requiring more power.

In rare applications where a completely new machine needs to be developed, the costs and risks must be considered. In some cases, intellectual property rights and available patents may a factor. Process compressor vendors usually need to supply a list of previous applications of their machines to show suitable references for all components in a particular application. This mitigates risk by showing the customer that they are not buying an untested prototype. Close collaboration between vendors and customers sometimes takes place to allow prototype machines to be installed and tested in new process applications in order to examine them closely in operation before eventual purchase and series production. In many situations, a further important issue is the environmental cost and sustainability, both in terms of the operation of the machine and the costs of disassembly. This particularly applies to compressors in subsea gas recovery applications, which are usually hermetically sealed, oil-free with magnetic bearings and direct-drive integrated motors (Fulton et al., 2001).

20.2.3 Preliminary Design of Centrifugal Compressors

Having completed the conceptual design, the next step is called the preliminary design. This phase begins with a definition of the velocity triangles of the stage in line with the target performance, together with estimation of the off-design behaviour for the machine. In the first instance, the velocity triangles of the impeller need to be selected as outlined in Chapter 11. The inlet angle at the casing is designed to minimise the casing inlet relative Mach number, and the outlet flow angle is selected to give the appropriate work coefficient and the necessary absolute flow angle at outlet if operating with a vaneless diffuser. This process will also determine the span of the blade at inlet, the outlet width ratio of the impeller and the meridional velocity at these positions; see Chapter 11. In a second step, a suitable axial length for the impeller on the hub is selected. This is typically $L/D_2 = 0.35$ for open impellers, but may be longer in mixed flow designs. Shorter impellers are used for multistage process applications to avoid the rotordynamic issues associated with long shafts. At this point, the decision between the different types of diffusers also needs to be made.

The preliminary design is almost invariably carried out using a 1D analytical approach working along a mean line. The theoretical background is little more than the Euler turbomachinery equation and velocity triangles, as outlined in earlier chapters, but usually the empirical basis and the correlations included in such mean-line codes are highly valuable aspects of the inhouse knowledge base of a turbomachinery manufacturer or consultant. It is a crucial part of the design process in the sense that it defines the geometrical envelope for the hardware within the available design space, taking into account previous experience, and determines the ultimate performance levels and performance map that can be expected.

Errors made at this stage, for example in the selection of inappropriate vector triangles, cannot be corrected by more sophisticated analysis in the later stages of the design process. No amount of CFD analysis and optimisation of the blade shape of a radial vaned impeller with no backsweep will, for instance, provide a wider operating margin between the peak efficiency point and the surge line, as this is a fundamental aspect of using backsweep; see Chapter 11. If a design is selected which is not close to the optimum flow coefficient, the penalty in performance due to an incorrect flow coefficient cannot be made good by better design. The need to keep the design as small, light and cheap as possible, ideally with no compromise in performance, is almost always present. As a result, higher flow coefficients than would achieve best efficiency are often used. The ability to assess the impacts of constraints quickly and accurately at this early stage in the process is of paramount importance. The great value of 1D analytical formulations is that they are rapid and can provide valuable insights to the trained eye. In this way, they simplify the design process by quickly establishing a first-pass geometry.

In some cases, it may be necessary to challenge the customer to refine the specifications in order to express legitimate concerns about the efficiency and duty targets and other issues. It is important to do this early on before too much time and energy has been spent on the detailed design. In turbochargers and gas turbines, the matching between the compressor and the turbine needs to be considered. For machines with several blade rows, component matching is important and special tools are needed to examine this. Component matching is especially important if the machine has high Mach numbers. Here the matching between the impeller and diffuser is of crucial significance, as outlined in Chapter 18.

In the development of a new design, it is imperative that such preliminary design tools do not lead to decisions that conflict with the results of the more complex CFD tools downstream in the design process. It is not intended that the 1D and 2D designs be revisited once these have been optimised. In an integrated design process, this is taken into account by adjusting classical 1D methods and correlations to take into account the experience obtained with CFD and with the test data available from earlier designs. Nevertheless, the preliminary design parameters and shapes may sometimes need to be revised in subsequent and more detailed design phases as more fundamental features of the flow patterns, such as flow separations and flow leakage, become available.

The principal of operation of such systems is for the user to specify the compressor duty (pressure ratio, mass flow, rotational speed), any fixed geometric constraints (inducer hub diameter, vane thickness, axial length) and some geometric and aerodynamic variables. The parameters used in this process are, to a certain extent, open for the designer to choose and may be changed during the iterative 1D design procedure. For example, one designer may prefer to work with a fixed value of impeller work coefficient, or of the backsweep, so that the diameter varies during the design, while another may prefer to fix the diameter and allow the backsweep to change with this fixed diameter. In addition, various parameters such as blade number, relative velocity ratio (the de Haller number along the casing contour), incidence and inlet and outlet blade angles need to be selected. The preliminary design code should allow for various combinations of these parameters to allow different designers to use different approaches.

When the preliminary design code is run, the parameters for the 1D geometry and aerodynamics are computed and a preliminary skeletal geometry of the impeller is produced. If efficiency correlations have been selected, it is useful to generate a chart which shows the correlation with the current design point, together with a preliminary estimate of the performance map. The user then assesses the results, the input data are revised accordingly and the code is rerun. This process is repeated until a satisfactory 1D design has been achieved.

20.2.4 Geometry Specification

The next step after the preliminary design is to expand the preliminary 1D mean-line geometry into a preliminary 3D geometry definition taking spanwise variations of parameters into account. Ideally this step will be automatic at the end of the preliminary 1D design. Normally, a complete preliminary 3D geometry is then generated, first with respect to the main components in the flow path but second for the key mechanical parts (shaft, casing, seals, etc.). These parts need early assessment to establish viability against the targets for life, integrity, weight and manufacturability, and also to estimate costs. The subsequent design steps refine this preliminary geometry in an iterative manner. The designer repeatedly adjusts the shape of the blades and of the flow channel until a suitable geometry is found that combines acceptable aerodynamic performance with low stress levels and is economic to manufacture.

The process is greatly simplified when a simple parameterised system of geometry definition for the components is available. Most geometry definition methods used for the design of radial impellers follow closely that published by Casey (1983) and make use of Bezier curves for the blade and flow channel surfaces as described in Chapter 14. Such data can easily be transferred to modern CAD systems for subsequent drafting and manufacture and passed on to automatic grid generation systems for CFD and FEM analysis.

20.2.5 Throughflow and Blade-to-Blade Design

Aerodynamically, the key tool used during this phase is a throughflow code which solves simplified circumferentially averaged equations of motion, as described in Chapter 15. The solution of the meridional flow field on the mean-stream surface is one of the most enduring elements of all turbomachinery design systems, as it dates back to the first CFD developments in the mid-1940s. It represents a very time-efficient route into the CFD- and FEM-dominated 3D refinement of the design. In most turbomachinery design systems, the meridional throughflow calculation is the backbone of the design process. It is fast, reliable and easy to understand; deals easily with multiple blade rows; and includes empirical loss, deviation and blockage correlations. The performance of and experience from earlier machines can then be taken into account in the design in a more straightforward way than with 3D CFD. It is useful if the throughflow code is also coupled to a blade-to-blade solver so that issues of blade loading and choking can also be identified during the 2D design stage. The general-purpose throughflow code used by the authors is specifically designed for single-stage or multistage radial turbomachinery applications. Details of the theoretical aspects of this code are described in Chapter 15. In the early stages of design, a throughflow code of quite low resolution and fidelity is appropriate, as what is needed is a converged solution of sufficient accuracy to show where any problems may occur.

The designer makes use of the throughflow code to examine the relative Mach number, the blade loading and the incidence at the inlet for each blade row. By optimising the shroud diameter and shroud blade inlet angle, as outlined in Chapter 11, the 1D design process minimises the inlet relative Mach number. The 2D throughflow code includes the hub-to-casing meridional velocity gradient due to curvature and the blade loading effects. The designer may need to make adjustments to the blade inlet angle from the 1D design to ensure low incidence across the span. The blade outlet angle from the 1D optimisation is not usually changed. The throughflow code enables the designer to adjust the impeller blade angle distribution along the casing and the hub from inlet to outlet, in line with considerations of local blade loading, throat area and blade lean.

20.2.6 Preliminary Structural Analysis

Before submitting the throughflow design to a complete 3D CFD flow analysis and an FE mechanical analysis, it is useful to ensure that the design meets the essential structural and vibrational requirements of the specification, as discussed in Chapter 19. An initial structural analysis is carried out employing a simplified finite element method on a coarse grid. This method conducts a stress and modal analysis using a structured mesh generated automatically from a 3D geometric description of the blade and a generic geometric description of the meridional section of the disc, together with data on material properties. The method analyses the blade alone or the blade together with a section of the blade disc and, if present, the shroud. A frequency assessment may be conducted to establish the frequency margins between the number of upstream and downstream blades and an assessment of blade flutter based on the reduced frequency. In a radial impeller, the stress distribution at the vane-to-hub interface is influenced significantly by the presence of the disc. In order to calculate a more realistic stress distribution in this region and to carry out a subsequent creep life assessment, this type of analysis is recommended rather than a vane only analysis. The analysis generates information that can be used to modify the blade thickness to reduce stresses and to move the natural frequency away from areas of risk.

20.2.7 3D CFD and FE Design

The design is then refined using 3D CFD and FE analytical tools. There are different levels of analysis that are possible with these tools. Generally, a balance must be struck between what can be achieved in a design iteration where time and resource constraints are relevant and what could theoretically be achieved with the highest levels of analysis.

The aim at this stage is to refine the simulation to provide details of the flow that could not be assessed in the 1D and 2D simulations. In this process, it is very useful to carry out CFD on similar impellers whose performance has previously been demonstrated in research and development (R&D) rig tests. The predicted velocity distributions and flow patterns of a new design can then be compared with earlier good designs to ensure that the best experience available from R&D on similar machines is considered. In this way, features of a particularly good design can be incorporated in a new design, while features that appear to lead to poor performance can be avoided even if the exact understanding of the good performance is not available.

The state-of-the-art CFD analysis in the design process comprises a multistage steady RANS calculation using mixing-plane interfaces between the blade rows, as outlined in Chapter 16. More sophisticated analyses tend to include as much detail of the full flow path as possible. Features such as fillet radii, tip-clearance flows, leakage flow paths over shrouded blade rows together with cavities and steps in the casing walls are now regularly taken into account in the analysis. Such systems require robust tools for automatic grid generation, not only of the blade rows, but also for the secondary flow paths. The ability to mix grids of different topologies and styles at general grid interfaces has hugely simplified the modelling task for analysts.

Transient CFD has also become more widely used. However, the literature suggests that, for most components, there is little benefit from a full transient over a steady-state solution for performance prediction at operating points close to design. The exception here is when unsteady effects need to be determined. The real benefit of transient analysis is where there is a transient boundary condition or where operation is far from the design point (towards stall in a compressor). Another application is in the calculation of flutter and forced response problems. Examples of these different styles of CFD analysis are given by Robinson et al. (2012), where the effect of the radial spacing of a vaned diffuser is examined using a mixing plane, a transient rotor–stator simulation of the full annulus and a computationally less expensive time transformation simulation.

The theoretical methods now available generally allow high levels of performance to be obtained with reasonable certainty. However, this is only true if the designer appreciates and considers the limitations of the techniques, and if the methods have been calibrated on similar units. Risks are always considerably greater for unfamiliar applications. Similarly, careful attention to mechanical aspects is essential – large performance penalties can occur all too easily if tip clearances are too high or there are protrusions or leakage flows into the primary flow path.

20.2.8 Optimisation and Inverse Methods

For aerodynamic design problems, an alternative to finding the optimum shape through individual iterations controlled by the designer is to use an advanced numerical multivariate optimisation method. Many such methods are available, and their use in design is increasing as computer power steadily improves. The computer Deep Blue that was programmed to play chess was already able to beat a chess grand master in 1996 and defeated the world chess champion in 1997. In 2016, the computer program AlphaGO trained itself to play Go, which is a more complex game than chess, and decisively defeated the world's greatest Go players. In 2017, a similar program called AlphaZero trained itself to play chess in 150 hours and defeated the world computer chess champion Stockfish 8 (Harari, 2018). Based on these achievements, it is not surprising that a numerical optimisation algorithm can deliver a better compressor design than that produced by an experienced designer and that such tools are set to become the state-of-the-art design techniques for centrifugal compressors.

Nonetheless, given that optimisation algorithms are mathematical interpretations of calculated data, they often have no explanatory power as such. There are two concerns with compressor design using advanced optimisation methods. The first is that, as described in Section 16.7, the numerical methods for the flow calculations are not perfect, especially with regard to the issue of instability and surge. This means an optimisation method might find an optimised design solely by exploiting the weaknesses of the simulation algorithms. Because of this, it is not satisfactory just to accept the results without critical examination. The second concern is how to interpret the results to gain a better understanding of the steps leading to the optimum solution. This allows the knowledge gained from the optimisation method to be passed on to novices or incorporated into improved preliminary design systems for future designs. The key to the use of such methods successfully is to use them to remove the donkey-work in the design process but to retain a mixture of human and artificial intelligence in making the final design decisions.

Many successful optimisation methods are available, including genetic algorithms, evolutionary algorithms, artificial neural networks, adjoint algorithms, simulated annealing and response surface methods. These methods have been reviewed by Sadrehaghighi (2018) for general aerodynamic problems and by Li and Zheng (2017) for turbomachinery applications. Many of the methods have been applied to centrifugal compressor design, for example, Pierret and Van den Braembussche (1999), Casey et al. (2008), Voss et al. (2014), Verstraete et al. (2010), Hehn et al. (2018) and Teichel (2018). An overview of the different optimisation methods as applied to centrifugal compressors is given by Van den Braembussche (2019). The basic principles of genetic algorithm optimisation processes are similar to the theory of natural selection of Darwin, whereby a population of individuals changes over several generations following laws of natural or artificial selection, involving the reproduction and mutation of the fittest individuals. Each individual is a different virtual compressor design and its chance of survival into the next generation (fitness) is related to how well it meets the design objectives. Genetic and evolutionary algorithms have the advantage over other optimisation methods of being very robust and not needing to compute gradients of parameters. They also ensure that a global, and not a local, optimum is found. Each method has advantages and disadvantages, and the methods can be combined to build hybridised algorithms. Currently, the major challenge with optimisation methods is to reduce the computation cost while improving the optimisation accuracy.

One way of reducing the computational effort of evolutionary algorithms is to couple them to the preliminary stress and flow analysis combined with rule-based assessment of the simulations (Casey et al., 2008; Cox et al., 2010). Combining the algorithms with the preliminary design tools delivers rapid simulations but has the disadvantage that the rules for assessing them are experience based and do not include accurate models of the losses. More complex methods based on neural networks coupled to CFD and FEM analyses enable the neural network to learn the effects of different design parameters from the calculations already carried out, thus reducing the computational effort. These methods have a promising future for centrifugal compressors, especially when they are used to improve preliminary design rules (Teichel, 2018) or are analysed to compare with previous designs (Hehn et al., 2018).

Inverse design methods are also available and are used for centrifugal compressor design (Zangeneh, 1994). In these, the designer specifies a pressure distribution or a swirl distribution, and the inverse method develops a profile shape by iterative modification of the blade shape. The computational cost is small, as little flow analysis is required and is therefore inexpensive. Inverse design methods can be combined with an optimisation method (Bonaiuti and Zangeneh, 2006; He and Shan, 2012).

This approach relies strongly on the experience of the designer who needs to specify realistic flow and pressure distributions which satisfy the various aerodynamic requirements. However, in some situations, it is unclear how to integrate mechanical design constraints into these methods.

20.2.9 Prototype Testing

At the end of the development process, the measurement of a prototype in a suitable test rig assesses the suitability and quality of the new components. For compressor applications, it is important both to achieve the required efficiency and volume flow to confirm that the required surge and choke margin has been attained. It is the authors' experience that even where great attention to detail has been paid during the design of a compressor stage, the final design optimisation may still require a development program on a suitable experimental rig. Issues that often need to be addressed during a test program are the conflicts between high-speed and low-speed performance; the effects due to small clearance gaps which were not included in the analysis (Spakovsky et al., 2009; Hazby et al., 2019); and compromises to balance the requirements of surge margin, choke margin and peak efficiency at different speeds. The final detail of the matching between a compressor impeller and a diffuser can also be adjusted at this stage by changes in the pinch or the throat area of the diffuser. In industrial compressors, the amount of design effort and testing is often reduced by using a system of adaptation or trimming of the meridional contour of existing stages to adapt a well tried and tested design to other flow conditions; see Section 11.11.

20.3 Compressor Testing

20.3.1 Types of Tests

Despite the well-established numerical methods that are currently available, a performance test of any new components or whole machines in a suitable experimental rig is the final arbiter of the design. Of relevance in compressor design is the fact that modern simulation methods are still not able to predict the onset of stall or surge in compressors with any precision; this phenomenon is usually of primary importance. The intrinsic unsteadiness of the flow near stall and the limitations of turbulence modelling restricts the capabilities of the steady-state CFD that is used for design, as outlined in Section 16.6. Every prototype centrifugal compressor will most likely be tested to verify its thermodynamic performance, and in some situations every compressor that is delivered will be tested.

Compressors play an integral role in the manufacturing processes of the chemical and petrochemical industries, and each machine is carefully audited before being commissioned in order to guarantee that it meets the performance and reliability standards that are agreed are needed for continuous operation. In industrial compressor applications, the testing of a new machine will be conducted either in the manufacturer's facility under strictly controlled conditions or in the field at actual operating conditions. Older compressors that have been placed in service after maintenance or have been operating for an extended period of time may require testing to verify the efficiency under normal operation and to identify the effects of erosion and fouling.

Such tests are regulated by the relevant international bodies for the different applications, such as the American Institute of Chemical Engineers (AIChE), the American Petroleum Institute (API), American Society of Heating, Refrigerating and Air-Conditioning Engineers (ASHRAE), the American Society of Mechanical Engineers (ASME), the British Standards Organisation (BSO), *Deutsches Institut für Normung* (DIN) (English: German Institute for Standardisation), the International Organization for Standardization (ISO), Society of Automotive Engineering (SAE) and *Verein Deutscher Ingenieure* (VDI) (Association of German Engineers). For industrial compressors, an exhaustive list of the relevant standards is provided by Brun and Kurz (2019), and the following are the most important of these codes:

- VDI 2045: Acceptance and performance test of dynamic and positive displacement compressors
- API 617: Centrifugal compressors for petroleum, chemical and gas industry services
- ASME PTC10: Performance test codes – compressors and exhausters

Similar testing procedures are defined for radial ventilator fans and for turbochargers, such as the following:

- VDI 2044: Acceptance and performance tests on fans
- ASHRAE 51: Laboratory method of testing fans
- SAE J1826_199503: Turbocharger gas stand test code

Additional guidelines on testing procedures and test methods for compressors are provided by Dimmock (1963), AGARD-AG-207 (1975) and Brun and Nored (2006).

The experimental studies are usually of two types. The first type ascertains the global operating characteristics of the compressor, and the second type determines flow conditions and performance, in particular components within the compressor. The second type allows the results of a new development to be incorporated in the further development of new components and improved design tools. The important categories of test are summarised in the following subsections.

20.3.2 Acceptance Tests for Industrial Compressors

Acceptance tests are carried out before the delivery of a new machine to the customer to demonstrate that the contractual obligations have been met. These are usually witnessed by the customer or by his or her delegated representatives. With industrial compressors, the objective of the acceptance test is to demonstrate that the compressor fulfils the guaranteed performance and mechanical stability at the rated operating conditions given in the specifications. The test confirms the shape of the compressor head-flow curve, the peak efficiency, the maximum and minimum flow limits at various speeds and the vibration signature of the compressor. Testing conducted during commissioning will establish a baseline of performance that can be used by the purchaser to assess deterioration and fouling over the lifetime of the machine and indicate whether the machine is suitable for a different application, one not originally envisaged in the specification.

The most importance test codes for regulating acceptance tests for industrial compressors are the ASME PTC-10 Performance Test Code (ASME, 1997) and the American Petroleum Institute code API standard 617 (API, 2016). The ASME Performance Test Code provides for two classes of testing which are known as type I, where the test gas is the same as the specified gas, and type II, where the test gas differs from the specified gas. API 617 refers to different tests, namely, full load, full pressure, full density and full speed, or a combination of these in parallel with the ASME PTC 10 type 1 tests. The full pressure condition permits the evaluation of the compressor for the mechanical and rotordynamic integrity of its components. Details can be found in Ishimoto et al. (2015) and Kocur et al. (2013).

For the type II tests, a detailed procedure is given for calculating and correcting results for differences in gas properties and test conditions based on similarity methods. Broadly, the tip-speed Mach number needs to be the same as for the substitute gas because of its effect on thermodynamic similarity. Otherwise, the volume ratio across the compressor will be affected and stages may be mismatched. A type 1 test is conducted with the specified gas at or very near the specified operating conditions. Because the actual and the test operating conditions may differ, the permissible deviations in the operating conditions are limited. Limitations on the volume ratio, machine Mach number and machine Reynolds number need to be specified.

However, because of physical limitations on the power available or limits on the pressure attainable in the test facility – and sometimes due to safety concerns – it may not be possible to conduct these tests in the manufacturer's own facility. A type II test may then be carried out using a carefully selected mixture of gases blended together to form a gas that has physical properties that closely resemble the specified gas. Principles of similarity are then applied to determine the speed and operating conditions. For example, a natural gas compressor (molecular weight 16) can be tested at a lower inlet pressure using inert CO_2 (molecular weight 44) leading to 20% of the mass flow, 50% of the rotational speed and approximately 6% of the power with the rated conditions (Almasi, 2018). The type II test is not run at the rated speed so that a separate mechanical test is required.

The process involves the measurement of the operating performance through the determination of the characteristic performance curves, i.e. the relationship between the pressure ratio, efficiency and power consumption as a function of the inlet volume flow and rotational speed. The determination of the safe operating limit and the maximum choke flow at each speed is an important aspect. In machines with variable guide vanes the performance at different guide vane setting needs also to be determined. The test usually includes measurements of casing and rotor vibrations, pressure fluctuations, bearing temperatures and noise.

A field test allows the compressor performance at the site of delivery to be compared with the factory performance test. Although the field test accuracy may be inherently lower, such tests provide a historical trend of possible degradation in performance and the need for upgrades (Brun and Nored, 2006).

20.3.3 Acceptance Tests for Turbochargers

Centrifugal compressors in turbochargers for road vehicles are not such a capital-intensive business as large industrial compressors. Turbochargers are produced in very large numbers, as roughly 75% of all automotive engines are now turbocharged, so that they have effectively become consumer products and replacements can be obtained off the shelf. Both the original equipment manufacturer (OEM) automotive companies and the turbocharger manufacturers have a large interest in running their own test programmes to verify performance. These tests enable the companies to acquire reliable data for the development of powertrain systems with a view to understanding current technology, investigating new innovations and improving charger and engine simulation models. The costs are sufficiently low that it is also quite usual for a turbocharger company to purchase a product from a competitor and to benchmark this to establish product competitiveness. The testing provides the compressor and turbine maps for use in 1D engine simulation codes to assess the matching of the turbocharger with specific engines.

To achieve high accuracy, most tests are carried out in a hot gas experimental facility for steady-flow turbocharger testing without a combustion engine. This removes the pulsation associated with intermittent cylinder operation. However, the absence of these pulsations may influence the stability of the compressor, such that the assessment of the surge margin may be incorrect (Watson and Janota, 1982). The hot gas test stands may be designed individually, but there is also a market for the purchase of standardised gas stands. Commonly, the stands use compressed dry air from a separate compressor. This is heated to between 300 K and 950 K before being channelled to drive the turbine. A valve at the turbine inlet controls the turbine expansion ratio and the air mass flow. At the compressor outlet, a valve is used to control the compression ratio and the compressor air mass flow rate.

The turbocharger test bench and the turbocharger are insulated to reduce heat transfers to the surroundings. Nevertheless, an issue associated with turbocharger testing, both in a hot gas stand or with an engine, is the internal heat transfer from the turbine to the compressor. This causes an increase in the temperature ratio across the compressor and hence an apparent increase in the power consumption, which is determined from the temperature rise. There is also an apparent drop in the efficiency of the compressor. As the turbine power and efficiency are derived from the apparent power of the compressor, the turbine appears better than it really is. Especially at low speeds, an apparent shift of efficiency from compressor to turbine results; see Section 6.7.2 and Sirakov and Casey (2012).

A further issue in testing on hot gas stands is that the inlet to the compressor is usually a simple optimal straight pipe. In a real engine application, the air inlet system is usually compromised by packaging constraints such that the boundary conditions are dictated by crash planes, pedestrian protection areas, vehicle styling, powerpack components and vehicle chassis components. These constraints influence the design and routing of the air inlet ducting, which may have to include elbows and abrupt changes. Elbows upstream of the compressor tend to reduce the pressure rise, but in some cases may reduce the volume flow at surge, leading to a wider range (Serrano et al., 2013). This effect is mainly a result of the stabilisation of the compression system through the additional pressure losses in the bend (Kerres et al., 2016). Abrupt changes in the inlet pipe, involving a plenum and generation of swirl, are also known to improve the surge performance, possibly by interaction with inlet recirculation at low flows; see Section 17.2.5.

A turbocharger has to operate reliably for as long as the engine on which it is fitted. Because of this, turbochargers undergo a number of stringent tests before being released for series production. This test programme includes tests of individual turbocharger components, tests on a turbocharger hot gas test stand and an engine test (Baines, 2005). One key test is the containment test, which ensures that if a compressor or turbine wheel bursts, the remaining parts of the wheel do not penetrate the compressor or turbine housing. Low cycle fatigue (LCF) tests are used to determine the wheel material load limits. A start-stop test identifies any issues with the high temperatures of the oil and components after the engine has been shut down. A cyclic endurance test is run on the engine for several hundred hours at varying load points to determine the wear on the individual components.

20.3.4 Prototype Tests

Prototype tests are usually carried out at the end of the development process in purpose-built experimental rigs in the manufacturer's test facility. The main objective is to confirm that the development objectives have been met before the start of serial production of a new machine. The tests may be carried out on a completely new machine configuration or on specific components. The tests show whether the behaviour of the machine meets expectations in terms of noise, efficiency, operating range and mechanical stability. They may also be used to make fine adjustments to the matching of components, to determine the schedule of vane angles for a inlet guide vane system or to validate global simulations of the machine.

20.4 Basic Research and Development Tests

20.4.1 Dedicated Research Rigs

The knowledge required to understand the performance of compressors is obtained from basic research and development tests. Centrifugal compressor designers rely heavily on such performance tests to validate new designs and to confirm that the required performance maps have been obtained. In addition, such tests can be used to demonstrate experience to customers and to provide data for calibration and for the development of prediction techniques. Although compressor vendors hope to collect useful information during production testing, in most cases, this testing covers only flange-to-flange performance. In a single-stage compressor, such testing may be very useful. However, in a multistage compressor, flange-to-flange data reflect the combined performance of all the stages and are of marginal value when the main interest is in the performance of one particular stage or stage component.

A far more practical means of collecting component performance information is in a dedicated single-stage test rig. The test rig can be fitted with detailed measurement planes allowing more information to be gained about the behaviour of the components and in order to validate simulations. Any weaknesses of the components can be identified and eliminated before serial production of the stages. Because of the longer timescales of such fundamental studies, single-stage test rigs are often sited in research institutes of companies, in government research institutions or in university environments. This leads to no disturbance of the production and delivery of new machines. Single-stage test stands for industrial compressors are often scale models of a particular stage in a multistage machine, whose full-size impeller diameter may be close to 1000 mm. Often the rig is designed around the smallest size impeller that is produced. The advantages of model tests are the fact that the test stand is small, which makes it easier to manipulate components, and requires a lower power, giving lower operating costs.

Many test rigs used for basic research and development tests are well described in easily accessible technical literature in journals, conferences and Ph.D. theses. A short survey with many references to some of these rigs is provided by Sitaram (2017). In many universities, master's and doctoral theses are freely available that describe such single-stage test stands and their instrumentation in great detail. Some examples are listed here, and readers are advised to examine these original sources for more detailed information.

20.4.2 Industrial Compressor Test Rigs

Several literature sources describe test stands for single stages of industrial compressors in an industrial environment (Benvenuti, 1978; Simon and Bülskämper, 1984; Dalbert et al., 1988; Sorokes and Welch, 1992; Kowalski et al., 2012). These sources detail the features and capabilities of industrial test rigs designed for the systematic testing of single-stage configurations both in open and closed test loops and also discuss the instrumentation and test data analysis systems. Closed-loop test rigs permit the operation of the stages with different pressure levels and with different gases, allowing the effect of the isentropic exponent and the Reynolds number to be evaluated. Such rigs can also be used to investigate the effect of high Mach numbers with a high molecular weight gas within the safe mechanical limits of the impellers. With reduced inlet pressures, the closed-loop rig allows experiments to be carried out at a high pressure ratio with low power consumption. Typical impeller diameters for such test rigs are between 250 mm and 450 mm.

The closed-loop rigs described by Dalbert et al. (1988) were used for published studies on the effect of Reynolds number by Casey (1985a), for studies on the design of low flow coefficient stages by Casey et al. (1990), for systematic studies of low-solidity diffusers by Kmecl and Dalbert (1999) and Kmecl et al. (1999) and loss prediction in diffusers by Ribi and Dalbert (2000). A single-stage rig used to examine the effects of diffuser and inlet guide vane adjustment on the stage performance characteristics is described by Simon et al. (1987). A different closed-loop test rig for studies of the effect of Reynolds number on low flow coefficient stages from the same industrial laboratory is detailed by Simon and Bülskämper (1984).

Newer research rigs for industrial stages developed with industrial partners are often located in university departments. Bianchini et al. (2015) describe an open-loop rig in the Department of Energy Engineering in the University of Florence, Italy. This rig is being used for fundamental studies of surge and stall in industrial compressor stages. Rossbach et al. (2015) describe a process compressor rig in the Institute of Jet Propulsion and Turbomachinery at the RWTH University of Aachen, Germany. The first experimental results from this rig include global stage performance data and stage efficiencies; based on these measurements, the negative impact of the return channel on the stage efficiency, especially at high flow rates, has been identified (Rube et al., 2016). Hazby et al. (2019) give another example of a test rig based in a university for an industrial company, as shown in Figure 20.2.

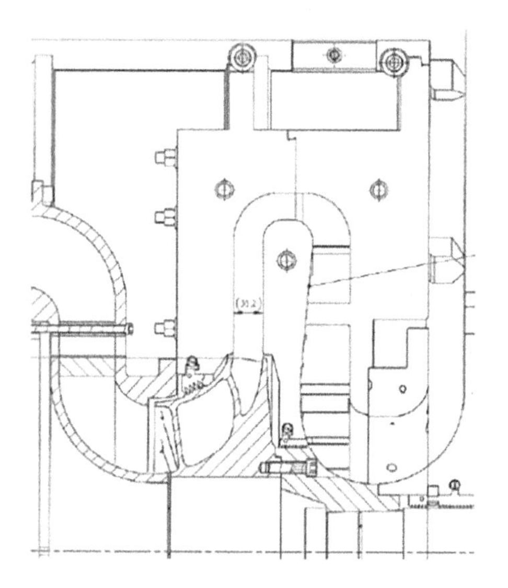

Figure 20.2 A test rig used for process compressors. (image courtesy of Howden CKD)

20.4.3 Rigs for Open Impellers

Single-stage compressors designed for specific applications may also be modified to produce a suitable test rig. These are most actively used for proprietary stage development, the results of which are usually not published. An example of an industrial rig being used for the development of new compressor stages and the validation of CFD simulations is provided by Hazby et al. (2018). A summary of published experimental work carried out in some of the most active large research test rigs is given in this section. Smaller rigs for turbocharger compressor testing are more or less ubiquitous in all turbomachinery and combustion engine laboratories and are not listed here. The major importance of this testing activity is demonstrated by the number of new compressor test rigs developed in the last 10 years.

During the period when the understanding of the complex flow field in centrifugal compressors was not as advanced as now, suitable research test rigs with open impellers provided the breakthrough in the understanding of the jet-wake flow pattern in real radial impellers. Important work was carried out in a test rig at the DFVLR, the forerunner of the DLR, where laser anemometry was used to investigate the flow patterns (Eckardt, 1975; Eckardt et. al., 1977). Subsequent publications by Eckardt's successors at the DLR research laboratory, for example Krain (1981) and more recently Elfert et al. (2017), show that this approach is still useful.

A test rig known as RADIVER in the institute for jet propulsion and turbomachinery in the RWTH Aachen, Germany, is used to examine the unsteady interaction between the impeller and a wedge diffuser including optical laser velocimetry (Ziegler et al., 2003a; 2003b). A further test stand in the Institute of Jet Propulsion and Turbomachinery in RWTH Aachen is used for testing aero-engine centrifugal compressors and examination of tip-clearance effects, bleed flows, Reynolds number variations and the axial alignment between impeller and diffuser. In addition, a variety of different diffuser and deswirler concepts have been experimentally tested and analysed numerically. There are many publications available covering such tests from this rig, for example Zachau et al. (2009), Kunte et al. (2013), Wilkosz et al. (2013) and Kaluza et al. (2017).

SuMa, which is short for Super Martin, is a closed-loop centrifugal compressor test facility with a full-size impeller at ABB Turbocharging in Baden, Switzerland (Wittwer and Küpfer, 1986). Recent experiments in this rig have been published by Spakovsky and Roduner (2009), Everitt et al. (2017), Gao et al. (2017) and Galloway et al. (2018). Experiments have examined different spike and modal stall precursors leading to instability, studied the pressure rise in different zones of vaned diffusers and investigated the effects of bleed flows.

A further centrifugal compressor test rig in Switzerland is in the laboratory for energy conversion in the Swiss Federal Institute of Technology, Zurich, as described by Hunziker and Gyarmathy (1994). Research activities have included extensive studies of instability (Ribi and Gyarmathy, 1993; Ribi, 1966), measurement of aerodynamic quantities with optical laser anemometry and fast response aerodynamic probes to study the flow structure within the compressor (Gossweiler, 1993; Roduner, 1999; Schleer, 2006), More recent work research work has concentrated on forced response or flutter of impeller blades, both from upstream excitation from flow distortion and from downstream excitation from the diffuser potential flow field (Kammerer, 2009; Zemp et al., 2010; Zemp, 2012).

Another active European research rig is a centrifugal compressor stage composed of a transonic backswept splittered unshrouded impeller and a radial vaned diffuser designed for the rear compression stage of a helicopter engine. This compressor is installed in the Department of Aerodynamics, Energetics and Propulsion (DAEP) in Toulouse, France, and is used for CFD validation studies of the effect of unsteadiness and control of flow separations in the diffuser (Trébinjac et al., 2009; Marsan et al., 2014).

A gas turbine–driven, high-speed, open-loop centrifugal compressor facility with a tip diameter of 216.5 mm is described by Cukurel et al. (2008) and Cukurel et al. (2012). This is installed in Purdue University, USA, and is equipped with windows suitable for particle image velocimetry (PIV) measurements of the flow patterns in the transonic diffuser flow. Another single-stage rig at the same university is known as CSTAR and was especially designed to investigate the flow physics of a centrifugal compressor as the last stage of an axial-centrifugal compressor in gas turbine applications (Methel, 2016). A third rig in the same establishment is known as SSCC (single-stage centrifugal compressor) and is designed for investigations emphasising inlet swirl distortion, tip-clearance effects, and impeller-diffuser interactions (Lou et al., 2016; Gooding et al., 2019).

20.4.4 Determination of Impeller and Diffuser Efficiencies

During the development testing of single-stage radial compressor stages, a static pressure measurement between the impeller and the diffuser should be included. This helps to determine and map the matching of the diffuser and the impeller for different conditions and is a great aid in making any development decisions. For example, it quickly identifies whether any deficit in performance is due to the impeller or the diffuser and whether the two components are well matched; see the examples and procedure described by Dalbert et al. (1988) and the example given in Figure 4.10.

The test rig should include a calibrated nozzle to measure the mass flow in the closed loop. The inlet total temperature and the outlet total temperature and pressure should also be measured. At impeller outlet, static pressure taps on the diffuser walls just downstream of the impeller (say at $r/r_2 = 1.02$) allow the static pressure just downstream of the impeller outlet to be determined. There should be several of these taps on each diffuser wall. Provided there are no circumferential distortions, several of these can be ganged together to give a single static pressure for each wall. The static pressure values from the two sides of the diffuser can be arithmetically averaged to give a mean static pressure at the impeller outlet. In reality, the measurement plane is not exactly at the impeller outlet, and the actual data analysis software needs to calculate the static pressure at the impeller outlet on the assumption of constant swirl using a dissipation loss model in the small space between the impeller outlet and the measurement plane.

In this procedure, a first guess is made of the density at the impeller outlet, and assuming no swirl in the inlet flow, the outlet velocity triangle can be determined from the Euler equation and the continuity equation:

$$\begin{aligned} c_{m2} &= \frac{\dot{m}}{\pi b_2 D_2 \rho_2} \\ c_{u2} &= c_p(T_{t2} - T_{t1})/u_2 \\ c_2 &= \sqrt{c_{m2}^2 + c_{u2}^2} \\ \tan\alpha_2 &= c_{m2}/c_{u2} \end{aligned} \tag{20.2}$$

The static pressure at the impeller outlet is measured. The static temperature at the impeller outlet can then be calculated from the impeller outlet velocity and the stage outlet total temperature, generating a new estimate for the impeller outlet density:

$$\begin{aligned} T_2 &= T_{t2} - \frac{c_2^2}{2c_p} \\ \rho_2 &= \frac{p_2}{RT_2} \end{aligned} \tag{20.3}$$

This iteration is continued until convergence, which is rapid. A recent example of the use of this method is given by Hazby et al. (2014). The method can be used to determine the total pressure at the impeller outlet from the measured static pressure and the total and static temperature at this position. Thus, all the data necessary for the determination of the mean velocity triangle, the loss coefficient of the diffuser components, the impeller total–static efficiency and diffuser static–total efficiency are made available. The method also supplies information on the outlet relative flow angle and allows the slip factor to be determined.

20.4.5 Simplified Research Rigs

High-speed, single-stage radial compressor test rigs are complex, and testing is time consuming, requires significant resources and is expensive. In axial compressors, some useful basic research can be carried out with simplified geometries, such as linear cascade wind tunnels, which are less cumbersome. Much of the work on such cascade rigs has a carryover to our understanding of radial compressor flows, such as the work on leading edges by Walreavens and Cumpsty (1995) and Wheeler and Miller (2008), as discussed in Section 7.3. This type of simple cascade rig is not suitable for radial impellers as there is no way to model the effect of a change in radius and the centrifugal and Coriolis forces related to this without a rotating rig.

Simpler rigs have been developed for fundamental studies on some aspects of the flow in radial stages. The effects of rotation in diffusing radial flow passages were studied by means of a simple straight-walled flow channel rotating at low speed by Moore (1973). The NASA Low-Speed Centrifugal Compressor (LSCC), a large, low-speed experimental rig for fundamental studies of secondary flows, was built at NASA (Hathaway et al., 1992). The flow field was investigated using laser anemometry and used to calibrate a 3D viscous CFD code. A specially designed rotating swirl generator was used by Filipenco et al. (2000) in order to generate a transonic swirling flow in an annular test rig for studying the effects of blockage in centrifugal compressor diffusers. A sector test rig has been used to evaluate return channels for multistage radial compressors (Simpson et al., 2008). In this test rig, a 90° sector of the return channel annulus can be examined by means of a blowdown flow through a preswirl section to generate the swirling flow at inlet to the return channel vanes. A similar concept has been applied by Dolle et al. (2018) with a test rig designed as a full 360° annulus with a crossover bend and a return channel. This has a stationary preswirler to set up the swirling flow whereby, to investigate different operating points of the stage, several different preswirlers are required to generate different velocity and flow angle profiles.

20.5 Instrumentation and Measurements

20.5.1 General

A detailed review of fluid dynamic measurement is outside the terms of reference of this book. This section provides a brief overview of the main items of relevance for compressor testing. For more detail, readers are referred to the various test codes outlined in Section 20.3.1 and to various reviews of testing methods and instrumentation (Tropea et al., 2007; Sitaram, 2017). Two problems occur in the measurement of overall compressor performance, that of correctly measuring the required quantity at a point in the fluid flow and that of obtaining an appropriate mean value from the measurements (Cumpsty and Horlock, 2006).

20.5.2 Measurement of Gas Mass-Flow Rate

An accurate measurement of the mass-flow rate through a compressor is essential if the performance is to be accurately determined. The most common flowmeter installed in compressor tests is an orifice meter. This is essentially a cylindrical tube containing a plate with a small concentric hole in the middle. Standard designs of such orifices are available, and they should be calibrated for best accuracy. The fluid accelerates through the orifice, and this kinetic energy is not completely recovered downstream. There is a loss in static pressure across the plate, and this is measured to determine the mass-flow rate. The point of maximum acceleration usually occurs slightly downstream of the actual physical orifice in a vena contracta. The correct sizing, installation, maintenance, adjustment and calibration are necessary to achieve the desired level of precision and repeatability. Issues related to the accuracy of such devices in compressor testing occur if the orifice is placed too close to an upstream throttle, or if the air is drawn from within a test cell with possible temperature stratification and swirl at the compressor intake. Other mass-flow meters that are also used are calibrated intake nozzles, venturi nozzles or an inlet bellmouth, where the measured quantity is the static pressure drop across the nozzle.

20.5.3 Measurement of Torque

Determination of torque with a torque meter is recommended to reduce test uncertainty as to the measured shaft power. Various torque measuring systems are available in the industry. In research tests, it is often possible to mount the stator components as a pendulum and to measure the reaction of this pendulum with a force transducer in order to determine the torque acting on the rotor. If a torque meter is used, this should be calibrated before and after the test. The measurement of torque also provides a good baseline for verifying the compressor performance determined from temperature and mass-flow measurements (Dalbert et al., 1988).

20.5.4 Measurement of Rotational Speed

Instruments for measuring rotational speed are called tachometers. Most tachometers operate by generating electric pulses – using magnetic or optical devices – and then counting these pulses. Magnetic pulse tachometers have a toothed wheel in front of a magnetic transducer, which is a coil around a permanent magnet. The inductance differs when a tooth or a void is in front of the magnetic transducer to generate the pulses that are counted. Optical pulse tachometers comprise a disc with holes on the rotating shaft; these holes allow the light from a lamp to reach a photocell. The resistance of the photocell depends on strength of the incident light and so provides suitable pulse for measurement of the rotating frequency.

20.5.5 Measurement of Steady Pressure

The stagnation pressure is usually used for global performance calculations, as described in Chapter 6. However, it is often more convenient to measure steady static pressures from pressure taps in the casing walls and then convert these pressures to total pressure using the dynamic head, as given in (2.26). The estimate of stagnation pressure from the static pressure measurement requires knowledge of the swirl in the flow. The pressure tap usually takes the form of straight-edged cylindrical orifice in the wall. This needs to be flush with the wall and small enough to have no influence on the flow (Tropea et al., 2007). The pressure tap is usually connected using small-bore tubing to a calibrated pressure transducer of appropriate sensitivity and range.

Whenever feasible, at least four pressure taps should be used, consistent with ASME PTC 10 recommendations. The accuracy of the static pressure measurement is dependent on the selected location. Four pressure sensors guarantee a reasonably accurate average measurement of pressure or temperature in the inlet or outlet duct, even in a nonuniform flow field. Additional pressure measurements can be carried out, if more than four sensors are insufficient to identify circumferential distortions of the flow. In some cases, it may be useful to gang the pressure measurement tubes together, which automatically generates a mean pressure level, and then make use of a single transducer for the measurement. In other situations, the use of multiple transducers provides redundancy, which allows the identification of outliers in the measurement data related to possible issues with the instrumentation.

Generally, the global performance is ascertained by measuring the pressures as close as possible to the compressor, using multiple pressure taps at suction and discharge. If suitable locations close to the compressor cannot be found, locations farther away may be used. In this case, the pressure measurement values must be corrected using empirical pressure losses for the ducting between the measurement location and the compressor inlet and discharge.

In development tests for new stages, it is advisable to measure the static pressure near the impeller outlet on both casing and hub sides, at the inlet to the diffuser, downstream of the diffuser and at the outlet of the volute or return channel. For open impellers, static pressure measurements are often made along the casing contour through the impeller. Abrupt changes in the slope of the pressure distribution may be used to identify a possible separation on the casing contour.

In closed impellers, pressure taps may be placed on the casing upstream and downstream of the seals to indicate the pressure drop across the seals. The seal mass flow can then be estimated using suitable correlations; Benvenuti et al., 1978a, 1978b; Dalbert et al., 1988). In vaned diffusers, an array of static pressure measurement points is selected upstream of the vanes and through the vane channel (see Figure 20.3, which is adapted from Lou et al., 2019). The distribution can be used to identify the static pressure rise in the different subcomponents of the diffuser (Gao et al., 2017; Galloway et al., 2018).

In some situations, it is useful to measure the total pressure directly, and for this purpose Pitot tubes are used, as invented by Henri Pitot (1695–1771) (Pitot, 1732). These are blunt-nosed tubes designed to decelerate the flow isentropically to the stagnation point on the nose of the tube, where a pressure tapping is placed to measure the stagnation pressure. Many forms of nose shape are available, and these can be combined with static pressure taps to create Pitot-static probes to estimate velocity. In some cases, a Kiel probe is used to measure the total pressure of a flow. The Kiel probe has an outer shroud which straightens the incident flow on to the Pitot tube within it and exhibits a high accuracy over a wide range of flow angles.

20.5.6 Rakes or Combs

Probes can be combined into rakes or combs of probes, which are an array of separate probes on a support. They are used to measure pressure or temperature profiles in the flow. The rake has the advantage that not only many separate readings can be made simultaneously but also that a simple average can be made of many readings. In some situations, where space is limited, probes on the leading edge of a single vane in a cascade may be deployed instead of a rake.

20.5.7 Measurement of Flow Angle

The measurement of flow angle can be determined by different forms of multihole probe configurations, including wedge probes, cylindrical probes and cobra probes.

These are able to determine the flow angle by turning the probe into the flow until the pressure difference sensor shows no difference, a process known as nulling. Alternatively, they are able to determine the flow direction by calibration of the pressure differences across the different holes.

20.5.8 Measurement of Temperature

In compressor measurements of temperature, it is important to allow adequate time for thermal equilibrium to be attained, that is, that the conditions have become steady. The outlet temperature takes the longest time to stabilise on account of the heat transfer to the metal of the compressor. For this reason, it is advisable not only to allow adequate settling time but also test from choke to surge and back in both directions along the characteristics. Lower speeds are usually tested first so the rig components can warm up steadily during the test procedure.

The approach to the measurement of temperature is similar to that of pressure, in that temperature should be measured very close to the compressor to ensure that the measured temperature is representative of the compressor temperature. Four temperature sensors should, however, be used to identify any inconsistent measurements which can be discarded. Whenever possible, the temperature should always be measured downstream of pressure. The pressure and temperature sensors should not be installed in the same line of sight. Thermocouples, thermistors and resistance

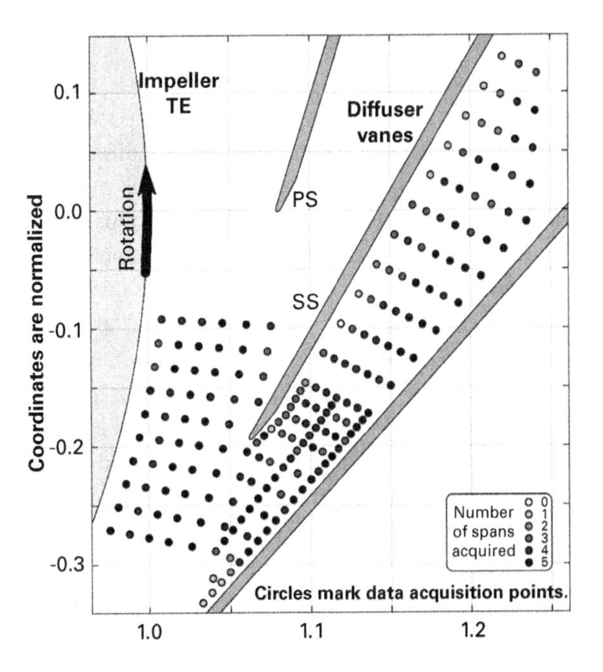

Figure 20.3 Static pressure measurements and traverse positions on the wall of a diffuser. (image courtesy of the compressor research laboratory of Purdue University)

temperature devices can be used to measure temperature, with thermocouples being the most popular. There are no practical methods to measure the static temperature, with the result that total temperature measurements are made with shielded thermocouples in Kiel probes.

20.5.9 Measurement of Unsteady Parameters

Turbomachinery flows are generally unsteady, and in many research situations it is necessary to use special techniques for the measurement of unsteady parameters. An overview of unsteady measurement techniques in turbomachinery is provided by Sieverding et al. (2000). In vaned stages, an array of unsteady static pressure measurement points is selected in the impeller, upstream of the vanes and through the vane channel (see Figure 20.3, adapted from Lou et al., 2019).

20.5.10 Tip Clearance Measurement Techniques

Compressor performance is strongly affected by the tip clearance between an open impeller and the casing, Factors affecting the clearance gap are manufacturing tolerances, individual blade asymmetries, rotor centrifugal growth, stress-related blade deflections, rotor thermal growth, stator thermal growth, stator casing gas loads, movement of the axial shaft due to thrust loads and wear and abrasion. Usually a measurement of the clearance is made by means of feeler gauges as part of the quality control process during cold assembly of the compressor components. Minimum cold clearances can be obtained if the casing and the impeller are manufactured together and the impeller tip dimensions are finished by honing with the associated casing so that both have similar production tolerances.

The measurement of tip clearances in operation is a more difficult task (Chivers, 1989; Janssen et al., 1994; Steiner, 2000; Sheard, 2011). It is made more difficult because both the radial and the axial clearances between the impeller and the casing are strongly dependent on the load cycle. In a cold start, the centrifugal force on the rotor initially reduces the radial clearance values until the running speed is reached. Depending on the thermal behaviour of the components, the clearance then changes. The blades are usually thinner than the casing and have a large surface area for heat transfer, so that these will also expand with the heating of the gas, meaning that a minimum clearance is reached shortly after the running speed is reached. A heavy casing warms up more slowly and ultimately expands away from the impeller, increasing the clearance again. Shutting down the compressor removes the centrifugal load immediately, and during cooling the tip clearance moves back towards the cold clearance.

There are several techniques available for measuring the clearance between the rotating and stationary components. The simplest are called rub probes. These are soft metal pins of lead or copper, for example, which are installed at the measuring point on the casing, and these become abraded during operation and in this way record the closest point of approach of the longest blade and the casing. Typically, two different probes at different positions along the meridional channel of an impeller are used to measure the axial and the radial clearance. An alternative method involves electromechanical pins which are driven by a motor to make contact with the blade tip and so measure the clearance at different operating conditions. Laser-optical and microwave probes are also available.

20.5.11 Typical Internal Instrumentation Assemblies

In most experimental studies, an array of different instruments is applied at different locations in the stage. Figure 20.4 shows an array of measurement locations applied in testing of a compressor stage with an open impeller by Lou et al. (2019). Typically, the pressure along the casing of the open impeller and the diffuser is measured by a series of static pressure taps for steady and unsteady pressures, and at various positions rakes of total pressure probes are included.

Similar details of the instrumentation and its location can be found in many other publications; recent examples are given in Lou et al. (2016), Zachau (2009), Elfert et al. (2017) and Galloway et al. (2018). In small turbochargers, it is not usually possible to include as much instrumentation. An example is given by Zheng et al. (2017). In this case, details are given of pressure probes around the circumference of the volute to assess the effect of volute-induced nonuniformities on the pressure distribution. Industrial compressors have similar requirements for instrumentation, but as they are usually shrouded there are no pressure taps above the impeller along the casing; examples are given by Dalbert et al. (1988).

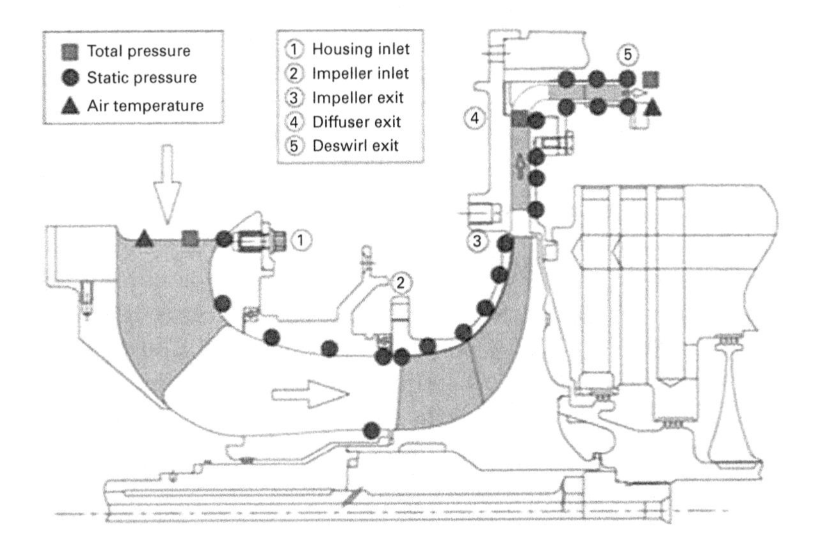

Figure 20.4 Internal stage measurements for a highly instrumented open impeller stage. (image courtesy of the compressor research laboratory of Purdue University)

20.6 Determination of Stall and Surge

The surge point of a compressor on a test rig is determined most simply by closing the throttle slowly in order to reduce the flow at constant speed and to increase the discharge pressure until audible surge is detected from the noise produced by the instability. This is usually unmistakable – a large industrial compressor going into deep surge at high speeds can be very frightening, as it produces a very loud repeating bang and leads to large low-frequency vibrations of the machine and the accompanying pipework and foundations. With smaller compressors, different audible phenomena may be identified, such as a burble related to mild surge in which the flow does not reverse but simply oscillates around a value near the surge flow. Since the objective is to determine the minimum stable flow and not the efficiency, it is not usual to wait for thermal equilibrium in surge measurements. A sweep through different speeds taking measurement points near the surge line may be carried out in order to identify any unexpected regions where a change of slope of the surge line may be used to identify other unstable phenomena at certain speeds which would require closer attention. In some situations, monitoring the vibration spectrum and the rotor orbit may provide additional information.

In compressor acceptance tests, additional measurements may also be used to obtain a more accurate determination of surge than is possible with an audible method and also to identify any regions of instability before the surge point (Colby, 2005). Pressure fluctuations are usually measured at the intake and discharge flanges with dynamic pressure transducers. Additional methods include the monitoring of rotor vibrations with accelerometers, the monitoring of changes in the acoustic signature and acoustic frequencies of the machine, identifying the location of the peak pressure rise and the monitoring any flow fluctuations with unsteady dynamic transducer measurements. Acoustic measurements are a satisfying low-cost and flexible solution that can capture many instability phenomena in a compressor in service.

Many different methodologies and instrumentation systems have been used to detect the instability of compressors in R&D laboratory tests. In these tests, the objective is usually to identify instabilities in the flow which represent the precursors to surge. The throttle valve is closed slowly, to reduce the mass flow, and the initiation of instability is identified as the point at which the amplitude of the unsteady pressure measurements starts to grow. Usually a distinct oscillation at a definite frequency can be identified, in the form of either spikes or modes, as described in Section 17.2. These may develop into either a rotating stall or a surge cycle. The unsteady pressure variations can be measured by several transducers around the circumference at a specific transverse cross section at a specific time. The number of low-pressure regions within the pressure field (over the compressor circumference) defines the number of stall cells.

References

Abbott, I. H. and von Doenhoff, A. E. (1949) *Theory of Wing Sections*. Dover, New York.

Abdel-Hamid, A. N. (1980) Analysis of Rotating Stall in Vaneless Diffusers of Centrifugal Compressors. ASME Paper 80-GT-184, ASME Gas Turbine Conference, 10–13 March, New Orleans. DOI:10.1115/80-GT-184

Abdel-Hamid, A. N. (1987) A New Technique for Stabilizing the Flow and Improving the Performance of Vaneless Radial Diffusers. *ASME J. Turbomach.*, 109(1):36–40. DOI:10.1115/1.3262067

Abdelwahab, A. and Gerber, G. (2008) A New Three-Dimensional Aerofoil Diffuser for Centrifugal Compressors. *I. Mech. Eng. Part A: Power Proc. Eng.*, 222(8):819–830. DOI:10.1243/09576509JPE579

Adkins, G. G. and Smith, L. H. (1982) Spanwise Mixing in Axial Flow Turbomachines. *ASME J. Eng. Power*, 104(1):97–110. DOI:10.1115/1.3227271

AGARD-AG-207 (1975) *Modern Methods of Testing Rotating Components of Turbomachinery*. AGARD Agardograph 207.

AGARD-AR-175 (1981) Through Flow Calculations in Axial Turbomachines. AGARD Advisory Report no. 175.

AGARD-CP-195 (1976) Throughflow Calculations in Axial Turbomachinery. AGARD Conference Proceeding No. 195.

AGARD-CP-282 (1980) Centrifugal Compressors, Flow Phenomena and Performance. AGARD Conference Proceeding No. 282.

Ahmad, M., Casey, M. V. and Sürken, N. (2009) Experimental Assessment of Droplet Impact Erosion Resistance of Steam Turbine Blade Materials. *J. Wear*, 267(9–10):1605–1618. DOI:10.1016/j.wear.2009.06.012

Akima, H. (1970) A New Method of Interpolation and Smooth Curve Fitting Based on Local Procedures. *JACM*, 17(4):589–602. DOI:10.1145/321607.321609

Aknouche, S. (2003) Impact of Tip Clearance Flow on Centrifugal Pump Impeller Performance. MSc Thesis, MIT.

Aldi, N., Casari, N., Pinelli, M. and Suman, A. (2018) A Statistical Survey on the Actual State-of-the-Art Performance of Radial Flow Fans Based on Market Data. Conference: Fan 2018, Darmstadt, 18–20 April.

Almasi, A. (2018) Shop Performance Test of Turbocompressors. 22 July, *Turbomachinery Magazine*.

Ambühl, H. and Bachmann, P. (1980) Bestimmung der Verlustanteile einzelner Stufenkomponenten in ein- und mehrstufigen Pumpturbinen radialer Bauart. *Escher Wyss Mitteilungen*, 53(1):82–91

Anish, S. and Sitaram, N. (2009) Computational Investigation of Impeller–Diffuser Interaction in a Centrifugal Compressor with Different Types of Diffusers *I. Mech. Eng. Part A: Power Proc. Eng.*, 223(2):167–178. DOI:10.1243/09576509JPE662

API standard 617 (2016) *Axial and Centrifugal Compressors and Expander-Compressors*, 8th Edition, August 2016. American Petroleum Institute, Washington.

ASHRAE (2016) *Laboratory Methods of Testing Fans for Aerodynamic Performance Rating*. ASHRAE/AMCA 51, ASHRAE, Atlanta.

ASME Performance test code PTC10 (1997) *Performance Test Code on Compressors and Exhausters*. ASME Standard PTC 10. ASME, New York.

Auchoybur, K. and Miller, R. (2018) The Sensitivity of 3D Separations in Multi-Stage Compressors. *J. Glob. Power Propuls. Soc.*, 2018(2):329–343. DOI:10.22261/jgpps.34c05t

Aungier, R. H. (1995a) Centrifugal Compressor Stage Preliminary Aerodynamic Design and Component Sizing. ASME Paper 95-GT-078, ASME Gas Turbine Congress, 5–8 June, Houston. DOI:10.1115/95-GT-078

Aungier, R. H. (1995b) Mean Streamline Aerodynamic Performance Analysis of Centrifugal Compressors. *ASME J. Turbomach.*, 117(3):360–366. DOI:10.1115/1.2835669

Aungier, R. H. (1995c) A Fast, Accurate Real Gas Equation of State for Fluid Dynamic Analysis Applications. *ASME J. Fluids Eng.*, 117(2):277–281. DOI:10.1115/1.2817141

Aungier, R. H. (2000) *Centrifugal Compressors – A Strategy for Aerodynamic Design and Analysis*. ASME Press, New York.

Backström, T. W. (2006) A Unified Correlation for Slip Factor in Centrifugal Impellers. *ASME J. Turbomach.*, 128(1):1–10. DOI:10.1115/1.2101853

Backström, T. W. (2008) The Effect of Specific Heat Ratio on the Performance of Compressible Flow Turbomachines. ASME Paper GT2008-50183, ASME Turbo Expo, 9-13 June, Berlin. DOI:10.1115/GT2008-50183

Baehr, H. D. and Kabelac, S. (2012) *Thermodynamik*. Springer Viehweg, Germany.

Baines, N., Wygant, K. D. and Dris, A. (2010) The Analysis of Heat Transfer in Automotive Turbochargers. *ASME. J. Eng. Gas Turbines Power*, 132(4):042301. DOI:10.1115/1.3204586

Baines, N. C. (2005) *Fundamentals of Turbocharging*. Concepts NREC, White River Junction.

Bakken, L.E., Jordal, K., Syverud, E. and Veer, T. (2004) Centenary of the First Gas Turbine to Give Net Power Output: A Tribute to Ægidius Elling. ASME paper GT2004-53211. ASME Turbo Expo, 14-17 June, Vienna. DOI:10.1115/GT2004-53211

Baldassarre, L., Bernocchi, A., Fontana, M. et al. (2014) Optimization of Swirl Brake Design and Assessment of Its Stabilizing Effect on Compressor Rotordynamic Performance. 43rd Turbomachinery Symposium, Texas A&M, Houston. DOI:10.21423/R15P94

Baldwin, W. S. and Lomax, H. (1978) Thin-Layer Approximation and Algebraic Model for Separated Turbulent Flows. AIAA Paper 87-257. DOI:10.2514/6.1978-257

Balje, O. E. (1962) A Study on Design Criteria and Matching of Turbomachines: Part A – Similarity Relations and Design Criteria of Turbines. *ASME J. Eng. Power*, 84(1):83–102. DOI:10.1115/1.3673386

Balje, O. E. (1970) Loss and Flow Path Studies on Centrifugal Compressors Part I and Part II. *ASME J. Eng. Power*, 92(3):287–300. DOI:10.1115/1.3445353

Balje, O. E. (1978) A Flow Model for Centrifugal Compressor Rotors. *ASME J. Eng. Power*, 100(1):148–158. DOI:10.1115/1.3446308

Balje, O. E. (1981) *Turbomachines, a Guide to Design, Selection and Theory*. John Wiley & Sons, Toronto.

Baltadjiev, N. D., Lettieri, C. and Spakovszky, Z. S. (2015) An Investigation of Real Gas Effects in Supercritical CO2 Centrifugal Compressors. *ASME J. Turbomach.*, 137 (9):091003. DOI:10.1115/1.4029616

Barbarin, V. I. and Mikirtichan, V. M. (1982) The Entropic Efficiency of Compressors and Turbines. *Fluid Mech. –Sov. Res.*, 11(3):36–47.

Baumann, U. (1999) *Rotordynamic Stability Tests on High-Pressure Radial Compressors*. 28th Turbomachinery Symposium, Texas A&M, Houston. DOI:10.21423/R1TD30

Behlke, R. F. (1986) The Development of a Second Generation of Controlled Diffusion Airfoils for Multistage Compressors. *ASME J. Turbomach.*, 108(1):32–40. DOI:10.1115/1.3262020

Beinecke, D. and Lüdtke, K. (1983) Die Auslegung von Turboverdichtern unter Berücksichtigung des realen Gasverhaltens. *VDI-Berichte* 487, 271–279.

Belardini, E, Pandit, R. et al. (2016) 2nd Quadrant Centrifugal Compressor Performance: Part II. ASME Paper GT2016-57124, ASME Turbo Expo 2016. 13–17 June, Seoul. DOI:10.1115/GT2016-57124

Bennett, I., Tourlidakis, A. and Elder, R. L. (2000) The design and analysis of pipe diffusers for centrifugal compressors. *I. Mech. Eng. Part A: Power and Energy*, 214(1):87–96. DOI:10.1243/0957650001537886

Benvenuti, E. (1978a) Aerodynamic Development of Stages for Industrial Centrifugal Compressors: Part 1 – Testing Requirements and Equipment – Immediate Experimental Evidence. ASME Paper 78-GT-4, ASME Gas Turbine Conference. 9–13 April, London. DOI:10.1115/78-GT-4

Benvenuti, E. (1978b) Aerodynamic Development of Stages for Industrial Centrifugal Compressors: Part 2 – Test Data Analysis Correlation and Use. ASME Paper 78-GT-5, ASME 1978 Gas Turbine Conference. 9–13 April, London. DOI:/10.1115/78-GT-5

Berdanier, R. A., Smith, N. R., Fabian, J. C. and Key, N. L. (2014) Humidity Effects on Experimental Compressor Performance – Corrected Conditions for Real Gases. *ASME J. Turbomach.*, 137(3):031011. DOI:10.1115/1.4028356

Berenyi, S. G. (2006) High Pressure Turbocharger for Solid Oxide Fuel Cells. ASME Paper GT2006-90222, ASME Turbo Expo 2006, Barcelona. DOI:10.1115/GT2006-90222

Berman, P. A. (1978) Compressed Air Energy Storage Turbomachinery. ASME Paper 78-GT-97, ASME Gas Turbine Congress, 9–13 April, London. DOI:10.1115/78-GT-97

Bianchini A., Carnele, E. A., Bilotti, D. et al. (2015) Development of a Research Test Rig for Advanced Analyses in Centrifugal Compressors. *Energy Procedia*, 82:230–236. DOI:10.1016/j.egypro.2015.12.027

Bidaut, Y. and Baumann, U. (2012) Identification of Eigenmodes and Determination of the Dynamical Behaviour of Open Impellers. ASME Paper GT2012-68182, ASME Turbo Expo, 11–15 June, Copenhagen. DOI:10.1115/GT2012-68182

Bidaut, Y. and Dessibourg, D. (2014) The Challenge for the Accurate Determination of the Axial Thrust in Centrifugal Compressors. 43rd Turbomachinery Symposium, Texas A&M, Houston. DOI:10.21423/R1QD7K

Biliotti, D., Bianchini, A., Vannini, G. et al. (2014) Analysis of the Rotordynamic Response of a Centrifugal Compressor Subject to Aerodynamic Loads Due to Rotating Stall. *ASME. J. Turbomach.*, 137(2):021002. DOI:10.1115/1.4028246

Bindon, J. P. (1989) The Measurement and Formation of Tip Clearance Loss. *ASME J. Turbomach.*, 111(3):257–263. DOI:10.1115/1.3262264

Bloch, H. P. (2006) *A Practical Guide to Compressor Technology*. Wiley, New York.

Bölcs, A. (2005) *Transonic Flow in Turbomachines*. Computerized Education Platform, Heat and Power Technology, Lecture series vol. 4, Stockholm.

Bolleter, U. (1980) European patent. EP0046173A1

Bommes, L., Fricke, J. and Grundmann, R. (2003) *Ventilatoren*. Vulkan Verlag Essen.

Bonaiuti, D. and Zangeneh, M. (2006) On the Coupling of Inverse Design and Optimization Techniques for Turbomachinery Blade Design. ASME Paper GT2006-90897, ASME Turbo Expo, 8–11 May, Barcelona. DOI:10.1115/GT2006-90897

Bosman, C. and Marsh, H. (1974) An Improved Method for Calculating the Flow in Turbo-Machines, Including a Consistent Loss Model. *I. Mech. Eng. Part C: J. Mech. Eng. Sci.*, 16 (1):25–31. DOI:10.1243/JMES_JOUR_1974_016_006_02

Bousquet, Y., Carbonneau, X., Dufour, G., Binder, N. and Trebinjac, I. (2014) Analysis of the Unsteady Flow Field in a Centrifugal Compressor from Peak Efficiency to Near Stall with Full-Annulus Simulations. *Int. J. Rotating Mach.*, 2014:ID 729629. DOI:10.1155/2014/729629

Boyce, M. (2002) *Centrifugal Compressors: A Basic Guide*. Pennwell Pub. Tulsa.

Bradshaw, P. (1994) Turbulence: The Chief Outstanding Difficulty of Our Subject. *Exp. Fluids*, 16:203–216.

Brandvik, T. and Pullan, G. (2011) An Accelerated 3D Navier–Stokes Solver for Flows in Turbomachines. *ASME J. Turbomach.*, 133(2):021025. DOI:10.1115/1.4001192

Braunscheidel, E. P., Welch, G. E., Skoch. G. J. et al. (2016) Aerodynamic Performance of a Compact, High Work-Factor Centrifugal Compressor at the Stage and Subcomponent Level, NASA/TM – 2015-218455.

Brebner, G. G. and Bagley, J. A. (1956) Pressure and Boundary Layer Measurements on a Two-Dimensional Wing at Low-Speed. ARC R&M 2886 HMSO.

Brown, L. E. (1972) Axial Flow Compressor and Turbine Loss Coefficients: A Comparison of Several Parameters. *ASME J. Eng. Power*, 94(3):195–201. DOI:10.1115/1.3445672

Brown, R. N., (2005) Compressors: Selection and Sizing, Gulf Professional Publishing, Houston.

Brown, W. B. and Bradshaw, G. R. (1947) Method of Designing Vaneless Diffusers and Experimental Investigation of Certain Undetermined Parameters. NACA TN1426.

Brun, K. and Kurz, R. (2019) Compression Machinery for Oil and Gas. Gulf Professional Publishing, Houston.

Brun, K. and Nored, M. G. (2006) *Guideline for Field Testing of Gas Turbine and Centrifugal Compressor Performance*. Gas Machinery Research Council, Southwest Research Institute, Dallas.

Bullock, R. O. and Johnsen, I. A. (1965) Aerodynamic Design of Axial Flow Compressors. NASA-SP-36.

Bulot, N. and Trebinjac, I. (2009) Effect of the Unsteadiness on the Diffuser Flow in a Transonic Centrifugal Compressor Stage. *Int. J. Rotating Mach.*, 2009:ID 932593. DOI:10.1155/2009/932593

Busemann, A. (1928) Das Forderhohenverhaltniss radialer Kreiselpumen mit logarithmischspir-aligen Schaufeln. *Z. Angew. Math. Mech.*, 8: 372–384.

Bygrave, J., Villanueva, A. and Enos, R. (2010) Upgrading the Performance of a Centrifugal Barrel Compressor Family. ASME Turbo Expo, 14–18 June, Glasgow. DOI:10.1115/GT2010-23767

Calvert, W. J. (1982) An Inviscid–Viscous Interaction Treatment to Predict the Blade-to-Blade Performance of Axial Compressors with Leading-Edge Normal Shock Waves. ASME Paper 82-GT-135, ASME Gas Turbine Conference, 18–22 April, London. DOI:10.1115/82-GT-135

Calvert, W. J. (1994) Inviscid–Viscous Method to Model Leading Edge Separation Bubbles. ASME Paper 94-GT-247, ASME Gas Turbine Congress, 13–16 June, The Hague. DOI:10.1115/94-GT-247

Calvert, W. J. and Ginder, R. B. (1999) Transonic Fan and Compressor Design. *I. Mech. Eng. Part C: J. Mech. Eng. Sci.*, 213(5):419–436. DOI:10.1243/0954406991522671

Came, P. M. (1978) The Development, Application and Experimental Evaluation of a Design Procedure for Centrifugal Compressors. *I. Mech. E.*, 192(1):49–67. DOI:10.1243/PIME_PROC_1978_192_007_02

Came, P. M. (1995) Streamline Curvature Throughflow Analysis. *Proceedings of First European Turbomachinery Conference*, VDI Berichte, Germany 1185:291.

Came, P. M. and Robinson, C. J. (1998) Centrifugal Compressor Design. *I. Mech. Eng. Part C: J. Mech. Eng. Sci.*, 213(2):139–155. DOI:10.1243/0954406991522239

Came, P. M., Connor, W. A., Fyles, A. and Swain, E. (1984) High Performance Turbochargers for Marine Diesel Engines. Paper 64. *Trans I Marine E.*, 96:1–12.

Camp, T. R., and Day, I. J. (1998) A Study of Spike and Modal Stall Phenomena in a Low-Speed Axial Compressor. *ASME. J. Turbomach.*, 120(3):393–401. DOI:10.1115/1.2841730

Campbell, W. (1924) Protection of Steam Turbine Disk Wheels from Axial Vibration. *Trans ASME* 46:31–160.

Carter, A. D. S. (1961) Blade Profiles for Axial Flow Fans, Pumps and Compressors. *I. Mech. Eng. Part C: J. Mech. Eng. Sci.*, 175(1):775–806. DOI:10.1243/PIME_PROC_1961_175_051_02

Casey, M. V. (1983) A Computational Geometry for the Blades and Internal Flow Channels of Centrifugal Compressors. *ASME J. Eng. Power.*, 105(2):288–295. DOI:10.1115/1.3227414

Casey, M. V. (1985a) The Effects of Reynolds Number on the Efficiency of Centrifugal Compressor Stages. *ASME J. Eng. Power.*, 107(2):541–548. DOI:10.1115/1.3239767

Casey, M. V. (1985b) The Aerodynamic Development of High-Performance Radial Compressor Stages for Industrial Turbocompressors. *VDI Berichte 572.1, Thermische Strömungsmaschinen 1985*. VDI Verlag, Düsseldorf.

Casey, M. V. (1994a) Computational Methods for Preliminary Design and Geometry Definition in Turbomachinery. AGARD-LS-195, AGARD Lecture Series on Turbomachinery Design Using CFD, NASA-Lewis.

Casey, M. V. (1994b) The Industrial Use of CFD in the Design of Turbomachinery. AGARD-LS-195, AGARD Lecture Series on Turbomachinery Design Using CFD, NASA-Lewis.

Casey, M. V. (2004) Third State of the Art Review for Thematic Area 6, CFD for Turbomachinery Internal Flows. EU Project QNET-CFD.

Casey, M. V. (2007) Accounting for Losses and Definitions of Efficiency in Turbomachinery Stages. *I. Mech. Eng. Part C: J. Mech. Eng. Sci.*, 221(6):735–743. DOI:10.1243/09576509JPE459

Casey, M. V. (2008) Manuscripts of Lecture Courses: Grundlagen der Thermischen Strömungsmaschinen (TSM), Turbochargers (TC), Turbokompressoren und Ventilatoren (TKV), on www.itsm.uni-stuttgart.de, University of Stuttgart.

Casey, M. V. and Fesich, T. M. (2010) The Efficiency of Turbocharger Components with Diabatic Flows. *ASME J. Eng. Power*, 132(7):072302. DOI:10.1115/1.4000300

Casey, M. V. and Roth, P. (1984) A Streamline Curvature Throughflow Method for Radial Turbocompressors. I. Mech. E. Conference on Computational Methods in Turbomachinery, April, Birmingham. Paper No. C57/84.

Casey, M. V. and Marty, F. (1985) Centrifugal Compressors – Performance at Design and Off-Design. *Proceedings of the Institute of Refrigeration, London, 1985–86*, 71–82.

Casey, M. V. and Robinson, C. J. (2006) A Guide to Turbocharger Compressor Characteristics. Paper 4, *Dieselmotorentechnik, 10th Symposium, 30–31 March*. TAE Esslingen, Esslingen.

Casey, M. V. and Robinson, C. J. (2010) A New Streamline Curvature Throughflow Code for Radial Turbomachinery. *ASME J. Turbomach.*, 132(3):03102. DOI:10.1115/1.3151601

Casey, M. V. and Robinson, C. J. (2011) A Unified Correction Method for Reynolds Number, Size and Roughness Effects on the Performance of Compressors. *I. Mech. E. Part A: J. Power Energy*, 225(7):864–876. DOI:10.1177/0957650911410161

Casey, M. V. and Robinson, C. J. (2013). A Method to Estimate the Performance Map of a Centrifugal Compressor Stage. *ASME J. Turbomach.*, 135(2):021034. DOI:10.1115/1.4006590

Casey, M. V. and Rusch, D. (2014) The Matching of a Vaned Diffuser with a Radial Compressor Impeller and Its Effect on the Stage Performance. *ASME J. Turbomach.*, 136 (12):121004. DOI:10.1115/1.4028218

Casey, M. V. and Schlegel M. (2010) Estimation of the Performance of Turbocharger Compressors at Extremely Low Pressure Ratios. *I. Mech. E. Part A: J. Power Energy*, 224(2):239–250. DOI:10.1243/09576509JPE810

Casey, M. V. and Wintergerste, T. (2000) ERCOFTAC Best Practice Guidelines for Industrial CFD. ERCOFTAC, January 2000.

Casey, M. V., Dalbert, P. and Schurter, E. (1990) Radial Compressor Stages for Low Flow Coefficients. Paper C403/004, I. Mech. Eng. International Conference, Machinery for the Oil and Gas Industries, Amsterdam.

Casey, M. V., Dalbert, P. and Roth, P. (1992) The Use of 3D Viscous Flow Calculations in the Design and Analysis of Industrial Centrifugal Compressors. *ASME J. Turbomach.*, 114 (1):27–37. DOI:10.1115/1.2927995

Casey, M. V., Gersbach, F. and Robinson, C. J. (2008) An Optimisation Technique for Radial Compressor Impellers. ASME Paper GT2008-50561, ASME Turbo Expo, 9–13 June, Berlin. DOI:10.1115/GT2008-50561

Casey, M. V., Krähenbuhl, D. and Zwyssig C. (2013) The Design of Ultra-High-Speed Miniature Compressors. 10th European Conference on Turbomachinery ETC10, 15-19 April, Lappeenranta.

Casey, M. V., Robinson, C. J. and Zwyssig, C. (2010) The Cordier Line for Mixed Flow Compressors. ASME Paper GT2010-22549, ASME Turbo Expo, 14–18 June, Glasgow. DOI:10.1115/GT2010-22549

Cengel, Y. and Boles, M. (2015) Thermodynamics, and Engineering Approach, 8th Edition. McGraw-Hill, India.

Chen, G. T., Greitzer, E. M., Tan, C. S. and Marble, F. E. (1991) Similarity Analysis of Compressor Tip Clearance Flow Structure. *ASME J. Turbomach.*, 113(2):260–269. DOI:10.1115/1.2929098

Chen, H. and Lei, V. (2013) Casing Treatment and Inlet Swirl of Centrifugal Compressors. *ASME J. Turbomach.*, 135(4):041010. DOI:10.1115/1.4007739

Chen, H. (2017) Noise of Turbocharger Compressors. ISROMAC 2017, Maui. https://hal.archives-ouvertes.fr/hal-02376818/document

Chen, N. (2010) *Aerothermodynamics of Turbomachinery*. Wiley, New York.

Cheshire, L. J. (1945). The Design and Development of Centrifugal Compressors for Aircraft Gas Turbines. *I. Mech. E.*, 153(1):426–440. DOI:10.1243/PIME_PROC_1945_153_048_02

Childs, D. (1993) *Turbomachinery Rotordynamics*. John Wiley & Sons, New York.

Childs, P. R. N. and Noronha, M. B. (1999/1997) The Impact of Machining Techniques on Centrifugal Compressor Impeller Performance. *ASME J. Turbomach.*, 121(4):637–643. DOI:10.1115/1.2836715

Chivers, J. W. R. (1989) A Technique for the Measurement of Blade Tip Clearance in a Gas Turbine. Ph.D. Thesis, Imperial College, London.

Christmann, R., Langler, F., Habermehl, M. et al. (2010) Low-Cycle Fatigue of Turbocharger Compressor Wheels Online Prediction and Lifetime Extension. I. Mech. E. 9th Int Conference on Turbocharging and Turbochargers, 2010. London.

Clark, C. J., Pullan, G., Curtis, E. and Goenaga, F. (2017) Secondary Flow Control in Low Aspect Ratio Vanes Using Splitters. *ASME J. Turbomach.*, 139(9):091003. DOI:10.1115/1.4036190

Clements, W. W. and Artt, D. W. (1987) The Influence of Diffuser Channel Geometry on the Flow Range and Efficiency of a Centrifugal Compressor. *I. Mech. E. Part A: J. Power Energy*, 201(2):145–152. DOI:10.1243/PIME_PROC_1987_201_016_02

Clements, W. W. and Artt, D. W. (1988) The Influence of Diffuser Channel Length-Width Ratio on the Efficiency of a Centrifugal Compressor. *P I Mech. Eng. A–J Pow*, 202(3):163–169. DOI:10.1243/PIME_PROC_1988_202_022_02

Colby, G. M. (2005) Hydraulic Shop Performance Testing of Centrifugal Compressors. 34th Turbomachinery Symposium, 2005, Texas A&M, Houston. DOI:10.21423/R18930

Coles, D. E. (1956) The Law of the Wake in the Turbulent Boundary Layer. *J. Fluid Mech.*, 1:191–226. DOI:10.1017/S0022112056000135

Conboy, T., Wright, S., Pasch, J. et al. (2012) Performance Characteristics of an Operating Supercritical CO2 Brayton Cycle. *ASME. J. Eng. Gas Turbines Power.*, 134(11):111703. DOI:10.1115/1.4007199

Coppage, J. E., Dallenbach, F., Eichenberger, H. P. et al. (1956) Study of Supersonic Radial Compressors for Refrigeration and Pressurisation systems. WADC Technical Report 55-257.

Coppinger, M. and Swain, E. (2000) Performance Prediction of an Industrial Centrifugal Compressor Inlet Guide Vane System. *I. Mech.E. Part A: J. Power Energy*, 214 (2):153–164. DOI:10.1243/0957650001538254

Cordes, G. (1963) Strömungstechnik der gasbeaufschlagten Axialturbine. Springer, Berlin.

Cordier, O. (1953) Ähnlichkeitsbedingungen für Strömungsmaschinen. *BWK Zeitschrift*, 5(10):337–340.

Cousins, W. T., Yu, L., Garofano, J. et al. (2014) Test and Simulation of the Effects of Surface Roughness on a Shrouded Centrifugal Impeller. ASME Paper GT2014-25480, ASME Turbo Expo, 16–20 June, Düsseldorf. DOI:10.1115/GT2014-25480

Cox, G. D., Fischer, C. and Casey, M. (2010) The Application of Throughflow Optimisation to the Design of Radial and Mixed Flow Turbines. C1302, I. Mech E. 9th International Conference on Turbochargers and Turbocharging, 19–20 May 2010, London.

Cumpsty, N. A. (2004) Compressor Aerodynamics. Krieger Publishing Company, Malabar.

Cumpsty, N. A. (2010) Some Lessons Learned. *ASME. J. Turbomach.*, 132(4):041018. DOI:10.1115/1.4001222

Cumpsty, N. A. and Heyes, A. (2018) Jet Propulsion, 3rd Edition, Cambridge University Press, Cambridge.

Cumpsty, N. A. and Horlock, J. H. (2006) Averaging Nonuniform Flow for a Purpose. *ASME J. Turbomach.*, 128(1):120–129. DOI:10.1115/1.2098807

Cukurel, B., Lawless, P. B. and Fleeter, S. (2008) PIV Investigations of a High Speed Centrifugal Compressor Diffuser: Spanwise Loading Variations. ASME Paper GT2008-51321, ASME Turbo Expo, 9–13 June, Berlin. DOI:10.1115/GT2008-51321

Cukurel, B., Lawless, P. B. and Fleeter, S. (2012) PIV Investigation of a High Speed Centrifugal Compressor Diffuser: Circumferential and Spanwise Variations. 43rd AIAA/ASME/SAE/ASEE Joint Propulsion Conference, 08–11 July, Cincinnati. DOI:10.2514/6.2007-5021

Dailey, J. W. and Nece R. E. (1960) Chamber Dimension Effects on Frictional Resistance of Enclosed Rotating Disks. *ASME J. Basic Eng.*, 82(1):217–230. DOI:10.1115/1.3662532

Dalbert, P., Casey, M. V. and Schurter, E. (1988) Development, Testing and Performance Prediction of Radial Stages for Multi-Stage Industrial Compressors. *J. Turbomach.*, 110 (3):283–292. DOI:10.1115/1.3262194

Dalbert, P., Ribi, B., Kmecl, T. and Casey, M. V. (1999). Radial Compressor Design for Industrial Compressors. *I. Mech. Eng. Part C: J. Mech. Eng. Sci.*, 213(1):71–83. DOI:10.1243/0954406991522194

Dallenbach, F. (1961) The Aerodynamic Design and Performance of Centrifugal and Mixed Flow Compressors. SAE Technical Paper 610160. DOI: 10.4271/610160.

Damle, S. V., Dang, T. Q. and Reddy, D. R. (1997) Throughflow Method for Turbomachines Applicable for All Flow Regimes. *ASME J. Turbomach.*, 119(2):256–262. DOI:10.1115/1.2841108

Dawes, W. N. (1987) Application of a Three-Dimensional Viscous Compressible Flow Solver to a High-Speed Centrifugal Compressor Rotor-Secondary Flow and Loss Generation. I. Mech. E. Conference on Turbomachinery – Efficiency Prediction and Improvement, 1–3 September, Robinson College, Cambridge.

Dawes, W. N. (1988) Development of a 3D Navier Stokes Solver for Application to All Types of Turbomachinery. ASME Paper 88-GT-70, ASME Gas Turbine Congress, 6–9 June, Amsterdam. DOI:10.1115/88-GT-70

Dawes, W. N. (1990) A Comparison of Zero and One Equation Turbulence Modelling for Turbomachinery Calculations. ASME Paper 90-GT-303 ASME Gas Turbine Congress, 11–14 June, Brussels. DOI:10.1115/90-GT-303

Dawes, W. N. (1992) Toward Improved Throughflow Capability: The Use of Three-Dimensional Viscous Flow Solvers in a Multistage Environment. *ASME J. Turbomach.*, 114(1):8–17. DOI:10.1115/1.2928002

Dawes, W. N. (1992) The Simulation of Three-Dimensional Viscous Flow in Turbomachinery Geometries Using a Solution-Adaptive Unstructured Mesh Methodology. *ASME J. Turbomach.*, 114(3):528–537. DOI:10.1115/1.2929176

Day, I. J. (1976) *Axial Compressor Stall*. Ph.D. Thesis, University of Cambridge

Day, I. J. (1993). Stall Inception in Axial Flow Compressors. *ASME. J. Turbomach.*, 115(1):1–9. DOI:10.1115/1.2929209

Day, I. J. (2016) Stall, Surge, and 75 Years of Research. *ASME J. Turbomach.*, 138(1):011001. DOI:10.1115/1.4031473

Day, I. J., Greitzer, E. M. and Cumpsty, N. A. (1978) Prediction of Compressor Performance in Rotating Stall. *ASME J. Eng. Power.*, 100(1):1–12. DOI:10.1115/1.3446318

De Haller, P. (1953) Das Verhalten von Tragflugelgittern in Axialverdichtern und in Windkanal. *BWK Zeitschrift*, 5(10):333–337.

Dean, R. C. Jr. (1959) On the Necessity of Unsteadiness in Fluid Machines. *J. Basic Eng.*, 81(1):24–28. DOI:10.1115/1.4008350

Dean, R. C. Jr. (1971) On the Unresolved Fluid Dynamics of the Centrifugal Compressor. In *Advanced Centrifugal Compressors*, ASME Gas Turbine Division, New York. 1–55.

Dean, R. C. Jr. and Senoo, Y. (1960) Rotating Wakes in Vaneless Diffusers. *J. Basic Eng.*, 82(3):563–570. DOI:10.1115/1.3662659

Del Greco, A. S., Biagi, F. R., Sassanelli, G. and Michelassi, V. (2007) A New Slip Factor Correlation for Centrifugal Impellers in a Wide Range of Flow Coefficients and Peripheral Mach Numbers. ASME Paper GT2007-27199, ASME Turbo Expo 2007, 14–17 May, Montreal. DOI:10.1115/GT2007-27199

Deniz, S. (1997) Effects of Inlet Conditions on Centrifugal Diffuser Performance. Report 225 GTL, MIT, Cambridge.

Deniz, S., Greitzer, E. M. and Cumpsty, N. A. (2000) Effects of Inlet Flow Field Conditions on the Performance of Centrifugal Compressor Diffusers: Part 2 – Straight Channel Diffuser. *ASME J. Turbomach.*, 122(1):11–21. DOI:10.1115/1.555424

Denton, J. D. (1975) A Time Marching Method for Two- and Three-Dimensional Blade to Blade Flows. ARC R&M. No. 3775 HMSO.

Denton, J. D. (1978) Throughflow Calculations for Transonic Axial Flow Turbines. *ASME J. Eng. Power*, 100(2):212–218. DOI:10.1115/1.3446336

Denton, J. D. (1983) An Improved Time-Marching Method for Turbomachinery Flow Calculation. *ASME J. Eng. Power*, 105(3):514–521. DOI:10.1115/1.3227444

Denton, J. D. (1986) The Use of a Distributed Body Force to Simulate Viscous Effects in 3D Flow Calculations. ASME Paper 86-GT-144, ASME Gas Turbine Conference, 8–12 June, Dusseldorf. DOI:10.1115/86-GT-144

Denton, J. D. (1992) The Calculation of Three-Dimensional Viscous Flow through Multistage Turbomachines. *ASME J. Turbomach.*, 114(1):18–26. DOI:10.1115/1.2927983

Denton, J. D. (1993) The 1993 IGTI Scholar Lecture: Loss Mechanisms in Turbomachines. *ASME J. Turbomach.*, 115(4):621–656. DOI:10.1115/1.2929299

Denton, J. D. (1997) Lessons from Rotor 37. *J. Thermal Science*, 6(1):1–13. DOI:10.1007/s11630-997-0010-9

Denton, J. D. (2010) Some Limitations of Turbomachinery CFD. ASME Turbo Expo 2010, 14–18 June. Glasgow. DOI:10.1115/GT2010-22540

Denton, J. D. (2017) Multall: An Open Source, CFD Based, Turbomachinery Design System. ASME Paper GT2017-63993, ASME Turbo Expo 2017, 26–30 June, Charlotte. DOI:10.1115/GT2017-63993

Denton, J. D. and Cumpsty, N. A. (1987) Loss Mechanisms in Turbomachines. I. Mech E Paper C260/87.

Denton, J. D. and Dawes, W. N. (1998) Computational Fluid Dynamics for Turbomachinery Design. *I. Mech. E. Part C: J. Mech. Eng. Sci.*, 213(2):107–124. DOI:10.1243/0954406991522211

Denton, J. D. and Xu, L. (2002) The Effects of Lean and Sweep on Transonic Fan Performance. ASME Paper GT2002-30327, ASME Turbo Expo, 3–6 June, Amsterdam. DOI:10.1115/GT2002-30327

Dick, E. (2015) *Fundamentals of Turbomachines*. Springer, Berlin.

Dick, E., Heirman, P. and Annerel, S. (2011) Optimization of the Deceleration Ratio in Impellers of Centrifugal Fans. *9th European Turbomachinery Conference, Conference Proceedings*, vol. 2, ed. M. Sen. Istanbul Technical University, Istanbul, 935–944.

Dickens, T. and Day, I. (2011) The Design of Highly Loaded Axial Compressors. *ASME J. Turbomach.*, 133(3):031007. DOI:10.1115/1.4001226

Dickmann, H. P. (2013) Shroud Contour Optimization for a Turbocharger Centrifugal Compressor Trim Family. Paper ETC2013-015, 10th European Turbomachinery Conference, 15–19 April, Lappeenranta.

Diehl, M. (2019) Mitigation of Tip Leakage Induced Phenomena in a Low Reynolds Number Centrifugal Compressor via Blade Loading Distribution. Thesis Nr. 9720 EPFL, Lausanne.

Dietmann, F. (2015) Zum Einfluss der Reynolds-Zahl und der Oberflächenrauigkeit bei thermischen Turbokompressoren. Ph.D. Thesis. ITSM, Stuttgart Shaker-Verlag.

Dietmann, F. and Casey, M. V. (2013) The Effects of Reynolds Number and Roughness on Compressor Performance. Paper ETC2013-052, 10th European Turbomachinery Conference, 15–19 April, Lappeenranta.

Dietmann, F., Casey, M. V. and Vogt, D. M. (2020) Reynolds Number and Roughness Effects on Turbocompressor Performance: Numerical Calculations and Measurement Data Evaluation. ASME Paper GT2020-14653, ASME Turbo Expo, 22–26 June, London.

Dietzen, F. J. and Nordmann, R. (1987) Calculating Rotordynamic Coefficients of Seals by Finite-Difference Techniques. *ASME J. Tribol.*, 109(3):388–394. DOI:10.1115/1.3261453

Dimmock, N. A. (1963) A Compressor Routine Test Code. ARC R&M 3337, HMSO.

Dixon, L. and Hall, C. (2010), Fluid Mechanics and Thermodynamics of Turbomachinery, 6th Edition. Butterworth Heinemann, Oxford.

Dolle, B., Brillert, D., Dohmen, H. J. and Benra, F. K. (2018) Investigation of Aerodynamic Effects in Stator Components of Multistage Centrifugal Compressors. ASME Paper GT2018-76783, ASME Turbo Expo, 11–15 June, Oslo.

Dong, L. and Cao, L. (2015) Effects of Residual Riblets of Impeller's Hub Surface on Aerodynamic Performance of Centrifugal Compressors. *Engi. Appli. Comput. Fluid Mech.*, 9(1):99–113. DOI:10.1080/19942060.2015.1004813

Dowson, P., Bauer, D. and Laney S. (2008) Selection of Materials and Material Related Processes for Centrifugal Compressors and Steam Turbines in the Oil and Petrochemical Industry. 37th Turbomachinery Symposium, Texas A&M, Houston. DOI:10.21423/R1SS8C

Drela, M., Giles, M. and Thompkins, W. T. (1986) Newton Solution of Coupled Euler and Boundary Layer Equations. In *Numerical and Physical Aspects of Aero-Dynamic Flow III*, ed. T. Cebeci. Springer-Verlag, New York, 143–154.

Drela, M. and Youngren, H. (1991) Viscous/Inviscid Method for Preliminary Design of Transonic Cascades. AIAA Paper No. 91-2364. DOI:10.2514/6.1991-2364

Dring, R. P. (1984) Blockage in Axial Compressors. *ASME J. Eng. Power*, 106(3):712–714. DOI:10.1115/1.3239628

Dubbel, H. (2001) *Taschenbuch für den Maschinenbau.* 20th Edition. Springer, Berlin, Germany.

Dunham, J. (1998) CFD Validation for Propulsion System Components. AGARD Advisory Report 355.

Eck, B. (1972) *Fans: Design and Operation of Centrifugal Axial Flow and Cross Flow Fans.* Pergamon Press, Oxford.

Eckardt, D. (1975) Instantaneous Measurements in the Jet-Wake Discharge Flow of a Centrifugal Compressor Impeller. *ASME J. Eng. Power*, 97(3):337–345. DOI:10.1115/1.3445999

Eckardt, D. (1976) Detailed Flow Investigations within a High-speed Centrifugal Compressor Impeller. *ASME J. Fluids Eng.*, 98(3):390–399. DOI:10.1115/1.3448334

Eckardt, D. (1978) Investigation of the Jet-Wake Flow of a Highly-Loaded Centrifugal Compressor Impeller. Translation of a doctoral thesis from Aachen, Germany. NASA TM -75232.

Eckardt, D. (1979) Flow Field Analysis of Radial and Backswept Centrifugal Compressor Impellers, Part 1: Flow Measurements Using Laser Velocimeter. In *Performance Prediction of Centrifugal Pumps and Compressors*, ed. Gopalakrishnan. ASME Publication, New York, 77–86

Eckardt, D. (2014) *Gas Turbine Powerhouse.* Oldenbourg, Munich

Eckert, B. and Schnell, E. (1961) *Axial- und Radial-Kompressoren.* Springer, Berlin.

Eckardt, D., Trülzsch, K. J. and Weimann, W. (1977) Vergleichende Strömungsuntersuchungen an den drei Radial-Verdichter Laufrädern mit konventionellen Messverfahren, *FVV Radialverdichter Vorhaben.* No. 182, Heft 237, FVV, Frankfurt.

Edminster, W. C. (1961) *Applied Hydrocarbon Thermodynamics*. Gulf Publishing Company, Houston.

Elfert, M., Weber, A., Wittrock, D., Peters, A., Voss, C. and Nicke, E. (2017) Experimental and Numerical Verification of an Optimization of a Fast-Rotating High-Performance Radial Compressor Impeller. *ASME. J. Turbomach.*, 139(10): 101007. DOI:10.1115/1.4036357

Eisele, K., Zhang, Z., Casey, M. V., Gülich, J. and Schachenmann, A. (1997) Flow Analysis in a Pump Diffuser – Part 1: LDA and PTV Measurements of the Unsteady Flow. *ASME J. Fluids Eng.*, 119(4):968–977. DOI:10.1115/1.2819525

Eisenlohr, G. and Chladek, H. (1994) Thermal Tip Clearance Control for Centrifugal Compressor of an APU Engine. *ASME J. Turbomach.*, 116(4):629–634. DOI:10.1115/1.2929453

Eisenlohr, G., Dalbert, P., Krain, H. et al. (1998) Analysis of the Transonic Flow at the Inlet of a High Pressure Ratio Centrifugal Impeller. ASME Paper 98-GT-024, ASMEGas Turbine Congress, 2–5 June, Stockholm. DOI:10.1115/98-GT-024

Elder, R. L. and Gill, M. E. (1985) A Discussion of the Factors Affecting Surge in Centrifugal Compressors. *ASME J. Eng. Power*, 107(2):499–506. DOI:10.1115/1.3239759

Elfert, M., Weber, A., Wittrock, D. et al. (2017) Experimental and Numerical Verification of an Optimization of a Fast-Rotating High-Performance Radial Compressor Impeller. *ASME. J. Turbomach.*, 139(10):101007. https://doi.org/10.1115/1.4036357

Emmons, H. W., Pearson, C. E. and Grant, H. P. (1955) Compressor Surge and Stall Propagation. ASME 53-A-65; *ASME J. Basic Eng.*, 77:455–469.

Engeda, A. (1998) Early Historical Development of the Centrifugal Impeller. ASME Paper 98-GT-22, ASME Gas Turbine Congress, 2–5 June, Stockholm. DOI:10.1115/98-GT-022

Engeda, A. (2001) The Design and Performance Results of Simple Flat Plate Low Solidity Vaned Diffusers. *I. Mech. E. Part A: J. Power Energy*, 215(1):109–118. DOI:10.1243/0957650011536471

Engeda, A. (2007) Effect of Impeller Exit Width Trimming on Compressor Performance. Proceedings of the 8th International Symposium on Experimental and Computational Aerothermodynamics of Internal Flows, July, Lyon. ISAIF8–00135.

Engels B. (1990) Verbesserung des Instationärverhaltens von Abgasturboladern, Seminar on Brennverfahrensentwicklung für direkeinspritzende Dieselmotoren, 25–26 January, Nurnberg.

Epple, P., Durst, F. and Delgado A. (2010) A Theoretical Derivation of the Cordier Diagram for Turbomachines, *I. Mech. E. Part C: J. Mech. Eng. Sci.*, 225(2):354-368. DOI:10.1243/09544062JMES2285

Epstein, A. H. (2004) Millimetre-Scale, Micro-Electro-Mechanical Systems for Gas Turbine Engines. *ASME J. Eng. Power.*, 126(2):205–226. DOI:10.1115/1.1739245

ESDU 74015 (2007) Performance in Incompressible Flow of Plane-Walled Diffusers with Single-Plane Expansion.

Everitt, J. N., Spakovszky, Z. S., Rusch, D. and Schiffmann, J. (2017) The Role Impeller Outflow Conditions on the Performance of Vaned Diffusers. *ASME J. Turbomach.*, 139 (4):041004. DOI:10.1115/1.4035048

Ewins, D. J. (1969) The Effects of Detuning upon the Forced Vibrations of Bladed Disks. *J. of Sound Vib.*, 9(1), 65–72. DOI:10.1016/0022-460X(69)90264-8

Falomi, S., Aringhieri, C., Iurisci, G. et al. (2016) Full Scale Validation of a High Pressure Ratio Centrifugal Compressor. 45th Turbomachinery Symposium, Texas A&M, Houston.

Farkas, F. (1977) The Development of Brown Boveri Gas Turbine Compressors. *Brown Boveri Review*, 64(1):52–59.

Faux, I. D. and Pratt, M. J. (1979) *Computational Geometry for Design and Manufacture*. Ellis Horwood, London.

Ferguson, T. B. (1963) *The Centrifugal Compressor Stage*. Butterworth, London.

Ferziger, J. H. and Peric, M. (2002) *Computational Methods for Fluid Dynamics*. Springer, Berlin

Filipenco, V. G., Deniz, S., Johnston, J. M., Greitzer, E. M. and Cumpsty, N. A. (2000) Effects of Inlet Flow Field Conditions on the Performance of Centrifugal Compressor Diffusers: Part 1 – Discrete Passage Diffuser. *ASME J. Turbomach.*, 122(1):1–10. DOI:10.1115/1.555418

Fink, D. A., Cumpsty, N. A. and Greitzer, E. M. (1992) Surge Dynamics in a Free-Spool Centrifugal Compressor System. *ASME J. Turbomach.*, 114(2):321–332. DOI:10.1115/1.2929146

Fisher, F. B. (1988) Application of Map Width Enhancement Devices to Turbocharger Compressor Stages. *SAE Trans.*, 1988(97):1303–1310.

Flathers, M. B., Bache, G. E. and Rainsberger, R. (1996) An Experimental and Computational Investigation of Flow in A Radial Inlet of an Industrial Pipeline Centrifugal Compressor. *ASME J. Turbomach.*, 118(2):371–384. DOI:10.1115/1.2836652

Fontana, M., Baldassarre, L., Bernocchi, A. et al. (2015) Axial Thrust in High Pressure Centrifugal Compressors: Description of a Calculation Model Validated by Experimental Data from Full Load Test. 44th Turbomachinery Symposium, Texas A&M, Houston.

Fox, R. W. and Kline, S. J. (1962) Flow Regimes in Curved Subsonic Diffusers. *ASME J. Basic Eng.*, 84(3):303–312. DOI:10.1115/1.3657307

Freeman, C. (1985) Effect of Tip Clearance Flow on Compressor Stability and Engine Performance. VKI Lecture series LS-1985-05, Von Karman institute, Brussels

Freeman, C. and Cumpsty, N. A. (1992) A Method for the Prediction of Supersonic Compressor Blade Performance. *J. Propul.*, 8(1):199. DOI:10.2514/3.23461

Frigne, P. and Van den Braembussche, R. (1979) *One Dimensional Design of Centrifugal Compressors Taking into Account Flow Separation in the Impeller*. Von Karman Institute for Fluid Mechanics, Sint-Genesius-Rode, TN 129.

Frigne, P. and Van den Braembussche, R. (1984) Distinction between Different Types of Impeller and Diffuser Rotating Stall in a Centrifugal Compressor with Vaneless Diffuser. *ASME J. Eng. Power*, 106(2):468–474. DOI:10.1115/1.3239589

Fritsch, G. and Giles, M. B. (1995) An Asymptotic Analysis of Mixing Loss. *ASME J. Turbomach.*, 117(3):367–374. DOI:10.1115/1.2835670

Fulton, J. W., Klein, J. M., Marriott, A. and Graham, D. A, (2001) Full Load Testing of an All-Electric Centrifugal Compressor for Miscible Gas Injection. 30th Turbomachinery Symposium, Texas A&M, Houston. DOI:10.21423/R1ZD2V

Galindo, J., Serrano, J. R., Margot, X., Tiseira, A., Schorn, N. and Kindl, H. (2007) Potential of Flow Pre-Whirl at the Compressor Inlet of Automotive Engine Turbochargers to Enlarge Surge Margin and Overcome Packaging Limitations, *Int. J. Heat Fluid Flow*, 28(3):374-387. DOI:10.1016/j.ijheatfluidflow.2006.06.002

Gallimore, S. J. (1986) Spanwise Mixing in Multistage Axial Flow Compressors: Part II. Throughflow Calculations Including Mixing. *J. Turbomach.*, 108(1):10–16. DOI:10.1115/1.3262009

Gallimore, S. J., Bolger, J. J., Cumpsty, N. A. et al. (2002) The Use of Sweep and Dihedral in Multistage Axial Flow Compressor Blading. Part I: University Research and Methods Development. *ASME J. Turbomach.*, 124(4):521–532. DOI:10.1115/1.1507333

Galloway, L. (2018) Enhancing Centrifugal Compressor Stability Using Porous Endwall Treatments in the Vaned Diffuser. Ph.D. Thesis, Queen's University, Belfast.

Galloway, L., Spence, S., In Kim, S., Rusch, D., Vogel, K. and Hunziker, R. (2018) An Investigation of the Stability Enhancement of a Centrifugal Compressor Stage Using a Porous Throat Diffuser. *ASME J. Turbomach.*, 140(1):011008. DOI:10.1115/1.4038181

Galpin, P. F., Broberg, R. B. and Hutchinson, B. R. (1995) Three-Dimensional Navier–Stokes Predictions of Steady State Rotor/Stator Interaction with Pitch Change. Third Annual Conference of the CFD Society of Canada, 25–27 June, Banff.

Galpin, P. F., Van Doormaal J. P. and Raithby G. D. (1985) Solution of the Incompressible Mass and Momentum Equations by Application of a Coupled Equation Line Solver. *Int. J. Numer. Methods in Fluids*, 5(7):615–625.

Galvas, M. (1973) Fortran Program for Predicting Off-Design Performance of Centrifugal Compressors. NASA-TND-7487. NASA Lewis Research Center, Cleveland.

Gao, C., Gu, C., Wang, T. and Yang, B. (2007) Analysis of Geometries' Effects on Rotating Stall in Vaneless Diffuser with Wavelet Neural Networks. *Int. J. Rotating Mach.*, 2007, Article ID 76476., DOI:10.1155/2007/76476.

Gao, R., Spakovszky, Z., Rusch, D. and Hunziker, R. (2017) Area Schedule-Based Design of High-Pressure Recovery Radial Diffusion Systems. *ASME J. Turbomach.*, 139(1):011012. DOI:10.1115/1.4034488

Garrison, L., and Cooper, N. (2009) Visualization and Post-Processing of Centrifugal Compressor Computational Fluid Dynamics Flow Fields. ASME Paper GT2009-60165, ASME Turbo Expo, Orlando, 8–12 June. DOI:10.1115/GT2009-60165

Gasch, R., Nordmann, R. and Pfützner, H. (2006) *Rotordynamik.* Springer Berlin.

Gaster, M. (1969) Structure and Behaviour of Laminar Separation Bubbles. ARC R&M 3595. HMSO.

Gault, D. E. (1957) A Correlation of Low-Speed, Airfoil-Section Stalling Characteristics with Reynolds Number and Airfoil Geometry. NACA-TN-3963.

Geller, W. (2015) *Thermodynamik für Maschinenbauer.* Springer, Berlin.

Gibson L., Galloway L., Kim, S. and Spence, S. (2017) Assessment of Turbulence Model Predictions for a Centrifugal Compressor Simulation. *J. GPPS.*, 1:142–156. DOI:10.22261/2II890

Giesecke, D., Stark, U., Garcia, R. H. and Friedrichs, J. (2017) Design and Optimisation of Compressor Airfoils by Using Class Function/Shape Function Methodology. Conference ISROMAC 2017 – 17th International Symposium on Transport Phenomena and Dynamics of Rotating Machinery, Maui.

Gill, A., von Backström, T. W. and Harms, T. M. (2007) Fundamentals of Four-Quadrant Axial Flow Compressor Maps. *I. Mech. E., Part A: J. Power Energy*, 221(7):1001–1010. https://doi.org/10.1243/09576509JPE354

Ginder, R. B. and Calvert, W. J. (1987) The Design of an Advanced Civil Fan Rotor. *ASME J. Turbomach.*, 109(3):340–345. DOI:10.1115/1.3262111

Godard, A., Trébinjac, L. and Roumeas, M. (2017) Experimental characterisation of the surge onset in a turbocompressor for a fuel cell application. Paper ID: ETC2017-040, 12th European Turbomachinery Conference, 3–7 April, Stockholm. DOI:10.29008/ETC2017-040

Goodhand, M. N. and Miller, R. J. (2011) Compressor Leading Edge Spikes: A New Performance Criterion. *ASME J. Turbomach.*, 133(2):021006. DOI:10.1115/1.4000567

Goodhand, M. N. and Miller, R. J. (2012) The Impact of Real Geometries on Three-Dimensional Separations in Compressors. *ASME J. Turbomach.*, 134(2):021007. DOI:10.1115/1.4002990

Gooding, W. J., Fabian, J. C. and Key, N. L. (2019) LDV Characterization of Unsteady Vaned Diffuser Flow in a Centrifugal Compressor. ASME Paper GT2019-90476, ASME Turbo Expo, 17–21 June, Phoenix.

Goodman, J. (1919) *Mechanics Applied to Engineering*. Longmans, Green and Co., London.

Gosman, A. D. and Johns, R. J. R. (1978) Development of a Predictive Tool for In-Cylinder Gas Motion in Engines. SAE International Congress, Detroit. Paper 780315.

Gostelow, J. P. (1984) *Cascade Aerodynamics*. Pergamon Press, Oxford.

Gossweiler, C. R. (1993) Sonden und Messsystem für schnelle aerodynamische Strömungsmessung mit piezoresistiven Druckgebern. Dissertation ETH Nr. 10253. Zürich

Goto, A. (1992) Study of Internal Flows in a Mixed-Flow Pump Impeller at Various Tip Clearances Using Three-Dimensional Viscous Flow Computations. *ASME J. Turbomach.*, 114(2):373–382. DOI:10.1115/1.2929154

Grates, D. R., Jeschke, P. and Niehuis, R. (2014) Numerical Investigation of the Unsteady Flow inside a Centrifugal Compressor Stage with Pipe Diffuser. *ASME. J. Turbomach.*, 136 (3):031012. DOI.org/10.1115w1.4024873

Gravdahl, J. D. and Egeland, O. (1998) Speed and Surge Control for a Low Order Centrifugal Compressor Model. *J. Modell. Identifi. Control*, 19(1):13–29. DOI:10.1109/87.784420

Green, J. E., Weeks, D. J. and Brooman, J. W. F. (1977) Prediction of Turbulent Boundary Layers and Wakes in Compressible Flow by a Lag-Entrainment Method. R.A.E. Technical Report, R.&M. 3791, 1973. HMSO.

Gregory-Smith, D. G. and Crossland, S. C. (2001) Prediction of Turbomachinery Flow Physics from CFD: Review of Recent Computations of APPACET Test Cases. *Task Quarterly*, 5 (4):407–432.

Greitzer, E. M. (1976) Surge and Rotating Stall in Axial Flow Compressors – Part II: Experimental Results and Comparison with Theory. *ASME J. Eng. Power.*, 98 (2):199–211. DOI:10.1115/1.3446139

Greitzer, E. M. (1981) The Stability of Pumping Systems. *ASME. J. Fluids Eng.*, 103(2):193–242. DOI:10.1115/1.3241725

Greitzer, E. M., Nikkanen, J. P., Haddad, D. E., Mazzawy, R. S. and Joslyn, H. D. (1979) A Fundamental Criterion for the Application of Rotor Casing Treatment. *ASME J. Fluids Eng.*, 101(2):237–243. DOI:10.1115/1.3448945

Greitzer, E. M., Tan, C. S. and Graf, M. B. (2004) *Internal Flow: Concepts and Applications*. Cambridge University Press, Cambridge.

Gresh, T. M. (2018) *Compressor Performance*. Butterworth Heinemann, Elsevier, Oxford.

Grieb, H. (2009) *Verdichter für Turbo-Flugtriebwerke*. Springer, Berlin and Heidelberg. DOI:10.1007/978-3-540-34374-5

Grigoriadis, P., Sens, M. and Müller, S. (2012) Variable Trim Compressor – A New Approach to Variable Compressor Geometry. Tenth International Conference on Turbochargers and Turbocharging I Mech E, London.

Grimaldi, A. and Michelassi, V. (2018) The Impact of Inlet Distortion and Reduced Frequency on the Performance of Centrifugal Compressors. *ASME J. Eng. Power*, 141(2):021012. DOI:10.1115/1.4040907

Grönman, A., Dietmann, F., Casey, M. V. and Backman, J. (2013) Review and Collection of Preliminary Design Rules for Low Solidity Diffusers. Tenth European Conference on Turbomachinery ETC10, 15–19 April, Lappeenranta.

Guasti, M. (1992) Analytic Geometry of Some Rectilinear Figures. *Int. J. Educ. Sci. Technol.*, 23:895–901.

Guidotti, E., Toni, L., Rubino, D. T. et al. (2014) Influence of Cavity Flows Modelling on Centrifugal Compressor Stages Performance Prediction across Different Flow Coefficient

Impellers. ASME Paper GT2014-25839, ASME Turbo Expo, 16–20 June, Düsseldorf. DOI:10.1115/GT2014-25830

Gülich, J. F. (2008) *Centrifugal Pumps*. Springer, Berlin.

Guo, G., Zhang ,Y., Xu, J., Zheng, X. and Zhuge W. (2008) Numerical Simulation of a Transonic Centrifugal Compressor Blades Tip Clearance Flow of Vehicle Turbocharger, ASME Paper GT2008-50957, ASME Turbo Expo, 9–13 June, Berlin. DOI:10.1115/GT2008-50957

Gyarmathy, G. (1996) Impeller-Diffuser Momentum Exchange during Rotating Stall. ASME Paper 96-WA/PID-6, ASME International Mechanical Engineering Congress, Atlanta.

Gyarmathy, G. (2003) *Personal Communication of Unpublished Turbomachinery Lecture Notes*, ETH, Zürich.

Haaland, S E. (1983) Simple and Explicit Formulae for the Friction Factor in Turbulent Pipe Flow, Trans. *ASME J. Fluids Eng*., 105:89–99. DOI:10.1115/1.3240948

Hah, C., Puterbaugh, S. L. and Wadia, A. R. (1998) Control of Shock Structure and Secondary Flow Field inside Transonic Compressor Rotors through Aerodynamic Sweep. ASME Paper 98-GT-561, ASME Gas Turbine Congress, 2–5 June, Stockholm. DOI:10.1115/98-GT-561

Hall, D. K., Greitzer, E. M., and Tan, C. S. (2012) Performance Limits of Axial Compressor Stages. ASME Paper GT2012-69709, ASME Turbo Expo, 11–15 June, Copenhagen. DOI:10.1115/GT2012-69709

Halstead, D. E., Wisler, D. C., Okishii, T. H., Walker, G. J., Hodson, H. P. and Shin H. W. (1997) Boundary Layer Development in Axial Compressors and Turbines – Part 1 of 4: Composite Picture. *ASME J. Turbomach*., 119(2):114–127. DOI:10.1115/1.2841105

Hama, F. R. (1954) Boundary Layer Characteristics for Smooth and Rough Surfaces. *SNAME*. 62:333–358.

Han, G., Gua, L., Yand, C. et al. (2017) Design and Analysis of a High Pressure Ratio Centrifugal Compressor with Three Diffusers. GPPS-2017-0104 Proceedings of Shanghai GPP Forum 2017, 30 October–1 November.

Hanimann, L., Mangani, L., Casartelli, E. et al. (2014) Development of a Novel Mixing Plane Interface Using a Fully Implicit Averaging for Stage Analysis. *ASME J. Turbomach*., 136 (8):081010. DOI:10.1115/1.4026323

Hanlon, P. C. (2001) *Compressor Handbook*. McGraw-Hill, New York.

Harada, H. (1985) Performance Characteristics of Shrouded and Unshrouded Impellers of a Centrifugal Compressor. *ASME J. Eng. Power*., 107(2):528–533. DOI:10.1115/1.3239765

Harari, Y. N. (2018) *21 Lessons for the 21st Century*. Jonathan Cape, London.

Hardin, J. R. (2002) A New Approach to Predicting Centrifugal Compressor Sideload Pressure. ASME Paper IMECE2002-39592, ASME Mechanical Engineering Congress, 17–22 November, New Orleans. DOI:10.1115/IMECE2002-39592

Harley, P., Spence, S., Filsinger, D., Dietrich, M. and Early, J. (2014) Meanline Modelling of Inlet Recirculation in Automotive Turbocharger Centrifugal Compressors. *ASME J. Turbomach*., 137(1):011007. DOI:10.1115/1.4028247

Harley, P., Starke, A. Bamba, T. and Filsinger D. (2017) Axial Groove Casing Treatment in an Automotive Turbocharger Centrifugal Compressor. *I. Mech. E. Part C: J. Mech. Eng. Sci*., 232(24):4472–4484. DOI:10.1177/0954406216688495

Harris, L. E. (1951) Some Factors in the Early Development of the Centrifugal Pump 1689 to 1851. *Trans. Newcomen Soc*., 28(1):187–202. DOI: 10.1179/tns.1951.014

Harvey, N. W., Rose, M. G., Taylor, M. D. et al. (1999) Non-Axisymmetric Turbine End Wall Design: Part I – Three-Dimensional Linear Design System. *ASME J. Turbomach*., 122(2):278–285. DOI:10.1115/1.555445

Hathaway, M. D. (2007) Passive Endwall Treatments for Enhancing Stability. NASA/TM-2007-214409.

Hathaway, M. D., Wood, J. R. and Wasserbauer, C. A. (1992) NASA Low-Speed Centrifugal Compressor for Three-Dimensional Viscous Code Assessment and Fundamental Flow Physics Research. *ASME J. Turbomach.*, 114(2):295–303. DOI:10.1115/1.2929143

Hausenblas, H. (1965) Trennung der Lauf- und Leitradverluste bei der Aufwertung von Versuchen an einstufigen Radialverdichtern, Forsch. *Ing. Wes.*, 31(1):11–13.

Hawthorne, W. R. (1974) Secondary Vorticity in Stratified Compressible Flows in Rotating Systems. C.U.E.D./A-Turbo/TR63.

Hayami, H., Senoo, Y. and Utsunomiya, K. (1990) Application of a Low-Solidity Cascade Diffuser to Transonic Centrifugal Compressor. *ASME J. Turbomach.*, 112(1):25–29. DOI:10.1115/1.2927416

Haywood, R. W. (1980) *Analysis of Engineering Cycles*. Pergamon Press, Oxford.

Hazby, H. R. (2010) Centrifugal Compressor Aerodynamics. Ph.D. Thesis, Cambridge University.

Hazby, H. R and Robinson C. J. (2018) Inducer Design of Centrifugal Impellers. Paper No: CON6516/161. The 13th International Conference on Turbochargers and Turbocharging, 16–17 May, I. Mech E.

Hazby, H. R. and Xu, L. (2009a) Numerical Investigation of the Effects of Leading Edge Sweep in a Small Transonic Impeller. Eighth European Turbomachinery Conference on Turbomachinery Fluid Dynamics and Thermodynamics, 23–27 March, Graz.

Hazby, H. R. and Xu, L. (2009b) Role of Tip Leakage in Stall of a Transonic Centrifugal Impeller. ASME Paper GT2009-59372, ASME Turbo Expo, 8–12 June, Orlando. DOI:10.1115/GT2009-59372

Hazby, H. R., Casey, M. V. and Březina, L. (2019) Effect of Leakage Flows on the Performance of a Family of Inline Centrifugal Compressors. *ASME J. Turbomach.*, 141(9):091006. DOI:10.1115/1.4043786

Hazby, H. R., Casey, M. V., Numakura, R. and Tamaki, H. (2014) Design and Testing of a High Flow Coefficient Mixed Flow Impeller. Eleventh International Conference on Turbochargers and Turbocharging, 13–14 May, London.

Hazby, H. R., Robinson, C. J., Casey, M. V., Rusch, D. and Hunziker, R. (2018) Free-Form versus Ruled Inducer Design in a Transonic Centrifugal Impeller. *ASME J. Turbomach.*, 140(1):011010. DOI:10.1115/1.4038176

Hazby, H. R., Xu, L. and Casey, M. V. (2017) Investigation of the Flow in a Small-Scale Turbocharger Centrifugal Compressor. *I. Mech. E. Part A: J. Power Energy*, 231(1):3–13. DOI:10.1177/0957650916671277

Hazby, H. R., Xu, L. and Schleer, M. (2013) Study of the Flow in a Vaneless Diffuser at Part Speed Operating Conditions. *ASME J. Turbomach.*, 136(3):031011. DOI:10.1115/1.4024693

Hazby, H., Woods, I., Casey, M., Numakura, R. and Tamaki, H. (2015). Effects of Blade Deformation on the Performance of a High Flow Coefficient Mixed Flow Impeller. *ASME J. Turbomach.*, 137(12):121005. DOI:10.1115/1.4031356

Hazby, H. R., O'Donoghue, R. and Robinson, C. J. (2020) Design and Modelling of Circular Volutes for Centrifugal Compressors. 14th International Conference on Turbochargers and Turbocharging,11–12 May, London.

He, L. and Shan, P. (2012) Three-Dimensional Aerodynamic Optimization for Axial-Flow Compressors Based on the Inverse Design and the Aerodynamic Parameters. *ASME. J. Turbomach.*, 134(3):031004. DOI:10.1115/1.4003252

He, X. and Zheng, X. (2019) Roles and Mechanisms of Casing Treatment on Different Scales of Flow Instability in High Pressure Ratio Centrifugal Compressors. *Aerosp. Sci. Technol.*, 84:734–746, DOI:10.1016/j.ast.2018.10.015

Head, M. R. (1960) Entrainment in the Turbulent Boundary Layer. ARC R&M 3152, HMSO.

Head, M. R. and Patel, V. C. (1968) Improved Entrainment Method for Calculating Turbulent Boundary Layer Development. ARC R&M 3643, HMSO.

Hederer, M., Editor (2011) *Radialverdichterforschung in der Forchungsvereinigung, Heft R555-2011*. Forschungsvereinigung Verbrennungskraftmaschinen, Frankfurt.

Hehn, A., Mosdzien, M., Grates, D. and Jeschke, P. (2018) Aerodynamic Optimization of a Transonic Centrifugal Compressor by Using Arbitrary Blade Surfaces. *ASME. J. Turbomach.*, 140(5):051011. DOI:.10.1115/1.4038908

Herbert, M. V. (1980) A Method of Performance Prediction for Centrifugal Compressors. ARC R&M 3843, HMSO.

Hetzer, T., Epple, P., and Delgado, A. (2010) Exact Solution of the Plane Flow and Slip Factor in Arbitrary Radial Blade Channels and Extended Design Method for Radial Impellers. ASME Paper IMECE2010-39081, ASME 2010 Mechanical Engineering Congress, 12–18 November, Vancouver. DOI:10.1115/IMECE2010-39081

Hirsch, C. (2007) *Numerical Computation of Internal and External Flows: The Fundamentals of Computational Fluid Dynamics*, 2nd Edition. Butterworth Heinemann, Oxford.

Hirsch, C. and Warzee, G. (1976) A Finite-Element Method for through Flow Calculations in Turbomachines. *ASME J. Fluids Eng.*, 98(3):403–414. DOI:10.1115/1.3448341

Hirsch, C. and Deconinck, H. (1985) Through Flow Models for Turbomachines: Streamsurface and Passage Averaged Representations. In *Thermodynamics and Fluid Dynamics of Turbomachinery*, eds. A. S. Üçer, P. Stow and Ch. Hirsch, vol. 1. Martinus Nijhoff Publishers, Dordrecht. 3–35.

Hirsch, Ch., Kang, S., and Pointel, G. (1996) A Numerically Supported Investigation of the 3D Flow in Centrifugal Impellers: Part II – Secondary Flow Structure. ASME Paper 96-GT-152, ASMEGas Turbine Congress, 10–13 June, Birmingham. DOI:10.1115/96-GT-152

Hirschmann, A., Casey, M. V. and Montgomery, M. (2013) A Zonal Calculation Method for Axial Gas Turbine Diffusers. ASME Turbo Expo, 3–7 June, San Antonio. DOI:10.1115/GT2013-94117.

Hobbs, D. E. and Weingold, H. D. (1984) Development of Controlled Diffusion Airfoils for Multistage Compressor Application. *ASME J. Eng. Power.*, 106(2):271–278. DOI:10.1115/1.3239559

Hodson, H. P., Hynes, T. P., Greitzer, E. M. and Tan, C. S. (2012) A Physical Interpretation of Stagnation Pressure and Enthalpy Changes in Unsteady Flow. *J. Turbomach.*, 134(6):060902. DOI:10.1115/1.4007208

Horlock, J. H. (1958) *Axial Flow Compressors*. Butterworth, Oxford, reprinted with additional material in 1973 by Krieger Publishing Co., Malabar, 1973.

Horlock, J. H. (1960) Losses and Efficiencies in Axial-Flow Turbines. *Int. J. Mech. Sci.*, 2:48–75. DOI:10.1016/0020-7403(60)90013-8

Horlock, J. H. (1966) *Axial Flow Turbines*. Butterworth, Oxford.

Horlock, J. H. (1971) On Entropy Production in Adiabatic Flow in Turbomachines. *ASME J. Basic Eng. Series D*, 93(4):587–593. DOI:10.1115/1.3425313

Horton, H. P. (1967) A Semi-Empirical Theory for the Growth and Bursting of Laminar Separation Bubbles. ARC Conf. Proc. 1073, HMSO.

Howard, J. H. G. and Ashrafizaadeh, M. (1994) A Numerical Investigation of Blade Lean Angle Effects on Flow in a Centrifugal Impeller. ASME Gas Turbine Congress, 13–16 June, The Hague. DOI:10.1115/94-GT-149

Howell, A. R. (1945) Fluid Dynamics of Axial Compressors. *I. Mech. E.*, 153 (1):441–452

Howell, A. R. and Calvert, W. J. (1978) A New Stage Stacking Technique for Axial Flow Compressor Performance Prediction. *ASME J. Eng. Power*, 100(4):698–703. DOI:10.1115/1.3446425

Hu, B., Brillert, D., Dohmen, H. J. and Benra, F. K. (2018) Investigation on Thrust and Moment Coefficients of a Centrifugal Turbomachine. *Int. J. Propuls. Power*, 3(2):9. DOI:10.3390/ijtpp3020009

Huntington, R. A. (1985) Evaluation of Polytropic Calculation Methods for Turbomachinery Performance. *ASME J. Eng. Power*, 107(4):872–876. DOI:10.1115/1.3239827

Hunziker, R. (1993) Einfluss der Diffusorgeometrie auf die Instabilitätsgrenze des Radialverdichters. ETH Dissertation Nr. 10252 Zürich.

Hunziker, R., and Gyarmathy, G. (1994) The Operational Stability of a Centrifugal Compressor and Its Dependence on the Characteristics of the Subcomponents. *ASME J. Turbomach.*, 116 (2):250–259. DOI:10.1115/1.2928359

Hunziker, R., Dickmann, H. P. and Emmrich, R. (2001) Numerical and Experimental Investigation of a Centrifugal Compressor with an Inducer Casing Bleed System. *I. Mech. E. Part A: J. Power Energy*, 215(6):783–791. DOI:10.1243/0957650011538910

Ibaraki, S., Tomita, S., Ebisu, M. and Takashi, S. (2012) Development of a Wide-Range Centrifugal Compressor for Automotive Turbochargers. *Mitsubishi Heavy Ind. Tech. Rev.*, 49(1):68–79.

Inoue, M. (1983) Radial Vaneless Diffusers: A Re-Examination of the Theories of Dean and Senoo and of Johnston and Dean. *ASME J. Fluids Eng.*, 105(1):21–27. DOI:10.1115/1.3240935

Inoue, M. and Cumpsty, N. A. (1984) Experimental Study of Centrifugal Impeller Discharge Flow in Vaneless and Vaned Diffusers. *ASME J. Eng. Power*, 106(2):455–467. DOI:10.1115/1.3239588

Ishimoto, L., Miranda, M. A. Silva, R. T. et al. (2015) Review of Centrifugal Compressors High Pressure Testing for Offshore Applications. 44th Turbomachinery Symposium, Texas A&M, Houston. DOI:10.21423/R1V04J

Issac, J. M., Sitaram, N. and Govardhan, M. (2004) Effect of Diffuser Vane Height and Position on the Performance of a Centrifugal Compressor. *I. Mech. E. Part A: J. Power Energy*, 218 (8):647–654. DOI:10.1243/0957650042584320.

ISO 1302 (2002) *Geometrical Product Specifications (GPS) – Indication of Surface Texture in Technical Product Documentation*. International Organisation for Standardisation, Geneva.

ISO 10439 (2015) *Petroleum, Petrochemical and Natural Gas Industries Axial and Centrifugal Compressors*. International Organisation for Standardisation, Geneva.

Itou, S., Oka, N., Furukawa, M., Yamada, K. et al. (2017) Optimum Aerodynamic Design of Centrifugal Compressor Using a Genetic Algorithm and an Inverse Method Based on Meridional Viscous Flow Analysis. ISROMAC 2017 International Symposium on Transport Phenomena and Dynamics of Rotating Machinery, 16–21 December, Hawaii.

Jaatinen-Värri, A., Tiainen, J., Turunen-Saaresti, T. et al. (2016) Centrifugal Compressor Tip Clearance and Impeller Flow. *J. Mech. Sci. Technol.*, 30(11):5029–5040. DOI:10.1007/s12206-016-1022-8

Jaatinen-Värri, A., Turunen-Saaresti, T., Grönman, A., Roytta, P. and Backman, J. (2013) The Tip Clearance Effects on the Centrifugal Compressor Vaneless Diffuser Flow Fields at Off-Design Conditions. ETC2013-065, 10th European Turbomachinery Conference, Lappeenranta.

Jansen, M. and Rautenberg, M. (1982) Design and Investigations of a Three Dimensionally Twisted Diffuser for Centrifugal Compressors. ASME Paper 82-GT-102, ASME Gas Turbine Conference, 18–22 April, London. DOI:10.1115/82-GT-102

Jansen, W. and Kirschner, A. (1974) *Impeller Blade Design Method for Centrifugal Compressors*. NASA Special Publication, Washington.

Janssen, M., Seume, J. and Zimmermann, H. (1994) The Model V84.3 Shop Tests: Tip Clearance Measurements and Evaluation. ASMEGas Turbine Congress, 13–16 June, The Hague. DOI:10.1115/94-GT-319

Japikse, D. (1985) Assessment of Single- and Two-Zone Modelling of Centrifugal Compressors, Studies in Component Performance: Part 3. ASME Paper 85-GT-73, ASME Gas Turbine Conference, 18–21 March, Houston. DOI:10.1115/85-GT-73

Japikse, D. (1990) *Centrifugal Compressor Design and Performance*. Concepts/NREC Publishing, White River Junction.

Japikse, D. and Baines N. C. (1984) *Introduction to Turbomachinery*. Concepts/NREC Publishing, White River Junction.

Japikse, D. and Baines N. C. (1998) *Turbomachinery Diffuser Design Technology*. Concepts/ NREC Publishing, White River Junction.

Jeffcott, H. H. (1919) Lateral Vibration of Loaded Shafts in the Neighbourhood of a Whirling Speed – the Effect of Want on Balance. *Phil. Mag.*, 37:304–314

Jennions, I. K. and Stow, P. (1985) A Quasi-Three-Dimensional Turbomachinery Blade Design System: Part II – Computerized System. *ASME J. Eng. Power*, 107(2):308–314. DOI:10.1115/1.3239716

Jenny, E. (1993) The BBC Turbocharger – a Swiss Success Story. Birkhaueser, Basel.

Jiang, P. M. and Whitfield, A. (1992) Investigation of Vaned Diffusers as a Variable Geometry Device for Application to Turbocharger Compressors. *I. Mech. Eng. Part D: J. Automob. Eng.*, 206(3):209–220. DOI:10.1243/PIME_PROC_1992_206_179_02

Johnson, M. W. (1978) Secondary Flow in Rotating Bends. *ASME J. Eng. Power*, 100(4):553–560. DOI:10.1115/1.3446393

Johnson, M. W. and Moore, J. (1980) The Development of Wake Flow in a Centrifugal Impeller. *ASME J. Eng. Power*, 102(2):382–389. DOI:10.1115/1.3230265

Johnson, M. W. and Moore, J. (1983) The Influence of Flow Rate on the Wake in a Centrifugal Impeller. *ASME J. Eng. Power*, 105(1):33–39. DOI:10.1115/1.3227395

Jung, S. and Pelton, R. (2016) Numerically Derived Design Guidelines of Self Recirculation Casing Treatment for Industrial Centrifugal Compressors. ASME Paper GT2016-56672, ASME Turbo Expo, 13–17 June, Seoul. DOI:10.1115/GT2016-56672

Kaluza, P., Landgraf, C., Schwarz, P., Jeschke, P. and Smythe, C. (2017) On the Influence of a Hub-Side Exducer Cavity and Bleed Air in a Close-Coupled Centrifugal Compressor Stage. *ASME. J. Turbomach.*, 139(7):071011. DOI:10.1115/1.4035606

Kammerer, A. (2009) Experimental Research into Resonant Vibration of Centrifugal Compressor Blades. ETH Dissertation No. 18587, Zurich.

Kang, S. and Hirsch, C. (1993) Experimental Study on the Three-Dimensional Flow within a Compressor Cascade with Tip Clearance: Part I – Velocity and Pressure Fields, and Part II – the Tip Leakage Vortex, *ASME J. Turbomach.*, 115(3):435–450. DOI:10.1115/ 1.2929271

Karstadt, S., Weiske, S. and Münz, S. (2018) Turbocharger with Variable Compressor Geometry – Another Contribution to Improved Fuel Economy by the Boosting System. 27th Aachen Colloquium Automobile and Engine Technology 2018.

Katsanis, T. (1966) Use of Arbitrary Quasi-Orthogonals for Calculating Flow Distribution in a Turbomachine. *ASME J. Eng. Power*, 88(2):197–202. DOI:10.1115/1.3678504

Kawakubo, T., Numakura, R. and Majima, K. (2008) Prediction of Surface Roughness Effects on Centrifugal Compressor Performance. ASME Paper FEDSM2008-55078, ASME Fluids Eng. Conference, 10–14 August, Jacksonville. DOI:10.1115/FEDSM2008-55078

Kenny, D. P. (1972) A Comparison of the Predicted and Measured Performance of High Pressure Ratio Centrifugal Compressor Diffusers. ASME Paper 72-GT-54, ASME 1972 Gas Turbine Conference, 26–30 March, San Francisco. DOI:10.1115/72-GT-54

Kerres, B., Cronhjort, A. and Mihaescu, M. (2016). Experimental Investigation of Upstream Installation Effects on the Turbocharger Compressor Map. 12th International Conference on Turbochargers and Turbocharging, 17–18 May, London.

Khalid, S. A., Khalsa, A. S., Waitz, I. A. et al. (1999) Endwall Blockage in Axial Compressors. *ASME J. Turbomach.*, 121(3):499–509. DOI:10.1115/1.2841344

Kim, Y. and Koch, J. (2004) Design and Numerical Investigation of Advanced Radial Inlet for a Centrifugal Compressor Stage. ASME Paper IMECE2004-60538, ASME Mech. Eng. Congress, 13–19 November, Anaheim. DOI:10.1115/IMECE2004-60538

Kim, Y., Engeda, A., Aungier, R. and Amineni, N. (2002). A Centrifugal Compressor Stage with Wide Flow Range Vaned Diffusers and Different Inlet Configurations. *I. Mech. E. Part A: J. Power Energy*, 216(4):307–320. DOI:10.1243/09576500260251156

Klein, S. and Nellis, G. (2012), *Thermodynamics*. Cambridge University Press, New York.

Kline, S. J., Abbott, D. E. and Fox, R. W. (1959) Optimum Design of Straight-Walled Diffusers. *ASME J. Basic Eng.*, 81(3):321–329. DOI:10.1115/1.4008462

Kmecl, T. and Dalbert, P. (1999) Optimization of a Vaned Diffuser Geometry for Radial Compressors: Part I – Investigation of the Influence of Geometry Parameters on Performance of a Diffuser. ASME Paper 99-GT-437. ASME Gas Turbine Congress, 7–10 June, Indianapolis. DOI:10.1115/99-GT-437

Kmecl, T., ter Harkel, R. and Dalbert, P. (1999) Optimization of a Vaned Diffuser Geometry for Radial Compressors: Part II – Optimization of a Diffuser Vane Profile in Low Solidity Diffusers. ASME Paper 99-GT-434, ASME Gas Turbine Congress, 7–10 June, Indianapolis. DOI:10.1115/99-GT-434

Koch, C. C. (1981) Stalling Pressure Rise Capability of Axial Flow Compressor Stages. *ASME J. Eng. Power*, 103(4):645–656. DOI:10.1115/1.3230787

Koch, C. C. and Smith, L. H. (1976) Loss Sources and Magnitudes in Axial-Flow Compressors. *ASME J. Eng. Power*, 98(3):411–424. DOI:10.1115/1.3446202

Koch, J., Sorokes, J. and Belhassan, M. (2011) Modelling and Prediction of Sidestream Inlet Pressure for Multistage Centrifugal Compressors. 40th Turbomachinery Symposium, Texas A&M, Houston. DOI:10.21423/R1SD29

Kocur, J. A. Jr. and Cloud, C. H. (2013) Shop Rotordynamic Testing – Options, Objectives, Benefits and Practices. 42nd Turbomachinery Symposium, Texas A&M, Houston. DOI:10.21423/R1GD2H

König, S., Petry, N. and Wagner, N. G. (2009) Aeroacoustic Phenomena in High Pressure Centrifugal Compressors – A Possible Root Cause for Impeller Failures. 38th Turbomachinery Symposium, Texas A&M, Houston. DOI:10.21423/R1735C

Kouremenos, D. A. and Antonopoulos, K. A. (1987) Isentropic Exponents of Real Gases and Application for the Air at Temperatures from 150 K to 450 K.A. *Acta Mech.*, 65:81–99. DOI:10.1007/BF01176874

Kowalski, S. C., Pacheco, J. E., Fakhri, S. and Sorokes, J. M. (2012) Centrifugal Stage Performance Prediction and Validation for High Mach Number Applications. 41st Turbomachinery Symposium, Texas A&M, Houston. DOI:10.21423/R1K05F

Krain, H. (1981) A Study on Centrifugal Impeller and Diffuser Flow. *ASME J. Eng. Power*, 103 (4):688–697. DOI:10.1115/1.3230791

Krain, H. (1987) Secondary Flow Measurements with L2F Technique in Centrifugal Compressors. *AGARD Proceedings, AGARD CP-421 (PEP Meeting, 69th Symposium, 4–8 May, Paris*, 34.1–34.10.

Krain, H. (2005) Review of Centrifugal Compressor's Application and Development. *ASME J. Turbomach.*, 127(1):25–34. DOI:10.1115/1.1791280

Krain, H. and Hoffman, W. (1989) Verification of an Impeller Design by Laser Measurements and 3D-Viscous Flow Calculations. ASME Paper 89-GT-159, ASME Gas Turbine Congress, 4–8 June, Toronto. DOI:10.1115/89-GT-159

Krain, H., Hoffmann, B. and Pak, H. (1995) Aerodynamics of a Centrifugal Compressor Impeller with Transonic Inlet Conditions. ASME Paper 95-GT-079, ASME 1995 Gas Turbine Congress, 5–8 June, Houston. DOI:10.1115/95-GT-079

Krain, H., Hoffmann, B., Rohne, K.-H., Eisenlohr, G. and Richter, F.-A. (2007) Improved High Pressure Ratio Centrifugal Compressor. ASME Paper GT2007-27100, ASME Turbo Expo, 14–17 May, Montreal. DOI:10.1115/GT2007-27100

Kucharski, W. (1918) Strömungen einer reibungsfreien Flüssigkeit bei Rotation fester Körper. Oldenbourg Technology and Engineering, Berlin.

Kumar, S. K., Kurz, R. and O'Connell, J. P. (1999) Equations of State for Gas Compressor Design and Testing. ASME Paper No: 99-GT-012, ASME Turbo Expo, 7–10 June, Indianapolis. DOI:10.1115/99-GT-012

Kunte, R., Jeschke, P. and Smythe, C. (2013) Experimental Investigation of a Truncated Pipe Diffuser with a Tandem De-Swirler in a Centrifugal Compressor Stage. *ASME. J. Turbomach.*, 135(3):031019. DOI:10.1115/1.4007526

Kurz, R. (2019). Optimization of Compressor Stations. *Journal of the Global Power and Propulsion Society*, 3:668–674. DOI:10.33737/jgpps/112399

Kurz, R., Brun, K. and Legrand, D. D. (1999) Field Performance Testing of Gas Turbine Driven Compressor Sets. 28th Turbomachinery Symposium, Texas A&M, Houston. DOI:10.21423/R1PM2F

Kurz, R., Marechale, R. K., Fowler, E. J. et al. (2016) Operation of Centrifugal Compressors in Choke Conditions. Asia Turbomachinery and Pump Symposium, 22–25 February, Singapore.

Kushner, F., Richard, S. J. and Strickland, R. A. (2000) Critical Review of Compressor Impeller Vibration Parameters for Failure Prevention. 29th Turbomachinery Symposium, Texas A&M, Houston. DOI:10.21423/R1F959

Lakshminarayana, B. (1996) *Fluid Dynamics and Heat Transfer in Turbomachinery*. Wiley, New York.

Langtry, R. B. and Menter, F. R. (2005) Transition Modelling for General CFD Applications in Aeronautics. AIAA Paper 2005-522, 10–13 January, Reno. DOI:10.2514/6.2005-522

Larosiliere, L. M., Skoch, G. J. and Prahst, P. S. (1997) Aerodynamic Synthesis of a Centrifugal Impeller Using CFD and Measurements. NASA Technical Memorandum, ARL-TR-1461.

Launder, B. E. and Spalding, D. B. (1974) The Numerical Computation of Turbulent Flow. *Comp. Meth. Appl. Mech. Eng.*, 3(2): 269–289. DOI:10.1016/0045-7825(74)90029-2

Lawless, P. B. and Fleeter, S. (1993) Rotating Stall Acoustic Signature in a Low Speed Centrifugal Compressor: Part 2 – Vaned Diffuser. ASME Paper 93-GT-254, ASME Gas Turbine Congress, 24–27 May, Cincinnati. DOI:10.1115/93-GT-254

Lee, B. I. and Kesler, M. G. (1975) A Generalized Thermodynamic Correlation Based on Three-Parameter Corresponding States. *AIChE*, 21:510–527. DOI:10.1002/aic.690210313

Lei, D. and Lixin, C. (2015) Effects of Residual Riblets of Impeller's Hub Surface on Aerodynamic Performance of Centrifugal Compressors. *Eng. Appl. Comp. Fluid Mech.*, 9(1):99–113, DOI:10.1080/19942060.2015.1004813

Lei, V., Spakovszky, Z. S. and Greitzer, E. M. (2008) A Criterion for Axial Compressor Hub-Corner Stall. *ASME J. Turbomach.*, 130(3):031006. DOI:10.1115/1.2775492

Leichtfuss, S., Bühler, J. Schiffer, H. P., Peters, P. and Hanna, M. (2019) A Casing Treatment with Axial Grooves for Centrifugal Compressors. *Int. J. Propuls. Power.* 4(3):27. DOI:10.3390/ijtpp4030027

Lenke, L. J. and Simon, H. (1999) Numerical Investigations on the Optimum Design of Return Channels of Multi-Stage Centrifugal Compressors. ASME Paper 99-GT-103, ASME Gas Turbine Congress, 7–10 June, Indianapolis. DOI:10.1115/99-GT-103

Lettieri, C., Baltadjiev, N., Casey, M. V. and Spakovszky, Z. (2014) Low-Flow-Coefficient Centrifugal Compressor Design for Supercritical CO2. *ASME J. Turbomach.*, 136 (8):081008. DOI:10.1115/1.4026322

Lewis, K. L. (1994) Spanwise Transport in Axial-Flow Turbines: Part 2 – Throughflow Calculations Including Mixing. *ASME J. Turbomach.*, 116(2):187–193. DOI:10.1115/1.2928352

Lewis, R. I. (1996), *Turbomachinery Performance Analysis*. John Wiley & Sons, New York.

Lewis, R. I., Fisher, E. H. and Saviolakis, A. (1972) Analysis of Mixed-Flow Rotor Cascades. ARC R&M 3703. HMSO.

Li, X., Zhao, Y., Liu, Z. and Chen, H. (2020) A New Methodology for Preliminary Design of Centrifugal Impellers with Pre-Whirl. *I. Mech. E., Part A: J. of Power and Energy*, 234 (3):251–262. DOI:10.1177/0957650919864193

Li, Z. and Zheng, X. (2017) Review of Design Optimization Methods for Turbomachinery Aerodynamics. *Prog. Aerosp. Sci.*, 93:1–23. DOI:10.1016/j.paerosci.2017.05.003

Lieblein S (1965) Experimental Flow in Two-Dimensional Cascades. *Aerodynamic Design of Axial-Flow Compressors*. NASA SP-36, ch. 6.

Liu, B., Chen, S. and Martin H. F. (2000) A Primary Variable Throughflow Code and Its Application to Last Stage Reverse Flow in LP Steam Turbine. Paper: IJPGC2000–5010, Int. Joint Power Generation Conference, 23–26 July, Miami Beach.

Lohmberg, A., Casey, M. V. and Ammann, S. (2003) Transonic Radial Compressor Inlet Design. *I. Mech. E. Part A: J. Power Energy*, 217(4):367–374. DOI:10.1243/095765003322315423

Longley, J. P. and Hynes, T. P. (1990) Stability of Flow through Multistage Axial Compressors. *ASME J. Turbomach.*, 112(1):126–132. DOI:10.1115/1.2927409

Lou, F. and Key, N. L. (2019) The Design Space for the Final-Stage Centrifugal Compressor in Aeroengines. AIAA Paper 2019-0944, AIAA Scitech, 2019 Forum, 7–11 January, San Diego. DOI:10.2514/6.2019-0944

Lou, F., Fabian, J. C. and Key, N. L. (2019) Design Considerations for Tip Clearance Sensitivity of Centrifugal Compressors in Aeroengines. *J. Propulsion and Power*, 35(3):666. DOI/10.2514/1.B37100

Lou, F., Harrison, H. M., Fabian, J. C. et al. (2016) Development of a Centrifugal Compressor Facility for Performance and Aeromechanics Research. ASME Paper GT2016-56188, ASME Turbo Expo, Seoul. DOI:10.1115/GT2016-56188

Lou, F. J., Fabian, H. M. and Key, N. L. (2014). The Effect of Gas Models on Compressor Efficiency Including Uncertainty. *ASME. J. Eng. Power*, 136(1):012601. DOI:10.1115/1.4025317

Lown, H. and Wiesner, F. J. (1959) Prediction of Choking Flow in Centrifugal Impellers. *ASME J. Basic Eng.*, 81(1):29–35. DOI:10.1115/1.4008351

Lüdtke, K. H. (2004) *Process Centrifugal Compressors*. Springer, Berlin.

Lyman, F. A. (1993) On the Conservation of Rothalpy in Turbomachines. *ASME. J. Turbomach.*, 115(3):520–526. DOI:10.1115/1.2929282

Macchi. E. (1985) The Use of Radial Equilibrium and Streamline Curvature Methods for Turbomachinery Design and Prediction. In *Thermodynamics and Fluid Mechanics of Turbomachinery*, vol. 1 , eds. A. S. Uecer, P. Stow and Ch. Hirsch. Springer, Netherlands, 33–66.

MacCormack, R. W. (1969) The Effect of Viscosity in Hypervelocity Impact Cratering. AIAA Paper, 69-354. DOI:10.2514/6.1969-354

Mack, R., Casey, M. V. et al. (1997) The Use of a Three-Dimensional Navier Stokes Code for the Sealing and Leakage Flows in Turbomachinery Applications. Second European Conference on Turbomachinery – Fluid Dynamics and Thermodynamics, 5–7 March, Antwerp.

Malik, A. and Zheng, Q. (2019) Effect of Double Splitter Blades Position in a Centrifugal Compressor Impeller. *I. Mech. E. Part A: J. Power Energy*, 233(6):689–701. DOI:10.1177/0957650918792462

Mallen, M. and Saville, G. (1977) Polytropic Processes in the Performance Prediction of Centrifugal Compressors. In *I. of Mech. E. Conference, C183/77, London*, 89–96.

Marcinowski, H. (1959) Einstufige Turboverdichter. *Chemie Ingieurtechnik* 31(4):237–247.

Marsan, A., Trébinjac, I., Coste, S. and Leroy, G. (2014) Influence of Unsteadiness on the Control of a Hub-Corner Separation within a Radial Vaned Diffuser. *ASME. J. Turbomach.*, 137(2):021008. DOI:10.1115/1.4028244

Marsh, H. (1968) A Digital Computer Program for the Through-Flow Fluid Mechanics in an Arbitrary Turbomachine Using a Matrix Method. ARC R&M, 3509, HMSO.

Mayle, R. E. (1991) The Role of Laminar-Turbulent Transition in Gas Turbine Engines. *ASME J. Turbomach.*, 113:509–537. DOI:10.1115/1.2929110

McDonald, G. B., Lennemann, E. and Howard, J. H. G. (1971) Measured and Predicted Flow Near the Exit of a Radial-Flow Impeller. *ASME. J. Eng. Power*, 93(4):441–446. DOI:10.1115/1.3445604

McDougall, N. M., Cumpsty, N. A. and Hynes, T. P. (1990). Stall Inception in Axial Compressors. *ASME. J. Turbomach.*, 112(1):116–123. DOI:10.1115/1.2927406

McKenzie, A. B. (1967) *Axial Flow Fans and Compressors: Aerodynamic Design and Performance*. Ashgate, Aldershot.

Medic, G., Sharma, O. P., Jongwook, J. et al. (2014) High Efficiency Centrifugal Compressor for Rotorcraft Applications. NASA/CR-2014-218114/REV1.

Mehldahl, A. (1941) Die Trennung der Rad- und Diffusorverluste bei Zentrifugalgebläsen, *Brown Boveri Mitteilung*, 28(8–9):203–206.

Meier, M., Gooding, W., Fabian, J. and Key, N. L. (2019) Considerations for Using Additive Manufacturing Technology in Centrifugal Compressor Research. *ASME. J. Eng. Power.*, 142(3):031018. DOI:10.1115/1.4044937

Melnik, R. E., Brook, J. W. and Del Guidice, P. (1986) Computation of Turbulent Separated Flow with an Integral Boundary Layer Method. In *Proceedings of the 10th International Conference on Numerical Methods in Fluid Dynamics, Beijing, and Lecture Notes in Physics*, 264:473–480.

Menter, F. R. (1994) Two-Equation Eddy-Viscosity Turbulence Models for Engineering Applications. *AIAA Journal*, 32(8):1598–1605. DOI:10.2514/3.12149

Menter, F. R. (1996) A Comparison of Some Recent Eddy-Viscosity Turbulence Models. *ASME J. Fluids Eng.*, 118(3):514–519. DOI:10.1115/1.2817788

Menter, F. R., Kuntz, M. and Langtry, R. (2003) Ten Years of Industrial Experience with the SST Turbulence Model. *Turbulence Heat and Mass Transfer*, 4:625–632.

Methel, J., Gooding, W. J., Fabian, J. C., Key, N. L. and Whitlock, M. (2016) The Development of a Low Specific Speed Centrifugal Compressor Research Facility. ASME Paper GT2016-56683, ASME Turbo Expo, 13–17 June, Seoul. DOI:10.1115/GT2016-56683

Michelassi, V. and Giachi, M. (1997) Experimental and Numerical Analysis of Compressor Inlet Volutes. ASME Paper 97-GT-481, ASME 1997 Gas Turbine Congress, 2–5 June, Orlando. DOI:10.1115/97-GT-481

Miller, D. S. (1990) *Internal Flow Systems*. BHRA Fluid Engineering, Cranfield.

Miller, R. J. and Denton J. D. (2018) *Loss Generation in Turbomachines*. Cambridge Turbomachinery Course, The Moller Centre, 25–29 June, Cambridge.

Mischo, B., Ribi, B., Seebass-Linggi, C. and Mauri, S. (2009) Influence of Labyrinth Seal Leakage on Centrifugal Compressor Performance. ASME Paper GT2009-59524, ASME Turbo Expo, 8–12 June, Orlando. DOI:10.1115/GT2009-59524

Mishina, H. and Gyobu, I. (1978) Performance Investigations of Large Capacity Centrifugal Compressors. ASME Paper 78-GT-3, ASME Gas Turbine Conference, 9–13 April, London. DOI:10.1115/78-GT-3

Moody, L. F. (1944) Friction Factors for Pipe Flow. *Trans. ASME*, 66(8):671–684.

Moore, G. E. (1965) Cramming More Components onto Integrated Circuits. *Electron. Mag.*, 38 (8):114–117. DOI:10.1109/N-SSC.2006.4785860

Moore, J. (1973) A Wake and an Eddy in a Rotating Radial Flow Passage – Part I: Experimental Observations; Part II: Flow Model. *ASME J. Eng. Power*, 95(3):205–212. DOI:10.1115/1.3445724

Moore, J. (1976) Eckhardt's Impeller – a Ghost from Ages Past. C.U.E.D./A-Turbo/TR83. Cambridge University Engineering Department.

Moore, J. and Moore, J. G. (1981) Calculations of Three-Dimensional, Viscous Flow and Wake Development in a Centrifugal Impeller. *ASME J. Eng. Power*, 103(2):367–372. DOI:10.1115/1.3230730

Moore, J. and Moore, J. G. (1983) Entropy Production Rates from Viscous Flow Calculations: Part I – a Turbulent Boundary Layer Flow. ASME Paper 83-GT-70, ASME Gas Turbine Conference, 27–31 March, Phoenix. DOI:10.1115/83-GT-70

Moore, F. K. and Greitzer, E. M. (1986). A Theory of Post-Stall Transients in Axial Compression Systems: Part I – Development of Equations. *ASME. J. Eng. Gas Turbines Power*, 108(1):68–76. DOI:10.1115/1.3239887

Moore, J., Moore, J. G. and Timmis, P. H. (1984). Performance Evaluation of Centrifugal Compressor Impellers Using Three-Dimensional Viscous Flow Calculations. *ASME J. Eng. Power*, 106(2):475–481. DOI:10.1115/1.3239590

Moore, M. J. (2002) *Micro-Turbine Generators*. PEP, London.

Moran, M. J. and Shapiro, H. N. (2007) *Fundamentals of Engineering Thermodynamics*. John, Wiley & Sons Inc., New York.

Morandin, M., Mercangoz, M., Hemrle, J. et al. (2013) Thermoeconomic Design Optimization of a Thermo-Electric Energy Storage System Based on Transcritical CO2 Cycles. *Energy*, 58:571–587 DOI:10.1016/j.energy.2013.05.038

Morris, R. E., and Kenny, D. P. (1968) High Pressure Ratio Centrifugal Compressors for Small Gas Turbine Engines. In *Aircraft Propulsion Systems AGARD Conference Proceedings No 31*, 10–14 June, Ottawa, Canada.

Mounier, V., Picard, C. and Schiffmann J. (2018) Data-Driven Predesign Tool for Small-Scale Centrifugal Compressor in Refrigeration. *ASME. J. Eng. Gas Turbines Power*, 140 (12):121011. DOI:10.1115/1.4040845

Neverov, V. and Liubimov, A. (2018) Design Optimization of a Multi-Stage Centrifugal Compressor. *Res. J.*, 1(2018):32–35.

Nichelson, B. J. (1988) Early Jet Engines and the Transition from Centrifugal to Axial Compressors. Ph.D. Thesis, University of Minnesota.

Nielsen, K. K., Childs, D. W. and Myllerup, C. M. (2001) Experimental and Theoretical Comparison of Two Swirl Brake Designs. *ASME J. Turbomach.*, 123(2):353–358. DOI:10.1115/1.1354140

Nikuradse, I. (1950) Laws of Flow in Rough Pipes. National Advisory Committee for Aeronautics. NACA TM1292.

Oakes, W. C., Lawless, P. B. and Fleeter, S. (1999) Instability Pathology of a High Speed Centrifugal Compressor. ASME Paper 99-GT-415, ASME Gas Turbine Congress, 7–10 June, Indianapolis. DOI:10.1115/99-GT-415

Oh, H. W., Yoon, E. S. and Chung, M. K. (1997) An Optimum Set of Loss Models for Performance Prediction of Centrifugal Compressors. *I. Mech. E. Part A: J. Power Energy*, 211(4):331–338. DOI:10.1243/0957650971537231

Pacciani R., Marconcini, M. and Arnone, A. (2017) A CFD-Based Throughflow Method with Three-Dimensional Flow Features Modelling. *Int. J. Propuls. Power*, 2(3):11. DOI:10.3390/ijtpp2030011

Paeng, K. S. and Chung, M. K. (2001) A New Slip Factor for Centrifugal Impellers. *I. Mech. E. Part A: J. Power Energy*, 215(5):645–649. DOI:10.1243/0957650011538776

Pakle, S. and Jiang, K. (2018) Design of a High-Performance Centrifugal Compressor with New Surge Margin Improvement Technique for High Speed Turbomachinery. *Journal of Propulsion and Power Research*, 7(1):19–29. DOI:10.1016/j.jppr.2018.02.004

Pampreen, R. C. (1972) The Use of Cascade Technology in Centrifugal Compressor Vaned Diffuser Design. *ASME J. Eng. Power*, 94(3): 187–192. DOI:10.1115/1.3445671

Pampreen, R. C. (1973) Small Turbomachinery Compressor and Fan Aerodynamics. *J. Eng. Power*, 95(3):251–256. DOI:10.1115/1.3445730

Patankar, S. (1980) *Numerical Heat Transfer and Fluid Flow*. CRC Press, Boca Raton.

Peng, D. Y. and Robinson D. B. (1976) A New Two Constant Equation of State. *Ind. Eng. Chem. Fundam.*, 15(1):59–64. DOI:10.1021/i160057a011

Peng, S., Li, T., Wang, X. et al. (2017) Toward a Sustainable Impeller Production: Environmental Impact Comparison of Different Impeller Manufacturing Methods. *J. Ind. Ecol.*, 21(1):216–229. DOI:0.1111/jiec.12628

Petry, N., Benra, F. K. and Koenig, S. (2010) Experimental Study of Acoustic Resonances in the Side Cavities of a High-Pressure Centrifugal Compressor Excited by Rotor/Stator

Interaction. ASME Paper GT2010-22054, ASME Turbo Expo, 14–18 June, Glasgow. DOI:10.1115/GT2010-22054

Pfleiderer, C. and Petermann, H. (1986) *Strömungsmaschinen*, 5th Edition. Springer, Berlin.

Pianko, M. and Wazelt F. (1983) Propulsion and Energetics Panel Working Group 14 on Suitable Averaging Techniques in Non-Uniform Internal Flows. AGARD Advisory Report No 182, Advisory Group for Aerospace Research and Development, Neuilly Sur Seine, France.

Pierret, S. and Van den Braembussche, R. A. (1999) Turbomachinery Blade Design Using a Navier–Stokes Solver and Artificial Neural Network. *ASME. J. Turbomach.*, 121 (2):326–332. DOI:10.1115/1.2841318

Pitot, H. (1732) Description d'une machine pour mesurer la vitesse des eaux et le sillage des vaisseaux. Histoire de l'Académie royale des sciences avec les mémoires de mathématique et de physique tirés des registres de cette Académie, 1732, 363–376.

Plöcker, U. and Knapp, H. (1978) Calculation of High-Pressure Vapor–Liquid Equilibria from a Corresponding-States Correlation with Emphasis on Asymmetric Mixtures. *Ind. Eng. Chem. Process Des. Dev.*, 17:324–332. DOI:10.1021/i260067a020

Podeur, M., Vogt, D. M., Mauri, S. and Jenny, P. (2019) Impeller Design and Multi-Stage Architecture Optimisation for Turbocompressors Operating with a Helium-Neon Gas Mixture. Paper IGTC-2019-153, International Gas Turbine Congress, 17–22 November, Tokyo.

Poling, B. E., Prausnitz, J. M. and O'Connell, J. P. (1977) Properties of Gases and Liquids, 5th Edition. McGraw-Hill Education, New York.

Polishuk, I. (2009) Generalized Cubic Equation of State Adjusted to the Virial Coefficients of Real Gases and Its Prediction of Auxiliary Thermodynamic Properties. *Ind. Eng. Chem. Res.*, 48(23):10708–10717. DOI:10.1021/ie900905p

Prandtl, L. (1938) Zur Berechnung der Grenzschichten. *J. Appli. Math. Mech.*, 18(1):77–82.

Pullan, G. (2017) A Web-Based Database Approach to CFD Post-Processing. AIAA 2017-0814. DOI:10.2514/6.2017-0814

Pullan, G., Young, A. M., Day, I. J., Greitzer, E. M. and Spakovszky, Z. S. (2015). Origins and Structure of Spike-Type Rotating Stall. *ASME. J. Turbomach.*, 137(5):051007. DOI:10.1115/1.4028494

Qiu, X., Japikse, D. and Anderson, M .R. (2008) A Meanline Model for Impeller Flow Recirculation. ASME Turbo Expo, 9–13 June, Berlin. DOI:10.1115/GT2008-51349

Qiu, X., Japikse, D., Zhao, J. and Anderson, M. R. (2011) Analysis and Validation of a Unified Slip Factor Model for Impellers at Design and Off-Design Conditions. *ASME J. Turbomach.*, 133(4):041018. DOI:10.1115/1.4003022

Radgen, P. and Blaustein, E. (2001) *Compressed Air Systems in the European Union.* Rondo Druck, Ebersbach-Röswälden.

Rahman, M. A., Chiba, A. and Tukao, T. (2004) Super High-Speed Electrical Machines. Summary. *Power Engineering Society General Meeting, 2004. IEEE, 10 June*, 2:1272–1275.

Rains, D. A. (1954) Tip Clearance Flows in Axial Flow Compressors and Pumps. California Institute of Technology Hydrodynamics Laboratory Report 5.

Raitor, T. and Neise, W. (2008) Sound Generation in Centrifugal Compressors. *J. Sound Vib.*, 314:738–756. DOI:10.1016/j.jsv.2008.01.034

Rankine, W. J. (1869) Centrifugal Whirling of Shafts. Engineer, XXVI, California Tech.

Raw, J. A. (1986) Surge Margin Enhancement by a Porous Throat Diffuser. *Can. Aeronaut. Space J.*, 32(1):54–60.

Raw, J. A. and Weir, G. C. (1980) The Prediction of Off-Design Characteristics of Axial and Axial/Radial Compressors. SAE Paper 800628, Turbine Powered Executive Aircraft Meeting, Phoenix. DOI:10.4271/800628

Reddy, T. C. S., Murty, G. V. R., Mukkavilli, P. and Reddy, D. N. (2004). Effect of the Setting Angle of a Low-Solidity Vaned Diffuser on the Performance of a Centrifugal Compressor Stage. *P I Mech. Eng. A -J Pow*, 218(8):637–646. DOI:10.1243/0957650042584294

Redlich, O. and Kwong, J. N. S. (1949) On the Thermodynamics of Solutions, an Equation of State. *Chem. Rev.*, 44:233–244. DOI:10.1021/cr60137a013

Reeves, G. B. (1977) Design and Performance of Selected Pipe-Type Diffusers. ASME Paper 77-GT-104, ASME Gas Turbine Conference, 27–31 March, Philadelphia. DOI:10.1115/77-GT-104

Reneau, L. R., Johnston, J. P. and Kline, S. J. (1967) Performance and Design of Straight, Two-Dimensional Diffusers. *ASME J. Basic Eng.*, 89(1):141–150. DOI:10.1115/1.3609544

Reynolds, O. (1895) On the Dynamical Theory of Incompressible Viscous Fluids and the Determination of the Criterion. *Phil. Trans R. Soc. London.* 186:123–164. DOI:10.1098/rsta.1895.0004

Ribaud, Y. (1987) Experimental Aerodynamic Analysis Relative to Three High Pressure Ratio Centrifugal Compressors. ASME Gas Turbine Conference, 31 May–4 June, Anaheim. DOI:10.1115/87-GT-153

Ribaut, M. (1968) Three-Dimensional Calculation of Flow in Turbomachines with the Aid of Singularities. *ASME J. Eng. Power*, 90(3):258–264. DOI:10.1115/1.3609184

Ribaut, M. (1977) On the Calculation of Three-Dimensional Divergent and Rotational Flow in Turbomachines. *ASME J. Fluids Eng.*, 99(1):187–196. DOI:10.1115/1.3448522

Ribaut, M. (1988) A Full Quasi-Three-Dimensional Calculation of Flow in Turbomachines. *ASME J. Turbomach.*, 110(3):401–404. DOI:10.1115/1.3262210

Ribi, B. (1966) Radialverdichter im Instabilitätsbereich. Dissertation ETH Nr. 11717, Zürich.

Ribi, B. (1996) *Flow in Radial Turbomachines, VKI Lecture Series 1996–01.* Von Karman Institute, Brussels.

Ribi, B. and Dalbert, P. (2000) One-Dimensional Performance Prediction of Subsonic Vaned Diffusers. *ASME J. Turbomach.*, 122(3):494–504. DOI:10.1115/1.1303816

Ribi, B. and Gyarmathy, G. (1993) Impeller Rotating Stall as a Trigger for the Transition from Mild to Deep Surge in a Subsonic Centrifugal Compressor. ASME Paper 93-GT-234, ASME Gas Turbine Congress, 24–27 May, Cincinnati. DOI:10.1115/93-GT-234

Riegels, F. W. (1961) *Airfoil Sections.* Butterworth, London

Roberts, S. K. and Sjolander, S. A. (2005) Effect of the Specific Heat Ratio on the Aerodynamic Performance of Turbomachinery. *ASME J. Eng. Power.* 127(4):773–780. DOI:10.1115/1.1995767

Robinson, C. J. (1991) Endwall Flows and Blading Design for Axial Flow Compressors, Ph.D. Thesis, Cranfield Institute of Technology.

Robinson, C. J., Casey, M. V. and Woods, I. (2011) An Integrated Approach to the Aero-Mechanical Optimisation of Turbocompressors. In Current Trends in Design and Computation of Turbomachinery. CKD NoveEnergo and TechSoft Engineering, Prague.

Robinson, C. J., Casey, M. V., Hutchinson, B. and Steed, R. (2012) Impeller-Diffuser Interaction in Centrifugal Compressors. ASME Paper GT2012-69151, ASME Turbo Expo, 11–15 June, Copenhagen. DOI:10.1115/GT2012-69151

Roduner, C. H. (1999) Strömungsstrukturen in Radialverdichtern, untersucht mit schnellen Sonden. Dissertation ETH Nr. 13428. Zürich.

Rolls-Royce Plc. (1986) *The Jet Engine*. BPCC Ltd, Dorset.

Rodgers, C. (1961) Influence of Impeller and Diffuser Characteristics and Matching on Radial Compressor Performance. SAE Tech. Prog. Ser.

Rodgers, C. (1978) A Diffusion Factor Correlation for Centrifugal Impeller Stalling. *ASME J. Eng. Power*, 100(4):592–601. DOI:10.1115/1.3446403.

Rodgers, C. (1980) Efficiency of Centrifugal Compressor Impellers. Paper 22 of AGARD Conference Proceedings No 282 Centrifugal Compressors, Flow Phenomena and Performance, Brussels, May.

Rodgers, C. (1982) The Performance of Centrifugal Compressor Channel Diffusers. ASME Paper 82-GT-10, ASME 1982 Gas Turbine Conference, 18–22 April, London. DOI:10.1115/82-GT-10

Rodgers, C. (1991) The Efficiencies of Single Stage Centrifugal Compressors for Aircraft Applications. ASME Paper 91-GT-77, ASME 1991 Gas Turbine Congress, 3–6 June, Orlando. DOI:10.1115/91-GT-077

Rodgers, C. (2000) Effects of Blade Number on the Efficiency of Centrifugal Compressor Impellers. ASME Turbo Expo, 8–11 May, Munich. DOI:10.1115/2000-GT-0455

Rodgers, C. (2001) Centrifugal Compressor Blade Trimming for a Range of Flows. ASME Turbo Expo, 4–7 June, New Orleans. DOI:10.1115/2001-GT-0316

Rodgers, C. (2005) Flow Ranges of 8.0:1 Pressure Ratio Centrifugal Compressor for Aviation Applications. ASME Paper GT2005-68041, ASME Turbo Expo, 6–9 June, Reno. DOI:10.1115/GT2005-68041

Roduner C. H. (1999) Strömungsstrukturen in Radialverdichter, untersucht mit schnellen Sonden. Dissertation ETH Nr. 13428, Zürich.

Rohne, K. and Banzhaf, M. (1991) Investigation of the Flow at the Exit of an Unshrouded Centrifugal Impeller and Comparison with the 'Classical' Jet-Wake Theory. *ASME J. Turbomach.*, 113(4):654–659. DOI:10.1115/1.2929131

Rossbach, T., Rube, C., Wedeking, M. et al. (2015) Performance Measurements of a Full-Stage Centrifugal Process Gas Compressor Test Rig. Paper ETC2015-084, 11th European Turbomachinery Conference, 23–27 March, Madrid.

Rothe, P. H. and Johnston, J. P. (1979) Free Shear Layer Behaviour in Rotating Systems. *ASME J. Fluids Eng.*, 101(1):117–120. DOI:10.1115/1.3448721

Rube, C., Rossbach, T., Wedeking, M., Grates, D. R. and Jeschke, P. (2016) Experimental and Numerical Investigation of the Flow inside the Return Channel of a Centrifugal Process Compressor. *ASME. J. Turbomach.*, 138(10):101006. DOI:10.1115/1.4032905

Runstadler, P. W. and Dean, R. C. Jr. (1969) Straight Channel Diffuser Performance at High Inlet Mach Numbers. *ASME J. Basic Eng.*, 91(3):397–412. DOI:10.1115/1.3571134

Runstadler, P. W. Jr., Dolan, F. X. and Dean, R. C. Jr. (1975) *Diffuser Data Book*. Technical Note TN-186. Creare Inc., Hanover.

Rusch, D. and Casey, M. V. (2013) The Design Space Boundaries for High Flow Capacity Centrifugal Compressors. *ASME J. Turbomach.*, 135(3):031035. DOI:10.1115/1.4007548

Sadrehaghighi, I. (2018) Aerodynamic Design and Optimisation. CFD Open Series Rev. 1.85.7 DOI:10.13140/RG.2.2.11383.73127/7

SAE (1995) J1826_199503. Turbocharger gas stand test code. Society of Automotive Engineers (SAE) Pennsylvania.

Sandberg, M. R. (2016) Centrifugal Compressor Configuration, Selection and Arrangement: A User's Perspective. 45th Turbomachinery Symposium, 12–15 September, Texas A&M, Houston. DOI:10.21423/R1QP46

Sandberg, M. R. and Colby, G. M. (2013) Limitations of ASME PTC 10 in Accurately Evaluating Centrifugal Compressor Thermodynamic Performance. 42nd Turbomachinery Symposium, 1–3 October, Texas A&M, Houston.

Sapiro, L. (1983) Effect of Impeller-Extended Shrouds on Centrifugal Compressor Performance as a Function of Specific Speed. *ASME J. Eng. Power*, 105(3):457–465. DOI:10.1115/1.3227437

Saravanamuttoo, H. I. H., Rogers G. F. C., Cohen, H. and Straznicky P. (2009) *Gas Turbine Theory*, 6th Edition, Pearson Education Limited, London.

Schaeffler, A. (1980) Experimental and Analytical Investigation of the Effects of Reynolds Number and Blade Surface Roughness on Multistage Axial Flow Compressors. *ASME J. Eng. Power*, 102(1):5–12. DOI:10.1115/1.3230232

Schiffmann, J. (2008) Integrated Design, Optimization and Experimental Investigation of a Direct Driven Turbocompressor for Domestic Heat Pumps. Thesis Nr. 4126 EPFL, Lausanne.

Schleer, M. (2006) Flow Structure and Stability of a Turbocharger Centrifugal Compressor. Dissertation ETH Nr. 16605, Zürich.

Schleer, M., Song, S. J. and Abhari, R. S. (2008) Clearance Effects on the Onset of Instability in a Centrifugal Compressor. *ASME J. Turbomach.*, 130(3):031002. DOI:10.1115/1.2776956

Schlichting, H. and Gersten, K. (2006) *Grenzschicht-Theorie*, 10th Edition. Springer Verlag, Berlin, Heidelberg, New York.

Schobeiri, M. (2005) Turbomachinery Flow Physics and Dynamic Performance. Springer, Berlin.

Schodl, R. (1980) A Laser-Two-Focus (L2F) Velocimeter for Automatic Flow Vector Measurements in the Rotating Components of Turbomachines. *ASME. J. Fluids Eng.*, 102(4):412–419. DOI:10.1115/1.3240713

Schmied, J. (2019) Application of MADYN 2000 to Rotor Dynamic Problems of Industrial Machinery. SIRM 2019 – 13th International Conference on Dynamics of Rotating Machines, 13–15 February, Copenhagen.

Scholz, N (1977) Aerodynamics of Cascades. AGARD AG 220.

Schreiber, C. (2017) Inlet Recirculation in Radial Compressors. Ph.D. Thesis, Cambridge University.

Schultz, J. M. (1962) The Polytropic Analysis of Centrifugal Compressors. *ASME J. Eng. Power*, 84(1):69–82. DOI:10.1115/1.3673381

Schweitzer, J. K. and Garberoglio, J. E. (1984) Maximum Loading Capability of Axial Flow Compressors. *J. Aircr.*, 21(8):593–600. DOI:10.2514/3.45028

Senoo, Y. and Ishida, M. (1986) Pressure Loss Due to the Tip Clearance of Impeller Blades in Centrifugal and Axial Blowers. *ASME J. Eng. Power*, 108(1):32–37. DOI:10.1115/1.3239882

Senoo, Y. and Kinoshita, Y. (1978) Limits of Rotating Stall and Stall in Vaneless Diffuser of Centrifugal Compressors. ASME Paper 78-GT-19, ASME Gas Turbine Conference, 9–13 April, London. DOI:10.1115/78-GT-19

Senoo, Y., Hayami, H. and Ueki, H. (1983) Low-Solidity Tandem-Cascade Diffusers for Wide-Flow-Range Centrifugal Blowers. ASME Paper 83-GT-3, ASME Gas Turbine Conference, 27–31 March, Phoenix. DOI:10.1115/83-GT-3

Senoo, Y., Kinoshita, Y., and Ishida, M. (1977) Asymmetric Flow in Vaneless Diffusers of Centrifugal Blowers. *ASME J. Fluids Eng.*, 99(1):104–111. DOI:10.1115/1.3448501

Seralathan, S. and Chowdhury D. G. (2013) Modification of Centrifugal Impeller and Effect of Impeller Extended Shrouds on Centrifugal Compressor Performance. *Procedia Eng.*, 64:1119–1128. DOI:10.1016/j.proeng.2013.09.190

Serrano, J. R., Margot, X., Tiseira Izaguirre, A. O. and García-Cuevas González, L. M. (2013) Optimization of the Inlet Air Line of an Automotive Turbocharger. *Int. J. Eng. Res.*, 14 (1):92–104. DOI:10.1177/1468087412449085.

Sheard, A. G. (2011) Blade by Blade Tip Clearance Measurement. *Int. J. Rot. Mach.*, 2011: Article ID 516128. DOI:10.1155/2011/516128

Shen, F., Yu, L., Cousins, W. T. et al. (2016) Numerical Investigation of the Flow Distortion Impact on a Refrigeration Centrifugal Compressor. ASME Turbo Expo, 13–17 June, Seoul. DOI:10.1115/GT2016-57063

Shepherd, D. G. (1956) *Principles of Turbomachinery*. Macmillan, London.

Sherstyuk, A. N. and Kosmin, V. M. (1969) Determining the Losses and Optimum Velocities in Centrifugal Compressor Volutes. *Therm. Eng.*, 2:70–72.

Shibata, T., Yagi, M., Nishida, H., Kobayashi, H. and Tanaka, M. (2012) Effect of Impeller Blade Loading on Compressor Stage Performance in a High Specific Speed Range. *ASME J. Turbomach.*, 134(4):041012. DOI:10.1115/1.4003659

Shum, Y. K. P., Tan, C. S. and Cumpsty, N. A. (2000) Impeller–Diffuser Interaction in a Centrifugal Compressor. *ASME J. Turbomach.*, 122(4):777–786. DOI:10.1115/1.1308570

Sieverding, C. H., Arts, T., Dénos, R. and Brouckaert J. F. (2000) Measurement Techniques for Unsteady Flows in Turbomachines. *Exp. Fluids*, 28(2000):285–321. DOI:10.1007/s003480050390

Simon, H. (1987) Design Concept and Performance of a Multistage Integrally Geared Centrifugal Compressor Series for Maximum Efficiencies and Operating Ranges. ASME Paper 87-GT-43, ASME Gas Turbine Conference, 31 May–4 June, Anaheim. DOI:10.1115/87-GT-43

Simon, H. and Bülskämper, A. (1984) On the Evaluation of Reynolds Number and Relative Surface Roughness Effects on Centrifugal Compressor Performance Based on Systematic Experimental Investigations. *ASME J. Eng. Power*, 106(2):489–498. DOI:10.1115/1.3239592

Simon, H. and Rothstein, E. (1983) On the Development of Return Passages of Multi-Stage Centrifugal Compressors. In *Return Passages of Multi-Stage Turbomachinery,* ed. P. Nykorowytsch. Applied Mechanics, Bioengineering and Fluids Engineering Conference, 20–22 June, Houston, 1–12

Simon, H., Wallmann, T. and Mönk, T. (1987) Improvements in Performance Characteristics of Single-Stage and Multistage Centrifugal Compressors by Simultaneous Adjustments of Inlet Guide Vanes and Diffuser Vanes. *ASME. J. Turbomach.*, 109(1):41–47. DOI:10.1115/1.3262068

Simpson, A., Aalburg, C., Schmitz, M. et al. (2008) Design, Validation and Application of a Radial Cascade for Centrifugal Compressors. ASME Paper GT2008-51262, ASME Turbo Expo, 9–13 June, Berlin. DOI:10.1115/GT2008-51262

Singh, M. P., Vargo, J. J., Schiffer, D. M. and Dello, J. D. (1988) Safe Diagram – a Design and Reliability Tool for Turbine Blading. 17th Turbomachinery Symposium, Texas A&M, Houston. DOI:10.21423/R1B673

Sirakov, B. and Casey, M. V. (2013). Evaluation of Heat Transfer Effects on Turbocharger Performance. *ASME J. Turbomach.*, 135(2):021011. DOI:10.1115/1.4006608

Sirakov, B., Gong, Y., Epstein, A. and Tan, Ch. (2004) Design and Characterization of Micro-Compressor Impellers. ASME Paper GT2004-53332, ASME Turbo Expo, 14–17 June, Vienna. DOI:10.1115/GT2004-53332

Sitaram, N. (2017) Survey of Available Techniques for High Speed Turbomachinery Testing. Paper ICTACEM-2017/360, Proc International Conference on Theoretical, Applied, Computational and Experimental Mechanics, 28–30 December, IIT Kharagpur, India.

Sivagnanasundaram, S., Spence, S., Early, J. and Nikpour, B. (2010) An Investigation of Compressor Map Width Enhancement and the Inducer Flow Field Using Various Configurations of Shroud Bleed Slot. ASME Paper GT2010-22154, ASME Turbo Expo, 14–18 June, Glasgow. DOI:10.1115/GT2010-22154

Skoch, G. J. (2003) Experimental Investigation of Centrifugal Compressor Stabilization Techniques. *ASME J. Turbomach.*, 125(4):704–713. DOI:10.1115/1.1624846

Smirnov, P. E., Hansen, T. and Menter, F. R. (2007) Numerical Simulation of Turbulent Flows in Centrifugal Compressor Stages with Different Radial Gaps. ASME Paper GT2007-27376, ASME Turbo Expo, 14–17 May, Montreal. DOI:10.1115/GT2007-27376

Smith L. H., Jr. (1955) Secondary Flow in Axial-Flow Turbomachinery. *Trans. ASME*, 77:1065–1076.

Smith, L. H., Jr. (1958) Recovery Ratio – a Measure of the Loss Recovery Potential of Compressor Stages. *Trans. ASME*, 80:517–524.

Smith, L. H., Jr. (1966) The Radial-Equilibrium Equation of Turbomachinery. *ASME J. Eng. Power*, 88:1–12. DOI:10.1115/1.3678471

Smith, L. H., Jr. (1970) Casing Boundary Layers in Multistage Axial-Flow Compressors. Brown Boueri Symposium. In *Flow Research in Blading*, ed. L. S. Dzung. Elsevier, Amsterdam. 275–300.

Smith, L. H., Jr. (2002) Axial Compressor Aerodesign Evolution at General Electric. *ASME J. Turbomach.*, 124(3):321–330. DOI:10.1115/1.1486219

Smith, S. F. (1965) A Simple Correlation of Turbine Efficiency. *J. Roy. Aeronaut. Soc.*, 69 (655):467–470. DOI:10.1017/S0001924000059108

Smith, D. J. L. and Merryweather, H. (1973) Representation of the Geometry of Centrifugal Impeller Vanes by Analytic Surfaces. *Intl. J. Num. Meth. Eng.*, 7:137–154.

Spakovszky, Z. S. and Roduner, C. H. (2009) Spike and Modal Stall Inception in an Advanced Turbocharger Centrifugal Compressor. *ASME. J. Turbomach.*, 131(3):031012. DOI:10.1115/1.2988166

Spasov, M. (2009) Private communication from Gamma technologies.

Spurr, A. (1980) The Prediction of 3D Transonic Flow in Turbomachinery Using a Combined Throughflow and Blade-to-Blade Time Marching Method. *Int. J. Heat Fluid Flow*, 2 (4):189–199. DOI:10.1016/0142-727X(80)90013-2

Soave, G. (1972) Equilibrium Constants from a Modified Redlich–Kwong Equation of State. *Chem. Eng. Sci.*, 72:1197–1203. DOI:10.1016/0009-2509(72)80096-4

Sorokes, J. M. and Welch, J. P. (1992) Experimental Results on a Rotatable Low Solidity Vaned Diffuser. ASME Paper G92-GT-019, ASME Gas Turbine Congress, 1–4 June, Cologne. DOI:10.1115/92-GT-019

Sorokes J. M. (1993) The Practical Application of CFD in the Design of Industrial Centrifugal Impellers. 22nd Turbomachinery Symposium, Texas A&M, Houston. DOI:10.21423/R1P074

Sorokes, J. M. (1994) A CFD Assessment of Entrance Area Distributions in a Centrifugal Compressor Vaneless Diffuser. ASME Paper 94-GT-090, ASME Gas Turbine Congress, 13–16 June, The Hague. DOI:10.1115/94-GT-090

Sorokes, J. M., Marshall, F. and Kuzdzal, M. (2018) A Review of Aerodynamically Induced Forces Acting on Centrifugal Compressors and Resulting Vibration Characteristics of Rotors. 47th Turbomachinery Symposium, Texas A&M, Houston. DOI:10.21423/R12366

Sovran, G. and Klomp, E. D. (1967) Experimentally Determined Optimum Geometries for Rectilinear Diffusers with Rectangular, Conical, or Annular Cross Section. In *Fluid Mechanics of Internal Flows*, ed. Sovran, G. Elsevier, Amsterdam, 270–319.

Sozer, E., Brehm C. and Kiris, C. C. (2014) Gradient Calculation Methods on Arbitrary Polyhedral Unstructured Meshes for Cell-Centered CFD Solvers. AIAA Paper 2014-1440, 52nd Aerospace Sciences Meeting, 13–17 January, National Harbor. DOI:10.2514/6.2014-1440

Spalart, P. R. and Allmaras, S. R. (1992) A One-Equation Turbulence Model for Aerodynamic Flows. AIAA Paper 92-0439. DOI:10.2514/6.1992-439

Spalding, B. (1998) Fluid Structure Interaction in the Presence of Heat Transfer and Chemical Reaction. Keynote lecture, ASME/JSME Joint Pressure Vessels and Piping Conference, July, Boston.

Span, R. and Wagner, W. (2003) Equations of State for Technical Applications. I. Simultaneously Optimized Functional Forms for Nonpolar and Polar Fluids. *Int. J. Thermophys.*, 24, 1–39. DOI/10.1023/A:1022390430888

Squire, H. B. and Winter, K. G. (1951) The Secondary Flow in a Cascade of Airfoils in a Nonuniform Stream. *J. Aero. Sci.*, 18(4):271–275. DOI:10.2514/8.1925

Stahlecker, D. (1999) Untersuchung der instationären Strömung eines beschaufelten Radialverdichterdiffusors mit einem Laser-Doppler-Anemometer. Dissertation ETH Nr. 13228. Zürich.

Stanitz, J. D. and Prian, V. D. (1951) A Rapid Approximate Method for Determining Velocity Distribution on Impeller Blades of Centrifugal Compressors. NACA TN 2421.

Starling, K. E. (1973) Fluid Thermodynamic Properties for Light Petroleum Systems. Gulf Publ., Houston.

Staubli, T., Gyarmathy, G. and Inderbitzen, A. (2001) Visualisation d'un décollement tournant dans un banc d'essais en eau d'un étage de compresseur centrifuge. *La Houille Blanche*, 2 (4):39–45.

Steiner, A. (2000) Techniques for Blade Tip Clearance Measurements with Capacitive Probes. *Meas. Sci. Technol.*, 11 (7):865–869. DOI:10.1088/0957-0233/11/7/303

Stephan, K. and Mayinger, F. (2000) Thermodynamik: Grundlagen und technische Anwendungen. Springer, Berlin.

Stodola, A. (1905) *Steam Turbines: With an Appendix on Gas Turbines and the Future of Heat Engines*. D. Van Nostrand Co., New York.

Storer, J. A. (1991) Tip Clearance Flow in Axial Compressors. Ph.D. Thesis, Cambridge University.

Storer, J. A. and Cumpsty, N. A. (1994) An Approximate Analysis and Prediction Method for Tip Clearance Loss in Axial Compressors. *ASME J. Turbomach.*, 116(4):648–656. DOI:10.1115/1.2929457

St. Peter, J. (1999) The History of Aircraft Gas Turbine Development in the United States. IGTI, Georgia.

Stranges, S. (2000) *Germany's Synthetic Fuel Industry, 1927–1945*. Springer, Dordrecht.

Stratford, B. S. (1959). An Experimental Flow with Zero Skin Friction throughout Its Region of Pressure Rise. *J. Fluid Mech.*, 5(1):17–35. DOI:10.1017/S0022112059000027

Stratford, B. S. and Tubbs, H. (1965) The Maximum Pressure Rise Attainable in Subsonic Diffusers. *J. Roy. Aeronauti. Soc.*, 69(652):275–278. DOI:10.1017/S0001924000059911

Strazisar, A. J. and Denton, J. D. (1995) CFD Code Assessment in Turbomachinery: A Progress Report. *Global Gas Turbine News*, May/June:12–14.

Strub, R. A. (1974) A New Axial-Radial 'Isotherm' Compressor. ASME Paper 74-GT-141, ASME 1974 Gas Turbine Conference, 30 March–4 April, Zürich. DOI:10.1115/74-GT-141

Strub, R. A. (1984) Rotors of Turbomachines – Power and Elegance. *Sulzer Tech. Revi.*, 1:29–36.

Stuart, C., Spence, S., Filsinger, D., Starke, A. and Kim, S. (2019) A Three-Zone Modelling Approach for Centrifugal Compressor Slip Factor Prediction. *ASME J. Turbomach.*, 141 (3):031008. DOI:10.1115/1.4042248

Sturmayr, A. and Hirsch, C. (1999) Throughflow Model for Design and Analysis Integrated in a Three-Dimensional Navier–Stokes Solver. *I. Mech. E. Part A: J. Power Energy*, 213 (4):263–273. DOI:10.1243/0957650991537608

Su, G.-J. (1946) Modified Law of Corresponding States for Real Gases. *Ind. Eng. Chem.*, 38 (8):803–806. DOI:10.1021/ie50440a018

Sun, Z., Sun, X., Zheng, X. and Linghu, Z. (2017) Flow Characteristics of a Pipe Diffuser for Centrifugal Compressors. *J. Appl. Fluid Mech.*, 10(1):143–155. DOI:10.18869/acadpub.jafm.73.238.26476

Swain E. (2005) Improving a One-Dimensional Centrifugal Compressor Performance Prediction Method. *I. Mech. E. Part A: J. Power Energy*, 219(8):653–659. DOI:10.1243/095765005X31351

Tain, L. and Cumpsty, N. A. (2000) Compressor Blade Leading Edges in Subsonic Compressible Flow. *I. Mech. E. Part C: J. Mech. Eng., Sci.*, 214(1):221–242. DOI:10.1243/0954406001522921

Tamaki, H. (2012) Effect of Recirculation Device with Counter Swirl Vane on Performance of High Pressure Ratio Centrifugal Compressor. *ASME J. Turbomach.*, 134(5): 051036. DOI:10.1115/1.4004820

Tamaki, H. (2019) A Study on Matching between Centrifugal Compressor Impeller and Low Solidity Diffuser and Its Extension to Vaneless Diffuser. *ASME. J. Eng. Gas Turbines Power*, 141(4):041026. DOI:10.1115/1.4041003

Tamaki, H., Nakao, H. and Saito, M. (1999) The Experimental Study of Matching between Centrifugal Compressor Impeller and Diffuser. *ASME J. Turbomach.*, 121(1):113–118. DOI:10.1115/1.2841218

Teichel, S. H. (2018) Optimised Design of Mixed Flow Compressors for an Active High-Lift System. Ph.D. Thesis, Gottfried Wilhelm Leibniz Universität Hannover.

Thompson, R. G. (1979) Performance Correlations for Flat and Conical Diffusers. ASME Paper 79-GT-52, ASME Gas Turbine Conference, 12–15 March, San Diego. DOI:10.1115/79-GT-52

Thomson, W. (1996) *Theory of Vibration with Applications*. CRC Press, Boca Raton.

Tobin, J. (2004) *To Conquer the Air: The Wright Brothers and the Great Race for Flight*. Free Press, New York.

Tognola, S. and Paranjpe, P. (1960) Theoretische Strömungsberechnungen und Modellversuche für Radialverdichter. *Escher-Wyss-Mitteilungen*, 33:58–66.

Traupel, W. (1962) Die Theorie der Strömung durch Radialmaschinen. G Braun, Karlsruhe.

Traupel, W. (1975) Grundzüge einer Theorie des radialen Verdichterlaufrades. *FVV Radialverdichter Vorhaben*, 83 and 115, Heft 237, FVV Frankfurt.

Traupel, W. (2000) *Thermische Turbomaschinen*, Springer, Berlin.

Trébinjac, I., Kulisa, P., Bulot, N. and Rochuon, N. (2009) Effect of Unsteadiness on the Performance of a Transonic Centrifugal Compressor Stage. *ASME. J. Turbomach.*, 131 (4):041011. DOI:10.1115/1.3070575

Tropea, C., Yarin, Y. L. and Foss, J. (2007) *Springer Handbook of Experimental Fluid Mechanics*. Springer, Berlin. DOI:10.1007/978-3-540-30299-5

Tsujimoto, Y., Hiroshi, T. and Doerfler, P. (2103) Effects of Acoustic Resonance and Volute Geometry on Phase Resonance in a Centrifugal Fan. *Int. J. Fluid Machi. Sys.*, 6(2):1–12. DOI:10.5293/IJFMS.2013.6.2.075

Vadasz, P. and Weiner, D. (1992) The Optimal Intercooling of Compressors by a Finite Number of Intercoolers. *ASME J. Energy Resour. Technol.* 114:155–260. DOI:10.1115/1.2905950

Van den Braembussche, R. A. (2006a) Flow and Loss Mechanisms in Volutes of Centrifugal Pumps. In *Design and Analysis of High Speed Pumps.* Educational Notes RTO-EN-AVT-143, Neuilly-sur-Seine, 12-1–12-26.

Van den Braembussche, R. A. (2006b) Optimization of Radial Impeller Geometry. In *Design and Analysis of High Speed Pumps.* Educational Notes RTO-EN-AVT-143, Neuilly-sur-Seine, 13:1–28.

Van den Braembussche, R. A. (2019) *Design and Analysis of Centrifugal Compressors.* ASME Press Series. ASME Press, Wiley, New York.

Van Gerner, H. J. (2014) *Heat Pump Conceptual Study and Design, NLR-CR-2014-009*, National Aerospace Laboratory, Netherlands.

Vance, J., Zeidan, F. and Murphy, B. (2010) *Machinery Vibration and Rotordynamics.* John Wiley & Sons, Inc., Hoboken.

Vavra, M. H. (1970a) Basic Elements for Advanced Designs of Radial Flow Compressors. In *AGARD Lecture Series 39 on Advanced Compressors*, June, Brussels.

Vavra, M. H. (1970b) Application of Throughflow to Radial Wheel Design. In *AGARD Lecture Series 39 on Advanced compressors*, June, Brussels.

VDI 2044 (2018) *Acceptance and Performance Tests on Fans.* VDI, Düsseldorf.

VDI 2045 Blatt 1 (1993) *Acceptance and Performance Tests on Turbo Compressors and Displacement Compressors; Test Procedure and Comparison with Guaranteed Values.* VDI, Düsseldorf.

VDI 2045 Blatt 2 (1993) *Acceptance and Performance Test on Dynamic and Positive Displacement Compressors; Theory and Examples.* VDI, Düsseldorf.

VDI 4675 (2012) *Balance Based Averaging of Inhomogeneous Flow Fields.* VDI, Düsseldorf.

Velasquez, E. I. G. (2017) Determination of a Suitable Set of Loss Models for Centrifugal Compressor Performance Prediction. *Chin. J. Aeronaut.*, 30 (5):1644–1650. DOI:10.1016/j.cja.2017.08.002

Verstraete, T., Alsalihi, Z. and Van den Braembussche, R. A. (2010) Multidisciplinary Optimization of a Radial Compressor for Micro-Gas-Turbine Applications. *ASME. J. Turbomach.*, 132(3):031004. DOI:10.1115/1.3144162

Von Backström, T. W. (2008) The Effect of Specific Heat Ratio on the Performance of Compressible Flow Turbomachines. ASME Paper GT2008-50183, ASME Turbo Expo 2008, 9–13 June, Berlin. DOI:10.1115/GT2008-50183

Von Ohain, H. P. (1989) The First Turbojet Flights and Other German Developments. In *Celebration of the Golden Anniversary of Jet Powered Flight*, ed. St. Peter, J. Dayton.

Voss, C., Aulich, M. and Raitor, T. (2014) Metamodel Assisted Aeromechanical Optimisation of a Transonic Centrifugal Compressor. 15th ISROMAC Symposium, Honolulu.

Wachter, J. and Woehrl, B. (1981) *Aufwertungen des Wirkungsgrades von Turbomaschinen der radialen Bauart in Abhängigkeit von Reynoldszahl und Geometrie*, Pfleiderer Gedächtnis Tagung, Braunschwieg, VDI 424:19–28.

Wagner, W. et al. (2000) The IAPWS Industrial Formulation 1997 for the Thermodynamic Properties of Water and Steam. *ASME J. Eng. Gas Turbines Power*, 122(1):150-184. DOI/10.1115/1.483186

Walreavens, R. E. and Cumpsty, N. A. (1995) Leading Edge Separation Bubbles on Turbomachine Blades. *ASME J. Turbomach.*, 117(1):115–125. DOI:10.1115/1.2835626

Walsh, P. P. and Fletcher, P. (2004) *Gas Turbine Performance.* Wiley, London.

Watson, N. and Janota, M. S. (1982) *Turbocharging the Internal Combustion Engine.* Macmillan, London.

Weber, C. R., and Koronowski, M. E. (1986) Meanline Performance Prediction of Volutes in Centrifugal Compressors. ASME Paper 86-GT-216, ASME Gas Turbine Conference, 8–12 June, Dusseldorf. DOI:10.1115/86-GT-216

Wheeler, A. P. S. and Miller, R. J. (2008) Compressor Wake/Leading-Edge Interactions at Off Design Incidences. ASME Paper GT2008-50177, ASME Turbo Expo, 9–13 June, Berlin. DOI:10.1115/GT2008-50177

Whitfield, A. (1978) Rationalization of Empirical Loss Coefficients and Their Application in One Dimensional Performance Prediction Procedures for Centrifugal Compressors. ASME Paper 78-GT-177, ASME Gas Turbine Conference, Wembley. DOI:10.1115/78-GT-177

Whitfield, A. and Baines, N. C. (1976) A General Computer Solution for Radial and Mixed Flow Turbomachine Performance Prediction. *Int. J. Mech. Sci.*, 18(4):179–184. DOI:10.1016/0020-7403(76)90023-0

Whitfield, A. and Baines, N. C. (1990) *Design of Radial Turbomachines.* Longman, John Wiley & Sons, London.

Whittle, F. (1945) The Early History of the Whittle Jet Propulsion Gas Turbine (the First James Clayton Lecture). *I. Mech E.*, 152:419–435.

Wiesner, F. J. (1960) Practical Stage Performance Correlations for Centrifugal Compressors. ASME Paper 60-HYD-17, Gas Turbine Power and Hydraulic Conference, March, Houston. DOI:10.1115/60-HYD-17

Wiesner, F. J. (1967) A Review of Slip Factors for Centrifugal Impellers. *Trans. ASME. J. Eng. Power*, 89:558–566. DOI:10.1115/1.3616734

Wilkosz, B., Schmidt, J., Guenther, C. et al. (2013) Numerical and Experimental Comparison of a Tandem and Single Vane De-Swirler Used in an Aeroengine Centrifugal Compressor. *ASME. J. Turbomach.*, 136(4):041005. DOI:10.1115/1.4024891

Wilcox, D. C. (1998) *Turbulence Modelling for CFD.* DCW Industries, Inc., Lake Arrowhead.

Wilson, D. G. and Korakianitis, T. (1998) *The Design of High Efficiency Turbomachinery and Gas Turbines.* Prentice Hall, Upper Saddle River.

Wittwer, D. and Küpfer, H. (1986) The New Radial Compressor Test Rig Super-Martin. *Brown Boveri Rev.*, 73(4):177–184.

Wilkinson, D. H. (1969) Streamline Curvature Methods for Calculating the Flow in Turbomachines, Report No W/M(3F), English Electric, Whetstone.

Wilkinson, D. H. (1970) Stability, Convergence, and Accuracy of Two-Dimensional Streamline Curvature Methods Using Quasi-Orthogonals. I. Mech. E. Convention, Glasgow, Paper 35.

Wittrock, D., Reutter, O., Nicke, E., Schmidt, T. and Klausmann, J. (2018) Design of a Transonic High Flow Coefficient Centrifugal Compressor by Using Advanced Design Methods. ASME Turbo Expo. 11–15 June, Oslo. DOI:10.1115/GT2018-75024

Wittrock, D., Junker, M., Beversdorff, M., Peters, A. and Nicke, E. (2019) A Deep Insight Into the Transonic Flow of an Advanced Centrifugal Compressor Design. ASME Turbo Expo, 17–21 June, Phoenix. DOI:10.1115/GT2019-90308.

Wolfe, T., Lee, Y. and Slipper, M. E. (2015) A Performance Prediction Model for Low-Speed Centrifugal Fans. *ASME J. Fluids Eng.*, 137(5):051106. DOI:10.1115/1.4029397

Woods, I. (2018) *Centrifugal Impeller Design Guidelines.* PCA Internal Report, Lincoln.

Wright, P. I. and Miller, D. C. (1991) An Improved Compressor Performance Prediction Model. Paper C423/028, I. Mech. Eng. Conference 1991–3: Turbomachinery: Latest Developments in a Changing Scene. Mechanical Engineering Publications, London.

Wu, C. H. (1952) A General Theory of Three-Dimensional Flow in Subsonic, and Supersonic Turbomachines of Axial, Radial and Mixed-Flow Types. NASA TN2604

Yoon, S. Y., Lin, Z. and Allaire, P. E. (2013) Control of Surge in Centrifugal Compressors by Active Magnetic Bearings. *Adv. Ind. Control.* DOI:10.1007/978-1-4471-4240-9_2

Yoshinaga, Y., Gyobu, I., Mishina, H., Koseki, F. and Nishida, H. (1980) Aerodynamic Performance of a Centrifugal Compressor with Vaned Diffusers. *ASME J. Fluids Eng.*, 102(4):486–493. DOI:10.1115/1.3240730

Yoshinaka, T. (1977) Surge Responsibility and Range Characteristics of Centrifugal Compressors. *Proc. Tokyo Joint Gas Turbine Conference*, 381–390.

Young, A. D. (1951) A Review of Some Stalling Research. ARC R&M No 2609, HMSO.

Younglove, B. and Ely, J. (1987) Thermophysical Properties of Fluids. II. Methane, Ethane, Propane, Isobutane and Normal Butane. *J. Phys. Chem. Ref. Data*, 16(4):577–798. DOI:10.1063/1.555785

Zachau, U., Niehuis, R., Hoenen, H. and Wisler, D. C. (2009). Experimental Investigation of the Flow in the Pipe Diffuser of a Centrifugal Compressor Stage under Selected Parameter Variations. ASME Paper GT2009-59320, ASME Turbo Expo, 8–12 June, Orlando. DOI:10.1115/GT2009-59320

Zangeneh, M. (1994) Inviscid-Viscous Interaction Method for Three-Dimensional Inverse Design of Centrifugal Impellers. *ASME J. Turbomach.*, 116(2):280–290. DOI:10.1115/1.2928362

Zangeneh, M., Goto, A. and Harada, H. (1998). On the Design Criteria for Suppression of Secondary Flows in Centrifugal and Mixed Flow Impellers. *ASME J. Turbomach.*, 120 (4):723–735. DOI:10.1115/1.2841783

Zemp, A. (2012) Experimental Investigation and Validation of High Cycle Fatigue Design Systems for Centrifugal Compressors. Dissertation ETH No. 20670, ETH Zürich.

Zemp, A., Abhari, R. S. and Ribi, B. (2011) Experimental Investigation of Forced Response Impeller Blade Vibration in a Centrifugal Compressor with Variable Inlet Guide Vanes: Part 1 – Blade Damping. ASMETurbo Expo, 6–10 June, Vancouver. DOI:10.1115/GT2011-46289

Zemp, A., Kammerer, A. and Abhari, R. S. (2010) Unsteady Computational Fluid Dynamics Investigation on Inlet Distortion in a Centrifugal Compressor. *ASME J. Turbomach.*, 132 (3):031015. DOI:10.1115/1.3147104

Zhang,Y., Xu, Y., Zhou, X. et al. (2017) Compressed Air Energy Storage System with Variable Configuration for Wind Power Generation. *Energy Proc.*, 142:3356–3362, DOI:10.1016/j.egypro.2017.12.470

Zheng, X., Jin, L., Du, T., Gan, B., Liu, F. and Qian, H. (2013) Effect of Temperature on the Strength of a Centrifugal Compressor Impeller for a Turbocharger. *I. Mech. E. Part C: J. Mech. Eng., Sci.*, 227(5):896–904. DOI:10.1177/0954406212454966

Zheng, X., Sun, Z., Kawakubo, T. and Tamaki, H. (2017) Experimental Investigation of Surge and Stall in a Turbocharger Centrifugal Compressor with a Vaned Diffuser. *Exp. Therm. Fluid Sci.*, 82:493–506. DOI:10.1016/j.expthermflusci.2016.11.036.

Ziada, S., Oengoeren, A. and Vogel, A. (2002) Acoustic Resonance in the Inlet Scroll of a Turbo-Compressor. *J. Fluids Struct.*, 16(3):361–373. DOI:10.1006/jfls.2001.0421

Ziegler, K. U., Gallus, H. E. and Niehuis, R. (2003a) A Study on Impeller–Diffuser Interaction – Part I: Influence on the Performance. *ASME. J. Turbomach.*, 125(1):173–182. DOI:10.1115/1.1516814

Ziegler, K. U., Gallus, H. E. and Niehuis, R. (2003b) A Study on Impeller–Diffuser Interaction – Part II: Detailed Flow Analysis. *ASME. J. Turbomach.*, 125(1):183–192. DOI:10.1115/1.1516815

Zucker, R. D. and Biblarz, O. (2002) *Fundamentals of Gas Dynamics*. Wiley, Hoboken.

Zweifel, O. (1941) The Determination of the Variation of State in Turbomachinery by Means of the Increase in Entropy. *Brown Boveri Rev.*, 28 (8/9):232–236.

Zweifel, O. (1945) The Spacing of Turbomachine Blading, Especially with Large Deflection. *Brown Boveri Rev.*, 32(12):436– 444.

Zwyssig, C., Round, S. D. and Kolar, J. W. (2008) Ultra-High-Speed Low Power Electrical Drive Systems. *IEEE Trans Ind. Electron.*, 55(2):577– 585. DOI:10.1109/TIE.2007.911950

Index

Printed in the United States
by Baker & Taylor Publisher Services